Physics, Geometry, and Topology

NATO ASI Series

Advanced Science Institutes Series

A series presenting the results of activities sponsored by the NATO Science Committee, which aims at the dissemination of advanced scientific and technological knowledge, with a view to strengthening links between scientific communities.

The series is published by an international board of publishers in conjunction with the NATO Scientific Affairs Division

A	**Life Sciences**	Plenum Publishing Corporation
B	**Physics**	New York and London
C	**Mathematical and Physical Sciences**	Kluwer Academic Publishers Dordrecht, Boston, and London
D	**Behavioral and Social Sciences**	
E	**Applied Sciences**	
F	**Computer and Systems Sciences**	Springer-Verlag
G	**Ecological Sciences**	Berlin, Heidelberg, New York, London,
H	**Cell Biology**	Paris, and Tokyo

Recent Volumes in this Series

Volume 234—Constructive Quantum Field Theory II
edited by G. Velo and A. S. Wightman

Volume 235—Disorder and Fracture
edited by J. C. Charmet, S. Roux, and E. Guyon

Volume 236—Microscopic Simulations of Complex Flows
edited by Michel Mareschal

Volume 237—New Trends in Nonlinear Dynamics and Pattern-Forming Phenomena: The Geometry of Nonequilibrium
edited by Pierre Coullet and Patrick Huerre

Volume 238—Physics, Geometry, and Topology
edited by H. C. Lee

Volume 239—Kinetics of Ordering and Growth at Surfaces
edited by Max G. Lagally

Volume 240—Global Climate and Ecosystem Change
edited by Gordon J. MacDonald and Luigi Sertorio

Volume 241—Applied Laser Spectroscopy
edited by Wolfgang Demtröder and Massimo Inguscio

Series B: Physics

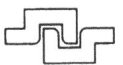

Physics, Geometry, and Topology

Edited by

H. C. Lee

Chalk River Laboratories
Chalk River, Ontario, Canada

Springer Science+Business Media, LLC

Proceedings of a NATO Advanced Study Institute
and Banff Summer School in Theoretical Physics
on Physics, Geometry, and Topology,
held August 14-25, 1989,
in Banff, Alberta, Canada

Library of Congress Cataloging-in-Publication Data

NATO Advanced Study Institute and Banff Summer School in Theoretical Physics on Physics, Geometry, and Topology (1989 : Banff, Alta.)
Physics, geometry, and topology/ edited by H.C. Lee.
p. cm. -- (NATO ASI series. Series B, Physics ; v. 238)
"Published in cooperation with NATO Scientific Affairs Division."
"Proceedings of a NATO Advanced Study Institute and Banff Summer School in Theoretical Physics on Physics, Geometry, and Topology, held August 14-25, 1989, in Banff, Alberta, Canada"--CIP t.p. verso.
Includes bibliographical references and index.
ISBN 978-0-306-43693-2 ISBN 978-1-4615-3802-8 (eBook)
DOI 10.1007/978-1-4615-3802-8
1. Geometry--Congresses. 2. Topology--Congresses.
3. Mathematical physics--Congresses. I. Lee, H. C. (Hoong-Chien), 1941- II. Title. III. Series.
QC20.7.G44N37 1989
516--dc20 90-48076
CIP
AC

Preface

The Banff NATO Summer School was held August 14-25, 1989 at the Banff Centre, Banff, Albert, Canada. It was a combination of two venues: a summer school in the annual series of Summer School in Theoretical Physics sponsored by the Theoretical Physics Division, Canadian Association of Physicists, and a NATO Advanced Study Institute. The Organizing Committee for the present school was composed of G. Kunstatter *(University of Winnipeg)*, H.C. Lee *(Chalk River Laboratories and University of Western Ontario)*, R. Kobes *(University of Winnipeg)*, D.J. Toms *(University of Newcastle Upon Tyne)* and Y.S. Wu *(University of Utah)*.

Thanks to the group of lecturers *(see Contents)* and the timeliness of the courses given, the school, entitled PHYSICS, GEOMETRY AND TOPOLOGY, was popular from the very outset. The number of applications outstripped the 90 places of accommodation reserved at the Banff Centre soon after the school was announced. As the eventual total number of participants was increased to 170, it was still necessary to turn away many deserving applicants.

In accordance with the spirit of the school, the geometrical and topological properties in each of the wide ranging topics covered by the lectures were emphasized. A recurring theme in a number of the lectures is the Yang-Baxter relation which characterizes a very large class of integrable systems including: many state models, two-dimensional conformal field theory, quantum field theory and quantum gravity in 2+1 dimensions.

Two separate workshops, one on "Geometrical Methods in Gauge Theories" and the other on "Physics, Braids and Links" were held during the school. These provided students and lecturers alike another venue to discuss recent developments in topics closely related to those lectured on in the formal part of the school. In addition, a heavily attended poster session gave every participant the opportunity to discuss her or his recent work.

The content of this book has two parts. Part One is the lectures; those by Fröhlich, Isham, De Vega and Seiberg are considerably expanded from the respective oral versions given at the school. Part Two contains some of the papers presented at the "Physics, Braids and Links" workshop. It is hoped that these will serve as a useful complement to Part One.

In the preparation of this book, invaluable editorial assistance was provided by Margaret Carey who, in addition to typing several manuscripts, formatted a good number of other pretyped manuscripts into the required camera ready form. Photographs of the lecturers were taken by Peter Leivo.

We are thankful to all of those who contributed to the success of the school: the lecturers, the question-asking students, the participants at the workshops, the authors and inspectors of the posters and, last but not least, the organizers and volunteers who helped in running the school.

The school was mainly funded by a NATO Advanced Study Institute Award and a Natural Science & Engineering Council of Canada Conference Grant. Additional funding came from AECL-Chalk River Laboratories, TRIUMF, Canadian Institute of Particle Physics, University of Winnipeg, and from the Canadian Association of Physicists in the form of a loan. These supports are gratefully acknowledged.

H.C. Lee
Chalk River

Contents

Part I

LECTURES - Banff NATO ASI
Physics, Geometry & Topology

Field Theory Methods and Strongly Correlated Electrons 1
Ian Affleck

Braid Statistics in Three-Dimensional Local Quantum Theory 15
Jurg Fröhlich

On the Algebraic Structure of the BRST Symmetry 81
Marc Henneaux

Black Hole Quantization and a Connection to String Theory 105
Gerard 't Hooft

An Introduction to General Topology and Quantum Topology 129
Chris Isham

Topics in Planar Physics 191
Roman Jackiw

Introduction to Conformal Field Theory and 241
Infinite Dimensional Algebras
David Olive

Lectures on RCFT 263
Nathan Seiberg

Chern-Simons Gauge Theory and the Spin-Statistics Connection 363
in Two Dimensional Quantum Mechanics
Gordon Semenoff

Yang-Baxter Algebras, Integrable Theories and Quantum Groups 387
Hector De Vega

Symmetry and Functional Integration 435
Claude-Michel Viallet

Topological Aspects of the Quantum Hall Effect 461
Yong-Shi Wu

PART II

SEMINARS - Workshop on Physics, Braids & Links

The Nonabelian Chern-Simons Term with Sources and Braid Source Statistics 493
Michelle Bourdeau

The Quantum Group Method of Quantising the Special Linear Group SL(2,C) 513
Nigel Burroughs

2+1 Dimensional Quantum Gravity and the Braid Group 541
Steven Carlip

Finite Renormalization of Chern-Simons Gauge Theory 553
Wei Chen

A New Family of N-State Representations of the Braid Group 573
Michel Couture

Link Polynomials and Solvable Models 583
Tetsuo Deguchi

Integrable Restrictions of Quantum Soliton Theory and Minimal Conformal Series 605
André LeClair

Tangles, Links and Twisted Quantum Groups 623
H.C. Lee

Participants *(Group photo, pg 656)* 659

Index 663

PART I

LECTURES
on
PHYSICS, GEOMETRY & TOPOLOGY

Field Theory Methods and Strongly Correlated Electrons

Ian Affleck†

Canadian Institute For Advanced Research
and
Physics Department
University of British Columbia
Vancouver, B.C., V6T2A6, Canada

Abstract

The one-dimensional Hubbard Model at arbitrary filling is reviewed using non-abelian bosonization and emphasizing the hidden O(4) symmetry at half-filling. This provides a one-dimensional warm-up for theories of high-T_c superconductivity involving "holons" and "spinons".

I. Introduction

There is currently enormous interest in the two dimensional Hubbard model and its generalizations, as theories for high-T_c superconductivity. Many open questions and controversies remain at this time. Many of the ideas and assumptions made about the two dimensional case are inspired by the one dimensional version which is much better understood. In particular, bosonization leads to a remarkable separation of the charge and spin excitations which have recently become known as "holons" and "spinons". It seems likely that many of these features are special to the one-dimensional case. Nonetheless, it is worhwhile to understand this case as well as possible, both as a warm-up for the two-dimensional problem and because of the application to quasi-one-dimensional systems. A very nice review of the 1D theory was given by V. Emery in 1979.[1] Discussion of its experimental relevance may be found in other articles in the same conference proceedings.[1]

A certain amount of progress has been made in bosonization techniques since that time thanks to non-abelian bosonization and the general development of conformal field theory. The aim of these notes is to re-examine the problem from a more modern perspective. A much more extensive discussion of the related problem of one-dimensional Heisenberg models was given in [2], where the Hubbard model was also discussed in passing. In those notes only the case of half-filling was covered, wherein spinons but not holons, occur. In the present notes I will

---

† Presented at the Nato Advanced Study Institute on *Physics, Geometry and Topology*, Banff, August 1989. Research supported by NSERC of Canada.

Physics, Geometry, and Topology
Edited by H. C. Lee
Plenum Press, New York, 1990

endeavour to avoid repeating the contents of [2] as much as possible and focus on the case not covered: spin-charge separation away from half-filling. I will also emphasize the hidden SO(4) symmetry of the Hubbard Model at half-filling and use it in bosonization. The reader is referred to [2] for original references and to other articles in the same volume for reviews of conformal field theory.

2. The Hubbard Model: Strong Coupling Limit

The Hubbard Hamiltonian is:

$$H_H = -t\Sigma_{<i,j>}(\psi^{\dagger\alpha}{}_i\psi_{\alpha,j} + h.c.) + U\Sigma_i n_{i\uparrow}n_{i\downarrow}.$$

Here $\psi_{i\alpha}$ annihilates an electron on lattice site i with spin α, $n_{i\downarrow}$ and $n_{i\uparrow}$ are the electron numbers $\psi^{\dagger\alpha}{}_i\psi_{\alpha i}$ where α=1 or 2 (not summed). A sum over nearest neighbours (each pair occuring *once*) on some lattice is implied in the first term. This is a standard model for electrons in well-localized atomic orbitals with a probablility t for transitions between neighbouring atoms. The constant U, which is normally positive, represents a highly screened Coulomb repulsion between electrons. It costs energy U to put 2 electrons on the same site. This very simplified form assumes that the screening length is on the order of the lattice spacing. In the extended Hubbard model a nearest neighour repulsion is also maintained. The on-site interaction can be written in a number of equivalent ways, including:

$$H' = (U/2)\Sigma_i n_i^2 \text{ or, } (-2U/3)\Sigma_i S_i^2.$$

Here n_i is the total electron number:

$$n_i = \psi^{\dagger\alpha}{}_i\psi_{\alpha i} \text{ (summed over } \alpha),$$

and S_i is the electron spin operator:

$$S_i = \psi^{\dagger\alpha}{}_i(1/2)\sigma_\alpha{}^\beta\psi_{\beta i}.$$

The various ways of writing H' differ by constants plus terms proportional to $\Sigma_i n_i$, the total electron number. This is a conserved quantity so such a term just shifts the chemical potential.

The strong-coupling limit of the Hubbard Model at half-filling (the total number of electrons equals the number of sites) is of special interest. Setting t=0, we see that any of the 2^V states (V is the volume, or the number of sites) with one particle per site is a ground state. Of course, this large degeneracy is broken by the hopping term. This problem may be studied in degenerate perturbation theory. A single application of the hopping term always produces a doubly occupied site and so increases the energy by U. Note that the system is an electric insulator in the strong-U limit. Transport of an electron by L sites only occurs in L^{th} order perturbation theory and is suppressed by a factor of $(t/U)^L$. Wave-functions are exponentially localized. This insulating behaviour at half-filling could not occur without interactions. Trivial insulators always have filled bands. This type of non-trivial insulator, known as a Mott-Hubbard insulator is often observed. It occurs, for example, in compounds closely related to high-T_c superconductors. Mixing of the ground states occurs in second-order degenerate perturbation theory in t. Let us consider the mixing induced by the hopping term on a particular link:

$$<\downarrow\uparrow|H_{int}(E_0-H_0)^{-1}H_{int}|\downarrow\uparrow> = -2t^2/U,$$
$$<\downarrow\uparrow|H_{int}(E_0-H_0)^{-1}H_{int}|\uparrow\downarrow> = 2t^2/U.$$

(The factor of 2 arises because there are two possible intermediate states with both electrons on either of the 2 sites. The minus sign occurs from anti-commuting the creation operators.) The matrix element is zero for parallel spins, since the 2 electrons cannot go onto the same site in the intermediate state. We may write down an effective Hamiltonian in the space of singly occupied states which has the same matrix elements.

$$H_{eff} = J\Sigma_{<i,j>}[S_i \cdot S_j - 1/4],$$

where,

$$J = 4t^2/U.$$

Thus Mott-Hubbard insulators are expected to be antiferromagnets. This is also frequently observed; in fact this may be the main mechanism responsible for antiferromagnetism. The magnetic interaction results from an exchange process in which neighbouring electrons share each others orbitals. This occurs more efficiently for anti-parallel spins due to the Coulomb repulsion and Fermi statistics.

The one-dimensional Heisenberg model is not trivial. In particular the naive antiferromagnetically ordered state with anti-parallel neighbouring spins (which occurs in higher dimension) is prevented from occuring by Coleman's theorem. It turns out that a convenient way of understanding the basic physics of the Hubbard model is to study weak coupling where the continuum limit can be taken. One can then investigate whether the behaviour is smooth right up to infinite U.

The Hubbard model has an obvious SU(2)xU(1) symmetry. There is actually a hidden SO(4) symmetry, or two commuting SU(2)'s, at half-filling which will be quite useful when we bosonize. This hidden symmetry can be seen by introducing real fermions:

$$\psi_1 = (\xi_1+i\xi_2)/2,\ \psi_2 = (\xi_3+i\xi_4)/2, \text{ on even sites, and}$$

$$\psi_1 = (\xi_2-i\xi_1)/2,\ \psi_2 = (\xi_4-i\xi_3)/2, \text{ on odd sites}$$

obeying:

$$\{\xi_{ar},\xi_{br'}\} = 2\delta_{r,r'}\delta_{ab}.$$

The hopping term takes the manifestly SO(4) invariant form:

$$H_0 = (-it/2)\Sigma_{<i,j>}\xi_{ai}\xi_{aj} \quad \text{(for i odd and j even).}$$

Noting that:

$$\psi^{\dagger\alpha}{}_i\psi_{\alpha i}-1 = (i/2)(\xi_1\xi_2+\xi_3\xi_4),$$

we see that the Hubbard interaction:

$$H' = (U/2)\Sigma_i(\psi^{\dagger\alpha}{}_i\psi_{\alpha i}-1)^2 = (-U/96)\Sigma_i\epsilon^{abcd}\xi_{ai}\,\xi_{bi}\,\xi_{ci}\,\xi_{di},$$

is also SO(4) invariant.

Alternatively, we may construct a matrix from the complex fermions:

$$\Psi_{\hat{\alpha}\beta} \equiv \begin{bmatrix} \psi_1 & \psi_2 \\ \psi^{\dagger 2} & -\psi^{\dagger 1} \end{bmatrix} \qquad (1a)$$

on even sites and:

$$\Psi_{\hat{\alpha}\beta} \equiv \begin{bmatrix} i\psi_1 & i\psi_2 \\ -i\psi^{\dagger 2} & i\psi^{\dagger 1} \end{bmatrix} \qquad (1b)$$

on odd sites.

An ordinary SU(2) transformation corresponds to right multiplication of Ψ: $\Psi \to \Psi U$, while the hidden SU(2) transformation is left multiplication. [The ordinary U(1) charge symmetry is a sub-group of the hidden SU(2)]. To see that these are indeed both symmetries of H note that the hopping term can be written:

$$H_0 = -it\Sigma_{<i,j>}\mathrm{tr}\Psi_i^\dagger\Psi_j \text{ (for i odd and j even)},$$

while the Hubbard interaction is proportional to $\Sigma_i \mathrm{Det}\Psi_i^\dagger\Psi_i$.

The number density and pair creation operators form a triplet under the hidden SU(2):

$$\hat{S}_j = (1/2)[(-1)^j(\psi_{j2}\psi_{j1}+\text{h.c.}), -i(-1)^j(\psi_{j1}\psi_{j2}-\text{h.c.}), \psi^{\dagger\alpha}{}_j\psi_{\alpha j}-1] = \mathrm{tr}\Psi_i^\dagger\sigma\Psi_i/4.$$

On the other hand the ordinary spin operators are:

$$S_j = \mathrm{tr}\Psi^\dagger{}_j\Psi_j\sigma/4.$$

Note that a chemical potential does not respect the hidden SU(2) since the number density forms part of a triplet. The O(4) invariant filling factor is seen to be half-filling since this corresponds to:

$$<\hat{S}^3>=0.$$

Note that a chemical potential term:

$$\mu\, \Sigma_i \psi^\dagger{}_i{}^{1\hat{\alpha}}\psi_{i\hat{\alpha}1},$$

breaks the SO(4) symmetry down to the obvious SU(2)xU(1) subgroup. Thus the hidden SU(2) symmetry is broken by moving away from half-filling.

The tranformation:

$$\psi_{i2} \rightarrow (-1)^i\psi^\dagger{}_i{}^2,$$

leaves the hopping term invariant and changes the sign of U. Thus the positive and negative U models are equivalent, with a redefinition of operators. This redefinition switches the charge and spin operators.

$$n_i - 1 \rightarrow \psi^\dagger{}_i\sigma^z\psi_i = 2S^z{}_i,$$

$$(-1)^i\psi_{j2}\psi_{j1} \rightarrow \psi^{\dagger 2}{}_j\psi_{j1} = \psi_j{}^\dagger\sigma^-\psi_j = 2S^-{}_j$$

ie. $\hat{S}_j \leftrightarrow S_j$.

The triplets of spin and charge operators are interchanged under the duality transformation.

3. Continuum Limit

The continuum limit is based on weak coupling. We begin by considering the model at U=0.

$$H = -t\Sigma_i[\psi_{i+1}{}^{\dagger\alpha}\psi_{i\alpha} + \text{h.c.}].$$

We may immediately find the ground state by Fourier transforming:

$$H \rightarrow -2t\Sigma_k \cos ak\, \psi^{\dagger\alpha}{}_k\psi_{\alpha k}.$$

The dispersion relation is drawn below. We find the groundstate by filling the Fermi sea up to the Fermi points $\pm k_F$. k_F is determined by fixing the total charge, or filling factor, k_F/π.

The low-energy excitations involve creating holes just below the Fermi surface and electrons just above it. Thus we are only concerned with the ψ_k operators for $k \approx \pm k_F$. Essentially we may truncate the Fourier expansion and only keep k in the range

$$|k \pm k_F| \leq \Lambda,$$

where Λ is an ultraviolet cut-off which can be taken to be $<<2\pi/a$, where a is the lattice spacing. This will be sufficient to study the physics at length scales >> $1/\Lambda$. Thus we write:

$$\psi(x)/\sqrt{a} \approx e^{ik_Fx}\psi_L(x) + e^{-ik_Fx}\psi_R(x),$$

where ψ_R and ψ_L are slowly varying on the lattice scale and contain the Fourier modes near $\pm\pi/2a$ respectively. It is convenient to define continuum Fourier modes:

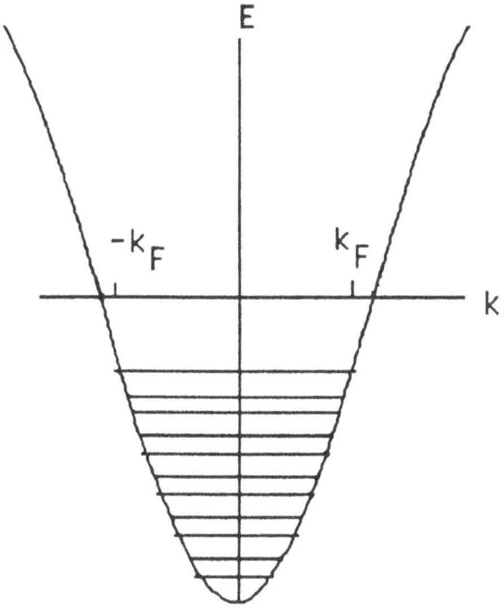

$a(k) = \psi(k_F+k)$, $b^\dagger(k) = \psi(k_F-k)$, for $k > 0$,

where a and b are electron and hole (or positron, in the relativistic limit) annihilation operators. Thus ψ_R annihilate right-moving electrons and creates right-moving holes:

$$\psi_R = (1/\sqrt{a})\Sigma_{k>0}[e^{-ikx}a_k + e^{ikx}b^\dagger_k].$$

Likewise, ψ_L is written in terms of the left-moving electrons and holes:

$$\psi_L = (1/\sqrt{a})\Sigma_{k<0}[e^{-ikx}a_k + e^{ikx}b^\dagger_k].$$

In the continuum approximation we linearize the dispersion relation near the Fermi surface. This leads to the continuum Hamiltonian

$$H = iv\int dx[\psi^{\dagger\alpha}{}_R(d/dx)\psi_{R\alpha} - \psi^\dagger{}_L{}^\alpha(d/dx)\psi_{L\alpha}],$$

where $v=2ta\sin k_Fa$, the Fermi velocity.

In what follows, we generally follow the convention of setting $v=1$; it can always be restored by dimensional analysis. We have obtained a Lorentz-invariant massless free Dirac fermion theory with 2 "flavours", the spin components. (Note that the electron spin appears as an internal quantum number in the (1+1) dimensional field theory; there is no intrinsic spin in one dimension.) The continuum free fermion theory has chiral U(1) and SU(2) symmetries. Operating on the left fermions we have a U(1) charge symmetry:

$$\psi_{Li\alpha} \rightarrow e^{i\theta}\psi_{Li\alpha},$$

and an SU(2) spin symmetry:

$$\psi_{Li\alpha} \rightarrow g_\alpha{}^\beta\psi_{Li\beta}.$$

Here, $g^\alpha{}_\beta$ is an SU(2) matrix. Likewise we may make *independent* transformations on the ψ_R's. The charge and spin of the left and right-moving fermions are separately conserved. Corresponding to these symmetries we have conserved currents. The light-cone components of the currents can be written:

$$J_{L,R} \equiv :\psi^\dagger{}_{L,R}{}^{i\alpha}\psi_{L,Ri\alpha}:,$$

$$\mathbf{J}_{L,R} \equiv \psi^\dagger{}_{L,R}{}^{i\alpha}\,(1/2)\sigma_\alpha{}^\beta\psi_{L,Ri\beta}.$$

(Note that the bold-face symbol **J**, denotes an SU(2) vector of currents.) The energy-momentum tensor can be written in a Sugawara form quadratic in currents:

$$T_L = (\pi/2)vJ_LJ_L + (2\pi/3)v\mathbf{J}_L\cdot\mathbf{J}_L$$

(and similarly for T_L). Here we have re-instated the velocity of light, v.

We now consider the interaction term. For small U (ie. U<<t), only states close to the Fermi surface are mixed with the groundstate (and low energy states) so we may use the continuum approximation. Thus we rewrite the interaction in terms of $\psi_{L,R}$. This gives a variety of different terms. Because of the assumption that $\psi_{L,R}$ vary slowly, terms with phase factors of the form $e^{\pm 2ik_F x}$ can be ignored. (More correctly they lead to higher derivitive operators which are irrelevant.) This will generally also be true of terms with phase factors $e^{\pm 4ik_F x}$, except in the special case $k_F=\pi/2a$, corresponding to half-filling, where these phases become zero since x=na. Upon rearranging the order of the operators (which only generates constants and terms proportional to the total conserved fermion number) we get five different non-derivitive 4-Fermi operators. These can be written:

$$(J_L J_L + J_R J_R),\ (\mathbf{J}_L \cdot \mathbf{J}_L + \mathbf{J}_R \cdot \mathbf{J}_R)$$

$$J_L J_R,\ \mathbf{J}_L \cdot \mathbf{J}_R$$

$$[e^{-4ik_F x}(\epsilon_{\alpha\beta}\psi^{\dagger}{}_L{}^{\alpha}\psi^{\dagger}{}_L{}^{\beta})(\psi_{R\gamma}\psi_{R\delta}\epsilon^{\gamma\delta}) + \text{h.c.}]$$

We see from the definition of the continuum fermi fields, that the single charge and spin symmetries of the Hubbard model correspond to the diagonal subgroup of the chiral symmetries of the free fermion theory, under which left and right fermions transform the same way. These five interaction terms are the most general operators respecting the diagonal U(1) and SU(2) symmetries of the Hubbard model and containing even numbers of left and right fields. Thus if we consider more general interactions in the lattice model we can only change the five continuum coupling constants; no other operators will appear. The first two terms are of the form of the free Hamiltonian and can be regarded as simply renormalizing the velocity. Since the coefficients of these two terms are not the same, we get two different velocities for the charge and spin terms. A theory with two different "velocities of light" could not be Lorentz invariant. Fortunately, as we shall see, the charge and spin sectors decouple and can be considered separately so we obtain two decoupled Lorentz invariant theories with different velocities of light.

The left-right terms correspond to Lorentz-invariant interaction terms in the Lagrangian (the same terms as in the Hamiltonian, but with a change of sign). This can be written:

$$L_{\text{int}} = \lambda_1(1/4)J_L J_R + \lambda_2(1/16)[(\epsilon_{\alpha\beta}\psi^{\dagger}{}_L{}^{\alpha}\psi^{\dagger}{}_L{}^{\beta})(\psi_{R\gamma}\psi_{R\delta}\epsilon^{\gamma\delta}) + \text{h.c.}] + \lambda_3 \mathbf{J}_L \cdot \mathbf{J}_R$$

These coupling constants have the values

$$-\lambda_1 = -\lambda_2 = \lambda_3 = U/t.$$

The λ_2 term is only present at half-filling where the phase vanishes. We have obtained a generalized, SU(2) invariant Thirring model.

The hidden SO(4) symmetry at half filling is present in the continuum theory due to the relations between the coupling constants. We make the hidden symmetry manifest by introducing left and right matrices Ψ_L and Ψ_R as was done for the lattice model in Eq. (1). The ordinary SU(2) currents become:

$$\mathbf{J}_L = (1/2)\text{tr}\Psi^{\dagger}{}_L\Psi_L\boldsymbol{\sigma}$$

while the hidden SU(2) current is:

$$\hat{\mathbf{J}}_L = (1/2)\text{tr}\Psi^{\dagger}{}_L\boldsymbol{\sigma}\Psi_L.$$

$\hat{J}_L{}^3$ is proportional to the charge current J_L. The two SO(4) and Lorentz invariant interactions are $\mathbf{J}_L \cdot \mathbf{J}_R$ and $\hat{\mathbf{J}}_L \cdot \hat{\mathbf{J}}_R$. This second interaction is proportional to

$$(1/4)J_L J_R - (1/16)[(\epsilon_{\alpha\beta}\psi^{\dagger}{}_L{}^{\alpha}\psi^{\dagger}{}_L{}^{\beta})(\psi_{R\gamma}\psi_{R\delta}\epsilon^{\gamma\delta}) + \text{h.c.}].$$

Thus SO(4) symmetry demands that $\lambda_1 = -\lambda_2$ as was obtained above. The duality symmetry of the Hubbard model also implies that λ_1 and λ_3 switch under $U \to -U$, as well as implying that the velocities in the spin and charge sector switch under $U \to -U$. This latter duality is made more manifest by observing that the charge part of T_L can be rewritten using standard point-splitting techniques as:

$$T_{Lc} = (\pi/2)vJ_L J_L = (2\pi/3)v\hat{J}_L \cdot \hat{J}_L$$

Note that the symmetries of the underlying lattice model imply a symmetry-preserving regularization of the field theory. Thus the symmetries and duality properties should be exactly true in the regulated interacting quantum field theory. Of course, once we move away from half-filling $\lambda_2 = 0$ due to the oscillating factor arising from the lattice theory and the hidden SU(2) symmetry is broken. We shall concentrate on the half-filled case for a while and then return to the general case in Section 4.

At this point we bosonize. The most elegant way of doing this is to use Witten's non-abelian bosonization. He showed that a theory of two free massless complex Dirac fermions (ie. four real fermions) is eqivalent to the SO(4) Wess-Zumino-Witten non-linear σ-model (WZW model) with central charge k=1. In terms of real fermions, ξ_a (a=1,...4), the bosonization formula is:

$$\xi_{La}\xi_{Rb} \propto O_{ab},$$

where O is an SO(4) matrix. Similarly, the left and right SO(4) currents of the fermion theory are equivalent to the currents in the WZW model. An equivalent and more convenient version of this bosonization formula is obtained for the complex fermions by rewriting the SO(4) matrix as a product of 2 SU(2) matrices $g^{\alpha}{}_{\beta}$ and $h^{\alpha}{}_{\beta}$ in the obvious and hidden SU(2) groups respectively:

$$\psi^{\dagger}{}_{L}{}^{\alpha}\psi_{R\beta} \propto g^{\alpha}{}_{\beta}h^{1}{}_{1} \qquad (2a)$$

$$\psi_{L\alpha}\psi_{R\beta} \propto \epsilon_{\alpha\gamma}g^{\gamma}{}_{\beta}h^{1}{}_{2}. \qquad (2b)$$

The ordinary and hidden SU(2) currents are written in terms of g or h only:

$$J_L = -(i/4\pi)\mathrm{tr}\partial_+ gg\tau\sigma, \quad \hat{J}_L = -(i/4\pi)\mathrm{tr}\partial_+ hh\tau\sigma.$$

The SU(2) matrices g and h are WZW fields with central charge k=1. We see that the continuum version of the Hubbard model decouples into separate sectors: g and h for spin and charge excitations. This is a consequence of bosonization and the group-theoretical equivalence:

$$SO(4) \Leftrightarrow SU(2)xSU(2).$$

It should be realized that this decoupling is not an exact property of the lattice model. The two sectors are coupled by higher dimension operators. Thus we should only expect it to appear asymptotically in the long-distance, low-energy properties of the Hubbard model.

Having separated charge and spin sectors, we are still faced with the problem of understanding the behaviour of each sector. We can understand the essential behaviour from the renormalisation group. The interaction in each sector is marginal (has dimension 2); it will be marginally relevant or irrelevant depending on the sign of the coupling. We see from the Appendix that, for U>0 the spin sector coupling, λ_3 is marginally irrelevant while the charge sector couplings, $\lambda_1 = \lambda_2$ is marginally relevant. Thus we expect to develop a gap in the charge sector of order:

$$m \propto \Lambda e^{-2\pi/\lambda_1},$$

where Λ is the ultraviolet cut-off. Inserting the original units this corresponds to an energy gap in the charge sector at weak coupling and half-filling of:

$$\Delta \propto te^{-4\pi t/U}.$$

Note that this exponentially small gap becomes O(U) in the strong coupling limit.

It is plausible that the gap increases monotonically with U (for fixed t). It is also plausible that the WZW field h develops an expectation value of the form:

$$<h^{\alpha}{}_{\beta}> \propto \delta^{\alpha}{}_{\beta}.$$

This is the unique form for this expectation value which is consistent with the hidden SU(2) symmetry of the model. This symmetry cannot be spontaneously broken due to Coleman's theorem. The bosonized model has an additional Z_2 symmetry $h \rightarrow -h$, which is a subgroup of the chiral (hidden) SU(2) symmetry of the WZW model. The above expectation value implies that this Z_2 is spontaneously broken. Note that this is not inconsistent with Coleman's theorem since it is only a discrete symmetry which is being broken. We will give an (essentially rigorous) argument below that this symmetry breaking occurs. The connected parts of h correlation functions should decay exponentially with a correlation length of $O(m^{-1})$. ie:

$$<h^{\dagger\alpha}{}_{\beta}(x)h^{\gamma}{}_{\delta}(0)> \rightarrow \text{const.}\cdot\delta^{\alpha}{}_{\delta}\delta^{\gamma}{}_{\beta} + O(e^{-mr}),$$

$$<\tilde{J}^{a}(x)\tilde{J}^{b}(0)> \rightarrow O(e^{-mr}), \text{ etc.}$$

On the other hand the spin coupling, λ_3 renormalizes to zero at long wavelengths. Thus the low energy correlations in the spin theory should be described by the WZW fixed point (up to logarithmic corrections coming from the marginally irrelevant operator). This implies that:

$$<g^{\dagger\alpha}{}_{\beta}(x)g^{\gamma}{}_{\delta}(0)> \rightarrow \text{constant}/r$$

(actually this is multiplicatively corrected by a power of a logarithm due to the marginal operator),

$$<J^{a}{}_{L}(x_{-})J^{b}{}_{L}(0)> \rightarrow -1/8\pi^2 x_{-}{}^2, \text{ etc.}$$

We may now read off the asymptotic behaviour of the Hubbard model correlation functions for the charge and spin operators introduced above, by using their bosonized forms. Thus, for any lattice operator bilinear in fermions, we first rewrite it in terms of $\psi_{L,R}$ and then bosonize. This gives, in general, two terms with phases 1, or $e^{\pm i(\pi/2a)x} = (-1)^{x/a}$.

$$S_x/a = (J_L + J_R) + \text{const}\cdot(-1)^{x/a}\,(\mathrm{tr}h)\cdot(\mathrm{tr}g\sigma)$$

$$\tilde{S}_x/a = (\tilde{J}_L + \tilde{J}_R) + \text{const}\cdot(-1)^{x/a}(\mathrm{tr}h\sigma)\cdot(\mathrm{tr}g)$$

Now using our assumption that $<\mathrm{tr}h> \neq 0$, whereas $<\mathrm{tr}h\sigma> = 0$ by the unbroken hidden SU(2) symmetry, we see that the $\tilde{S}$ correlation function (charge density and pair creation operator) decays exponentially, whereas the S (ordinary spin) correlation function decays with power laws determined by the dimensions of J and g:

$$<S^{a}(0,0)S^{b}(x,t)> = a^2\delta^{ab}[-(1/8\pi^2)(1/x_{+}{}^2+1/x_{-}{}^2) + \text{const}\cdot(-1)^{x/a}(x^2-t^2)^{-1/2}].$$

(The coefficent in from of the non-alternating term is actually determined completely by the condition that the conserved total spin obey the same commutation relations in the lattice and continuum theories. On the other hand the coefficient in front of the alternating term is non-universal and is not determined by any general argument. Finally, it should be remembered that we have set v=1 in this formula.)

Of course, these results apply equally well to the U<0 theory at half-filling, due to the duality transformation. We simply interchange the $\tilde{S}$ and S operators so that the ordinary spin correlation function decays exponentially whereas the charge density and pair creation operator correlation functions decay with these power laws. This can again be seen to be consistent with a physical picture of the large (negative) U limit.

4. Away From Half-Filling

Our continuum analysis goes through essentially the same way at general k_F except for the crucial difference that $\lambda_2=0$, ie the term: $(\varepsilon_{\alpha\beta}\psi^{\dagger}{}_L{}^{\alpha}\psi^{\dagger}{}_L{}^{\beta})(\psi_{R\gamma}\psi_{R\delta}\varepsilon^{\gamma\delta})$ no longer appears in the continuum Lagrangian due to the alternating phase factor. We see from the renormalization group equations in the Appendix that λ_1 does not renormalize so that adjusting U simply moves λ_1 along the fixed critical line. Thus we we expect to get gapless spin *and* charge excitations, in agreement with the large U limit.

The most convenient way of handling the charge sector once the hidden SU(2) is broken down to the obvious U(1) subgroup, is to replace the WZW field h by an eqivalent free boson, φ. We adopt standard string theory normalization so that the Lagrangian is written:

$$L = (1/2\pi)(\partial_\mu\varphi)^2.$$

The boson is defined as an angular variable, that is φ and $\varphi+2\pi R$ are regarded as equivalent. It is well known that the k=1 SU(2) WZW model is equivalent to a free boson with radius $R=1/\sqrt{2}$. The currents and the field h can be rewritten in terms of φ:

$$\hat{J}_L{}^3 = (1/\pi\sqrt{8})\partial_+\varphi,\ \hat{J}_L{}^{\pm} \equiv \hat{J}_L{}^1 \pm i\hat{J}_L{}^2 \propto e^{\pm i\sqrt{8}\,\varphi_L},$$

$$g \propto \begin{bmatrix} e^{i\sqrt{2}\,\varphi} & e^{i\sqrt{2}\,\tilde{\varphi}} \\ e^{-i\sqrt{2}\,\tilde{\varphi}} & e^{-i\sqrt{2}\,\varphi} \end{bmatrix}$$

where $\varphi = \varphi_L + \varphi_R$, $\tilde{\varphi} = \varphi_L - \varphi_R$.

The charge part of the Lagrangian becomes:

$$L = (1/2\pi)(1-\lambda_1/8\pi)(\partial_\mu\varphi)^2 + \text{const}\cdot\lambda_2\cdot\cos\sqrt{8}\,\varphi$$

(We temporarily include λ_2 in order to reanalyse the half-filling case in terms of φ.) We may rescale φ by $\sqrt{1-\lambda/8\pi}$ in order to give the kinetic energy the canonical normalization. This means the radius changes to

$$R = [(1-\lambda_1/8\pi)/2]^{1/2}.$$

The cosine interaction is now $\cos 2\varphi/(1-\lambda_1/8\pi)^{1/2}$. This has scaling dimension:

$$x = 2/(1-\lambda_1/8\pi).$$

Since $\lambda_1 = -U/t$ in the Hubbard model, we see that this cosine interaction is relevant for U>0. In fact we have just obtained the sine-Gordon model [with $\beta^2=8\pi/(1-\lambda_1/8\pi)$ using a conventionally (ie. non string theory) normalized boson field]. We reobtain the previous conclusion that the model develops a mass gap. We in fact learn more using the spectrum of the sine-Gordon model which is based on the semi-classical analysis and known to be correct in the quantum theory. The symmetry of translation $\varphi \rightarrow \varphi + \pi/R$ is spontaneously broken. Note that due to the periodic *definition* of φ this is only a Z_2 symmetry. The spectrum for small λ_1 consists of a massive soliton doublet. This Z_2 symmetry corresponds to $h \rightarrow -h$ in the WZW representation and the known results on the sine-Gordon model substantiate the claim made earlier that this Z_2 symmetry is spontaneously broken.

We may now easily analyze the model away from half-filling. Since the sine-Gordon interaction is now absent, we simply obtain a free boson critical theory in the charge sector, but with a U dependent radius. All fermion bilinear correlation functions may be read off by rescaling φ. Requiring preservation of the canonical commutation relations implies:

$$\varphi \to (2/R)^{1/2}\varphi,\ \tilde{\varphi} \to (R/2)^{1/2}\tilde{\varphi}.$$

Thus:

$$S/a = (J_L + J_R) + \mathrm{const}[e^{ik_F x} \cdot e^{i\varphi/R} + \mathrm{h.c.}]\mathrm{tr}g\sigma$$

$$n/a = (1/\sqrt{R})\partial_1\varphi + i\cdot\mathrm{const}[e^{ik_F x} \cdot e^{i\varphi/R} - \mathrm{h.c.}]\mathrm{tr}g$$

$$\varepsilon^{\alpha\beta}\psi_\alpha\psi_\beta \propto \exp(i2R\tilde{\varphi})\{[e^{2ik_F x}e^{-i\varphi/R} + \mathrm{h.c.}] + \mathrm{const}\cdot\mathrm{tr}g\}$$

We can now read off the scaling dimensions of these operators (half the correlation exponent). Note that in each case the first term contains purely spin or charge operators while the other contains a product of both.

	unform	2kF
spin & charge	1	$1/2+1/4R^2$
pair	$R^2+1/4R^2$	$1/2+R^2$

Note that all dimensions are 1 for U=0, $R=1/\sqrt{2}$, the free fermion case. With increasing U and R the pair dimensions get larger while the $2k_F$ spin and charge dimensions get smaller. Standard arguments[3] which treat the 1D model exactly and the 3D couplings in mean field theory suggest that long-range order will occur with the addition of weak 3D inter-chain couplings only if the dimension is less than 2. This suggests that the quasi-1D positive U Hubbard model may show spin-density or charge-density wave long-range-order but not superconductivity. Although the quasi-1D case is not directly relevant, this represents some difficulty for proponents of Hubbard model-based theories of high-T_c superconductivity.

To complete our renormalization group analysis of the Hubbard model and to understand what may happen if we add more general lattice interactions (nearest neighbour repulsion, for example) it is essential to examine all possible relevant operators which may be added to the critical theory. In general, any interaction not forbidden by symmetry should be generated (even if it does not appear in a tree level analysis such as performed above). Let us begin with SO(4) invariant Hamiltonians. (Generalizations of the Hubbard model which are SO(4) invariant at half filling can be easily be written down.) The only non-trivial primary field in the k=1 SU(2) WZW model is g (or h) itself. The only Lorentz invariant and diagonal SO(4) invariant operators which could be added to the Lagrangian are the current-current terms already considered and trg, trh, trg·trh. Even allowing for non-Lorentz invariant operators which respect the limited spatial symmetries of the lattice model (eg. parity) permits no further relevant operators besides these two(except for renormalization of the velocities already considered). Fortunately, there are discrete symmetries of the lattice theory which forbid these operators. Consider the symmetry of translation by 1 site. In the continuum fermion formulation this is an *internal* discrete symmetry:

$$\psi_L \to i\psi_L,\ \psi_R \to -i\psi_R.$$

This is actually a Z_2 subgroup of the chiral U(1) symmetry of the free fermion continuum theory. From the bosonization formulas of Eq. (2) we see that it corresponds to:

$$g \to g,\ h \to -\sigma^3 h\sigma^3.$$

This forbids the interactions trh or trh·trg. There is a second symmetry of the form:

$$g \to -g \text{ and } h \to -h,$$

which is considerably more subtle in the fermionic theory. Note that *all* fermion bilinears are invariant under this transformation. Consequently if we begin with an arbitrary fermionic Lagrangian (including all irrelevant local fermion operators) its bosonized form will respect this symmetry. Since we only generate local fermionic operators in perturbation theory in the lattice fermionic theory, it follows that the symmetry holds to all orders in perturbation theory. This additional symmetry forbids the interaction term trg.

A related fact is that, even though the first symmetry is spontaneously broken in the Hubbard model at half-filling, as we discussed above, there is no non-zero order parameter *local in fermion operators*. Local order parameters in the Hubbard model can only be non-zero if both symmetries are broken. Spin-Peierls, dimer order occurs if:

$$(-1)^i S_i \cdot S_{i+1} \propto \text{trg} \cdot \text{trh}$$

is non-zero. (This equation is obtained by using the operator product expansion to eliminate current operators.) This requires both trg and trh to have expectation values, breaking *both* discrete symmetries. Thus although, in some formal sense, there is a broken discrete symmetry in the half-filled Hubbard model, there is no corresponding order parameter local in fermion operators. (This is perhaps somewhat analogous to a situation in the 2D Ising model. One can define a dual order parameter which is non-zero in the *high-T phase*. However, this order parameter is non-local in the original spin variables; all local order parameters are zero. Whether or not one should consider the high-T phase to have a broken symmetry is largely a matter of definition.) An alternative point of view might be that the bosonic theory is only equivalent to the fermionic one if we make the identification:

$$(g,h) \Leftrightarrow (-g,-h).$$

The same discrete symmetries persist away from half-filling with the SO(4) symmetry broken. $h \to -h$ becomes $\varphi \to \pi R \varphi$, as discussed above. We also need to worry about possibly generating $\text{trh}\sigma^3$ or $\text{trh}\sigma^3\text{trh}$, since they are permitted by the remaining U(1) subgroup of the hidden SU(2). However, these ioperators are forbidden by the discrete symmetries.

We now are in a position to describe the general phase diagram for generalized Hubbard models with arbitrary interactions respecting the same symmetries as the ordinary Hubbard model. There are two possible phases for the spin sector, depending on the sign of the single marginal operator one with a gap and one without. There are three phases in the charge sector. There is a gapless phase with no broken symmetries and two phases with gaps in which either trh or $\text{trh}\sigma^3$ is non-zero. These order parameters correspond to $\cos\varphi/R$ and $\sin\varphi/R$. Which is non-zero depends on the sign of λ_2: ie. the sign of the $\cos 2\varphi/R$ interaction. Altogether we have six phases. However there are only only non-zero local fermionic order parameters in the two phases where there are both charge *and* spin gaps as discussed above. These phases have either spin-Peierls order or charge-density-wave order depending on whether trh or $\text{trh}\sigma^3$ has an expectation value. The latter follows from the fact that $\text{trg}\cdot\text{trh}\sigma^3$ is the bosonized form of the alternating part of the charge density. While only these six phases should occur at weak coupling, at sufficiently strong coupling in some models both trh and $\text{trh}\sigma^3$ might be simultaneously non-zero, corresponding to simultaneous Ising and dimer order. In sine-Gordan language this means that both $\cos\varphi/R$ and $\sin\varphi/R$ are non-zero. This could occur when R is sufficiently large that a $\cos 4\varphi/R$ term is relevant since the classical minimum can then occur at arbitrary φ.

Away from half-filling translation symmetry forbids the $\cos 2\varphi/R$ term and inhibits the development of a charge gap. Translation by one site now corresponds to:

$$\psi_L \to e^{ik_F}\psi_L,\ \psi_R \to e^{-ik_F}\psi_R,$$

or

$$g \to g,\ \varphi \to \varphi + 2k_F R.$$

This symmetry would only permit interaction of the type $\cos(n\pi\varphi/k_F R)$ for integer n. However, the periodic definition of φ implies that only $\cos n\varphi/R$ is single-valued. Hence cosine interactions can only occur for commensurate filling factors: $k_F/\pi = p/q$, p and q integer. For this filling factor the most relevant operator is $\cos q\varphi/R$ with dimension $q^2/4R^2$. This will only be relevant for $R \geq q/2\sqrt{2}$. Since $R=1/\sqrt{2}$ at weak coupling, only the q=2, half-filling instability occurs at weak coupling. The other ones correspond to other types of order with larger unit cells and should occur in some models at sufficiently strong coupling. (They probably don't occur in the ordinary Hubbard model.)

VI. Conclusions

Field theory methods, in particular bosonization, are very useful in understanding the behaviour of the one-dimensional Hubbard model. At half-filling the low energy spin-degrees of freedom are described by the k=1 SU(2) WZW model (with a marginally irrelevant interaction added) for any positive U; as has been well verifed by numerical work. Away from half-filling there are also gapless charge excitations. Field theory methods predict that two decoupled free boson theories, one for spin (equivalent to the WZW model) and one for charge, describe the asymptotic behaviour.

In the two dimensional case the existence of independent spin and charge excitations remains an open question.

Appendix: Renormalization Group Equations:

In a general 2D conformal field theory, the one-loop β-function can be obtained from the operator product expansion. For a set of marginal interactions, O_i with (Minkowski Space) Lagrangian:

$$L' = \Sigma_i \lambda_i O_i,$$

the β-function:

$$d\lambda_i/d\ln L = \beta_{ijk}\lambda_j\lambda_k,$$

with:

$$O_i(x)O_j(y) \to \beta_{ijk}O_k(x)/\pi(x-y)^2.$$ This follows from expanding e^{-S} to quadratic order in L', using the operator product expansion and the integral:

$$\int d^2x/x^2 \approx 2\pi \ln L,$$

where L is the infra-red cut-off. We consider an anisotropic interaction:

$$L' = \Sigma_a \lambda_a J_L^a J_R^a.$$

(Don't confuse these λ^a's with the three couplings defined in the Hubbard model.) In general the operator product expansion for a current and *any* operator (of arbitrary dimension) is:

$$J_R^a(z)O(z') = S_R^a O(z')/2\pi i(z-z'),$$

where S_R^a is the generator of $SU(2)_R$ transformations in the representation under which O transforms. (This follows from doing a contour integral over z which gives the commutator of the two operators.) Thus:

$$J_R^a(z)J_R^b(z') = \varepsilon^{abc}J_R^c/2\pi i(z-z'),$$

$$d\lambda_1/d\ln L = -\lambda_2\lambda_3/2\pi$$

$$d\lambda_2/d\ln L = -\lambda_1\lambda_3/2\pi$$

$$d\lambda_3/d\ln L = -\lambda_2\lambda_1/2\pi$$

Let us now specialize to a general U(1) invariant interaction:

$$L' = \lambda_{+-}(1/2)(J_L^+J_R^- + J_L^-J_R^+) + \lambda_3 J_L^z J_R^z:$$

$$d\lambda_{+-}/d\ln L = -\lambda_{+-}\lambda_3/2\pi$$

$$d\lambda_3/d\ln L = -\lambda_{+-}^2/2\pi.$$

These are very well-known RG equations which occur in a number of problems, including Kosterlitz's analysis of the 2D classical xy model. Noting that

$$d(\lambda_{+-}^2-\lambda_3^2)/d\ln L = 0,$$

we see that the RG flow lines are hyperbolas:

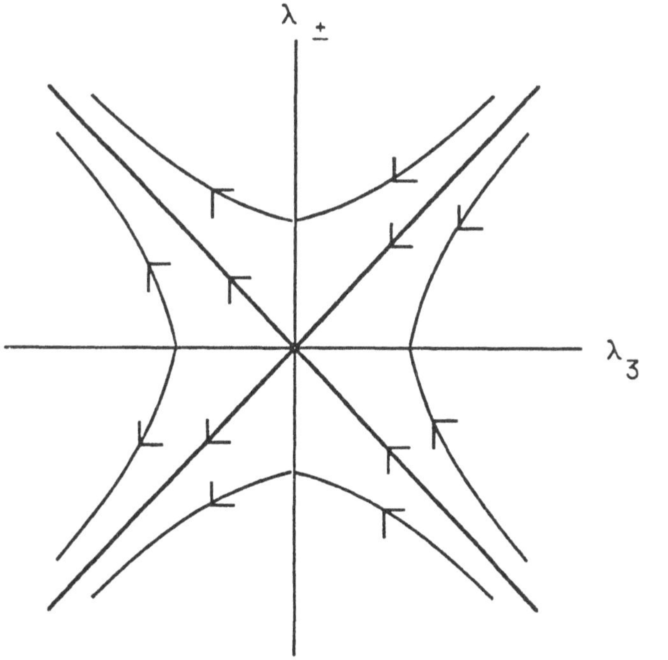

For $\lambda_3 \geq 0$ and $|\lambda_\pm| \leq \lambda_3$, the couplings flow to the fixed line on the positive λ_3 axis, but otherwise they flow to infinity. Finally, let us consider the SU(2) invariant case, $\lambda_{+-}=\lambda_3$. This condition determines the separatrix: the straight line RG flow to (and from) zero coupling. For λ initially positive, it renormalizes to zero.

References

1. V.J. Emery, Highly Conducting One-Dimensional Solids (ed. J.T. Devreese, R.P. Evrard and V.E. van Doren, Plenum Press, New York, 1979), p. 247.

2. I. Affleck, lectures notes, Les Houches Summer School Fields, Strings and Critical Phenomena, 1988.

3. D.J. Scalapino, M. Sears and R.A. Ferrell, Phys. Rev. B6, 3409 (1972).

BRAID STATISTICS IN THREE-DIMENSIONAL LOCAL QUANTUM THEORY*

J. Fröhlich[1], F. Gabbiani[1] and P.-A. Marchetti[2]

[1] Theoretical Physics, ETH-Hönggerberg, CH-8093 Zürich

[2] Dipartimento di Fisica, Universitá di Padova
I. N. F. N., I-35131 Padova

Abstract. The general theory of superselection sectors and their statistics in three-dimensional local quantum theory is outlined. It is shown that abelian and non-abelian braid statistics can arise, provided that reflections at lines in two-dimensional space (parity) are not symmetries of the theory. Braid statistics is completely described by a family of braid- and fusion matrices satisfying polynomial equations. These braid- and fusion matrices have properties very similar to those of the corresponding matrices in two-dimensional conformal field theory and determine invariants for coloured links in S^3. The role of quantum group theory in three-dimensional local quantum theory is elucidated. Excitations with braid statistics are believed to play an important role in fractional quantum Hall systems and two-dimensional high-T_c superconductors. Three-dimensional gauge theories with Chern-Simons term exhibiting such excitations are briefly described.

---

* *for the proceedings of the Banff Summer School in Theoretical Physics "Physics, Geometry and Topology", August 1989; lectures presented by J. Fröhlich.*

Physics, Geometry, and Topology
Edited by H. C. Lee
Plenum Press, New York, 1990

1. Introduction

1.1 Some remarks on the history of fractional statistics

Since the beginnings of quantum theory the notion of identical particles and their statistics played a key role in the study of quantum systems. Implicitly, Bose-Einstein statistics already appeared in Planck's law of black body radiation and more explicitly in the study of monatomic gases carried out by Bose and Einstein. Pauli discovered his famous exclusion principle in the context of the old quantum theory. After the discovery of quantum mechanics, Heisenberg showed that the statistics of identical particles is described by the symmetry properties of n-particle wave functions under permutations of their arguments. For bosons, the complete wave functions are totally symmetric, for fermions they are totally anti-symmetric.

In local, relativistic quantum field theory, statistics was cast in the form of commutation relations between pairs of local fields at space-like separated arguments: Two Bose fields at space-like separated arguments commute, while Fermi fields anti-commute. The principles of local, relativistic quantum field theory led to the discovery of another basic fact: Bosons have integral spin, while fermions have half-integral spin. This connection between spin and statistics, though previously known in examples, was first shown to be a general feature of local, relativistic quantum field theory by Fierz, in 1939, in the context of free-field theory [1]. It was later shown to be a general consequence of the basic principles of local quantum theory [2].

After the advent of the quark model in strong interaction physics [3], the statistics problem became important again: For hadrons to have the observed spin, quarks had to have half-integral spin. Assuming that the spin-statistics connection is correct for unobservable particles (quarks), one concluded that quarks had to be fermions. This, however, led to difficulties in constructing phenomenologically viable quark wave functions for hadrons, in particular for the proton. A possible way out of these difficulties appeared to be to view quarks as particles with parastatistics: The symmetry properties of n-particle wave functions would then be described by higher-dimensional representations of the permutation group, S_n, of n elements. However, ultimately the statistics problem in the quark model of hadrons turned out to be one motivation (among several ones) for the introduction of colour and was neatly resolved by it: By introducing additional internal degrees of freedom the apparent parastatistics of quarks could be reinterpreted as ordinary Fermi statistics, and the standard connection between spin and statistics was saved. [The possibility of converting parastatistics into ordinary Bose- or Fermi statistics through the introduction of additional internal degrees of freedom had apparently been suggested by Fierz, Glaser and others.]

A deep analysis of statistics based on fundamental postulates of local quantum theory was carried out by Doplicher, Haag and Roberts [4], at the beginning of the seventies. They classified all possible statistics (para-Bose and para-Fermi statistics of order $d = 1, 2, 3, \ldots$) compatible with locality and certain general assumptions on the nature of physical states, for theories in four or more dimensions. The starting point of their analysis was reconsidered and given a better and more general foundation that includes gauge theories by Buchholz and Fredenhagen [5]. In an awsome effort, Doplicher and Roberts finally succeeded in proving that the parastatistics of "charged" particles in local, relativistic quantum theory could always be reinterpreted as ordinary Bose- or Fermi statistics by introducing additonal, internal degrees of freedom on which a global, compact internal symmetry group acts [6].

It has been known for some time that in quantum theory in two and three space-time dimensions the statistics of particles and fields is not in general described by representations of the permutation groups. In two-dimensional space-time, the concept of particle statistics becomes meaningless. Particle positions form an ordered set in a one-dimensional space, and hence the symmetry properties of wave functions under exchanging particles are not physically relevant. [For example, the Pauli principle can be understood as a consequence of local hard-core interactions in a system of bosons.] However, the statistics of fields in local quantum theory in two-dimensional Minkowski space is an interesting concept, at least for a certain class of theories. In the early seventies, the quantum theory of solitons in simple models of two-dimensional, relativistic quantum field theory became topical among theorists. The commutation relations of what people nowadays call vertex operators of the massless free field in two space-time dimensions were studied by Streater and Wilde [7] and found to be different, in general, from local commutators or anti-commutators.

More subtle examples of "exotic" commutation relations between soliton fields and meson fields in two-dimensional, interacting scalar theories with soliton sectors ($\lambda\varphi_2^4$ - and, more generally, $P(\varphi)_2$-models exhibiting quantum kinks) were studied by one of us in [8]. It was recognized there that the key facts behind the appearance of soliton sectors and exotic field statistics in two space-time dimensional theories are vacuum degeneracy and the property of two-dimensional Minkowski space that the causal complement of a bounded double cone consists of two disconnected wedges. It was recognized only fairly recently by several people that the new field statistics encountered in two-dimensional models with quantum kinks [8] may lead to representations of the braid groups. Earlier, one of us (J.F.) noted that, in the Euclidean description of two-dimensional quantum field theory, exotic statistics manifests itself as non-trivial monodromy of Euclidean Green functions. This was a model-independent interpretation of the monodromy properties found in the study of order-disorder correlation functions of the two-dimensional Ising model by Kadanoff and Ceva [9] and exploited in the work of Jimbo, Miwa and Sato [10]. For a brief sketch of such results see [11], and [12] for interesting related results.

While, during the late seventies, there was little interest in two-dimensional models, the situation changed after a resurge of interest in string theory and the appearance of the fundamental paper of Belavin, Polyakov and Zamolodchikov on two-dimensional conformal field theory [13], in 1984. Soon, it was recognized that the chiral fields of two-dimensional conformal field theories provide interesting examples of exotic statistics, or "exchange algebras" [14]. This was conceptualized in [15]. In this paper, a general theory of exotic statistics in two-dimensional theories, not limited to conformal theories, was sketched, and it was suggested that the proper framework for a more rigorous analysis was algebraic field theory [16,4], in combination with the theory of Yang-Baxter representations of the braid groups [17]. Subsequently, this point of view was developed in [18]. It should be emphasized, however, that there are examples of field statistics in two-dimensional theories <u>not</u> covered by the framework of [18]. Similar themes in the context of conformal field theory were studied in [19].

In these notes, we focus on the study of field- and particle statistics in <u>three-dimensional local quantum theory</u> which is of considerable interest in condensed matter theory. It appears to be a rather old observation that, for quantum-mechanical systems in two-dimensional space, particle statistics is described by representations of the braid groups, [20]. In fact, the right story to tell about statistics in quantum mechanics is that it is described by unitary representations of the fundamental group of the classical configuration space of identical particles on the Hilbert space of quantum mechanical wave functions. In two space dimensions the classical n-particle configuration space is

$$M_n = [(\mathsf{E}^2)^{\times n} \backslash D_n]/S_n, \tag{1.1}$$

where

$$D_n = \{(\underline{x}_1, \cdots, \underline{x}_n) \in (\mathsf{E}^2)^{\times n} : \underline{x}_i = \underline{x}_j \quad ; \quad \text{for some } i \neq j\},$$

and S_n is the permutation group of n elements. It is easily seen that the fundamental group of M_n is given by

$$\pi_1(M_n) = B_n,$$

where B_n is the braid group on n strands [21,17].

Quantum mechanical systems in two-dimensional space with abelian braid statistics were first proposed and studied by Leinaas and Myrheim [22]. Consider an array of identical point particles carrying an electric charge q and a magnetic flux (vorticity) ϕ. Such particles have been termed "anyons" by Wilczek [23]. Naturally, such a system exhibits an Aharonov-Bohm effect: If one particle moves around a positively oriented loop enclosing k particles the total wave function picks up a phase factor $exp(2ikq\phi)$. If two particles are exchanged along paths whose composition forms a positively oriented loop enclosing k particles the wave function gets multiplied by a factor $exp(iq\phi + 2ikq\phi)$.

Elements of the braid group B_n label homotopy classes of loops in M_n of which the two loops just described are examples. The braid group B_n has generators $\tau_i, i = 1, \cdots, n-1$; the generator $\tau_i^{\pm 1}$ describes the exchange of particle i with particle $i+1$ (in some order chosen on the points of E^2) along paths whose composition forms a $\binom{positively}{negatively}$ oriented loop in the plane not enclosing any other particle. An element $b \in B_n$ can be written as a word in the generators $\{\tau_1, \ldots, \tau_{n-1}\}$ modulo the relations

$$\begin{aligned} \tau_i\, \tau_{i+1}\, \tau_i &= \tau_{i+1}\, \tau_i\, \tau_{i+1}, \\ \tau_i\, \tau_j &= \tau_j\, \tau_i, \quad \text{for } \mid i-j \mid \geq 2. \end{aligned} \tag{1.2}$$

The unitary representation of B_n carried by the space of n-anyon wave functions is defined by assigning a phase factor $exp(\pm i\, q\, \phi)$ to $\tau_j^{\pm 1}$, for all $j = 1, \ldots, n-1$. This is a one-dimensional (abelian) representation of B_n. We shall see that there are, in principle, other non-abelian representations of B_n that could describe the statistics of non-abelian anyons. Let ρ be a representation of B_n. If

$$\rho(\tau_i^2) \;=\; \rho(\tau_i)^2 \;= \mathbb{I}$$

then ρ factors through a representation of S_n, and the corresponding particles have ordinary permutation group statistics. Thus if

$$q\phi/2\pi \;\in\; \frac{1}{2}\mathbb{Z} \tag{1.3}$$

anyons are bosons or fermions. More precisely, if $q\phi/2\pi$ is an integer anyons are bosons, while if $q\phi/2\pi$ is a half-integer anyons are fermions. For other values of $\theta \equiv q\phi/2\pi$, anyons have what has been called fractional-, or intermediate-, or θ-statistics. We shall speak of braid (group) statistics, as opposed to permutation (group) statistics.

It should be emphasized that it is not necesary to think of anyons as particles carrying electric charge and vorticity. By a gauge transformation the vector potential can be gauged away. But then if $\theta \notin \frac{1}{2}\mathbb{Z}$ the Hilbert space of anyon wave functions must be chosen to be a space of multi-valued functions with half-monodromies given by the phase factors $exp(\pm 2\pi i\theta)$. Such wave functions can be viewed as single-valued functions on the universal cover, $\widetilde{M}_n$, of M_n. This description of n-anyon systems is more natural if there are no electrostatic interactions between anyons, i.e. if the Coulomb interactions are absent or nearly completely screened.

In this picture, an n-anyon wave function will have the form

$$\psi(\underline{x}_1, \ldots, \underline{x}_n) \;=\; \prod_{i<j}(z_i - z_j)^{2\theta}\; g(\underline{x}_1, \ldots, \underline{x}_n), \tag{1.4}$$

where $g(\underline{x}_1, \ldots, \underline{x}_n)$ is a single-valued, symmetric function on $(\mathsf{E}^2)^{\times n}$, and z_j is the complex number $(x_j^1 + i\, x_j^2)$ corresponding to $\underline{x}_j$. An example is a Laughlin-type wave function [25]

$$\psi(\underline{x}_1, \ldots, \underline{x}_n) \;=\; const.\; \prod_{i<j}(z_i - z_j)^{2\theta} \prod_{k=1}^{n}\; exp(-\mid z_k \mid^2 /4)\,. \tag{1.5}$$

To conclude this section, it might be mentioned that a somewhat systematic analysis of gauge theories with Chern-Simons term (and related 0(3) non-linear σ-models with a Hopf term) in three space-time dimensions describing particles with braid statistics was initiated in [23,26-28]. It was noted in [15,28] that the braid statistics of anyons is closely related to the 't Hooft commutation relations between Wilson loops and vortex creation operators. Some general comments on three-dimensional theories are also contained in [18].

1.2 Physical realizations of braid statistics

One should ask why, as physicists, we should care about anyons and braid statistics? Is there experimental evidence for excitations with braid statistics in two-dimensional systems of condensed matter physics? By now, the standard answer is that excitations with braid statistics appear to play an important role in systems exhibiting a fractional quantum Hall effect. Moreover, pure anyon gases, with $\theta \in \mathbb{Q}\backslash\frac{1}{2}\mathbb{Z}$, appear to be superconductors of a new type [29]. It has been speculated that anyon superconductivity may describe (essentially two-dimensional) high-T_c superconductors. This would be fairly plausible if one could exhibit doped anti-ferromagnets that admit a flux phase breaking parity and time reversal invariance.

Let us briefly recall Laughlin's argument explaining the role of anyons in the fractional quantum Hall effect: An idealized experimental set-up is sketched in Fig.1

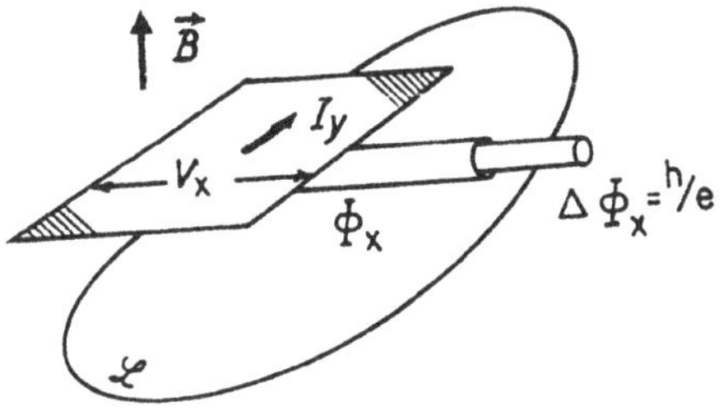

Fig. 1

The transverse conductivity is

$$\sigma_{xy} = {}^{I_y}/\,V_x \tag{1.6}$$

Experimentally, one finds that in films of $GaAl_x\,As_{1-x}$ at low temperatures and in strong magnetic fields, σ_{xy}, plotted as a function of the electron density ρ, has plateaux at the values

$$\nu\, e^2/h\,, \text{ with } \nu = \begin{cases} 1,2,3,\ldots \\ \frac{1}{5},\frac{2}{7},\frac{1}{3},\frac{2}{5},\frac{2}{3},\ldots\,, \end{cases} \tag{1.7}$$

more generally, for $\nu = p/r$, where p and r do not have a common divisor, and r is odd. [More recently, plateaux appear to have been observed for even values of r, too.] Fractional values of ν could be understood as an effect of fractionally charged excitations in such a system: Suppose there are charge carriers of charge $q = f\,e$, $f \in \mathbb{Q}$, in the conducting rectangle depicted in Fig. 1. Imagine that the total magnetic flux through the loop $\mathcal{L}$ is increased adiabatically by one quantum of flux from ϕ_x to $\phi_x + \Delta\phi_x$, $(\Delta\phi_x = h/e)$. The microscopic quantum-mechanical Hamiltonians $H(\phi_x)$ and $H(\phi_x + \Delta\phi_x)$ of the system are then gauge-equivalent, hence have the same spectrum. If the Fermi energy of the system is in a mobility gap, the adiabatic process described above will therefore not change the total energy of the system. Suppose that during that process n charge carriers of charge fe move from one to another edge in the x-direction. This changes the total energy by an amount

$$\Delta\, U = n\,f\,e\,V_x. \tag{1.8}$$

By Faraday's induction law the work of the current in the y-direction, during the same adiabatic process, is given by

$$\Delta\, W = -\int_{-\infty}^{\infty} dt\; I_y \frac{d\phi_x}{dt} = -I_y\,\frac{h}{e}. \tag{1.9}$$

Since the total energy remains unchanged,

$$\Delta\, U + \Delta\, W = 0.$$

By (1.6), (1.8) and (1.9),

$$\sigma_{xy} = n f\; e^2/h. \tag{1.10}$$

Comparison with experimental data, (1.7), shows that if the naïve argument just sketched is correct then f must be fractional, for fractional values of ν. For strongly correlated many-electron systems, one might imagine mechanisms giving rise to "solitons" which correspond to a fraction, $1/r$, of an electron. [Solitons of this kind are well known in one-dimensional systems such as polyacethylene.] Solitons corresponding to a fraction of an electron carry fractional charge and fractional spin. Excitations with fractional spin in two dimensional systems necessarily obey fractional statistics, as we shall see in Sect. 3. Phenomenological wave functions for an assembly of n such excitations have the form (1.4), with $\theta = \frac{1}{2r}$, [25].

We believe that a detailed understanding of how anyons emerge in two-dimensional, highly correlated many-body systems is still missing.

1.3 Three-dimensional gauge theories with braid statistics

Next, we give a mini-review of three-dimensional gauge theories with Chern-Simons term describing particles with braid statistics. The simplest examples are abelian gauge theories. They have an action given by

$$S[A] = \frac{1}{2} \int_{M^3} [e^{-2} F^2 + \frac{\theta}{2\pi} \varepsilon^{\mu\nu\lambda} A_\mu \partial_\nu A_\lambda - 2 j^\mu A_\mu + \text{(pure matter terms)}] \, . \tag{1.11}$$

The term proportional to θ is the Chern-Simons term, breaking parity, j^μ is the matter current. By varying the action with respect to A we find the modified Maxwell equations. In particular, one has that

$$\frac{1}{e^2} \, div \, \vec{E} = j^0 - \frac{\theta}{2\pi} B. \tag{1.12}$$

This equation shows that if the electric field is screened vortices carry an electric charge $\frac{\theta}{2\pi} \phi$, ($\phi \equiv \int d^2x \, B(\vec{x}, t)$ is their vorticity), and particles with an electric charge q carry vorticity q/θ . The abelian braid statistics of charged particles and vortices in such theories can be understood as a consequence of the Aharonov-Bohm effect. This can be substantiated in models by means of non-perturbative calculations with functional integrals [26], (or in an operator formalism, using 't Hooft commutation relations [28]). Abelian gauge theories with Chern-Simons term have been studied in detail in [30], and their particle spectrum and statistics are analyzed in [23, 26-28, 31].

We now ask whether non-abelian gauge theories with Chern-Simons term in three dimensions might describe particles with non-abelian braid statistics? Here we only sketch a preliminary answer to this question.

Consider a simply connected, compact gauge group $G(\stackrel{e.g.}{=} SU(N))$. Let A be a vector potential (connection) with values in $Lie(G)$. Following Witten [32], we consider the pure Chern-Simons theory with action

$$S_{C.S.}[A] = \frac{k}{4\pi} \int_{M^3} tr(A \wedge dA + \frac{2}{3} A \wedge A \wedge A) \tag{1.13}$$

This theory can be quantized using geometrical quantization [32] or functional integrals [33]. Gauge-invariance and unitarity constrain the coupling constant of the theory to be quantized,

$$k_{eff.} = n + h_G, \; n = 1, 2, 3, \ldots, \tag{1.14}$$

where h_G is the dual Coxeter number of G; (e.g. $h_G = N$, for $G = SU(N)$).

Static colour sources in this theory have non-abelian braid statistics described by Yang-Baxter matrices that are identical to the braid matrices of the Wess-Zumino-Witten models corresponding to the group G, at level n. These braid matrices can be understood as holonomy matrices of the Knizhnik-Zamolodchikov connection. These results follow from [32-34]. Although pure Chern-Simons theory has interesting applications to pure mathematics [32], it is uninteresting for physics. It is a purely topological theory, and hence its Hamilton opertor vanishes on physical states. The physical state spaces are finite-dimensional.

An idea for constructing non-topological gauge theories with non-abelian braid statistics is to add non-topological terms to $S_{C.S.}$. Consider a theory with Euclidean action

$$S[A] \stackrel{e.g.}{=} g^{-2} \int tr(F^2) - \frac{ik}{4\pi} \int tr(A \wedge dA + \frac{2}{3} A \wedge A \wedge A) + \lambda \int \bar{\psi}(\not{D}_A + m)\psi + \ldots, \tag{1.15}$$

where g, λ and m are positive constants, and ψ is a two-component spinor field in the fundamental representation of G; (there may be further matter fields, e.g. Higgs fields). The conjectured properties of this theory are as follows.

1) The first two terms in $S[A]$ on the r.h.s. of (1.15) make the gluon massive [30]. Hence all interactions mediated by gluons are expected to be of short range. In particular, one expects that there is no confinement of colour in this theory.

2) Long-range interactions in this theory are purely topological, just like in pure Chern-Simons theory. The effective action at very large distance scales is essentially a pure Chern-Simons action with a renormalized value of k, the renormalization depending only on the number and nature of matter fields [30].

3) Since particle statistics can be determined at arbitrarily large distance scales, one expects, on the basis of 1) and 2), that the statistics of coloured particles in this theory is the same as the statistics of static colour sources in pure Chern-Simons theory, for a renormalized value of k, which we have described above.

4) Coloured one-particle states in this theory are expected to be created by applying Mandelstam string operators to the physical vacuum, Ω. The Mandelstam string operators, denoted by $\psi(C_x)$, are smeared out versions of

$$\psi(y)\ P(exp \int_{\gamma_y} A_\mu(\xi)\ d\xi^\mu), \tag{1.16}$$

where γ_y is a space-like path starting at $y \in M^3$ and reaching out to space-like ∞. By averaging the operators (1.16) over Poincaré transformations in the vicinity of $(\mathbb{1}, 0)$, we obtain operators, $\psi(C_x)$, localized in some space-like cone C_x with apex at some point $x \in M^3$ (and with arbitrarily small opening angle) which are densely defined, closed operators on the physical Hilbert space. [By polar decomposition, these operators could then be replaced by colour-carrying, bounded operators; see also Sect. 2.]

A more careful definition of the operators $\psi(C_x)$ reveals that they are multi-valued. They are elements of fibres of a certain vector bundle of operators over the circle of asymptotic directions in two-dimensional space. In order to construct a section of operators in this bundle, we choose a "reference cone", $\overset{\circ}{\mathcal{C}}$, (corresponding to a boundary condition at ∞), as sketched in Fig. 2.

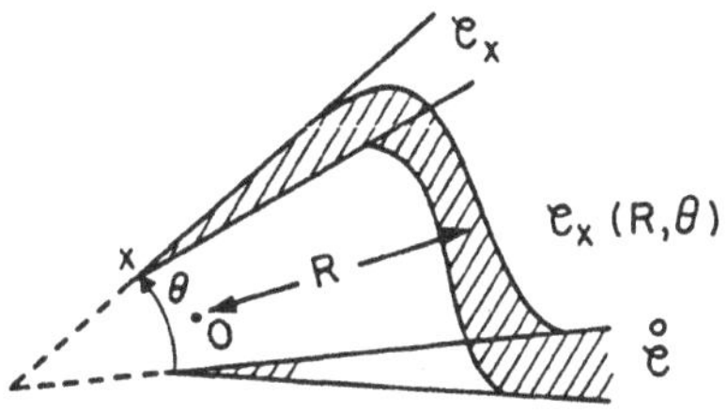

Fig. 2

The angle θ through which $\overset{\circ}{C}$ is rotated to have the direction of C_x is called the asymptotic direction, as (C_x), of C_x. We define the operator $\psi(C_x, \theta)$ as a limit

$$\psi(C_x, \theta) = \lim_{R \to \infty} \psi(C_x(R, \theta)). \tag{1.17}$$

The space-like cone C_x can also be reached by rotating $\overset{\circ}{C}$ through an angle $\theta - 2\pi$; see Fig. 3.

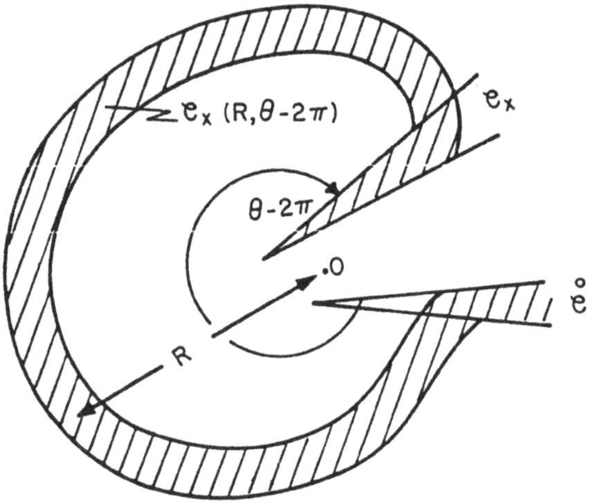

Fig. 3

We define

$$\psi(C_x, \theta - 2\pi) = \lim_{R \to \infty} \psi(C_x(R, \theta - 2\pi)). \tag{1.18}$$

The point is that $\psi(C_x, \theta)$ and $\psi(C_x, \theta - 2\pi)$ are distinct:

$$\psi(C_x, \theta - 2\pi) = V_{-2\pi}\, \psi(C_x, \theta), \tag{1.19}$$

where the operator $V_{-2\pi}$ commutes with all local observables of the theory and can be expressed in terms of the fractional spins of coloured particles.

The operators $\{\psi(C_x, \theta)\}$ are expected to obey (non-abelian) braid statistics: If C_{x_1} and C_{x_2} are space-like separated then

$$\psi(C_{x_1}, \theta)\, \psi(C_{x_2}, \varphi) = R^{\pm}\, \psi(C_{x_2}, \varphi)\, \psi(C_{x_1}, \theta), \tag{1.20}$$

for $\theta \gtrless \varphi$. The operators R^+ and R^- are unitary operators commuting with all local observables of the theory. Braid statistics (as opposed to permutation statistics) arises if $R^+ \neq R^-$ - which is expected for the gauge theories discussed here, unless the Chern-Simons term in the effective (large-scale) gauge field action vanishes.

One may wonder whether the gauge theories discussed here are of purely academic interest? It is conceivable, though far from established, that such theories arise as large-scale effective theories in highly correlated two-dimensional quantum many-body systems. It is a challenge to find physically plausible model systems of this type.

From now on, we outline a general analysis of braid statistics in three-dimensional local quantum theory which is mathematically rigorous [28]. It lends support to the idea that systems with infinitely many degrees of freedom in two space dimensions, with broken parity (and time reversal invariance), will generically exhibit excitations with braid statistics.

2. The algebraic formulation of local quantum theory

In the algebraic formulation of local, relativistic quantum theory [16,4], the basic object is an algebra of local observables. Physical properties of a system are extracted from the representation theory of that algebra.

The construction of algebras of local observables might proceed as follows: We imagine that we are given a local, relativistic quantum field theory, in the sense of Wightman [2], in the vacuum representation. The vacuum Hilbert space is denoted by $\mathcal{H}_1$. It contains a Poincaré-invariant vector Ω, the physical vacuum. Of particular interest for our purposes are gauge theories. The gauge-invariant, local observables of such a theory are Wilson loop operators $W(\mathcal{L})$, where $\mathcal{L}$ is a space-like loop in M^3, Mandelstam string operators, denoted $\psi(\gamma_{xy})$, where γ_{xy} is a space-like curve starting at x and ending at y, and local, gauge-invariant currents, $J^{\alpha}_{\mu_1,\ldots,\mu_k}(\vec{x},t)$, where $\mu_1,\ldots,\mu_k$ are Lorentz indices, and α labels different currents with the same tensorial properties under Lorentz transformations. In order to obtain densely defined, closed operators on $\mathcal{H}_1$, the distributional fields introduced above must be smeared out with test functions. Let $\mathcal{O}$ be a bounded open region in M^3, e.g. a double cone, and let f be a test function with $supp\, f \subset \mathcal{O}$. If $J^{\alpha}_{\mu_1,\ldots,\mu_k}$ is a real current, and f is a real test function one expects that the operator

$$J^{\alpha}_{\mu_1,\dots,\mu_k}(f) = \int d^3x \, J_{\mu_1,\dots,\mu_k}(\vec{x},t)\, f(\vec{x},t) \tag{2.1}$$

is selfadjoint on the vacuum sector $\mathcal{H}_1$. Similarly, by averaging Wilson loops and Mandelstam operators over (finite-dimensional) families of loops or curves, respectively, contained in $\mathcal{O}$, we can hope to construct further selfadjoint operators on $\mathcal{H}_1$. All these operators have the common feature that they are localized in the space-time region $\mathcal{O}$.

We now define the local algebra $\mathcal{A}(\mathcal{O})$ to be the von Neumann algebra generated by all bounded functions of all the gauge-invariant, selfadjoint operators localized in $\mathcal{O}$ introduced above. The algebra $\mathcal{A}(\mathcal{O})$ is closed in the weak operator topology determined by the scalar product on $\mathcal{H}_1$.

If $\mathcal{S}$ is an unbounded space-time region in M^3 we define the algebra $\mathcal{A}(\mathcal{S})$ by setting

$$\mathcal{A}(\mathcal{S}) = \overline{\bigcup_{\substack{\mathcal{O} \subset \mathcal{S} \\ \mathcal{O} \text{ bounded}}} \mathcal{A}(\mathcal{O})}^{\,n}, \tag{2.2}$$

where the closure is taken in the operator norm. In particular, the "algebra of all (quasi-) local observables", $\mathcal{A}$, is defined to be

$$\mathcal{A} = \mathcal{A}(\mathcal{S} = \mathsf{M}^3). \tag{2.3}$$

The algebras $\mathcal{A}(\mathcal{S}), \mathcal{A}$ are C^*-algebras. We define the "relative commutant", $\mathcal{A}^c(\mathcal{S})$, of $\mathcal{A}(\mathcal{S})$ in $\mathcal{A}$ by

$$\mathcal{A}^c(\mathcal{S}) = \{A \in \mathcal{A} : [A,B] = 0, \ \forall B \in \mathcal{A}(\mathcal{S})\}. \tag{2.4}$$

Let $\mathcal{C}_0$ be a wedge in two-dimensional space. The causal completion, $\mathcal{C}$, of $\mathcal{C}_0$ is defined as follows. Let $\mathcal{C}_0'$ (the causal complement of $\mathcal{C}_0$) be given by

$$\mathcal{C}_0' = \{x \in \mathsf{M}^3 : (x-y)^2 < 0, \ \forall y \in \mathcal{C}_0\}. \tag{2.5}$$

Then one sets

$$\mathcal{C} = (\mathcal{C}_0')'. \tag{2.6}$$

The causal completion of a wedge $\mathcal{C}_0$ is called a <u>simple domain</u>. If the opening angle of $\mathcal{C}_0$ is smaller than π $\mathcal{C}$ is called a <u>space-like cone</u>.

Let U_1 be the unitary representation of the quantum mechanical Poincaré group $\tilde{\mathcal{P}}_+^\uparrow$ on the vacuum sector $\mathcal{H}_1$. We define a representation, α, of $\mathcal{P}_+^\uparrow$ on $\mathcal{A}$ by setting

$$\alpha_{(\Lambda,a)}(A) = U_1[\Lambda,a]\, A\, U_1[\Lambda,a]^{-1}. \tag{2.7}$$

This is a representation of $\mathcal{P}_+^\uparrow$ as a group of *automorphisms on $\mathcal{A}$.

Next, we recall some basic properties of the net $\{\mathcal{A}(\mathcal{O})\}_{\mathcal{O}\subset \mathsf{M}^3}$ of local observable algebras and the representation α of $\mathcal{P}_+^\uparrow$ on $\mathcal{A}$ which are believed to be true in every "reasonable" local, relativistic QFT.

(1) Locality: For all $A \in \mathcal{A}(\mathcal{S})$ and all $B \in \mathcal{A}(\mathcal{S}')$,

$$[A,B] = 0, \tag{2.8}$$

i.e. $\mathcal{A}(\mathcal{S}') \subseteq \mathcal{A}^c(\mathcal{S})$; ($\mathcal{S}'$ is the causal complement of $\mathcal{S}$, see (2.5)).

(2) For $\mathcal{S}_1 \subseteq \mathcal{S}_2$, $\mathcal{A}(\mathcal{S}_1) \subseteq \mathcal{A}(\mathcal{S}_2)$, and, for arbitrary $\mathcal{S}_1$ and $\mathcal{S}_2$,

$$\mathcal{A}(\mathcal{S}_1 \cup \mathcal{S}_2) \supseteq \mathcal{A}(\mathcal{S}_1) \vee \mathcal{A}(\mathcal{S}_2). \tag{2.9}$$

(3) Duality [4,5]: Let $\mathcal{B}$ be an algebra of bounded operators on $\mathcal{H}_1$. By $\mathcal{B}'$ we denote the algebra of all bounded operators on $\mathcal{H}_1$ commuting with all operators in $\mathcal{B}$, (the "commutant of $\mathcal{B}$"). It is reasonable to expect that

$$\mathcal{A}^c(\mathcal{S})' = \overline{\mathcal{A}(\mathcal{S})}^{w}, \tag{2.10}$$

where $\overline{(\cdot)}^{w}$ denotes closure in the weak operator topology. See [5] for more discussion of (2.10).

(4) Poincaré covariance: Let

$$\mathcal{O}_{(\Lambda,a)} = \{x \in \mathsf{M}^3 :\ \Lambda^{-1}(x-a) \in \mathcal{O}\},\ (\Lambda,a) \in \mathcal{P}_+^\uparrow\}.$$

Then

$$\alpha_{(\Lambda,a)}(\mathcal{A}(\mathcal{O})) = \mathcal{A}(\mathcal{O}_{(\Lambda,a)}). \tag{2.11}$$

The basic objects in the algebraic approach to local quantum theory are

$$\left\{ \{\mathcal{A}(\mathcal{O})\}_{\mathcal{O}\subset \mathsf{M}^3},\ \alpha \right\} \tag{2.12}$$

satisfying properties (1)-(4), above. For a more precise description of this structure see [4-6, 28].

The physics of a system described by a given pair $\{\{\mathcal{A}(\mathcal{O})\}_{\mathcal{O}\subset M^3}, \alpha\}$ can be inferred from the representation theory of $\{\mathcal{A}, \alpha\}$.

Definition 2.1 A *representation, j, of $\mathcal{A}$ on a separable Hilbert space $\mathcal{H}_j$ is called a covariant positive-energy representation iff there exists a unitary representation, U_j, of $\tilde{\mathcal{P}}_+^\uparrow$ on $\mathcal{H}_j$ such that

- $j(\alpha_{(\Lambda,a)}(A)) = U_j[\Lambda,a]\, j(A)\, U_j[\Lambda,a]^{-1}$, for all $A \in \mathcal{A}$ and all $(\Lambda,a) \in \mathcal{P}_+^\uparrow$; ($[\Lambda,a]$ denotes an element in $\tilde{\mathcal{P}}_+^\uparrow$ projecting onto (Λ,a)).
- $U_j(\mathbb{1},a) = e^{ia\cdot P_j}$, with $spec(P_j) \subseteq \bar{V}_+$, where $\bar{V}_+$ denotes the closure of the forward light cone. This is the relativistic spectrum condition.

The superselection sectors of a system described by $\{\mathcal{A}, \alpha\}$ are the representation spaces, $\mathcal{H}_j$, corresponding to irreducible covariant positive-energy representations, j, of $\mathcal{A}$.

The physical state space of the theory, $\mathcal{H}$, is defined to be

$$\mathcal{H} = \bigoplus_j \mathcal{H}_j \tag{2.13}$$

where the direct sum extends over all inequivalent, irreducible covariant positive-energy representations of $\mathcal{A}$. It carries a unitary representation

$$U = \bigoplus_j U_j \tag{2.14}$$

of $\tilde{\mathcal{P}}_+^\uparrow$ satisfying the relativistic spectrum condition.

Buchholz and Fredenhagen have analyzed the covariant positive-energy representations of $\mathcal{A}$ for systems which admit a complete particle interpretation and without zero-mass particles, [5]. For simplicity, we may suppose that there is only one vacuum sector, $\mathcal{H}_1$, containing a unique vacuum Ω. Then the results in [5] show that (under suitably precise hypotheses on the particle structure of the theory) a covariant positive-energy representation, j, of $\mathcal{A}$ has the property that, for an arbitrary space-like cone $C \subset M^3$,

$$j\,|_{\mathcal{A}^c(C)} \simeq 1\,|_{\mathcal{A}^c(C)}\,, \tag{2.15}$$

where $\simeq$ denotes "unitarily equivalent". This implies that there exists a unitary operator, V_C, from $\mathcal{H}_j$ to the vacuum sector $\mathcal{H}_1$ such that

$$j(A) = V_C^*\, A\, V_C\,, \tag{2.16}$$

for all $A \in \mathcal{A}^c(C)$; (we are identifying the abstract element $A \in \mathcal{A}$ with the operator $1(A)$ on $\mathcal{H}_1$). The proof of (2.16) involves using the Reeh-Schlieder theorem [2].

We now define a representation, $\rho_C^j \equiv \rho_C$, of $\mathcal{A}$ on $\mathcal{H}_1$ by setting

$$\rho_C(A) = V_C\, j(A)\, V_C^* . \tag{2.17}$$

For $A \in \mathcal{A}$, $\rho_C(A)$ is a bounded operator on $\mathcal{H}_1$, and by (2.16),

$$\rho_C(A) = A, \text{ for all } A \in \mathcal{A}^c(C). \tag{2.18}$$

Let C_a be some auxiliary space-like cone of arbitrarily small opening angle, and let

$$C_a + x = \{y \in \mathsf{M}^3 : y - x \in C_a\}.$$

We define an enlarged C^*algebra, $\mathcal{B}^{C_a}$, containing $\mathcal{A}$, by setting

$$\mathcal{B}^{C_a} = \overline{\bigcup_{x \in \mathsf{M}^3} \overline{\mathcal{A}^c(C_a + x)}^{w}}^{n} . \tag{2.19}$$

It has been shown in [5] that ρ_C has a continuous extension to $\mathcal{B}^{C_a}$, and if C is space-like separated from $C_a + x$, for some x, then ρ_C is a *morphism on $\mathcal{B}^{C_a}$, i.e., ρ_C is a linear map from $\mathcal{B}^{C_a}$ into $\mathcal{B}^{C_a}$ such that

$$\rho_C(A \cdot B) = \rho_C(A) \cdot \rho_C(B) \text{ and } \rho_C(A^*) = \rho_C(A)^* . \tag{2.20}$$

We may often keep the choice of C_a fixed and then write $\mathcal{B}$ for $\mathcal{B}^{C_a}$.

Next, we introduce the notion of "charge-transport operators" [4,5] which plays quite a basic role: Consider two *morphisms, ρ_{C_1} and ρ_{C_2}, of $\mathcal{B}$ equivalent to a given covariant positive-energy representation, j, of $\mathcal{B}$. Then there exists a unitary operator $\Gamma_{\rho_{C_1},\rho_{C_2}}$, a "charge-transport operator", such that

$$\rho_{C_1}(A) = \Gamma_{\rho_{C_1},\rho_{C_2}}\, \rho_{C_2}(A)\, \Gamma^*_{\rho_{C_1},\rho_{C_2}} \tag{2.21}$$

on $\mathcal{H}_1$. Let $\mathcal{S}$ be any simple domain containing $C_1 \cup C_2$. Then, since

$$\rho_{C_1}(A) = \rho_{C_2}(A) = A, \text{ for } A \in \mathcal{A}^c(C_1) \cap \mathcal{A}^c(C_2), \tag{2.22}$$

it follows that

$$\begin{aligned} \Gamma_{\rho_{C_1},\rho_{C_2}} &\in (\mathcal{A}^c(C_1) \cap \mathcal{A}^c(C_2))' \\ &\subseteq (\mathcal{A}^c(\mathcal{S}))' = \overline{\mathcal{A}(\mathcal{S})}^{w} . \end{aligned} \tag{2.23}$$

The last equality in (2.23) follows from duality, (2.10). Thus if $\mathcal{S}$ is in the causal complement of $C_a + x$, for some x, $\Gamma_{\rho_{C_1},\rho_{C_2}} \in \mathcal{B}$, and hence $i(\Gamma_{\rho_{C_1},\rho_{C_2}})$ is defined, for an arbitrary covariant positive-energy representation, i, of $\mathcal{B}$.

Note that it does not follow from (2.23) that $\Gamma_{\rho_{C_1},\rho_{C_2}} \in \overline{\mathcal{A}(C_1)}^w \vee \overline{\mathcal{A}(C_2)}^w$. In fact, theories in which

$$\Gamma_{\rho_{C_1},\rho_{C_2}} \in \overline{\mathcal{A}(C_1)}^w \vee \overline{\mathcal{A}(C_2)}^w \tag{2.24}$$

have ordinary permutation statistics. In other words, braid statistics is tied to a failure of (2.24), [28]. The operators $\psi(C_x, \theta)$ introduced in (1.17) are likely to determine charge transport operators for which (2.24) does not hold, as discussed in [28] for abelian gauge theories.

Summary According to [5], the class of representations of $\{\mathcal{A}, \alpha\}$ describing the physics of a system (at zero temperature) consists of all covariant positive-energy representations, j, localizable in space-like cones, in the sense of Eqs. (2.15), (2.16). These representations are unitarily equivalent to representations of $\mathcal{A}$ on $\mathcal{H}_1$ determined by *morphisms, ρ_C^j, of the extended algebra $\mathcal{B}$ localized in space-like cones C. If $\rho_{C_1}^j$ and $\rho_{C_2}^j$ are both unitarily equivalent to j, and C_1 and C_2 are space-like separated from $C_a + x$, for some x, then there is a unitary intertwiner, $\Gamma_{\rho_{C_1},\rho_{C_2}}$, (a charge-transport operator), such that

$$\rho_{C_1}(A)\,\Gamma_{\rho_{C_1},\rho_{C_2}} = \Gamma_{\rho_{C_1},\rho_{C_2}}\,\rho_{C_2}(A), \quad \text{with } \Gamma_{\rho_{C_1},\rho_{C_2}} \in \mathcal{B}. \tag{2.25}$$

We denote by L the complete list of all inequivalent, irreducible, covariant positive-energy representations of $\{\mathcal{A}, \alpha\}$ localizable in space-like cones.

The fact that ρ_C is a *morphism of $\mathcal{B}$ (if $C \subset (C_a + x)'$, for some x) and (2.23) permit us to define a composition of representations in L: For $C_1 \subset (C_a + x)'$ and $C_2 \subset (C_a + x)'$, for some x, $\rho_{C_1}^j(A) \in \mathcal{B}$, for $A \in \mathcal{A} \subseteq \mathcal{B}$, $j \in L$; hence, for $i \in L$,

$$\rho_{C_2}^i \circ \rho_{C_1}^j(A) \equiv \rho_{C_2}^i(\rho_{C_1}^j(A)) \tag{2.26}$$

is well defined. We define $i \times j$ to be the representation of $\mathcal{A}$, unique up to unitary equivalence, unitarily equivalent to the vacuum representation, 1, of $\rho_{C_2}^i \circ \rho_{C_1}^j(\mathcal{A})$.

If C_1 and C_2 are chosen to be space-like separated then $\rho^i_{C_2} \circ \rho^j_{C_1}(A) = \rho^j_{C_1} \circ \rho^i_{C_2}(A)$, hence

$$i \times j = j \times i . \tag{2.27}$$

Clearly, $i \times j$ is localizable in cones. One easily deduces from (2.23) that, for i and j in L, $i \times j$ is again a covariant, positive-energy representation of $\{\mathcal{A}, \alpha\}$; see [28] for details.

In [4,5,35] natural hypotheses on $\{\mathcal{A}, \alpha\}$ have been isolated which imply the following

<u>Property 2.2</u>

(P1) Every covariant positive-energy representation of $\{\mathcal{A}, \alpha\}$ is completely reducible into a direct sum of <u>irreducible</u>, covariant positive-energy representations.

(P2) There is a unique involution $^-$ ("charge conjugation") on $L : j \in L \mapsto \bar{j} \in L$, such that $j \times \bar{j}$ contains the vacuum representation, 1, of $\mathcal{A}$ precisely <u>once</u> as a subrepresentation.

These properties are deep properties, and it is a non-trivial task to derive them from the structure of $\{\mathcal{A}, \alpha\}$; see [5,35]. Henceforth, they will be assumed to hold.

As a corollary of the fact that, for i and j in L, $i \times j$ is a covariant positive-energy representation of $\{\mathcal{A}, \alpha\}$, and of (P1), we have that $i \times j$ can be decomposed into a direct sum of irreducible representations belonging to L:

$$i \times j = \bigoplus_{k \in L} \bigoplus_{\mu=1}^{N_{kij}} k^{(\mu)} , \tag{2.28}$$

where $k^{(\mu)}$ is unitarily equivalent to $k \in L$, and $N_{kij} \in \{0, 1, 2, \dots\}$ is the multiplicity of k in $i \times j$. By property (P2), N_{kij} can also be interpreted as the multiplicity of 1 in $\bar{k} \times i \times j$. This and (2.27) show that

$$N_{kij} = N_{kji} = N_{\bar{j}\bar{k}i} . \tag{2.29}$$

We define $|L| \times |L|$ matrices, N_j, $j \in L$, by setting

$$(\mathsf{N}_j)_{ki} = N_{kji} \in \{0, 1, 2, \dots\} , \tag{2.30}$$

($|L|$ is the cardinality of L). Clearly

$$1 \times j = j \times 1 = j,$$

so that

$$\mathsf{N}_1 = \mathbb{1} . \tag{2.31}$$

As shown in [28], (2.27) implies that

$$\mathsf{N}_i \cdot \mathsf{N}_j = \mathsf{N}_j \cdot \mathsf{N}_i, \text{ for all } i, j \in L . \tag{2.32}$$

Properties (2.29)-(2.32) identify the matrices $\{\mathsf{N}_j\}$ as matrices of fusion rules. It is an outstanding open problem to classify all possible fusion rules. Standard examples of fusion rules are those derived from the representation theory of a compact group $G, (N_{k\,j\,i} =$ multiplicity of irrep. k in the tensor product, $j \otimes i$, of two irreps. j and i of G; in this connection see also [6]), or from the representation theory of a quantum group, $U_q(\mathcal{G})$, $q = exp(2\pi i/m)$. For example, if $\min\limits_{j\neq 1} \|\mathsf{N}_j\| < 2$ then $\min\limits_{j\neq 1} \|\mathsf{N}_j\| = 2\cos(\frac{\pi}{n+1})$, where $n \equiv \mid L \mid$, and

$$\begin{aligned} N_{k\,j\,i} &= 1, \text{ for } \mid i-j \mid < k < \min\{i+j,\ 2(n+1)-i-j\} \\ &= 0, \text{ otherwise.} \end{aligned} \tag{2.33}$$

These fusion rules can be derived from the representation theory of $U_q(sl(2))$, $q = exp(2\pi i/n+2)$.

Suppose now that $N_{k\,j\,i} \neq 0$. Then k appears as a subrepresentation of $j \times i$. By the definition of composition, $\times$, this implies that the representation $\rho^j_{C_1} \circ \rho^i_{C_2}$ contains $N_{k\,j\,i}$ subrepresentations $\rho^k_{\tilde{C},\mu}$, $\mu = 1, \ldots, N_{k\,j\,i}$; (here C_1, C_2 and $\tilde{C}$ are space-like cones space-like separated from $C_a + x$, for some x). Equivalently, the representation i of $\rho^j_C(\mathcal{A})$ contains $N_{k\,j\,i}$ subrepresentations $k^{(\mu)}, \mu = 1, \ldots, N_{k\,j\,i}$, unitarily equivalent to the representation k of $\mathcal{A}$. Hence the superselection sector $\mathcal{H}_i$ can be decomposed into a direct sum of spaces

$$\mathcal{H}_i = \bigoplus_{k\in L} \bigoplus_{\mu=1}^{N_{k\,j\,i}} \mathcal{H}_i(k, j; \mu) \tag{2.34}$$

with the property that the representation i of $\rho^j_C(\mathcal{A})$ on $\mathcal{H}_i(k, j; \mu)$ is unitarily equivalent to the representation k of $\mathcal{A}$.

We now wish to construct a complex vector bundle, $\mathcal{J} \equiv \mathcal{J}_{k\,j\,i}$, of intertwining operators from $\mathcal{H}_k$ to $\mathcal{H}_i$. The base space of this bundle is the space, $\mathcal{M}_j$, of all *morphisms, ρ_C^j, of an extended algebra $\mathcal{B}^{C_a}$ localizable in some space-like cone C space-like separated from C_a, for some choice of an auxiliary cone C_a, and with the property that the representation $1(\rho_C^j(\cdot))$ of $\mathcal{A}$ is unitarily equivalent to the representation j of $\mathcal{A}$. The fibre space, $\mathcal{V}_k(\rho_C^j)_i$, above a point $\rho_C^j \in \mathcal{M}_j$ is a complex vector space of operators

$$V \; : \; \mathcal{H}_k \; \to \; \mathcal{H}_i \, , \tag{2.35}$$

satisfying the intertwining relations

$$i(\rho_C^j(A))V \;=\; V\,k(A), \text{ for all } A \in \mathcal{A}. \tag{2.36}$$

By (2.34), the range of V is contained in the subspace $\bigoplus_{\mu=1}^{N_{kji}} \mathcal{H}_i(k,j;\mu)$ of $\mathcal{H}_i$.

The space $\mathcal{V}_k(\rho_C^j)_i$ is equipped with a scalar product: For V and W in $\mathcal{V}_k(\rho_C^j)_i$, V^*W is an operator from $\mathcal{H}_k$ to $\mathcal{H}_k$ which, by (2.36), satisfies

$$k(A)\,V^*W \;=\; V^*W\,k(A). \tag{2.37}$$

Since k is an irreducible representation of $\mathcal{A}$, it follows from Schur's lemma that

$$V^*W \;=\; c\cdot \mathbb{1}, \; c \in \mathbb{C}. \tag{2.38}$$

The complex number c depends anti-linearly on V and linearly on W. Moreover, for $V = W \neq 0$, c is strictly positive. Hence c defines a scalar product,

$$\langle \, V, \, W \, \rangle, \tag{2.39}$$

on $\mathcal{V}_k(\rho_C^j)_i$.

Clearly, the multiplicity, N_{kji}, of k in $j \times i \simeq i(\rho_C^j(\cdot))$ does not depend on the choice of the *morphism $\rho_C^j \in \mathcal{M}_j$. Hence all fibres $\mathcal{V}_k(\rho_C^j)_i$ are isomorphic, as complex Hilbert spaces, to $\mathbb{C}^{N_{kji}}$, equipped with the usual scalar product; in particular,

$$dim\; \mathcal{V}_k(\rho)_i \;=\; N_{kji}, \text{ for all } \rho \in \mathcal{M}_j. \tag{2.40}$$

Physically speaking, local sections of $\mathcal{J}_{kji}$, (operators $V(\rho) : \mathcal{H}_k \to \mathcal{H}_i, \rho \in \mathcal{M}_j$, satisfying the intertwining relations (2.36)), are interpreted as the unobservable "charged" fields of the theory which play an important rôle in the construction of scattering theory [4,5,28].

In order to describe the bundle $\mathcal{J}_{kji}$ more explicitly, we propose to construct an atlas of local coordinate charts, along with its transition functions, for $\mathcal{J}_{kji}$. The details of this construction are given in [28]. Here we just outline some basic ideas.

First, we define the manifold $\mathcal{M}_j$, of *morphisms of type j more precisely. We choose two space-like separated, space-like auxiliary cones, C_a^I and C_a^{II}, where C_a^{II} is obtained from C_a^I by a Euclidean motion. The corresponding enlarged C^* algebras,

$$\mathcal{B}^I \equiv \mathcal{B}^{C_a^I}, \quad \text{and} \quad \mathcal{B}^{II} \equiv \mathcal{B}^{C_a^{II}}, \tag{2.41}$$

are defined as in (2.19). We pick two reference morphisms, ρ^I and ρ^{II}, with the properties that $1(\rho^{\#}(\cdot)) \simeq j$, and $\rho^{\#}$ is localized in a space-like cone $C_{\#}$, for $\# = I, II$, such that C_I and C_{II} are space-like separated from $C_a^I \cup C_a^{II}$. We could choose $\rho^I = \rho^{II} \equiv \rho_0$, where ρ_0 is localized in a space-like cone C_0 space-like separated from $C_a^I \cup C_a^{II}$.

Next, we define two groups, $\mathcal{U}^I$ and $\mathcal{U}^{II}$, of unitary operators, as follows:

$$\mathcal{U}^{\#} := \{\Gamma : \Gamma \in \mathcal{B}^{\#},\ \Gamma^* = \Gamma^{-1}\}.$$

If ρ is a *morphism in $\mathcal{M}_j$ localized in a space-like cone C such that C is space-like separated from $C_a^{\#} + x$, for some $x \in \mathrm{M}^3$ — this is written, for short, as $\rho \times C_a^{\#}$ — then there exists an operator $\Gamma^{\#} \in \mathcal{U}^{\#}$ such that

$$\rho(A)\,\Gamma^{\#}_{\rho,\rho^{\#}} = \Gamma^{\#}_{\rho,\rho^{\#}}\,\rho^{\#}(A), \quad \text{for all } A \in \mathcal{A}; \tag{2.42}$$

see (2.21) and (2.23). If $\Gamma'_{\rho,\rho^{\#}}$ is another element of $\mathcal{U}^{\#}$ for which (2.42) holds then $(\Gamma^{\#}_{\rho,\rho^{\#}})^*\,\Gamma'_{\rho,\rho^{\#}}$ commutes with $\rho^{\#}(\mathcal{A})$, and, since $\mathcal{M}_j$ consists of irreducible morphisms,

$$(\Gamma^{\#}_{\rho,\rho^{\#}})^*\,\Gamma'_{\rho,\rho^{\#}} = e^{i\theta}, \quad \text{for some } \theta \in [0, 2\pi). \tag{2.43}$$

Thus $\Gamma^{\#}_{\rho,\rho^{\#}}$ is unique up to a phase factor. We define

$$\left[\Gamma^{\#}_{\rho,\rho^{\#}}\right] = \left\{\Gamma'_{\rho,\rho^{\#}} \in \mathcal{U}^{\#} : \Gamma'_{\rho,\rho^{\#}} = e^{i\theta}\,\Gamma^{\#}_{\rho,\rho^{\#}}\right\} \in P\mathcal{U}^{\#}. \tag{2.44}$$

It is straightforward to verify the following properties: Let ρ_1, ρ_2, and ρ_3 be three *morphisms in $\mathcal{M}_j$, with $\rho_i \times C_a^{\#}$, for $i = 1,2,3$. Then

$$\text{(a)} \qquad \left[\Gamma^{\#}_{\rho_i,\rho_i}\right] = \text{identity};$$

$$\text{(b)} \qquad \left[\left(\Gamma^{\#}_{\rho_i,\rho_k}\right)^*\right] = \left[\Gamma^{\#}_{\rho_k,\rho_i}\right];$$

$$\text{(c)} \qquad \left[\Gamma^{\#}_{\rho_1,\rho_2}\right]\left[\Gamma^{\#}_{\rho_2,\rho_3}\right] = \left[\Gamma^{\#}_{\rho_1,\rho_3}\right];$$

$$\text{(d)} \qquad \left[\Gamma^{I}_{\rho_i,\rho_k}\right] = \left[\Gamma^{II}_{\rho_i,\rho_k}\right]$$

if ρ_i and ρ_k are localized in a simple domain $\mathcal{S}$ space-like separated from $C_a^I \cup C_a^{II}$.

Local coordinates on $\mathcal{M}_j$ in the vicinity of a reference morphism $\rho^{\#}$ are given by the coordinate map

$$\phi^{\#}_{\rho^{\#}} : \rho \in \mathcal{M}_j \mapsto \left[\Gamma^{\#}_{\rho,\rho^{\#}}\right] \in P\mathcal{U}^{\#}. \tag{2.45}$$

By (b) and (c), the transition function $\phi^I_{\rho^I} \circ \left(\phi^{II}_{\rho^{II}}\right)^{-1}$ is given by right multiplication by

$$\left[\Gamma^{I}_{\rho^{II},\rho^I}\right].$$

This defines $\mathcal{M}_j$ as an infinite-dimensional topological manifold modelled on the projective unitary group $P\mathcal{U}$, where $\mathcal{U}$ is the group of unitary operators in a C^*algebra $\mathcal{B}$ isomorphic to $\mathcal{B}^{\#}$; (note that, as abstract C^*algebras, $\mathcal{B}^I$ and $\mathcal{B}^{II}$ are isomorphic). It is not hard to see that the fundamental group of $\mathcal{M}_j$ is given by

$$\pi_1(\mathcal{M}_j) = \mathbb{Z}. \tag{2.46}$$

The geometrical fact underlying (2.46) is that the manifold of asymptotic directions of the space-like cones in which the *morphisms $\rho \in \mathcal{M}_j$ are localized is a circle, the circle of points at infinity in two-dimensional space. For additional details see [28].

Next, we shall construct coordinate charts for the bundle $\mathcal{J}_{kji}$. For any *morphism $\rho \in \mathcal{M}_j$, with $\rho \times C_a^{\#}$, we shall choose a representative $\Gamma^{\#}_{\rho,\rho^{\#}} \in \left[\Gamma^{\#}_{\rho,\rho^{\#}}\right]$, in a definite way explained in [28]. By (2.23), $\Gamma^{\#}_{\rho,\rho^{\#}} \in \mathcal{B}^{\#}$. Let $V(\rho^{\#})$ be an intertwiner satisfying (2.36). Then, by (2.42) and (2.36), the operator

$$V(\rho) = i\left(\Gamma^{\#}_{\rho,\rho^{\#}}\right) V(\rho^{\#}) \tag{2.47}$$

is an element of the fibre $\mathcal{V}_k(\rho)_i$ above ρ. [Note that $i(\Gamma^{\#}_{\rho,\rho^{\#}})$ is a well-defined, unitary operator on $\mathcal{H}_i$, because $\Gamma^{\#}_{\rho,\rho^{\#}} \in \mathcal{B}^{\#}$, and i is a representation of $\mathcal{B}^{\#}$.] Let us choose an orthonormal basis

$$\{V_\mu^{ik}(\rho^{\#})\}_{\mu=1}^{N_{kji}} \subset \mathcal{V}_k(\rho^{\#})_i \,. \tag{2.48}$$

Then (2.47) shows that

$$\left\{ i\left(\Gamma^{\#}_{\rho,\rho^{\#}}\right) V_\mu^{ik}\left(\rho^{\#}\right) \right\}_{\mu=1}^{N_{kji}} \tag{2.49}$$

is an orthonormal basis in $\mathcal{V}_k(\rho)_i$.

Two coordinate charts on $\mathcal{J}_{kji}$ are now constructed as follows: A pair $\{\rho, V(\rho)\}$, $\rho \in \mathcal{M}_j$, belongs to the chart $\mathcal{N}^{\#}$ of $\mathcal{J}_{kji}$ iff $\rho \times \mathcal{C}^{\#}_a$, (i.e. ρ is localized in a space-like cone $\mathcal{C}$ with the property that $\mathcal{C}$ is space-like separated from $\mathcal{C}^{\#}_a + x$, for some $x \in \mathsf{M}^3$); $\# = I$, or II. The $\mathcal{N}^{\#}$-coordinates of $\{\rho, V(\rho)\}$ are given by

$$\left\{ [\Gamma^{\#}_{\rho,\rho^{\#}}] \in P\mathcal{U}^{\#},\ \langle V(\rho),\ i(\Gamma^{\#}_{\rho,\rho^{\#}})\ V_\mu^{ik}(\rho^{\#}),\rangle\ \mu = 1,\ldots,N_{kji} \right\}, \tag{2.50}$$

where $\langle\cdot,\cdot\rangle$ is the scalar product on $\mathcal{V}_k(\rho)_i$ constructed in (2.39).

An atlas for $\mathcal{J}_{kji}$ consists of the two coordinate charts $\mathcal{N}^I$ and $\mathcal{N}^{II}$, together with transition functions on $\mathcal{N}^I \cap \mathcal{N}^{II}$. In order to calculate the transition functions, we consider the geometrical situation sketched in Fig. 4:

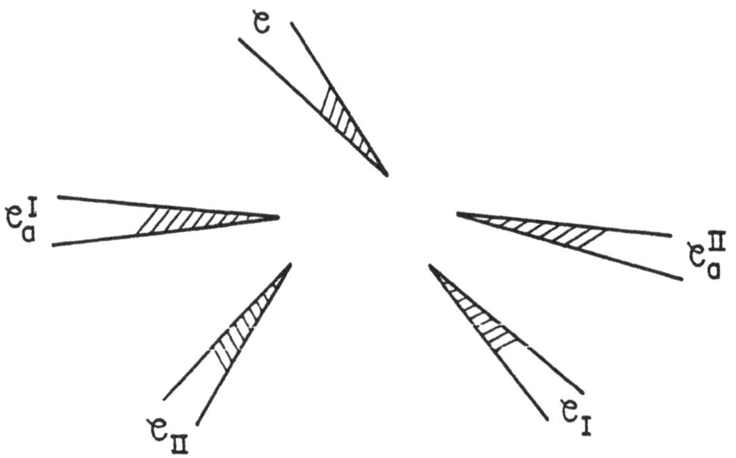

Fig. 4

The reference morphisms ρ^I and ρ^{II} are localized in the space-like cones C_I and C_{II}; the morphism ρ is localized in C. The cones C, C_I and C_{II} are space-like separated from the two auxiliary cones C_a^I and C_a^{II}. The $\mathcal{N}^{\#}$-coordinates of $\{\rho,\, V(\rho)\}$ are given by

$$\left\{ [\Gamma^{\#}_{\rho,\rho^{\#}}],\ \langle V(\rho),\, i(\Gamma^{\#}_{\rho,\rho^{\#}})\ V^{ik}_{\mu}(\rho^{\#})\rangle,\ \mu = 1,\ldots,N_{kji} \right\}, \tag{2.51}$$

with $\Gamma^{\#}_{\rho,\rho^{\#}} \in \mathcal{U}^{\#} \subset \mathcal{B}^{\#}$. The transition functions are calculated by comparing (2.51), for $\# = I$, with (2.51), for $\# = II$. This requires some work [28] which we now sketch.

(1) Since $\rho^{II} \in \mathcal{N}^I$, (see Fig. 4), we can express $\{\rho^{II},\, V^{ik}_{\nu}(\rho^{II})\}$ in $\mathcal{N}^I$-coordinates. They are given by

$$\left\{ [\Gamma^{I}_{\rho^{II},\rho^{I}}],\ a_{\nu\mu}(\rho^{II},\rho^{I}),\ \mu = 1,\ldots,N_{kji} \right\}, \tag{2.52}$$

where

$$a_{\nu\mu}(\rho^{II},\rho^{I}) = \langle V^{ik}_{\nu}(\rho^{II}),\, i(\Gamma^{I}_{\rho^{II},\rho^{I}})\ V^{ik}_{\mu}(\rho^{I})\rangle. \tag{2.53}$$

(2) The calculation of transition functions on $\mathcal{N}^I \cap \mathcal{N}^{II}$ is complicated by the circumstance that there is no abstract C^*algebra containing $\Gamma^{I}_{\rho,\rho^{I}}$, $\Gamma^{I}_{\rho^{II},\rho^{I}}$ <u>and</u> $\Gamma^{II}_{\rho,\rho^{II}}$, for arbitrary $\rho \in \mathcal{N}^I \cap \mathcal{N}^{II}$. The operators $\Gamma^{I}_{\rho,\rho^{I}}$ and $\Gamma^{I}_{\rho^{II},\rho^{I}}$ are elements of $\mathcal{B}^I$, while $\Gamma^{II}_{\rho,\rho^{II}}$ is an element of $\mathcal{B}^{II}$. Though not disjoint, the algebras $\mathcal{B}^I$ and $\mathcal{B}^{II}$ are distinct, so that, a priori, multiplication of $\left(\Gamma^{I}_{\rho,\rho^{I}}\right)^*$ with $\Gamma^{II}_{\rho,\rho^{II}}$ is not defined. However, $i(\mathcal{B}^I)$ and $i(\mathcal{B}^{II})$ are both naturally imbedded in $B(\mathcal{H}_i)$, the algebra of all bounded operators on the representation space $\mathcal{H}_i$. Hence $i(\Gamma^{I}_{\rho,\rho^{I}})^*\, i(\Gamma^{II}_{\rho,\rho^{II}})$ is defined as multiplication of operators in $B(\mathcal{H}_i)$. We define

$$V^{i}_{-2\pi} = i(\Gamma^{I}_{\rho^{II},\rho^{I}})\ i(\Gamma^{I}_{\rho,\rho^{I}})^*\ i(\Gamma^{II}_{\rho,\rho^{II}}), \tag{2.54}$$

where ρ is localized in a space-like cone C located as sketched in Fig. 4.* From Fig. 4 we see that $V^{i}_{-2\pi}$ has the interpretation of rotating the *morphism ρ^{II} through an angle -2π. One might therefore expect that $V^{i}_{-2\pi}$ can be expressed in terms of <u>spins</u>.

---

* *This definition is unambiguous only if the phases of the charge transport operators are chosen in a definite way; see [28].*

(3) In order to compute $V^i_{-2\pi}$, we compare an arbitrary intertwiner

$$V(\rho^{II}) = \sum_{\mu=1}^{N_{kji}} b_\mu \, V_\mu^{ik}(\rho^{II}) \in \mathcal{V}_k(\rho^{II})_i$$

with the operator $V^i_{-2\pi}\, V(\rho^{II})$. We note that $V(\rho^{II})$ and $V^i_{-2\pi}\, V(\rho^{II})$ both satisfy the intertwining relations

$$\begin{aligned} i(\rho^{II}(A))\, V(\rho^{II}) &= V(\rho^{II})\, k(A), \text{ and} \\ i(\rho^{II}(A))\, V^i_{-2\pi}\, V(\rho^{II}) &= V^i_{-2\pi}\, V(\rho^{II})\, k(A), \end{aligned} \tag{2.55}$$

for all $A \in \mathcal{A}$; (the second equation in (2.55) follows from the first one and from (2.54)). Hence $V^i_{-2\pi}\, V(\rho^{II})$ belongs to $\mathcal{V}_k(\rho^{II})_i$, too, and by Schur's lemma, there exists an $N_{kji} \times N_{kji}$-matrix, $\left(V^-_{\nu\mu}(i,k)\right)_{\nu,\mu=1}^{N_{kji}}$, such that

$$V^i_{-2\pi}\, V(\rho^{II}) = \sum_\nu c_\nu \, V_\nu^{ik}(\rho^{II}),$$

with

$$c_\nu = \sum_\mu V^-_{\nu\mu}(i,k)\, b_\mu \, . \tag{2.56}$$

We propose to calculate the matrices $V^-(i,k)$.

(4) We recall that all representations in L are irreducible, covariant positive-energy representations of $\{\mathcal{A}, \alpha\}$. Thus, for $k \in L$, there is a representation U_k of $\tilde{\mathcal{P}}_+^\uparrow$ on $\mathcal{H}_k$. Let $U_k(2\pi)$ be the unitary operator representing the space rotation through an angle 2π. Clearly $U_k(2\pi)$ commutes with $k(\mathcal{A})$, and, since k is irreducible, it follows that

$$U_k(2\pi) = e^{2\pi i s_k}, \tag{2.57}$$

where s_k is called the <u>spin of the representation</u> k. Since the little group in $\tilde{\mathcal{P}}_+^\uparrow$ of a time-like vector is isomorphic to the covering group, $\widetilde{SO}(2) = \mathsf{R}$, of the subgroup of space rotations in M^3, the spin s_k can be an arbitrary real number in the interval $[0,1)$.

Consider the following loop of intertwiners:

$$\{V(\rho^{II}, \theta) \, : \, -2\pi \le \theta \le 0\},$$

with

$$V(\rho^{II},\theta) := U_i(\theta)\,V(\rho^{II})\,U_k(-\theta), \tag{2.58}$$

where $V(\rho^{II}) \in \mathcal{V}_k(\rho^{II})_i$, and $U_k(\theta)$ represents a space rotation through an angle θ on $\mathcal{H}_k$. One easily checks that

$$V(\rho^{II},\theta) \in \mathcal{V}_k(\alpha_\theta \circ \rho^{II} \circ \alpha_{-\theta})_i,$$

where α_θ is the *automorphism of $\mathcal{A}$ representing the space rotation through an angle θ. Hence $V(\rho^{II},-2\pi) \in \mathcal{V}_k(\rho^{II})_i$. Given the geometrical situation sketched in Fig. 4, it is not surprizing that one can choose the charge-transport operators $\{\Gamma^I_{\rho,\rho^I},\ \rho \in \mathcal{N}^I\}$ and $\{\Gamma^{II}_{\rho,\rho^{II}},\ \rho \in \mathcal{N}^{II}\}$ such that

$$V^i_{-2\pi}\,V(\rho^{II}) = V(\rho^{II},-2\pi). \tag{2.59}$$

A complete analysis of this point is non-trivial, and we refer the reader to [28] for details.

From (2.57), (2.58) and (2.59) we conclude the following theorem proven in [28].

Theorem

The matrix $V^-(i,k)$ introduced in (2.56) is given by

$$V^-(i,k) = e^{2\pi i(s_k - s_i)}\,\mathbb{1}; \tag{2.60}$$

equivalently,

$$V^i_{-2\pi}\,V(\rho^{II}) = e^{2\pi i(s_k - s_i)}\,V(\rho^{II}), \tag{2.61}$$

for all $V(\rho^{II}) \in \mathcal{V}_k(\rho^{II})_i$.

□

(5) We are now ready to calculate the transition functions on $\mathcal{N}^I \cap \mathcal{N}^{II}$. By (2.52), left multiplication by $i(\Gamma^I_{\rho^{II},\rho^I})^*$ maps $\mathcal{V}_k(\rho^{II})_i$ onto $\mathcal{V}_k(\rho^I)_i$. By (2.54) and (2.61), we have that, for an arbitrary intertwiner $V(\rho^{II}) \in \mathcal{V}_k(\rho^{II})_i$,

$$\begin{aligned} i(\Gamma^{II}_{\rho,\rho^{II}})\,V(\rho^{II}) &= i(\Gamma^I_{\rho,\rho^I})\,i(\Gamma^I_{\rho^{II},\rho^I})^*\,e^{2\pi i(s_k - s_i)}\,V(\rho^{II}) \\ &= i(\Gamma^I_{\rho,\rho^{II}})\,e^{2\pi i(s_k - s_i)}\,V(\rho^{II}), \end{aligned} \tag{2.62}$$

where $\Gamma^{I}_{\rho,\rho^{II}} \equiv \Gamma^{I}_{\rho,\rho^{I}} \left(\Gamma^{I}_{\rho^{II},\rho^{I}}\right)^{*}$. Using (2.62) and (2.52), we find that

$$\begin{aligned}
&\langle V(\rho), i\,(\Gamma^{II}_{\rho,\rho^{II}})\, V^{ik}_{\nu}(\rho^{II})\rangle \\
&\quad = \langle V(\rho),\, i(\Gamma^{I}_{\rho,\rho^{II}})\, V^{ik}_{\nu}(\rho^{II})\rangle\, e^{2\pi\, i(s_k - s_i)} \\
&\quad = \sum_{\mu} \bar{a}_{\nu\mu}\, e^{2\pi\, i(s_k - s_i)} \langle V(\rho),\, i(\Gamma^{I}_{\rho,\rho^{I}})\, V^{ik}_{\mu}(\rho^{I})\rangle,
\end{aligned} \tag{2.63}$$

where $a_{\nu\mu} = a_{\nu\mu}(\rho^{II}, \rho^{I})$ is given by (2.53). Eq. (2.45) and (2.63) show that the transition function on the component of $\mathcal{N}^{I} \cap \mathcal{N}^{II}$ described in Fig. 4 is given by

$$\left\{ R[\Gamma^{I}_{\rho^{I},\rho^{II}}],\ \left(\bar{a}_{\nu\mu}\, e^{2\pi\, i(s_k - s_i)}\right)^{N_{kji}}_{\nu,\mu=1} \right\}, \tag{2.64}$$

where $R[\Gamma]$ denotes right-multiplication by $[\Gamma]$ and corresponds to the transformation

$$[\Gamma^{I}_{\rho,\rho^{I}}] \ \mapsto\ [\Gamma^{II}_{\rho,\rho^{II}}] \ =\ [\Gamma^{I}_{\rho,\rho^{I}}]\,[\Gamma^{I}_{\rho^{I},\rho^{II}}], \tag{2.65}$$

and $(\bar{a}_{\nu\mu}\, e^{2\pi\, i(s_k - s_i)})$ describes the transformation (2.63). The transition function on the component of $\mathcal{N}^{I} \cap \mathcal{N}^{II}$ shown in Fig. 4', below, is given by

$$\left\{ R[\Gamma^{I}_{\rho^{I},\rho^{II}}],\ \left(\bar{a}_{\nu\mu}\right)^{N_{kji}}_{\nu\mu=1} \right\}. \tag{2.66}$$

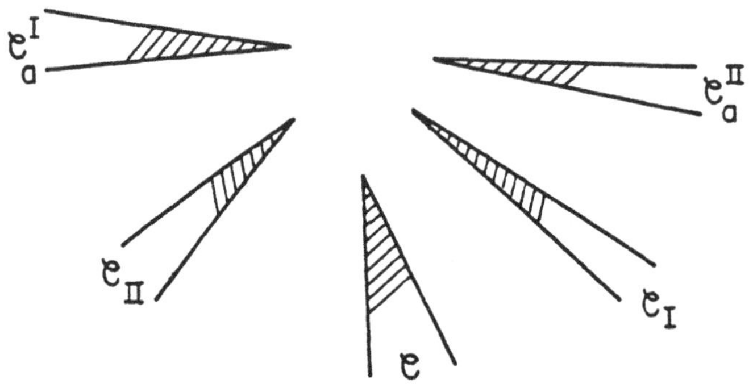

Fig. 4'

If we choose $\rho^{I} = \rho^{II}$ then

$$\begin{aligned}
a_{\nu\mu} &= \delta_{\nu\mu}, \text{ and} \\
V^{I}_{-2\pi} &= i(\Gamma^{I}_{\rho,\rho^{I}})^{*}\, i(\Gamma^{II}_{\rho,\rho^{I}}).
\end{aligned} \tag{2.67}$$

This completes our sketch of the construction of the vector bundles $\mathcal{J}_{kji}$ of intertwiners ('charged fields') from $\mathcal{H}_k$ to $\mathcal{H}_i$.

Formulas (2.64) and (2.66) reflect the non-trivial topology of $\mathcal{M}_j$ which, in turn, reflects the topology of the manifold of space-like asymptotic directions of space-time.

This concludes our brief review of the algebraic approach to local, relativistic quantum theory [4,5,6,28]. In the next section, we shall discuss the structure of the algebra of intertwiners.

3. Statistics and fusion of intertwiners

Let C_a be some space-like cone in three-dimensional Minkowski space M^3, and let $\mathcal{S}$ be a simple domain (region) contained in the space-like complement of C_a. [Space-like cones and simple domains were defined in (2.5),(2.6).] Let $C \subset \mathcal{S}$ be some space-like cone. With C we associate an angle $\theta(C)$ as follows: We choose polar coordinates (r, θ) in two-dimensional space, $\{(\vec{x}, t) \in \mathsf{M}^3 : t = 0\}$. Let $\sigma(C)$ be the half-line in space bisecting the wedge C_0 whose causal completion is the cone C; see (2.5),(2.6). Let $\theta(C)$ be the asymptotic angle of $\sigma(C)$; $\theta(C)$ is called the asymptotic direction of C. If ρ is some *morphism of $\mathcal{B}^{C_a}$ localized in C then $\theta(C)$ is called the *asymptotic direction* of ρ and is denoted by $as(\rho)$. We may choose our coordinates such that $\theta(C_a) = \pi$, and require that

$$| \, as(\rho) \, | < \pi \, , \tag{3.1}$$

for all *morphisms ρ localized in space-like cones contained in $\mathcal{S}$.

Let ρ_1 and ρ_2 be two *morphisms of $\mathcal{B}^{C_a}$ localized in space-like cones C_1 and C_2, respectively. We say that ρ_1 and ρ_2 are *causally independent*, denoted by $\rho_1 \times \rho_2$, iff C_1 and C_2 are space-like separated.

For every irreducible, covariant, positive-energy representation $j \in L$, we choose a *reference morphism* $\rho_0^j \in \mathcal{M}_j$, localized in a space-like cone $C_0^j \subset \mathcal{S}$, and a basis of intertwiners

$$V_\mu^{ik}(\rho_0^j) \, : \, \mathcal{H}_k \to \mathcal{H}_i \, , \tag{3.2}$$

satisfying the intertwining relations

$$i(\rho_0^j(A))\, V_\mu^{ik}(\rho_0^j) \;=\; V_\mu^{ik}(\rho_0^j)\, k(A)\,, \tag{3.3}$$

for $\mu = 1, \ldots, N_{kji}$. [We recall that $\mathcal{M}_j$ is the space of all *morphisms, ρ^j, localized in space-like cones with the property that the representation j of $\mathcal{A}$ is unitarily equivalent to the vacuum representation, 1, of $\rho^j(\mathcal{A})$.]

Let ρ^j be some other *morphism of $\mathcal{B}^{C_a}$ contained in $\mathcal{M}_j$ and localized in a space-like cone contained in $\mathcal{S}$. As explained in Sect. 2, (2.21), (2.23) there is then a unitary operator $\Gamma_{\rho^j,\rho_0^j} \equiv \Gamma^{\mathcal{S}}_{\rho^j,\rho_0^j} \in \overline{\mathcal{A}(\mathcal{S})}^w$ such that

$$\rho^j(A)\, \Gamma^{\mathcal{S}}_{\rho^j,\rho_0^j} \;=\; \Gamma^{\mathcal{S}}_{\rho^j,\rho_0^j}\, \rho_0^j(A)\,, \tag{3.4}$$

for all $A \in \mathcal{A}$. Moreover, if ρ_1, ρ_2 and ρ_3 are *morphisms in $\mathcal{M}_j$ localized in space-like cones $\subset \mathcal{S}$

$$\left[\Gamma^{\mathcal{S}}_{\rho_i,\rho_i}\right] \;=\; \mathbb{1}, \;\; \left[(\Gamma^{\mathcal{S}}_{\rho_i,\rho_j})^*\right] \;=\; \left[\Gamma^{\mathcal{S}}_{\rho_j,\rho_i}\right], \tag{3.5}$$

for $i, j = 1, 2, 3$, and

$$\left[\Gamma^{\mathcal{S}}_{\rho_1,\rho_2}\right] \left[\Gamma^{\mathcal{S}}_{\rho_2,\rho_3}\right] \;=\; \left[\Gamma^{\mathcal{S}}_{\rho_1,\rho_3}\right]. \tag{3.6}$$

See Sect. 2, (a)-(c), after (2.44).

A basis of intertwiners, $\left\{V_\mu^{ik}(\rho^j)\right\}_{\mu=1}^{N_{kji}}$, associated with ρ^j is obtained by setting

$$V_\mu^{ik}(\rho^j) \;=\; i(\Gamma^{\mathcal{S}}_{\rho^j,\rho_0^j})\, V_\mu^{ik}(\rho_0^j)\,; \tag{3.7}$$

see (2.49). They satisfy the intertwining relations

$$i(\rho^j(A))\, V_\mu^{ik}(\rho^j) \;=\; V_\mu^{ik}(\rho^j)\, k(A)\,. \tag{3.8}$$

3.1. The statistics of intertwiners

The structure of the algebra of intertwiners is described in the following basic result.

<u>Theorem 1</u> [28]

<u>For p and q in L, let $\rho^p \in \mathcal{M}_p$ and $\rho^q \in \mathcal{M}_q$ be two *morphisms of $\mathcal{B}^{C_a}$ localized in space-like cones contained in $\mathcal{S}$. Let the intertwiners $\left\{V_\mu^{ik}(\rho^{p,q})\right\}$ be defined as in (3.7). Then there are matrices, called statistics matrices,</u>

$$\left(R^{\pm}(j,p,q,k)^{l\kappa\lambda}_{i\mu\nu}\right)$$

only depending on the classes $\mathcal{M}_p$ and $\mathcal{M}_q$, such that

$$\begin{aligned} & V^{ji}_{\mu}(\rho^p)\, V^{ik}_{\nu}(\rho^q) \\ & \qquad = \sum_{l,\alpha,\beta} R^{\pm}(j,p,q,k)^{l\alpha\beta}_{i\mu\nu}\, V^{jl}_{\alpha}(\rho^q)\, V^{lk}_{\beta}(\rho^p)\,, \end{aligned} \tag{3.9}$$

provided $\rho^p \times \rho^q$ and $as(\rho^p) \gtrless as(\rho^q)$.

□

Remarks The proof of Theorem 1 is given in [28]. That there is a relation of the form of (3.9) is not difficult to see. It is a straightforward consequence of Schur's lemma: Consider the operator

$$V \equiv V^{lk}_{\beta}(\rho^p)^*\, V^{jl}_{\alpha}(\rho^q)^*\, V^{ji}_{\mu}(\rho^p)\, V^{ik}_{\nu}(\rho^q)\,. \tag{3.10}$$

From the intertwining relations (3.8) and their adjoint it follows that

$$k(A)V = V k(A), \quad \text{for all } A \in \mathcal{A}\,. \tag{3.11}$$

Since k is an irreducible representation of $\mathcal{A}$, it follows from Schur's lemma that

$$V = \lambda\, 1\!\!1,\ \lambda \in \mathbb{C}\,. \tag{3.12}$$

We denote λ by $R(j,\rho^p,\rho^q,k)^{l\alpha\beta}_{i\mu\nu}$. Next, we note that

$$\left\{V^{jl}_{\alpha}(\rho^q)\, V^{lk}_{\beta}(\rho^p) \,:\, l \in L,\ \alpha = 1,\ldots,N_{lqj},\ \beta = 1,\ldots,N_{kpl}\right\} \tag{3.13}$$

is a basis of intertwiners from $\mathcal{H}_k$ to $\mathcal{H}_j$ intertwining the representations $j(\rho^q \circ \rho^p(\,\cdot\,))$ and $k(\,\cdot\,)$ of the algebra $\mathcal{A}$; see e.g. [28]. From (3.10),(3.12) and (3.13) we conclude that

$$V^{ji}_{\mu}(\rho^p)\, V^{ik}_{\nu}(\rho^q) = \sum_{l,\alpha,\beta} R(j,\rho^p,\rho^q,k)^{l\alpha\beta}_{i\mu\nu}\, V^{jl}_{\alpha}(\rho^q)\, V^{lk}_{\beta}(\rho^p)\,. \tag{3.14}$$

Next, one shows, using an argument invented in the proof of Lemma 2.6 of [4], that if $\rho^p \times \rho^q$ then

$$R(j,\rho^p,\rho^q,k)^{l\alpha\beta}_{i\mu\nu} = R^{\pm}(j,p,q,k)^{l\alpha\beta}_{i\mu\nu}\,, \tag{3.15}$$

for $as(\rho^p) \gtrless as(\rho^q)$, where the matrices $\left(R^{\pm}(j,p,q,k)^{l\alpha\beta}_{i\mu\nu}\right)$ only depend on the classes $\mathcal{M}_p$ and $\mathcal{M}_q$ of *morphisms to which ρ^p and ρ^q belong, but are independent of the specific choice of ρ^p and ρ^q; they are also independent of the choice of the auxiliary cone $\mathcal{C}_a$, (as long as $\mathcal{C}_a$ is space-like separated from the localization cones of ρ^p and ρ^q). Although these facts are not difficult to prove, technically, they are somwhat more subtle than (3.14). For proofs see [28].

Next, we investigate the properties of the statistics matrices $\left(R^{\#}(j,p,q,k)^{l\alpha\beta}_{i\mu\nu}\right)$ somewhat systematically. For that purpose we introduce a graphical notation:

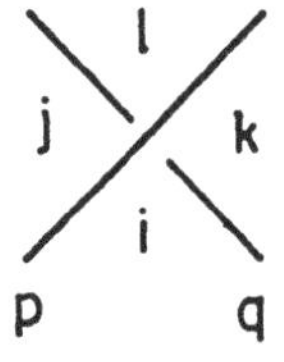

$$\longleftrightarrow \qquad R^{+}(j,p,q,k)^{l}_{i} \tag{3.16}$$

$$\longleftrightarrow \qquad R^{-}(j,p,q,k)^{l}_{i}\,. \tag{3.17}$$

We have dropped the Greek multiplicity indices μ, ν, α and β; (a more complete notation would be

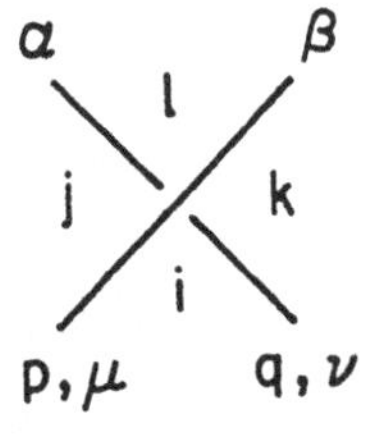

, etc.; repeated indices are to be summed over).

Iterating equ. (3.9), we find that

(1)

$\sum_{l}$ 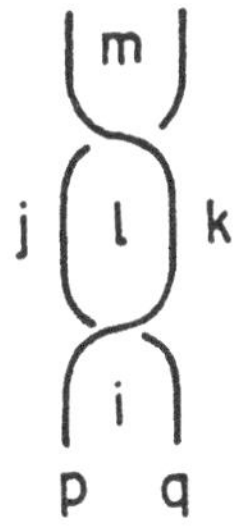

$= \delta_i^m$ 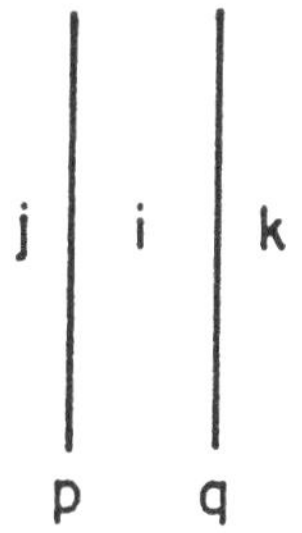

(3.18)

where

$\longleftrightarrow \quad \mathbb{1}\,\big|_{\mathbb{C}^{N_{ipj}}}\,,$ (3.19)

or

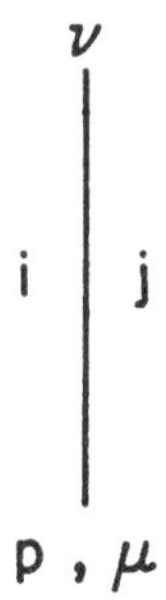

$\longleftrightarrow \quad \delta^{\nu}_{\mu}\,, \quad \mu, \nu = 1, \ldots, N_{ipj}\,.$ (3.19′)

By considering

$$V_{\mu}^{ji}(\rho^p)\, V_{\nu}^{ik}(\rho^q)\, V_{\kappa}^{kl}(\rho^r), \tag{3.20}$$

and assuming that $as(\rho^p)$, $as(\rho^q)$ and $as(\rho^r)$ are ordered in some way and ρ^p, ρ^q and ρ^r are pairwise causally independent, (i.e. $\rho^p \times \rho^q$, $\rho^p \times \rho^r$ and $\rho^q \times \rho^r$), we find by permuting the order of the factors in (3.20) to

$$V_{\alpha}^{jm}(\rho^r)\, V_{\beta}^{mn}(\rho^q)\, V_{\gamma}^{nl}(\rho^p) \tag{3.21}$$

in two distinct ways that

(2)

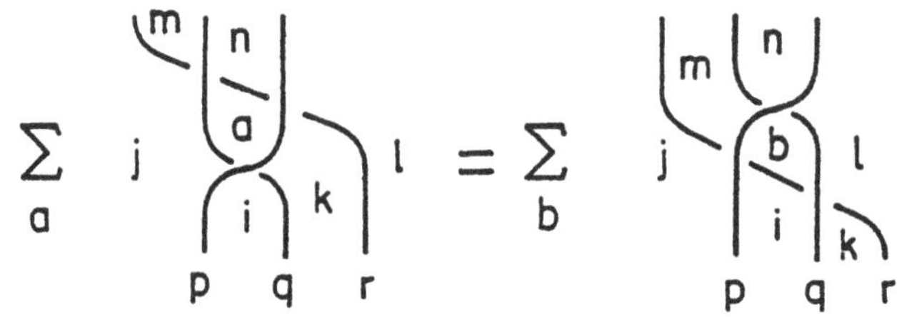

(3.22)

here it is assumed that $as(\rho^p) > as(\rho^q) > as(\rho^r)$. Other related identities are found for other orderings of $as(\rho^p)$, $as(\rho^q)$ and $as(\rho^r)$. The equations (3.22) are homogeneous, cubic equations in the matrices $R^{\pm}$. They represent the sos-form of the Yang-Baxter equations (YBE) without spectral parameter. The derivation of (3.22) from (3.20) and (3.21) was first given in [15].

From (1) and (2) we conclude that the matrices $(R^{\pm}(j,p,q,k)_i^l)$ generate representations of the groupoid, B_n^c, of coloured braids on n strands.

Next, we derive a basic relation between R^+ and R^-: We consider two *morphisms ρ^p and ρ^q, localized in space-like cones C^p and C^q whose projection onto two-dimensional space is shown in Fig. 5. We suppose that the reference morphisms, ρ_0^p and ρ_0^q are localized in a space-like cone C_0. The cones C^p, C^q and C_0 are assumed to be contained in a simple region S whose space-like complement, S', is the auxiliary cone C_a.

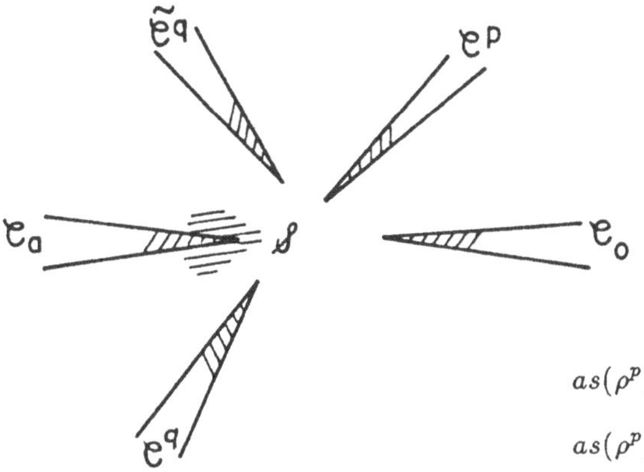

$$as(\rho^p) > as(\rho^q),$$
$$as(\rho^p) < as(\tilde{\rho}^q).$$

Fig. 5

We also consider a *morphism, $\tilde{\rho}^q$, localized in a space-like cone $\tilde{C}^q$, as shown in Fig. 5. Then we have from Theorem 1 that

$$V^{ij}(\rho^p)\, V^{jk}(\rho^q) \;=\; \sum R^+(i,p,q,k)^l_j \; V^{il}(\rho^q)\, V^{lk}(\rho^p), \tag{3.23}$$

and

$$V^{ij}(\rho^p)\, V^{jk}(\tilde{\rho}^q) \;=\; \sum R^-(i,p,q,k)^l_j \; V^{il}(\tilde{\rho}^q)\, V^{lk}(\rho^p). \tag{3.24}$$

[We omit the multiplicity indices $\mu, \nu, \ldots$ everywhere.] Since $R^-(i,p,q,k)^l_j$ is independent of the choice of the auxiliary cone, C_a, (3.24) does not change if C_a is replaced by a new auxiliary cone, $\hat{C}_a$, chosen as indicated in Fig. 6.

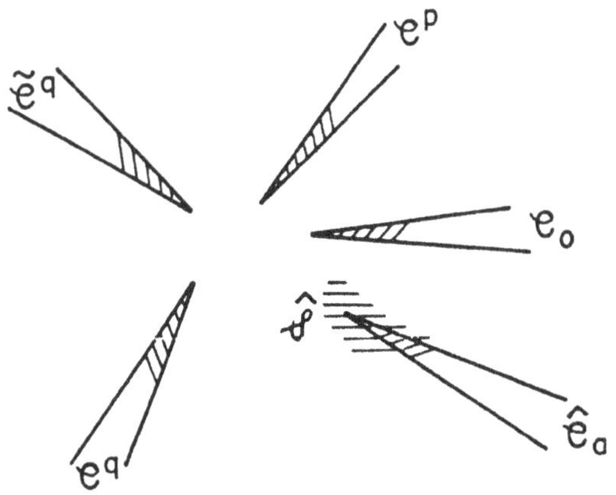

Fig. 6

In the situation shown in Fig. 6,

$$as(\rho^q) \geq as(\tilde{\rho}^q) > as(\rho^p). \tag{3.25}$$

We define

$$V^{ij}_{\hat{S}}(\rho^{\#}) = i(\Gamma^{\hat{S}}_{\rho^{\#},\rho_0^{\#}})\, V^{ij}(\rho_0^{\#}), \tag{3.26}$$

for $\# = p, q$. We also recall that

$$V^{ij}(\rho^{\#}) = i(\Gamma^{S}_{\rho^{\#},\rho_0^{\#}})\, V^{ij}(\rho_0^{\#}); \tag{3.27}$$

see (3.7). Thus we conclude from (3.24) and the remark that R^- does not change if C_a is replaced by $\hat{C}_a$ that

$$V^{ij}_{\hat{S}}(\rho^p)\, V^{jk}_{\hat{S}}(\tilde{\rho}^q) = \sum R^-(i,p,q,k)^l_j\, V^{il}_{\hat{S}}(\tilde{\rho}^q)\, V^{lk}_{\hat{S}}(\rho^p). \tag{3.28}$$

But in the situation shown in Fig. 6, $\tilde{C}^q$ can be rotated to C^q in the positive direction inside $\hat{S}$, and, since $R^-(i,p,q,k)^l_j$ only depends on $\mathcal{M}_p$ and $\mathcal{M}_q$, but not on $\tilde{\rho}^q$, we can replace $\tilde{\rho}^q$ by ρ^q and conclude that

$$V^{ij}_{\hat{S}}(\rho^p)\, V^{jk}_{\hat{S}}(\rho^q) = \sum R^-(i,p,q,k)^l_j\, V^{il}_{\hat{S}}(\rho^q)\, V^{lk}_{\hat{S}}(\rho^p). \tag{3.29}$$

From (3.26) and (3.27) we obtain that

$$\begin{aligned} V^{ij}(\rho^q) &= i(\Gamma^{S}_{\rho^q,\rho_0^q})\, i(\Gamma^{\hat{S}}_{\rho^q,\rho_0^q})^*\, V^{ij}_{\hat{S}}(\rho^q) \\ &= i(\Gamma^{S}_{\rho^q,\rho_0^q})\, i(\Gamma^{\hat{S}}_{\rho_0^q,\rho^q})\, V^{ij}_{\hat{S}}(\rho^q). \end{aligned} \tag{3.30}$$

But from the calculations in Sect. 2, (2.54),(2.56),(2.60), (2.61) and (2.67), we infer that

$$\begin{aligned} & i(\Gamma^{S}_{\rho^q,\rho_0^q})\, i(\Gamma^{\hat{S}}_{\rho_0^q,\rho^q})\, V^{ij}_{\hat{S}}(\rho^q) \\ &= V^{i}_{-2\pi}\, V^{ij}_{\hat{S}}(\rho^q) \\ &= e^{2\pi i(s_j - s_i)}\, V^{ij}_{\hat{S}}(\rho^q), \end{aligned}$$

where s_j is the spin of representation j. Hence

$$V^{ij}(\rho^q) = e^{2\pi i(s_j - s_i)}\, V^{ij}_{\hat{S}}(\rho^q). \tag{3.31}$$

From Fig. 5 and Fig. 6 we also learn that

$$V^{ij}(\rho^p) = V^{ij}_{\hat{S}}(\rho^p). \tag{3.32}$$

Inserting (3.31) and (3.32) into (3.29) we have that

$$\begin{aligned}V^{ij}(\rho^p)\,V^{jk}(\rho^q) &= e^{2\pi i(s_j-s_k)}\,V^{ij}_{\hat{S}}(\rho^p)\,V^{jk}_{\hat{S}}(\rho^q)\\ &= e^{2\pi i(s_j-s_k)}\sum_l R^-(i,p,q,k)^l_j\,V^{il}_{\hat{S}}(\rho^q)\,V^{lk}_{\hat{S}}(\rho^p)\\ &= \sum_l e^{2\pi i(s_j-s_k+s_l-s_i)}\,R^-(i,p,q,k)^l_j\,V^{il}(\rho^q)\,V^{lk}(\rho^p)\,.\end{aligned} \tag{3.33}$$

Comparing (3.33) with (3.23) we arrive at the following fundamental identity:

(3) $$\boxed{R^+(i,p,q,k)^{l\alpha\beta}_{j\mu\nu} = e^{2\pi i(s_j+s_l-s_i-s_k)}\,R^-(i,p,q,k)^{l\alpha\beta}_{j\mu\nu}} \tag{3.34}$$

where s_j is the spin of representation j.

<u>Remark</u> If s_j is reinterpreted as the conformal dimension of a representation j of some chiral algebra then identities (1),(2) and (3), (see (3.18),(3.22) and (3.34)), become well known identities for the braid matrices of conformal field theory [19].

Equ. (3.34) has the following obvious, but important corollary: If all representations $j \in L$ have integer spins, i.e.

$$s_j = 0 \ mod.\,\mathbb{Z}, \quad \text{for all } j \in L,$$

then

$$R^+(i,p,q,k)^l_j = R^-(i,p,q,k)^l_j \equiv R(i,p,q,k)^l_j\,, \tag{3.35}$$

for arbitrary i,p,q,k,j and l in L. In this case, (3.18) and (3.22) imply that the matrices $(R(i,p,q,k)^l_j)$ define representations of the <u>permutation groups</u>, S_n, of n elements. Hence, <u>in a theory in which all representations have integer spin, the statistics of the intertwiners</u> $\{V^{ij}(\rho^p)\}$ <u>is ordinary permutation group statistics</u>, as analyzed by Doplicher, Haag and Roberts in [4].

Next, we prove a <u>connection between spin and statistics</u>. It is based on the following simple, but basic result: Given a representation $j \in L$, $\bar{j}$ denotes its conjugate representation; $\bar{j}$ is the unique representation of $\rho^j(\mathcal{A})$, $\rho^j \in \mathcal{M}_j$, containing precisely one subrepresentation unitarily equivalent to the vacuum representation, 1, of $\mathcal{A}$.

<u>Lemma 2</u> [28]

(1) $$R^-(l,\bar{q},\bar{p},j)^{m\alpha\beta}_{k\mu\nu} = \overline{R^+(j,p,q,l)^{m\beta\alpha}_{k\nu\mu}}$$

(2) $R^{\pm}(j,p,q,1)^{l\alpha\beta}_{k\mu\nu} = \delta^{\bar q}_{k}\,\delta^{l}_{\bar p}\,\delta^{1}_{\nu}\,\delta^{\beta}_{1}\; R^{\pm}(j,p,q,1)^{\bar p\alpha 1}_{\bar q\mu 1}$

(3) $R^{\pm}(1,p,q,j)^{l\alpha\beta}_{k\mu\nu} = \delta^{p}_{k}\,\delta^{l}_{q}\,\delta^{1}_{\mu}\,\delta^{\alpha}_{1}\; R^{\pm}(1,p,q,j)^{q1\beta}_{p1\nu}\,.$

□

We omit the proof of this lemma.

We now note that, by Lemma 2, parts (2) and (3), the only non-zero matrix elements of the matrices $R_{\pm}(1,p,\bar p,1)^{m\alpha\beta}_{k\mu\nu})$ are $R^{\pm}(1,p,\bar p,1)^{\bar p 1 1}_{p 1 1}$. By Lemma 2, part (1), and since $\bar{\bar p} = p$,

$$R^{-}(1,p,\bar p,1)^{\bar p 1 1}_{p 1 1} = \overline{R^{+}(1,p,\bar p,1)^{\bar p 1 1}_{p 1 1}}\,. \tag{3.36}$$

If one chooses the intertwiners to be partial isometries then one sees that

$$|\,R^{\pm}(1,p,\bar p,1)^{\bar p 1 1}_{p 1 1}\,| = 1.$$

We may therefore introduce the notation

$$R^{+}(1,p,\bar p,1)^{\bar p 1 1}_{p 1 1} = e^{2\pi i\,\theta_p}\,. \tag{3.37}$$

By (3.36)

$$R^{-}(1,p,\bar p,1)^{\bar p 1 1}_{p 1 1} = e^{-2\pi i\,\theta_p}\,. \tag{3.38}$$

Next, we apply (3.34) to conclude that

$$e^{2\pi i\,\theta_p} = e^{2\pi i(s_p+s_{\bar p})}\; e^{-2\pi i\,\theta_p}\,, \tag{3.39}$$

and we have used that $s_1 = 0 \; mod.\, \mathbb{Z}$. Finally, we note that

$$s_p = s_{\bar p}. \tag{3.40}$$

Thus, combining (3.39) and (3.40) we have that

$$\boxed{s_p = \theta_p \; mod.\, \frac{1}{2}\mathbb{Z}\,.} \tag{3.41}$$

This is the simplest connection between spin and statistics. More precise results of a similar nature will be proven below.

3.2 Fusion of intertwiners

For p and q in L, we consider *morphisms $\rho^p \in \mathcal{M}_p$ and $\rho^q \in \mathcal{M}_q$ localized in space-like cones $\mathcal{C}^p$ and $\mathcal{C}^q$, respectively, which are contained in the interior of a simple region $\mathcal{S} \subset \mathsf{M}^3$. The space-like complement of $\mathcal{S}$ is assumed to contain a non-empty, space-like auxiliary cone, $\mathcal{C}_a$. Then ρ^p and ρ^q are *morphisms of the extended algebra $\mathcal{B}^{\mathcal{C}_a}$, defined in (2.19). In particular, the composition, $\rho^p \circ \rho^q$, of ρ^p with ρ^q is well defined on the algebra $\mathcal{A}$ of quasi-local observables. Property 2.2, (P1), (Sect. 2, after (2.27)) guarantees that the product representation $p \times q$ can be decomposed into a direct sum of irreducible, localizable, covariant positive-energy representations, i.e.,

$$p \times q \;=\; \bigoplus_{r \in L} \bigoplus_{\mu=1}^{N_{rpq}} r^{(\mu)}\,, \tag{3.42}$$

see (2.28). Let $\mathcal{C}^r$ be a space-like cone contained in the interior of $\mathcal{S}$, and let $\rho^r \in \mathcal{M}_r$ be a *morphism of $\mathcal{B}^{\mathcal{C}_a}$ localized in $\mathcal{C}^r$ with the property that the representation r of $\mathcal{A}$ is unitarily equivalent to the vacuum representation, 1, of $\rho^r(\mathcal{A})$. Then there exist N_{rpq} partial isometries,

$$\Gamma^{\mathcal{S}}_{\rho^p \circ \rho^q, \rho^r}(\mu) \in \overline{\mathcal{A}(\mathcal{S})}^{w} \subset \mathcal{B}^{\mathcal{C}_a} \tag{3.43}$$

such that

$$\rho^p \circ \rho^q(A)\, \Gamma^{\mathcal{S}}_{\rho^p \circ \rho^q, \rho^r}(\mu) \;=\; \Gamma^{\mathcal{S}}_{\rho^p \circ \rho^q, \rho^r}(\mu)\, \rho^r(A)\,, \tag{3.44}$$

for all $A \in \mathcal{A}$, $\mu = 1, \ldots, N_{rpq}$; see [4,5].

If $\mathcal{S}_0 \subseteq \mathcal{S}$ is a simple domain containing the cones $\mathcal{C}^p, \mathcal{C}^q$ and $\mathcal{C}^r$ in the interior then, actually,

$$\Gamma^{\mathcal{S}}_{\rho^p \circ \rho^q, \rho^r}(\mu) \;\in\; \overline{\mathcal{A}(\mathcal{S}_0)}^{w}\,. \tag{3.45}$$

Next, we consider a product of intertwiners $V^{ij}_{\alpha}(\rho^p)\, V^{jk}_{\beta}(\rho^q)$. They satisfy the intertwining relations

$$\begin{aligned} i(\rho^p \circ \rho^q(A))\, V^{ij}_{\alpha}(\rho^p)\, V^{jk}_{\beta}(\rho^q) \\ = V^{ij}_{\alpha}(\rho^p)\, V^{jk}_{\beta}(\rho^q)\, k(A). \end{aligned} \tag{3.46}$$

We wish to compare the properties of $V^{ij}_{\alpha}(\rho^p)\, V^{jk}_{\beta}(\rho^q)$ to those of the operators

$$i\big(\Gamma^{\mathcal{S}}_{\rho^p \circ \rho^q, \rho^r}(\mu)\big)\; V^{ik}_{\nu}(\rho^r) \tag{3.47}$$

which, by (3.44), satisfy the same intertwining relations

$$
\begin{aligned}
&i(\rho^p \circ \rho^q(A))\, i(\Gamma^S_{\rho^p \circ \rho^q, \rho^r}(\mu))\; V^{ik}_\nu(\rho^r) \\
&\qquad = i(\Gamma^S_{\rho^p \circ \rho^q, \rho^r}(\mu))\; V^{ik}_\nu(\rho^r)\; k(A),
\end{aligned}
\tag{3.48}
$$

for all $A \in \mathcal{A}$, $\mu = 1, \ldots, N_{rpq}$. Equs. (3.46),(3.48) and Schur's lemma suggest that $V^{ij}_\alpha(\rho^p)\, V^{jk}_\beta(\rho^q)$ can be expanded in a sum over the operators $i(\Gamma^S_{\rho^p \circ \rho^q, \rho^r}(\mu))\, V^{ik}_\nu(\rho^r)$. This expansion will be called fusion.

In order to make these ideas precise, we start with a special case of fusion: Consider the operators $V^{\bar{r}\bar{q}}_\alpha(\rho^p)\, V^{\bar{q}1}(\rho^q)$ and $V^{\bar{r}1}(\rho^r)$. [We recall that to every representation $r \in L$ there exists a unique conjugate representation $\bar{r} \in L$ such that $\bar{r} \times r$ contains the vacuum representation, 1, precisely once; see Property 2.2, (P2), Sect. 2, after (2.27). From this one can conclude that

$$
V^{j1}(\rho^r) \;=\; 0, \text{ unless } j \;=\; \bar{r}, \tag{3.49}
$$

and that there exists precisely one partial isometry $V^{\bar{r}1}(\rho^r) : \mathcal{H}_1 \to \mathcal{H}_{\bar{r}}$ which is unique up to a phase.] By (3.42) and (3.44), there exist complex numbers, $\sigma_\alpha(r; p, q)$, $\alpha = 1, \ldots, N_{rpq}$, such that

$$
V^{\bar{r}\bar{q}}_\alpha(\rho^p)\; V^{\bar{q}1}(\rho^q) \;=\; \sigma_\alpha(r;p,q)\, \bar{r}(\Gamma^S_{\rho^p \circ \rho^q, \rho^r}(\alpha))\; V^{\bar{r}1}(\rho^r). \tag{3.50}
$$

Since we choose the operators $\{V^{ij}_\alpha(\rho^p)\}$ to be partial isometries, it follows that

$$
\|\bar{r}(\Gamma^S_{\rho^p \circ \rho^q, \rho^r}(\alpha))\; V^{\bar{r}1}(\rho^r)\,\Omega\| \;=\; 1,
$$

where $\Omega \in \mathcal{H}_1$ is the vacuum vector. Hence

$$
\sigma_\alpha(r;p,q) \;=\; \big\langle \bar{r}\,(\Gamma^S_{\rho^p \circ \rho^q, \rho^r}(\alpha))\; V^{\bar{r}1}(\rho^r)\,\Omega,\; V^{\bar{r}\bar{q}}_\alpha(\rho^p)\; V^{\bar{q}1}(\rho^q)\,\Omega \big\rangle. \tag{3.51}
$$

This formula is quite useful: Suppose that ρ^p and ρ^q are causally independent, i.e. the cones C^p and C^q are space-like separated. Then $\rho^p \circ \rho^q = \rho^q \circ \rho^p$, and hence we may choose the intertwining opertors $\Gamma^S_{\rho^p \circ \rho^q, \rho^r}(\alpha)$ and $\Gamma^S_{\rho^q \circ \rho^p, \rho^r}(\alpha)$ to be equal. Moreover, by Theorem 1, (3.9), and Lemma 2, part (2),

$$
\begin{aligned}
&V^{\bar{r}\bar{q}}_\alpha(\rho^p)\; V^{\bar{q}1}(\rho^q)\,\Omega \\
&\qquad = \sum_\mu R^{\pm}(\bar{r}, p, q, 1)^{\bar{p}\mu 1}_{\bar{q}\alpha 1}\; V^{\bar{r}\bar{p}}_\mu(\rho^q)\; V^{\bar{p}1}(\rho^p)\,\Omega
\end{aligned}
\tag{3.52}
$$

if $as(\rho^p) \gtrless as(\rho^q)$. Hence

$$\sigma_\alpha(r;p,q) = \sum_\mu R^\pm(\bar{r},p,q,1)^{\bar{p}\mu 1}_{\bar{q}\alpha 1}\,\sigma_\mu(r;q,p)\,, \tag{3.53}$$

provided $\rho^p \times \rho^q$ and $as(\rho^p) \gtrless as(\rho^q)$.

We now state our basic result on the fusion of intertwiners.

Theorem 3 [28]

There exist matrices $\left(F(i,p,q,k)^{r\mu\nu}_{j\alpha\beta}\right)$ only depending on the representations i,p,q,k,j and r, but not on the specific choice of ρ^p, ρ^q and ρ^r, such that

$$\begin{aligned} &V^{ij}_\alpha(\rho^p)\,V^{jk}_\beta(\rho^q) \\ &= \sum_{r,\mu,\nu} F(i,p,q,k)^{r\mu\nu}_{j\alpha\beta}\,\sigma_\mu(r;p,q)\,i(\Gamma^S_{\rho^p\circ\rho^q,\rho^r}(\mu))\,V^{ik}_\nu(\rho^r)\,. \end{aligned} \tag{3.54}$$

The matrices $\left(F(i,p,q,k)^{r\mu\nu}_{j\alpha\beta}\right)$ can be expressed in terms of the matrices $\left(R^\pm(l,a,b,m)^{k\gamma\delta}_{n\lambda\kappa}\right)$.

□

We shall outline the main ideas going into the proof of Theorem 3: Let $\rho^{\bar{k}}$ be a *morphism of $\mathcal{B}^{C_a}$ localized in a space-like cone $C^{\bar{k}} \subset \mathcal{S}$ and suppose that C^p, C^q and $C^{\bar{k}}$ are pairwise space-like separated. Let us suppose, for example, that $as(\rho^p),\ as(\rho^q) > as(\rho^{\bar{k}})$. We consider the operator $V^{ij}_\alpha(\rho^p)\,V^{jk}_\beta(\rho^q)\,k(A)\,V^{k1}(\rho^{\bar{k}})$, where A is an arbitrary element of $\mathcal{A}$, and apply the intertwining relations (3.8) and the commutation relations (3.9) between intertwiners. Then

$$\begin{aligned} &V^{ij}_\alpha(\rho^p)\,V^{jk}_\beta(\rho^q)\,k(A)\,V^{k1}(\rho^{\bar{k}}) \\ &= i(\rho^p\circ\rho^q(A))\,V^{ij}_\alpha(\rho^p)\,V^{jk}_\beta(\rho^q)\,V^{k1}(\rho^{\bar{k}}) \\ &= \sum R^+(j,q,\bar{k},1)^{\bar{q}\mu 1}_{k\beta 1}\,i(\rho^p\circ\rho^q(A))\,V^{ij}_\alpha(\rho^p)\,V^{j\bar{q}}_\mu(\rho^{\bar{k}})\,V^{\bar{q}1}(\rho^q) \\ &= \sum R^+(j,q,\bar{k},1)^{\bar{q}\mu 1}_{k\beta 1}\,R^+(i,p,\bar{k},\bar{q})^{\bar{r}\gamma\nu}_{j\alpha\mu}\,i(\rho^p\circ\rho^q(A)) \\ &\qquad\times V^{i\bar{r}}_\gamma(\rho^{\bar{k}})\,V^{\bar{r}\bar{q}}_\nu(\rho^p)\,V^{\bar{q}1}(\rho^q) \\ &= \sum R^+(j,q,\bar{k},1)^{\bar{q}\mu 1}_{k\beta 1}\,R^+(i,p,\bar{k},\bar{q})^{\bar{r}\gamma\nu}_{j\alpha\mu}\,\sigma_\nu(r;p,q) \\ &\qquad\times i(\rho^p\circ\rho^q(A))\,V^{i\bar{r}}_\gamma(\rho^{\bar{k}})\,\bar{r}(\Gamma^S_{\rho^p\circ\rho^q,\rho^r}(\nu))\,V^{\bar{r}1}(\rho^r)\,, \end{aligned} \tag{3.55}$$

and we have used (3.50). Let $\mathcal{S}_{p,q} \subset \mathcal{S}$ be a simple domain containing the cones C^p and C^q and space-like separated from $C^{\bar{k}}$. We may choose ρ^r to be localized in $\mathcal{S}_{p,q}$.

Then $\Gamma^S_{\rho^p \circ \rho^q, \rho^r}(\nu) \in \overline{\mathcal{A}(S_{p,q})}^w$, and hence

$$V^{i\bar{r}}_\gamma(\rho^{\bar{k}})\, \bar{r}(\Gamma^S_{\rho^p \circ \rho^q, \rho^r}(\nu)) = i(\Gamma^S_{\rho^p \circ \rho^q, \rho^r}(\nu))\, V^{i\bar{r}}_\gamma(\rho^{\bar{k}}),$$

by (3.8). We therefore derive from (3.55) that

$$\begin{aligned} &V^{ij}_\alpha(\rho^p)\, V^{jk}_\beta(\rho^q)\, k(A)\, V^{k1}(\rho^{\bar{k}}) \\ &\quad = \sum R^+(j,q,\bar{k},1)^{\bar{q}\mu 1}_{k\beta 1}\, R^+(i,p,\bar{k},\bar{q})^{\bar{r}\gamma\nu}_{j\alpha\mu}\, R^-(i,\bar{k},r,1)^{k\delta 1}_{\bar{r}\gamma 1} \\ &\quad \times \sigma_\nu(r;p,q)\, i(\rho^p \circ \rho^q(A))\, i(\Gamma^S_{\rho^p \circ \rho^q, \rho^r}(\nu))\, V^{ik}_\delta(\rho^r)\, V^{k1}(\rho^{\bar{k}}). \end{aligned} \tag{3.56}$$

From the intertwining relations (3.44) and (3.8) we derive that

$$\begin{aligned} &i(\rho^p \circ \rho^q(A))\, i(\Gamma^S_{\rho^p \circ \rho^q, \rho^r}(\nu))\, V^{ik}_\delta(\rho^r) \\ &\quad = i(\Gamma^S_{\rho^p \circ \rho^q, \rho^r}(\nu))\, i(\rho^r(A))\, V^{ik}_\delta(\rho^r) \\ &\quad = i(\Gamma^S_{\rho^p \circ \rho^q, \rho^r}(\nu))\, V^{ik}_\delta(\rho^r)\, k(A). \end{aligned} \tag{3.57}$$

Combining (3.56) and (3.57) we find that

$$\begin{aligned} &V^{ij}_\alpha(\rho^p)\, V^{jk}_\beta(\rho^q)\, k(A)\, V^{k1}(\rho^{\bar{k}}) \\ &\quad = \sum R^+(j,q,\bar{k},1)^{\bar{q}\mu 1}_{k\beta 1}\, R^+(i,p,\bar{k},\bar{q})^{\bar{r}\gamma\nu}_{j\alpha\mu}\, R^-(i,\bar{k},r,1)^{k\delta 1}_{\bar{r}\gamma 1} \\ &\quad \times \sigma_\nu(r;p,q)\, i(\Gamma^S_{\rho^p \circ \rho^q, \rho^r}(\nu))\, V^{ik}_\delta(\rho^r)\, k(A)\, V^{k1}(\rho^{\bar{k}}). \end{aligned} \tag{3.58}$$

Next, we note that

$$\{k(A)\, V^{k1}(\rho^{\bar{k}})\xi \,:\, A \in \mathcal{A},\ \xi \in \mathcal{H}_1\}$$

is dense in $\mathcal{H}_k$. Thus (3.58) proves (3.54), with

$$\begin{aligned} F(i,p,q,k)^{r\nu\delta}_{j\alpha\beta} &= \sum_{\mu,\gamma} R^+(j,q,\bar{k},1)^{\bar{q}\mu 1}_{k\beta 1}\, R^+(i,p,\bar{k},\bar{q})^{\bar{r}\gamma\nu}_{j\alpha\mu} \\ &\quad \times R^-(i,\bar{k},r,1)^{k\delta 1}_{\bar{r}\gamma 1}. \end{aligned} \tag{3.59}$$

By choosing $as(\rho^{\bar{k}}) > as(\rho^p), as(\rho^q)$, we find that

$$\begin{aligned} F(i,p,q,k)^{r\nu\delta}_{j\alpha\beta} &= \sum_{\mu,\gamma} R^-(j,q,\bar{k},1)^{\bar{q}\mu 1}_{k\beta 1}\, R^-(i,p,\bar{k},\bar{q})^{\bar{r}\gamma\nu}_{j\alpha\mu} \\ &\quad \times R^+(i,\bar{k},r,1)^{k\delta 1}_{\bar{r}\gamma 1}. \end{aligned} \tag{3.60}$$

We must ask whether (3.59) and (3.60) are consistent? The reader verifies without the slightest difficulty that the consistency of (3.59) and (3.60) follows from the fundamental identity (3.34).

In order to discuss further properties of the fusion matrices $\left(F(i,p,q,k)^{r\nu\delta}_{j\alpha\beta}\right)$ it is helpful to introduce a graphical notation:

We denote $F(i,p,q,k)^{r\nu\delta}_{j\alpha\beta}$ by

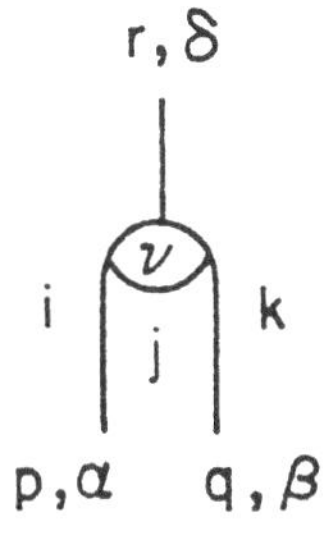

(3.61)

As in conformal field theory [19,36] it is easy to derive the following "polynomial equations":

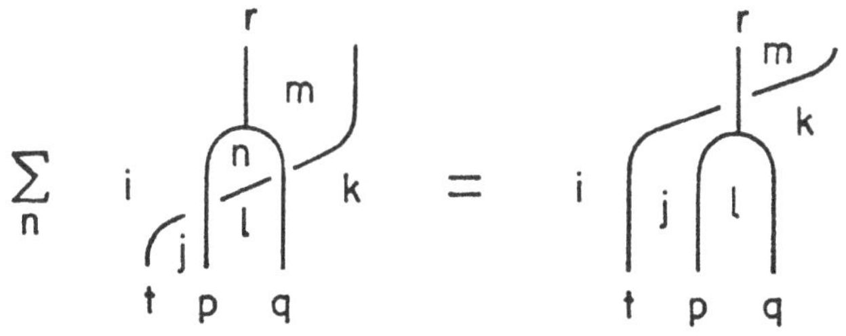

(3.62)

where we have used notation (3.17). Similarly,

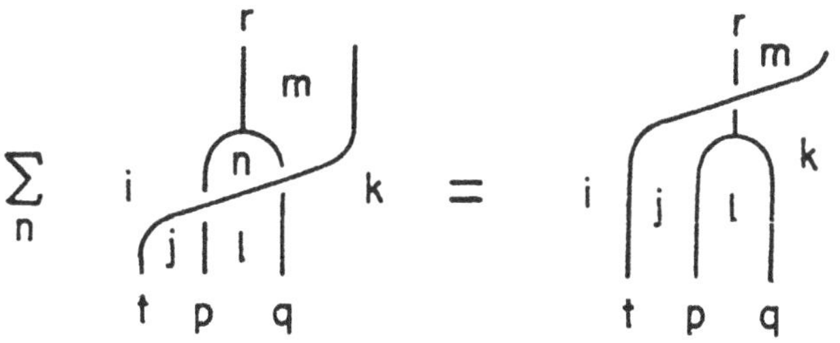

(3.63)

where we have used (3.16), and

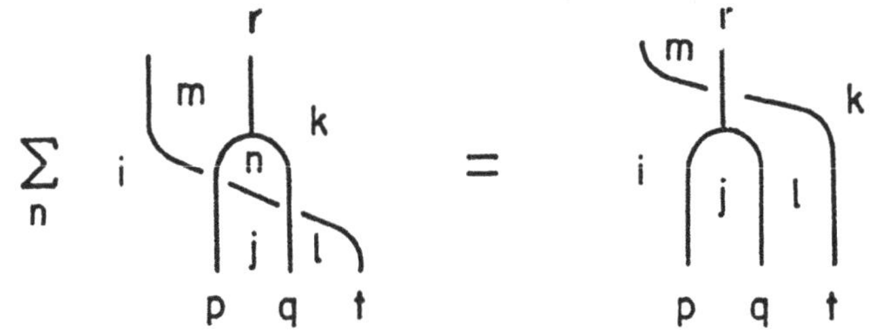

(3.64)

etc.. In (3.62)-(3.64) and henceforth we omit the Greek indices $\alpha, \beta, \nu, \delta, \ldots$.

Furthermore, from (3.9),(3.53) and (3.54) one easily derives [28,37] that

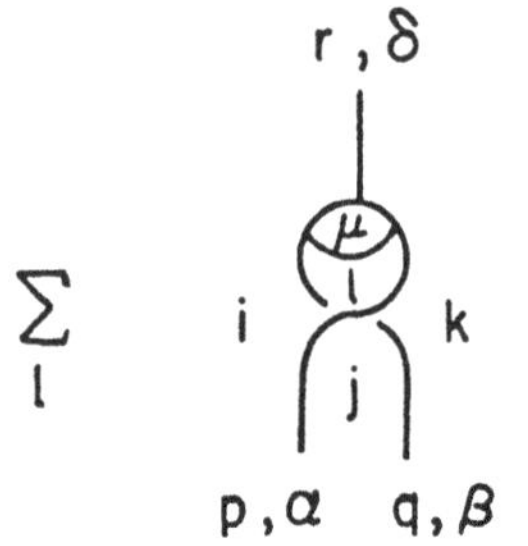

$$= \sum_{\nu} R^{+}(\bar{r}, p, q, 1)^{\bar{p}\nu 1}_{\bar{q}\mu 1}$$

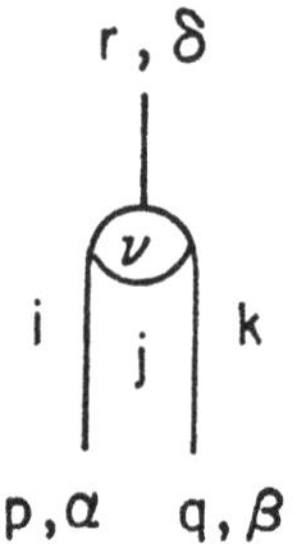

(3.65)

A similar equation holds with R^- replacing R^+.

Next, we introduce the monodromy matrices

$$M(i,p,q,k)^{l\gamma\delta}_{j\alpha\beta} = \sum_{n\mu\nu} R^{+}(i,p,q,k)^{n\mu\nu}_{j\alpha\beta} \, R^{+}(i,q,p,k)^{l\gamma\delta}_{n\mu\nu} \, . \tag{3.66}$$

Graphically,

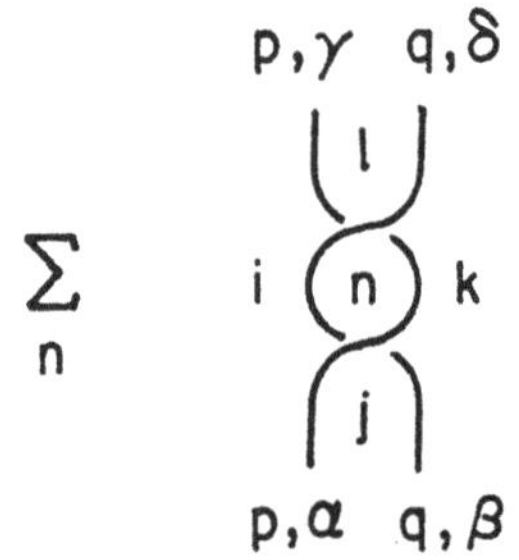

$$\longleftrightarrow \quad M(i,p,q,k)^{l\gamma\delta}_{j\alpha\beta} \, . \tag{3.67}$$

Iterating (3.65) we find that

$$\sum_{n,l}$$

$$= \sum_{n,\nu} R^+(\bar{r},q,p,1)^{\bar{q}\nu 1}_{\bar{p}\mu 1}$$

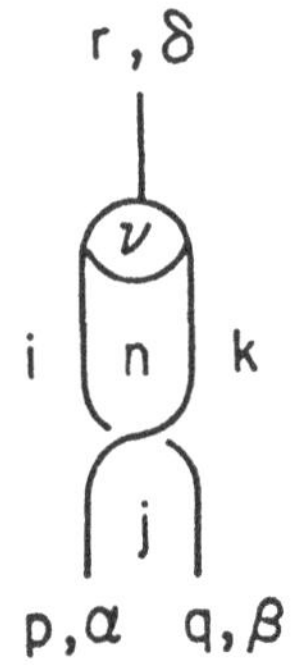

$$= \sum_{n,\nu,\kappa} R^+(\bar{r},q,p,1)^{\bar{q}\nu 1}_{\bar{p}\mu 1}\, R^+(\bar{r},p,q,1)^{\bar{p}\kappa 1}_{\bar{q}\nu 1} \qquad (3.68)$$

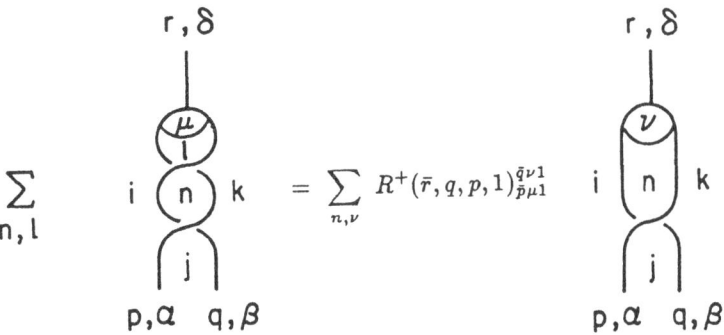

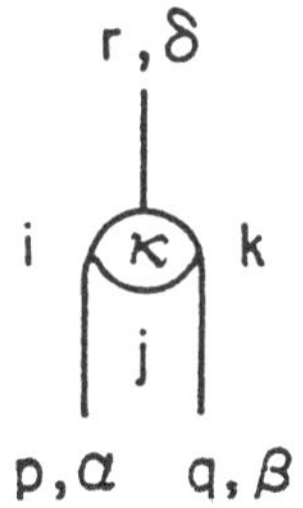

Our fundamental identity (3.34) says that

$$R^+(\bar{r},p,q,1)^{\bar{p}\kappa 1}_{\bar{q}\nu 1} = e^{2\pi i(s_q+s_{\bar{p}}-s_r)}\, R^-(\bar{r},p,q,1)^{\bar{p}\kappa 1}_{\bar{q}\nu 1}\,, \qquad (3.69)$$

where we have used that $s_1 = 0 \; mod.\, \mathbf{Z}$, and (3.18) says that

$$\sum_{\nu} R^+(\bar{r},q,p,1)^{\bar{q}\nu 1}_{\bar{p}\mu 1}\, R^-(\bar{r},p,q,1)^{\bar{p}\kappa 1}_{\bar{q}\nu 1} = \delta^{\kappa}_{\mu}\,. \qquad (3.70)$$

By combining (3.68), (3.69) and (3.70) we find that

$$\boxed{\sum_{l\gamma\delta} M(i,p,q,k)^{l\gamma\delta}_{j\alpha\beta}\, F(i,p,q,k)^{r\mu\nu}_{l\gamma\delta} = e^{2\pi i(s_p+s_q-s_r)}\, F(i,p,q,k)^{r\mu\nu}_{j\alpha\beta}\,,} \qquad (3.71)$$

where we have also used that $s_j = s_{\bar{j}}$, for all $j \in L$.

3.3 Spin spectrum, spin addition rules, spin-statistics

As shown in [37] and refs. given there, equ. (3.71) has rather interesting consequences:

(1) The fusion matrices $F(i,p,q,k)^{r\mu\nu}_{l\gamma\delta}$ diagonalize the monodromy matrices $\big(M(i,p,q,k)^{l\gamma\delta}_{j\alpha\beta}\big)$.

(2) The spectrum of $M(i,p,q,k)$ is given by $\{e^{2\pi i(s_p+s_q-s_r)} : r \in L,\ N_{rpq} \neq 0\}$.

(3) Let us assume that we are dealing with a theory which has only finitely many distinct superselection sectors, i.e., $|L| < \infty$. Then we have the following result.

Theorem 4 [37]

If the number, $|L|$, of superselection sectors is finite all the spins $s_j, j \in L$, are rational numbers.

In analogy with conventional jargon in conformal field theory [38], field theories in three space-time dimensions with only finitely many distinct superselection sectors, $|L| < \infty$, are called rational theories.

(4) Next, we consider a three-dimensional theory with permutation group statistics, i.e.,

$$R^+(j,p,q,k) = R^-(j,p,q,k). \tag{3.72}$$

Then, using (3.18) one concludes that all monodromy matrices are trivial, i.e.,

$$M(i,p,q,k)^{l\gamma\delta}_{j\alpha\beta} = \delta^l_j\ \delta^\gamma_\alpha\ \delta^\delta_\beta\,, \tag{3.73}$$

and hence all their eigenvalues are equal to 1,

$$e^{2\pi i(s_p+s_q-s_r)} = 1\,, \tag{3.74}$$

for all p, q and r for which $N_{rpq} \neq 0$.

If $q = \bar{p}$ then $s_q = s_p$ and $N_{1pq} \neq 0$. In this case (3.74) implies that

$$e^{4\pi i s_p} = 1, \text{ for all } p\,, \tag{3.75}$$

or, equivalently,

$$\boxed{s_p \in \frac{1}{2}\,\mathbb{Z}, \text{ for all } p.} \tag{3.76}$$

We define "spin parity", σ_p, by setting

$$\sigma_p = e^{2\pi i s_p},\ p \in L. \tag{3.77}$$

Since, by (3.74), (3.72) implies that

$$s_p + s_q - s_r \in \mathbb{Z}\,,$$

for all p, q and r satisfying the fusion rules, i.e., $N_{rpq} \neq 0$, we conclude, using (3.76), that

$$\sigma_p \sigma_q = \sigma_r , \tag{3.78}$$

for p, q and r such that $N_{rpq} \neq 0$; in words: spin parity is conserved under fusion.

Conversely, let us assume that we consider a theory with the property that

$$s_p \in \frac{1}{2} \mathbb{Z}, \text{ for all } p \in L, \text{ and} \tag{3.79}$$

$\sigma_p \sigma_q = \sigma_r$, for all p, q and r for which $N_{rpq} \neq 0$.

Then, by (3.74)

$$M(i,p,q,k)^{l\gamma\delta}_{j\alpha\beta} = \delta^l_j \, \delta^\gamma_\alpha \, \delta^\delta_\beta \tag{3.80}$$

for all i, p, q, k, j and l satisfying the fusion rules. Hence

$$R^+(i,p,q,k) = R^-(i,p,q,k) , \tag{3.81}$$

i.e. the theory has standard permutation group statistics.

We summarize these findings in a theorem.

Theorem 5 [28]

The conditions

(i) $R^+(i,p,q,k) = R^-(i,p,q,k)$, for all i, p, q, k, and

(ii) $s_p \in \frac{1}{2}\mathbb{Z}$, for all $p \in L$, and $\sigma_p \cdot \sigma_q = \sigma_r$, for all p, q, r for which $N_{rpq} \neq 0$,

are equivalent.

(5) The analysis developed in Sects. 2 and 3 can also be applied to theories in $d \geq 4$ space-time dimensions or to theories in three space-time dimensions with superselection sectors generated by *morphisms of an algebra $\mathcal{A}$ of quasi-local observables localized in bounded space-time regions, as described in [4]. In both cases, one shows easily (see also [4,5]) that

$$R^+(i,p,q,k) = R^-(i,p,q,k),$$

for all i, p, q, k. Hence all sectors of such theories have integral or half-integral spin and spin parity is conserved in such theories. [The fact that, in such theories, $s_j \in \frac{1}{2}\mathbb{Z}$, for all $j \in L$, can also be derived from the structure of the Poincaré group ($d \geq 4$)

and from locality and the relativistic spectrum condition $(d = 3)$.]

A fundamental theorem due to Doplicher and Roberts [6] says that, in a theory with standard permutation group statistics, the fusion rules (N_{rpq}) can be derived from the representation theory of some compact group, G , and, by introducing internal degrees of freedom on which the representations of G act, the permutation group statistics can be reduced to ordinary Bose- and Fermi statistics.

(6) It follows from (3.66) and (3.71) that the 1×1 matrix $M(1,p,\bar{p},1)^p_p = (M(1,p,\bar{p},1)^{p11}_{p11})$ is given by

$$M(1,p,\bar{p},1)^p_p = R^+(1,p,\bar{p},1)^{\bar{p}}_p \; R^+(1,\bar{p},p,1)^p_{\bar{p}} = e^{4\pi i s_p} , \qquad (3.82)$$

and we have used that $s_{\bar{p}} = s_p$. In a relativistically covariant theory, a field-theoretic argument suggests that *

$$R^+(1,p,\bar{p},1)^{\bar{p}}_p = R^+(1,\bar{p},p,1)^p_{\bar{p}} = e^{2\pi i s_p} . \qquad (3.83)$$

By definition of the angle θ_p , see (3.37),

$$R^+(1,p,\bar{p},1)^{\bar{p}}_p = e^{2\pi i \theta_p} .$$

Thus

$$s_p = \theta_p, \; mod. \; \mathbb{Z}, \qquad (3.84)$$

for all p . This connection between spin and statistics strengthens the one found in (3.41). Equ. (3.84) is well known for theories with standard permutation group statistics [2]. For more details see [28].

In all examples known to us, the matrices $R^{\pm}(i,p,q,k)^l_j$ and $F(i,p,q,k)^r_j$ can be obtained from the representation theory of some group $(R^+ = R^-$, [6]) or some quantum group $(R^+ \neq R^-)$ with the help of the socalled *vertex – sos* transformation: see Sect. 4. By analogy with the results of Doplicher and Roberts [6], one might conjecture that this is always true. In these cases, the spectrum $\{e^{2\pi i(s_p+s_q-s_r)} : N_{rpq} \neq 0\}$ of the monodromy matrices, $M(i,p,q,k)^l_j$, is com-

* *Under somewhat stronger hypotheses on the structure of the theory, (3.83) can be proven.*

pletely determined by the representation theory of the quantum group. This observation and equ. (3.84) permit us to calculate all spins s_p *mod.* $\mathbb{Z}$, for all $p \in L$.

Let us briefly consider the example of a quantum deformation, $U_q(\mathcal{G})$, of the universal enveloping algebra of a classical Lie algebra $\mathcal{G}$, with $q^N = 1$, for some integer $N \geq 2$.

Then one has the formula

$$s_p = \pm\frac{1}{2N}\left(\langle\nu_p,\nu_p\rangle + \langle\rho,\nu_p\rangle\right) + s_p^{(a)} \text{ mod. } \mathbb{Z}\,, \tag{3.85}$$

where ν_p is the highest weight of some finite-dimensional, unitarizable, irreducible highest-weight representation of $U_q(\mathcal{G})$, ρ is the sum of positive roots of $\mathcal{G}$, and $s_p^{(a)}$ is a contribution to s_p that comes from an abelian factor in the braid group statistics of the theory. (For example, $s_p^{(a)} = \frac{1}{2}n_p, n_p = 0, 1$, would describe a contribution to s_p due to an ordinary Fermi field.) Below, the structure of $s_p^{(a)}$ is described completely. The proof of (3.85) is obtained by comparing equ. (3.71) with a formula in [45]. See Sect. 4 for more details concerning connections between our theory and quantum group theory.

The example of abelian braid group statistics is elementary and can be analysed completely. It has interesting applications in quantum field theory and condensed matter physics. The results reviewed below have been conjectured in [26] on the basis of an analysis of concrete models describing anyons. As shown in [28], abelian braid group statistics implies that all representations $p \in L$ are unitarily equivalent to representations $1(\rho^p(\cdot))$ of $\mathcal{A}$, where ρ^p is a *automorphism of $\mathcal{B}^{C_a}$. Then $\bar{p}$ is equivalent to $1((\rho^p)^{-1}(\cdot))$, hence $\bar{p} \times p = 1$, and every power $p^{\times n} = p \times \cdots \times p$ of the representation p belongs to L. The set $\{p^{\times n} : n \in \mathbb{Z}\}$ is a subset of L invariant under composition, whose fusion rules are described by $\mathbb{Z}$. The braid matrices of such theories have the form [28].

$$R^{\pm}(j,p,q,k)_l^m = \begin{cases} e^{\pm 2\pi i \theta_{p,q}}, & \text{if all fusion rules are satisfied;} \\ 0, & \text{otherwise.} \end{cases} \tag{3.86}$$

It then follows from equ. (3.64), (with $q = \bar{p}$, $t = p$, $r = 1$, $k = 1$, $l = \bar{p}$, $j = 1$, $i = \bar{p}$ and $m = 1$) that

$$R^+(1,\bar{p},p,1)_{\bar{p}}^{p}\, R^+(\bar{p},p,p,p)_1^1 = 1,$$

i.e.

$$e^{2\pi i \theta_{\bar{p},p}} \, e^{2\pi i \theta_{p,p}} = 1 , \tag{3.87}$$

for all $p \in L$. Previously, $\theta_{\bar{p},p}$ has been denoted by θ_p which was shown to be given by $s_p \; mod.\, \mathbb{Z}$; see (3.84). Hence (3.87) and (3.84) yield

$$\boxed{\theta_{p,p} = -\theta_{\bar{p},p} = -s_p \; mod.\, \mathbb{Z} ,} \tag{3.88}$$

for all $p \in L$. By (3.86),

$$R^{+}(\bar{p} \times \bar{p}, p, p, 1)^{\bar{p}}_{\bar{p}} = e^{2\pi i \theta_{p,p}} . \tag{3.89}$$

Furthermore (3.66) and (3.71) show that

$$\left(R^{+}(\bar{p} \times \bar{p}, p, p, 1)^{\bar{p}}_{\bar{p}}\right)^2 = e^{2\pi i (2 s_p - s_{p \times p})} . \tag{3.90}$$

Combining (3.88)-(3.90), we conclude that

$$s_{p \times p} = 4 s_p \; mod.\, \mathbb{Z} , \tag{3.91}$$

for all $p \in L$. Iterating these arguments, one finds that

$$\boxed{s_{p \times n} = n^2 \, s_p \; mod.\, \mathbb{Z} ,} \tag{3.92}$$

for all $p \in L$.

Equs. (3.86), (3.88) and (3.92) represent a completely general, model-independent proof of relations conjectured in [26] on the basis of an analysis of specific models describing anyons.

In the non-abelian case, results analogous to (3.92) can be proven by using (3.39) and the polynomial equations (3.64), provided the fusion matrices $\{F(i,p,q,k)^r_j\}$ can be calculated without using equs. (3.59) or (3.60). This is the case if, for example, the matrices $R^{\pm}$ and F can be derived from the representation theory of some quantum group via the vertex-sos transformation; see [28,37].

Finally, we wish to point out that the norms, $\|\mathsf{N}_p\|$, of the multiplicity matrices, N_p, defined by $(\mathsf{N}_p)_{rq} = N_{rpq}$ (fusion rules) define "statistical dimensions" which can be interpreted in terms of indices of subfactors in the sense of Jones, [18,39].

Further results are discussed in [28].

4. Connections to knot theory and to quantum group theory

Let us first draw attention to a deep connection between the theory developed in Sect. 3 and the theory of knots and links in $S^3 = \dot{\mathbb{R}}^3$. This connection is the same as one described in detail in [40] in the context of conformal field theory; (see also [37]). The point is that from a family of matrices $\{R^{\pm}(i,p,q,k)^l_j,\ F(i,p,q,k)^r_j\}$ satisfying

(a) equs. (3.18) and (3.22) (YBE);

(b) equs. (3.62) - (3.65); equs. (3.83);

one can derive a family of invariants for oriented knots and links in S^3, by an explicit construction described in detail in [40]. Furthermore, as was sketched in [40], one can derive a family of invariants for tri-valent ribbon graphs.

We briefly sketch the construction of invariants for knots and links from $\{R^{\pm}, F\}$: In $\dot{\mathbb{R}}^3$ we choose a two-dimensional plane, π, a unit vector $\vec{n}$ normal to π and a unit vector $\vec{e} \in \pi$. Given an oriented link, $\mathcal{L}$, in $\mathbb{R}^3$, we choose a representative of $\mathcal{L}$, (i.e. a system of loops in $\dot{\mathbb{R}}^3$ representing $\mathcal{L}$), which has a non-degenerate projection along $\vec{n}$ onto π. Such a projection is called a shadow of $\mathcal{L}$. If under- and overcrossings in the ordering fixed by $\vec{n}$, are recorded on the crossings of lines in the shadow of $\mathcal{L}$ we speak of a diagram, $D(\mathcal{L})$, of $\mathcal{L}$. The link $\mathcal{L}$ can obviously be reconstructed from $\vec{n}$ and $D(\mathcal{L})$. The diagram $D(\mathcal{L})$ is marked as follows: If the coordinate function in the direction of $\vec{e}$ on π has a local maximum or a local minimum at a point $p \in D(\mathcal{L})$ then p is marked by a dot:

Next, a marked diagram, $D(\mathcal{L})$, is decorated as follows: The shadow of $\mathcal{L}$ on π decomposes π into disjoint regions $\Omega_1, \ldots, \Omega_N$, where Ω_1 is defined to be the region containing the point at infinity, and $N \geq 2$ is determined by $D(\mathcal{L})$. Then we assign to every region Ω_l a representation k_l of $\mathcal{A}$ belonging to the list L introduced in Sect. 2. Without loss of generality, we may assign the vacuum representation,

1, to Ω_1. A <u>component</u> of $D(\mathcal{L})$ is the projection of an <u>oriented</u> loop (connected component) of the link in $\mathring{R}^3$ onto π. Every component, C_i, of $D(\mathcal{L})$ is assigned a representation $j_i \in L$. The resulting marked, decorated diagram of $\mathcal{L}$ is denoted by $D(\mathcal{L}; j_1, \ldots, j_n; k_2, \ldots, k_N)$; ($n = \#$ of components of $\mathcal{L}$). The <u>elements</u> of a decorated diagram of $\mathcal{L}$ are defined in Fig. 7 and are assigned matrices R^+, R^- or F, as shown in Fig. 7, (a)-(f):

(a)

(b)

(c) $\longleftrightarrow R^+(i, p.q, k)_j^l \; e^{i\varphi_{pq}}$

(d) $\longleftrightarrow R^-(i, p, q, k)_j^l \; e^{-i\varphi_{pq}}$

(e) $\longleftrightarrow F(k, p, \bar{p}, k)_l^1 \equiv F_{pl}^k$

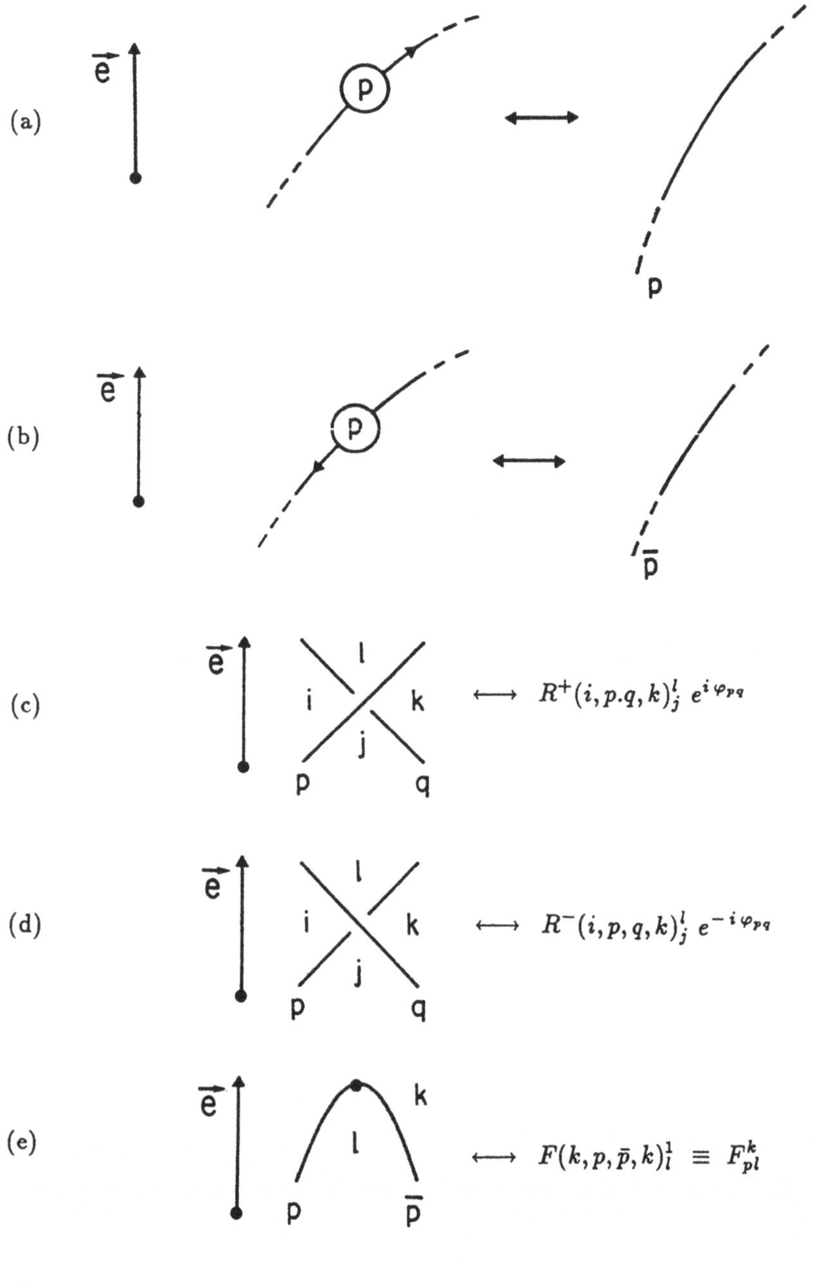

(f)

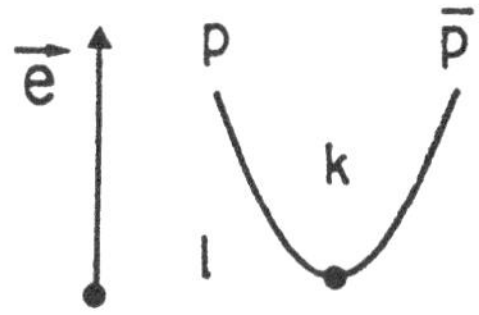

$\longleftrightarrow \quad (F^k_{\bar{p}l})^{-1}$

Fig. 7

The angles φ_{pq} are given by

$$\varphi_{pq} \;=\; 2\pi\sqrt{s_p \cdot s_q}\,. \tag{4.1}$$

By $(F^k_{pl})^{-1}$ is meant the inverse of the matrix F^k_{pl} with matrix elements

$$\left(F^k_{pl}\right)_{\alpha\beta} \;=\; F\left(k,p,\bar{p},k\right)^{1\,1\,1}_{l\alpha\beta}\,. \tag{4.2}$$

With $\mathcal{D}(\mathcal{L};j_1,\dots j_n;k_2,\dots k_N)$ we associate a complex number $i(\mathcal{L};\ j_1,\dots,j_n;\ k_2,\dots,k_N)$ by calculating the sum of products of matrix elements of the matrices $(R^\pm,F,F^{-1})$ associated to all the elements of $\mathcal{D}(\mathcal{L};j_1,\dots,j_n;\ k_2,\dots,k_N)$ by the rules specified in Fig. 7, (a) - (f).

We may now quote one of the main results of [40] (Theorem 6.1): Let $\mathcal{L}$ be an oriented link in S^3 with diagram $D(\mathcal{L})$. The component, i, of $\mathcal{L}$ projected onto C_i is assigned the "colour" $j_i \in L$. With an oriented, coloured link $(\mathcal{L};j_1,\dots,j_n)$ we associate a complex number

$$I(\mathcal{L};j_1,\dots,j_n) \;=\; \sum_{k_2,\dots,k_N\in L} i(\mathcal{L};j_1,\dots,j_n;k_2,\dots,k_N)\,. \tag{4.3}$$

Then we have the following result

Theorem 6 The numbers $I(\mathcal{L};j_1,\dots,j_n)$ define invariants for oriented, coloured links in S^3.

Remark. The proof is identical to the proof of Theorem 6.1 in [40]. Since it is somewhat lengthy, we shall not repeat it here.

It has been outlined in [40] how, from the same data $\{R^\pm,F\}$ one can, in principle, construct invariants for oriented, coloured links embedded in general three-dimensional manifolds. See also [42] for related results.

Next, we wish to briefly address the question of connections between the theory developed in Sects. 2 and 3 and quantum group theory. This question has been discussed in the context of two-dimensional conformal field theory in [40,37,43]. The results, or better: speculations, found there carry over to the present framework essentially without change.

Let $\mathcal{K}$ be a Hopf algebra with co-multiplication $\Delta : \mathcal{K} \to \mathcal{K} \otimes \mathcal{K}$ and universal R-matrix $\mathcal{R} \in \mathcal{K} \otimes \mathcal{K}$. Let I be a list of finite-dimensional, unitarizable, irreducible highest-weight representations of $\mathcal{K}$. Comultiplication, Δ, is what is needed to define tensor product representations of $\mathcal{K}$: For i and j in I we define the representation $i \otimes j$ of $\mathcal{K}$ by

$$i \otimes j : \; A \in \mathcal{K} \longrightarrow i \otimes j(\Delta(A)) \in \; End(V_i \otimes V_j), \tag{4.4}$$

where V_i is the representation space of $i \in I$. Let $P_{ij} : V_i \otimes V_j \to V_j \otimes V_i$ be the transposition operator. We define

$$R_{ij} \;=\; P_{ij}\, i \otimes j\,(\mathcal{R})\,. \tag{4.5}$$

The matrices $\{R_{ij}\}$ are Yang-Baxter matrices intertwining $i \otimes j$ with $j \otimes i$, i.e.

$$j \otimes i(\Delta(A))\, R_{ij} \;=\; R_{ij}\, i \otimes j\,(\Delta(A))\,. \tag{4.6}$$

Since $\mathcal{K}$ will in general not be a semi-simple algebra, it need not be possible to decompose a tensor product representation $i \otimes j$ of $\mathcal{K}$ into a direct sum of representations in I. But $i \otimes j$ may contain some $k \in I$ as a subrepresentation. We let N_{kij} denote the multiplicity of k in $i \otimes j$. There is then a basis of Clebsch-Gordan matrices

$$P_{k,ij}(\mu) : \; V_i \otimes V_j \;\longrightarrow\; V_k,\; \mu = 1, \ldots, N_{kij}\,, \tag{4.7}$$

intertwining k with $i \otimes j$, i.e.,

$$k(A)\, P_{k,ij}(\mu) \;=\; P_{k,ij}(\mu)\, i \otimes j\,(\Delta(A))\,. \tag{4.8}$$

Experience with the representation theory of some simple quantum groups suggests to introduce the following assumption on the structure of $\{\mathcal{K}, I\}$: Define a representation $i \otimes j \otimes l, i, j, l \in I$, by

$$i \otimes j \otimes l : \; A \mapsto i \otimes j \otimes l((\Delta \otimes \mathbb{1})\, \Delta(A))\,. \tag{4.9}$$

Let N_{kijl} be the multiplicity of $k \in I$ in $i \otimes j \otimes l$. Then

$$\text{(A1)} \quad N_{kijl} \;=\; \sum_{r \in I} N_{kir}\, N_{rjl}\,. \tag{4.10}$$

Let $P_{k,ijl}(\lambda)$, $\lambda = 1, \ldots, N_{kijl}$, be a basis of Clebsch-Gordan matrices intertwining k with $i \otimes j \otimes l$.

(A2) We assume that every $P_{k,ijl}(\lambda)$ can be represented as

$$P_{k,ijl}(\lambda) = \sum_{r \in I, \mu, \nu} c(\lambda, \mu, \nu; r) \, P_{k,ir}(\mu) \, P_{r,jl}(\nu), \tag{4.11}$$

for some coefficients $c(\lambda, \mu, \nu; r) \in \mathbb{C}$.

Now we can describe the vertex-sos transformation mentioned at the end of Sect. 3: We consider two further families of N_{kijl} Clebsch-Gordan-matrices intertwining k with $i \otimes j \otimes l$:

$$\text{(i)} \quad P_{k,jr}(\alpha) \, P_{r,il}(\beta) \, (R_{ij}^{\pm 1} \otimes \mathbb{1}_l),$$

$$\tag{4.12}$$

$$\text{(ii)} \quad P_{k,rl}(\gamma) \, P_{r,ij}(\delta),$$

with $r \in I$. By assumption (A2), these matrices can be expanded in the basis (4.11):

$$P_{k,jr}(\alpha) \, P_{r,il}(\beta) \, (R_{ij}^{\pm 1} \otimes \mathbb{1}_l) = \sum_{p\mu\nu} \rho^{\pm}(k,i,j,l)_{p\mu\nu}^{r\alpha\beta} \, P_{k,ip}(\mu) \, P_{p,jl}(\nu), \tag{4.13}$$

and

$$P_{k,rl}(\gamma) \, P_{r,ij}(\delta) = \sum_{p\mu\nu} \varphi(k,i,j,l)_{p\mu\nu}^{r\gamma\delta} \, P_{k,ip}(\mu) \, P_{p,jl}(\nu), \tag{4.14}$$

for complex numbers $\rho^{\pm}(k,i,j,l)_{p\mu\nu}^{r\alpha\beta}$ and $\varphi(k,i,j,l)_{p\mu\nu}^{r\gamma\delta}$. By arguments very similar to those used in Sect. 3, one can show that the matrices $\rho^{\pm}(k,i,j,l)$ and $\varphi(k,i,j,l)$ satisfy polynomial equations analogous to (3.18),(3.22),(3.62)-(3.65); see [45,40]. Thus a Hopf algebra $\mathcal{K}$ and the set I of finite-dimensional, unitarizable, irreducible highest-weight representations, subject to assumptions (A1) and (A2), determine "six-index symbols" $\rho^{\pm}(k,i,j,l)_p^r$ and $\varphi(k,i,j,l)_p^r$ which have the same properties as $R^{\pm}(k,i,j,l)_p^r$ and $F(k,i,j,l)_p^r$.

The general problem one would like to solve is to find conditions on families of matrices $\{\rho^{\pm}(k,i,j,l)_p^r, \; \varphi(k,i,j,l)_p^r\}$, satisfying equs. (3.18),(3.22) and (3.62)-(3.65), which guarantee that these matrices are derivable from a Hopf algebra via the vertex-sos transformations (4.13),(4.14). This problem appears to be open.

Let us now suppose that there is a bijection from L to I taking the vacuum representation, 1, of $\mathcal{A}$ to the trivial representation of $\mathcal{K}$, also denoted by 1, and preserving the fusion rules $\{N_{kij}\}$. A point in L and its image in I will then be denoted by the same lower-case Latin letter. Let us further suppose that

$$R^{\pm}(k,i,j,l)_p^r \;=\; \rho^{\pm}(k,i,j,l)_p^r\,, \tag{4.15}$$

and

$$F(k,i,j,l)_p^r \;=\; \varphi(k,i,j,l)_p^r\,. \tag{4.16}$$

Then the algebra $\mathcal{K}$ plays the rôle of a "global symmetry algebra" of the quantum theory described by $\{\mathcal{A}, L\}$, which acts trivially on all observables in $\mathcal{A}$ but can be made to act non-trivially on the unobservable, charged fields of the theory. This is analogous to what has been found in [6]. In order to make these remarks more concrete, we define "$\mathcal{K}-vertices$" $v_\mu^{ki}(x^j)\,:\,V_i \to V_k$, for $x^j \in V_j$, by setting

$$v_\mu^{ki}(x^j)\,x \;\equiv\; P_{k,ji}(\mu)\,(x^j \otimes x)\,, \tag{4.17}$$

for all $x \in V_i,\;\; \mu = 1,\ldots,N_{kji}$.

Let D_{ipj} be the "structure constants" discussed in Sect. 4 of [40], and refs. given there. Then we may introduce new charged fields, $\psi(\rho^p, x^p)$, $\rho^p \in \mathcal{M}_p$, $x^p \in V_p$, by setting

$$\psi(\rho^p, x^p) \;=\; \sum_{ij\mu\nu} (D_{ipj})^{\mu\nu}\; V_\mu^{ij}(\rho^p)\; v_\nu^{ij}(x^p)\,. \tag{4.18}$$

Let ρ^p and ρ^q be two *morphisms localized in space-like separated, space-like cones, C^p and C^q, respectively, contained in a simple domain $S \subset M^3$. Let $\{e_\alpha^p\}$ and $\{e_\beta^q\}$ be bases for V_p, V_q, respectively, and set

$$\psi_\alpha(\rho^p) \;\equiv\; \psi(\rho^p, e_\alpha^p),\;\; \psi_\beta(\rho^q) \;\equiv\; \psi(\rho^q, e_\beta^q).$$

Then the commutation relations between $\psi_\alpha(\rho^p)$ ans $\psi_\beta(\rho^q)$ are given by

$$\psi_\alpha(\rho^p)\,\psi_\beta(\rho^q) \;=\; \sum_{\gamma\delta} (R_{pq}^{\pm 1})_{\alpha\beta,\gamma\delta}\;\psi_\gamma(\rho^q)\,\psi_\delta(\rho^p)\,, \tag{4.19}$$

if $as(\rho^p) \gtrless as(\rho^q)$. Here $(R^{\pm 1}_{pq})_{\alpha\beta,\gamma\delta}$ are the matrix elements of $R^{\pm 1}_{pq}$, defined in (4.5), in the bases $\{e^p_\alpha\}$ and $\{e^q_\beta\}$. See [37,43] for details. It will be shown elsewhere how to construct a *algebra out of the fields $\{\psi_\alpha(\rho^p)\}$.

Remarks

(1) It is clear how to define an action of $\mathcal{K}$ on the fields $\psi(\rho^p, x^p)$.

(2) If we work with the fields $\psi(\rho^p, x^p)$ it is natural to define the physical Hilbert space of the quantum theory described by $\{\mathcal{A}, L\}$ as follows:

$$\mathcal{H}_{phys.} = \bigoplus_{p \in L} \mathcal{H}_p \otimes V_p .$$

Let π be the representation of $\mathcal{A}$ on $\mathcal{H}_{phys.}$. Then we have that

$$\pi\big(\rho^p(A)\big)\ \psi(\rho^p, x^p) = \psi(\rho^p, x^p)\ \pi(A), \tag{4.20}$$

for all $A \in \mathcal{A}$.

See [6] for a complete theory in the case where $R^+ = R^-$, where $\mathcal{K}$ can be replaced by a compact group. An ansatz of the form (4.19) for commutation relations between unobservable, charged fields was first discussed in [15], (for theories in two space-time dimensions).

5. Back to physics

It is now time to ask how the braid statistics of charged fields (intertwiners), found in Sect. 3 and further discussed in the second half of Sect. 4, will manifest itself physically? One answer to this question is found by studying collision (scattering) theory in quantum theories with charged fields obeying braid statistics. A form of collision theory in the algebraic formulation of local, relativistic quantum theory, inspired by Haag-Ruelle theory [2], has been developed in [4,5] and can be adapted to theories with fields obeying braid statistics, as sketched in [26,28]. The result of the analysis presented in these papers is that the momentum-space wave functions describing incoming or outgoing states of charged particles have symmetry properties under exchanging the momenta (and spins, internal quantum numbers) of charged particles along oriented paths in momentum space that can be described in terms of the braid matrices $\{R^+, R^-\}$. Thus, the statistics of charged fields (intertwiners) determines the statistics of asymptotic charged particles which manifests itself in symmetry properties of scattering amplitudes and cross sections. For all details we must refer the reader to [4,5,26,28].

Next, we wish to reconsider the question of what kinds of Lagrangian models of local, relativistic quantum theories in three space-time dimensions have charged fields obeying braid statistics? Heuristically, a rather clear-cut answer can be given.

It is known from the work of Doplicher, Haag and Roberts [4] that if the charged fields (intertwiners) can be localized in bounded space-time regions and the dimension of space-time is $d \geq 3$ then $R^+(i,p,q,k)^l_j = R^-(i,p,q,k)^l_j$, and, by Theorem 5, Sect. 3,

$$s_p \in \frac{1}{2}\mathbb{Z}, \text{ for all } p \in L,$$

and "spin parity" is conserved under fusion, (i.e, $\sigma_p \cdot \sigma_q = \sigma_r$ if $N_{rpq} \neq 0$) .

Thus, in order to find examples of theories with charged fields satisfying braid group statistics we must look for theories whose charged fields cannot be localized in bounded space-time regions, but only in space-like cones. Charged fields only localizable in space-like cones are a typical feature of gauge theories.

Consider a three-dimensional theory with charged fields only localizable in space-like cones. Let P denote space-reflection at a line $l \subset \{(\vec{x},t) \in \mathsf{M}^3 : t = 0\}$. Let $\mathcal{O}$ be a region in M^3. We define

$$\mathcal{O}^P = \{(\vec{x},t) \in \mathsf{M}^3 : (P\vec{x},t) \in \mathcal{O}\}. \tag{5.1}$$

We suppose that P is represented on $\mathcal{A}$ by a *automorphism α_P with the property that

$$\alpha_P(\mathcal{A}(\mathcal{O})) = \mathcal{A}(\mathcal{O}^P). \tag{5.2}$$

Let ρ^p be a *morphism of an extended algebra $\mathcal{B}^{C_a}$, with $p \in L$. We define ρ^p_P by setting

$$\rho^p_P(A) = \alpha_P \circ \rho^p \circ \alpha_P(A). \tag{5.3}$$

Let ρ^p and ρ^q be *morphisms of $\mathcal{B}^{C_a}$ localized in space-like separated, space-like cones, C^p and C^q, (space-like separated from the auxiliary cone C_a). It follows easily from (5.2), (5.3) and (2.18) that

$$as(\rho^p) \gtrless as(\rho^q) \iff as(\rho^p_P) \lessgtr as(\rho^q_P). \tag{5.4}$$

Next, suppose that, for all $p \in L$, the *morphisms ρ^p and ρ^p_P are unitarily equivalent and that α_P can be implemented unitarily on the total physical Hilbert space of states, $\mathcal{H}_{phys.}$, of the theory, i.e.,

$$\rho^p \sim \rho^p_P, \text{ for all } p \in L, \ \pi(\alpha_P(A)) = U(P)\pi(A)U(P)^*, \tag{5.5}$$

for all $A \in \mathcal{A}$, where $U(P)$ is a unitary involution on $\mathcal{H}_{phys.}$. Let us recall the commutation relations (3.9) between charged fields established in Theorem 1 of Sect. 3:

$$\begin{aligned} &V^{ji}_{\mu}(\rho^p)\, V^{ik}_{\nu}(\rho^q) \\ &\quad = \sum_{l\alpha\beta} R^{\pm}(j,p,q,k)^{l\alpha\beta}_{i\mu\nu}\, V^{jl}_{\alpha}(\rho^q)\, V^{lk}_{\beta}(\rho^p), \end{aligned} \tag{5.6}$$

if $\rho^p \times \rho^q$ and $as(\rho^p) \gtrless as(\rho^q)$.

By applying (5.5) and (5.4) to (5.6) we obtain the following

<u>**Theorem 7**</u> <u>If space reflection at a line,</u> P, <u>is a symmetry of the theory, in the sense of equ.</u> (5.5), <u>then</u>

$$R^{+}(j,p,q,k)^{l}_{i} = R^{-}(j,p,q,k)^{l}_{i},$$

<u>i.e., the theory has ordinary permutation group statistics,</u> $s_p \in \frac{1}{2}\mathbb{Z}$, <u>and "spin parity" is conserved under fusion.</u>

Thus, in order for a three-dimensional, local quantum theory to exhibit <u>non-trivial braid group statistics</u> it must have charged fields which <u>cannot</u> be localized in bounded space-time regions and it must <u>break</u> the symmetry of space-reflection at a line.

Let us suppose that we look for a relativistic theory with these features. Then it must likely be a gauge theory without confinement which breaks space-reflection at a line. Thinking in terms of Lagrangian theories, we conclude that it must be a gauge theory with a Chern-Simons term in the effective gauge field Langrangian, as discussed in Sect. 1, or a theory which can be reformulated as a Chern-Simons theory, such as an $O(3)$ non-linear σ-model with a Hopf term in the effective Lagrangian.

In view of applications of our theory to condensed matter physics, it is useful to recall the essential assumptions on which the general theory developed in Sects. 2, 3 and 5 rests. They are as follows:

(1) The algebra of observables, $\mathcal{A}$, of the theory has a local structure

$$\mathcal{A} = \overline{\cup \mathcal{A}(C)}^{n}, \tag{5.7}$$

where $\mathcal{A}(C)$ is an algebra of observables corresponding to measurements in a localized region, C, of space-time; (C might be a wedge contained in a time slice). One requires some suitable form of locality, duality and the Reeh-Schlieder theorem; (see Sect. 2).

(2) Existence of a space-time translation- and rotation covariant "ground state representation", with the property that the generator of time translations, (the Hamiltonian), H satisfies the spectrum condition

$$H \geq 0. \tag{5.8}$$

(3) One considers then the class of all space-time translation- and rotation-covariant representations, p, of $\mathcal{A}$ satisfying the spectrum condition (5.8) and requires that p be localizable in an arbitrary "wedge", with respect to the local structure (5.7), in the sense of (2.18).

We emphasize that full relativistic covariance is not needed; covariance under the projective group of Euclidean motions in space, and time translations is already a little more than what we need for our analysis.

From these remarks we conclude that braid statistics can, in principle, be encountered in non-relativistic systems of condensed matter theory with broken space-reflection symmetry! Archetypal examples are correlated electronic systems in a strong external magnetic field, such as quantum Hall systems. Other systems are two-dimensional systems with flux phases, (perhaps of relevance in high T_c superconductivity).

These matters will be discussed in more detail in a future publication. Our approach to these problems is somewhat comparable in its spirit to the topological approach to classifying defects in ordered media [46]: Very general arguments based on symmetry and topological considerations yield a considerable amount of insight.

Acknowledgements The ideas in Sect. 3 are intimately related to ideas worked out in collaborations by one of us (J.F.) with G. Felder, G. Keller and C. King. Their help is gratefully acknowledged.

References

1. M. Fierz, Helvetica Physica Acta 12, 3 (1939).
2. R. Jost, "The General Theory of Quantized Fields", Providence, R.I.: American Math. Soc. Publ. 1965.
   R.F. Streater and A.S. Wightman, "PCT, Spin and Statistics, and All That", Reading, Mass.: Benjamin 1978.
3. J.J.J. Kokkedee, "The Quark Model", New York: Benjamin, 1969.
   R.N. Mohapatra and C.H. Lai (eds.) "Gauge Theories of Fundamental Interactions", Singapore: World Scientific 1981.
   H.S. Green, Phys. Rev. 90, 270 (1953);
   O.W. Greenberg and A.M.L. Messiah, Phys. Rev. 136, B 248 (1964).
4. R. Haag and D. Kastler, J. Math. Phys. 5, 848 (1964);
   S. Doplicher, R. Haag, and J.E. Roberts, Commun. Math. Phys. 23, 199 (1971); Commun. Math. Phys. 35, 49 (1974).
   J.E. Roberts, "Statistics and the Intertwiner Calculus", in "C*-Algebras and Their Applications to Statistical Mechanics and Quantum Field Theory", Proc. of Int. School of Physics "Enrico Fermi", Course LX, North Holland 1976.
5. D. Buchholz and K. Fredenhagen, Commun. Math. Phys. 84, 1 (1982).
6. S. Doplicher and J.E. Roberts, "C*-Algebras and Duality for Compact Groups ..." in: Proc. of VIIIth Int. Congress on Math. Phys., K. Mebkhout and R. Sénéor (eds.), Singapore: World Scientific 1987; preprint, Rome 1989.
7. R.F. Streater and I. Wilde, Nuclear Physics B 24, 561 (1970).
8. J. Fröhlich, Commun. Math. Phys. 47, 269 (1976);
   Acta Phys. Austriaca Suppl. XV, 133 (1976); unpublished manuscript (1976).
   J. Bellissard, J. Fröhlich and B. Gidas, Commun. Math. Phys. 60, 37 (1978).
9. L. Kadanoff and H. Ceva, Phys. Rev. B 11, 3918 (1971).
10. M. Sato, T. Miwa and M. Jimbo, Publ. RIMS, Kyoto University, 15, 871 (1979).
11. J. Fröhlich, in "Recent Developments in Gauge Theories" (Cargèse 1979), G. 't Hooft et al. (eds.), New York: Plenum Press, 1980.

12. E.C. Marino and J.A. Swieca, Nucl. Phys. B 170 [FS1], 175 (1980).

E.C. Marino, B. Schroer and J.A. Swieca, Nucl. Phys. B 200 [FS4], 473 (1982).

R. Köberle and E.C. Marino, Phys. Lett. 126 B, 475 (1983).

J. Fröhlich and P.-A. Marchetti, Commun. Math. Phys. 116, 127 (1988).

13. A.A. Belavin, A.M. Polyakov and A.B. Zamolodchikov, Nucl. Phys. B 247, 83 (1984).

14. V.S. Dotsenko and V.A. Fateev, Nucl Phys. B 240, [FS12], 312 (1984), B 251 [FS13], 691 (1985).

A. Tsuchiya and Y. Kanie, Lett. Math. Phys. 13, 303 (1987).

K.H. Rehren and B. Schroer, Phys. Lett. B 198, 480 (1987).

15. J. Fröhlich, "Statistics of Fields, the Yang-Baxter Equation, and the Theory of Knots and Links", in: "Non-Perturbative Quantum Field Theory", G. 't Hooft et al. (eds.), New York: Plenum 1988.

16. R. Haag and D. Kastler, J. Math. Phys. 5, 848 (1964).

17. V.F.R. Jones, Invent. Math. 72, 1 (1983);

"Braid Groups, Hecke Algebras and Type II_1 Factors", in: "Geometric Methods in Operator Algebrs", Proc. US-Japan Seminar, 242 (1986);

V.F.R. Jones, Ann. Math. 126, 335 (1987);

H. Wenzl, Invent. Math. 92, 349 (1988);

A. Ocneanu, "Quantized Groups, String Algebra, and Galois Theory for Algebras", Penn. State Univ., preprint 1988.

18. K. Fredenhagen, K.H. Rehren and B. Schroer, "Superselection Sectors with Braid Group Statistics and Exchange Algebras, I: General Theory", F.U. Berlin, preprint 1988.

19. K.H. Rehren, Commun. Math. Phys. 116, 675 (1988).

J. Fröhlich, "Statistics and Monodromy in Two- and Three-Dimensional Quantum Field Theory", in: "Differential Geometrical Methods in Theoretical Phy--sics", K. Bleuler and M. Werner (eds.), Dordrecht: Kluwer Academic Publ. 1988.

G. Moore and N. Seiberg, Phys. Lett. B 212, 451 (1988); Nucl. Phys. B 313, 16 (1989); Commun. Math. Phys. 123, 177 (1989).

G. Felder, J. Fröhlich and G. Keller, Commun. Math. Phys. 124, 417 (1989);

"On the Structure of Unitary Conformal Field Theory II: Representation Theoretic Approach", to appear in Commun. Math. Phys.

20. L.S. Schulman, J. Math. Phys. 12, 304 (1971);
M.G.G. Laidlaw and C. De Witt-Morette, Phys. Rev. D 3, 1375 (1971); J.S. Dowker, J. Phys. A 5, 936 (1972); J. Phys. A 18, 3521 (1985); E.C.G. Sudarshan, "Topology, Quantum Theory and Dynamics", University of Texas, preprint 1989.
21. J. Birman, "Braids, Links and Mapping Class Groups", Ann. Math. Studies 82, Princeton: Princeton University Press 1974.
22. J.M. Leinaas and J. Myrheim, Il Nuovo Cimento 37 B, 1 (1977).
23. F. Wilczek, Phys. Rev. Lett. 48, 1144 (1982); Phys. Rev. Lett. 49, 957 (1982); F. Wilczek and A. Zee, Phys. Rev. Lett. 51, 2250 (1983).
24. Y.S. Wu, "Fractional Quantum Statistics in Two-Dimensional Systems", in: Proc. 2nd Int. Symp. Foundations of Quantum Mechanics, Tokyo 1986, pp. 171 - 180.
25. R.E. Prange and S.M. Girvin (eds.), "The Quantum Hall Effect", Berlin-New York: Springer-Verlag 1987.
T. Chakrabarty, P. Pietiläinen, "The Fractional Quantum Hall Effect", Berlin-New York: Springer-Verlag 1988.
26. J. Fröhlich and P.-A. Marchetti, Lett. Math. Phys. 16, 347 (1988); Commun. Math. Phys. 121, 177 (1989).
27. J. Fröhlich, "Statistics and Monodromy in Two- and Three-Dimensional Quantum Field Theory"; see ref. 19.
28. J. Fröhlich, F. Gabbiani and P.-A. Marchetti, "Superselection Structure and Statistics in Three-Dimensional Local Quantum Theory", Proc. 12th John Hopkins Workshop on "Current Problems in High Energy Particle Theory", Florence 1989, G. Lusanna (ed.).
J. Fröhlich, F. Gabbiani and P.-A. Marchetti, "Braid Statistics in Three-Dimensional Local Quantum Theory", in preparation.
29. R.B.Laughlin, Science 242, 525 (1988);
V. Kalmeyer and R.B. Laughlin, Phys. Rev. Lett. 59, 2095 (1987).
A. Fetter, C. Hanna and R.B. Laughlin, Stanford University, preprint 1988.

Y.-H. Chen, F. Wilczek, E. Witten and B.I. Halperin, IASSNS-HEP-89/27, preprint 1989.

30. W. Siegel, Nucl. Phys. B 156, 135 (1979);
R. Jackiw and S. Templeton, Phys. Rev. D 23, 2291 (1981);
J.F. Schönfeld, Nucl. Phys. B 185, 157 (1981);
S. Deser, R. Jackiw and S. Templeton, Phys. Rev. Lett. 48, 975 (1982); Ann. Phys. (N.Y.) 140, 372 (1982);
I. Affleck, J. Harvey and E. Witten, Nucl. Phys. B 206, 413 (1982);
A.N. Redlich, Phys. Rev. Lett. 52, 18 (1984); Phys. Rev. D 29, 2366 (1984);
R.D. Pisarski and S. Rao, Phys. Rev. D 32, 2081 (1985), and refs. given there.

31. A. Coste and M. Lüscher, "Parity Anomaly and Fermion Boson Transmutation in 3-Dimensional Lattice QED", DESY 89-017, preprint 1989.
M. Lüscher, "Bosonization in 2+1 Dimensions", DESY, preprint 1989.
G.W. Semonoff and P. Sodano, preprint 1989.

32. E. Witten, Commun. Math. Phys. 121, 351 (1989).

33. J. Fröhlich and C. King, "The Chern-Simons Theory and Knot Polynomials", Commun. Math. Phys., in press.

34. A. Tsuchiya and Y. Kanie, Lett. Math. Phys. 13, 303 (1987);
T. Kohno, Ann. Inst. Fourier 37, 139 (1987); Adv. Studies in Pure Math. 12, 189 (1987).

35. K. Fredenhagen, Commun. Math. Phys. 79, 141 (1981).

36. G. Moore and N. Seiberg, Phys. Lett. B 212, 451 (1988).

37. G. Felder, J. Fröhlich and G. Keller, "On the Structure of Unitary Conformal Field Theory II ...", see ref. 19.

38. D. Friedan and S. Shenker, Nucl. Phys. B 281, 509 (1987);
G. Anderson and G. Moore, Commun. Math. Phys. 117, 441 (1988);
C. Vafa, Phys. Lett. B 206, 421 (1988).

39. R. Longo, "Index of Subfactors and Statistics of Quantum Fields", Commun. Math. Phys., in press; "Index of Subfactors and Statistics ... II: Correspondences, Braid Group Statistics and Jones Polynomial", Rome, preprint 1989.

40. J. Fröhlich and C. King, "Two-Dimensional Conformal Field Theory and Three-Dimensional Topology", J. Mod. Phys. A, in press.

41. D. Rolfsen, "Knots and Links", Publish or Perish, Math. Lecture Series, 1976;
G. Burde and H. Zieschang, "Knots", de Gruyter Studies in Mathematics 5.
V.F.R. Jones, Bulletin AMS 12, 103 (1985).
C.N. Yang and M.L. Ge (eds.), "Braid Group, Knot Theory and Statistical Mechanics", Adv. Series in Math. Physics, vol. 9, Singapore: World Scientific 1989.
42. N.Yu. Reshetikhin and V.G. Turaev, "Invariants of 3-Manifolds via Link Polynomials and Quantum Groups" MSRI 04008-89, preprint 1989.
43. L. Alvarez-Gaumé, C. Gomez and G. Sierra, "Quantum Group Interpretation of Some Conformal Field Theories", CERN-TH-5267/88, preprint 1988;
Refs. 40 and 37;
G. Moore and N.Yu. Reshetikhin, "A Comment on Quantum Group Symmetry in Conformal Field Theory", IASSNS-HEP-89/18;
D. Buchholz, G. Mack and I. Todorov, unpublished.
44 . V. Pasquier, Commun. Math. Phys. 118, 355 (1988).
45. N.Yu. Reshetikhin, "Quantized Universal Enveloping Algebras, The Yang-Baxter Equation and Invariants of Links I, II", LOMI preprints 1987.
46. N.D. Mermin, Rev. Mod. Phys. 51, 591 (1979);
L. Michel, Rev. Mod. Phys. 52, 617 (1980).

ON THE ALGEBRAIC STRUCTURE OF THE BRST SYMMETRY

Marc Henneaux*

Faculte des Sciences, Universite Libre de Bruxelles
Campus Plaine C.P. 231, B-1050 Bruxelles, Belgium
and
Centro de Estudios Cientificos de Santiago
Casilla 16443, Santiago 9, Chile

ABSTRACT

The algebraic structure of the BRST symmetry is explained in both the Lagrangian and the Hamiltonian cases.

1. INTRODUCTION

Three important features of the BRST symmetry have been gradually recognized in the last years. These are:

(i) The definition of the BRST symmetry does not involve any gauge choice, even though it was historically discovered only

---

* Maitre de recherches au Fonds National de la Recherche Scientifique (Belgium).

Physics, Geometry, and Topology
Edited by H. C. Lee
Plenum Press, New York, 1990

after fixing the gauge. Hence, the BRST symmetry possesses a manifestly intrinsic significance.
(ii) The BRST symmetry can be constructed for arbitrary gauge theories with either closed or "open" algebras. It can be used as a substitute for the original gauge symmetry, whatever the structure of this gauge symmetry is. The group structure (if any) is thus not a fundamental ingredient of the BRST construction, which relies on more primitive concepts.
(iii) The BRST symmetry does not depend on how one chooses to represent the gauge symmetry.

The purposes of these notes is to explain the rationale behind the BRST symmetry with an emphasis on these important aspects. We will stress here the main ideas, without giving the explicit proofs. We refer to the original literature for more details.

It turns out that the appropriate algebraic framework for discussing the BRST symmetry is provided by the "Homological Perturbation Theory" of Gugenheim and Stasheff [1].

The elements of this subject of algebraic topology necessary for defining the BRST symmetry were rediscovered independently by physicists and we will follow here the physicists'approach, notations and terminology.

These notes are based on the papers [2-6] where references to related works may be found. Let us simply indicate here that the algebraic approach analyzed in these notes has also been discussed along different lines and in more specialized situations in [7, 8](irreducible Lagrangian case) and [9, 10] (irreducible Hamiltonian case). Finally, the whole construction owes much to the pioneering works by Batalin, Fradkin and Vilkovisky [11, 12] .

2. SOME RESULTS OF ALGEBRAIC TOPOLOGY

We will be interested in graded derivations that are either even or odd and which (conventionally) act from the right. The elements of the graded algebra on which they act will be denoted by capital Latin letters, while the derivations themselves will be denoted by small Greek or Latin letters. With a right action, the Leibnitz rule reads

$$\sigma(AB) = A(\sigma B) + (-)^{\varepsilon_\sigma \varepsilon_B} (\sigma A) B \tag{1}$$

where ε_σ and ε_B are respectively the parity of σ and B.

Given two derivations σ , ρ , one defines their graded commutator as

$$[\sigma, \rho] = \sigma\rho - (-)^{\varepsilon_\sigma \varepsilon_\rho} \rho\sigma \tag{2}$$

It is a derivation of parity $\varepsilon_\sigma + \varepsilon_\rho$. Note that $[\ ,\ ]$ is actually the anticommutator when σ and ρ are both odd. The graded commutator obeys a graded version of the Jacobi identity,

$$[[\sigma,\rho],\nu] + (-)^{\varepsilon_\sigma(\varepsilon_\rho+\varepsilon_\nu)} [[\rho,\nu],\sigma] + (-)^{\varepsilon_\nu(\varepsilon_\sigma+\varepsilon_\rho)} [[\nu,\sigma],\rho] = 0 \tag{3}$$

Of particular interest are odd derivations that are "nilpotent of order two" -or, as one also more simply says, just "nilpotent"-,

$$d^2 \equiv \frac{1}{2}[d,d] = 0 \tag{4}$$

These will be called here "differentials". The cohomology groups $(\mathrm{Ker}d/\mathrm{Im}d)^k$ of d will be denoted by $H^k(d)$, where k is the grading associated with d. So, $H^k(d)$ is the set of equivalence classes of elements A of degree k annihilated by d ($dA = 0$) modulo the elements of degree k in the image of d ($A \sim A+dB$). Because of the Leibnitz rule, a product structure is well defined in Kerd/Imd.

One can also define a cohomology in the space of derivations. A derivation α is d-invariant if it commutes with d, $[\alpha, d] = 0$. If β is a derivation, $[\beta, d]$ is d-invariant by the Jacobi identity. Thus, one defines $\mathcal{H}^k(d)$ as the set of d-invariant derivations at degree k modulo the d-exact ones, $[\alpha, d] = 0$, $\alpha \sim \alpha + [\beta, d]$. The degree of α is such that $\deg(\alpha A) = \deg \alpha + \deg A$ and can be positive or negative even if the degree of the elements A of the algebra is non negative.

The main result of algebraic topology needed in the BRST construction concerns differential complexes with two gradings. One of the gradings will be called the "resolution degree" - or "r-degree" - and will be denoted by r. For reasons that will become clear later on, the second grading will be called "total ghost number" - or just "ghost number" - and will be denoted by gh. The r-degree r(A) of any element in the algebra is required to be non-negative, $r(A) \geqslant 0$.

Let us assume that there are two odd derivations δ and $\overset{(0)}{s}$ with the following properties,

$$\delta^2 = 0 \tag{5a}$$

$$r(\delta) = -1 \Longleftrightarrow \quad r(\delta A) = r(A) - 1 \quad \text{for } r(A) \geqslant 1, \tag{5b}$$

$$\delta A = 0 \quad \text{for} \quad r(A) = 0 \tag{5b'}$$

$$gh\,\delta = 1 \tag{5c}$$

$$H_k(\delta) = 0\ ,\ \mathcal{H}_k(\delta) = 0\ ,\ k \neq 0 \quad (k = \text{r-degree}) \tag{5d}$$

$$\overset{(0)}{s}\delta + \delta\overset{(0)}{s} = 0 \quad ,\quad 2\overset{(0)}{s}{}^2 = -[\delta, \overset{(1)}{s}] \tag{5e}$$

$$r(\overset{(0)}{s}) = 0\ , \qquad gh(\overset{(0)}{s}) = 1 \tag{5f}$$

In (5e), $\overset{(1)}{s}$ is a derivation of r-degree one and ghost number one. So, δ is a differential with trivial homology for $k \neq 0$, while $\overset{(0)}{s}$ is "almost" a differential. It fails to be nilpotent by a term that is δ-exact. One says that $\overset{(0)}{s}$ is a "differential modulo δ".

One can define the "cohomology of $\overset{(0)}{s}$ modulo δ" as

$$H^k(\overset{(0)}{s}) = \{\overset{(0)}{s}A = \delta M,\ A \sim A + \overset{(0)}{s}B + \delta C\} \tag{6}$$

where A is of ghost number k and of r-degree <u>zero</u>. In other words, one considers elements in the algebra that are $\overset{(0)}{s}$-closed modulo δ, and one identifies two such objects that differ by $\overset{(0)}{s}$- or δ-exact terms. This makes sense because of (5a), (5e) and of the fact that δB vanishes for B of r-degree zero, as there is no element of negative r-degree.

Working with a "differential modulo δ" is somewhat awkward, and one idea of homological perturbation theory is to replace $\overset{(0)}{s}$ by a true differential s without changing the cohomology : one constructs a "model" for (6).

<u>Theorem</u> : given (5), there exists an odd derivation s of total ghost number one with the following properties :

$$s = \delta + \overset{(0)}{s} + \overset{(1)}{s} + \overset{(2)}{s} + \ldots \tag{7a}$$

$$r(\overset{(k)}{s}) = k \quad ,\quad gh(\overset{(k)}{s}) = 1 \tag{7b}$$

$$s^2 = 0 \tag{7c}$$

$$H^k(s) = H^k(\overset{(0)}{s}) \tag{7d}$$

This theorem is a straightforward consequence of the Jacobi identity and of $\mathcal{H}_k(\delta) = 0$ for $k \neq 0$. It is proven recursively as follows. First, the equations (5a) and (5e) imply that the components of s^2 of r-degrees -2, -1 and 0 vanish. So, let us suppose that we have constructed s recursively up to order $n \geqslant 1$, and let us set

$$\overset{(n)}{\sigma} = \delta + \overset{(0)}{s} + \ldots + \overset{(n)}{s} \tag{8a}$$

By construction, $\overset{(n)}{\sigma}{}^2$ starts at order n,

$$2\overset{(n)}{\sigma}{}^2 = [\overset{(n)}{\sigma}, \overset{(n)}{\sigma}] = \overset{(n)}{\rho} + \overset{(n+1)}{\rho} + \ldots \tag{8b}$$

where $\overset{(k)}{\rho}$ is of r-degree k. We want to show that it is possible to add to $\overset{(n)}{\sigma}$ a term $\overset{(n+1)}{s}$ such that $(\overset{(n)}{\sigma} + \overset{(n+1)}{s})^2$ starts at order n+1, i.e., that one can kill $\overset{(n)}{\rho}$ by a suitable choice of $\overset{(n+1)}{s}$.

One has

$$[\overset{(n)}{\sigma} + \overset{(n+1)}{s}, \overset{(n)}{\sigma} + \overset{(n+1)}{s}] = 2[\delta, \overset{(n+1)}{s}] + \overset{(n)}{\rho} + \text{"higher order"}. \tag{8c}$$

To remove $\overset{(n)}{\rho}$, one must take $\overset{(n+1)}{s}$ solution of $2[\delta, \overset{(n+1)}{s}] + \overset{(n)}{\rho} = 0$. This equation possesses a solution if and only if $[\delta, \overset{(n)}{\rho}] = 0$. This condition is clearly necessary ($\delta^2 = 0$) but it is also sufficient as $\mathcal{H}_k(\delta) = 0$, $k \neq 0$. To check $[\delta, \overset{(n)}{\rho}] = 0$, one observes that $[\delta, \overset{(n)}{\rho}]$ is just the component of r-degree n-1 of $[[\overset{(n)}{\sigma}, \overset{(n)}{\sigma}], \overset{(n)}{\sigma}]$ and so, it vanishes by the graded Jacobi identity.

Once the existence of s is established, one proves $H^k(s) = H^k(\overset{(0)}{s})$ by expanding A, ghA = k, according to the resolution degree,

$$A = \sum_{m \geq 0} \overset{(m)}{A} \quad , \qquad r(\overset{(m)}{A}) = m \, , \; gh \overset{(m)}{A} = k \tag{9}$$

If sA = 0, then $\overset{(0)}{s}\overset{(0)}{A} + \delta \overset{(1)}{A} = 0$ and $\overset{(0)}{A}$ is thus $\overset{(0)}{s}$-closed modulo δ. The addition of sB to A modifies $\overset{(0)}{A}$ as $\overset{(0)}{A} \longrightarrow \overset{(0)}{A} + \overset{(0)}{s}\overset{(0)}{B} + \delta \overset{(1)}{B}$. Therefore, cohomological classes of s of ghost number k defines cohomological classes of $\overset{(0)}{s}$ modulo δ with same ghost number and r-degree zero. The mapping between these cohomological spaces is easily shown to be bijective by the same perturbative techniques that led to the proof of the existence of s : given $\overset{(0)}{A}$ solution of $\overset{(0)}{s}\overset{(0)}{A} + \delta \overset{(1)}{A} = 0$, one can improve $\overset{(0)}{A}$ by higher order terms, $\overset{(0)}{A} \longrightarrow A = \overset{(0)}{A}$ + "more" so that sA = 0. Furthermore, the extensions associated with $\overset{(0)}{A}$ and $\overset{(0)}{A} + \overset{(0)}{s}\overset{(0)}{B} + \delta \overset{(1)}{B}$ differ by a s-exact term. The formula (7d) is thereby proven. [This proof of (7d) is actually just a rephrasing of standard spectral sequence arguments in the special case where the spectral sequence collapses after the second step].

It should be noticed that s is not unique since one has the possibility of adding a δ-exact term at each resolution degree level.

Remark : Given the r-degree and the total ghost number, one can define a third grading ("filtration degree") by

$$\deg A = gh A + r(A) \tag{10a}$$

One has,

$$\deg \delta = 0 \quad , \quad \deg \overset{(0)}{s} = 1, \quad \deg \overset{(m)}{s} = m+1. \tag{10b}$$

In the main applications discussed below, it is actually this third grading and the r-degree that come naturally. They are known in those cases as the "pure ghost number" and the "antighost number" respectively. The ghost number is a derived concept defined by (10a), i.e.,

$$gh A = pure\,gh A - antigh A \tag{10c}$$

Furthermore, it turns out that the pure ghost number is non negative, so that (7d) becomes

$$H^k(s) = \begin{cases} 0 \quad , \quad k<0 & (11a) \\ H^k(s^{(0)}) \quad , \quad k \geq 0 & (11b) \end{cases}$$

3. A GEOMETRIC APPLICATION

3.1 Geometric ingredients

The theorem proved above can be applied to the following geometrical setting.

Consider the following ingredients :
(1) A surface Σ in a manifold I, defined by the equations

$$\Sigma : \quad f_{a_0} = 0 \quad , \quad a_0 = 1, \ldots, A_0 \tag{12}$$

(2) A set of vector fields $\vec{X}_{\beta_0}$ ($\beta_0 = 1, \ldots, B_0$) that
(i) are tangent to Σ,

$$\vec{X}_{\beta_0} f_{a_0} = t_{\beta_0 a_0}^{b_0} f_{b_0} \left(\Leftrightarrow \vec{X}_{\beta_0} f_{a_0} = 0 \text{ on } \Sigma \right); \tag{13a}$$

and (ii) close on Σ,

$$[\vec{X}_{\alpha_0}, \vec{X}_{\beta_0}] = C_{\alpha_0\beta_0}^{\gamma_0} \vec{X}_{\gamma_0} + f_{a_0} \vec{M}_{\alpha_0\beta_0}^{a_0} . \tag{13b}$$

The equations (12) describing Σ may or may not be independent, i.e., there may not or may be some relations on the f_{a_0}'s

$$R_{a_1}^{a_0} f_{a_0} = 0 \text{ (identically)}, \quad a_1 = 1, \ldots, A_1 \tag{14}$$

The need to represent Σ by an overcomplete set of equations may follow from covariance requirements or may be dictated by topological considerations (the normal bundle to Σ may be non trivial).

Similarly, we allow, if necessary, for relations among the $\vec{X}_{\alpha_0}$'s,

$$Z^{\alpha_0}_{\alpha_1} \vec{X}_{\alpha_0} = f_{a_0} \vec{\mu}^{a_0}_{\alpha_1} \quad (\Leftrightarrow Z^{\alpha_0}_{\alpha_1} \vec{X}_{\alpha_0} = 0 \text{ on } \Sigma). \tag{15}$$

By (13b), the vectors $\vec{X}_{\alpha_0}$ generate integrable orbits on Σ (figure 1)

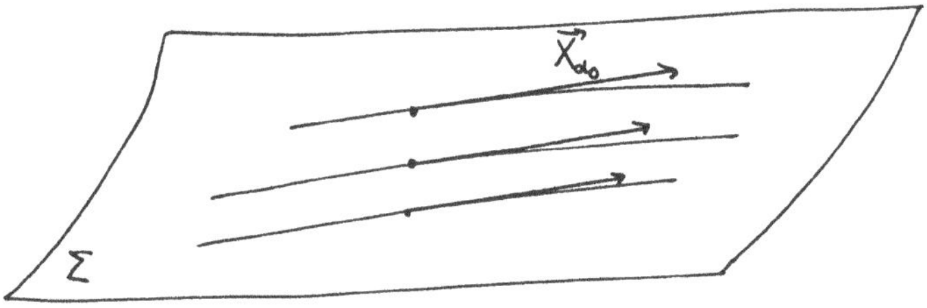

Fig. 1

Off Σ, however, the bracket $[\vec{X}_{\alpha_0}, \vec{X}_{\beta_0}]$ is in general no longer a combination of the $\vec{X}_{\alpha_0}$'s.

At this stage, we do not assume any additional relation among f_{a_0}, $R^{a_0}_{a_1}$, $\vec{X}_{\alpha_0}$ and $Z^{\alpha_0}_{\alpha_1}$, besides (12)-(15).

3.2 Exterior derivative along the orbits

One can define on Σ an "exterior derivative operator d along the orbits" that measures how the functions vary along the orbits. This operator obeys $d^2 = 0$ on Σ and is such that $H^o(d)$ is given by the functions on Σ that are constant along the orbits.

If the vector fields $\vec{X}_{\alpha_0}$ are independent on Σ, the "p-forms along the orbits" on which d acts can be represented as polynomials in the elements of the dual basis,

$$A = \frac{1}{p!} A_{\alpha_0 \ldots \gamma_0} \eta^{\alpha_0} \ldots \eta^{\gamma_0} \tag{16}$$

The differential d can then be defined by

$$dF = (\vec{X}_{\alpha_0} F)\, \eta^{\alpha_0} \tag{17a}$$

for the functions, and

$$d\eta^{\alpha_0} = \frac{1}{2}\, C^{\alpha_0}{}_{\beta_0\gamma_0}\, \eta^{\beta_0}\, \eta^{\gamma_0} \tag{17b}$$

for the 1-forms η^{α_0}. One easily verifies $d^2 = 0$ on Σ.

If the vector fields $\vec{X}_{\alpha_0}$ are not independent on Σ, there is no dual basis. One can nevertheless still represent the forms along the orbits as polynomials in abstract objects η^{α_0}, as in (16), with

$$A_{\alpha_0 \cdots \gamma_0} \equiv A(\vec{X}_{\alpha_0}, \ldots, \vec{X}_{\gamma_0}) \tag{18}$$

However, the coefficient functions $A_{\alpha_0 \ldots \gamma_0}$ are now no longer free as they must obey on Σ the conditions

$$Z_{\alpha_1}^{\beta_0}\, A_{\alpha_0 \cdots \beta_0 \cdots \gamma_0} = 0 \tag{19}$$

The exterior derivative d can still be computed according to the rules (17). One verifies again $d^2 = 0$ on Σ for polynomials in η^{α_0} obeying (19).

One can relax the condition (19) by introducing further formal objects η^{α_1} in degree two, whose aim is precisely to enforce (19). In this more convenient description, the p-forms are identified with the polynomials in η^{α_0} and η^{α_1}. One defines, for functions,

$$dF = (\vec{X}_{\alpha_0} F)\, \eta^{\alpha_0} \tag{20a}$$

as in (17a), but one replaces (17b) by

$$d\eta^{\alpha_0} = \frac{1}{2}\, C^{\alpha_0}{}_{\beta_0\gamma_0}\, \eta^{\beta_0}\, \eta^{\gamma_0} + Z_{\alpha_1}^{\alpha_0}\, \eta^{\alpha_1} \tag{20b}$$

The definition of $d\eta^{\alpha_1}$ is given in [4] and we refer to that reference for the details. Again one finds $d^2 = 0$ on Σ. The advantage of this new alternative description is that the coefficients of η^{α_0} and η^{α_1} can be treated as <u>free</u> and are not subject to a condition of the type (19). This condition arises when one imposes the closedness condition, $dA = 0$ (see [4]).

The grading of d will be called from now on the pure ghost number. It is non negative. The 1-forms η^{α_0} will be identified with the ghosts below. The forms η^{α_1} in degree 2 that appear when the vector fields are not independent on Σ ("reducible case") will become the "ghosts of ghosts".

3.3 From Σ to I

The exterior differential d has been defined so far on the surface Σ. In practice, however, it turns out to be more convenient to work with functions on the whole manifold I. So, the problem is to "lift" d from Σ to I. This can be done by assuming that the coefficients of the p-forms are functions on I. One than still uses the same rules (20) for computing dA. This makes sense because the vector fields $\vec{X}_{\alpha_0}$ and the functions $C^{\alpha_0}_{\beta_0\gamma_0}$, $Z^{\alpha_0}_{\alpha_1}$, are defined everywhere and not just on the surface Σ.

With these rules, one finds $d^2 \approx 0$ where "$\approx$" means "equal on Σ to" -i.e., "weakly equal to". In general, however, $d^2 \neq 0$ off Σ. So, nilpotency is lost. Furthermore, the cohomology of d is given by the set of weakly closed p-forms modulo weakly exact ones.*

The situation is very similar to that encountered in the theorem of the previous section. One has an operator that is not exactly nilpotent, and whose cohomology is given by a further identification besides the one implied by the mere addition of an exact term (see formulas (5e) and (6)).

---

*Even when $d^2 = 0$ off Σ, the identification of forms that coincide on Σ is necessary in order to get the cohomology of the exterior derivative d along the orbits on Σ.

We will show that the situation is not only similar, but can be made to be identical.

To that end, we introduce a second complex, the "Koszul-Tate resolution" [13] , whose purpose is to enforce the restriction to the surface Σ through its homology. The differential of this complex is denoted by δ and is just the δ of the previous section. Its grading, the r-degree, will be called here the antighost number (see remark at the end of section 2).

The definition of δ is given by

$$\delta F = 0 \tag{21a}$$

$$\delta t^*_{a_0} = f_{a_0} , \quad \text{antigh}\, t^*_{a_0} = 1 \tag{21b}$$

$$\delta t^*_{a_1} = R^{a_0}_{a_1} \; t^*_{a_0} , \quad \text{antigh}\, t^*_{a_1} = 2 \tag{21c}$$

where F is an arbitrary function on I, and where $t^*_{a_0}$, $t^*_{a_1}$ are respectively the generators in degrees one and two. We assume no relation on $R^{a_0}_{a_1}$. Also, if there is no $R^{a_0}_{a_1}$, i.e., if the functions f_{a_0} are independent, there is no $t^*_{a_1}$ and (21c) is absent. The complex is then generated by just $t^*_{a_0}$.

One can show [4],

$$H_k(\delta) = 0 \qquad k \neq 0 \tag{22a}$$

$$H_0(\delta) = \{ \text{functions on } \Sigma \} \tag{22b}$$

Also,

$$\mathcal{H}_k(\delta) = 0 \qquad k \neq 0 \tag{23a}$$

This follows from the existence of a contracting homotopy γ in degree k ≠ 0 that acts as a derivation. Therefore, any derivation of antighost number k ≠ 0 obeying $[t, \delta] = 0$ can be written as $[u, \delta]$ for some u.

If one extends δ to the ghosts as

$$\delta \eta^{\alpha_0} = 0 \tag{24a}$$

$$\delta \eta^{\alpha_1} = 0 \tag{24b}$$

and d to $t^*_{a_0}$ and $t^*_{a_1}$, as

$$dt^*_{a_0} = 0 \ , \quad dt^*_{a_1} = 0 \tag{24c}$$

one finds $d^2 = [\delta, \overset{(1)}{s}]$ for some $\overset{(1)}{s}$, but one does not get $[\delta, d]\, t^*_{a_0} = 0$ or $[\delta, d]\, t^*_{a_1} = 0$. This indicates that (24c) is not quite the right extension of d when acting on $t^*_{a_0}$ or $t^*_{a_1}$. The appropriate extension is, however, easy to find and exists because the vector fields $\vec{X}_{\alpha_0}$ are tangent to Σ. One defines

$$\overset{(0)}{s} F = dF \ , \quad \overset{(0)}{s} \eta^{\alpha_0} = d\eta^{\alpha_0}, \quad \overset{(0)}{s} \eta^{\alpha_1} = d\eta^{\alpha_1} \tag{25a}$$

and

$$\overset{(0)}{s} t^*_{a_0} = t_{\beta_0 a_0}^{\ b_0}\, t^*_{b_0} \eta^{\beta_0} \tag{25b}$$

One easily checks $(\overset{(0)}{s}\delta + \delta \overset{(0)}{s})\, t^*_{a_0} = 0$. Using $H^1(\delta) = 0$, one then shows the existence of $\overset{(0)}{s} t^*_{a_1}$ such that $\overset{(0)}{s}\delta + \delta\overset{(0)}{s}$ vanishes on $t^*_{a_1}$ as well. The definite expression will not be needed here and involves some manipulations of the structure functions and vectors f_{a_0}, $\vec{X}_{\alpha_0}$, $R^{a_0}_{a_1}$, $t_{\beta_0 a_0}^{\ b_0}$, $C^{\gamma_0}_{\alpha_0\beta_0}$ and $\vec{M}^{a_0}_{\alpha_0\beta_0}$. Similarly, using $d^2 \approx 0$ and $H^k(\delta) = 0$ for $k = 1,2$, one also checks that $2\overset{(0)}{s}{}^2$ is equal to $-[\overset{(1)}{s}, \delta]$ for some $\overset{(1)}{s}$.

One can therefore apply the theorem of the previous section. As $H^k(\overset{(0)}{s}) = H^k(d)$, one has proved the existence of a nilpotent operator s graded by the total ghost number, such that

$$H^k(s) = \begin{cases} 0 \ , & k < 0 \quad (26a) \\ H^k(d) \ , & k \geq 0 \quad (26b) \end{cases}$$

This nilpotent operator s exists whenever the above geometrical ingredients are available.

We will now indicate why and how this construction yields the BRST generator of gauge theories. We will see that the differential s gives actually the BRST symmetry.

4. HAMILTONIAN BRST CONSTRUCTION

The Hamiltonian description of gauge systems contains the previous geometric ingredients.

Let I be the phase space. The system is classically constrained to lie on the constraint surface Σ ,

$$\Sigma : G_{a_0} = 0 \tag{27}$$

where G_{a_0} are the "constraints" and are identified with the f_{a_0}'s of section 3. If the constraints all originate from the gauge invariance, as we assume for simplicity, they are "first class" [14]

$$[G_{a_0}, G_{b_0}] = C^{c_0}_{a_0 b_0} G_{c_0} \tag{28}$$

The constraints functions G_{a_0} play two roles. Not only do they constrain the system to be on Σ , but they also generate the gauge transformations through the Poisson bracket [14],

$$\delta_\varepsilon F \equiv \varepsilon^{a_0} (X_{a_0} F) = [F, G_{a_0}] \varepsilon^{a_0}. \tag{29}$$

The vector fields X_{a_0} are tangent to Σ and close on Σ by the first class property. So, all the previous conditions are indeed met, with $X_{a_0} \to X_{\alpha_0}$.

Furthermore, because of (29), there is a complete symmetry between, on the one hand, the differential

δ associated with the constraint aspect of the problem ($G_{a_0} = 0$) and, on the other hand, the exterior derivative d associated with the gauge orbits generated by X_{a_0} ($= [\ , G_{a_0}]$).

There are as many generators $\mathcal{P}_{a_0}$, $\mathcal{P}_{a_1}$ ... ($\equiv t^*_{a_0}$, $t^*_{a_1}$ ... of previous section) of the Koszul-Tate complex as there are generators η^{a_0}, η^{a_1}, ... of the exterior derivative complex. If one declares that these generators are conjugate,

$$[\mathcal{P}_{a_0}, \eta^{b_0}] = -\delta^{b_0}_{a_0}, \quad [\mathcal{P}_{a_1}, \eta^{b_1}] = -\delta^{b_1}_{a_1} \quad \text{etc...} \tag{30}$$

one finds that the "BRST symmetry" s whose existence is guaranteed by the previous theorem can be taken to be a canonical transformation,

$$sF = [F, \Omega] \tag{31}$$

The grading and nilpotency properties of s imply for the BRST generator Ω,

$$gh\, \Omega = 1 \quad , \quad \varepsilon(\Omega) = 1 \tag{32a}$$

$$[\Omega, \Omega] = 0 \tag{32b}$$

The fact that s is a canonical transformation is not implied by the theorem of section 2. This property can be checked, however, along identical lines : one directly proves the existence of Ω by expanding it again according to antighost number [2, 4, 6].

The existence of a canonical structure in the extended phase space including the ghosts η^{a_0}, η^{a_1}, ... and their momenta $\mathcal{P}_{a_0}$, $\mathcal{P}_{a_1}$, ... follows from the presence of a canonical structure in phase space to begin with, as well as from the fact that the two different complexes hidden in s are both generated by the same functions G_{a_0}, respectively through (27) and (29).

As a consequence of the above general theorem, one finds $H^0(s) = H^0(d) = \{$gauge invariant functions$\}$. So, BRST invariance completely captures gauge invariance.

Remarks : 1. In the expansions

$$\Omega = \sum_k \overset{(k)}{\Omega}$$

$$s = \delta + \overset{(0)}{s} + \sum_{k \geq 1} \overset{(k)}{s}$$

of Ω and s, one does not have

$$\overset{(k)}{s} A = [A, \overset{(k)}{\Omega}]$$

even though the equality

$$s A = [A, \Omega]$$

is true for the whole sums. This is because the antighost number of $[A, \overset{(k)}{\Omega}]$ is in general not equal to the antighost number of A plus k. One has for instance

$$\text{antigh}\, [F(q,p), \overset{(k)}{\Omega}] = k = \text{antigh}\, F + k$$

but

$$\text{antigh}\, ([\mathcal{P}_{a_0}, \overset{(k)}{\Omega}]) = \text{antigh}\, (\mathcal{P}_{a_0}) + k - 1$$

$$\text{antigh}\, ([\eta^{a_0}, \overset{(k)}{\Omega}]) = \text{antigh}\, (\eta^{a_0}) + k - 1$$

$$\text{antigh}\, ([\mathcal{P}_{a_1}, \overset{(k)}{\Omega}]) = \text{antigh}\, (\mathcal{P}_{a_1}) + k - 2$$

etc...

2. A first class constraint surface is also called a "coisotropic surface" in mathematical terminology. This is because the induced 2-form on Σ has minimum rank (equal to $\dim \Sigma - \text{codim}\, \Sigma$). The null directions are just spanned by the vector fields X_{a_0} tangent to the gauge orbits.

3. Different representations of the constraint surface and of the gauge transformations, $G_{a_0} \rightarrow \bar{G}_{a_0} = M^{b_0}_{a_0} G_{b_0}$, can be shown to lead to canonically related Ω's [2, 4].

5. LAGRANGIAN BRST CONSTRUCTION

The general construction of the BRST symmetry within the Lagrangian formalism requires the antifields of Batalin and Vilkovisky [8, 12]. It contains again the same two key geometrical ingredients : a surface Σ, and a set of vector fields tangent to Σ that close on Σ. Furthermore, there is also a symmetry between these two ingredients. That symmetry allows foı a canonical description of the BRST symmetry. However, the symmetry is shifted by one unit as compared with the Hamiltonian description. Accordingly, the required canonical structure is odd ("antibracket")*.

To see explicitly how these features are realized, consider the case when the gauge transformations are independent. We denote the fields/field histories by ϕ^i and the gauge transformations by

$$\delta_\varepsilon \phi^i = R^i_\alpha \varepsilon^\alpha \tag{33a}$$

These transformations leave the action $S_0(\phi)$ invariant,

$$\frac{\delta S_0}{\delta \phi^i} R^i_\alpha = 0 \tag{33b}$$

and close "on-shell",

$$\begin{aligned}[R_\alpha, R_\beta]^i &\equiv R^j_\alpha \frac{\delta R^i_\beta}{\delta \phi^j} - R^j_\beta \frac{\delta R^i_\alpha}{\delta \phi^j} \\ &= C^\gamma_{\alpha\beta}(\phi) R^i_\gamma + M^{ij}_{\alpha\beta} \frac{\delta S_0}{\delta \phi^j},\end{aligned} \tag{34a}$$

$$M^{ij}_{\alpha\beta} = - M^{ji}_{\alpha\beta} \tag{34b}$$

Let I be the space of all the field histories. Let Σ be the stationary surface where the equations of motion hold,

*The general theory of odd canonical structures may be found in [15].

$$\Sigma : \quad \frac{\delta S_0}{\delta \phi^i} = 0 \tag{35}$$

The vector fields X_α defined by $X_\alpha F \equiv (\delta F/\delta\phi^i) R^i_\alpha$ are tangent to Σ and close on Σ by virtue of (34a). Hence, they generate orbits, the "gauge orbits". The gauge invariant functions are defined on the quotient space of the stationary surface Σ by the gauge orbits.

These are again the geometric ingredients described in section 3. One can hence construct the Lagrangian "BRST symmetry" s along the lines of the general theorem proved there.

To that end, one needs the following spectrum of variables (besides the fields ϕ^i)

(i) for δ :

$$\phi^*_i, \quad \text{antigh}(\phi^*_i) = 1, \quad \delta\phi^*_i = \frac{\delta S_0}{\delta\phi^i} \tag{36a}$$

$$\phi^*_\alpha, \quad \text{antigh}(\phi^*_\alpha) = 2, \quad \delta\phi^*_\alpha = R^i_\alpha \phi^*_i \tag{36b}$$

(see 21) ;

(ii) for d :

$$C^\alpha, \quad \text{pure gh}\, C^\alpha = 1, \quad dC^\alpha = \frac{1}{2} C^\alpha{}_{\beta\gamma} C^\beta C^\gamma \tag{37}$$

The BRST transformation s acts on ϕ^i, ϕ^*_i, C^α and ϕ^*_α. The variables ϕ^*_i and ϕ^*_α of the Koszul-Tate complex are known as the "antifields", the variables C^α are the "ghosts". There is again a symmetry between δ and d :

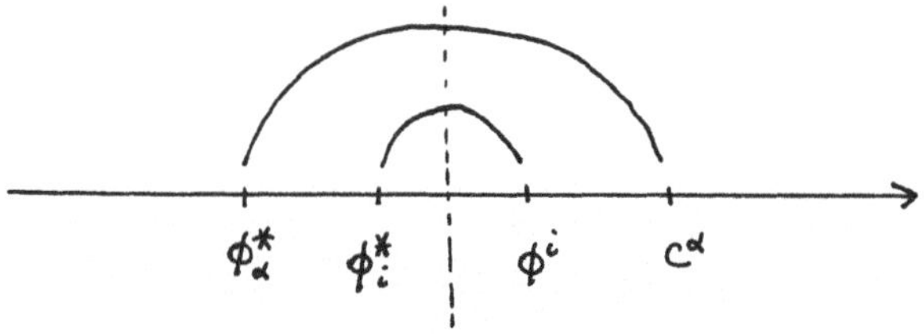

(38)

This symmetry follows from the fact that it is the same action S_0 that determines both the stationary surface through (35) and the gauge symmetries through (33).

It is therefore tempting to declare that (ϕ^i, ϕ^*_i) and $(C^\alpha, \phi^*_\alpha)$ form conjugate pairs,

$$(\phi^i, \phi^*_j) = \delta^i_j \tag{39a}$$

$$(C^\alpha, \phi^*_\beta) = \delta^\alpha_\beta \tag{39b}$$

However, the Grassmann parities and the ghost numbers do not match. This is in contrast with what happens in the Hamiltonian description. The "antibracket" (,) must accordingly carry an odd Grassmann parity and a ghost number equal to +1.

It turns out that the Lagrangian BRST symmetry whose existence is guaranteed by the above theorem is again a canonical transformation, but this time in the antibracket,

$$sF = (F, S) \tag{40}$$

The grading and nilpotency properties of s imply for the "BRST generator" S, taking the features of the antibracket into account,

$$gh\, S = 0 \qquad \varepsilon(S) = 0 \tag{41a}$$

$$(S, S) = 0 \tag{41b}$$

The fact that s is a canonical transformation is not implied by the theorem of section 2. This property can be checked, however, along identical lines : one directly proves the existence of the generator S by expanding it according to antighost number [8, 5],

$$S(\phi, \phi^*, C) = \overset{(0)}{S} + \sum_{k \geq 1} \overset{(k)}{S} \tag{42a}$$

$$\text{antigh}\, \overset{(k)}{S} = k \tag{42b}$$

The zeroth order piece $\overset{(0)}{S}$ of S is simply the classical action S_o,

$$\overset{(0)}{S} = S_o \tag{42c}$$

This is required if s is to contain δ (see(36a)).

Remark : a detailed exposition of the antifield formalism with a discussion of the Lagrangian path integral may be found in [16]. It is also explained there that different representations of the gauge symmetry lead to canonically related generators S's.

6. GAUGE FIXED BRST COHOMOLOGY

The theorem of section 2 can be used in reverse as follows.

Let s be a differential which can be decomposed according to some "resolution degree" as

$$s = \delta + \overset{(0)}{s} + \sum_{k \geq 1} \overset{(k)}{s} \tag{43a}$$

$$r(\delta) = -1, \quad r(\overset{(0)}{s}) = 0, \quad r(\overset{(k)}{s}) = k \tag{43b}$$

Assume that δ is cohomologically trivial for $r \geq 1$. Then, from $s^2 = 0$, one knows that

$$\overset{(0)}{s}\delta + \delta\overset{(0)}{s} = 0 \tag{44a}$$

$$2\overset{(0)}{s}{}^2 + [\delta, \overset{(1)}{s}] = 0 \tag{44b}$$

and furthermore, by the theorem, the cohomology of $\overset{(0)}{s}$ modulo δ is equal to the cohomology of s.

This application of the theorem is of interest for analyzing the so-called "gauge fixed BRST cohomology", which we now describe.

In order to construct the Lagrangian path integral, it is necessary to eliminate the antifields ϕ^*_A. [We collectively denote the fields ϕ^i and the ghosts C^α by ϕ^A ; the antifields ϕ^*_i and ϕ^*_α are denoted by ϕ^*_A]. This is done as follows.

(i) First one makes an appropriate canonical transforma- in the antibracket. This transforms $S(\phi, \phi^*)$ into $S_\psi(\phi, \phi^*)$, where ψ is a function that characterizes the canonical transformation.

(ii) Second, one sets the antifields equal to zero to get the gauge fixed action $S_\psi(\phi)$,

$$S_\psi(\phi) = S_\psi(\phi, \phi^* = 0) \tag{45}$$

The canonical transformation is "appropriate" if $S_\psi(\phi)$ has no gauge invariance.

In the gauge fixation procedure, all the antifields are put on the same footing, independently of their antighost number. Let us introduce a r -degree that gives equal weight to each antifield,

$$r(\phi^*_A) = 1 \tag{46a}$$

$$r(\phi^A) = 0 \tag{46b}$$

So, r(A) just counts the number of antifields in A, irrespectively of their antighost number.

Let s'_ψ be the form of the BRST symmetry after canonical transformation,

$$s'_\psi \phi^A = (\phi^A, S_\psi) \tag{47a}$$

$$s'_\psi \phi^*_A = (\phi^*_A, S_\psi) \tag{47b}$$

with $S_\psi \equiv S_\psi(\phi, \phi^*)$. One can expand s'_ψ according to the r-degree,

$$s'_\psi = \delta' + \overset{(0)}{s}{}'_\psi + \sum_{k \geq 1} \overset{(k)}{s}{}' \tag{48}$$

One gets

$$\delta' \phi^*_A = \frac{\delta S_\psi(\phi)}{\delta \phi^A} \tag{49}$$

so that the antifields ϕ^*_A can be viewed as the generators of the Koszul complex associated with the gauge-fixed stationary surface. The Koszul differential δ' is cohomologically trivial for $r \geqslant 1$ because the equations of motion $\frac{\delta S_\psi(\phi)}{\delta \phi^A} = 0$ are independent as the action is gauge fixed.

Furthermore,

$$\overset{(0)}{s}{}'_\psi \phi^A = (s'_\psi \phi^A)(\phi, \phi^* = 0) \tag{50}$$

The operator $\overset{(0)}{s}{}'_\psi$ defines what is known as the gauge fixed BRST symmetry. It is a differential modulo δ'. Using (6) and (49), one finds that its cohomology modulo δ' is given by

$$\overset{(0)}{s}{}'_\psi A = \lambda^A \frac{\delta S_\psi}{\delta \phi^A} \tag{51a}$$

$$A \sim A + \overset{(0)}{s}{}'_\psi B + \mu^A \frac{\delta S_\psi}{\delta \phi^A} \tag{51b}$$

with $A(\phi)$, $B(\phi)$ and $\mu^A(\phi)$. By the above theorem, this gauge fixed BRST cohomology is guaranteed to be equal to the BRST cohomology of the non-gauge fixed s.

7. FINAL COMMENTS

In this paper, we have explained what we believe is the algebraic structure of the BRST symmetry. This structure, rediscovered by the physicists, had actually been studied by mathematicians before and is part of "homological perturbation theory". The only slight subtlety in applying the theorems of homological perturbation theory is to identify the appropriate gradings appearing in the problem.

Besides the study of the algebraic structure of the BRST symmetry, the lectures given at Banff also covered some aspects of the operator quantization. It was in particular stressed that one could define a sensible scalar product in the space of the ghosts and of the gauge modes if one used the (pseudo-) Fock representation of the (graded) commutation relations. The scalar product in the "big" Hilbert space reduces then to the appropriate, finite scalar product in the space of the physical states. This approach is equivalent to the "weak Dirac" quantization in which only the negative frequency components of the constraints are imposed on the physical states.

The details can be found in [17].

Acknowledgements

The author is grateful to the organizers of the school in Banff for their kind invitation. He also acknowledges enlightening discussions with Jim Stasheff.

REFERENCES

[1] J.D. Stasheff, Trans. Amer. Math. Soc. 18 (1963) 215, 293 ;
V.K.A.M. Gugenheim and J.P. May, Mem. AMS 142 (1974) ;
V.K.A.M. Gugenheim, J. Pure Appl. Alg. 25 (1982) 197 ;
V.K.A.M. Gugenheim and J.D. Stasheff, Bull. Soc. Math. Belg. 38 (1986) 237;
V.K.A.M. Gugenheim and L. Lambe, "Perturbation Theory in Differential Homological Algebra I", Il. J. Math. (to appear) ;
J. Huebschmann and T. Kadeishvili, "Minimal Models for Chain Algebras over a Local Ring" (Heidelberg preprint).

[2] M. Henneaux, Phys. Rep. 126 (1985) 1.

[3] M. Henneaux and C. Teitelboim, Commun. Math. Phys. 115 (1988) 213.

[4] J. Fisch, M. Henneaux, J. Stasheff and C. Teitelboim, Commun. Math. Phys. 120 (1989) 379.

[5] J. Fisch and M. Henneaux, "Homological Perturbation Theory and the Algebraic Structure of the Antifield-Antibracket Formalism for Gauge Theories", U.L.B. preprint TH2/89-02, to appear in Commun. Math Phys.

[6] J.D. Stasheff, "Bull. Amer. Math. Soc. 19 (1988) 287; see also "Homological Reduction of Constrained Poisson Algebras, UNC-Math preprint 1988-1989.

[7] B.L. Voronov and I.V. Tyutin, Theor. Math. Phys. USSR 50 (1982) 218.

[8] I.A. Batalin and G.A. Vilkovisky, J. Math. Phys. 26 (1985) 172.

[9] M. Dubois-Violette, Ann. Inst. Fourier 37 (1987) 45 ;

J.M. Figueroa - O'Farrill and T. Kimura, ITP-SB-88-81 preprint (revised version January 1989).

[10] A.D. Browning and D. Mc Mullan, J. Math. Phys. 28 (1987) 438.

[11] E.S. Fradkin and G.A. Vilkovisky, Phys. Lett. 55B (1975) 229.

I.A. Batalin and G.A. Vilkovisky, Phys. Lett. 69B (1977) 309.

E.S. Fradkin and T.E. Fradkina, Phys. Lett. 72B (1978) 343.

[12] I.A. Batalin and G.A. Vilkovisky, Phys. Lett. 102B (1981) 27 ; 120B (1983) 166.

[13] J.L. Koszul, Bull. Soc. Math. France 78 (1950) 5 ;

A. Borel, Ann. Math. 57 (1953) 115 ;

J. Tate, Ill. J. Math. 1 (1957) 14.

[14] P.A.M. Dirac, Can. J. Math. 2 (1950) 129 ; "Lectures on Quantum Mechanics", Yeshiva University (1964).

M. Henneaux and C. Teitelboim, "Classical and Quantum Mechanics of Constrained Hamiltonian Systems", book in preparation (Princeton University Press).

[15] D.A. Leites, Sov. Math. Dokl. 18 (1977) 1277.

[16] M. Henneaux - "Lectures on the Antifield- BRST Formalism for Gauge Theories", to appear in Proceedings of 1989 GIFT Meeting, North Holland Delta Series, and in "Quantum Mechanics of Fundamental Systems III" C. Teitelboim and J. Zanelli eds, Plenum Press (New York: 1990), ULB preprint TH2/89-07.

[17] M. Henneaux, Ann. Phys. 194 (1989) 281, see also contribution in "Quantum Mechanics of Fundamental Systems 1", pp. 117-143, ed. C. Teitelboim, Plenum Press, New York 1988.

BLACK HOLE QUANTIZATION AND A CONNECTION TO STRING THEORY

Gerard 't Hooft

Institute for Theoretical Physics
Princetonplein 5, P.O. Box 80.006
3508 TA UTRECHT, The Netherlands

1. INTRODUCTION

Gravitation is the weakest and the strongest force known in physics. When considered as a force between two single electrons, it is nearly 43 orders of magnitude weaker than the electro-magnetic force. But gravity works collectively: when an amount of matter, somewhat more than the mass of our Sun, is allowed to cool and compress under its own weight, then sooner or later a complete collapse has to take place. No other force can then overcome the gravitational one. The process of collapse can be computed using well-established laws of physics, and few physicists doubt on the final outcome: a black hole[1].

Black holes form an extremely interesting theoretical laboratory. A quite surprising result was found by S. Hawking[2] in 1975. Applying quantum field theory to the region surrounding a black hole he discovered that a black hole must radiate matter of all sorts, behaving like an ideal radiating black body with a temperature T given by*

$$k_B T = 1/8\pi M \,, \tag{1.1}$$

where k_B is Boltzmann's constant and M the black hole mass, in units where $\hbar=c=\kappa=1$ (κ is Newton's gravitational constant.)

This result implies that black holes must loose energy, becoming ever lighter. The mass loss per unit of time is proportional to T^4 (the energy density of the radiation), and the surface area, which goes like M^2. One thus expects

$$\frac{dM}{dt} = -\,C\,T^4 M^2 = -\,C/M^2 \,, \tag{1.2}$$

where C is a constant of order one in natural units. Consequently, the

*Although this is the most widely accepted value, its derivation is not free from assumptions. An alternative theory yielding a different value can be formulated[3].

Physics, Geometry, and Topology
Edited by H. C. Lee
Plenum Press, New York, 1990

mass will decrease as

$$M(t) \cong C'(t_0-t)^{\frac{1}{3}} \, , \qquad (1.3)$$

and the natural lifetime of a black hole is proportional to M^3, so that black holes of astronomical sizes are extremely stable.

The average time between two Hawking emissions is of order of the black hole size, which is M. If these emissions could be seen as transitions from one state into another, these states would be resonances off the real axis, so that

$$\mathrm{Im}(M) \cong 1/M \; ; \qquad (1.4)$$

$$s \cong M^2 - i\Gamma \, , \qquad (1.5)$$

with $\Gamma = \mathcal{O}(1)$ in natural units.

Is there a lower limiting value for their mass? If so this can only be when either quantum field theory or general relativity, or both, cease to be valid. The fundamental principles of these theories are unobjectionable as long as the gravitational force can be quantized perturbatively, and this is the case when all distance scales used are considerably larger than

$$\sqrt{\frac{\kappa\hbar}{c^3}} = 1.6\cdot 10^{-33}\mathrm{cm} \, , \qquad (1.6)$$

called the *Planck length*. At this length scale we have energies of the order

$$\sqrt{\frac{\hbar c^5}{\kappa}} = 1.22 \cdot 10^{22} \text{ MeV} \, , \qquad (1.7)$$

the Planck energy. One could expect that the lightest black hole has a mass of this order, but at this energy black holes may well become indistinguishable from elementary particles. After all, elementary particles with such a mass would be surrounded by the same gravitational field, so they also can be seen as "collapsed objects", and it may well be that all particles at such energies are unstable, emitting radiation much like black holes do while they decay.

Plausible as this picture may seem (and indeed we will adopt it), there are problems with it. It suggests namely that black holes should form a discrete spectrum just like elementary particles, to be characterized by quantum numbers. Also larger black holes could then, at least in priciple, be characterized by discrete quantum numbers, and this is not what one gets applying Hawking's techniques.

Either something is wrong with our conventional picture of what matter is like at the Planck scale, or something is missing in the theory leading to Hawking radiation (or both).

To find out what the possible resolution of this dilemma is we will study Hawking radiation. Assuming that quantum mechanics in some sense will continue to be exactly valid at the Planck scale, we will concentrate on the assumptions that went into the derivation of the Hawking radiation, and ask ourselves how one can improve the procedure so that no contradiction arises. According to the standard treatment of black holes they "have no hair", that is, surface details are fundamentally unobservable. Our present picture is that this changes dramatically in a complete quantum description: there is lots of hair, one follicle (Boolian yes or no) per Planckian unit of its surface area. In the classical limit this hair becomes unobservable.

As we will see some very fundamental aspects of the quantum theory are at stake; we will also see that black holes are fairly closely related to string theory. Indeed, string theory can perhaps be reformulated in terms of black holes, or more precisely: it could be that the Veneziano or Koba-Nielsen amplitudes, commonly attributed to an underlying theory of strings, may be given an alternative explanation in terms of dynamical features of a horizon.

2. THE BLACK HOLE

A black hole is a solution of the classical Einstein equations for the space-time metric, and the equation of state for matter, such that space-time can be seen to be divided in two regions, "region *I*" and "region *III*" (called that way for no particularly good reason), which are defined as follows:

i) All points in *I* can be connected to the outside world by a timelike geodesic directed into the future, and

ii) None of the points in *III* can be connected that way to the outside world.

We have to explain what is meant by "outside world". This concept only means something if we have a non-compact asymptotically flat region surrounding the "black hole" solution at $t \to \infty$. In contrast, region *III* can at all times be enclosed inside a closed surface with an area never exceeding a certain number Σ. For a black hole with an infinite lifetime the question whether or not a signal will be able to escape from region *III* will be unambiguous. The (lightlike) surface dividing the two regions is called the future event horizon, or horizon for short†. Within the horizon, in region *III*, all light cones are directed inwards, so that no information can escape. In the center we usually have some sort of *singularity*. The singularity is essentially a divergence, where presumably our known laws of physics break down. Remarkably however, the singularity will not play a significant role in our discussion. This is because it is well shielded from the observable world; we don't see it, so its precise nature is irrelevant. In these lectures we'll mainly focus on the horizon.

As can be seen from Figure 1, a black hole is expected to rapidly settle for a stable, stationary configuration. In a particular coordinate frame the space-time metric can then be written as

$$\begin{aligned} ds^2 &= -\left(1-\frac{2M}{r}\right)dt^2 + \left(1-\frac{2M}{r}\right)^{-1}dr^2 + r^2 d\Omega^2 \ ; \\ d\Omega^2 &= d\vartheta^2 + \sin^2\vartheta\, d\varphi^2 \ . \end{aligned} \tag{2.1}$$

where M is the mass of the black hole in natural units. This we refer

†The well-informed reader, knowing that we expect black holes to have only a finite lifetime, might wonder how this affects our definition of the horizon. Indeed, it makes the horizon somewhat "fuzzy". However, the uncertainty this induces to the location of the horizon will be far less than other fluctuations that we will discuss.

to as the Schwarzschild solution.[♪]

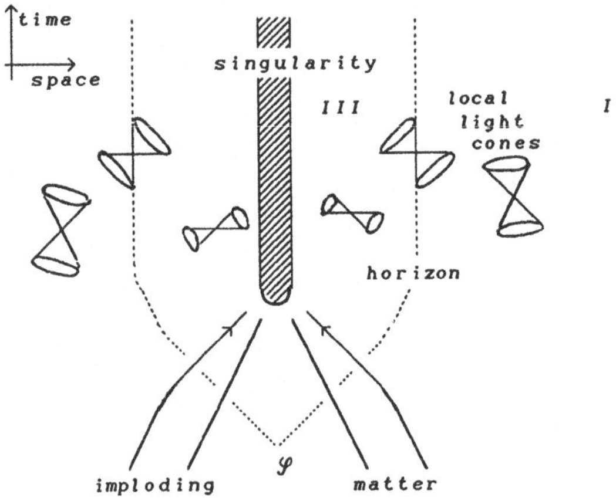

Fig. 1. The black hole.

In this representation there may seem to be a singularity at $r = 2M$, but this is a coordinate artefact. There is a better set of coordinates, called "Kruskal coordinates", in which we see that (2.1) in fact describes part of a larger space-time[4]. Replace r and t by x and y, defined by

$$xy = \left(1 - r/2M\right) e^{r/2M} ; \qquad (2.2)$$
$$x/y = - e^{t/2M} .$$

Then in terms of x, y, ϑ and φ we have

$$ds^2 = -2A(x,y)dxdy + r^2 d\Omega^2 , \qquad (2.3)$$

with $d\Omega^2$ as in eq. (2.1) and

$$A(x,y) = \frac{16M^3}{r} e^{-r/2M} , \qquad (2.4)$$

which is not singular at $r = 2M$. Notice now that for every r and t we have in general *two* solutions for x and y, differing from each other by a sign. This inplies that our universe is smoothly connected to another, equal, universe, and we obtain regions *II* and *IV*, see Figure 2.

[♪]For simplicity we limit ourselves to the non-rotating chargeless Schwarzschild solution. The more general Kerr-Newman solution (having charge and angular momentum) has a different structure at $r=0$ and a somewhat different horizon structure[4]. Since we are mainly concerned about the horizon our arguments can be extended to this more general case.

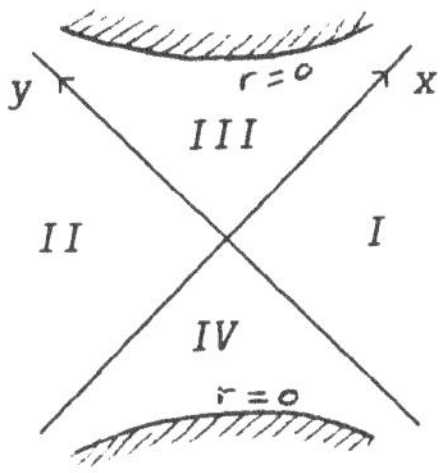

Fig. 2. The Kruskal coordinates.

If the Kruskal coordinates are used to describe the black hole of Fig. 1 then the regions II and IV in there are unphysical. This is because the infalling matter that created the black hole was ignored. Region IV would be at the infinite past only if the black hole is "eternal": it existed already before the universe was there[f]. Region II can only be reached by travelling faster than the speed of light. In realistic black holes (the ones formed by a collapsing object) regions II and IV are replaced by flat space. On the y-axis (x=0) one then has the imploding material which gives a right-hand side to Einstein's equations such that this hybrid configuration (regions I and III glued onto flat space) is a solution.

3. FIELD THEORY IN RINDLER SPACE

Let us now concentrate on the region $r \cong 2M$. Write

$$x = t_1+z_1 \ ; \quad y = t_1-z_1 \ ; \quad r/2M -1 = \zeta^2 \ ; \quad t/4M = \tau \quad . \tag{3.1}$$

We look at a small angular region so that ϑ and φ can be replaced by two transverse coordinates $\tilde{x}_1 = (x_1 \ , y_1)$. Then close to the horizon ($r=2M$) we have the coordinate mapping

$$z_1 = \zeta \cosh \tau \quad ; \quad t_1 = \zeta \sinh \tau \quad . \tag{3.2}$$

We call $\tilde{x}_1$, z_1 and t_1 the locally regular coordinates. ζ and τ are the Rindler coordinates[5]. We see that they are directly related to the Schwarzschild coordinates r and t. In terms of the regular coordinates space-time is not at all singular near the horizon.

Note now that a shift in the τ parameter (or equivalently in the Schwarzschild time t) corresponds to a Lorentz boost in x_1 space. All equations of physics are invariant under such boosts and therefore invariant under shifts in Rindler (or Schwarzschild) time.

A field theory – any local field theory – can be defined by giving the Hamiltonian as the integral of a Hamilton density over 3-space. In ordinary flat coordinates this is

$$H_{reg} = \int \mathcal{H}(\mathbf{x}) \ d^3\mathbf{x} \quad . \tag{3.3}$$

The Hamiltonian is the operator that generates shifts in time t_1 . Now at τ=0 an infinitesimal shift $\delta\tau$ in τ corresponds to a shift

[f]More about "eternal" black holes in sect. 7.

$$\delta t_1 = \zeta\delta\tau \quad , \tag{3.4}$$

in t_1, as one can easily see from (3.2). Thus, the generator for an infinitesimal translation in τ is

$$H_{Rin} = \int\mathcal{H}(\mathbf{x})z\,d^3\mathbf{x} \quad = \quad H_I - H_{II} \quad , \tag{3.5}$$

with

$$H_I = \int_{z>0} H(\mathbf{x})z\,d^3\mathbf{x} \quad ; \quad H_{II} = \int_{z<0} H(\mathbf{x})|z|\,d^3\mathbf{x} \quad . \tag{3.6}$$

We recognise (3.5) as the generator of a Lorentz transformation.

In most field theories it is not difficult to check that

$$[H_I, H_{II}] = 0 \quad , \tag{3.7}$$

because the integrands in (3.6) vanish at $z=0$.♪ Indeed, one expects (3.7) because no signal can be transmitted between regions I and II.

To describe physical processes at the visible part of a black hole one needs only H_I. Would particles described by H_I alone "bounce" against the horizon? To see what happens it is instructive to consider a field theory first in its Lagrange form. The simplest Lagrangian is

$$\mathcal{L} = -\tfrac{1}{2}(\partial_z\varphi)^2 + \tfrac{1}{2}(\partial_t\varphi)^2 - \tfrac{1}{2}(\tilde{\partial}_x\varphi)^2 - \tfrac{1}{2}m^2\varphi^2 \quad , \tag{3.8}$$

and if we write $\zeta = e^{\sigma}$ then this becomes in the Rindler coordinates

$$\mathcal{L}_{Rin} = \mathcal{L}\,\frac{dz\,dt}{d\sigma\,d\tau} = -\tfrac{1}{2}(\partial_\sigma\varphi)^2 + \tfrac{1}{2}(\partial_\tau\varphi)^2 + e^{2\sigma}\left(-\tfrac{1}{2}(\tilde{\partial}_x\varphi)^2 - \tfrac{1}{2}m^2\varphi^2\right). \tag{3.9}$$

We see that at a given transverse momentum $\tilde{k}$ all wavelike solutions will satisfy

$$\partial_\sigma^2\varphi = \partial_\tau^2\varphi \quad , \quad \text{at} \quad \sigma \to -\infty \, , \tag{3.10}$$

$$\varphi = \varphi_{out}(\tau-\sigma) + \varphi_{in}(\tau+\sigma) \quad . \tag{3.11}$$

This means that the boundary $\sigma = -\infty$ is open! An infinite world of particles, on their way in or out, exists in the region $\zeta \cong 0$.

One would conclude from this that black holes are in a fundamental way different from other soliton like configurations in gauge theories such as magnetic monopoles: even if we would enclose them in a finite space surrounding the black hole, particles near the black hole will form a continuous spectrum because they occupy a non-compact space.

♪The only contribution to (3.7) could come from the origin. Now the commutator $[\mathcal{H}(\mathbf{x}),\mathcal{H}(\mathbf{y})]$ contains at most one derivative of a Dirac deltafunction, whereas there are two factors z in eq. (3.7). This is why, after partial integration, one gets zero.

It should be clear however that particles near $\sigma \to -\infty$ probe the infinitely small distance regime. Therefore, since quantum gravity is non-renormalizable, there will be values for σ beyond which the above analysis fails. Can gravitational forces turn the spectrum into a discrete one? The author believes this to be the case[6]. But first we will explain what Hawking's result implies for this boundary.

4. THE HAWKING EFFECT

In this chapter we briefly review the derivation of the Hawking effect[2]. Everything can be understood as a feature of the central region in Kruskal space, and indeed all we need is the Rindler coordinate transformation (3.2). Only non-rotating, non-interacting particles are considered; other cases are more complicated but yield the same results.

A scalar field in the regular coordinates r_1 can be written as

$$\Phi(\mathbf{r}_1, t_1) = \int \frac{d^3\mathbf{k}}{\sqrt{2k^0(\mathbf{k})V}} \left(a(\mathbf{k})e^{i\mathbf{k}\mathbf{r}_1 - ik^0 t_1} + a^\dagger(\mathbf{k})e^{-i\mathbf{k}\mathbf{r}_1 + ik^0 t_1}\right),$$
$$\dot{\Phi}(\mathbf{r}_1, t_1) = \int \frac{d^3\mathbf{k}\ (ik^0)}{\sqrt{2k^0(\mathbf{k})V}} \left(-a(\mathbf{k})e^{i\mathbf{k}\mathbf{r}_1 - ik^0 t_1} + a^\dagger(\mathbf{k})e^{-i\mathbf{k}\mathbf{r}_1 + ik^0 t_1}\right). \quad (4.1)$$

Here, $V = (2\pi)^3$, and we have

$$[a(\mathbf{k}), a^\dagger(\mathbf{k}')] = \delta^3(\mathbf{k}-\mathbf{k}') , \quad (4.2)$$

and

$$[\dot{\Phi}(\mathbf{r}), \Phi(\mathbf{r}')] = -i\delta^3(\mathbf{r}-\mathbf{r}') , \quad (4.3)$$

etc. We see that $a^\dagger(\mathbf{k})$ and $a(\mathbf{k})$ are not only the Fourier components of the observable field $\Phi(\mathbf{r}, t)$ but also play the role of creation and annihilation operators. They create or annihilate objects with momentum $\mathbf{k}$ and energy k^0. That they serve these two purposes at once is no coincidence. It is easy to see that, just *because* they are the Fourier components of the fields, they may act on any state $|\psi\rangle$ with momentum $\mathbf{k}_1$ and energy $k^0_{,1}$ to give a state $a(\mathbf{k}, k^0)|\psi\rangle$ that has momentum $\mathbf{k}_1-\mathbf{k}$ and energy $k^0_{,1}-k^0$.

We now turn to Rindler space, which can be seen to represent any small section of the region in the immediate neighborhood of a black hole horizon (see (3.1) and (3.2)). The trick is now to construct the Fourier components of the fields Φ in the two regions I and II with respect to Rindler time τ . Let us use the light cone coordinates

$$u = \tfrac{1}{2}(t_1 - z_1) , \quad v = \tfrac{1}{2}(t_1 + z_1) ,$$
$$k_+ = k^0 + k_3 , \quad k_- = k^0 - k_3 . \quad (4.4)$$

In Rindler time these evolve as

$$v \to ve^{\tau} , \quad u \to ue^{-\tau} . \quad (4.5)$$

For the Fourier transform with respect to τ we need the following function,

$$K(\omega,\alpha,\beta) = \int_0^\infty \frac{dx}{x} x^{i\omega} e^{-ix\alpha - i\beta/x} , \tag{4.6}$$

which will relate $a(\mathbf{k},k^0)$ to the Rindler annihilation operators $a_I(\tilde{k},\omega)$ and $a_{II}(\tilde{k},\omega)$, where $\tilde{k}$ is the transverse component of the momentum. One finds that, in region I,

$$\Phi(\mathbf{r}_1,t_1) = \int_0^\infty d\omega \int \frac{d^2\tilde{k}}{\sqrt{4\pi V}} e^{i\tilde{k}\tilde{r}} \sqrt{1-e^{-2\pi\omega}} \Big[K(-\omega,\mu u,\mu v)\; a_I(\tilde{k},\omega) + K^*(-\omega,\mu u,\mu v)\; a_I{}^\dagger(-\tilde{k},\omega) \Big] , \tag{4.7}$$

where a_I and $a_I{}^\dagger$ are normalized such that

$$[a_I(\tilde{k},\omega), a_I{}^\dagger(\tilde{k}',\omega')] = \delta^2(\tilde{k}-\tilde{k}')\delta(\omega-\omega') . \tag{4.8}$$

In a similar way we have operators a_{II} and $a_{II}{}^\dagger$, commuting with a_I and $a_I{}^\dagger$, describing the fields Φ in region II.

We now observe that both $a_{I,II}$ and $a^\dagger{}_{I,II}$ depend linearly on both $a(\mathbf{k},k^0)$ and $a^\dagger(\mathbf{k},k^0)$. Transformations that mix creation and annihilation operators are called Bogolyubov transformations. It turns out that[*]

$$\begin{aligned} a_{I,II}(\omega) &\propto a(k) + e^{-\pi\omega} a^\dagger(k) ; \\ a^\dagger{}_{I,II}(\omega) &\propto a(k) + e^{\pi\omega} a^\dagger(k) . \end{aligned} \tag{4.9}$$

Inversely, $a(k)$ is the sum of operators proportional to $a_I(\omega) - e^{-\pi\omega} a^\dagger{}_{II}(\omega)$ and $a_{II}(\omega) - e^{-\pi\omega} a^\dagger{}_I(\omega)$.

Since $\mathcal{H}$ in region I depends only on Φ in region I , which only depends on a_I , it is not surprising to see that H_I only depends on a_I and H_{II} on a_{II} :

$$\begin{aligned} H_I &= \int_0^\infty d\omega \int d^2\tilde{k}\; \omega\; a_I{}^\dagger a_I + C ; \\ H_{II} &= \int_0^\infty d\omega \int d^2\tilde{k}\; \omega\; a_{II}{}^\dagger a_{II} + C , \end{aligned} \tag{4.10}$$

where C is a common, irrelevant constant coming from the ordering process. It cancels in H_R , eq. (3.5).

In the previous chapter we stated that an observer in region I only works with H_I . Therefore he has only a_I to his disposal, not

[*]The exponent can be seen to be directly related to a rotation over 180° in Euclidean space[7])

a_{II} . Now suppose that in terms of the *regular coordinates* $\mathbf{r}_1$, t_1 an observer would see a vacuum in the region around the origin. This would be a state $|\Omega\rangle$ in Hilbert space defined by

$$a|\Omega\rangle = a_{\ell}|\Omega\rangle = a_R|\Omega\rangle = 0 \ , \quad \text{for all } \tilde{k},\ \omega \ . \tag{4.11}$$

Now let us choose a new basis of states for the Hilbert space which for each $\tilde{k}$, ω have given values for the quanta

$$n_I = a_I{}^{\dagger} a_I \quad , \quad n_{II} = a_{II}{}^{\dagger} a_{II} \ . \tag{4.12}$$

Let us call these basis elements $|n_I, n_{II}\rangle$. Clearly,

$$\prod_{\tilde{k},\omega} |0,0\rangle \neq |\Omega\rangle \ . \tag{4.13}$$

To express $|\Omega\rangle$ in our Rindler basis we use, from eqs (4.9) and their inverse,

$$\begin{aligned} a_I(\tilde{k},\omega)|\Omega\rangle - e^{-\pi\omega} a_{II}{}^{\dagger}(-\tilde{k},\omega)|\Omega\rangle &= 0 \ ; \\ a_{II}(\tilde{k},\omega)|\Omega\rangle - e^{-\pi\omega} a_I{}^{\dagger}(-\tilde{k},\omega)|\Omega\rangle &= 0 \ , \end{aligned} \tag{4.14}$$

so that, when acting on $|\Omega\rangle$, we have

$$a_I{}^{\dagger} a_I = e^{-\pi\omega} a_I{}^{\dagger} a_{II}{}^{\dagger} = e^{-\pi\omega} a_{II}{}^{\dagger} a_I{}^{\dagger} = a_{II}{}^{\dagger} a_{II} \ . \tag{4.15}$$

Consequently, $|\Omega\rangle$ consists only of states with $n_I = n_{II}$:

$$|\Omega\rangle = \sum_n f_n |n,n\rangle \ . \tag{4.16}$$

We find f_n from eq. (4.14):

$$\sum_n f_n \sqrt{n}\, |n-1,n\rangle = e^{-\pi\omega} \sum_n f_n \sqrt{n+1}\, |n,n+1\rangle \ ; \tag{4.17}$$

$$f_{n+1} = e^{-\pi\omega} f_n \quad , \tag{4.18}$$

Conclusion:

$$|\Omega\rangle = \prod_{\tilde{k},\omega} \sqrt{1-e^{-2\pi\omega}} \sum_{n=0}^{\infty} e^{-n\pi\omega} |n,n\rangle_{\pm\tilde{k},\omega} \ . \tag{4.19}$$

where the square root is a normalization factor. Notice that eq. (4.16) implies

$$H_R|\Omega\rangle = 0 \ , \tag{4.20}$$

which just means that $|\Omega\rangle$ is Lorentz-invariant.

If $\mathbb{H}_I$ is the Hilbert space in region I and $\mathbb{H}_{II}$ in region II then we see that $|\Omega\rangle$ is a superposition of states in $\mathbb{H}_I \otimes \mathbb{H}_{II}$. The expectation value of any operator $\mathcal{O}$ in $\mathbb{H}_I$ is

$$\langle \mathcal{O} \rangle = \sum_{\{n_I\},\{n_{II}\},\{n_I'\}} \langle \Omega | \{n_I, n_{II}\} \rangle \, \mathcal{O}_{\{n_I\},\{n_I'\}} \, \langle \{n_I', n_{II}\} | \Omega \rangle =$$

$$= \left[\prod_{\tilde{k},\omega} \sum_{n_I(\tilde{k},\omega)} \right] \mathcal{O}_{\{n_I(\tilde{k},\omega)\},\{n_I(\tilde{k},\omega)\}} \prod_{\tilde{k},\omega} (1-e^{-2\pi\omega}) e^{-2\pi\omega n_I(\tilde{k},\omega)} . \tag{4.21}$$

This one could write as

$$\langle \mathcal{O} \rangle = \mathrm{Tr}\, \rho\, \mathcal{O} \quad , \tag{4.22}$$

where ρ is a density matrix:

$$\rho = \prod_{\tilde{k},\omega} \rho_{n,n'} \quad ; \quad \rho_{n,n'} = (1-e^{-2\pi\omega}) e^{-2\pi\omega n} \, \delta_{n,n'} \; . \tag{4.23}$$

This is the density matrix for a thermal system at temperature given by

$$k_B T = 1/2\pi \; . \tag{4.24}$$

Thus, assuming that the most energetic particles near the point $\zeta = 0$ are absent when measured by a freely falling observer, the Rindler observer who uses τ as his time parameter sees particles emerging from his horizon corresponding to radiation with this temperature. Noticing the relation (3.1) between τ and Schwarzschild time t, one derives that in proper units the temperature with which the black hole radiates is given by

$$k_B T = 1/8\pi M \; , \tag{4.25}$$

the Hawking temperature[β]. In our way of dealing with Rindler space one sees that this result corresponds to information concerning the boundary condition at $\sigma=-\infty$, or $r=2M$.

It is important to try to understand this boundary condition. The physics generated by the Lagrangian (3.9) is not altered. So one still seems to have a continuous spectrum. All we get is the statement that particles come back from the horizon at random, with weight factors equal to the Boltzmann factor corresponding to the Hawking temperature. This random behavior is different from what one would get if there were some given scattering matrix reflecting particles against the horizon. How different?

[β]See footnote of Sect. 1.

5. THE BLACK-HOLE SPECTRUM AND THERMODYNAMICS

The existence of an infinite spectrum of plane waves to and from the horizon, described by eq. (3.11), suggests that the spectrum of black holes is continuous. Yet, a more compelling reason exists to suggest that the spectrum should be discrete. A continuous spectrum would imply that an infinite number of mutually orthogonal states exists that can be stored within the volume defined by the horizon, with energies E between M and $M+\mathrm{d}M$, where M is the black hole mass. A small region of space would allow an infinite amount of information to be stored in there, at the cost of virtually no energy. This, we think, sounds unlikely.

Suppose for a moment that the density of black holes at mass M (and charge Q and angular momentum L), is given by some finite number $\rho(M)$ (or $\rho(M,Q,L)$). Let us now compare the absorption process of an object with energy ΔE by a black hole with mass M, with the Hawking emission process for the same object by a hole with mass $M+\Delta E$:

$$(M) + (\Delta E) \Longleftrightarrow (M+\Delta E) . \tag{5.1}$$

The absorption process has a cross section σ given approximately by

$$\sigma \simeq \pi R^2 , \tag{5.2}$$

where R is the black hole radius, $R=2M$. The emission probability W is approximately

$$W \simeq \pi R^2 \rho_1(\Delta E)\, \mathrm{e}^{-\beta_H \Delta E} , \tag{5.3}$$

where β_H is the inverse Hawking temperature, $\beta_H = 8\pi M$ (putting the gravitational constant κ and Boltzmann's constant k_B equal to one). $\rho_1(\Delta E)$ is the density of states for the objects radiated out with energy ΔE.

If there were a quantum mechanical theory for the black hole, the same quantities could be expressed in terms of transition amplitudes, using the "golden rule":

$$\sigma = |\langle M+\Delta E|\mathcal{T}|M,\Delta E\rangle|^2 \rho(M+\Delta E) ; \tag{5.4}$$

where $\mathcal{T}$ is the transition matrix, and

$$W = |\langle M,\Delta E|\mathcal{T}|M+\Delta E\rangle|^2 \rho(M)\rho_1(\Delta E) . \tag{5.5}$$

By virtue of PCT invariance, the matrix elements in (5.4) and (5.5) would be each other's conjugates, and therefore we find

$$\rho(M+\Delta E)/\rho(M) = \rho_1(\Delta E)\, \sigma/W = \mathrm{e}^{\beta_H \Delta E} ; \tag{5.6}$$

this should hold for a range of values for ΔE as long as $\Delta E \ll M$, and with $\beta_H = 8\pi M$ we find

$$\rho(M) = C\, \mathrm{e}^{4\pi M^2} , \tag{5.7}$$

where the universal constant C is the only unknown. Note that the exponent is one quarter of the area of the horizon, this is what one also finds in more general cases.

C could be finite, in which case we indeed have a finite spectrum density, or C is infinite, but in this case equations (5.4) and (5.5) could be considered at the lower end of the continuum. Let $|M+\Delta E\rangle$ be in the continuum but $|M\rangle$ one of the discrete states directly underneath. Either $|M+\Delta E\rangle$ would be an absolutely stable thing ($W=0$), in which case *virtual pair creation* of these things would give infinite contributions to graviton self-energy diagrams, or the cross section σ for collisions against $|M\rangle$ would tend to infinity. Neither of these latter options sound physically very attractive, which is why we suspect C to be finite. We must realize however that very large numbers are indigenous in quantum gravity. It could be that C is of the order of 10^{40} or 10^{-40}.

It is unlikely that C is exactly constant. There may well be subdominant corrections either in the exponent, or in the form of powers in front of the exponent.

There is a different way to derive the same expression (5.7) for $\rho(M)$. This is by applying *thermodynamics* to the black hole[8]. The free energy F is defined by

$$e^{-\beta F} = \int_{M}^{M+\delta M} \rho(M)\mathrm{d}M\, e^{-\beta_H M} \simeq \rho(M)\, e^{-8\pi M^2} , \tag{5.8}$$

where the integral could be suppressed[£] because the leading term is given by the exponent anyhow. We inserted Hawking's value for the temperature and used natural units for the mass (the unit is the Planck mass, 21.7 μg).

The expectation value for the energy M is

$$\langle M\rangle = \frac{\partial}{\partial\beta}(\beta F) = \beta/8\pi \ ; \tag{5.9}$$

$$\beta F = \beta^2/16\pi = 4\pi M^2 , \tag{5.10}$$

and we conclude that

$$\rho = C\, e^{4\pi M^2} . \tag{5.11}$$

The philosophy used is that black holes behave just like little containers with some gas or liquid inside. If these communicate thermally with the outside world, one can deduce information concerning the total number of states from these reactions. The only thing slightly unusual about the thermal black holes is their negative specific heat:

$$c = \frac{\partial}{\partial T}\langle M\rangle = -\beta^2\frac{\partial}{\partial\beta}(\beta/8\pi) = -\beta^2/8\pi = -8\pi M^2 . \tag{5.12}$$

[£] In this argument it was necessary to introduce *bounds* M and $M+\delta M$ for the integral, where δM is some small number. The reason for this is the fundamental instability of the black hole caused by its negative specific heat, eq. (5.12).

This implies that black holes cannot be completely in equilibrium with their environment. Martinez and York[9] speculate that this may have deep implications for the probabilistic interpretation of quantum mechanical expressions concerning black holes.

If indeed $\rho(M)$ is finite then this has also consequences for the symmetry aspects of a quantum gravity theory: we cannot have any absolute global conservation laws such as baryon number conservation![10] The reason for this is that one could imagine dropping an unlimited amount of baryons into the hole, waiting each time for an equal amount of energy to reemerge in the form of Hawking radiation. Our derivation implies that equal numbers of baryons and anti-baryons should come out, on the average. So one could increase the hole's baryon number indefinitely. If ρ is finite there is however only a finite number of black hole states so that baryon number sooner or later becomes an ill-specified quantum number. Any preference of the hole to emit baryons rather than antibaryons would contradict our derivations of sect. 4.

Now this remark implies that applying quantum field theory alone to the black hole horizon will not yield us further details concerning these spectral states. We could have global symmetries in this quantum field theory, and they will be standing in the way! something fundamental is missing.

Our preliminary standpoint will be the following: our present understanding of the laws of physics is imprecise[11]; applying them gives us only statistical information about the black hole states. Remember that for large black holes (5.11) is an enormous number. Each of these states could be described by a slightly different Hamiltonian. Since our statistical answer was a consequence of our applying quantum mechanics, this may well mean that the usual statistical interpretation of quantum mechanics may be a consequence of an incomplete description.

Imagine just a slight "uncertainty" in the Hamiltonian,

$$H \rightarrow H + \delta H , \tag{5.13}$$

where δH has some *probabilistic* distribution. In that case the solutions to the Schrödinger equation,

$$\frac{d\psi}{dt} = -iH\psi , \tag{5.14}$$

will have a "thermal" contribution, to be described by a density matrix, just as what one gets as Hawking radiation, eq. (4.23).

It is an important observation that our seemingly innocent application of the quantum mechanical rules gives us a Hamiltonian for the black hole with a built in uncertainty. Only one Hamiltonian in this distribution will correctly describe our world. The methods of Sect. 4 will not tell us which. It seems that quantum mechanics here is incomplete, and that the missing information concerning our Hamiltonian could only be provided by some sort of hidden variables. A similar situation occurs when one tries to incorporate *wormholes* in quantum gravity: these give rise to uncomputable renormalizations of the physical coupling constants[12]).

6. THE HORIZON'S DISPLACEMENTS

Our problem can be traced back to the plane wave solutions (3.11) of eqs. (3.10): there will be mutually independent states of ingoing and outgoing particles. In reality we expect that the outgoing radiation will be *determined* by what comes in *via* an *S*-matrix. The first thing to suspect is that it was wrong to neglect gravitational interactions between ingoing and outgoing radiation. What will these interactions be like? Fortunately, they can be computed rather precisely.

Consider again Rindler space, described by the coordinates $\tilde{x}$, ζ and τ . We ignore matter that fell in long ago. At time $\tau=\tau_1$ a light particle is dropped in. See Figure 3.

Now a translation in τ corresponds to a Lorentz boost in the regular coordinates. Therefore, at Rindler time $\tau=\tau_2$, $\tau_2 \gg \tau_1$, the infallen particle appears to be boosted to an enormous energy. Sooner or later the gravitational fields due to this energy become important.

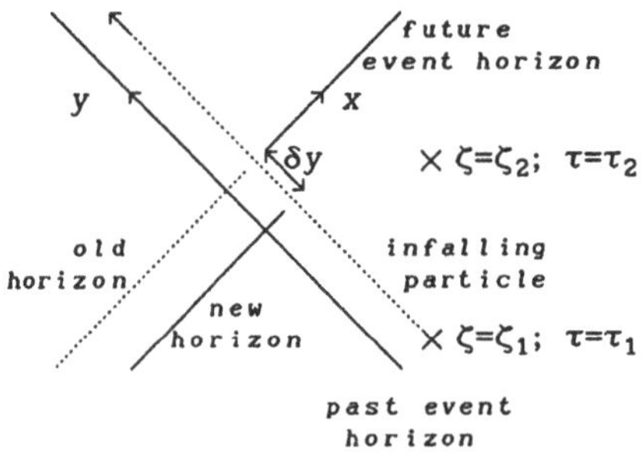

Fig. 3. The horizon displacement.

To describe its gravitational field the particle may be considered massless, moving with light velocity. The field is easily found[13] by first taking a light particle at rest and then boosting it together with its Schwarzschild field. The result is a field that takes the form of a "shock wave" (called "inpulsive wave" in the literature,) not unlike the sonic boom of an airplane moving with the speed of sound. Both in front of the particle and behind it we have flat space-time, and at the space-time points x that satisfy

$$p_1 \cdot (x-x_1(t)) = 0 \quad , \tag{6.1}$$

where p_1 is the particle's 4-momentum and $x_1(t)$ its trajectory, the two flat spaces are seamed together, shifted by an amount δx that depends on the transverse distance $\tilde{x}$:

$$\delta x = 2\kappa p_1 \log(1/\tilde{x}^2) \quad , \tag{6.2}$$

which, because of its nonlinear $\tilde{x}$ dependence produces a delta-distributed curvature.

Now (6.1) and (6.2) hold for a particle in a flat space-time background. How do they generalise when the particle moves in(to) a black hole? This turns out to be straightforward[14], if at finite τ_1 the particle had little energy. The situation at late τ_2 is sketched in Figure 3, which must now be seen as representing a black hole in its Kruskal coordinates. Two regular Schwarzschild solutions, described in their Kruskal coordinates, are glued together at the line $x=0$, shifted by an amount δy in the y direction, where δy now depends on the angles $\Omega = (\vartheta, \varphi)$. The functional dependence of δy on ϑ and φ is determined by the equation

$$(1-\Delta_\Omega)\ \delta y(\Omega) \quad = \quad 4\pi\kappa p_{in}\ \delta^2(\Omega,\Omega_1) \quad , \tag{6.3}$$

where Δ_Ω is the angular Laplacian and Ω_1 is the set of angles at which the particle dropped in. The solution of eq. (6.3) can be written as

$$\delta y(\Omega) \quad = \quad f(\Omega,\Omega_1)\ p_{in} \quad , \tag{6.4}$$

where $f(\Omega,\Omega_1)$ is a Green function approaching the logarithm of eq. (6.2) for large black holes. It is uniquely determined by (6.3) because $1-\Delta_\Omega$ has a unique inverse.

Eq. (6.3) is found by writing the space-time metric obtained from glueing the two Schwarzschild solutions together as an *Ansatz* and then imposing Einstein's equations. The Green function f can be given in an integral form but this is not very illuminating. One finds that $f(\Omega,\Omega_1) > 0$ for all Ω, Ω_1, and f diverges logarithmically when $\Omega \to \Omega_1$.

A consequence of these observations is that if we drop a particle into the black hole, the position of the horizon at times τ *before* the particle fell in, changes, as drawn in Fig. 3. This change is barely perceptible at times $\tau \lesssim \tau_1$, but at times $\tau_2 \gg \tau_1$ the change is large. An observer there sees Hawking radiation that now originated in a *different* region of space-time than it would have if the particle had not been thrown in.

Is this consequence of any importance? What does it matter if the Hawking radiation originated somewhere else? It will certainly look the same as before.

We will argue in the next section that only in a quantum theory that is detailed enough to give us a scattering matrix instead of a density matrix, shifting horizons will be relevant. Indeed, important constraints on the scattering matrix will be found.

7. THE SCATTERING MATRIX: BLACK HOLE - WHITE HOLE DUALITY

Suppose a scattering matrix exists. This means that if we have completely specified the state $\{p_1, p_2, \ldots\}$ of all particles that ever went into the black hole, the outgoing matter should be in one well-specified state $|\psi\rangle_{out}$. A basis for $|\psi\rangle_{out}$ is the set of states where all outgoing particles have well-specified momenta at a certain time $\tau=\tau_0$. Now at $\tau=\tau_1>\tau_0$ we drop a light particle into the hole, with momentum p_{in} (in regular coordinates) at solid angle Ω_1. The change this induces for the outgoing wave is now determined primarily by the horizon shift (if other, non-gravitational interactions may be ignored). Thus, the new state will now be

$$|\psi\rangle_{out} \to e^{-i\int P_{out}(\Omega)\delta y(\Omega)\mathrm{d}^3\Omega}\ |\psi\rangle_{out} \ , \tag{7.1}$$

where $P_{out}(\Omega)$ is the operator that generates a shift in the configurations at the solid angle Ω. It is, of course, also the total momentum emerging at solid angle Ω. In here, we can now substitute eq. (6.4) for δy .

Now this means that if we know $|\psi\rangle_{out}$ at one stage, then $|\psi\rangle_{out}$ can, in principle, be determined after allowing any number of particles to fall in. If we may ignore non-gravitational interactions, we see that all states $|\psi\rangle_{out}$ ever to be produced by the black hole are generated by the operator $P_{out}(\Omega)$ from one single state. Therefore, $|\psi\rangle_{out}$ must be generated by the *algebra* of these operators. Similarly, the *ingoing* particles are only distinguished by the *total* momentum $p_1(\Omega_1)$ at each solid angle Ω_1 .

We find the following important result[15]. For the incoming wave functions one may diagonalize the operators $P_{in}(\Omega_1) = p_{in}(\Omega_1)$, and for the outgoing states we diagonalize $P_{out}(\Omega)$. Eq. (7.1) then tells us how a change in p_{in} affects the outgoing state. Up to a proportionality factor, the complete transformation rule for ingoing states into outgoing states should be generated by this equation. This rule is not difficult to find:

$$\langle\{p_{out}(\Omega)\}|\{p_{in}(\Omega')\}\rangle = N\, e^{-i\int d^2\Omega\, d^2\Omega'\; p_{out}(\Omega) f(\Omega,\Omega') p_{in}(\Omega')} , \tag{7.2}$$

where N is a normalization factor[#].

Eq. (7.2) is the *S*-matrix we wanted. *If* an *S*-matrix exists, and if we may ignore other than the longitudinal gravitational forces, it must be this one. It is here that we see a strikingly close resemblance to string theory[22]. As in string theory, eq. (7.2) should be universal. The amplitude is nearly the same as the one used as a starting point in string theory; only the string coupling constant comes out being imaginary[15,23]). If we replace the usual string amplitudes (for which after all no direct physical motivation can be found) by (7.2) or possible refinements[17]) of (7.2), then the *spectrum of massless states at the zero-slope limit* will remain the same. Thus we imagine that the qualitative successes of string phenomenology can also be attributed to this amplitude.

However, our amplitudes were directly motivated by consistency requirements for black holes, and in our implementation of these requirements a number of approximations were made. And it is not hard to argue that (7.2) cannot be exactly correct.

The problem with it is the algebra that generated the basis in which it is defined. We have the following commutation rules

$$[p_{in}(\Omega), p_{in}(\Omega')] = 0 \quad ; \quad [p_{in}(\Omega),\ x_{in}(\Omega')] = -i\delta^2(\Omega,\Omega') \quad ; \tag{7.3}$$

$$[p_{out}(\Omega), p_{out}(\Omega')] = 0 \;;\; [p_{out}(\Omega),\ y_{out}(\Omega')] = -i\delta^2(\Omega,\Omega') \quad , \tag{7.4}$$

and we have the relation

$$y_{out}(\Omega) \;=\; \int d^2\Omega_1\; f(\Omega,\Omega_1)\; p_{in}(\Omega_1) \quad . \tag{7.5}$$

[#] If other properties of the in- and outgoing particles are taken into account besides their momenta, (e.g. electric charge) then N becomes a unitary matrix. In the case of electric charge this matrix represents the contribution of a fifth, compactified, dimension[17].

This implies

$$[x_{in}(\Omega), y_{out}(\Omega')] \quad = \quad if(\Omega,\Omega') \quad , \tag{7.6}$$

so that we have also

$$x_{in}(\Omega) \quad = \quad -\int d^2\Omega' \; f(\Omega,\Omega') \; p_{out}(\Omega') \quad . \tag{7.7}$$

The operators x_{in} and y_{out} could be interpreted as coordinates of "particles", but then there should be *exactly one particle at every value of* Ω . This is where this Hilbert space differs from ordinary Fock space, where we may have any number of particles (mostly this will be zero) at every mode. This is also why it will be difficult to interpret our *S*-matrix directly as a matrix describing scattering of familiar particles.

But in spite of the unusual way in which the dynamical variables are represented we do believe that this description of the Hilbert space surrounding the black hole can be defended. Imagine a lattice-like cut-off on the horizon, where the lattice length is of the order of the Planck length. At every lattice site Ω there is exactly one particle. This leaves us more than enough particles to reconstruct ordinary Hilbert space. Ordinary particle physics is at the low-energy limit, where we never need to know what happens when two or more particles sit at exactly the same site Ω .

Thus, on the one hand we have ordinary particles, but alternatively, x_{in} and y_{out} can be seen as the position operators for the *past* and the *future* horizon. We then recognise an important consequence of our description of Hilbert space: past and future horizons cannot both be localized accurately; these obey an uncertainty relation. Indeed, they are each other's dual conjungates, much in the same way as coordinates are dual to momenta.

We claim that this also does away with a question considered often in the literature: does the time-reversed black hole (called "white hole") exist? Does the "eternal black hole" (one with a past white hole that was already there before the universe began) exist? The questions are not appropriate if our *S*-matrix exists: white hole coordinates are ill-specified once a black hole was localised.

The difficult but perhaps exciting picture that emerges is that the exact shape of either the past or the future horizon may completely determine the particle content of the black hole's vicinity. It should be possible to refine this picture by incorporating gravitational forces in the transverse direction, and non-gravitational forces. It is not hard to take the electromagnetic force into account. Here, the electric charge density operator $\rho(\Omega)$ and the gauge phase operator $\phi(\Omega)$ are each other's duals. Electromagnetic shock waves (Čerenkov radiation) surrounding charged massless particles are very similar to gravitational shock waves[16].

As will be explained in the next section, we expect a cut-off in Ω space. Probably the transverse forces are responsible for that. Surely, if coordinates in the transverse direction would be specified with accuracies better than the Planck length (implying $\delta\Omega \lesssim M_{Pl}/M$, where M is the black hole mass), then momenta in the transverse direction exceed the Planck energy so that also shifts in the transverse direction will arise that are bigger than $\delta\Omega$. The simplest cut-off would be a lattice in Ω space, but reality will probably be more complicated. What we expact actually to happen is that the Hilbert space algebra itself will produce a cut-off. A glimpse of a possible idea will be given in Sect. 9.

8. DISCRETE PHYSICS

Imagine a volume $V = L^3$ in three-dimensional space. Suppose that one could define a quantum (field) theory of all phenomena (particles, black holes) inside this volume, in terms of a Hilbert space. We ask: how many dimensions (basis elements) $\mathcal{D}$ does this Hilbert space have as V grows? In ordinary field theories the answer is strictly infinite. Even a single particle can occupy infinitely many states, so surely a second-quantized theory will have an infinite-dimensional Hilbert space. Only if we would introduce a rigid lattice cut-off, and accept only fundamental fermions in our theory, the total dimensionality would be finite. It would grow like

$$\mathcal{D} \cong \mathcal{O}(e^{\Lambda^3 L^3}) \quad , \tag{8.1}$$

where Λ is the inverse lattice size.

But this changes if the gravitational force is taken into account. In that case we cannot allow the total energy to exceed a certain value depending on V :

$$E_{tot} \leq L/2\kappa \quad , \tag{8.2}$$

because otherwise a black hole would form with a horizon that stretches beyond the edges of V .

Most quantum field theories have only

$$\mathcal{D} \cong \mathcal{O}(e^{C\kappa^{-3/4} L^{3/2}}) \tag{8.3}$$

basis elements with energy less than that (C is some constant)*). If we allow black holes we expect more states. Suppose there are N black holes with labels $i=1, \dots, N$. Each black hole has a density of states of the order of

$$\mathcal{D}_i \cong e^{4\pi M_i^2} \quad , \tag{8.4}$$

(leaving again $M_{Pl}^2 = \kappa = 1$), and the total system has

$$\mathcal{D} = \prod_i \mathcal{D}_i = e^{4\pi \sum_i M_i^2} \leq e^{4\pi (\sum_i M_i)^2} \quad ; \tag{8.5}$$

whereas the total energy is

$$E_{tot} = \sum_i M_i \leq \tfrac{1}{2} L \quad . \tag{8.6}$$

---

*) Eq. (8.3) holds for a theory with free massless particles. It can be derived most easily by counting methods familiar in statistical physics.

Therefore, allowing black holes we find

$$\mathcal{D} \lesssim e^{L^2\pi} , \qquad (8.7)$$

which, surprisingly, grows exponentially with the surface area L^2 rather than the volume L^3 .

We claim that in any "complete" quantum theory of the world the total dimensionality of Hilbert space inside a volume L^3 must be finite and approximately given by eq. (8.7). Theories where this dimensionality does not grow exponentially with the volume but with the surface area are not easy to construct. It could mean that there is some sort of additional constraint to be imposed on all states in the "physical part" of Hilbert space, whose solutions would have the dimensionality given by (8.7).

There is an interesting class of finite-dimensional theories. These are the so-called cellular automata. A cellular automaton is a system containing a definite number of completely discrete and limited variables located at "cells". The contents of each cell are continuously updated according to a given arithmetic rule depending on the previous values and the values inside neighboring cells. They are ideal for computer simulations.

A cellular automaton may either be completely deterministic or quantum mechanical. In a quantum mechanical cellular automaton the updating is prescribed by some unitary operator U defined in Hilbert space. At each tick of the clock we have

$$|\psi(t+1)\rangle = U\,|\psi(t)\rangle \ . \qquad (8.8)$$

This equation is the direct discrete analogon of the Schrödinger equation. The model is *deterministic* if a basis exists in terms of which the operator U happens to be also a permutation operator:

$$U\,|n_1,\ n_2,\ \ldots\rangle = |\mathrm{P}(n_1,\ n_2,\ \ldots)\rangle \ . \qquad (8.9)$$

If ever we find the operator U describing the real world it might not be easy to establish whether or not it can be seen to be a deterministic one, because the search for the corresponding basis may be difficult. It is very tempting to *suspect* that U should be deterministic[18]. It then remains to be seen how we can understand that nevertheless features typical for quantum mechanics govern the macroscopic world. Experiences with computer simulations with cellular automata show that they often behave *chaotically*. This means that there is no simple way to describe the macroscopic world in terms of deterministic laws in spite of local determinism. Perhaps the *only* way to describe the macroscopic world is *via* the Schrödinger equation (8.8) or in other words: perhaps our world is a cellular automaton but it allows only a statistical description. The mathematics of this statistics may happen to be that of conventional quantum mechanics[19].

There are numerous difficulties with such a picture but we think it cannot be ruled out. It would be a neat way to resolve the philosphical difficulties usually caused by quantum mechanics. In particular we will have to face the Einstein Podolski Rosen paradox[20] and the question how one can understand the violation of Bell's inequalities[21] by quantum mechanical interference effects. We stress however that strictly speaking there is no logical contradiction here at all because the *vacuum state* $|\emptyset\rangle$ will have to be a superposition of all basis elements, in particular if we use the deterministic evolution operator P of eq. (8.9). Since all our experimental set-ups are always surrounded by a vacuum we should not be surprised to find "interference effects". To be more precise: we expect that what is referred to as the 'vacuum state' is neither a single deterministically described state nor a completely random statistical mixture of all states, but a very special statistical mixture with many long-distance correlations. These correlations occur also between spacelike seperated points, yet cannot transmit any information.

In one attempt to improve our calculation of the S matrix elements (7.2) we found that the horizon can probably not be kept topologically as simple as an S_2 sphere. Indeed at the point $\mathcal{S}$ in Fig. 1 the topology must be very involved. One might speculate that the topological structures at $\mathcal{S}$ are denumerable and this could be a starting point for a completely discrete theory for the black hole horizon. But we did not see how to continue along this line. The following section shows a more promising approach.

9. CONFORMAL OPERATOR ALGEBRA FOR THE BLACK HOLE HORIZON

Our theory describes states in Hilbert space by giving the coordinates in space *and* time for the horizon intersection surface $x^{\mu}(\sigma_1,\sigma_2)$. We can identify, following the equations of Section 7,

$$\begin{aligned} \tilde{x} &= (x_1,x_2) \simeq (\vartheta,\varphi); \\ x^+ &= x_{in}; \qquad x^- = x_{out} \quad . \end{aligned} \tag{9.1}$$

For simplicity we limit ourselves to large black holes so that the horizon is approximately flat and $\tilde{x}$ can be treated as two flat coordinates. However, let us rewrite the equations of Section 7 in such a way that they become covariant under general two dimensional coordinate transformations. Equation (7.6) can be seen to be in the special gauge

$$\sigma_1 = x_1, \qquad \sigma_2 = x_2. \tag{9.2}$$

We have

$$[x^+(\tilde{\sigma}),\ x^-(\tilde{\sigma}')] = if(\tilde{\sigma},\tilde{\sigma}'), \tag{9.3}$$

with

$$\partial^2_{\sigma} f(\tilde{\sigma},\tilde{\sigma}') = -4\pi\kappa\delta^2(\tilde{\sigma}-\tilde{\sigma}'). \tag{9.4}$$

Now this is probably only an approximation that holds when

$$x^{\mu}(\sigma_1,\sigma_2) \simeq (\sigma_1,\sigma_2,0,0), \tag{9.5}$$

so (9.3) can be rewritten covariantly as

$$[x^\mu(\tilde\sigma),\ x^\nu(\tilde\sigma')] = if^{\mu\nu}(\tilde\sigma,\tilde\sigma')\ ;$$
$$f^{\mu\nu}(\tilde\sigma,\tilde\sigma') \cong \frac{2\kappa}{\sqrt{g}}\,\varepsilon^{\mu\nu\rho\lambda}\varepsilon^{ab}\partial_a x^\rho\partial_b x^\lambda\ \log|\tilde\sigma-\tilde\sigma'|, \tag{9.6}$$

where g is the determinant of the induced metric in σ space:

$$g_{ab} = \partial_a x^\mu \partial_b x^\mu. \tag{9.7}$$

The first part of eq. (9.6) and eq. (9.7) is entirely covariant under general coordinate transformations in $\tilde\sigma$ space and *special* coordinate transformations in x space**.

Let us now make the transition to a conformal gauge,

$$g_{ab} = \lambda\delta_{ab}\ , \tag{9.9}$$

and try to construct an algebra that should form the basis of a fundamental conformal theory for black holes. The resulting theory will be characterized by the fact that the x^μ do not commute (when x is large the commutators become negligible, so we are dealing with a small-distance effect here.

Writing

$$A^{\mu\nu}(\tilde\sigma) = -8\pi\kappa\ \varepsilon^{\mu\nu\kappa\lambda}\ \frac{\partial_1 x^\kappa(\tilde\sigma)\ \partial_2 x^\lambda(\tilde\sigma)}{\sqrt{g(\tilde\sigma)}}\ , \tag{9.10}$$

one derives

$$[\tilde\partial_\sigma x^\mu(\tilde\sigma),\tilde\partial_{\sigma'} x^\nu(\tilde\sigma')] = A^{\mu\nu}(\tilde\sigma)\delta^2(\tilde\sigma-\tilde\sigma')\ . \tag{9.11}$$

We *derived* this commutator equation only for the case that the surface $x^\mu(\tilde\sigma)$ deviates only infinitesimally from a flat surface. It is tempting to assume (9.11) to have a more general validity in any set of conformal coordinates. We can then solve the differential equation to find the commutators for $x^\mu(\sigma)$, by Fourier transforming in $\tilde\sigma$ space:

$$[x^\mu(\tilde k),x^\nu(\tilde p)] = -\frac{A^{\mu\nu}(\tilde k+\tilde p)}{(\tilde k\cdot\tilde p)}\ , \tag{9.12}$$

and Fourier transforming back:

$$[x^\mu(\tilde\sigma),x^\nu(\tilde\sigma')] = \frac{1}{(2\pi)^2}\ \mathcal{P}\!\int\frac{A^{\mu\nu}(\tilde\sigma'')\,d^2\tilde\sigma''}{(\tilde\sigma-\tilde\sigma'')\cdot(\tilde\sigma'-\tilde\sigma'')}\ , \tag{9.13}$$

**Indeed we described effects close to the horizon in a space-time metric $g_{\mu\nu}$ that is locally normalized to stay close to $\delta_{\mu\nu}$. The transition to general $g_{\mu\nu}(x)$ may well be as in string theories.

where $\mathcal{P}$ stands for a priciple value integration over the poles in $\tilde{\sigma}''$ space (which form a circle through $\tilde{\sigma}$ and $\tilde{\sigma}'$).

Unfortunately, this algebra becomes very complicated because of the highly non-linear form (9.10) of $A^{\mu\nu}$. The Hilbert space structure of our surfaces would become easier to fathom if we had linear commutation rules. Clearly, (9.13) suggests that we should look at the composite fields

$$W^{\mu\nu}(\tilde{\sigma}) = \varepsilon^{ab}\frac{\partial x^\mu}{\partial\sigma^a}\frac{\partial x^\nu}{\partial\sigma^b} . \tag{9.14}$$

From (9.13) it is not possible to derive simple commutation rules for the W fields. But we could start again from the beginning. For approximately flat surfaces in the transverse direction, and in the gauge (9.2), we have (if μ and ν are 3 or 4):

$$W^{12} = 1 \quad ; \quad W^{1\mu} = \frac{\partial x^\mu}{\partial\sigma^2} \quad ; \quad W^{2\mu} = -\frac{\partial x^\mu}{\partial\sigma^1} \quad ; \quad W^{34} = O(\partial x^\mu)^2 . \tag{9.15}$$

The commutation rule (9.11) can then be rewritten in the form

$$\sum_\lambda [W^{\lambda\mu}(\tilde{\sigma}), W^{\lambda\nu}(\tilde{\sigma}')] = \tfrac{1}{2}T\,\varepsilon^{\mu\nu\kappa\lambda}W^{\kappa\lambda}(\tilde{\sigma})\delta^2(\tilde{\sigma}-\tilde{\sigma}') , \tag{9.16}$$

which again is written in such a way that it remains true in all coordinate frames. T is a constant ('string constant') of order one in Planck units. In stead of (9.13) we can take this to be the equation that generalizes to arbitrary surfaces. It has the advantage of being linear (the factor $\sqrt{g}$ cancels out).

The algebra (9.16) is not complete, because the left hand side still contains a summation. What we could do about this is the following*†. Let $K^{\mu\nu}$ be i times the self dual part of $W^{\mu\nu}$:

$$K^{\mu\nu} = i(W^{\mu\nu} + \tfrac{1}{2}\varepsilon^{\mu\nu\kappa\lambda}W^{\kappa\lambda}) . \tag{9.17}$$

It has three independent components:

$$K_1 = i(W^{23} + W^{14}) \quad ; \quad K_2 = i(W^{31} + W^{24}) \quad ; \quad K_3 = i(W^{12} + W^{34}) . \tag{9.18}$$

Now from (9.16) we derive that these obey a complete commutator algebra,

$$[K_a(\tilde{\sigma}), K_b(\tilde{\sigma}')] = iT\varepsilon_{abc}K_c(\tilde{\sigma})\delta^2(\tilde{\sigma}-\tilde{\sigma}') . \tag{9.19}$$

Apart from a complication to be mentioned shortly, this is a local and complete algebra of the kind we were looking for. At first sight it

---

*†This part of these notes was written early October 1989, enabling us to add an argument here that was not yet known when the lectures were given.

seems to generate an infinite dimensional Hilbert space because we have these operators at every $\tilde{\sigma}$. But then we should remember that the operators K , like the W , are distributions*). Let us introduce test functions $f(\sigma)$, $g(\sigma)$ and define operators

$$L_a^{(f)} = \int K_a(\tilde{\sigma})f(\tilde{\sigma})\mathrm{d}^2\tilde{\sigma} \quad , \tag{9.20}$$

then these satisfy commutation rules:

$$[L_a^{(f)}, L_a^{(g)}] = i\varepsilon_{abc}L_c^{(fg)} \quad . \tag{9.21}$$

Let us now restrict to test functions $f(\tilde{\sigma})$ that can only take the values 0 or 1 . Then $L_a^{(f)}$ satisfy the commutation rules of ordinary angular momentum operators. Note that for such an f the integral (9.20) is nothing but a boundary integral:

$$L_1^{(f)} = i \oint_{\delta f} (x^2\mathrm{d}x^3 + x^1\mathrm{d}x^4) \quad , \quad \text{etc.}, \tag{9.22}$$

where δf stands for the boundary of the support of f . We conclude that for every closed curve δf on $\tilde{\sigma}$ space we have three 'angular momentum' operators $L_a^{(f)}$ that satisfy the usual commutation rules and addition rules for angular momenta. Given such a bunch of closed curves f_i we can characterize the contribution of that part of the horizon to Hilbert space by the usual quantum numbers l_i and m_i . These are discrete and so, in some sense, we seem to come close to our aim of realizing a discrete Hilbert space for black holes. As anticipated in Sect 8, there is a set of quantum numbers whose number and values are limited per unit of surface area (see eq. 9.22).

Unfortunately, there is a snag. The operators L_a are not Hermitean. From the definition (9.14) we see that W^{ij} are Hermitean and W^{i4} anti-Hermitean. Therefore, $L_a^{\dagger}$ correspond to the *anti*-self dual parts of $W^{\mu\nu}$. The commutation rules between L_a and $L_a^{\dagger}$ are non-local (they may perhaps follow from (9.13)). The operators L^2 are Hermitean, but not necessarily positive (they are only nonnegative for *time-like* surface elements). If we may assume the smallest surface elements to be timelike we can still build our surface using quantum numbers l_i and m_i but the states we get are *not properly normalized* (it is for finding the norms of the states that we need Hermitean conjugation). If

$$\psi\{l_i, m_i\}$$

are the basis elements constructed using the self dual operators L_i , and

$$\phi\{l_i, m_i\}$$

the basis elements generated by the anti-self dual $L_i^{\dagger}$, then we have

---

*) $W^{\mu\nu}$ is actually the two-form $W^{\mu\nu} = \mathrm{d}x^{\mu}\wedge\mathrm{d}x^{\nu}$.

$$\langle \phi\{l_i',m_i'\} | \psi\{l_i,m_i\} \rangle = \prod_i \delta_{l_i,l_i'} \delta_{m_i,m_i'} \quad , \tag{9.23}$$

but the ψ themselves, or the ϕ themselves, are not orthonormal. Therefore it is far from clear whether or not we actually obtained a complete representation of our Hilbert space.

REFERENCES

1. K.S. Thorne, "Black Holes: the Membrane Paradigm", Yale Univ. press, New Haven, 1986; S. Chandrasekhar, "The Mathematical Theory of Black Holes", Clarendon Press, Oxford University Press
2. S.W. Hawking, Comm. Math. Phys. **43** (1975) 199
   J.B. Hartle and S.W. Hawking, Phys. Rev. **D13** (1976) 2188
   W.G. Unruh, Phys. Rev. **D14** (1976) 870
   S.W. Hawking and G. Gibbons, Phys. Rev. **D15** (1977) 2738
   S.W. Hawking, Comm. Math. Phys. **87** (1982) 395
3. G. 't Hooft, J. Geom. and Phys. **1** (1984) 45
4. S.W. Hawking and G.F.R. Ellis, "The large-scale structure of space-time", Cambridge Univ. Press 1973
   C.W. Misner, K.S. Thorne and J.A. Wheeler, "Gravitation", Freeman, San Francisco, 1973
5. W. Rindler, Am. J. Phys. **34** (1966) 1174
6. G. 't Hooft, Nucl. Phys. **B256** (1985) 727
7 R. Friedberg, T.D. Lee and Y. Pang. Nucl.Phys. **B264** (1986) 437 and Nucl. Phys. **B276** (1986) 549
8. J.D. Bekenstein, Nuovo Cim. Lett. **4** (1972) 737; Phys. Rev. **D7** (1973) 2333; **D9** (1974) 3292
9. E.A. Martinez and J.W. York, Jr., Chapel Hill preprint (1989)
10 J.D. Bekenstein, Phys. Rev. **D5** (1972) 1239, 2403
11 S.W. Hawking, Comm. Math. Phys. **87** (1982) 395
   S.W. Hawking and R. Laflamme, Phys. Lett **B209** (1988) 39
12 S. Coleman, Nucl. Phys. **B306** (1988) 643; **B307** (1988) 864
   S.B. Giddings and A. Strominger, Nucl. Phys. **B321** (1989) 481
13 P.C. Aichelburg and R.U. Sexl, J. Gen. Rel. Grav. **2** (1971) 303
14 T. Dray and G. 't Hooft, Nucl Phys. **B253** (1985) 173
15. G. 't Hooft, *in* Proceedings of the 4th seminar on Quantum Gravity, May 25-29, 1987, Moscow, USSR, ed. by M.A. Markov et al, World Scientific 1988, pp. 551-567
16 G. 't Hooft, Phys. Lett. **B198** (1987) 61; Nucl. Phys. **B304** (1988) 867
17 G. 't Hooft, to be Publ.
18. E. Fredkin and T. Toffoli, Int. J. Theor. Phys. **21** (1982) 219; T. Toffoli, Int. J. Theor. Phys. **21** (1982) 165; Physica **10D** (1984) 117; R.P. Feynman, Int. J. Theor. Phys. **21** (1982) 467; S. Wolfram, Physica **10D** (1984) 1; N.H. Margolus, Doctoral thesis, M.I.T., June 1987; id., Physica **10D** (1984) 81
19. G. 't Hooft, J. Stat. Phys. **53** (1988) 323
20. A. Einstein, B. Podolski and N. Rosen, Phys. Rev. **47** (1935) 777; M. Jammer, "The conceptual Development of Quantum Mechanics (Mc. Graw-Hill, 1966)
21. J.S. Bell, Physics **1** (1964) 195
22 G. 't Hooft, Physica Scripta **T15** (1987) 143
23 G. 't Hooft, talk presented at the 10th Workshop on Grand Unification, Chapel Hill, April 1989, to be publ.

AN INTRODUCTION TO GENERAL TOPOLOGY AND QUANTUM TOPOLOGY

C.J. Isham

Blackett Laboratory
Imperial College
South Kensington
London SW7 2BZ

§1 INTRODUCTION

The aim of this course is to give a short introduction to the classical theory of general topology and to consider some ways in which one might attempt to formulate a genuine theory of 'quantum topology'. These days, it is hardly necessary to motivate speaking to theoretical physicists about topology — the title of this summer school speaks for itself! However, much of the topology used in theoretical physics is in fact *differential geometry*. For example, the spacetime of classical general relativity is modelled by a smooth four-dimensional manifold, and infinite-dimensional manifolds of maps between various finite-dimensional manifolds play an important rôle in a number of branches of modern non-linear field theory (the non-linear σ-model, Yang-Mills theory, string and membrane theory,...).

Notwithstanding the current popularity of differential geometry, my strong belief is that its days are numbered, at least so far as the subject of quantum gravity is concerned. Smooth manifolds and local differential equations belong primarily to the world of classical physics and I do not believe that these are appropriate tools with which to probe the structure of spacetime (in so far as this is a meaningful concept at all) near the Planck length. At best, they are likely to be applicable in the semi-classical limit of the quantum theory of gravity (whatever that might be) and a lot more thought needs to be given to the question of which mathematical structures are really relevant for discussing the concepts of space and/or time in the "deep" quantum region. In particular, I shall be concerned in this course with the idea that the topology of space or spacetime is itself subject to the laws of quantum theory, and that this necessitates moving outside the realm of smooth manifolds.

Many problems arise when trying to identify the most appropriate mathematical framework in which to discuss the ideas of quantum topology. These include severe

Physics, Geometry, and Topology
Edited by H. C. Lee
Plenum Press, New York, 1990

conceptual difficulties, not least in regard to the applicability of the concepts of 'space' and 'time' and the overarching question of whether *any* idea of quantum topology can be made compatible with conventional quantum theory in which normally:

(i) the topology of space or spacetime is assumed to be that of a differentiable manifold;

(ii) this manifold is part of the fixed, classical background structure. Thus different topological structures for space would be expected to generate only different superselection sectors; in which case the problem reduces to seeing what effect these various background topologies have on systems that are otherwise to be quantized in a conventional way.

However, within the context of quantum gravity, the concept of 'quantum topology' contains two ideas that are considerably more radical. The first is that the topological structure of space and/or spacetime should be regarded as a quantum variable rather than as part of the fixed background within which the theory is to be formulated; the second is that the quantum variables describing the topology of space may change in time via some sort of quantum (tunelling?) process. But what does this really mean? In particular, how literally can the word "topology" be taken? In practice, most discussions do not involve general topological spaces at all but are locked firmly into the framework of differential geometry. For example, in the "Euclidean" approach to quantum gravity, the basic entity of interest is the transition probability amplitude $K(g_1, \Sigma_1 ; g_2, \Sigma_2)$ for going from a metric g_1 on a three-manifold Σ_1 to a metric g_2 on a three-manifold Σ_2. This is expressed as the functional integral

$$K(g_1, \Sigma_1 ; g_2, \Sigma_2) = \sum_M \int e^{-S(g)}\, dg \tag{1.1}$$

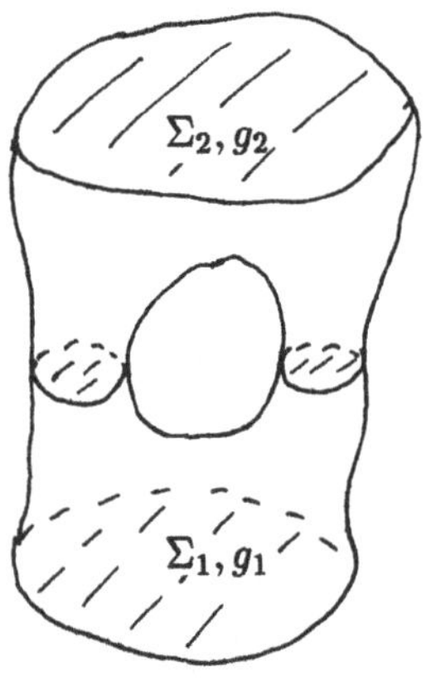

in which $S(g)$ is the classical action, the sum is over every four-manifold M whose boundary is the disjoint union of Σ_1 and Σ_2, and the integral is over all *Riemannian* four-metrics on M which induce the given three-metrics g_1 and g_2 on Σ_1 and Σ_2 respectively (Hawking, 1979).

To appreciate the significance of (1.1) we must first review very briefly some of the central ideas in the canonical quantization of gravity. This starts with an arbitrary decomposition of the (Lorentzian) spacetime manifold M of classical general relativity into a one-parameter family of spacelike three-manifolds. The spacetime metric of M has ten independent components ${}^4g_{\mu\nu}(x,t)$, $\mu,\nu = 0\ldots3$ but the components ${}^4g_{0\nu}$ are purely "gauge" degrees of freedom and are concerned only with how the spatial coordinates are related from one spacelike slice to another and how the spacelike slices are situated within the four-manifold (for a comprehensive review see Kuchař (1981)). Thus an initial canonical analysis apparently yields a collection of six canonical variables $g_{ij}(x)$, $i,j = 1\ldots3$ (the metric on the three-manifold) with an associated set $p^{kl}(x)$ of conjugate variables. However, it transpires that not all the Einstein equations $G_{\mu\nu}({}^4g) = 0$ relate to dynamical evolution. More precisely, once the time derivative of ${}^4g$ has been replaced with the appropriate combination of g and p, the four equations $G_{0\nu} = 0$ become *constraints* on the canonical variables. These constraints play a fundamental rôle in the theory. In fact, they essentially *determine* the dynamical evolution in the sense that any four-metric on M with the property that the constraints are satisfied on all spacelike slices necessarily satisfies the remaining dynamical equations $G_{ij} = 0$.

A naive analysis of the canonical commutation relations

$$[\hat{g}_{ij}(x), \hat{p}^{kl}(y)] = i\hbar\delta^k_{(i}\delta^l_{j)}\delta(x,y) \tag{1.2}$$

suggests a "Schrödinger representation" [1] in which states are functionals $\Psi(g)$ with $(\hat{g}_{ij}(x)\Psi)(g) = g_{ij}(x)\Psi(g)$ and $(\hat{p}^{kl}(y)\Psi)(g) = -i\hbar(\frac{\delta}{\delta g_{kl}(y)}\Psi)(g)$. The constraints are first-class and, following the usual Dirac procedure, are imposed as constraints on the state vector. Furthermore, because of the relation between the constraints and the dynamics mentioned above, these quantum constraint equations are deemed to constitute the *entire* content of the canonical quantization of gravity — there is no additional time-dependent Schrödinger equation. Three of this set of four constraints merely affirm that Ψ depends only on the intrinsic geometry of the spacelike surface (that is, the state vector is invariant under spatial diffeomorphisms) and hence that $\Psi(g)$ is a function of $6-3=3$ degrees of freedom per spatial point. However, the final equation (the famous Wheeler-DeWitt equation) has the highly non-trivial form $\big(\mathcal{H}(\hat{g},\hat{\pi})\Psi\big)(g) = 0$ where

$$\mathcal{H}(g,p) = \frac{1}{2}(\det g)^{-\frac{1}{2}}(g_{ij}g_{kl} + g_{il}g_{kj} - g_{ik}g_{jl})p^{ik}p^{jl} - (\det g)^{\frac{1}{2}}R(g) \tag{1.3}$$

[1] Representations of this type can be very useful in conventional quantum field theory. This has been especially emphasised by Jackiw in the context of Yang-Mills theory and is developed by him in his article in this volume.

in which $R(g)$ is the curvature scalar of the three-metric g.

Of the many difficult questions that can be asked about this formalism one of the most subtle concerns the status of 'time'. There is no external time label attached to $\Psi(g)$. [1] Instead, one must try to identify time as some function of the metric variables themselves (e.g. the volume of the three-space) and then reinterpret the Wheeler-DeWitt equation as describing the evolution of the remaining physical variables (two degrees of freedom per spatial point) with respect to this "internal time" (Kuchar, 1981). However, in practice it does not seem possible to make a precise identification of this type and the evidence suggests strongly that our conventional notion of time can be sustained at best in the semi-classical limit of the quantum theory — in the "deep" quantum region the concept of time breaks down entirely. This has many important implications for the quantum gravity programme and we will return to it later in the context of quantum topology.

After this discursion into conventional canonical quantum gravity let us return to (1.1) and to the meaning which can be attached to $K(g_1, \Sigma_1\,; g_2, \Sigma_2)$. Although this was computed using an integral over Riemannian metrics it can nevertheless be shown that, if g_1 is fixed, it satisfies the Wheeler-DeWitt equation in g_2 appropriate for a *Lorentzian* signature spacetime. In particular, if the Σ_2 manifold is absent (so that the manifold M has a single, connected, three-boundary Σ_1) we get the remarkable "creation *ex nihilo*" theory of Hartle and Hawking (1983). This affords a most elegant way of relating the spatial (canonical) picture to an underlying spacetime structure (albeit Riemannian) but of course it poses many questions in its own right. One particularly contentious issue is the use of Riemannian rather than Lorentzian metrics for this purpose. However, from the viewpoint of quantum topology, the most relevant feature of (1.1) is undoubtedly the sum over M which, if taken literally, implies that the ideas of quantum topology are to be developed purely within the category of differential geometry. In particular, the initial and final spaces Σ_1 and Σ_2 are differentiable manifolds and the topology change is implemented with the aid of the interpolating four-manifolds M.

But it is very debatable whether this picture is at all reliable. The well-known failure of (1.1) at the level of perturbative renormalizability means we have to work hard if the theory is to be retained in this form. Some of the more obvious options are:

1. Try and "patch" the classical action with the addition of terms to the Lagrangian which will cancel the infinities. Supergravity was a much-studied theory of

[1] I am assuming here that the spatial three-manifold is compact. In the non-compact case the possibility arises of selecting a time variable using the fixed background geometry at spatial infinity. In this case there *is* an additional time-dependent Schrödinger equation.

this type, but not many people now believe that the problem of quantum gravity can be resolved in this way.

2. Accept the fact that general relativity provides a phenomenological description of the gravitational field that is valid only at low energies. The "true" theory of quantum gravity is something quite different (for example, superstrings?) from which the classical theory (including perhaps space and time themselves) emerges only in some suitable limit. But it then seems most unlikely that (1.1) will have any fundamental meaning. In particular, if the semi-classical computation of (1.1) breaks down at the Planck length then any picture of topology changes at this scale must be equally suspect. Thus the elegant picture of quantum topology mediated by interpolating manifolds is only likely to be correct if the theory contains a scale at which such effects could occur which is significantly larger than the Planck length itself.

3. Continue to hope that there may exist some new, non-perturbative, way of evaluating the functional integral (1.1) which will give well-defined, finite answers. One of the most promising approaches of this sort is the work of Rovelli and Smolin (1988, 1989) which exploits the considerable simplification obtained for the expression for $\mathcal{H}$ using Ashtekar's new variables (Ashtekar, 1988). Note that even if it is position 2. that turns out ultimately to be the correct one, it is still possible that a non-perturbative picture of this type will provide a more reliable guide to what the complete theory looks like in the "intermediate" quantum regions than is afforded by the older methods.

4. Argue that the problem of reconciling general relativity and quantum theory is so acute that the structure of quantum theory itself must be radically revised.

Many workers in quantum gravity have entertained ideas of this final type and, at the very deepest level, this may well be the correct way to proceed. (Penrose is an especially articulate advocate of this view; see his (1987) paper for a review of his current position). However, I have always been intrigued by the question of how far conventional quantum theory can be pushed and, in particular, of the extent to which the geometrical and topological ideas in classical general relativity can still be maintained in the quantum régime (or, if not, with what they should be replaced). If the question of quantum topology is pursued from the viewpoint of this third option, it becomes natural to ask whether the objects summed over in (1.1) can really be just differentiable manifolds. One of the important properties of the functional integral measures that arise in conventional quantum field theory is their tendency to be supported on spaces of *distributions* rather than smooth functions. This "roughening up" process is typical of quantum theory in general [1] and suggests that the sum in (1.1) might have to be extended to include topological spaces that are more singular than

[1] In the context of the canonical Schrödinger approach to quantum field theory, this roughening is reflected in the feature that the domain of the state vectors $\Psi(\phi)$ is typically forced to be a space of distributional fields ϕ rather than smooth functions.

differentiable manifolds. The critical issue then becomes the selection of the appropriate mathematical concepts with which to model spacetime structure in this deep quantum régime.

As remarked already, I support this view in the strong sense of doubting the validity at the Planck length of the entire framework of differential geometry. This iconoclastic urge forms part of the motivation for the present course of lectures in which I wish to extend the ideas of quantum topology from differential geometry to more general types of topological space. In particular, I wish to consider the possibility of state vectors that are functions $\psi(\tau)$ of arbitrary topologies τ, with differentiable manifolds appearing only in the semi-classical limit in which conventional canonical quantum gravity is presumably valid.

Many difficult questions arise *a priori* when contemplating such a programme. For example

1. Can we talk about "all" topologies in a meaningful way?

2. If so, do they form a "nice" space which can be considered as the "configuration space" of a system that is to be quantized? How is this quantization to be performed? Can it be done without introducing some background manifold?

3. What is the status of 'time' in such a theory? Is the attachment of a continuous time label t to a state vector $\psi(\tau)$ compatible with our depreciation of differential geometry, or should one develop a purely topological version of the differentiable cobordism picture afforded by (1.1)? In the latter case, the interpretation of the state vectors $\psi(\tau)$ would presumably have to include some sort of "internal time", at least in a semi-classical limit, just as in conventional quantum gravity an internal time is extracted from part of the three-metric g appearing in the state vector $\psi(g)$.

4. A topology on a set X is a specific type of structure with a precise mathematical definition. But what is special about "topology" rather than any of the other related mathematical structures? That is, what is the most appropriate quantum generalization of the idea of a smooth manifold?

5. What is the ontological status of the points in the set X? In particular, what is the cardinality of X (i.e. "how many" points are there) and should this number itself be subject to quantum fluctuations?

Some of these issues will be addressed in what follows. However, one of the aims of this course is to give a simple introduction to the *classical* theory of general topology as well as to the more exotic quantum ideas, and much of the material is intended for this purpose. Thus section 2 contains an introduction to the theory of metric spaces. These provide a natural generalization of differentiable manifolds in which the topological properties are still defined in terms of real numbers via the use of distance functions. In so far as quantum theory ascribes fundamental significance to the real or complex numbers, it could be argued that it is metric topologies which are likely to be most compatible with quantum ideas. The theory of quantum metric topology is not much developed but some preliminary ideas are contained in section 3. The remaining sections deal with more general topological spaces and start with a discussion of partially ordered sets and lattices. This is in line with my general aim of presenting the ideas of topology in a fairly algebraic way on the grounds that this is likely to be particularly appropriate when one's ultimate aims are quantum mechanical. Section 5 deals with the general idea of a 'topology' via the intermediate concept of a 'neighbourhood space'. I have deliberately taken this detour (rather than plunging immediately into the axioms for open sets etc.) as I feel it is helpful when debating the central question of which particular set of axioms for topological-like systems is most relevant in the construction of a mathematical model for spacetime. In section 6 we will consider a theory of quantum topology involving wave-functions $\psi(\tau)$, eigenstates of topology, etc., and the paper concludes with a discussion of some of the open issues and possible directions for future research.

Some set theory notation

$\forall$	"for all"
$\exists$	"there exists"
$\iff$	"if and only if"
$\alpha := \beta$	The entity α is *defined* to be the entity β.
$\{x \mid P(x)\}$	The set of all x such that the proposition $P(x)$ is true.
$x \in X$	x belongs to the set X
$A \subset B$	A is a subset of B (this includes the possibility that $A = B$).
$A \cap B$	The *intersection* of A and $B := \{x \in X \mid x \in A \text{ and } x \in B\}$.
$A \cup B$	The *union* of A and $B := \{x \in X \mid x \in A \text{ or } x \in B\}$
$\lvert X \rvert$	The cardinal number of the set X.
$\mathcal{F}(X,Y)$	The set of all functions/maps from the set X to the set Y.
$C(X,Y)$	All *continuous maps* between the topological spaces X and Y.
$C^\infty(X,Y)$	All *infinitely differentiable* functions between the manifolds X and Y.
$\mathbb{R}$	The set of real numbers.
$\mathbb{C}$	The set of complex numbers.

§2 METRIC SPACES

§2.1 Main Ideas

A key idea in any topological-type structure on a set X is the sense in which a point $x \in X$ can be said to be "near" to another point $y \in X$; without such a concept, the points in X are totally disconnected from each other. In particular, we would like to say that a sequence $\{x_n\}$ of points in X *converges* to a point $x \in X$ if the elements of the sequence get arbitrarily near to x in an appropriate way. We will use the idea of the convergence of sequences to develop the theory of metric spaces and, in §6, general topological spaces.

In the case of complex numbers, the "nearness" of z_1 to z_2 is measured by the value of the modulus $|z_1 - z_2|$, and to say that the sequence $\{z_n\}$ converges to z means that

$$\forall \epsilon > 0, \; \exists n_0 \text{ such that } n > n_0 \text{ implies } |z_n - z| < \epsilon \tag{2.1.1}$$

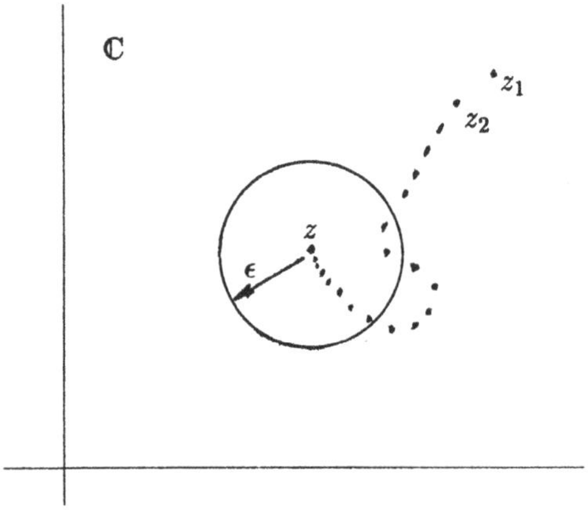

Thus the disks $B_\epsilon(z) := \{ z' \in \mathbb{C} \mid |z - z'| < \epsilon \}$ 'trap' the sequence. That is

$$\forall \epsilon > 0, \; \exists n_0 \text{ such that } n > n_0 \text{ implies } z_n \in B_\epsilon(z) \tag{2.1.2}$$

or, in terms of the *tails* $T_n := \{ z_k \mid k > n \}$ of the sequences,

$$\forall \epsilon > 0, \; \exists n_0 \text{ such that } T_{n_0} \subset B_\epsilon(z) \tag{2.1.3}$$

This notion of convergence can be generalised at once to the space $\mathbb{R}^n$ of all n-tuples of real numbers with the aid of the distance function

$$d(\mathbf{x}, \mathbf{y}) := \sqrt{(\mathbf{x} - \mathbf{y}) \cdot (\mathbf{x} - \mathbf{y})} \tag{2.1.4}$$

and the associated balls

$$B_\epsilon(\mathbf{x}) := \{\mathbf{y} \in \mathbb{R}^n \mid d(\mathbf{x}, \mathbf{y}) < \epsilon\} \tag{2.1.5}$$

Then a sequence of points $\mathbf{x}_n \in \mathbb{R}^n$ is said to *converge* to $\mathbf{x} \in \mathbb{R}^n$ (denoted $\mathbf{x}_n \to \mathbf{x}$) if

$$\forall \epsilon > 0, \ \exists n_0 \text{ such that } n > n_0 \text{ implies } \mathbf{x}_n \in B_\epsilon(\mathbf{x}) \tag{2.1.6}$$

The concept of a distance function can be generalized to an arbitrary set X by extracting the crucial properties (*vis a vis* convergence) of the Euclidean distance $d(\mathbf{x}, \mathbf{y})$ defined in (2.1.4). Thus a *metric* on a set X is a map $d : X \times X \to \mathbb{R}_+$ (the positive real numbers) satisfying the three conditions

$$d(x, y) = d(y, x) \tag{2.1.7}$$

$$d(x, y) \geq 0 \text{ and } = 0 \iff x = y \tag{2.1.8}$$

$$d(x, y) \leq d(x, z) + d(z, y) \tag{2.1.9}$$

for all $x, y, z \in X$. If (2.1.8) is replaced by the weaker condition $d(x, y) \geq 0$ (i.e., there may be $x \neq y$ such that $d(x, y) = 0$) then X is said to be a *pseudo-metric*. Once again, convergence of a sequence can be defined in terms of the tails of the sequence being trapped by the balls surrounding a point. That is, $x_n \to x$ means

$$\forall \epsilon > 0, \ \exists n_0 \text{ such that } T_{n_0} \subset B_\epsilon(x) \tag{2.1.10}.$$

where $B_\epsilon(x) := \{y \in X \mid d(x, y) < \epsilon\}$. Note that any given sequence of points may not converge at all but, if it does, it converges to only one point (exercise!). In a more general type of topological space a sequence can converge to more than one point — see later.

It is important to know when two metrics can be regarded as being equivalent. For example, metrics $d^{(1)}(x, y)$ and $d^{(2)}(x, y)$ are said to be *isometric* if there exists a bijection $i : X \to X$ such that, for all $x, y \in X$

$$d^{(1)}(x, y) = d^{(2)}(i(x), i(y)) \tag{2.1.11}$$

However, of greater interest to us is when two metrics lead to the same set of convergent sequences (and with each sequence converging to the same point in both metrics). This motivates the definition that $d^{(2)}$ is *stronger* than $d^{(1)}$ (or $d^{(1)}$ is *weaker* than $d^{(2)}$) if

$$\forall x \in X, \ \forall \epsilon > 0, \ \exists \epsilon' > 0 \text{ such that } B^{(2)}_{\epsilon'}(x) \subset B^{(1)}_\epsilon(x) \tag{2.1.12}$$

(as we shall see later, this means that the topology associated with $d^{(2)}$ is stronger than that associated with $d^{(1)}$). A pair of metrics is said to be *equivalent* if each is stronger than the other. Note that (i) a $d^{(2)}$-convergent sequence is automatically $d^{(1)}$-convergent, and (ii) equivalent metrics admit the same set of convergent sequences. A result of considerable importance is the converse to (ii). That is, it can be shown that if two metrics induce the same set of convergent sequences (with the same limits) then they are necessarily equivalent. Some of the material needed to prove this will be introduced later.

§2.2 Examples of Metric Spaces

1. If a differentiable manifold Σ is equipped with a Riemannian metric g, the distance between a pair of points $x, y \in \Sigma$ is defined to be

$$d(x,y) := \inf_{\gamma} \int \left(g_{ab}(\gamma(t))\dot{\gamma}^a(t)\dot{\gamma}^b(t)\right)^{\frac{1}{2}} dt \tag{2.2.1}$$

where the *infimum* is over all piece-wise differentiable curves $t \mapsto \gamma(t)$ in Σ which pass through the points x and y.

2. A metric can be defined on any set X by

$$d(x,y) := \begin{cases} 1, & \text{if } x \neq y; \\ 0, & \text{if } x = y. \end{cases} \tag{2.2.2}$$

3. On $\mathbb{R}^n$, some equivalent metrics are

$$d(\mathbf{x},\mathbf{y}) := \sqrt{(\mathbf{x}-\mathbf{y})\cdot(\mathbf{x}-\mathbf{y})} \tag{2.2.3}$$

$$d(\mathbf{x},\mathbf{y}) := \max_{i=1\ldots n} |\mathbf{x}_i - \mathbf{y}_i| \tag{2.2.4}$$

$$d(\mathbf{x},\mathbf{y}) := \left(\sum_{i=1}^{n} |\mathbf{x}_i - \mathbf{y}_i|^p\right)^{\frac{1}{p}} \qquad p \geq 1 \tag{2.2.5}$$

4. Let $C([a,b],\mathbb{R})$ denote the space of all real-valued, continuous, functions defined on the closed interval $[a,b] := \{r \in \mathbb{R} \mid a \leq r \leq b\}$. A metric can be defined on $C([a,b],\mathbb{R})$ by

$$d(f,g) := \int_a^b |f(t) - g(t)|\, dt \tag{2.2.6}$$

An inequivalent metric is

$$d(f,g) := \sup_{t\in[a,b]} |f(t) - g(t)| \tag{2.2.7}$$

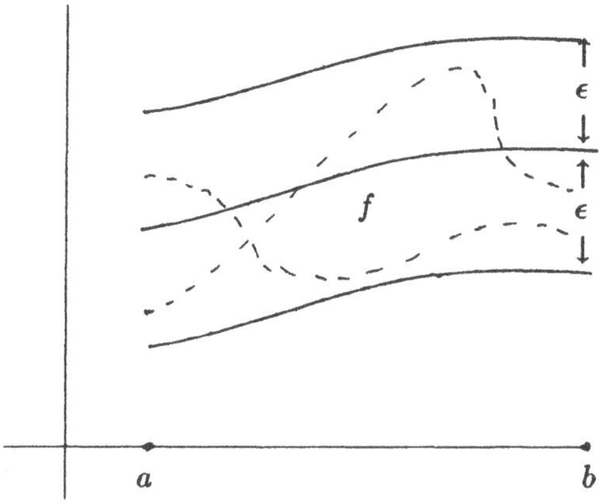

Yet another inequivalent metric is

$$d(f,g) := \left\{ \int_a^b |f(t) - g(t)|^2 \, dt \right\}^{\frac{1}{2}} \tag{2.2.8}$$

5. On any set X let $\ell^2(X)$ denote the set of all real-valued functions f on X such that

   (i) $f(x) = 0$ for all but a countable set of $x \in X$

   (ii) $\sum_{x \in X} \left(f(x)\right)^2$ converges.

Then a distance function can be defined on $\ell^2(X)$ by

$$d(f,g) := \sqrt{\sum_{x \in X} |f(x) - g(x)|^2} \tag{2.2.9}$$

§2.3 Operations on Metrics

There a number of ways in which metrics on a set X may be combined to form a new metric. Operations of this type are of potential interest in any quantization scheme as the algebraic structure they reflect may become encoded in the structure of the quantum operators. Some specific examples of operations on metrics are as follows.

1. If d_i, $i = 1 \ldots n$ is a finite set of metrics on X then

$$d(x,y) := \sum_{i=1}^{n} a_i d_i(x,y) \tag{2.3.1}$$

defines a metric on X if a_i is any set of real numbers, greater than or equal to zero and with at least one of them non-zero.

2. If d_1 and d_2 are a pair of metrics on X then a new metric, called the *join* of d_1 and d_2, can be defined by

$$d_1 \vee d_2(x,y) := \max(d_1(x,y), d_2(x,y)) \tag{2.3.2}$$

Another metric, called the *meet* of d_1 and d_2, is defined by

$$d_1 \wedge d_2(x,y) := \inf_{x=x_1 \ldots y=x_r} \sum_{k=2}^{r} \min(d_1(x_{k-1},x_k), d_2(x_{k-1},x_k)). \tag{2.3.3}$$

where the *infimum* is taken over all finite subsets $\{x = x_1, x_2 \ldots x_r = y\}$ of X. It is interesting to note (Birkhoff, 1967) that the set of all metrics on X forms a lattice under these two operations (see §4 for a short introduction to lattices).

3. If d is any metric on X, define $d_b(x,y) := \min(1, d(x,y))$. Then d_b is a *bounded* metric which can be shown to be equivalent to d. Thus if we are only interested in metrics up to equivalence, nothing is lost by requiring them to be bounded functions on $X \times X$.

§2.4 Some Topological Concepts in Metric Spaces

In the present context, by 'topological' concepts I mean those dealing with the relations of points and sets (the word "topological" comes from the Greek τοπος meaning "place"). In particular, if A is a subset of the metric space X, every point in X belongs to just one of three categories in respect of its relation to $A \subset X$.

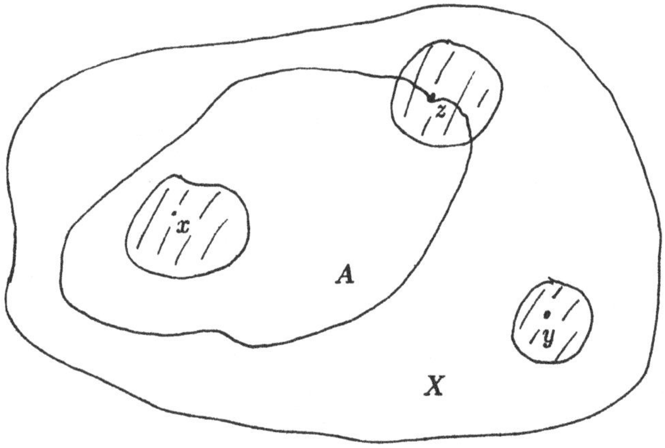

1. x is an *interior* point of A if there exists a ball $B_\epsilon(x)$ such that $B_\epsilon(x) \subset A$.
2. y is an *exterior* point of A if there exists a ball $B_\epsilon(y)$ such that $B_\epsilon(y) \cap A = \emptyset$.
3. z is a *boundary* point of A if every ball $B_\epsilon(z)$ intersects both A and its complement $A^c := \{x \in X \mid x \notin A\}$

In addition a point $x \in X$ is said to be a *limit* point of A if $B_\epsilon(x) \cap A \neq \emptyset$ for all $\epsilon > 0$. Thus a limit point is either an interior point or a boundary point of A.

Comments

(a) The *interior*, *exterior*, *boundary* of A are defined respectively to be the set of all interior, exterior and boundary points of A and are denoted $Int(A)$, $Ext(A)$ and $Bnd(A)$. It is easy to see that

$$
\begin{aligned}
Int(A) &= Ext(A^c) \\
Ext(A) &= Int(A^c) \\
Bnd(A) &= Bnd(A^c)
\end{aligned}
$$

$$
\begin{aligned}
Int(A) &\subset A \\
A \cap Ext(A) &= \emptyset
\end{aligned}
$$

(b) A set A is said to be *open* if it contains none of its boundary points. It is *closed* if it contains all its boundary points. Note that

$$
\begin{aligned}
A \text{ is open} &\iff A = Int(A) \\
A \text{ is open} &\iff A^c \text{ is closed}
\end{aligned}
$$

In the example of the real line $\mathbb{R}$ with its metric $d(x, y) := |x - y|$, the interval $\{ x \in \mathbb{R} \mid a < x < b \}$ (for any $a < b$) is an open set; similarly $\{ x \in \mathbb{R} \mid a \leq x \leq b \}$ is an example of a closed set. On the other hand, $\{ x \in \mathbb{R} \mid a \leq x < b \}$ is neither open nor closed.

(c) The collection of all open sets in any metric space is called the *topology* associated with the space and possesses the following very important set of properties (exercise!):

1. The union of an *arbitrary* family of open sets is open.
2. The intersection of any *finite* family of open sets is open.
3. The empty set $\emptyset$ and X itself are open.

The analogous properties for closed sets are:

1. The intersection of an *arbitrary* family of closed sets is closed.
2. The union of any *finite* family of closed sets is closed.
3. The empty set $\emptyset$ and X itself are closed.

(d) The topology associated with a metric space is determined equally by either the collection of all open sets or the collection of all closed sets. In the latter context it is therefore significant that

1. $A \subset X$ is closed if and only if it contains all its limit points.

2. A point $x \in X$ is a limit point of a subset A if and only if there exists a sequence $\{x_n\}$ in A which converges to x.

Thus a subset A is closed if and only if the limit of every convergent sequence x_n of points in A itself lies in A. That is, the closed sets (and hence the topology) associated with a metric are uniquely determined by its collection of convergent sequences. This is the key to proving the result mentioned earlier that two metrics with the same set of convergent sequences are equivalent.

(e) We will see later the precise sense in which a metric space is a special case of a general topological space. Thus the topological differences between, for example, a 2-sphere, a 2-torus and a 1264-sphere are entirely coded into their respective distance functions, which could all be considered to be defined on some common abstract set X with the cardinality of the continuum.

§3 QUANTUM THEORY OF METRIC FUNCTIONS

§3.1 Riemannian-Geometry Driven Quantum Topology

One way of constructing a theory of quantum topology might be to consider the set Metric(X) of all distance functions on a set X and try to quantize it by looking for a collection of hermitian operators $\hat{d}(x, y)$, $x, y \in X$, which are compatible in some way with the fundamental classical relations (2.1.7-9). A striking property of these relations is that two of them are *inequalities*, the triangle inequality (2.1.9) being particularly important. It is always a non-trivial problem to quantize a system defined in terms of inequalities — for example, there is no analogue of the Dirac bracket that allows, at least formally, a quantization of a system specified by a collection of *equalities* — and in the present case we are looking for a sort of "two-point" scalar quantum field theory on X which is consistent with these relations.

Ideas of quantizing the distance go back to the early work of Wheeler (1964,1967) who conceived the possibility of quantum fluctuations in the topology being induced by large fluctuations in the canonically-quantized metric of three-space. Thus the starting point for this "Riemannian-geometry driven" quantum topology is the classical expression (2.2.1) for a distance function on a given three-manifold Σ

$$d(x, y) := \inf_{\gamma} \int \left(g_{ab}(\gamma(t))\dot{\gamma}^a(t)\dot{\gamma}^b(t)\right)^{\frac{1}{2}} dt \tag{3.1.1}$$

where we note that since $g_{ab}(x)$ becomes an operator in the canonical quantization of general relativity, so does $d(x, y)$. An expectation value $< \psi, \hat{d}(x, y)\psi >$ is a symmetric function of x and y and will satisfy (2.1.7) and (2.1.9) provided that

$$\hat{g}_{ab}(\gamma(t))\dot{\gamma}^a(t)\dot{\gamma}^b(t) \tag{3.1.2}$$

is a well-defined positive operator so that the square root can be taken.

An important result in classical differential geometry is the demonstration that, for all choices of the Riemannian metric g, the distance function defined by (3.1.1) generates the same set of convergent sequences. In fact, the associated topology is automatically equivalent to the original manifold topology on Σ (Helgason, 1962). However, the singular nature of quantum field theory suggests that an operator version of (3.1.1) may lead to a distance function that no longer satisfies the strong form of (2.1.8) because of the existence of states $|\psi>$ for which $<\psi, \hat{d}(x,y)\psi>$ is only a pseudo-metric, that is, it may vanish for certain $x \neq y$. This is one sense in which the topology imparted to the set Σ by the quantized $d(x,y)$ may become state dependent. Thus we obtain Riemannian-metric driven quantum fluctuations of topology around the background manifold Σ.

Unfortunately, many problems arise in attempting to implement this idea. For example:

1. The conventional canonical commutation relations [1]

$$[\hat{g}_{ab}(x), \hat{p}^{cd}(y)] = i\hbar\delta^c_{(a}\delta^d_{b)}\delta(x,y) \tag{3.1.3}$$

are not compatible with the positivity condition on (3.1.2). This is analogous to the problem which arises when trying to construct a quantum theory for a system whose configuration space is the positive real numbers $\mathbb{R}_+$: the commutation relation $[\hat{x}, \hat{p}] = i\hbar$ implies that

$$e^{ia\hat{p}}\hat{x}e^{-ia\hat{p}} = \hat{x} + \hbar a \tag{3.1.4}$$

and hence that the (generalised) eigenvectors of $\hat{x}$ have eigenvalues which can be any real number. One way of resolving this difficulty is to employ the *affine* commutation relations (Klauder, 1970a,b; Isham, 1984) $[\hat{x}, \hat{\pi}] = i\hbar\hat{x}$. In a non-trivial irreducible representation of this algebra, the eigenvalue spectrum of $\hat{x}$ is either strictly positive or strictly negative and a consistent quantum theory on $\mathbb{R}_+$ can therefore be constructed by using only representations of the former type. The analogous affine commutation relations for general relativity are

$$[\hat{g}_{ab}(x), \hat{\pi}^c_d(y)] = i\hbar\delta^c_{(a}\hat{g}_{b)d}(x)\delta(x,y) \tag{3.1.5}$$

and

$$[\hat{\pi}^a_b(x), \hat{\pi}^c_d(y)] = i\hbar(\delta^a_d\hat{\pi}^c_b(x) - \delta^c_b\hat{\pi}^a_d(x))\delta(x,y). \tag{3.1.6}$$

---

[1] The components $g_{ab}(x)$ of the metric are taken with respect to a globally-defined basis for the tangent spaces of Σ. Such a frame always exists for a compact three-manifold and avoids the problems that can arise if the (only locally-defined) coordinate components $g_{ij}(x)$ are employed.

2. More seriously, even if (formal) positivity of (3.1.2) has been achieved by using (3.1.5-6), (3.1.1) involves smearing the operator $\hat{g}_{ab}(x)$ with a tensor field that is concentrated on a one-dimensional curve. It is likely that the resulting operator is still very singular, in which case taking the square root becomes problematical. [1]

3. This also raises the question of whether we should not be looking for some *smeared* version $\hat{d}(f,g)$ of $\hat{d}(x,y)$? If so, we must confront the difficult question of the precise vector space to which the test functions f and g should belong. In conventional quantum field theory the functions employed for this purpose are usually required to be continuous, but this notion is only meaningful if a topology is present on the space on which the functions are defined. So it looks as if a smearing operation might require the introduction of some background topological structure. This applies in principle to all attempts to quantize the distance functions on a set and presumably produces a theory of quantum fluctuations of topology around this background. In any event, even at the classical level it is interesting to ask what (if any) type of "generalized topology" can be associated with a distributional distance function.

These difficulties (especially 2.) have proved intractable and no one has yet managed to produce a theory of quantum topology based just on the canonical quantization of general relativity augmented with an operator version of (3.1.1). Thus other ways must be sought for tackling the quantization of distance functions. One possibility is to "invert" (3.1.1) by using $d(x,y)$ as a canonical variable instead of the metric tensor $g_{ab}(x)$. The idea would be to find a complete set of canonical variables involving $d(x,y)$ plus some suitable conjugates which are canonically equivalent at the classical level to the original pair (g_{ab}, p^{cd}) but which are (by construction) adapted to the problem in hand and which can be quantized without reference to the traditional commutation relations. The idea of using $d(x,y)$ as "half" of the set of canonical variables seems plausible but it has not yet been developed. Therefore, for the rest of this section I will consider a more ambitious approach in which all references to background manifolds and differential geometry are dropped and an attempt is made to quantize directly a system whose "configuration space" is the set Metric(X) of all metric functions on a given set X.

§3.2 Quantization On Metric(X)

The idea here would be to try to construct directly operators $\hat{d}(x,y)$ which commute with each other

$$[\hat{d}(x_1,y_1), \hat{d}(x_2,y_2)] = 0 \tag{3.2.1}$$

[1] But note that Colombeau has recently developed an approach to generalized functions in which there exists such an exotic entity as the square root of the Dirac delta-function (Phil Parker, private communication). It would be interesting to see if this novel approach has any application in the context of quantum gravity/topology.

and then to find a collection of conjugate variables which can maintain consistency with the defining conditions (2.1.7-9) for a classical distance function. By far the most significant of these is the "triangle inequality" (2.1.9) which is a non-local analogue of the situation in Riemannian geometry whereby, for each $x \in X$, the Riemannian metric $g_{ab}(x)$ is required to be a positive-definite matrix. It is this condition that is guaranteed (at least formally) by replacing the canonical commutation relations with the affine relations (3.1.5-6). The heart of the group-theoretical quantization scheme employed in the construction of (3.1.5-6) lies in the existence of a vector space W in which the configuration space Q of the system can be embedded as an orbit of a group G which acts linearly on this vector space (Isham, 1985). The general theory then shows that a suitable canonical group is the semi-direct product of G with the dual of W. In the case of $Riem(\Sigma)$, the vector space can be chosen to be the set of all covariant rank-2 symmetric tensor fields on Σ and the appropriate group is the set $C^\infty(\Sigma, GL(3,\mathbb{R}))$ of $GL(3,\mathbb{R})$-valued differentiable functions on Σ — this is the group whose (non-abelian) algebra is represented by (3.1.6). The action of $\Lambda \in C^\infty(\Sigma, GL(3,\mathbb{R}))$ on a tensor field t_{ab} is

$$t_{ab}(x) \to \Lambda_a{}^c(x)\Lambda_b{}^d(x)t_{cd}(x). \tag{3.2.2}$$

In the case of the space Metric(X), a natural choice for the vector space would be the set $\mathcal{F}(X \times X, \mathbb{R})$ of all $\mathbb{R}$-valued functions on $X \times X$. [1] The challenge then is to find a Lie group that acts on the space $\mathcal{F}(X \times X, \mathbb{R})$ such that the subspace Metric(X) appears as one of the orbits. The generators of such a group could then be used as the canonical conjugates to the variables $d(x,y)$ and would preserve the classical conditions on a distance function in the same sense that the affine relations (3.1.5-6) are compatible with the restrictions on a metric tensor. However, I do not know of any direct way of constructing such a group and therefore a different approach must be found.

One possibility is to consider some fixed Banach space $\mathcal{B}$ and to use the set of all injections $\rho : X \to \mathcal{B}$ to generate distance functions according to the formula

$$d_\rho(x,y) := N\big(\rho(x) - \rho(y)\big) \tag{3.2.3}$$

where N is the (fixed) norm on $\mathcal{B}$. The idea now would be to quantize the vector space of maps ρ so that the quantum distance function becomes

$$\hat{d}(x,y) := N\big(\hat{\rho}(x) - \hat{\rho}(y)\big). \tag{3.2.4}$$

It can be shown (Isham, 1988) that there exists a $\mathcal{B}$ such that every bounded metric can be obtained in this way (which entails no real loss of generality since every metric is

---

[1] A better choice might be the subspace of all such functions that are bounded. This has the advantage of possessing a natural Banach space structure obtained from the sup-norm. This space can be employed if we restrict our attention to bounded metrics.

equivalent to one that is bounded). The main difficulty with this approach is the extra "gauge" degrees of freedom which arise from the fact that many different embeddings of X in $\mathcal{B}$ yield the same distance function via (3.2.3). It is possible to fix this gauge in some simple model situations in which X is a finite set, but it is not clear how to proceed in general.

Another approach, "dual" to the above, is to choose a vector space $\mathcal{B}$ and a fixed embedding ρ_0 of X in $\mathcal{B}$, and then to quantize the set of all norms on $\mathcal{B}$. Thus the quantized operator $\hat{d}(x,y)$ is now obtained from (3.2.3) in the form

$$\hat{d}(x,y) := \hat{N}\big(\rho_0(x) - \rho_0(y)\big) \tag{3.2.5}$$

and the problem reduces to finding a vector space $\mathcal{B}$ such that all distance functions can be obtained in this way and then to construct the quantum theory of norms on such a space. In the latter context, we recall that a norm is a real-valued function N on $\mathcal{B}$ such that

$$N(rv) = |r|N(v), \quad \forall v \in \mathcal{B},\ \forall r \in \mathbb{R} \tag{3.2.6}$$

$$N(v) \geq 0 \text{ and } = 0 \iff v = 0 \tag{3.2.7}$$

$$N(u+v) \leq N(u) + N(v), \quad \forall u, v \in \mathcal{B} \tag{3.2.8}$$

Of course, the problem of constructing a quantum operator $\hat{N}$ which is compatible with these conditions is similar in many respects to the general metric problem we are trying to solve. However, (3.2.6-8) are simpler than (2.1.7-9) in several significant ways and the subject of "quantum norm theory" should be more tractable than tackling the metrics on X directly. A concrete attempt in this direction is contained in a forthcoming paper with Renteln and Kubyshin (1989).

The idea of quantizing the set of all distance functions on a set X is attractive because it bears at least some resemblance to a more conventional quantum field theory with $\hat{d}(x,y)$ regarded as a two-point quantum field. However, in practice this idea has not been taken very far and I will turn now to the most ambitious programme of all: the quantization of the collection of *all* topologies on a set X, not just those that can be obtained from a distance function.

§4 PARTIALLY ORDERED SETS AND LATTICES

§4.1 Partially Ordered Sets

In developing the general theory of topology it is useful to emphasise certain algebraic properties that arise naturally in this context and which are also central to the quantum programme to be discussed later. The relevant concepts are 'partially ordered sets' and 'lattices' which play an important rôle in many branches of mathematics. The classic reference for the latter is Birkhoff (1967); another useful source is Grätzer (1978).

Definitions

1. A *relation* R on a set X is a subset of $X \times X$, and $x \in X$ is said to be *R-related* to $y \in X$ (denoted xRy) if the pair $(x, y) \in R \subset X \times X$. Note that a function $f : X \to X$ defines a relation $\{(x, f(x)) \mid x \in X\}$ but there are many relations that are not derived from functions.

2. A *partially ordered set* (or *poset*) is a set X and a relation $\preceq$ on X which is:

(P1)	*Reflexive*:	for all $x \in X$, $\quad x \preceq x$
(P2)	*Antisymmetric*:	for all $x, y \in X$, $x \preceq y$ and $y \preceq x$ implies $x = y$
(P3)	*Transitive*:	for all $x, y, z \in X$, $x \preceq y$ and $y \preceq z$ implies $x \preceq z$.

The notation $x \prec y$ will be used if it is necessary to emphasise that $x \preceq y$ but $x \neq y$. Note that any particular pair of elements $x, y \in X$ may not be related either way. However, if it is true that for any $x, y \in X$ either $x \preceq y$ or $y \preceq x$ then X is said to be *totally* ordered.

3. An element y in a poset X *covers* another element x if $x \prec y$ and there is no $z \in X$ such that $x \prec z \prec y$. This is denoted diagramatically by

A finite partially ordered set is determined uniquely by its diagram of covering elements.

4. For our later use we recall also the definition of an *equivalence relation* on a set X. This is a relation R that is:

(E1)	*Reflexive*:	for all $x \in X$, xRx
(E2)	*Symmetric*:	for all $x, y \in X$, xRy implies yRx
(E3)	*Transitive*:	for all $x, y, z \in X$, xRy and yRz implies xRz.

It should be noted that any equivalence relation R on a set X partitions X into disjoint equivalence classes in which all the elements in any class are equivalent to each other. The set of all such equivalence classes is denoted X/R. An important example in theoretical physics is the set of gauge orbits of the action of a gauge group on the space of connections in a Yang-Mills theory.

Examples

1. The real numbers $\mathbb{R}$ are totally ordered with respect to the usual ordering $\leq$. Note than no $r \in \mathbb{R}$ possesses a cover since given any pair of real numbers there always exists a third one which lies between them.

2. The set Metric(X) of all metric functions on a set X can be partially ordered by saying that $d^{(1)} \preceq d^{(2)}$ if the open balls for the two metrics satisfy (2.1.12). This means that the topology associated with $d^{(2)}$ is stronger than that associated with $d^{(1)}$ and, as we shall see later, topologies can be partially ordered by the relation of one being stronger than the other. However, this relation is *not* a partial ordering on Metric(X) since $d^{(1)} \preceq d^{(2)}$ and $d^{(2)} \preceq d^{(1)}$ does not imply that $d^{(1)} = d^{(2)}$ but only that the two metrics are equivalent (that is, they admit the same set of convergent sequences).

3. If X is any set, the set of all subsets of X is denoted $\mathrm{P}(X)$ and is sometimes known as the *power set* of X. Thus, $A \subset X \iff A \in \mathrm{P}(X)$. In the general theory of topological convergence (to be developed later) the central idea is to associate with each $x \in X$ a collection $\mathcal{N}(x)$ of subsets of X (the "neighbourhoods" of the point x) which determine whether or not a sequence converges to x. Thus $\mathcal{N}(x) \subset \mathrm{P}(X)$ or, equivalently, $\mathcal{N}(x) \in \mathrm{P}(\mathrm{P}(X))$. Similarly, the collection $\mathcal{N} := \{\mathcal{N}(x) \mid x \in X\}$ can be regarded as a subset of $\mathrm{P}(\mathrm{P}(X))$ or as an element of $\mathrm{P}(\mathrm{P}(\mathrm{P}(X)))$.

The set $\mathrm{P}(X)$ has a natural partial ordering defined by

$$A \preceq B \text{ means } A \subset B \tag{4.1.1}$$

where $A, B \subset X$. Note that the notation $A \subset B$ includes the possibility that $A = B$. The covers of a subset A are all subsets of X obtained by adding a single point to A.

The simplest non-trivial example is the partial ordering diagram for the two-element set $X = \{a, b\}$:

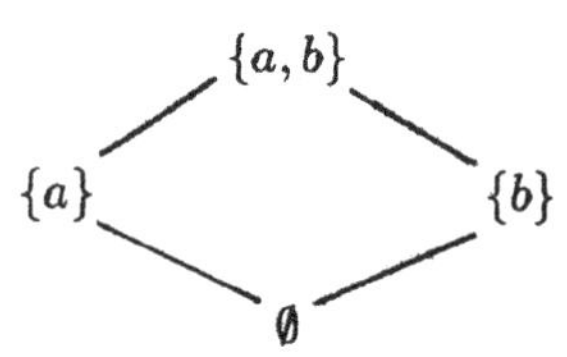

while the diagram for $X = \{a, b, c\}$ is

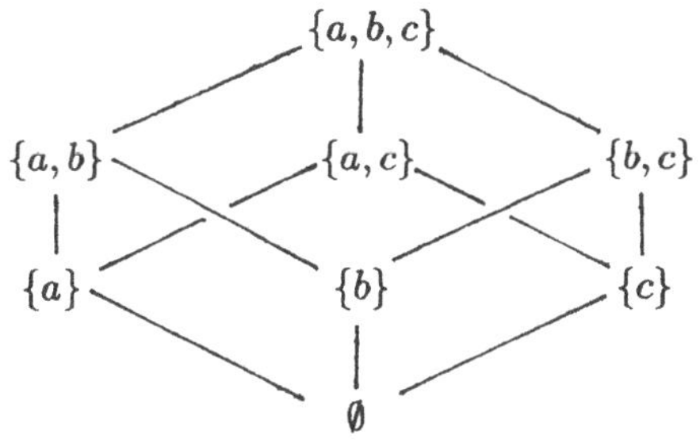

4. Partially ordered sets play an important rôle in classical general relativity. Let M be a spacetime manifold equipped with a Lorentzian metric. Then if $q, p \in M$, define $q \preceq p$ if p lies in the causal future (Kronheimer and Penrose, 1967) of q. This is a partial ordering and, rather remarkably, it can be shown that the entire metric (up to an overall conformal factor) can be recovered from this ordering (Hawking et al, 1976; Malament, 1977). This feature has been exploited in a variety of suggestions that spacetime should be regarded as a discrete set but still with a causal structure/partial ordering (for example: Finkelstein, 1969; t'Hooft, 1979; Bombelli et al, 1987a,b).

§4.2 Lattices

Definitions

1. In any poset P, a *join* (or *least upper bound*) of $a, b \in P$ is an element $a \vee b \in P$ such that:

(i) $a \vee b$ is an *upper bound* of a and b. That is, $a \preceq a \vee b$ and $b \preceq a \vee b$.

(ii) If there exists $c \in P$ such that $a \preceq c$ and $b \preceq c$ then $a \vee b \preceq c$.

2. A *meet* (or *greatest lower bound*) of $a, b \in P$ is an element $a \wedge b \in P$ such that:

(i) $a \wedge b$ is a *lower bound* of a and b. That is, $a \wedge b \preceq a$ and $a \wedge b \preceq b$.

(ii) If there exists $c \in P$ such that $c \preceq a$ and $c \preceq b$ then $c \preceq a \wedge b$.

Note that, if it exists, a join or meet is necessarily unique (exercise!).

3. A *lattice* is a poset $\mathcal{L}$ in which every pair of elements possesses a join and a meet.

A *unit* element 1 is such that, for all $a \in \mathcal{L}$, $a \preceq 1$.

A *null* element 0 is such that, for all $a \in \mathcal{L}$, $0 \preceq a$.

4. The lattice is said to be *complete* if a greatest lower bound and a least upper bound exist for *every* subset S of $\mathcal{L}$ (all that is guaranteed by the definition of a lattice is that these bounds will exist for all *finite* subsets of $\mathcal{L}$). These bounds will be denoted $\bigwedge S$ and $\bigvee S$ respectively.

5. A lattice is *distributive* if, for all $a, b, c \in \mathcal{L}$

$$a \wedge (b \vee c) = (a \wedge b) \vee (a \wedge c) \tag{4.2.1}$$

This is equivalent to

$$a \vee (b \wedge c) = (a \vee b) \wedge (a \vee c). \tag{4.2.2}$$

6. In any lattice one always has (exercise!)

$$a \vee (b \wedge c) \preceq (a \vee b) \wedge c \tag{4.2.3}$$

for all $a, b, c \in \mathcal{L}$ such that $a \preceq c$. If equality holds for all such a, b, c then the lattice is said to be *modular*. Every distributive lattice is modular but there exist modular lattices that are not distributive.

7. A lattice is *complemented* if it contains both a unit element and a null element and if to each $a \in \mathcal{L}$ there exists $a' \in \mathcal{L}$ (not necessarily unique) such that

$$a \vee a' = 1 \tag{4.2.4}$$

$$a \wedge a' = 0 \tag{4.2.5}$$

8. A *Boolean algebra* is a complemented, distributive lattice.

Comments

1. All the lattices we will be considering will have both a unit element and a null element.

2. $a \preceq b \iff a \wedge b = a \iff a \vee b = b$ (4.2.6)

3. For all $a \in \mathcal{L}$, $1 \wedge a = a$. Thus $\mathcal{L}$ is a semigroup with respect to the $\wedge$-operation with 1 as the unit element; it is not a group since no element other than 1 has an inverse. Similarly, $0 \vee a = a$ and hence $\mathcal{L}$ is also a semigroup with respect to the $\vee$-operation with 0 as the unit element. In addition we have, for all $a \in \mathcal{L}$,

$$1 \vee a = 1, \text{ and } 0 \wedge a = 0. \tag{4.2.7}$$

Thus 1 and 0 are *absorptive* elements for the $\vee$-semigroup and the $\wedge$-semigroup respectively.

4. If $\mathcal{L}$ is distributive then any complement a' of an element $a \in \mathcal{L}$ is unique.

Examples

1. An example of a Boolean algebra. $a \wedge b = 0$ and $a \vee b = 1$.

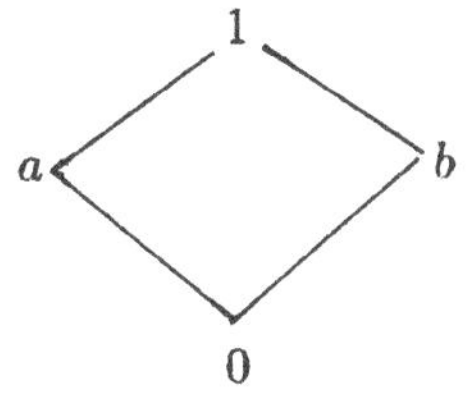

(4.2.8)

2. Two non-distributive lattices with 5 elements.

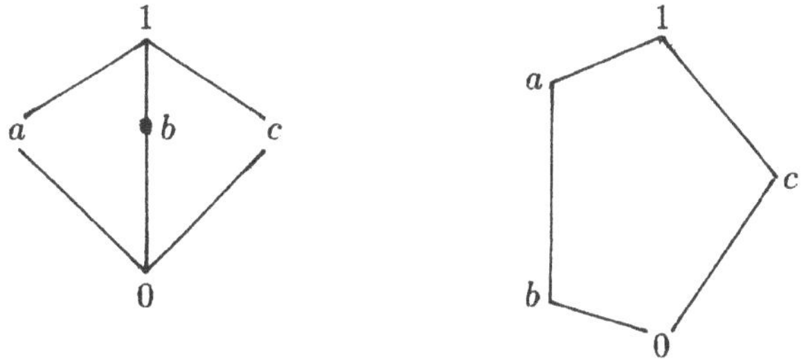

(4.2.9)

3. The set P(X) of all subsets of X is a Boolean algebra with

$$A \wedge B := A \cap B$$
$$A \vee B := A \cup B \tag{4.2.10}$$

and $A \preceq B$ if and only if $A \subset B$. The unit and null elements are

$$1 = X$$
$$0 = \emptyset \tag{4.2.11}$$

The lattice theory complement A' of $A \subset X$ is the set-theoretic complement A^c. Clearly this lattice is complete.

4. The set of all metrics on a set X can be partially ordered by $d^{(1)} \preceq d^{(2)}$ if, for all $x, y \in X$, $d^{(1)}(x, y) \leq d^{(2)}(x, y)$. It becomes a lattice under the join and meet operations defined in (2.3.2-3). Note however that there is no null or unit element since, given any metric $d(x, y)$, a larger (resp. smaller) metric can always be constructed by multiplying $d(x, y)$ by a positive real number that is greater than (resp. less than) 1.

5. If V is a vector space, the set of all linear subspaces of V is a lattice with

$$W_1 \wedge W_2 := W_1 \cap W_2 \tag{4.2.12}$$

and with $W_1 \vee W_2$ defined to be the smallest subspace containing the pair of subspaces W_1 and W_2.

Note that if V is a Hilbert space, the lattice of linear subspaces is complemented with the complement of a subspace W being defined as its orthogonal complement with respect to the Hilbert space inner product $<,>$:

$$W' := W_\perp = \{v \in V \mid \forall w \in W, \ <v, w>= 0\} \tag{4.2.13}$$

Correspondingly, the set of all hermitian projection operators on V also forms a complemented lattice. This particular lattice has been used extensively in investigations into the axiomatic foundations of general quantum theory (Jauch, 1973; Varadarajan, 1968). It differs strikingly from the analogous lattice of propositions in classical mechanics. The basic type of yes-no question that can be asked there is whether the point in phase space representing the state of the system does, or does not, lie in any particular subspace of the phase space. Thus the propositional lattice of classical physics is essentially the lattice of subsets of phase space. This distinction between the quantum and classical lattices has given rise to the interesting subject of "quantum logic".

A lattice satisfies several very important algebraic relations:

(L1)	*Idempotency*:	$a \vee a = a, \quad a \wedge a = a$
(L2)	*Commutativity*:	$a \vee b = b \vee a, \quad a \wedge b = b \wedge a$
(L3)	*Associativity*:	$(a \vee b) \vee c = a \vee (b \vee c)$
		$(a \wedge b) \wedge c = a \wedge (b \wedge c)$

In addition, any lattice satisfies the *absorptive* laws:

(L4) $a \wedge (a \vee b) = a, \qquad a \vee (a \wedge b) = a$

Conversely, there is the important theorem:

Theorem.

A non-empty set $\mathcal{L}$ equipped with binary operations (L1)–(L4) can be given a partial ordering by defining

$$a \preceq b \iff a = a \wedge b$$

The resulting structure is a lattice in which the meet and join operations are $a \wedge b$ and $a \vee b$ respectively.

§5 TOPOLOGICAL SPACES

§5.1 Examples of non-metric convergence

The general theory of topology may be approached in a number of different ways which are reflected in the variety of styles to be found in the many textbooks that are available on the subject. We will study the theory of general topological spaces in terms of the convergence of sequences and generalizations thereof. A selection of particularly useful references in this context is Bourbaki (1966), Császár (1978), Dugundji (1966) and Kelly (1955).

In a non-metric space X, it is no longer possible to define "nearness" using a real number. Instead we attempt to "trap" the tails of a sequence with subsets of X which are defined as far as possible to be analogues of the balls $B_\epsilon(x)$ in a metric space. A generalized structure of this type will consist of a suitable collection $\mathcal{N} = \{\mathcal{N}(x) \mid x \in X\}$ of families $\mathcal{N}(x)$ ("neighbourhoods" of x) of subsets of X and convergence is defined purely in terms of these subsets. Specifically, a sequence $\{x_n\}$ in X is defined to *converge* to x with respect to $\mathcal{N}(x)$ (denoted $x_n \overset{\mathcal{N}(x)}{\longrightarrow} x$) if

$$\forall N \in \mathcal{N}(x),\ \exists\, n_0 \text{ such that } n > n_0 \text{ implies } x_n \in N \tag{5.1.1}$$

(note that this does not rule out the possibility that a sequence may converge to many different points at once).

In order that the sequence $x_n := x$ for all n should always converge to x (which seems a minimal requirement for the concept of 'convergence' to be useful) it is necessary that each neighbourhood of x should contain the point x. We could start to consider other *a priori* requirements on these neighbourhoods but let us first give some simple examples of convergence which are not associated with any metric or pseudo-metric.

1. In $\mathbb{C} \cup \{\infty\}$, if $z_n \in \mathbb{C} \subset \mathbb{C} \cup \{\infty\}$ we define $z_n \to \infty$ to mean that, for all $\epsilon > 0$, there exists some n_0 such that $n > n_0$ implies $|z_n| > \epsilon$. A relevant family of neighbourhoods of the symbol ∞ is therefore the collection of sets

$$N_\epsilon(\infty) := \{\infty\} \cup \{\, z \in \mathbb{C} \mid |z| > \epsilon \,\} \tag{5.1.2}$$

where ϵ is any positive real number. (Strictly speaking, this can be derived from a metric topology on the Riemann sphere.)

2. Let $\mathcal{F}([a,b], \mathbb{R})$ denote the set of all real-valued functions on the closed interval $[a,b] \subset \mathbb{R}$ A sequence of functions f_n is said to converge *pointwise* to a function f if, for all $t \in [a,b]$, the sequence of real numbers $f_n(t)$ converges to the real number $f(t)$ in the usual way. That is

$$\forall t \in [a,b],\ \forall \epsilon > 0,\ \exists n_0(\epsilon, t) \text{ such that } n > n_0 \text{ implies } |f_n(t) - f(t)| < \epsilon. \tag{5.1.3}$$

A suitable family of neighbourhoods of $f \in \mathcal{F}([a,b],\mathbb{R})$ is clearly all sets of the form

$$N_{t,\epsilon}(f) := \{g \in \mathcal{F}([a,b],\mathbb{R}) \mid |g(t) - f(t)| < \epsilon\} \tag{5.1.4}$$

where t in any real number in the closed interval $[a,b]$ and ϵ is any positive number.

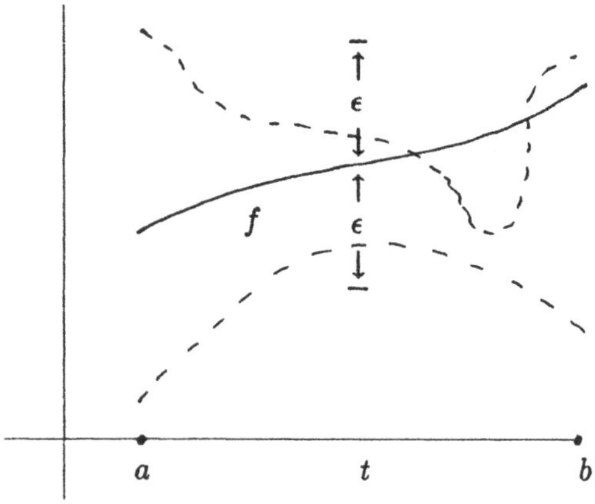

§5.2 Neighbourhood Spaces

Two crucial questions are

1. What properties must be possessed by the collections $\mathcal{N}(x)$ of subsets of X in order to give a "useful" notion of convergence?

2. When do two different families of neighbourhoods lead to the same sets of convergent sequences?

It turns out to be more useful to start with the second question, in which case the first important concept is the following:

Definition

If $\alpha, \beta \subset \mathrm{P}(X)$, the collection β of subsets of X is said to be *finer* that the collection α if, for each $A \in \alpha$, there exists a subset $B \in \beta$ such that $B \subset A$. This will be denoted by $\alpha \vdash \beta$.

Comments

1. If $\alpha \subset \beta$ then it is trivial that β is finer than α.

2. The definition (5.1.1) of convergence with respect to $\mathcal{N}(x)$ is equivalent to the statement that the set $T := \{T_n\}$ of tails of the sequence is finer than the family $\mathcal{N}(x)$.

3. If $\alpha \vdash \beta$ and $\beta \vdash \gamma$ then $\alpha \vdash \gamma$. In particular, if $\mathcal{N}^{(2)}(x)$ is finer than $\mathcal{N}^{(1)}(x)$ then any sequence $\{x_n\}$ that converges to x with respect to $\mathcal{N}^{(2)}(x)$ necessarily converges with respect to $\mathcal{N}^{(1)}(x)$.

The families $\mathcal{N}^{(2)}(x)$ and $\mathcal{N}^{(1)}(x)$ are defined to be *equivalent* (denoted $\mathcal{N}^{(2)}(x) \cong \mathcal{N}^{(1)}(x)$) if each is finer than the other. This implies that both sets of neighbourhoods of $x \in X$ produce the same collection of sequences that converge to x. This defines an equivalence relation on subsets of $\mathrm{P}(X)$. That is

$$\mathcal{N}(x) \cong \mathcal{N}(x) \tag{5.2.1}$$

$$\mathcal{N}^{(1)}(x) \cong \mathcal{N}^{(2)}(x) \text{ implies } \mathcal{N}^{(2)}(x) \cong \mathcal{N}^{(1)}(x) \tag{5.2.2}$$

$$\mathcal{N}^{(1)}(x) \cong \mathcal{N}^{(2)}(x) \text{ and } \mathcal{N}^{(2)}(x) \cong \mathcal{N}^{(3)}(x) \text{ implies } \mathcal{N}^{(1)}(x) \cong \mathcal{N}^{(3)}(x) \tag{5.2.3}$$

As far as convergence is concerned, we are only interested in neighbourhoods up to equivalence. However, the situation is not totally dissimilar to that in a gauge theory and, in the present context, it is useful to find a natural "gauge choice"; that is, a set of conditions on the elements of $\mathcal{N}(x)$ that select a unique representative from the class of equivalent collections of neighbourhoods. We will achieve this as follows.

Firstly, given any family of subsets $\mathcal{N}(x)$, define $\mathcal{N}'(x)$ to be the union of $\mathcal{N}(x)$ with the collection of all finite intersections of sets belonging to $\mathcal{N}(x)$. Clearly $\mathcal{N}'(x)$ is finer than $\mathcal{N}(x)$ (since $\mathcal{N}(x) \subset \mathcal{N}'(x)$) and hence if x_n converges to x with respect to $\mathcal{N}'(x)$ it also converges with respect to $\mathcal{N}(x)$.

Conversely, if $x_n \xrightarrow{\mathcal{N}(x)} x$ then, for any finite collection $A_1, A_2 \ldots A_m \in \mathcal{N}(x)$, there exists $n_1, n_2 \ldots n_m$ such that

$$\begin{aligned} &n > n_1 \text{ implies } x_n \in A_1 \\ &n > n_2 \text{ implies } x_n \in A_2 \\ &\quad\cdot \\ &\quad\cdot \\ &n > n_m \text{ implies } x_n \in A_m. \end{aligned}$$

Thus $n > \max(n_1, n_2 \ldots n_m)$ implies that $x_n \in A_1 \cap A_2 \cap \ldots \cap A_m$ and so $x_n \xrightarrow{\mathcal{N}'(x)} x$. Hence there is no loss of generality in choosing $\mathcal{N}(x)$ to be closed under the formation of finite intersections of its members. Thus $\mathcal{N}(x)$ can be taken to be a subsemigroup of the $\wedge$-semigroup structure on the lattice $\mathrm{P}(X)$ (with the unit X if this is included in $\mathcal{N}(x)$).

It is clear that the convergence properties of the family of neighbourhoods $\mathcal{N}(x)$ is not affected if we add to $\mathcal{N}(x)$ any subset of X that is a superset of an element of this family. In any lattice $\mathcal{L}$, a subset U of $\mathcal{L}$ is said to be an *upper set* if $a \in U$ implies that $b \in U$ for all $b \in \mathcal{L}$ satisfying $a \preceq b$. If A is any subset of $\mathcal{L}$, the upper set *generated* by A is defined to be

$$\uparrow(A) := \{ b \in \mathcal{L} \mid \exists a \in A \text{ with } a \preceq b \} \tag{5.2.4}$$

In particular, if $a \in \mathcal{L}$,

$$\uparrow(a) := \uparrow(\{a\}) = \{\, b \in \mathcal{L} \mid a \preceq b \,\} \tag{5.2.5}$$

and, for the lattice $\mathrm{P}(X)$, if $\alpha \subset \mathrm{P}(X)$,

$$\uparrow(\alpha) = \{\, B \subset X \mid \exists A \in \alpha \text{ with } A \subset B \,\}. \tag{5.2.6}$$

Note that

a) In $\mathrm{P}(X)$, $\alpha \cong \uparrow(\alpha)$

b) If $\alpha \vdash \beta$, and β is upper, then $\alpha \subset \beta$. Therefore, if α and β are both upper it follows that $\alpha \cong \beta$ if and only if $\alpha = \beta$. Thus each equivalence class contains precisely one upper family and hence there is no loss in generality in requiring the families $\mathcal{N}(x)$ to be

(i) closed under finite intersections;

(ii) upper families.

These two conditions constitute our "gauge choice" for the elements in $\mathcal{N}(x)$ and it is interesting to note that they are characteristic of an important type of algebraic object in a general lattice:

Definitions

1. An *ideal* in a lattice $\mathcal{L}$ is a subset $I \subset \mathcal{L}$ such that

   (i) $a, b \in I$ implies $a \vee b \in I$

   (ii) $a \in I$ and $b \preceq a$ implies $b \in I$.

2. A *dual ideal* in a lattice $\mathcal{L}$ is a subset $D \subset \mathcal{L}$ such that

   (i) $a, b \in D$ implies $a \wedge b \in D$

   (ii) $a \in D$ and $a \preceq b$ implies $b \in D$.

3. An ideal I is *proper* if $1 \notin I$, that is, $I \neq \mathcal{L}$.

   A dual ideal D is *proper* if $0 \notin D$, that is, $D \neq \mathcal{L}$.

Comments

1. $\uparrow(a) := \{b \in \mathcal{L} \mid a \preceq b\}$ is a dual ideal (exercise!) called the *principal* dual ideal generated by a.

$\downarrow(a) := \{b \in \mathcal{L} \mid b \preceq a\}$ is an ideal (exercise!) called the *principal* ideal generated by a.

2. If I_1 and I_2 are ideals then so is $I_1 \cap I_2$. Then if S is any subset of $\mathcal{L}$, $\mathcal{I}_S := \cap\{I \mid S \subset I\}$ is an ideal called the ideal *generated* by S. There is a similar construction for dual ideals.

Now we apply these concepts to the lattice $\mathrm{P}(X)$. In particular, a *filter* $\mathcal{F}$ on X is defined to be a proper dual ideal in the lattice $\mathrm{P}(X)$. Thus $\mathcal{F}$ is a family of subsets of X such that

(i) $\emptyset \notin \mathcal{F}$

(ii) $\mathcal{F}$ is closed under finite intersections

(iii) $\mathcal{F}$ is an upper family.

Note that a *principal* filter is therefore the set of all supersets of some $A \subset X$.

It follows from the above that our final form of the idea of convergence can be cast in terms of filters on X:

Definition

1. A *neighbourhood structure* $\mathcal{N}$ on a set X is an assignment to each $x \in X$ of a filter $\mathcal{N}(x)$ on X all of whose elements contain the point x. The pair $(X, \mathcal{N})$ (or simply X if $\mathcal{N}$ is understood) is called a *neighbourhood space* (Császár, 1978).

2. A sequence $\{x_n\}$ *converges* to x with respect to $\mathcal{N}$ if

$$\forall N \in \mathcal{N}(x),\ \exists n_0 \text{ such that } n > n_0 \text{ implies } x_n \in N. \tag{5.2.7}$$

This is about the most general notion of convergence of sequences that one could conceive and forms the foundation for a variety of special structures in which the filters $\mathcal{N}(x)$ are restricted in some way. As we shall see shortly, a 'topology' is one such example.

3. A *filter base* $\mathcal{B}$ is a family of non-empty subsets of X such that if $A, B \in \mathcal{B}$ then there exists $C \in \mathcal{B}$ such that $C \subset A \cap B$.

It is easy to show that

1. Every filter is a filter base.

2. If $\mathcal{B}$ is a filter base then $\uparrow(\mathcal{B}) = \{B \mid \exists A \in \mathcal{B} \text{ such that } A \subset B\}$ is a filter equal to the filter generated by $\mathcal{B}$. $\mathcal{B}$ is said to be a *base* for this filter. In practice, it is often very convenient to deal with filter bases rather than the (frequently much larger!) filters which they generate. Nothing is lost in so doing since a sequence converges with respect to a filter if and only if it converges with respect to any associated filter base (convergence with respect to a filter base is defined in the obvious way).

3. If $\mathcal{B}$ is a filter base and if $\alpha \subset P(X)$ is such that $\alpha \cong \mathcal{B}$, then α is also a filter base.

4. A non-empty collection $\mathcal{B}$ of subsets of X is a base for a specific filter $\mathcal{F}$ on X if and only if

(i) $\mathcal{B} \subset \mathcal{F}$

(ii) If $A \in \mathcal{F}$, there exists $B \in \mathcal{B}$ such that $B \subset A$

5. The collection T of tails of a sequence $\{x_n\}$ form a filter base and the condition for convergence can be rewritten as x_n converges to x with respect to $\mathcal{N}$ if and only if T is finer than $\mathcal{N}(x)$. The significance of this version is that it admits a very important generalization to an arbitrary filter base $\mathcal{B}$:

Definition

A filter base $\mathcal{B}$ ***converges*** to $x \in X$ if $\mathcal{B}$ is finer than $\mathcal{N}(x)$.

Note that this is true if and only if $\mathcal{B}$ is finer than any filter base $\mathcal{B}(x)$ for $\mathcal{N}(x)$. This is frequently useful in practice. Note also that collections of neighbourhoods $\mathcal{N}^{(1)}$ and $\mathcal{N}^{(2)}$ that are equivalent admit the same set of convergent filters.

Examples

1. In a metric space, the set of balls $B_\epsilon(x)$, $\epsilon > 0$, form a filter base $\mathcal{B}(x)$ for the filter $\mathcal{N}(x)$ of all neighbourhoods of x.

2. In $\mathbb{C} \cup \{\infty\}$, the sets $N_\epsilon(\infty)$, $\epsilon > 0$, form a filter base for the filter of all neighbourhoods of ∞.

3. On the space of functions $\mathcal{F}([a,b],\mathbb{R})$, we defined a collection of neighbourhoods of f as $N_{t,\epsilon}(f) := \{g \mid |g(t) - f(t)| < \epsilon\}$. These form what is known as a filter ***subbase***. That is, the set of all finite intersections of sets of this type forms a filter base.

Many of the topological concepts introduced in the context of metric spaces possess precise analogues in a neighbourhood space. For example, an ***interior point*** of a set $A \subset X$ is any point x such that there exists $N \in \mathcal{N}(x)$ with $N \subset A$. The definitions of ***exterior*** point, ***boundary*** point and ***limit*** point generalize in a similar way and the definitions of ***open*** and ***closed*** sets are as before. It is important to note that none of these concepts change if the filter $\mathcal{N}(x)$ is replaced by any filter base equivalent to it.

As in the case of metric spaces, A is open if and only if A^c is closed and if and only if $A = Int(A)$. Similarly, (i) $\emptyset$ and X are open and closed; (ii) the collection of all open sets is closed under finite intersections and arbitrary unions; (iii) the collection of all closed sets is closed under arbitrary intersections and finite unions. It is also the case that A is closed if and only if it contains all its limit points.

§5.3 Topological Spaces

The concept of a neighbourhood space is a considerable generalization of that of a metric space and, as we have seen, it allows meaningful, set-theoretic based, ideas of 'nearness' and 'convergence'. However, the collective experience of the mathematical community is that it needs to be supplemented with an additional requirement in order to yield a really useful tool. The problem with a general neighbourhood space is the absence of any *a priori* relation between the filters $\mathcal{N}(x)$ at different points x in X. The crucial extra requirement which has emerged over the years is that any neighbourhood of a point x should also be a neighbourhood of all points "sufficiently near" to x. It is fascinating to ponder whether or not this is also relevant for the mathematical model which is to describe *physical* spacetime. This is certainly what is assumed in, for example, classical general relativity but it is by no means obvious that it should also hold sway in the quantum realm. In any event, the precise mathematical definition is as follows:

Definition

A *topological space* is a neighbourhood space $(X, \mathcal{N})$ in which, for all $x \in X$, it is true that, for all $N \in \mathcal{N}(x)$ there exists $N_1 \in \mathcal{N}(x)$ such that, for all $y \in N_1$, $N \in \mathcal{N}(y)$.

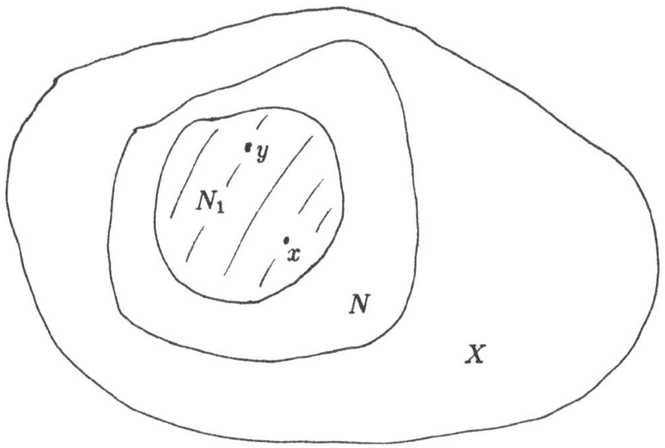

The following theorem (exercise!) is of considerable importance:

Theorem

A neighbourhood space $(X, \mathcal{N})$ is a topological space if and only if each filter $\mathcal{N}(x)$ has a filter base consisting of open sets.

Thus, in a topological space, the neighbourhoods are essentially determined by the open sets alone; this would not be true in a more general neighbourhood space.

Examples

1. Every metric or pseudo-metric space is a topological space since the balls $B_\epsilon(x) := \{y \in X \mid d(x,y) < \epsilon\}$ are open (exercise!) and also form a basis for the neighbourhoods of x. These spaces have many important properties. For example, every metric space is ***first countable***, that is, there exists a countable basis for the neighbourhoods of each point (simply choose the set of all balls with rational radii). Note however that finite metric spaces are rather uninteresting: any such space automatically possesses the ***discrete*** topology (see below) defined as the topology in which, for each $x \in X$, $\mathcal{N}(x)$ is the set of all subsets of X containing the point x (exercise!).

2. The neighbourhoods of a point $x \in \mathbb{C} \cup \{\infty\}$ are generated by the usual open discs if $x \in \mathbb{C} \subset \mathbb{C} \cup \{\infty\}$ and by the sets $N_\epsilon(\infty)$ if $x = \infty$. These sets are open, and hence $\mathbb{C} \cup \{\infty\}$ is a topological space.

3. In the neighbourhood structure associated with pointwise convergence on the space of functions $\mathcal{F}([a,b],\mathbb{R})$, the sets $N_{t,\epsilon}(f)$ are open (exercise!) and therefore so are their finite intersections. But these form a filter base for $\mathcal{N}(f)$ and hence this function space is a topological space.

4. The number of topologies that can be placed on a set X is much smaller than the number of neighbourhood structures. For example, on a finite set X with $|X| = n$ there are 2^{n^2-n} different neighbourhood structures (exercise!) but the number of topologies is less. Thus on $X = \{a,b,c\}$ there are 64 such structures whereas (see later) there are only 29 topologies. An example of a neighbourhood structure that is not a topology is $\mathcal{N}(a) = \{\{a,b\},X\}$, $\mathcal{N}(b) = \{\{b,c\},X\}$ and $\mathcal{N}(c) = \{\{a,c\},X\}$; the only open sets are $\emptyset$ and X, so they cannot form a basis for the neighbourhoods.

We have seen that the collection of all open sets in a neighbourhood space satisfies the three conditions:

$\emptyset$ and X are open

An arbitrary union of open sets is open

Any finite intersections of open sets is open

and one of the central properties of a topological space (closely related to the theorem mentioned above) is that the *converse* is true:

Theorem

Let τ be any family of subsets of a set X satisfying the three conditions:

(τ1) $\emptyset$ and X belong to τ

(τ2) An arbitrary union of elements of τ belongs to τ

(τ3) Any finite intersection of elements of τ belongs to τ

Then τ is the family of open sets of a topology on X with a neighbourhood base $\mathcal{B}(x) := \{ O \in \tau \mid x \in O \}$ for all $x \in X$.

Comments

1. Let $\mathcal{N}(X)$ and $\tau(X)$ denote respectively the set of all neighbourhood structures on X and the set of all topologies on X. Then $\tau(X) \subset \mathcal{N}(X)$; that is, there is an injection $i : \tau(X) \to \mathcal{N}(X)$. Now the open sets of an *arbitrary* neighbourhood structure $\mathcal{N}$ on X obey the conditions in the theorem above and therefore generate a topological space associated with $\mathcal{N}$. This defines a map $k : \mathcal{N}(X) \to \tau(X)$ and, in the diagram

$$\tau(X) \xrightarrow{i} \mathcal{N}(X) \xrightarrow{k} \tau(X), \tag{5.3.1}$$

it can be shown that, for any topology τ, $k \circ i(\tau) = \tau$. Thus the axioms $(\tau 1) - (\tau 3)$ constitute a complete, alternative, way of defining what is meant by a 'topology' on a set X. In fact, many introductions to general topology start at this point by defining a topology on a space X to be a collection τ of subsets of X which satisfies the conditions $(\tau 1) - (\tau 3)$. For this reason, we will denote a topological space by (X, τ) rather than $(X, \mathcal{N})$.

2. The set $\tau(X)$ of all topologies on a set X is a partially ordered set with $\tau_1 \preceq \tau_2$ defined to mean that every τ_1-open set is automatically τ_2-open (that is, τ_2 has "more" open sets than τ_1). We say that τ_1 is *weaker* or *coarser* than τ_2 and that τ_2 is *stronger* or *finer* than τ_1. The strongest topology is $P(X)$ (that is, every subset of X is open) and is called the *discrete* topology. The weakest topology (the *indiscrete* topology) is just $\{\emptyset, X\}$. This notation is compatible with the use in §2.1 of the words "stronger" and "weaker" in relation to metrics. We will make considerable use of this ordering when discussing the quantization of general topologies.

3. As in the case of neighbourhood structures, it is convenient to introduce the notion of a base, or subbase, for a topology. Thus a collection $\mathcal{B}$ of subsets of X is said to be a *base* for a topology τ if every τ-open set can be written as a union of members of $\mathcal{B}$. A collection is said to be a *subbase* if the set of all finite intersections of elements of the collection forms a base for the topology.

4. In practice, topologies are almost always defined in terms of their open sets or of a base or subbase for the open sets. For example, an interesting topology on any infinite set X is the *cofinite* topology defined to be the one whose open sets are $\emptyset$ and the complements of all finite subsets of X.

5. Partially ordered sets possess several natural topologies related to their ordering structure. One of the simplest is the collection of all upper sets which is clearly a topology since it is closed under arbitrary unions and intersections. The collection of all lower sets yields another example.

6. An equivalent way of defining a topological structure on a set X is as a collection $\mathcal{C}$ of subsets of X that (i) include $\emptyset$ and X; (ii) are closed under arbitrary

intersections; and (iii) are closed under finite unions. The complements of the elements in this family then form the collection of open sets for a unique topology on X in which the original collection C is the family of closed sets. Thus a topology is also determined by its collection of closed sets, that is, sets which contain all their limit points. In this context it is important to note that if x_n is a sequence in $A \subset X$ that converges to x then x is a limit point of A. However, unlike the situation for metric spaces, the converse may not be true. That is, a subset A of a general topological space may have a limit point to which no sequence of elements in A converges. But what *is* true is that there will always be a *filter* on A that converges to the limit point. Thus a general topological structure is determined by its collection of convergent filters. A necessary and sufficient condition for a topology to be determined by the set of convergent sequences alone is that it be first countable.

It is worth remarking that generalised convergence can also be discussed using what are called *nets*. First we must define a *directed set*. This is any partially ordered set D with the additional property that if $\alpha, \beta \in D$ then there exists $\gamma \in D$ such that $\alpha \preceq \gamma$ and $\beta \preceq \gamma$. A *net* on a set X is any function $f : D \to X$ for some directed set D and we say that the net *converges* to a point $x \in X$ with respect to a neighbourhood structure $\mathcal{N}$ on X if

$$\forall N \in \mathcal{N}(x),\ \exists \alpha \in D \text{ such that } \gamma \succeq \alpha \text{ implies } f(\gamma) \in N. \tag{5.3.2}$$

The set of all positive integers is a special example of a directed set and so it is clear that (5.3.2) is a far-reaching generalization of the definition (5.2.7) of a convergent sequence. One can prove that there is a one-to-one correspondence between filters and nets, and therefore there is no loss in using the latter. Many authors prefer this (for example, Kelly 1955) because of the intuitive similarity between nets and sequences. I have elected to concentrate on filters because their definition as dual ideals in the lattice $P(X)$ is in line with my desire to emphasise the algebraic aspects of general topology.

7. A most important concept in topology is that of a "compact" space which means a space that is, in some sense, of "finite size". The classic examples of compact spaces are spheres, tori, or any other subspaces of euclidean space $\mathbb{R}^n$ that are closed and bounded (a subspace A is bounded if $\sup\{\, d(x,y) \mid x, y \in X\,\} < \infty$). One characteristic feature of such a set is that any infinite subset of points must necessarily cluster together in some way. More precisely, it can be shown that every sequence $\{x_n\}$ in a closed and bounded subset of $\mathbb{R}^n$ necessarily has at least one *accumulation* point. That is, a point x such that any neighbourhood of x is visited infinitely many times by the sequence:

$$\forall N \in \mathcal{N}(x),\ \forall n,\ \exists n' > n \text{ such that } x_{n'} \in N \tag{5.3.3}$$

or, in terms of the tails T_n of the sequence,

$$\forall N \in \mathcal{N}(x),\ \forall T_n,\ N \cap T_n \neq \emptyset. \tag{5.3.4}$$

One might try and reverse this result and *define* a general compact space to be any topological space in which (5.3.4) is true. However, it turns out that this is too broad and the most useful definition is to strengthen (5.3.4) by including *all* filter bases, not just the tails of sequences. More precisely, a topological space X is said to be *compact* if every filter base $\mathcal{B}$ on X has an accumulation point. That is, there exists $x \in X$ such that

$$\forall N \in \mathcal{N}(x),\ \forall A \in \mathcal{B},\ N \cap A \neq \emptyset \tag{5.3.5}$$

§5.4 Morphisms in Topology

The concept of a 'morphism' appears in many branches of mathematics as a structure-preserving map between two sets equipped with the same type of mathematical structure. The first relevant question in the present context is whether a map $f : X \to Y$ between a pair of sets X and Y induces any maps between $\mathrm{P}(X)$ and $\mathrm{P}(Y)$ that respect the lattice structure. From a purely set-theoretic perspective, there are two natural maps, one from $\mathrm{P}(X)$ to $\mathrm{P}(Y)$ and the other from $\mathrm{P}(Y)$ to $\mathrm{P}(X)$:

1. The induced map from $\mathrm{P}(X)$ to $\mathrm{P}(Y)$ is defined on a subset $A \subset X$ by

$$f(A) := \{\, f(x) \in Y \mid x \in A \,\} \subset Y \tag{5.4.1}$$

Then

$$\text{(i) } f(A \cup B) = f(A) \cup f(B) \tag{5.4.2}$$

and

$$\text{(ii) } f(A \cap B) \subset f(A) \cap f(B). \tag{5.4.3}$$

Note that the equality may not hold in (5.4.3) and hence the induced map from $\mathrm{P}(X)$ to $\mathrm{P}(Y)$ does *not* preserve the lattice structure.

2. The second induced map is from $\mathrm{P}(Y)$ to $\mathrm{P}(X)$ and is defined on $A \subset Y$ by the inverse set map

$$f^{-1}(A) := \{\, x \in X \mid f(x) \in A \,\} \tag{5.4.4}$$

which, it should be noted, is well-defined even if $f : X \to Y$ is not one-to-one. This induced map satisfies

$$\text{(i) } f^{-1}(A \cup B) = f^{-1}(A) \cup f^{-1}(B) \tag{5.4.5}$$

$$\text{(ii) } f^{-1}(A \cap B) = f^{-1}(A) \cap f^{-1}(B) \tag{5.4.6}$$

with generalizations to arbitrary families. Hence it *does* preserve the lattice operations.

This result motivates (and renders consistent) the definition of a continuous map in the following way.

Definition

A map $f : (X, \tau) \to (Y, \tau')$ is said to be *continuous* if, for all $O \in \tau'$, $f^{-1}(O) \in \tau$.

Comments

1. It follows from (5.4.5-6) that a continuous map induces a homomorphism from the lattice τ' into the lattice τ. It is in this sense that a continuous function is a morphism in general topology when the theory is viewed from the perspective of the lattice of open sets. The significance of this will be touched on later in the context of the theory of frames and locales.

2. A more intuitive idea of continuity is that a small variation in x produces only a small variation in the value $f(x)$ of the function. In the absence of a metric, the concept of "small" must be defined in terms of the neighbourhoods of the points x and $f(x)$, and in fact it can be shown that a function $f : (X, \tau) \to (Y, \tau')$ is continuous if and only if

$$\forall x \in X,\ \forall M \in \mathcal{N}(f(x)), \exists N \in \mathcal{N}(x) \text{ such that } f(N) \subset M \tag{5.4.7}$$

or, equivalently, that $\mathcal{N}(f(x)) \vdash f(\mathcal{N}(x))$. Note that in the case where $X = Y = \mathbb{R}$, this reduces to the familiar definition

$$\forall x \in \mathbb{R},\ \forall \epsilon > 0,\ \exists \delta > 0 \text{ such that } |x - y| < \delta \text{ implies } |f(x) - f(y)| < \epsilon. \tag{5.4.8}$$

3. The "small" variation is often phrased in terms of sequences, and indeed it is true that if a function f is continuous and if $x_n \to x$ then $f(x_n) \to f(x)$. If X is a metric space, the converse also holds. That is, if $f : X \to Y$ is such that, for all points $x \in X$ and for any convergent sequence $x_n \to x$, $f(x_n)$ converges to $f(x)$, then f is continuous. For more general spaces this is false, but what *is* true is that a function $f : (X, \tau) \to (Y, \tau')$ is continuous if and only if for all $x \in X$ and for any filter base $\mathcal{B}$ on X which converges to x it is true that $f(\mathcal{B})$ converges to $f(x)$. (Exercise: show that if $\mathcal{B}$ is a filter base on X and f is any map from X to Y, then $f(\mathcal{B}) := \{ f(A) \mid A \in \mathcal{B} \}$ is a filter base on Y).

4. One of the important practical problems in the theory of topology is to find ways of actually constructing topologies on a given set. Two of the most useful techniques involve placing a topology on a set X with the aid of maps to or from X and some other topological space Y. For example, if τ is a topology on Y and f is a map from X to Y, the *induced* topology on X is defined to be

$$f^{-1}(\tau) := \{f^{-1}(O) \mid O \in \tau\}. \tag{5.4.9}$$

This is the coarsest topology on X such that f is continuous. A special case is when X is a subset of Y with an injection $i : X \hookrightarrow Y$. The induced topology on X is then called the *subspace* topology and consists of all sets of the form $X \cap O$ where O is open in the topology τ on Y. Note that the results (5.4.5-6) are crucial for this construction.

Another important example arises when (Y, τ) is a topological space and there is a surjective map $p : Y \to X$. The *identification* topology on X is defined as

$$p(\tau) := \{A \subset X \mid p^{-1}(A) \in \tau\}. \qquad (5.4.10)$$

This is the finest topology on X such that p is continuous. Topologies of this type frequently occur when some equivalence relation R is defined on Y with X being the space Y/R of equivalence classes and the map p being the canonical map of an element of Y onto its equivalence class. In fact, this example is universal since if p is *any* surjective map from a space Y onto a set X we can define an equivalence relation on Y by saying that two points y_1 and y_2 are equivalent if $p(y_1) = p(y_2)$. It is then easy to see that a bijection can be established between X and Y/R by mapping the point $p(y) \in X$ to the equivalence class of y in Y. Note that the points in Y which are mapped into the same point in Y/R are precisely those that are equivalent to each other. Hence one says that Y/R is obtained by "identifying equivalent points", which explains the origin of the name 'identification topology'.

Another crucial question is when two topological spaces can be regarded as being essentially equivalent. Thus we are looking for the appropriate meaning of an "isomorphism" in the topological case. Generally speaking, an isomorphism between two structures of the same type involves a bijective map between the underlying sets with the property that both it and its inverse are morphisms. In the context of topology, working from the idea of a continuous map as a morphism, we have

Definition

A map $f : (X, \tau) \to (Y, \tau')$ is a *homeomorphism* (an isomorphism in the context of general topology) if

(i) f is a bijection

(ii) f and f^{-1} are continuous.

We shall write this as $(X, \tau) \simeq (Y, \tau')$. Note that the symbol f^{-1} refers here to the map from Y to X that is the actual inverse of the map f from X to Y. It should not be confused with the inverse *set* map defined in (5.4.4). When f is invertible the two maps are related by $\{f^{-1}(y)\} = f^{-1}\{y\}$ for all $y \in Y$.

Comments

1. f is a homeomorphism if and only if (i) for all $O \in \tau$, $f(O)$ is τ'-open; and (ii) for all $O \in \tau'$, $f^{-1}(O)$ is τ-open. Thus f induces a bijective map between the two collections of open sets which preserves the algebraic operations of forming unions and

intersections; that is, an isomorphism of the lattice structures associated with the two topologies.

2. The set Perm(X) of all bijections ("permutations") of X onto itself is a group. If τ is a topology on X and $\phi \in \text{Perm}(X)$, $\phi(\tau)$ is defined to be the topology whose open sets are $\{\, \phi(O) \mid O \in \tau \,\}$. By construction, $\tau \simeq \phi(\tau)$ (that is, they are homeomorphic) and conversely, if τ_1 and τ_2 are a pair of topologies on the same set X that are homeomorphic, then there exists $\phi \in \text{Perm}(X)$ such that $\tau_2 = \phi(\tau_1)$. Thus the set $\tau(X)$ of all topologies on X decomposes under the action of Perm(X) as a disjoint union of orbits which are the homeomorphism classes of topology. This will be an important ingredient in our later discussions of the quantum theory of general topologies.

§5.5 Separation Axioms

An important question in any topological space X is the extent to which points can be separated or distinguished from each other by listing the collection of open sets to which each belongs. From the viewpoint of conventional physics it is important to note that if X is the manifold of three-space then any real "object" needs an open subset in which to exist. More precisely, it cannot exist as a subset of a closed subset unless this has a non-trivial interior. [1] In the context of quantum field theory this is related to the Bohr and Rosenfeld analysis (1933,1950) of the need to smear quantum fields with test functions which are non-vanishing on an open set. It thus seems plausible to argue that it is meaningless to distinguish physically between two points in X if the collections of open sets to which they belong are identical. The relevant mathematical definitions for handling this type of consideration are as follows.

Definition

1. A topological space X is T_0 if, given any pair of points $x, y \in X$, at least one of them is contained in an open set that excludes the other. This is equivalent to saying that, for all $x, y \in X$, $\mathcal{N}(x) \neq \mathcal{N}(y)$.

2. The space is T_1 if, given any pair of points x, y each one is contained in an open set that excludes the other.

3. The space is T_2 or *Hausdorff* if for any pair of points $x, y \in X$ there exist open sets O_1 and O_2 such that $x \in O_1$, $y \in O_2$ and $O_1 \cap O_2 = \emptyset$.

Comments

1. The *closure* $\overline{A}$ of any subset A of a topological space X is defined to be the

[1] One could say that all open sets are "fat" whereas closed sets come in both thin and fat varieties. For example, a segment of a line in the plane is thin whereas a closed disc is fat.

smallest closed set containing A (it can be constructed as the intersection of all closed sets containing A). It is easy to see that $\mathcal{N}(x) = \mathcal{N}(y)$ if and only if $\overline{\{x\}} = \overline{\{y\}}$.

2. If X is T_0, a partial ordering can be defined by

$$x \preceq y \iff \overline{\{x\}} \subset \overline{\{y\}} \tag{5.5.1}$$

3. From the remarks made above it could be asserted that a topological space must be at least T_0 if all its points are to have "physical meaning" in the sense of being distinguishable by objects located in open sets. It is important therefore to note that to *any* topological space X there is a canonically associated T_0 space. This is constructed by defining the equivalence relation R on X

$$xRy \iff \mathcal{N}(x) = \mathcal{N}(y) \tag{5.5.2}$$

and then equipping X/R with the identification topology of (5.4.10). The resulting space is T_0 and can be regarded as what is obtained from the original space X once points that cannot be "physically separated" have been identified. It is interesting to note that if this procedure is applied to a space with a pseudo-metric ρ, the resulting space is in fact T_2 with a metric topology induced by the distance function $d([x], [y]) := \rho(x, y)$ where $[x]$ denotes the equivalence class of the point x. This is clearly relevant to our discussion in §3 of possible quantum topology effects arising from a quantized distance function whose expectation value in certain states becomes a pseudo-metric.

4. A space is T_1 if and only if, for every point $x \in X$, the subset $\{x\}$ of X is closed. Note that there exist topological spaces that are T_0 but not T_1. An example on the set $X = \{a, b\}$ is the topology $\{\emptyset, X, \{a\}, \{a, b\}\}$. Indeed, it is easy to see that the only topology on a finite set that is T_1 is the discrete topology $P(X)$.

5. An example of a topology that is T_1 but not T_2 is afforded by the cofinite topology on any infinite set X. Note also that every metric space is Hausdorff but the only Hausdorff topology on a finite set is the discrete topology.

6. A subspace of a T_i topology, $i = 0, 1, 2$ is also a T_i topology.

7. There exist more refined notions of separation which involve, for example, the extension of the Hausdorff axiom to include the ability to distinguish between *arbitrary* closed sets (not merely single points) with the aid of non-intersecting open sets which contain them. However, the ideas introduced above will suffice for our present purposes.

8. The question of the *uniqueness* of the limits of sequences (or, more generally, filters) in a topological space has a precise answer in terms of the separation properties of the space. Specifically, it can be shown that a necessary and sufficient condition for a topological space X to be Hausdorff is that every filter on X converges to at most one point in X.

§5.6 Frames and Locales

Because our ultimate interest in topology lies in quantization, it is useful to discuss briefly certain algebraic structures which are associated with any topological space and from which the topology can be largely reconstructed. One well known example is that the topology of a compact Hausdorff space X is uniquely specified by the ring structure of its set of real-valued continuous functions. However, of more relevance to our present investigation is the fact that the collection of subsets of X which constitutes a topology forms a *sublattice* of $P(X)$. The interesting question is then the extent to which this lattice structure determines the topology. In particular, given the lattice, can we reconstruct the topological space (up to homeomorphisms)? From the discussion in §5.5 it seems unlikely that the topology can be reproduced completely if X contains points which cannot be separated by specifying the open sets to which they belong; that is, if the topology is not T_0. For example, the topologies $\tau_1 := \{\emptyset, X, \{a\}\}$ and $\tau_2 := \{\emptyset, X, \{b, c\}\}$ on the set $X = \{a, b, c\}$ have isomorphic lattices of open sets but τ_1 is not homeomorphic to τ_2.

The basic step is to try to reconstruct the points of X from the lattice associated with a topology τ on X. Since the only question we can ask of a point is whether or not it belongs to any particular open set, we are lead to consider the collection of mappings $h_x : \tau \to \{0, 1\}$ defined on open sets O by

$$h_x(O) = \begin{cases} 1, & \text{if } x \in O; \\ 0, & \text{otherwise.} \end{cases} \tag{5.6.1}$$

We see at once that each h_x is a homomorphism from the lattice τ onto the lattice $\{0, 1\}$ of two points. This inspires an attempt to define a "generalized point" associated with the lattice as *any* homomorphism from the lattice into $\{0, 1\}$. A topology can be constructed on the set $\mathrm{pt}(\tau)$ of all such homomorphisms by defining the open sets to be all subsets of the form $\{h \in \mathrm{pt}(\tau) \mid h(O) = 1\}$ where O in any τ-open subset of X. The natural map $h : X \to \mathrm{pt}(\tau)$, defined by $x \mapsto h_x$, is clearly continuous with respect to this topology. A number of important statements can be made concerning this construction (Johnstone, 1986):

1. The topology on $\mathrm{pt}(\tau)$ is T_0.

2. x and y determine the same homomorphism if and only if $\mathcal{N}(x) = \mathcal{N}(y)$. Thus the map $h : X \to \mathrm{pt}(\tau)$, is one-to-one if and only if the topology τ on X is T_0. If a relation R is defined on X as in (5.5.2) it is clear that the map $[x] \mapsto h_x$ is a continuous injection of X/R into $\mathrm{pt}(\tau)$.

3. The map $x \mapsto h_x$ may not be surjective; that is, there may exist homomorphisms that are *not* of the form h_x for any $x \in X$. Spaces for which this map is both one-to-one and onto (it is then necessarily a homeomorphism) are called *sober*—they are the spaces whose topology is completely captured by the lattice structure of their

open sets. For example, all Hausdorff spaces are sober; the cofinite topology on an infinite set X is not.

At this point it is important to observe that there is no reason why the constructions above cannot be applied to lattices that are not *a priori* lattices of open sets in any topology! To see what type of lattice is appropriate for such a treatment we note that the lattice of open sets associated with a topology is closed under arbitrary unions (the join operation) and, if the meet of an arbitrary family of open sets is defined to be the interior of their intersection, it becomes a complete sublattice of $P(X)$. Furthermore, it obeys the infinite distributive law

$$A \wedge \bigvee S = \bigvee \{ A \wedge B \mid B \in S \} \tag{5.6.2}$$

where S is any collection of open sets. It is this collection of properties which is "axiomatised" to construct a purely algebraic definition of a topology-like structure. More precisely, a *frame* or *locale* [1] is defined to be any complete lattice which satisfies the infinite distributive law. [2] Many of the ideas in topology generalise to this situation and this has given rise to the interesting subject of "pointless" topology (Johnstone, 1982, 1983). In particular, the "points" associated with any frame ℓ are defined to be the homomorphisms from ℓ into $\{0, 1\}$ and are given the topology in which the open sets are all subsets of $\mathrm{pt}(\ell)$ of the form $\{h \in \mathrm{pt}(\ell) \mid h(a) = 1\}$ for some $a \in \ell$.

If locales/frames are to replace pointset topology there has to be some analogue of the important idea of a continuous map $f : (X, \tau_1) \rightarrow (Y, \tau_2)$ from one topological space to another. The key step here is the result mentioned in §5.4 that such a map induces a homomorphism from the lattice τ_2 into the lattice τ_1. Thus homomorphisms between locales replace the idea of continuous maps. In this context it is relevant to note that

1. If f is a one-to-one map, the associated lattice map is surjective; the converse holds if τ_1 is a T_0 topology.

2. If f is surjective, the associated lattice map is injective.

It is clear from these results how one might set about constructing the appropriate generalizations to frames/locales of the ideas of subspace and quotient space.

This algebraic generalization of topology is rather fascinating and it is attractive to speculate that structures of this type might one day form an important ingredient in a proper understanding of the quantum theory of space and time. I will return to this possibility in the concluding section of these notes.

---

[1] There is a technical difference between these two concepts which comes into play only when morphisms are being considered. See the literature for more details.

[2] See Vickers, 1989, for an interesting discussion of how such lattices arise naturally in computer science.

§6 QUANTUM TOPOLOGY ON $\tau(X)$

§6.1 The Main Problem

Our aim is to construct a quantum theory for a system whose "configuration space" is the class of all topologies. The family of *all* topologies would consist of all such collections τ on all possible sets X. However, this family is so big that it is a "class" (in the technical sense), not a set. We will avoid this logical problem by restricting our attention to the family $\tau(X)$ of all topologies on a *fixed* set X, which is a proper set. Note that if we do not wish to distinguish between homeomorphic topologies a more appropriate configuration space is the quotient space $\tau(X)/\mathrm{Perm}(X)$ of homeomorphism classes of topology. We will return to this possibility later.

The problem with which we are faced is non-trivial. The set of neighbourhood structures on a set X is equal to the set of all functions from X into the space of filters—a "non-linear σ-model" type situation which, conceivably, might even be quantized as such. However, the requirements on a neighbourhood structure to be a topology are complex and it is not at all clear how to proceed once this condition has been imposed. The main problem is the absence of any unique *a priori* way of quantizing a system with a given configuration space Q, not even if it is a differentiable manifold (which $\tau(X)$ certainly is not). It is tempting perhaps to postulate that quantum states can be represented by complex-valued functions on Q (or cross-sections of a vector bundle over Q) whose Hilbert space structure is defined by integrating with respect to some measure μ on Q. However, one implication of such a scheme is that any suitably bounded continuous function $f: Q \to \mathbb{C}$ can be made into a self-adjoint operator by defining

$$(\hat{f}\psi)(q) := f(q)\psi(q) \tag{6.1.1}$$

and even when Q is simply a vector space E we know that this does not always work. For example, in quantum field theory it is usually impossible to define an operator version of $(\phi(x))^2$: we must instead restrict our attention to polynomials (or suitable limits of polynomials) in the smeared fields $\phi(f)$ where the test-function f belongs to E.

This is related to a very important result which can be derived from the spectral theory of the abelian algebra generated by these smeared fields. This shows that the Hilbert space on which the fields are defined is always isomorphic to one of the form $L^2(E', d\mu)$ for some probability measure μ on the topological dual E' of E, with the operator fields $\hat{\phi}(f)$ acting as

$$(\hat{\phi}(f)\psi)(\chi) = <\chi, f> \psi(\chi) \tag{6.1.2}$$

Thus the domain space of the quantum states is not the configuration space E itself but rather the ***distributional dual*** E' of E. This is the "canonical" version of the functional-integral problem alluded to in §1.

As yet we have said nothing about how to construct a quantum theory on $\tau(X)$. However, the analogue of conventional quantum field theory suggests that the following questions are of considerable interest:

1. Is there a preferred minimal *abelian algebra* of functions on $\tau(X)$ which can play a rôle analogous to that of the smeared fields $\phi(f)$ in quantum field theory?

2. Is it necessary to use a space $\tau(X)'$ of *distributional* topologies? If so:

   What is a "distributional" version of a topology?

   Is there an appropriate topology on $\tau(X)'$?

3. Is there a *spectral theorem* for the minimal algebra which justifies choosing the Hilbert space to be $L^2(\tau(X), d\mu)$ [or $L^2(\tau(X)', d\mu)$]?

4. What are the variables *conjugate* to the functions on $Q = \tau(X)$ selected in 1.? When Q is a manifold, the conjugate variables are associated with generators of diffeomorphisms of Q and are essentially vector fields which act on wave functions on Q as partial differential operators. However, this does not provide much guidance in handling a space like $\tau(X)$ which is not a manifold; we certainly do not expect to be able to write literally a "momentum" operator as $-i\partial\psi(\tau)/\partial\tau$!

It is clear that at the heart of the quantization programme lies the problem of finding a suitable preferred algebra of observables on $Q = \tau(X)$. The q-variables normally arise as generators of transformations along the p-directions on the classical phase space T^*Q while the p-variables are associated with transformations of Q. In our case, since Q is not a manifold, there is no cotangent bundle. However, it is still reasonable to ask if there are any "natural" transformations of $\tau(X)$ which might provide a foundation for a quantization scheme.

One possibility which suggests itself is that, given a particular topology τ, one might try to increase or decrease the number of open sets it contains. For example, if a subset A of X does not already belong to the collection of subsets which constitutes τ, can a new topology be formed by appending it to this collection? In general, the answer is "no" since the ensuing collection will not be closed under finite intersections or arbitrary unions. However, as we shall see shortly, it is possible to add the "missing sets" in a minimal way so that we *do* obtain a genuine topology and thus generate a transformation of the set $\tau(X)$. Our intention is to base the quantization of $\tau(X)$ on operations of this type plus the analogous technique for removing open sets. What is involved technically is a natural lattice structure possessed by $\tau(X)$ and the first step is to spell this out in detail.

§6.2 Lattice Structure on $\tau(X)$

The first step in placing a lattice structure on $\tau(X)$ is to recall from §5.3 that topologies can be partially ordered by the relation

$$\tau_1 \preceq \tau_2 \text{ if } \tau_1 \subset \tau_2. \tag{6.2.1}$$

The use in the quantum theory of such an ordering can be motivated in part by considering the Riemannian-geometry driven situation discussed in §3. If $\hat{d}(x,y)$ is a quantized metric we anticipate that for certain states $|\psi>$ the expectation value $<\psi, \hat{d}(x,y)\psi>$ may be a pseudo-metric. The existence of points with vanishing distance implies the loss of certain sets that were open in the original manifold topology and hence, potentially, to a topology that is coarser than the original one. In order to be finer than the original topology, the expectation value must be discontinuous (although the converse is not true—the two topologies might not be related at all).

However, a more potent motivation is the existence on $\tau(X)$ of the algebraic structure of a lattice. Lattice operations on $\tau(X)$ can be defined by

$$\begin{aligned} \tau_1 \wedge \tau_2 &:= \tau_1 \cap \tau_2 = \{A \subset X \mid A \text{ is open in both } \tau_1 \text{ and } \tau_2\} \\ \tau_1 \vee \tau_2 &:= \text{ coarsest topology containing } \{A_1 \cap A_2 \mid A_1 \in \tau_1, A_2 \in \tau_2\} \end{aligned} \tag{6.2.2}$$

and are compatible with the partial ordering in the sense of (4.2.6). The null and unit element are respectively the weakest topology $\{\emptyset, X\}$ and the strongest topology $P(X)$.

It is instructive to study a few simple examples where X is a finite set. The number of topologies that can be placed on a given set X has been calculated for the cases $|X| = 1-7$ and is equal to 1,4,29,355,6942,209527 and 9535241 respectively. In general, if $|X|$ is a finite integer n it is known that $2^n \leq |\tau(X)| \leq 2^{n(n-1)}$. When X is infinite it can be shown (Fröhlich, 1964) that the cardinality of $\tau(X)$ is two orders of infinity higher than that of X (that is, $|\tau(X)| = 2^{2^{|X|}} \equiv |P(P(X))|$). The simplest case is when X has one element, $\{a\}$ say, for which there is just the single topology $\{\emptyset, \{a\}\}$. If $X = \{a,b\}$ there are four topologies arranged in a lattice isomorphic to that in (4.2.8):

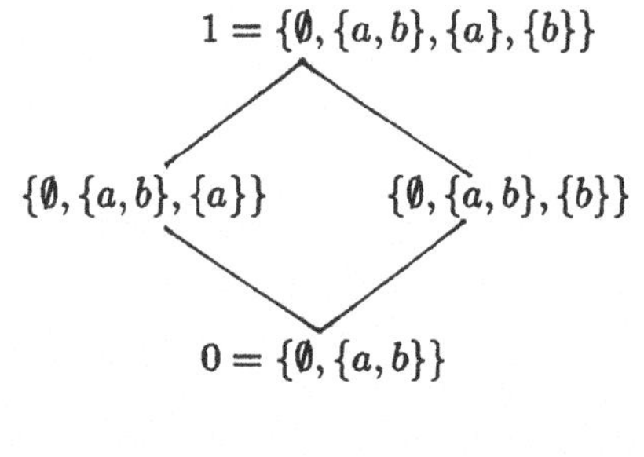

The first really interesting example is when X is a set $\{a, b, c\}$ of cardinality 3. The lattice diagram for this case is shown below using a notation which has been chosen for maximum typographical simplicity. For example, $ab(ab)(ac)$ means the topology whose open sets other than $\emptyset$ and X are the subsets $\{a\}$, $\{b\}$, $\{a, b\}$ and $\{a, c\}$.

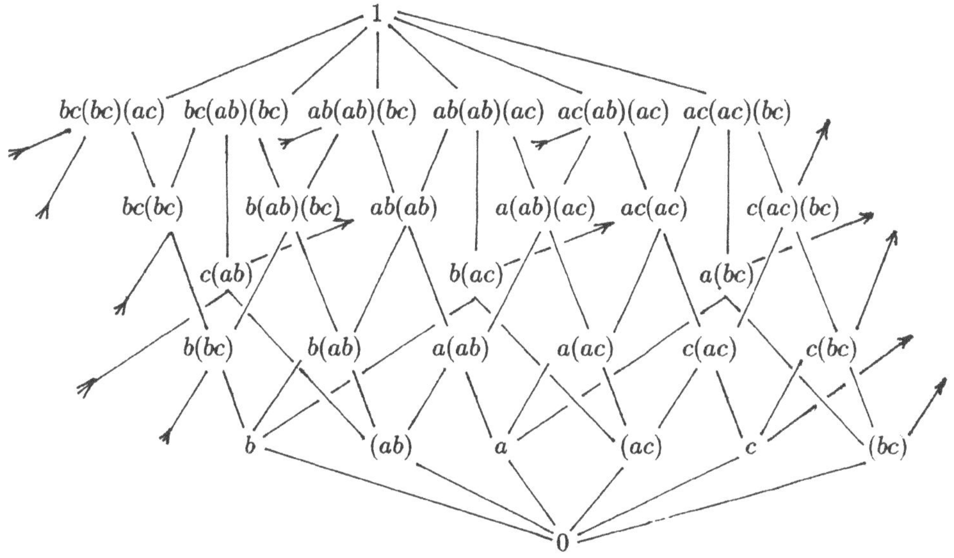

The lattice of all topologies on X possesses many interesting properties (Larson and Andima, 1975). For example, it is complete and *atomic*. Thus for each $A \subset X$, $\tau_A = \{\emptyset, X, A\}$ is an atom (that is, τ_A covers the trivial topology 0) and every topology τ is determined by these atoms in the sense that

$$\tau = \bigvee\{\tau_A \mid \tau_A \preceq \tau\} \tag{6.2.3}$$

In the example $X = \{a, b, c\}$, the atoms are the six topologies b,(ab),a,(ac),c and (bc). Note that the action referred to earlier of "adding" a subset A to a topology τ in a minimal way consists of forming the join $\tau_A \vee \tau$. However, contrary to what one might expect, $\tau_A \vee \tau$ does *not* necessarily cover τ.

The lattice $\tau(X)$ is also *anti-atomic*. That is, there exist topologies τ_A with the properties that (i) the maximal topology 1 covers τ_A, and (ii) every topology is uniquely determined by the anti-atoms that lie above it. Note that the minimal way of "weakening" a topology τ is to form $\tau_A \wedge \tau$ for some anti-atom τ_A. In the example $X = \{a, b, c\}$, the anti-atoms are the topologies $bc(bc)(ac)$, $bc(ab)(bc)$, $ab(ab)(bc)$, $ab(ab)(ac)$, $ac(ab)(ac)$ and $ac(ac)(bc)$. In general, the anti-atoms are topologies of the form

$$\tau_{(x,\mathcal{U})} := \{A \subset X \mid x \notin A \text{ or } A \in \mathcal{U}\} \tag{6.2.4}$$

where $\mathcal{U}$ is any ultra-filter (a maximal element with respect to the natural partial ordering of filters) not equal to the principal ultra-filter of all subsets of X containing the point $x \in X$.

The lattice structure also has important properties in relation to the action of the group Perm(X). The induced maps on $\tau(X)$ preserve all the lattice operations and are therefore lattice *automorphisms*. Furthermore, Hartmanis (1958) and Fröhlich (1964) showed that if $|X| = \infty$, the group of automorphisms Aut($\tau(X)$) is actually equal to Perm(X) — that is, the topological properties of an element of $\tau(X)$ are determined solely by its position in the lattice. (For finite X there is an additional automorphism induced by the transformation on the atoms $\tau_A \to \tau_{A^c}$ where A^c is the set-theoretic complement of the subset A of X.) This equality of Aut($\tau(X)$) with Perm(X) reinforces the idea that Perm(X) is the natural "gauge group" of the theory and therefore has important implications for the quantum theory.

We come now to the critical question of selecting a set of functions on $\tau(X)$ to use as a basis for a quantization scheme. One natural family of functions associated with the lattice structure is

$$R_\tau(\tau') := \begin{cases} 1, & \text{if } \tau \preceq \tau' \\ 0, & \text{otherwise.} \end{cases} \tag{6.2.5}$$

In particular,

$$R_{\tau_A}(\tau') = \begin{cases} 1, & \text{if } A \text{ is open in } \tau' \text{ (ie } \tau_A \preceq \tau') \\ 0, & \text{otherwise.} \end{cases} \tag{6.2.6}$$

and since $\tau(X)$ is atomic, this set of variables is clearly large enough to distinguish between the different topologies.

We recall that we are looking for an *algebra* of functions on $Q = \tau(X)$, and therefore it is important that

$$\begin{aligned} R_{\tau_1}(\tau')R_{\tau_2}(\tau') &= 1 \text{ if } \tau_1 \preceq \tau' \text{ and } \tau_2 \preceq \tau' \\ &= 0 \text{ otherwise.} \end{aligned} \tag{6.2.7}$$

However, $\tau_1, \tau_2 \preceq \tau'$ if and only if $\tau_1 \vee \tau_2 \preceq \tau'$ and hence

$$R_{\tau_1} R_{\tau_2} = R_{\tau_1 \vee \tau_2} \tag{6.2.8}$$

Thus we do indeed have closure, with the set of functions generating a representation of the $\vee$-semigroup operation in the lattice. One consequence of (6.2.8) is

$$(R_\tau)^2 = R_\tau \text{ for all } \tau \in \tau(X) \tag{6.2.9}$$

and hence, in the quantum theory, $\{\hat{R}_\tau \mid \tau \in \tau(X)\}$ can be expected to be a family of hermitian projection operators satisfying

$$\hat{R}_{\tau_1} \hat{R}_{\tau_2} = \hat{R}_{\tau_1 \vee \tau_2}. \tag{6.2.10}$$

Note that quantizing a system in terms of families of projection operators played a central rôle in Mackey's (1963) axiomatization of quantum theory.

When $|\tau(X)|$ is finite, it is easy to produce representations of this algebra by defining states to be vectors $\psi(\tau)$ with

$$(\hat{R}_\tau \psi)(\tau') := \begin{cases} \psi(\tau'), & \text{if } \tau \preceq \tau' \\ 0, & \text{otherwise.} \end{cases} \tag{6.2.11}$$

However, when $|\tau(X)|$ is infinite, account must be taken of the afore-mentioned possibility that the domain of the state vectors is a space of "distributional" topologies. In the analogous case in quantum field theory an important property of the space of distributions E' is that any distribution can be approximated arbitrarily closely by a smooth function from E. Topologically speaking, E is *dense* in E': that is, any neighbourhood of any element of E' always contains an element of E. Therefore we might expect/hope that whatever the space of "distributional" topologies might be, it will carry a topology such that $\tau(X)$ is embedded as a dense subset.

The key to finding such a space lies in the the *spectral theory* of the abelian algebra generated by the operators $\{\hat{R}_\tau \mid \tau \in \tau(X)\}$. The equation $(\hat{R}_\tau)^2 = \hat{R}_\tau$ implies that the eigenvalues of $\hat{R}_\tau$ are 0 or 1. Also, (6.2.10) gives

$$[\hat{R}_{\tau_1}, \hat{R}_{\tau_2}] = 0 \tag{6.2.12}$$

and hence there should be a complete set of simultaneous eigenvectors $|h\!>$ such that

$$\hat{R}_\tau |h\!> = h(\tau)|h\!> \tag{6.2.13}$$

where $h(\tau) = 0$ or 1. The maps $h : \tau(X) \to \{0,1\} \subset \mathbb{R}$ are of considerable interest since they form the *domain* space of the state vectors — in Dirac notation, $\psi(h) = \,<\!h \mid \psi\!>$. Equation (6.2.10) implies that

$$h(\tau_1)h(\tau_2) = h(\tau_1 \vee \tau_2) \quad \forall \tau_1, \tau_2 \in \tau(X) \tag{6.2.14}$$

so that h is a $\vee$-homomorphism (or *semi-character*) from $\tau(X)$ into $\{0,1\} \subset \mathbb{R}_+$.

Now define $I_h := \{\tau \mid h(\tau) = 1\}$. Then

(a) $\tau_1, \tau_2 \in I_h$ implies $\tau_1 \vee \tau_2 \in I_h$

(b) If $\tau' \preceq \tau$, then $\tau' \vee \tau = \tau$ and hence (6.2.14) implies that $h(\tau')h(\tau) = h(\tau)$. Then if $\tau \in I_h$, $h(\tau') = 1$ and so $\tau' \in I_h$.

These are precisely the properties for I_h to be an *ideal* in the lattice $\tau(X)$. Conversely, if I is an ideal, define $h_I : \tau(X) \to \mathbb{R}$ by

$$h_I(\tau) := \begin{cases} 1, & \text{if } \tau \in I \\ 0, & \text{otherwise} \end{cases} \tag{6.2.15}$$

But the definition of an ideal is equivalent to the statement that $\tau_1, \tau_2 \in I$ if and only if $\tau_1 \vee \tau_2 \in I$ (exercise!). Hence h_I satisfies (6.2.14). Furthermore, $h_{I_h}(\tau) = h(\tau)$ so that $h \rightarrow I_h$ is a bijection and therefore the domain space of the quantum states is the set $I(\tau(X))$ of all ideals in the lattice.

The important conclusion of this analysis is that the space of quantum states is expected to be equivalent to a Hilbert space of the form $L^2(I(\tau(X)), d\mu)$ for some measure μ on $I(\tau(X))$. It is useful to emphasise this analogy of the space of ideals with the vector space of distributions by writing $h_I(\tau)$ as $< I, \tau >$. Thus the basic eigenvalue equation (6.2.13) becomes $\hat{R}_\tau|I> = < I, \tau > |I>$ which translates into the action on $\psi \in L^2(I(\tau(X)), d\mu)$

$$(\hat{R}_\tau\psi)(I) = < I, \tau > \psi(I) \tag{6.2.16}$$

and which should be compared with the quantum field theoretic analogue (6.1.2). Note that the classical function R_τ in (6.2.5) can be written simply as $R_\tau(\tau') = < \downarrow(\tau'), \tau >$.

Comments

1. The set of ideals in any lattice is itself a lattice in which the meet operation is just set intersection $\cap$ (exercise!)

2. The rigorous version of the discussion above on the simultaneous eigenvectors of the operators $\{\hat{R}_\tau \mid \tau \in \tau(X)\}$ employs the the Gel'fand spectral theorem (Rudin, 1973). An integral part of this theory is the construction of a natural topology on $I(\tau(X))$ in which a subbasis for the open sets is all subsets of $I(\tau(X))$ of the form

$$Q_\tau := \{I \in I(\tau(X)) \mid \tau \in I\} \tag{6.2.17}$$

plus their set-theoretic complements. This topology has the property that $I(\tau(X))$ is both compact and Hausdorff.

3. The automorphism of the lattice $\tau(X)$ induced by a bijection of X passes to an automorphism of the lattice $I(\tau(X))$ which can be shown to be a *homeomorphism* with respect to the spectral topology on $I(\tau(X))$.

4. $\tau(X)$ can be embedded as a subset of $I(\tau(X))$ via the injection

$$\tau(X) \rightarrow I(\tau(X)), \quad \tau \mapsto \downarrow(\tau) := \{\tau' \in \tau(X) \mid \tau' \preceq \tau\} \tag{6.2.18}$$

which maps each topology τ into the principal ideal which it generates. This map is *surjective* if $\tau(X)$ is finite since then every ideal is principal:

$$I \equiv \downarrow(\bigvee\{\tau \in I\}) \tag{6.2.19}$$

Thus, for finite X, our expectation that state vectors can be written as functions on $\tau(X)$ is justified and (6.2.16) reproduces (6.2.11). In a general infinite lattice, ideals exist which are *not* principal but, in the present case, it can be shown that $\tau(X)$ is a *dense* subset of $I(\tau(X))$, analogous to the situation in quantum field theory where the smooth functions are dense in the space of distributions.

5. A key step in building a quantum theory is the construction of an inner product on the space of states which, in our case, means studying the existence of suitable measures on the space $I(\tau(X))$. In this context it is fortunate that the spectral theory of the algebra generated by the operators $\{\hat{R}_\tau \mid \tau \in \tau(X)\}$ can be extended well beyond the Gel'fand results. In particular, there is a precise analogue for semi-groups of the well-known theory for abelian groups which expresses all representations in terms of characters. First, if μ is any regular measure on the compact Hausdorff space $I(\tau(X))$ then (6.2.16) defines a cyclic representation of the algebra with cyclic vector $\Omega(I) := 1$ and in which

$$\kappa(\tau) := <\Omega, \hat{R}_\tau \Omega> = \int_{I(\tau(X))} <I, \tau> \, d\mu(I) \tag{6.2.20}$$

is a *positive semi-definite* function on $\tau(X)$. That is

$$\sum_{j,k=1}^{n} c_j c_k^* \kappa(\tau_j \vee \tau_k) \geq 0 \tag{6.2.21}$$

for all finite sets $c_1 \ldots c_n \in \mathbb{C}$.

The crucial result is the converse statement drawn from the general spectral theory of semigroups (Berg *et al* 1984) which affirms that to each such positive semi-definite function κ, there exists a regular measure on $I(\tau(X))$ such that (6.2.20) is true. The analogous statement in conventional quantum field theory is that any such function on the topological vector space E of smooth test functions generates a unitary representation $f \mapsto e^{i\hat{\phi}(f)}$ of E with $\kappa(f) = <\Omega, e^{i\hat{\phi}(f)}\Omega>$ and with this expectation value being expressible as a "Fourier transform" over E'. Thus, at least in principle, these spectral-theorem results on the algebra of operators $\{\hat{R}_\tau \mid \tau \in \tau(X)\}$ provide a way of placing the lattice-based theory of quantum topology on a respectable footing.

§6.3 The Complementary Variables

We must consider now the vital question of the construction of complementary "momentum" variables. In a conventional quantum theory of a system whose configuration space Q is a manifold, these variables are usually identified with the self-adjoint generators of a unitary representation of a Lie group of transformations of Q. However, a more relevant case for our present purposes is when Q is a discrete space. For example, if Q is equal to the integers, we can define

$$\begin{aligned}(\hat{T}\psi)(n) &= \psi(n+1) \\ (\hat{T}^\dagger\psi)(n) &= \psi(n-1).\end{aligned} \tag{6.3.1}$$

Thus $\hat{T}$ is unitary with respect to the usual l^2 scalar product, but there is no corresponding generator. Instead, there are the hermitian operators

$$\hat{q} := (\hat{T} + \hat{T}^\dagger)/2, \qquad \hat{\pi} := (\hat{T} - \hat{T}^\dagger)/2i \tag{6.3.2}$$

so that

$$(\hat{\pi}\psi)(n) = \big(\psi(n+1) - \psi(n-1)\big)/2i \tag{6.3.3}$$

which is the nearest we get to "differentiation" in this case. Note that $\hat{q}$ is not an independent observable since $\hat{T}^\dagger\hat{T} = \hat{T}\hat{T}^\dagger = 1$ implies that

$$\hat{q}^2 + \hat{\pi}^2 \equiv 1 \tag{6.3.4}$$

In the case of interest with $Q = \tau(X)$ we note that the operation

$$\begin{aligned} m_a : \tau(X) &\to \tau(X) \\ \tau &\mapsto a \wedge \tau \end{aligned} \tag{6.3.5}$$

is *continuous* with respect to the spectral topology on $\tau(X)$, as is its extension to $I(\tau(X))$ defined by $m_a(I) := \downarrow(a) \cap I$. Having already obtained a representation of the $\vee$-operation (from the $\hat{R}_\tau$-operators) it seems natural to complement this by choosing for the analogue of a group action on $Q = \tau(X)$, the action of the semi-group $\tau(X)$ with respect to the $\wedge$-operation. Thus, on the Hilbert space $L^2(I(\tau(X)), d\mu)$, we define

$$(\hat{M}_\tau\psi)(I) := \psi(\downarrow(\tau) \cap I). \tag{6.3.6}$$

This gives an algebra

$$\begin{aligned} \hat{R}_\tau \hat{R}_{\tau'} &= \hat{R}_{\tau\vee\tau'} \\ \hat{M}_\tau \hat{M}_{\tau'} &= \hat{M}_{\tau\wedge\tau'} \\ \hat{M}_\tau \hat{R}_{\tau'} &= <\downarrow(\tau), \tau'> \hat{R}_{\tau'} \hat{M}_\tau \end{aligned} \tag{6.3.7}$$

which is reminiscent of the canonical commutation relations in conventional quantum theory. We will "axiomatise" (6.3.7) as being the basic algebra for quantum topology on the lattice $\tau(X)$.

Note that $\hat{M}_\tau$ is not unitary but is instead more like a *creation* or *annihilation* operator. The associated hermitian operators are

$$\hat{\pi}_\tau := (\hat{M}_\tau - \hat{M}_\tau^\dagger)/i, \quad \hat{q}_\tau := (\hat{M}_\tau + \hat{M}_\tau^\dagger) \tag{6.3.8}$$

which, by virtue of the relation $(\hat{M}_\tau)^2 = \hat{M}_\tau$, satisfy the constraint

$$(\hat{q}_\tau - 1)^2 - (\hat{\pi}_\tau)^2 = 1 \tag{6.3.9}$$

so that $\hat{q}_\tau$ and $\hat{\pi}_\tau$ are not independent variables.

Example

Let us consider the example of a finite set X. The quantum states lie in the vector space $L^2(\tau(X), d\mu) \simeq \mathbb{C}^{|\tau(X)|}$ in which the inner product between a pair of vectors is the finite sum

$$<\psi, \phi> = \sum_\tau \mu(\tau)\psi(\tau)^*\phi(\tau) \tag{6.3.10}$$

The eigenstates of the operators $\hat{R}_\tau$ can be written as $| \tau' >$, so that

$$\hat{R}_\tau \mid \tau' > = \begin{cases} \mid \tau' >, & \text{if } \tau \preceq \tau' \\ 0, & \text{otherwise.} \end{cases} \tag{6.3.11}$$

The adjoint operator $\hat{M}_\tau^\dagger$ acts to *weaken* the topology:

$$\hat{M}_\tau^\dagger |\tau' > = (\mu(\tau')/\mu(\tau \wedge \tau'))^{\frac{1}{2}} |\tau \wedge \tau' > \tag{6.3.12}$$

whilst the "creation" operator $\hat{M}_\tau$ acts to *strengthen* the topology as

$$\hat{M}_\tau |\tau' > = \begin{cases} \frac{1}{\sqrt{\mu(\tau')}} \sum_{\{a | \tau \wedge a = \tau'\}} \sqrt{\mu(a)} |a >, & \text{if } \tau \geq \tau' \\ 0, & \text{otherwise.} \end{cases} \tag{6.3.13}$$

where the sum over the set $m_\tau^{-1}(\tau') = \{a \in \tau(X) \mid \tau \wedge a = \tau'\}$ will typically contain *more* than one element. This contrasts sharply with the more familiar examples of annihilation and creation operators and has its origin in the fact that the map $m_\tau : \tau(X) \to \tau(X)$ is *many*-to-one. Thus $\hat{M}_\tau$ not only strengthens a topology, it also *broadens* the state in the sense that an eigenstate of topology becomes a linear superposition of eigenstates.

§6.4 The Rôle of Perm(X)

We turn now to the important question of what rôle the permutation group Perm(X) is to play in the quantum theory. If we do not wish to distinguish between homeomorphic topologies then it is necessary to think of Perm(X) as a sort of gauge-group of the theory and to regard the quotient space $\tau(X)/\text{Perm}(X)$ as the true "configuration space". However it should be noted that the action of Perm(X) on $\tau(X)$ is not free and many topologies have a non-vanishing "little group". Indeed, the little group of a topology τ is just the group of all homeomorphisms of (X, τ) with itself.

As far as the quantum theory is concerned, there appear to be three different ways in which one might proceed:

1. Quantize on $\tau(X)$ and then impose constraints on "physical" state vectors and/or observables in the well-known manner advocated by Dirac.

2. Fix a gauge for Perm(X) and then quantize the remaining "physical" degrees of freedom.

3. Try and quantize directly on the space $\tau(X)/\mathrm{Perm}(X)$ of homeomorphism classes of topology.

For most gauge theories, the second "brute-force" method is often employed as a consistency check on the other, more elegant, approaches. However, in our case it is not clear how to set about specifying a gauge for the group Perm(X) and we are obliged to consider one of the other two schemes. In many ways the third is perhaps the most attractive but it is not obvious how to proceed in this direction either since although each $\phi \in \mathrm{Perm}(X)$ induces an automorphism of the lattice structure on $\tau(X)$, $\tau(X)/\mathrm{Perm}(X)$ does *not* inherit any lattice structure from $\tau(X)$ — that is, if $\tau_1 \simeq \tau_1'$ and $\tau_2 \simeq \tau_2'$ it is not necessarily the case that $\tau_1 \wedge \tau_2$ (resp. $\tau_1 \vee \tau_2$) is homeomorphic to $\tau_1' \wedge \tau_2'$ (resp. $\tau_1' \vee \tau_2'$). For example, when $X = \{a, b, c\}$ the topology $b(ac)$ is homeomorphic to $c(ab)$ via the permutation that exchanges the points b and c. However, $b(ac) \wedge a(ab) = 0$ whereas $c(ab) \wedge a(ab) = (ab)$.

On the other hand, $\tau(X)/\mathrm{Perm}(X)$ does possess a natural partial ordering which might be of use. This is the ordering defined between a pair of homeomorphism classes of topology α and β as

$$\alpha \preceq \beta \text{ iff } \exists \tau_1 \in \alpha \text{ and } \tau_2 \in \beta \text{ such that } \tau_1 \preceq \tau_2. \tag{6.4.1}$$

For example, the ordering diagram for the case $X = \{a, b, c\}$ is given below where $[\tau]$ denotes the equivalence class (orbit of Perm(X)) containing the topology τ.

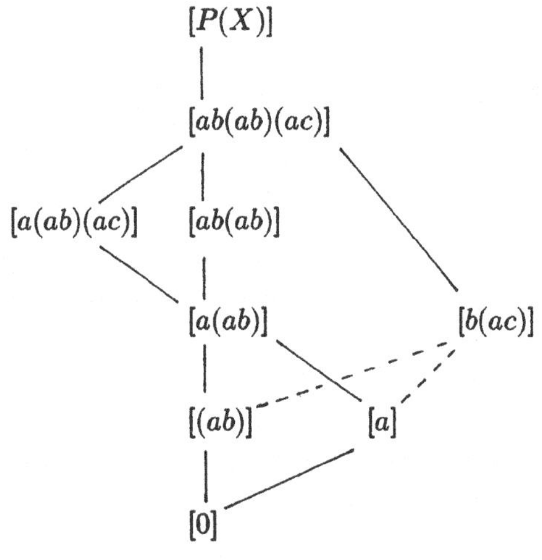

Note that $\tau_1 \preceq \tau_2$ implies $[\tau_1] \preceq [\tau_2]$ but the converse is not true in general and there may exist "extra" links between a pair of equivalence classes $[\tau_1]$ and $[\tau_2]$. This occurs when

(i) $\tau_1 \not\prec \tau_2$

and

(ii) there exist topologies τ_1' and τ_2' such that $\tau_1' \prec \tau_2'$ with $\tau_1 \simeq \tau_1'$ and $\tau_2 \simeq \tau_2'$.

For example, in the case above, we have $(ab) \prec a(ab)$ and hence $[(ab)] \prec [a(ab)]$. However we also have

$$(ab) \prec c(ab) \tag{6.4.2}$$

and $c(ab) \simeq b(ac)$ — that is $[c(ab)] = [b(ac)]$ — via the permutation of $X = \{a, b, c\}$ which exchanges the points b and c. This induces the additional ordering relation

$$[(ab)] \prec [b(ac)] \tag{6.4.3}$$

as shown by the dotted line in the diagram. Of course, which links are deemed to be "extra" depends on the choices that are made for a representative topology in each equivalence class of topologies. In an intrinsic sense, all links have equal status and there is no real distinction between them.

The non-lattice nature of this partially-ordered set is illustrated by the sub-block

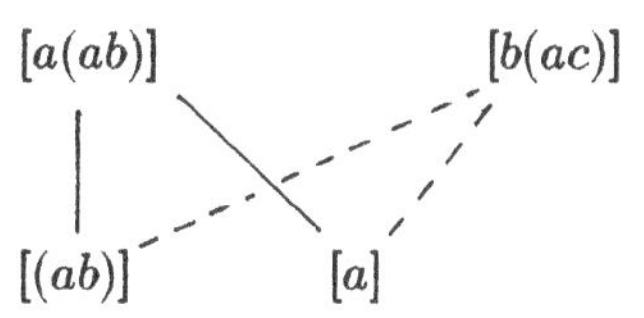

which shows, for example, that the "join" of the pair of equivalence classes $[(ab)]$ and $[a]$ would have to be both $[a(ab)]$ and $[b(ac)]$. We shall see below how this partial ordering can be used in the quantum theory.

Let us consider now the imposition of Dirac-style constraints on the Hilbert space $L^2(I(\tau(X)), d\mu)$ associated with the quantum theory on $\tau(X)$. If the measure μ is *quasi-invariant* under the action of Perm(X) (that is, if the transformed measure μ_Λ and the original measure μ have the same sets of measure zero), a unitary representation of this group can be obtained in the usual way by defining

$$(\hat{U}(\Lambda)\psi)(I) = (d\mu_\Lambda(I)/d\mu)^{\frac{1}{2}}\psi(\Lambda^{-1}(I)) \tag{6.4.4}$$

where $d\mu_\Lambda(I)/d\mu$ is the Jacobian (Radon-Nikodym derivative) of μ_Λ with respect to μ.

A simple calculation shows that

$$\hat{U}(\Lambda)\hat{R}_\tau\hat{U}(\Lambda)^{-1} = \hat{R}_{\Lambda(\tau)} \tag{6.4.5}$$

so that the $\hat{R}_\tau$-variables transform covariantly with respect to Perm(X). The same is not automatically true of the $\hat{M}_\tau$ operators unless the Jacobian satisfies an appropriate condition. We will assume from now on that such a measure has been found so that

$$\hat{U}(\Lambda)\hat{M}_\tau\hat{U}(\Lambda)^{-1} = \hat{M}_{\Lambda(\tau)} \tag{6.4.6}$$

To illustrate the basic idea let us consider the case where $|X|$ is finite. Then from (6.4.5-6) it follows that gauge-invariant operators can be constructed by defining

$$\begin{aligned}\hat{R}_O &:= \sum_{\tau\in O} \hat{R}_\tau,\\ \hat{M}_O &:= \sum_{\tau\in O} \hat{M}_\tau\end{aligned} \tag{6.4.7}$$

where O is an orbit of Perm(X) on $\tau(X)$. For example, when $X = \{a, b, c\}$ we have

$$\begin{aligned}\hat{R}_{[(ab)]} &= \hat{R}_{(ab)} + \hat{R}_{(bc)} + \hat{R}_{(ac)}\\ \hat{R}_{[b(ac)]} &= \hat{R}_{b(ac)} + \hat{R}_{a(bc)} + \hat{R}_{c(ab)}\end{aligned} \tag{6.4.8}$$

The crucial question for us is whether or not these new operators satisfy any sort of algebra. By explicit calculation we find that, whereas $\hat{R}_{(ab)}\hat{R}_{b(ac)} = \hat{R}_{ba(ab)(ac)}$, the new operators satisfy the relation

$$\hat{R}_{[(ab)]}\hat{R}_{[b(ac)]} = \hat{R}_{[ba(ab)(ac)]} + \hat{R}_{[b(ac)]} \tag{6.4.9}$$

so we do indeed get algebraic closure. This is in fact a special case of the more general theorem (Isham, 1989a):

Theorem

If O and O' are two orbits of Perm(X) on $\tau(X)$, then there exist positive integers $n_j(O, O')$ such that

$$\hat{R}_O\hat{R}_{O'} = \sum_j n_j(O, O')\hat{R}_{O_j} \tag{6.4.10}$$

where the sum is over all orbits O_j that pass through topologies $\tau \vee \tau'$ for some $\tau \in O$ and $\tau' \in O'$.

A similar result holds for products of $\hat{M}_{[\tau]}$. However, the situation is more complicated for mixed products. For example, when $X = \{a, b, c\}$

$$\hat{M}_{[b(ac)]}\hat{R}_{[(ab)]} = \hat{R}_{(ab)}\hat{M}_{c(ab)} + \hat{R}_{(bc)}\hat{M}_{a(bc)} + \hat{R}_{(ca)}\hat{M}_{b(ca)} \tag{6.4.11}$$

and the right hand side is not of the form $\hat{R}_O\hat{M}_{O'}$ or sums of such. On the other hand, it *is* of the form $\sum \hat{R}_\tau\hat{M}_{\tau'}$ where (τ, τ') belongs to an orbit of Perm(X) on the

Cartesian product $\tau(X) \times \tau(X)$. This suggests strongly that gauge-invariant operators should be of the form

$$\hat{T}_O = \sum_{(\tau,\tau')\in O} \hat{R}_\tau \hat{M}_{\tau'} \tag{6.4.12}$$

where O is an orbit in $\tau(X) \times \tau(X)$. Indeed, one can prove the result:

Theorem

$$\hat{T}_O \hat{T}_{O'} = \sum n_i(O, O') \hat{T}_{O_i} \tag{6.4.13}$$

for some integers $n_i(O, O')$ where i labels an orbit of Perm(X) in $\tau(X) \times \tau(X)$.

By these means we have succeeded in finding a set of gauge-invariant operators which satisfy a well-defined algebra. This is the outcome of the Dirac-constraint method as applied to the observables in the original $Q = \tau(X)$ theory. A similar analysis can be used to study "physical" states, although there a possibility here of the analogue of "θ-factors" — that is, physical states might only be invariant up to an overall phase factor.

Note that although this approach has been phrased in the context of the first of our three *a priori* ways of handling the gauge-group, it is clear that the procedure can be reversed and Perm(X)-invariant quantization *defined* to be the construction of representations of the algebra (6.4.13), where the numbers $n_i(O, O')$ are determined from the partial-ordering diagram for $\tau(X)$/Perm(X). This could be regarded as a *bona fide* quantization of the system whose configuration space is the space $\tau(X)$/Perm(X) of homeomorphism classes of topology.

§7 Conclusions

We have presented a short introduction to the general theory of topological spaces in which special emphasis has been placed on the more algebraic aspects of the subject. In particular: (i) a topology on a set X is a complete sublattice of the lattice P(X) with a certain distributive property, and (ii) the set $\tau(X)$ of all topologies on X is a lattice. We have seen how the latter may be exploited to develop a possible structure for a quantum theory of topology and we have also emphasised the importance of studying a quantum theory of metric topologies. However many difficulties remain, both at the technical and at the conceptual levels.

For example, the ability to handle spaces X that are infinite is based on the spectral theory in §6.2 and the introduction of the space of ideals $I(\tau(X))$. However, as it stands, the development in §6.4 of the quantum theory on $\tau(X)$/Perm(X) only works for a finite set X: an extension to the infinite case presumably requires integrating over the orbits of Perm(X). One way of doing this would be to exhibit Perm(X) as a topological group equipped with something like a Haar measure. This in turn might be achieved by embedding the group as a closed (and therefore compact) orbit in the compact Hausdorff space $I(\tau(X))$. The minimum needed for this is a point in $I(\tau(X))$ whose isotropy group is trivial—that is, a topology on X with no non-trivial homeomorphisms. When $X = \{1, 2 \ldots n\}$ is finite, an example of such a topology is

$$\tau := \{\emptyset, \{1\}, \{1,2\}, \ldots \{1, 2, \ldots n-1\}, X\} \tag{7.1}$$

since any non-trivial permutation of X must affect at least one of these subsets, thus giving rise to a trivial isotropy group. It might be possible to extend this construction to the infinite case with the aid of a well-ordering on X.

One may also wonder about the relation between the metric topology approach and that based on the lattice of all topologies. In this respect it is worth noting that the set of all T_1 topologies forms a complete sublattice of $\tau(X)$ (Larson and Andima, 1975). This lattice is only non-trivial when $|X| = \infty$ and has the cofinite topology as its null element. The lattice is antiatomic (but not atomic) and shares with $\tau(X)$ the property that its group of automorphisms is equal to Perm(X). This lattice might form an alternative candidate for quantization; it has the advantage of being related to metrics in the sense that every (first countable) T_1 topology can be written as the lattice product of all finer metric topologies (Raghavan and Reilly, 1986).

There are a number of other technical issues which could be considered [1] but many of the most important problems contain conceptual ingredients as well, especially those concerned with what the formalism means physically. In particular, we must say something about the (i) rôle of 'time' in these theories, and (ii) the status of the points in the set X.

The treatment of quantum topology presented in these notes is based on canonical-type ideas with the points in X being associated in some way with the points of "physical space". It might seem natural therefore to attach a time label to the states $\psi(\tau)$ (or, more generally, functions of ideals) and to consider a Schrödinger-type evolution equation. This involves constructing a suitable Hamiltonian which, one might hope, will have some natural expression in terms of the lattice structure of $\tau(X)$. In this context it is noteworthy that the states $|\tau>$ are actually eigenstates of the operators $\hat{M}^{\dagger}_{\tau'}\hat{M}_{\tau'}$ (shades of the simple harmonic oscillator!) and therefore a family of model "free" Hamiltonians can be constructed by taking linear combinations of such operators. The addition of other, non-diagonal, operators yields Hamiltonians that produce genuine quantum-mechanical changes in topology (Isham, 1989).

However, the use of a continuous (and, indeed, differentiable) time label seems at odds with our policy to eschew the concepts of differential geometry—a more natural

[1] For example, it is interesting to consider possible relations between the quantum ideas developed here and the work of probability theorists on "random topology". See Frank (1971), Schweizer and Sklar (1983).

approach might be to invoke a discrete picture of time, with the Schrödinger equation being replaced with an appropriate difference equation. The idea that time should be discretized is one that recurs frequently and has appeared in several of the more adventurous approaches to quantum gravity. One example is the work of Noyes and McGoveran on "programme universe" and the general research effort of the ANPA members (Noyes, 1989). Other good examples are Finkelstein's early work (1969) on spacetime code, his more recent ideas on "quantum net dynamics" (1989), and the discretization of causal structure in t'Hooft (1979) and in Bombelli et al, (1987a,b) and Bombelli and Meyer (1989).

Yet another possibility is to follow the lead of one of the more conventional schools of quantum gravity and construct an object $K(\tau_1, \tau_2)$ which, in analogy with (1.1), might be obtained by summing/integrating over all topological spaces which induce the topologies τ_1 and τ_2 on a subset consisting of the disjoint union of sets X_1 and X_2. This would involve the construction of some sort of action principle, [1] which raises the very general question of how classical general relativity is supposed to fit into this quantum scheme. This in turn requires an understanding of what is perhaps the deepest question of all: what is the status of the set X and its 'points', and how do smooth differentiable manifolds enter the picture?

Several different approaches can be taken to this issue, depending on one's view on the nature of space and time. One possibility is to proclaim that the points of space are in some sense "real" and that, furthermore, they have the cardinality c of the real line; to emphasise this point let us denote the set of points by X_c. It is still necessary to make some decision concerning the concept of 'time' (which, as suggested already by conventional canonical quantum gravity, may well be valid only in some semi-classical limit) but it is clear that, in these circumstances, we are genuinely interested in the lattice $\tau(X_c)$ of all topologies on X_c, which of course contains all differentiable manifolds of all possible dimensions. The crucial technical questions then are (i) how do these manifolds fit into the lattice $\tau(X_c)$?, and (ii) what does the the typical neighbourhood of a manifold look like? Armed with this knowledge, one could aspire to append some type of topological action to one's favourite action for ordinary quantum gravity. Note that similar considerations apply to "Riemannian-metric" driven quantum topology in which quantum fluctuations in topology occur about some background manifold.

A quite different approach (and the one which I favour) is to argue that the assignment of cardinality c to the points in space and time is purely a mathematical trick (similar perhaps to the way the rationals are completed to the reals) and that, in so far as the concept of 'point' has any meaning at all, there is only a finite number of

[1] See p499 in Wheeler (1964) for an early, but most interesting, reflection on this possibility.

them. [1] Thus the fact that we apparently "see" a continuous spacetime is an illusion which arises because our spectacles are too coarse-grained to distinguish the fine detail. Thoughts of this type lay behind Penrose's spin network theory (Penrose, 1971) and Sorkin's work (1983) on approximating a manifold by a finite covering of open sets. As we shall see, the latter idea is particularly relevant to the programme being explored here.

It is clear that, in the classical limit, the topology on our set X must "approximate" that of a manifold Σ. But what does this really mean? One possibility might be to consider the set X as a subset of Σ and to argue that the topology we apparently see on the latter is really a reflection of a topology on X which is a subspace of the manifold topology. However, this does not work because the Hausdorff topology on Σ always induces the *discrete* topology on any finite subset of points.

From a mathematical point of view, the other natural approach is to consider a *surjective* map from Σ onto X and argue that it is the associated identification topology τ on X which should be regarded as an approximation to the manifold. The key to understanding the potential physical significance of this construction is the observation (§5.6) that such an surjection induces an *injection* of the lattice τ into the lattice of open sets on the manifold. This suggests two somewhat different pictures.

The first scheme would be to use the lattice $\tau(X_c)$ but argue that the only topologies on X_c which have any physical significance are those that possess a *finite* number of open sets. The collection $\tau_F(X_c)$ of all such finite topologies is closed under meets and joins and includes the trivial topology $0 = \{\emptyset, X_c\}$. Note that the vector subspace spanned by eigenstates $|\tau>$ of topologies of this type is invariant under the action of $M_{\tau'}$ and $(M_{\tau'})^{\dagger}$ with $\tau' \in \tau_F(X_c)$. Manifolds fit into the scheme as follows. Each smooth manifold Σ belongs to $\tau(X_c)$, but of course the topology is not finite. However, each finite *covering* of Σ by a collection of Σ-open subsets (i) generates a sublattice of the topology of Σ, and (ii) forms a subbase for a finite topology on X_c which is coarser than the manifold topology (that is, it lies in the subset $\downarrow(\Sigma)$). The finer the covering, the closer the topology on X_c approximates that of the manifold Σ. Thus we arrive at a quantization scheme which involves just the finite topologies on X_c and the subalgebra of (6.3.7) generated by such topologies.

A somewhat different point of view (or perhaps an extension to the above) is to follow Sorkin (1983) and argue that points in X_c which cannot be distinguished by elements of a covering of the space should be identified. The resulting set, constructed

[1] If the universe is truly infinite in size it might be necessary to employ a countably infinite set when considering genuine cosmological questions. However, to simplify the exposition, I will assume in what follows that the universe is finite in size and that, correspondingly, any differentiable manifold that serves as a model for physical space is compact.

as in (5.5.2), is finite and carries the T_0 identification topology. The suggestion then is that it is *this* set X which should be identified with the set of "real" spatial points. The basic idea therefore is that:

1. There are really only a finite number of points in the universe (neglecting cosmological considerations). These can carry a variety of topologies which are subject to quantum laws. Only topologies that are at least T_0 are physically meaningful; T_1 and T_2 topologies have no rôle in this scheme since the only topology of these types on a finite set is the discrete topology.

2. There exist "classical" states in which the average lattice of open sets on X looks like a *sublattice* of the lattice of open sets on a manifold. This is why we assume (erroneously) that we live in a differentiable manifold.

Note however that, seen from this viewpoint, there is no reason for the number of points in X to be fixed. Indeed, even if we start with a single finite set X, the existence of non-T_0 topologies will lead us to pass to the corresponding quotient topology, which exists on a space of smaller cardinality. There are several ways in which one might attempt to extend the earlier quantum discussion to include a varying cardinality for the set X. However, reflection on the discussion above suggests a more radical possibility. It is clear that what is really being asserted is that the most important feature of space is not the points which it contains but rather the open subsets and the lattice relations between them. But then, since, physically speaking, a "point" is a most peculiar concept anyway, why not drop it altogether and deal directly with frames/locales?

Thus the final picture is one in which the fundamental space and/or spacetime concept is of a 'region', and the important property is the relation between these regions—the way they overlap as coded in the lattice structure. Both the number and the interrelations of such regions are subject to quantum fluctuations which might be handled with some extension to frames/locales of the formalism presented in these notes. The classical limit of such a theory will yield a collection of regions that imitate some (fairly refined) covering by open sets of a differentiable manifold and hence return us safely to the classical world as and when it becomes necessary. There is even a possibility of emulating the theory of quantum groups (much discussed at this meeting) by considering deformations of the locale lattices to non-commutative algebras and hence to "quantum topologies" rather than simply a quantization of collections of classical topologies. By these means we would truly arrive at a theory of "points without points"—a concept of which John Wheeler, the inventor of quantum topology, would surely have approved!

Acknowledgements

I most grateful to the organizers of the Banff Advanced Summer Institute for arranging such a pleasant and stimulating meeting, and for inviting me to participate in it.

REFERENCES

Ashtekar A. 1988 (With invited contributions) *New Perspectives in Canonical Gravity* (Bibliopolis, Naples).

Birkhoff G. 1967 *Lattice Theory* (Amer. Math. Soc., New York).

Berg C., Christensen J.P.R. and Ressel P. 1984 *Harmonic Analysis on Semigroups* (Springer-Verlag, New York).

Bohr N. and Rosenfeld L. 1933 *Mat.-fys. Medd. Dan. Vid. Selsk.* **12**; 1950 *Phys. Rev.* **78** 794. Both papers (including an English translation of the first) are conveniently reprinted in *Quantum Theory and Measurement*, J.A. Wheeler and W.H. Zurek eds., 1983 (Princeton University Press, Princeton).

Bombelli L., Lee L., Meyer D. and Sorkin R.D. 1987 *Phys. Rev. Letts.* **59** 521; *Phys. Rev. Letts.* **60** 656.

Bombelli L. and Meyer D.A. 1989 *The origin of Lorentzian geometry*, preprint.

Bourbaki N. 1966 *General Topology: Part 1* (Addison-Wesley, London).

Császár A. 1978 *General Topology* (Adam Hilger, Bristol).

Dugundji J. 1966 *Topology* (Allyn and Bacon Inc., Boston).

Finkelstein D. 1969 *Phys. Rev.* **184** 1261; 1989 *Quantum net dynamics*, *Inter. Jour. Theor. Phys.* , in press.

Frank M. J. 1971 *Jour. Math. Anal. Appl.* **34** 67.

Fröhlich O. 1964 *Math. Ann.* **156** 79.

Grätzer G. 1978 *General Lattice Theory* (Birkhäuser, Basel).

Hartle J.B. and Hawking S.W. 1983 *Phys. Rev.* **D28** 2960.

Hartmanis J. 1958 *Canad. Jour. Math.* **10** 547.

Hawking S.W., King A.R. and McCarthy P.J. 1976 *Jour. Math. Phys.* **17** 171.

Hawking S.W. 1979 in *General Relativity: An Einstein Centenary Survey*, S.W. Hawking and W. Israel eds., (Cambridge University Press, Cambridge).

Helgason S. 1962 *Differential Geometry and Symmetric Spaces* (Academic Press, New York).

't Hooft G. 1979 in *Recent Developments in Gravitation. Cargèse 1978* M. Levy and S. Deser eds. (Plenum, New York).

Isham C.J. 1984 in *Relativity, Groups and Topology II*, B.S. DeWitt and R. Stora eds. (North-Holland, Amsterdam).

Isham C.J. 1988 in *Proceedings of the Osgood Hill Conference on Conceptual Problems in Quantum Gravity*, in press.

Isham C.J. 1989 *Quantum topology and quantization on the lattice of topologies*, *Class. Qu. Grav.*, in press.

Jauch J.M. 1973 *Foundations of Quantum Mechanics* (Addison-Wesley, Massachusetts)

Johnstone P.T. 1983 *Bull. Amer. Math. Soc.* **8** 41.

Johnstone P.T. 1986 *Stone Spaces* (Cambridge University Press, Cambridge).

Kelley J.L. 1955 *General Topology* (Van Nostrand, New York).

Klauder J.R. 1970a in *Relativity*, M.S. Carmeli, S.I. Flicker and L. Witten eds. (Plenum Press, New York); 1970b *Comm. Math. Phys.* **18** 307.

Kronheimer E.H. and Penrose R. 1967 *Proc. Camb. Phil. Soc.* **63** 481.

Kuchař K. 1981 in *Quantum Gravity 2: A Second Oxford Symposium*, C.J. Isham, R. Penrose and D.W. Sciama eds. (Oxford University Press, Oxford).

Larson R.L. and Andima S.L. 1975 *Rocky Mount. Jour. Math.* **5** 177.

Mackey G.W. 1963 *Mathematical Foundations of Quantum Mechanics* (Benjamin, New York).

Malament D. 1977 *Jour. Math. Phys.* **18** 1399.

Noyes H.P. 1977 (ed.) *Discrete and Combinatorial Physics: Proceedings of ANPA 9* (ANPA WEST, 25 Buena Vista Way, Mill Valley, CA 94941). (ANPA stands for the "Alternative Natural Philosophy Association".)

Penrose R. 1971 in *Quantum Theory and Beyond*, Ted Bastin ed. (Cambridge University Press, Cambridge).

Penrose R. 1987 in *Three Hundred Years of Gravitation*, S.W. Hawking and W. Israel eds. (Cambridge University Press, Cambridge).

Raghavan T.G. and Reilly I.L. 1986 *Mat. Vesnik* **38** 91.

Rovelli C. and Smolin L. 1988 *Phys. Rev. Lett.* **61** 1155; 1989 *Loop space representation of quantum general relativity*, *Nucl. Phys.* . In press.

Rudin W. 1973 *Functional Analysis* (McGraw-Hill, London).

Schweizer B. and Sklar A. 1983 *Probabilistic Metric Spaces* (North-Holland, New York).

Sorkin R.D. 1983 in *General Relativity and Gravitation: Proceedings of the GR10 Conference*, F. de Felice and A. Pascolini eds. (Consiglio Nazionale delle Ricerche, Rome).

Steiner A.K. 1966 *Trans. Amer. Math. Soc.* **122** 379.

Varadarajan V.S. 1968 *Geometry of Quantum Theory* (Van Nostrand, Princeton).

Vickers S. 1989 *Topology via Logic* (Cambridge University Press, Cambridge).

Wheeler J.A. 1964 in *Relativity, Groups and Topology*, C. DeWitt and B.S. DeWitt eds. (Gordon and Breach, London); 1967 in *Batelle Rencontres 1967*, C. DeWitt and J.A. Wheeler eds. (Benjamin, New York).

TOPICS IN PLANAR PHYSICS *

R. Jackiw †

Department of Physics
Columbia University
538 West 120th Street
New York, NY 10027 U.S.A.

I. OVERVIEW

While the evident goal of physics is to explain phenomena in four-dimensional space-time, where physical Nature resides, and perhaps even to explain why Nature resides in four dimensions, the means that we have come to employ in reaching this goal are sufficiently intricate that it has proven useful to make a detour from the direct path, and to wander into lower-dimensional worlds, with the hope that in the simpler setting we can learn useful things about the agreed upon four-dimensional problem. This indeed has happened, initially in two dimensions, where we first encountered spontaneous gauge symmetry breaking, anomalies, the soliton phenomenon, to name

---

* This work is supported in part by funds provided by the U. S. Department of Energy (D.O.E.) under contract #DE-AC02-76ER03069.

† On sabbatical leave from the Center for Theoretical Physics, Laboratory for Nuclear Science and Department of Physics, Massachusetts Institute of Technology, Cambridge, MA 02139 U.S.A.

Physics, Geometry, and Topology
Edited by H. C. Lee
Plenum Press, New York, 1990

three important examples. Moreover, when it was appreciated that there exist physical environments — not in particle physics but in condensed matter and statistical systems — which are properly described by two-dimensional field theories — *e.g.* linear chains whose time evolution gives rise to two-dimensional dynamics, or planar arrays in equilibrium whose static properties are governed by two-dimensional Euclidean field theory — a physical application of the pedagogical investigations could be made, for example to solitons and fractional charge in polyacetylene or to conformally invariant critical phenomena. Additionally, mathematical and speculative uses for two-dimensional field theories were found in the string program.

Thus the foray into two dimensions proved very useful and it was natural to seek a repetition of these successes in three dimensions. My research in this area began in the early 1980's, when conversations with G. 't Hooft during the 1980 Schladming school[1] persuaded me that three-dimensional gauge theories were very interesting and little understood. Three-dimensions provided an unexplored terrain where discoveries could be made, because most physicists were populating the vast expanses in dimensions greater than four.

In my lectures, I shall describe some of the interesting things that we have found in the intervening decade. The subject has become very large, because many higher- and lower-dimensional colleagues have descended/ascended to three dimensions. Here there is time only for a selection of topics, drawn from the research by my collaborators and by me on geometrical planar models: gauge[2] and gravitational[3] theories. Regrettably I cannot acquaint you with the many interesting results in this area by the Princeton group,[4] nor with the investigations of the Texas group on non-geometrical planar field theories.[5]

The reasons for studying planar theories in three-dimensional space-time are pretty much the same as those put forward above for studying two-dimensional models. First, there is the pedagogical motive: there is still much to learn about quantum field theory whose analysis is more accessible in three dimensions than in four; also there are interesting structures to explore that are peculiar to three dimensions [more generally to odd dimensions]. Second, there are possible physical applications: the high-temperature behavior of four-dimensional field theories is governed by their three-dimensional analogs; interesting condensed matter phenomena like the quantum Hall effect and high-T_c superconductivity appear to involve planar gauge theoretic dynamics; motion in the presence of cosmic strings is adequately described by planar gravity. Third, there are mathematical and speculative applications: field theoretic construction of mathematically interesting three-dimensional characteristics and invariants; a fresh perspective on conformal two-dimensional field theories; description of membranes, which for some represent the next step beyond strings.

II. PLANAR GAUGE THEORIES

A. Topologically Massive Gauge Theories

Gauge theoretic dynamics in any dimension can be governed by the Maxwell/Yang–Mills Lagrange density.

$$\mathcal{L}_{\mathrm{YM}} = \frac{1}{2} \operatorname{tr} F^{\mu\nu} F_{\mu\nu} \qquad (2.A.1)$$

$$F_{\mu\nu} = \partial_\mu A_\nu - \partial_\nu A_\mu + [A_\mu, A_\nu] \qquad (2.A.2)$$

Here A_μ and $F_{\mu\nu}$, the gauge connection and curvature, are anti-Hermitian matrices that belong to the Lie algebra of the gauge group, which is generated by matrices T^a satisfying

$$[T^a, T^b] = f^{abc}T^c \tag{2.A.3a}$$

The T^a are normalized by

$$\operatorname{tr} T^a T^b = -\frac{1}{2}\delta^{ab} \tag{2.A.3b}$$

and provide a basis for expanding A_μ and $F_{\mu\nu}$ in components.

$$A_\mu = A^a_\mu T^a \tag{2.A.4a}$$

$$F_{\mu\nu} = F^a_{\mu\nu} T^a \tag{2.A.4b}$$

Our metric for flat three-dimensional space-time is $\eta_{\mu\nu} = \eta^{\mu\nu} = \operatorname{diag}(1, -1, -1)$.

However, in three dimensions another structure is available that can supplement/replace (2.A.1): the Chern–Simons term.

$$\Omega(A) = -\frac{1}{8\pi^2}\epsilon^{\alpha\beta\gamma} \operatorname{tr}\left(\partial_\alpha A_\beta A_\gamma + \frac{2}{3}A_\alpha A_\beta A_\gamma\right) \tag{2.A.5}$$

For dimensional balance with (2.A.1) the strength κ with which the Chern–Simons term enters dynamics must have dimensionality of mass [in our units where $\hbar, c$, and the gauge coupling are set to unity]. Thus we are led to consider the Lagrange density[2a]

$$\mathcal{L} = \frac{1}{2}\operatorname{tr} F^{\mu\nu}F_{\mu\nu} + 8\pi^2\kappa\Omega(A) = \mathcal{L}_{\mathrm{YM}} + \mathcal{L}_{\mathrm{CS}} \tag{2.A.6a}$$

$$\mathcal{L}_{\mathrm{CS}} = -\kappa\epsilon^{\alpha\beta\gamma} \operatorname{tr}\left(\partial_\alpha A_\beta A_\gamma + \frac{2}{3}A_\alpha A_\beta A_\gamma\right) \tag{2.A.6b}$$

The field equation that follows from (2.A.6) is

$$D_\mu F^{\mu\nu} + \frac{\kappa}{2}\epsilon^{\nu\alpha\beta}F_{\alpha\beta} = 0 \tag{2.A.7}$$

The covariant derivative D_μ acts by differentiation and commutation: $D_\mu \equiv \partial_\mu + [A_\mu, \;\;]$.

$\mathcal{L}_{\mathrm{CS}}$ is not gauge invariant, but changes under the gauge transformation g.

$$A_\mu \to g^{-1}A_\mu g + g^{-1}\partial_\mu g \tag{2.A.8}$$

$$\mathcal{L}_{\mathrm{CS}} \to \mathcal{L}_{\mathrm{CS}} + \kappa\epsilon^{\alpha\beta\gamma}\partial_\alpha \operatorname{tr}\left(\partial_\beta g\, g^{-1}A_\gamma\right) + 8\pi^2\kappa w(g) \tag{2.A.9}$$

$$w(g) \equiv \frac{1}{24\pi^2}\epsilon^{\alpha\beta\gamma} \operatorname{tr}\left(g^{-1}\partial_\alpha g\, g^{-1}\partial_\beta g\, g^{-1}\partial_\gamma g\right) \tag{2.A.10}$$

Since the field equation (2.A.7) is gauge covariant, the change in the Lagrange density must be a total derivative. This is seen explicitly in the next-to-last term of (2.A.9). However, for the last term the identity

$$w(g) = \partial_\alpha w^\alpha(g) \tag{2.A.11}$$

can be established only locally in group space. For example, for the $SU(2)$ gauge group, with g parametrized as $g = e^\lambda$, where λ is in the Lie algebra, one verifies that

$$w^\alpha(g) = \frac{1}{4\pi^2}\epsilon^{\alpha\beta\gamma} \operatorname{tr} \lambda\partial_\beta\lambda\partial_\gamma\lambda\left(\frac{|\lambda| - \sin|\lambda|}{|\lambda|^3}\right) \qquad |\lambda|^2 \equiv -2\operatorname{tr}\lambda^2 \;. \tag{2.A.12}$$

The Chern–Simons action $I_{CS} = \int d^3x \mathcal{L}_{CS}$ remains gauge non-invariant: although the next-to-last term in (2.A.9) integrates to zero [for g tending to the identity at infinity and for sufficiently well-behaved vector potentials], we recognize that the integral of the last term is proportional to the winding number $W(g)$ of g.

$$W(g) = \int d^3x\, w(g) = \frac{1}{24\pi^2} \int d^3x\, \epsilon^{\alpha\beta\gamma}\, \mathrm{tr}\, \left(g^{-1}\partial_\alpha g\, g^{-1}\partial_\beta g\, g^{-1}\partial_\gamma g\right) \qquad (2.A.13)$$

When the gauge group is compact and non-Abelian, and thus has Π_3 equal to $\mathbf{Z}$, $W(g)$ takes integer values. Hence, gauge invariance of the quantum theory [defined by the functional integral of the phase exponential of the action] requires quantizing the coupling constant κ.[2b]

$$\kappa = \frac{n}{4\pi}\ , \qquad n \in \mathbf{Z} \qquad (2.A.14)$$

For a canonical, Hamiltonian description, we work in the Weyl $[A_0 = 0]$ gauge, where the canonical variables are $\mathbf{A}$, while the canonically conjugate momenta $\mathbf{\Pi}$ possess a contribution from the Chern-Simons term.

$$\Pi^i = \dot{A}^i + \frac{\kappa}{2}\epsilon^{ij}A^j \qquad (2.A.15)$$

The Hamiltonian H, when expressed in terms of the electric and magnetic fields, $\mathbf{E} = -\dot{\mathbf{A}}$, and $B = -\frac{1}{2}\epsilon^{ij}F_{ij}$ respectively, does not see the Chern–Simons coupling.

$$H = -\int_{\mathbf{x}} \mathrm{tr}\, \left(E^2 + B^2\right) \qquad (2.A.16)$$

This is a consequence of the topological nature of the Chern–Simons term: even in curved space-time, the generally covariant generalization of $\mathcal{L}_{CS}$ does not make use of the space-time metric tensor $g_{\mu\nu}$ — in contrast to $\mathcal{L}_{YM}$. Therefore, when $g_{\mu\nu}$ is varied to produce the energy-momentum tensor $T^{\mu\nu}$ no contribution arises from the Chern–Simons term and $T^{\mu\nu}$ as well as the energy density and the Hamiltonian retain their Yang–Mills form, when expressed in terms of configuration space variables: $\mathbf{A}$ and its derivatives. Of course the Chern–Simons term reappears when H is expressed in canonical variables $\mathbf{A}$ and $\mathbf{\Pi}$ from (2.A.15).

The Hamiltonian equations that follow from (2.A.16) must be supplemented by a subsidiary condition which coincides with the time component of the field equation (2.A.7) *i.e.* Gauss' law.

$$\mathbf{D} \cdot \mathbf{E} - \kappa B = 0 \qquad (2.A.17a)$$

In terms of canonical variables this reads

$$\mathbf{D} \cdot \mathbf{\Pi} + \frac{\kappa}{2}\nabla \times \mathbf{A} = 0 \qquad (2.A.17b)$$

In the quantum theory, (2.A.17) is imposed as a condition on states. This is most transparent in a Schrödinger representation,[2g,h] where states are functionals of the dynamical variable $\mathbf{A}(\mathbf{x})$,

$$|\Psi\rangle \longleftrightarrow \Psi(\mathbf{A}) \qquad (2.A.18a)$$

on which the operator $\mathbf{A}$ acts by multiplication,

$$\mathbf{A}^a(\mathbf{x})|\Psi\rangle \longleftrightarrow \mathbf{A}^a(\mathbf{x})\Psi(\mathbf{A}) \qquad (2.A.18b)$$

and the canonical momentum operator, by functional differentiation.

$$\boldsymbol{\Pi}^a(\mathbf{x})|\Psi\rangle \longleftrightarrow \frac{1}{i}\,\frac{\delta}{\delta \mathbf{A}^a(\mathbf{x})}\Psi(\mathbf{A}) \tag{2.A.18c}$$

[The Schrödinger representation is at fixed time, hence the time argument of all operators is omitted.] Then (2.A.17) is realized as a functional differential equation that is obeyed by all physical states.

$$\left\{\left(\mathbf{D}\cdot\frac{\delta}{\delta\mathbf{A}}\right)^a + i\frac{\kappa}{2}\nabla\times\mathbf{A}^a\right\}\Psi(\mathbf{A}) = 0 \tag{2.A.19}$$

Gauss' law expresses gauge invariance in a quantum theory: the quantity that must vanish — the left-hand side of (2.A.17) — forms the generator of fixed-time gauge transformations that remain invariances of the theory in the Weyl gauge, and the condition (2.A.19) demands that states be annihilated by this generator. Without the Chern–Simons term, (2.A.19) translates into the statement that physical states are gauge invariant: $\Psi(\mathbf{A}^g) = \Psi(\mathbf{A})$. However, the Chern–Simons term alters the result: states in the quantum theory of (2.A.6) and (2.A.16), satisfying (2.A.19), respond to a gauge transformation with a 1-cocycle.[2g]

$$\Psi(\mathbf{A}^g) = e^{2\pi i\alpha_1(\mathbf{A};g)}\Psi(\mathbf{A}) \tag{2.A.20a}$$

$$\alpha_1(\mathbf{A};g) = -\frac{\kappa}{2\pi}\int_{\mathbf{x}}\epsilon^{ij}\,\mathrm{tr}\,\partial_i g\,g^{-1}A^j + 4\pi\kappa\int_{\mathbf{x}} w^0(g) \tag{2.A.20b}$$

Here $w^0(g)$ is given by (2.A.10) and (2.A.11); $\int d^2\mathbf{x}\,w^0(g)$ is globally ill-defined, owing to an integer ambiguity: it changes by $4\pi\kappa W(g) = 4\pi\kappa\times$ (integer) when g is taken through a smooth closed loop of group elements depending on the spatial two-vector $\mathbf{x}$ and on a homotopy parameter. This multivaluedness does not affect the exponentiated form in (2.A.20a), provided κ is quantized according to (2.A.14). Thus we obtain another perspective on (2.A.14): κ must be quantized so that physical states, which necessarily satisfy (2.A.20a), be single-valued.

Note also from (2.A.9) that $\frac{d}{dt}2\pi\alpha_1$ is precisely the change under a gauge transformation of our Lagrangian $L = \int_{\mathbf{x}}\mathcal{L}$; this examplifies a general relation between a 1-cocycle in the action of a symmetry transformation on states and the non-invariance of the Lagrangian against the symmetry transformation in question.[2g]

The Abelian theory can be analyzed completely and it is established that κ provides a mass for the excitations. We are dealing with a massive "photon," which nevertheless respects gauge invariance. One may explicitly construct the states of this non-interacting, but nevertheless interesting model. For example, the vacuum state Ψ_0, *i.e.* the lowest eigenstate of

$$H = \frac{1}{2}\int_{\mathbf{x}}\left(\left(i\frac{\delta}{\delta A^i(\mathbf{x})} + \frac{\kappa}{2}\epsilon^{ij}A^j(\mathbf{x})\right)^2 + B^2(\mathbf{x})\right) \tag{2.A.21}$$

is

$$\Psi_0(\mathbf{A}) = \left(\exp i\frac{\kappa}{2}\int BA_L\right)\left(\exp -\frac{1}{2}\int A_T^i\sqrt{-\nabla^2+\kappa^2}\,A_T^i\right) \tag{2.A.22a}$$

where $\mathbf{A}$ has been decomposed into its transverse and longitudinal parts.

$$\mathbf{A} = \nabla A_L + \mathbf{A}_T \tag{2.A.22b}$$

The first factor in (2.A.22a) gives rise to the [Abelian] 1-cocycle after a gauge transformation is performed.

$$\Psi_0(\mathbf{A} - \nabla\lambda) = e^{-i\frac{\kappa}{2}\int B\lambda}\Psi_0(\mathbf{A}) \tag{2.A.23}$$

The kernel in the second factor,

$$\left(\sqrt{-\nabla^2 + \kappa^2}\right)(\mathbf{x}, \mathbf{y}) = \int \frac{d^2\mathbf{k}}{(2\pi)^2} e^{-i\mathbf{k}\cdot(\mathbf{x}-\mathbf{y})}\sqrt{k^2 + \kappa^2} \tag{2.A.24}$$

exhibits the massive nature of the excitations. [Higher states are obtained by multiplying Ψ_0 by polynomials in $\mathbf{A}$.]

The spin of the excitation is ± 1, the sign being correlated with the sign of κ.

While the non-Abelian theory cannot be similarly solved, its linear approximation coincides of course with the Abelian model discussed above, and there is no reason to doubt that here too the excitations are massive. For this reason we call the model (2.A.6) a ***topologically massive gauge theory.***

Let us further examine the Chern–Simons modified Gauss law (2.A.17). Specifically in the Abelian case, and also with matter couplings arising from a charge density ρ, (2.A.17a) reads

$$\nabla \cdot \mathbf{E} - \kappa B = \rho \tag{2.A.25}$$

Integrating this over all space gives zero for the integral of $\nabla \cdot \mathbf{E}$, because the gauge invariant electric field is short-range owing to long-distance damping caused by the mass κ; the integral of B is the flux of Φ through the plane and the integral of ρ is the total charge Q. Hence we get

$$\Phi = -\frac{1}{\kappa}Q \tag{2.A.26}$$

This means that particles carrying charge Q also carry magnetic flux $-Q/\kappa$. Since $B = \nabla \times \mathbf{A}$, we further see that the gauge-variant vector potential $\mathbf{A}$ is long-range, so that $\int d^2\mathbf{x}\, \nabla \times \mathbf{A} \neq 0$, while the gauge-invariant magnetic field is short-range, so that $\int d^2\mathbf{x}\, B$ converges. In other words, we are dealing with a vortex-like object.[2a,b]

We conclude this discussion of topologically massive gauge theories with the following observations.

(a) The Chern–Simons term violates P and T, and conserves C and PT.

(b) When fermions couple to the gauge field, a Chern–Simons term is induced by fermion radiative corrections.[2b,4a,6] Fermi fields in three-dimensional space-time are described by two-component spinors, and the three "Dirac" matrices can be chosen to be the 2×2 Pauli matrices. A fermion mass term, constructed from these two-component spinors, also violates P and T; indeed it is the supersymmetric partner of the Chern–Simons mass term.[7] Thus it is natural that massive fermions radiatively generate the Chern–Simons term. However, also massless fermions do so, owing to the mass term that is present in Pauli–Villars regularization, which is needed to preserve gauge invariance against "small" gauge transformations with vanishing winding number. The technical mechanism that induces the Chern–Simons term through fermion loops relies on the trace of three Pauli matrices being non-zero, but proportional to the three-index Levi–Civita anti-symmetric epsilon tensor. Moreover, in the non-Abelian theory the coefficient of the Chern–Simons term, induced by a minimal set of fermions, is not properly quantized. To preserve gauge invariance against "large" gauge transformations with non-zero winding number, while retaining a minimal fermion content, it is necessary to include a "bare" Chern–Simons term so that the total coefficient is properly quantized. As a consequence there is no gauge invariant, parity preserving, non-Abelian gauge theory interacting with a minimal set of fermions. This is the "parity anomaly" of three-dimensions — the lower dimensional analog of the four-dimensional chiral anomaly.

(c) It is to be emphasized that there is no topological quantization of κ in the Abelian theory: Π_3 is trivial. This fact may also be established through a gauge invariant formulation of the Abelian model.[2c] Consider the Lagrange density

$$\mathcal{L} = \frac{1}{2}\epsilon^{\alpha\beta\gamma}\partial_\alpha F_\beta F_\gamma - \frac{\kappa}{2}F_\alpha F^\alpha \tag{2.A.27}$$

leading to the field equation,

$$\epsilon^{\mu\alpha\beta}\partial_\alpha F_\beta - \kappa F^\mu = 0 \tag{2.A.28a}$$

which also implies transversality of F^μ.

$$\partial_\mu F^\mu = 0 \tag{2.A.28b}$$

In three dimensions, a vector is dual to an antisymmetric tensor.

$$F^\mu = \frac{1}{2}\epsilon^{\mu\alpha\beta}F_{\alpha\beta} \quad , \quad F^{\alpha\beta} = \epsilon^{\alpha\beta\mu}F_\mu \tag{2.A.29}$$

Substituting (2.A.29) into (2.A.28a) shows that $F_{\mu\nu}$ satisfies [the Abelian version of] (2.A.7). We recognize that $F_{\alpha\beta}$ is just the gauge curvature, F^μ is its dual, both are gauge-invariant. In the absence of dynamical charged matter [external conserved matter currents can be coupled through $j_\mu F^\mu$], there is no need for a gauge-variant vector potential, which, as a consequence of (2.A.28b) and (2.A.29), can be introduced in topologically simple spaces, where a transverse vector can be written as a curl.

$$F^\mu = \epsilon^{\mu\alpha\beta}\partial_\alpha A_\beta \quad , \quad F_{\alpha\beta} = \partial_\alpha A_\beta - \partial_\beta A_\alpha \tag{2.A.30}$$

Of course A_μ is gauge-variant. However, the gauge symmetry acts trivially on (2.A.27) and does not constrain κ.

B. Non-Abelian Chern–Simons Gauge Theories

Because Eq. (2.A.26) encapsulates the physically novel and important consequences of the Chern–Simons term, it is natural to consider a truncation where (2.A.26) holds locally in space. This is achieved when the kinetic action for the gauge field is just the Chern–Simons term, with no Maxwell/Yang–Mills term. Such a model can be viewed as the $\kappa \to \infty$ limit of (2.A.6) and (2.A.7); it is a physically meaningful truncation at low energy, or at large distance, where the lower-derivative Chern–Simons term dominates the higher-derivative Maxwell term.

The *Chern–Simons theory* [without matter interactions] is governed by the Lagrange density[8]

$$\mathcal{L}_{\text{CS}} = 8\pi^2\kappa\Omega(A) = -\kappa\epsilon^{\alpha\beta\gamma}\,\text{tr}\left(\partial_\alpha A_\beta A_\gamma + \frac{2}{3}A_\alpha A_\beta A_\gamma\right) \tag{2.B.1}$$

with field equation

$$\kappa\epsilon^{\alpha\beta\gamma}F_{\beta\gamma} = 0 \tag{2.B.2}$$

which implies that $F_{\alpha\beta}$ vanishes, and so A_α must be pure gauge, at least locally. Nevertheless, the quantum theory retains non-trivial, interesting features. Of course, the gauge properties of the Chern–Simons theory are the same as those of the topologically massive model; in particular for non-Abelian gauge groups κ is quantized as in (2.A.14).

The canonical formulation in the Weyl gauge begins with (2.B.1) at $A_0 = 0$.

$$\mathcal{L}_{\mathrm{CS}} = \frac{\kappa}{2}\epsilon^{ij}\dot{A}_i^a A_j^a \tag{2.B.3}$$

The Hamiltonian vanishes. The Euler–Lagrange equation which follows from (2.B.3)

$$\dot{A}_i^a = 0 \tag{2.B.4}$$

coincides with the spatial component of (2.B.2) when $A_0 = 0$, while the time component — the Chern–Simons Gauss law — is imposed as a constraint.

$$G^a \equiv -\frac{\kappa}{2}\epsilon^{ij}F_{ij}^a = 0 \tag{2.B.5}$$

In the Weyl gauge, the theory is invariant under static gauge transformations,

$$\delta A_i^a = \partial_i\lambda^a + f^{abc}A_i^b\lambda^c \quad , \tag{2.B.6}$$

which are generated by

$$G = \int_{\mathbf{x}} \lambda^a G^a \quad . \tag{2.B.7}$$

Thus the constraint sets the generator to zero.

$\mathcal{L}_{\mathrm{CS}}$ is first-order in time derivatives, and the quantization of the corresponding symplectic structure leads to non-trivial equal-time commutation relations between vector potentials.[2c]

$$\left[A_i^a(\mathbf{x}), A_j^b(\mathbf{y})\right] = \frac{i}{\kappa}\epsilon_{ij}\delta^{ab}\delta^2(\mathbf{x}-\mathbf{y}) \tag{2.B.8}$$

Consequently, the generators follow the group's Lie algebra,

$$i\left[G^a(\mathbf{x}), G^b(\mathbf{y})\right] = f^{abc}G^c(\mathbf{x})\delta^2(\mathbf{x}-\mathbf{y}) \tag{2.B.9}$$

and there is no apparent obstruction to demanding that the constraint (2.B.5) be met by requiring that $G^a(\mathbf{x})$ annihilate physical states.

$$G^a(\mathbf{x})|\Psi\rangle = 0 \tag{2.B.10}$$

However, we show that (2.B.10) in fact cannot be satisfied unless κ is quantized. The Gauss law is all there is to this theory; since the Hamiltonian vanishes, (2.B.4) is trivially satisfied. Equation (2.B.10) is most readily analyzed on the Schrödinger representation.

In order to give a Schrödinger representation for the canonical algebra (2.B.8), we have to decide which operator is realized by multiplication, which by [functional] differentiation, and on what function(s) the state functionals depend. This procedure of dividing phase space into "coordinates" and "momenta" is called choosing a *polarization*.

We choose a *Cartesian polarization*: the state functionals depend on A_1^a, which we call φ^a, and so A_2^a is realized by functional differentiation with respect to φ^a.

$$|\Psi\rangle \longleftrightarrow \Psi(\varphi) \tag{2.B.11a}$$

$$A_1^a(\mathbf{x})|\Psi\rangle \longleftrightarrow \varphi^a(\mathbf{x})\Psi(\varphi) \ , \tag{2.B.11b}$$

$$A_2^a(\mathbf{x})|\Psi\rangle \longleftrightarrow \frac{1}{i\kappa}\,\frac{\delta}{\delta\varphi^a(\mathbf{x})}\Psi(\varphi) \ . \tag{2.B.11c}$$

Other polarizations are available. For example, A_i^a may be decomposed into longitudinal and transverse parts, which serve as coordinates and momenta. This choice has the advantage of being rotationally invariant, unlike the Cartesian polarization, and also permits a very simple treatment of the Abelian theory — a fact that we shall exploit later. However, non-Abelian gauge transformations are represented very awkwardly in this polarization, because they do not respect the longitudinal/transverse decomposition.

Another possible choice is a holomorphic polarization that uses the non-Hermitian pair: $A^a \equiv \frac{1}{\sqrt{2}}(A_1^a + iA_2^a)$ and $A^{a*} \equiv \frac{1}{\sqrt{2}}(A_1^a - iA_2^a)$. This too will be briefly described.

We now show that the action of the gauge group on states is realized with a 1-cocycle.[2d] The exponential of the generator G is the unitary operator $U(g)$ that implements a finite gauge transformation g,

$$\begin{aligned} U(g) &= \exp\left(i\int_{\mathbf{x}}\lambda^a G^a\right) \\ g &= e^{\lambda} \end{aligned} \tag{2.B.12}$$

and according to (2.B.9), the composition law follows that of the group.

$$U(g_1)U(g_2) = U(g_1 g_2) \tag{2.B.13}$$

In the Cartesian polarization (2.B.11), the generator is a functional differential operator, acting on functionals of φ,

$$\begin{aligned} \int_{\mathbf{x}}\lambda^a G^a &= -\int_{\mathbf{x}}\lambda^a(\mathbf{x})\left(\partial_1\frac{1}{i}\,\frac{\delta}{\delta\varphi^a(\mathbf{x})} + f^{abc}\varphi^b(\mathbf{x})\frac{1}{i}\,\frac{\delta}{\delta\varphi^c(\mathbf{x})}\right) - \kappa\int_{\mathbf{x}}\varphi^a(\mathbf{x})\partial_2\lambda^a(\mathbf{x}) \\ &\equiv G_\varphi + 2\kappa\int_{\mathbf{x}}\mathrm{tr}\,(\varphi\partial_2\lambda) \ . \end{aligned} \tag{2.B.14}$$

G_φ generates infinitesimal gauge transformations on $\varphi^a = A_1^a$; the last term, needed to generate the transformation on $\delta/i\kappa\delta\varphi^a = A_2^a$, is responsible for the 1-cocycle.

$$\begin{aligned} U(g)\Psi(\varphi) &= e^{iG}e^{-iG_\varphi}\Psi(\varphi^g) \\ \varphi^g &\equiv g^{-1}\varphi g + g^{-1}\partial_1 g \end{aligned} \tag{2.B.15}$$

The prefactor $e^{iG}e^{-iG_\varphi}$ is evaluated by introducing a homotopy parameter τ, $\tau \in [0,1]$, and solving a differential equation in τ.

$$\begin{aligned} -i\frac{\partial}{\partial\tau}\left(e^{i\tau G}e^{-i\tau G_\varphi}\right) &= \left(e^{i\tau G}e^{-i\tau G_\varphi}\right)\left(2\kappa\int_{\mathbf{x}}\mathrm{tr}\,(\varphi^{g_\tau}\partial_2\lambda)\right) \\ g_\tau &= e^{\tau\lambda} \end{aligned} \tag{2.B.16}$$

The last factor in (2.B.16) reads

$$\begin{aligned}2\kappa\int_{\mathbf{x}} \operatorname{tr}\,(\varphi^{g_\tau}\partial_2\lambda) &= 2\kappa\int_{\mathbf{x}} \operatorname{tr}\,(g_\tau^{-1}\varphi g_\tau\partial_2\lambda) + 2\kappa\int_{\mathbf{x}} \operatorname{tr}\,(g_\tau^{-1}\partial_1 g_\tau\partial_2\lambda) \\ =&2\kappa\int_{\mathbf{x}} \operatorname{tr}\,(\varphi g_\tau\partial_2\lambda g_\tau^{-1}) + \kappa\int_{\mathbf{x}} \operatorname{tr}\,(g_\tau^{-1}\partial_1 g_\tau\partial_2\lambda + g_\tau^{-1}\partial_2 g_\tau\partial_1\lambda) \\ &+\kappa\int_{\mathbf{x}} \epsilon^{ij}\operatorname{tr}\,(g_\tau^{-1}\partial_i g_\tau\partial_j\lambda)\end{aligned} \tag{2.B.17}$$

Since $\partial_\tau g_\tau = g_\tau\lambda$, the first term in the last equality is recognized as the τ derivative of $2\kappa\int \operatorname{tr}\,(\varphi\partial_2 g_\tau g_\tau^{-1})$, and the second as the τ derivative of $\kappa\int_{\mathbf{x}} \operatorname{tr}\,(g_\tau^{-1}\partial_1 g_\tau g_\tau^{-1}\partial_2 g_\tau)$. The last term in (2.B.17), after an integration by parts and use of (2.A.10) and (2.A.11), is seen to equal

$$\begin{aligned}\kappa\int_{\mathbf{x}} \epsilon^{ij}\operatorname{tr}\,(g_\tau^{-1}\partial_i g_\tau\partial_j\lambda) &= -\kappa\int_{\mathbf{x}} \operatorname{tr}\,(g_\tau^{-1}\partial_0 g_\tau g_\tau^{-1}\partial_i g_\tau g_\tau^{-1}\partial_j g_\tau) \\ &= -\frac{\kappa}{3}\int_{\mathbf{x}} \epsilon^{\alpha\beta\gamma}\operatorname{tr}\,(g_\tau^{-1}\partial_\alpha g_\tau g_\tau^{-1}\partial_\beta g_\tau g_\tau^{-1}\partial_\gamma g_\tau) \\ &= -8\pi^2\kappa\int_{\mathbf{x}} w(g_\tau) \\ &= -8\pi^2\kappa\frac{d}{d\tau}\int_{\mathbf{x}} w^0(g_\tau)\ .\end{aligned} \tag{2.B.18}$$

Thus $e^{iG}e^{-iG_\varphi} = e^{-2\pi i\alpha_1(\varphi,g)}$ where

$$\alpha_1(\varphi,g) = -\frac{\kappa}{2\pi}\int_{\mathbf{x}}\left\{\operatorname{tr}\,(2\varphi\partial_2 g\,g^{-1} + g^{-1}\partial_1 g\,g^{-1}\partial_2 g)\right\} + 4\pi\kappa\int_{\mathbf{x}} w^0(g) \tag{2.B.19}$$

and (2.B.15) becomes

$$U(g)\Psi(\varphi) = e^{-2\pi i\alpha_1(\varphi;g)}\Psi(\varphi^g)\ . \tag{2.B.20}$$

The last term in (2.B.19), which appears also in the 1-cocycle of topologically massive gauge theories [see (2.A.20b)], is multivalued for the same reason.

The Gauss law constraint (2.B.10) requires that physical states $\Psi(\varphi)$ be left unchanged by the action of $U(g)$, since the generator annihilates them.

$$U(g)\Psi(\varphi) = \Psi(\varphi) \tag{2.B.21}$$

Therefore in this theory, as in topologically massive gauge theories, functionals describing physical states are not gauge invariant; rather, according to (2.B.20), they satisfy

$$\Psi(\varphi^g) = e^{2\pi i\alpha_1(\varphi;g)}\Psi(\varphi)\ . \tag{2.B.22}$$

Only when $4\pi\kappa$ is an integer can this condition be met with single-valued functionals.

As indicated earlier, it is generally true that when a symmetric theory is described by a Lagrangian that changes by a total time derivative under a finite symmetry transformation, $L \to L + \frac{d}{dt}2\pi\alpha$, the 1-cocycle is just α, evaluated at fixed time. To verify this for the Chern–Simons theory, we must cast the Lagrangian in phase space form: the kinetic term should involve $p\dot{q}$, *i.e.* $\kappa A_2^a\dot{A}_1^a$ rather than $\frac{\kappa}{2}\epsilon^{ij}\dot{A}_i^a A_j^a$. This is achieved by subtracting $\kappa\frac{d}{dt}\operatorname{tr}\,(A_1A_2)$ from (2.B.3).

$$\tilde{L}_{\mathrm{CS}} = L_{\mathrm{CS}} - \kappa\frac{d}{dt}\int_{\mathbf{x}} \operatorname{tr}\,(A_1A_2) \tag{2.B.23}$$

Then from (2.A.8), (2.A.9), (2.A.10) and (2.A.11) it follows that $\tilde{L}_{\mathrm{CS}}$ transforms as

$$
\begin{aligned}
&\tilde{L}_{\mathrm{CS}} \to \tilde{L}_{\mathrm{CS}} \\
&+ \frac{d}{dt}\kappa \int_{\mathbf{x}} \Big\{ \mathrm{tr}\left(\epsilon^{ij}\partial_i g\, g^{-1} A_j - \partial_1 g\, g^{-1} A_2 - A_1 \partial_2 g\, g^{-1} - g^{-1}\partial_1 g\, g^{-1}\partial_2 g \right) \\
&+ 8\pi^2 \omega^0(g) \Big\} \\
&= \tilde{L}_{\mathrm{CS}} + \frac{d}{dt}\kappa \int_{\mathbf{x}} \{ -\,\mathrm{tr}\ (2\varphi \partial_2 g\, g^{-1} + g^{-1}\partial_1 g\, g^{-1} \partial_2 g) + 8\pi^2 w^0(g) \} \quad ,
\end{aligned}
\tag{2.B.24}
$$

in agreement with the general result and with (2.B.19).

We see that the 1-cocycle for the Chern–Simons theory is essentially the same as the one in(2.A.20) for the topologically massive gauge theory, once differences in polarization are taken into account: in the former there is a single field variable $\varphi^a = A_1^a$, while the latter is described by a pair A_i^a, $i = 1, 2$. In particular, the multivalued contribution to each is the same.

Next we construct explicitly states that obey (2.B.22), thus solving the Gauss law constraint. To this end we write

$$
\Psi(\varphi) = e^{2\pi i \alpha_0(\varphi)} \psi(\varphi)
\tag{2.B.25}
$$

and seek a quantity $\alpha_0(\varphi)$, called a *cochain*, that satisfies

$$
\alpha_0(\varphi^g) - \alpha_0(\varphi) = \alpha_1(\varphi; g)
\tag{2.B.26}
$$

Then (2.B.25) solves (2.B.22) with gauge invariant $\psi(\varphi)$.

If Eq. (2.B.26) holds the 1-cocycle α_1 is *trivial* — it is a *coboundary*. It is known that α_1 is non-trivial in *local* cohomology, but a *spatially non-local* functional that trivializes α_1 can be constructed. It is easy to verify that

$$
\alpha_0(\varphi) = 4\pi\kappa \int_{\mathbf{x}} w^0(h) - \frac{\kappa}{2\pi}\int_{\mathbf{x}} \mathrm{tr}\ (\varphi h^{-1}\partial_2 h) \quad ,
\tag{2.B.27}
$$

where h is defined by the non-local relation

$$
\varphi \equiv h^{-1}\partial_1 h \quad ,
\tag{2.B.28}
$$

solves (2.B.26) and therefore trivialized the cocycle (2.B.19).

It is to be remembered that the multivalued contribution to the trivializing functional (2.B.27) is related to the effective action of chiral fermions coupled to an external gauge field in two [Euclidean] dimensions. This connection arises because the fermion determinant is not gauge invariant; under a gauge transformation its change is related to the Chern–Simons 1-cocycle.[9]

The gauge invariant functional $\psi(\varphi)$ in (2.B.25) is formed solely from $\varphi = A_1$. It must be constructed from path-ordered exponential integrals of φ along x^1 at fixed x^2; *e.g.* closed Wilson loops $\Phi(C_{x^1}; x^2) = \mathrm{tr}\, P \exp \int_{C_{x^1}} dx^1 \varphi(x^1, x^2)$ where P denotes path ordering. [In two-dimensional, sourceless electrodynamics analogous holonomies around closed loops in the single spatial direction are the only surviving degrees of freedom in the quantized theory; they give rise to the vacuum angle and probe a possible vacuum electric field.[2h]] Whether such one-dimensional closed loops, or other gauge invariant constructions, exist depends on the topology of the two-dimensional space-like manifold on which the fixed-time canonical formalism is defined. Apart from this gauge-invariant functional, physical states of the non-Abelian Chern–Simons theory are given by

$$\Psi(\varphi) = N\, e^{2\pi i \alpha_0(\varphi)} \tag{2.B.29}$$

where N provides normalization and the other factor is related to the two-dimensional chiral fermion determinant — a non-local functional, as is normal for a physical wave functional. The necessary quantization of κ, (2.A.14), is again evident: from (2.B.25) and (2.B.29) we see that $\alpha_0(\varphi)$ contains the multi-valued term $4\pi\kappa \int_{\mathbf{x}} \omega^0(h)$, which with unquantixed κ would render $\Psi(\varphi)$ in (2.B.25) or (2.B.29) multi-valued.

Since the Hamiltonian vanishes, the physical states (2.B.25) or (2.B.29) solve the Chern–Simons theory for all times. It is instructive to demonstrate explicitly that the vector potential A_i acting on the state (2.B.29) is a pure gauge. To exhibit the action of $A_2^a = \delta / i\kappa\delta\varphi^a$ on $\Psi(\varphi)$ we need $\delta\alpha_0(\varphi)/i\kappa\delta\varphi^a$. This can be found from the definition (2.B.28), which implies $\partial_1(\delta h\, h^{-1}) = h\delta\varphi h^{-1}$, and from the formula (2.B.27) for $2\pi\alpha_0$, which has the consequence that $\delta(2\pi\alpha_0) = -2\kappa \int_{\mathbf{x}} \mathrm{tr}\,\{\partial_2 h\, h^{-1}\partial_1(\delta h\, h^{-1})\}$. Thus

$$\begin{aligned} A_2(\mathbf{x})\Psi(\varphi) &= \frac{2\pi}{\kappa} T^a \frac{\delta\alpha_0(\varphi)}{\delta\varphi^a(\mathbf{x})}\Psi(\varphi) \\ &= h^{-1}(\mathbf{x})\partial_2 h(\mathbf{x})\Psi(\varphi) \ . \end{aligned} \tag{2.B.30a}$$

Together with (2.B.28)

$$A_1(\mathbf{x})\Psi(\varphi) = h^{-1}(\mathbf{x})\partial_1 h(\mathbf{x})\Psi(\varphi) \tag{2.B.30b}$$

the desired result is obtained.

$$A_i\Psi(\varphi) = h^{-1}\partial_i h\Psi(\varphi) \tag{2.B.30c}$$

The wave-functional (2.B.29) has φ-independent norm $|N|^2$, hence it cannot be normlized by [funcitonal] integration over φ. This is to be expected of states that satisfy Gauss' law, because the Gauss law operator has a continuous spectrum. The resolution of course is that the integration measure $\mathcal{D}\varphi$ must be gauge fixed — $\delta(\varphi)$ is a natural choice leading to trivial integrals.

Note that it is possible to formulate the theory in terms of a gauge invariant, spatially non-local Lagrangian. From (2.B.24) it follows that the gauge transform of $\tilde{L}_{\mathrm{CS}}(A)$ is

$$\tilde{L}_{\mathrm{CS}}(A^g) = \tilde{L}_{\mathrm{CS}}(A) + \frac{d}{dt} 2\pi\alpha_1(\varphi; g) \tag{2.B.31a}$$

With (2.B.26), this can be presented as

$$\tilde{L}_{\mathrm{CS}}(A^g) - \frac{d}{dt}2\pi\alpha_0(\varphi^g) = \tilde{L}_{\mathrm{CS}}(A) - \frac{d}{dt}2\pi\alpha_0(\varphi) \tag{2.B.31b}$$

This means that the equivalent Lagrangian

$$\begin{aligned} L_{\mathrm{CS}}^{\mathrm{invariant}}(A) &\equiv \tilde{L}_{\mathrm{CS}}(A) - \frac{d}{dt}2\pi\alpha_0(\varphi) \\ &= \tilde{L}_{\mathrm{CS}}(A) - 2\pi \int_{\mathbf{x}} \dot{\varphi}^a \frac{\delta\alpha_0(\varphi)}{\delta\varphi^a} \\ &= -2\kappa \int_{\mathbf{x}} \mathrm{tr}\ (A_2 - h^{-1}\partial_2 h)\,\dot{A}_1 \end{aligned} \tag{2.B.31c}$$

is invariant against time-independent gauge transformations ($h^g \equiv hg$). In (2.B.31c), $h^{-1}\partial_2 h$ is the non-local functional of $\varphi = A_1$, defined by (2.B.28); $A_2 - h^{-1}\partial_2 h$ does *not* vanish here, only when acting on physical states.

The same results may be presented in the holomorphic representation, wherein states are functionals of a complex function $\mathcal{A}^a$ and $A^{a*} \equiv \frac{1}{\sqrt{2}}(A_1^a - iA_2^a)$ is realized by multiplication by $\mathcal{A}^a$ while $A^a \equiv \frac{1}{\sqrt{2}}(A_1^a + iA_2^a)$ acts by functional differentiation.

$$|\Psi\rangle \longrightarrow \Psi(\mathcal{A}) \tag{2.B.32a}$$

$$A^{a*}(\mathbf{x})|\Psi\rangle \longrightarrow \mathcal{A}^a(\mathbf{x})\Psi(\mathcal{A}) \tag{2.B.32b}$$

$$A^a(\mathbf{x})|\Psi\rangle \longrightarrow \frac{\delta}{\delta\mathcal{A}^a(\mathbf{z})}\Psi(\mathcal{A}) \tag{2.B.32c}$$

This action reproduces the commutator between A and A^*.

$$\left[A^a(\mathbf{x}), A^{*b}(\mathbf{y})\right] = \frac{1}{\kappa}\delta^{ab}\delta(\mathbf{x}-\mathbf{y}) \tag{2.B.33}$$

The adjoint relationship between the two operators is maintained, provided inner products involve a non-trivial measure in the functional integral

$$\langle\Psi_1|\Psi_2\rangle = \int \mathcal{D}\mathcal{A}\mathcal{D}\mathcal{A}^*\, e^{-\kappa\int_{\mathbf{x}} \mathcal{A}^{c*}(\mathbf{x})\mathcal{A}^c(\mathbf{x})}\Psi_1^*(\mathcal{A})\Psi_2(\mathcal{A}) \tag{2.B.34a}$$

where

$$\mathcal{D}\mathcal{A}\mathcal{D}\mathcal{A}^* = \mathcal{D}A_1^a\,\mathcal{D}A_2^a / \det(2\pi i) \tag{2.B.34b}$$

A development paralleling the previous discussion in the Cartesian polarization results in a wave functional in the holomorphic polarization analogous to (2.B.27) and (2.B.28).

$$\Psi(\mathcal{A}) = N\, e^{2\pi i\alpha_0(\mathcal{A})} \tag{2.B.35a}$$

$$\alpha_0(\mathcal{A}) = \frac{i\kappa}{2\pi}\int_{\mathbf{x}} \operatorname{tr}\left(\mathcal{A}h^{-1}\partial_+ h\right) + 4\pi\kappa\int_{\mathbf{x}}\omega^0(h) \tag{2.B.35b}$$

Here h is defined through

$$\mathcal{A} \equiv h^{-1}\partial_- h \tag{2.B.36}$$

and $\partial_\pm \equiv \frac{1}{\sqrt{2}}(\partial_1 \pm i\partial_2)$. The multivalued phase is of course encountered once again, while the integration measure $\exp -\kappa\int_{\mathbf{x}}\mathcal{A}^{c*}\mathcal{A}^c$ insures convergence of the functional integrals — no further gauge fixing is needed.

We conclude this presentation of the quantum Chern–Simons theory by noting that the Gauss law constraint is here solved after quantization. One may alternatively solve it classically, and quantize the remaining degrees of freedom. In general the two procedures do not commute, as is seen from the following example.[2d]

Consider the quantum mechanics for planar motion of a point particle described by the two-vector $\mathbf{q} = (q^1, q^2)$. The Lagrangian

$$L = \frac{1}{2}\dot{\mathbf{q}}^2 - V(q) \tag{2.B.37}$$

is invariant under rotations through the angle θ,

$$\delta q^i = -\theta\epsilon^{ij}q^j \tag{2.B.38}$$

when $V(q)$ depends only on the magnitude of $\mathbf{q}$. Here q^i is a function of time t, and the overdot denotes differentiation with respect to t. However, θ is time-independent — the rotation in (2.B.38) is "global." In the usual way, one knows that invariance under (2.B.38) implies conservation of angular momentum $J = \mathbf{q} \times \mathbf{p}$, where $\mathbf{p} \equiv \frac{\partial L}{\partial \dot{\mathbf{q}}} = \dot{\mathbf{q}}$, and the familiar Hamiltonian operator for fixed angular momentum ℓ is

$$H_\ell = -\frac{\hbar^2}{2}\frac{\partial^2}{\partial r^2} + \frac{\hbar^2\left(\ell^2 - \frac{1}{4}\right)}{2r^2} + V(r) \quad , \qquad r \geq 0 \ . \tag{2.B.39}$$

[We temporarily restore Planck's constant.] One may promote the global symmetry under rotations (2.B.38) to a "local" gauge symmetry by introducing into (2.B.37) a "gauge potential" $a(t)$,

$$L_a = \frac{1}{2}\left(\dot{q}^i + a\epsilon^{ij}q^j\right)\left(\dot{q}^i + a\epsilon^{ik}q^k\right) - V(q) \tag{2.B.40}$$

that transforms under time-dependent rotations.

$$\delta a = \dot{\theta} \tag{2.B.41}$$

In the Weyl gauge, $a = 0$, L_a coincides with (2.B.37), but now there is an additional constraint that captures the Lagrangian equation of motion obtained from (2.B.40) by varying a: the rotation generator J must annihilate physical states, which therefore are only s-states. If the constraint is solved classically, the classical Hamiltonian for rotationally invariant motion with vanishing angular momentum is

$$H_{\ell=0}^{\text{classical}} = \frac{p_r^2}{2} + V(r) \quad , \tag{2.B.42a}$$

whose quantized form, obtained with the naive replacement $p_r^2 \to -\hbar^2\partial_r^2$,

$$H_{\ell=0}^{\text{classical}} \longrightarrow -\frac{\hbar^2}{2}\frac{\partial^2}{\partial r^2} + V(r) \tag{2.B.42b}$$

does not reproduce the quantum s-wave Hamiltonian that survives from (2.B.39) if the constraint is imposed *after* quantization.

$$H_{\ell=0} = -\frac{\hbar^2}{2}\frac{\partial^2}{\partial r^2} - \frac{\hbar^2}{8r^2} + V(r) \tag{2.B.43}$$

The $\mathcal{O}(\hbar^2)$ difference between (2.B.42b) and (2.B.43) can be viewed as an ordering ambiguity *i.e.* (2.B.43) follows from (2.B.42a) if p_r^2 is taken to be $-\frac{\hbar^2}{\sqrt{r}}\partial_r r \partial_r \frac{1}{\sqrt{r}}$, but without further information there is no way to justify the "correct" choice. No such ambiguity arises if one imposes the constraint after quantization. [Henceforth we return $\hbar$ to unity.]

This is not to say that there will always be a discrepancy when phase space is reduced before or after quantization. For example, in many-body quantum mechanics, the passage to the center-of-mass rest frame [which may be formulated as a gauge principle for translations[28]] produces the same quantum theory whether it is carried out before or after quantization. Similarly, enforcing the Gauss law in quantum electrodynamics does not involve ordering ambiguities. It is therefore surprising that we find non-commutativity of quantization and phase-space reduction already for the Abelian Chern–Simons theory in flat space, which I shall now describe.

C. Abelian Chern–Simons Gauge Theory with Sources

We begin with the Lagrange density

$$\mathcal{L}_{CS} = \frac{\kappa}{2}\epsilon^{\alpha\beta\gamma}\partial_\alpha A_\beta A_\gamma - A_\mu j^\mu \tag{2.C.1}$$

where $j^\mu = (\rho, \mathbf{j})$ is the conserved matter current with time-independent charge.

$$Q = \int_{\mathbf{x}} \rho(t, \mathbf{x}) \tag{2.C.2}$$

Here the fields are real functions, and the coupling constant is absorbed in the definition of the current j^μ. We leave the matter Lagrangian unspecified; indeed we take the current to be an external, conserved, c-number source. The previous canonical development holds in this theory, except that the Hamiltonian does not vanish,

$$H = \int_{\mathbf{x}} A_i j^i \tag{2.C.3}$$

and the Gauss law constraint acquires an inhomogeneous term.

$$\epsilon^{ij}\partial_i A_j = -B = \frac{1}{\kappa}\rho \tag{2.C.4}$$

This constraint on the magnetic field B implies that particles with charge Q are also flux-tubes for A-flux, $\Phi = -Q/\kappa$, and leads to exotic statistics and angular momentum of the charge- and flux-carrying particles.[10]

In contrast to the non-Abelian theory, here the gauge field contribution to the constraint (2.C.4) is linear, and may be chosen to be the momentum conjugate to a coordinate θ. This is achieved by decomposing A_i into its longitudinal and transverse components,

$$\begin{aligned} A_i &= \partial_i\theta + \epsilon^{ij}\partial_j^{-1}B \\ \partial_j^{-1} &\equiv \partial_j/\nabla^2 \end{aligned} \tag{2.C.5}$$

The decomposition (2.C.5) is unique and well-defined provided there are no zero modes of the two-dimensional Laplacian ∇^2. This we assume here; indeed, we consider space-time to be Minkowskian.

The commutation relation (2.B.8) implies that B and θ form a canonical pair.

$$[\theta(\mathbf{x}), B(\mathbf{y})] = \frac{i}{\kappa}\delta^2(\mathbf{x}-\mathbf{y}) \tag{2.C.6}$$

In the Schrödinger representation we realize B as a functional derivative with respect to the coordinate, θ, $B = \delta/i\kappa\delta\theta$. This is the *rotationally invariant polarization*.

The constraint (2.C.4) reduces to

$$\left[\frac{1}{i}\frac{\delta}{\delta\theta(\mathbf{x})} + \rho(\mathbf{x})\right]\Psi = 0 \ , \tag{2.C.7}$$

and is solved by

$$\Psi(\theta; t) = N(t)\exp\left[-i\int_{\mathbf{x}}\rho\theta\right] \ , \tag{2.C.8}$$

where $N(t)$ is a θ-independent, but possibly time-dependent normalization factor. $N(t)$ is then determined by requiring that $\Psi(\theta;t)$ satisfies the time-dependent Schrödinger equation.

$$i\partial_t \Psi(\theta;t) = \left[\int_{\mathbf{x}} j^i A_i\right] \Psi(\theta;t) \tag{2.C.9}$$

Inserting the decomposition (2.C.5) and using the continuity equation for the matter current, we find

$$N(t) = \exp -\frac{i}{\kappa}\int_0^t dt' \int_{\mathbf{x}} \rho(t',\mathbf{x}) j(t',\mathbf{x}) \ , \tag{2.C.10}$$

where we have written j^i in terms of its longitudinal and transverse parts.

$$j^i = -\partial_i^{-1}\dot{\rho} + \epsilon^{ij}\partial_j j \tag{2.C.11}$$

$N(t)$ is normalized to 1 at $t = 0$. Note that for static matter the state Ψ is an eigenstate of the Hamiltonian with energy eigenvalue

$$E = \frac{1}{\kappa}\int_{\mathbf{x}} \rho j \ . \tag{2.C.12}$$

We see that the theory admits a unique physical state, which in the absence of sources is described by the wave functional $\Psi = 1$.

Of course, the above development can alternatively be presented in the Cartesian polarization $A_1 = \varphi$, $A_2 = \delta/i\kappa\delta\varphi$, which we used for the non-Abelian theory. In the absence of external sources, the unique physical state that satisfies Gauss's law is

$$\Psi(\varphi) = N \exp\left[\frac{i\kappa}{2}\int \varphi\frac{\partial_2}{\partial_1}\varphi\right] \ , \tag{2.C.13}$$

in agreement with (2.B.25), (2.B.27) and (2.B.29). Ψ responds to a gauge transformation by

$$\begin{aligned} \Psi\left(\varphi^g\right) &= \Psi\left(\varphi + \partial_1\lambda\right) = e^{i2\pi\alpha_1(\varphi;\lambda)}\Psi(\varphi) \ , \\ \alpha_1\left(\varphi;\lambda\right) &= \frac{\kappa}{2\pi}\int_{\mathbf{x}}\left[\varphi\partial_2\lambda + \frac{1}{2}\partial_1\lambda\partial_2\lambda\right] \ , \end{aligned} \tag{2.C.14}$$

in agreement with (2.B.22).

The unitary transformation functional that connects the Cartesian polarization to the rotationally invariant one is

$$\langle\theta|\varphi\rangle = \det{}^{1/2}\left(\frac{\kappa}{2\pi}\frac{\Delta}{\partial_2}\right)\exp\left[\frac{i\kappa}{2}\int\left\{\left(\partial_i\theta - \frac{\partial_i}{\partial_1}\varphi\right)\frac{\partial_1}{\partial_2}\left(\partial_i\theta - \frac{\partial_i}{\partial_1}\varphi\right) - \varphi\frac{\partial_2}{\partial_1}\varphi\right\}\right] \tag{2.C.15}$$

$$\Psi(\theta) = \langle\theta|\Psi\rangle = 1 \ , \qquad \Psi(\varphi) = \langle\varphi|\Psi\rangle = N\exp\left[\frac{i\kappa}{2}\int\varphi\frac{\partial_2}{\partial_1}\varphi\right]$$

In the rotationally invariant polarization the physical state, in the absence of external sources, $\Psi = 1$, is obviously gauge invariant. This fact realizes the gauge invariant formulation described previously: in the rotationally invariant polarization the Lagrangian, apart from a total time derivative, is invariant against time-independent gauge transformations.

$$\tilde{L}_{CS} = \kappa\int_{\mathbf{x}} B\dot{\theta} - \frac{\kappa}{2}\frac{d}{dt}\int_{\mathbf{x}} B\theta \tag{2.C.16a}$$

$$L_{CS}^{\text{invariant}} = \kappa\int_{\mathbf{x}} B\dot{\theta} \tag{2.C.16b}$$

Finally in the holomorphic polarization (2.B.32) the state of the Abelian theory is

$$\Psi(\mathcal{A}) = N \exp \frac{\kappa}{2} \int \mathcal{A} \frac{\partial_+}{\partial_-} \mathcal{A} \tag{2.C.17}$$

We have already remarked that the Chern–Simons theory may be viewed as the $\kappa \to \infty$ limit of the topologically massive model. It is interesting to examine in detail how one theory passes to the other, and this we can do explicitly for the wave functionals of the Abelian theory. For $\mathcal{L}$ of (2.A.6a) to pass into $\mathcal{L}_{CS}$ of (2.A.6b) it suffices [in the Abelian case] to rescale the potential by $\sqrt{\frac{\kappa'}{\kappa}}$ and set κ to infinity, with κ' remaining as the coefficient of the Chern–Simons Lagrange density. Performing this limit on the ground state wave functional (2.A.22a) leaves

$$\Psi_0(\mathbf{A}) \underset{\kappa\to\infty}{\longrightarrow} \left(\exp i\frac{\kappa'}{2} \int BA_L\right) \left(\exp -\frac{\kappa'}{2} \int \mathbf{A}_T \cdot \mathbf{A}_T\right) \tag{2.C.18a}$$

In terms of complex variables $\mathcal{A} \equiv \frac{1}{\sqrt{2}}(A_1 - iA_2)$ and $\mathcal{A}^* = \frac{1}{\sqrt{2}}(A_1 + iA_2)$, (2.C.18a) reads

$$\Psi_0(\mathbf{A}) \underset{\kappa\to\infty}{\longrightarrow} \exp\left(\frac{\kappa'}{2} \int \mathcal{A} \frac{\partial_+}{\partial_-} \mathcal{A}\right) \exp -\frac{\kappa'}{2} \int \mathcal{A}^* \mathcal{A} \tag{2.C.18b}$$

Comparison with (2.C.17) shows that in the limit, the ground state wave functional of the topologically massive theory tends to the unique state of the Chern–Simons theory, times the square root of the measure factor, so that the probability measures correctly pass into each other.[2e] Other details on this limit can be found in the literature.[2e]

D. Quantum Holonomy

In the Abelian Chern–Simons theory defined on the topologically trivial plane there is very little structure. Indeed when the constraints are solved before quantization, there is no structure at all if there are no sources. However, by quantizing first, and computing the quantum holonomy around closed loops at fixed time, we encounter non-trivial results that show an explicit difference between solving the constraints before and after quantization.[2d]

The holonomy operator $\Phi(C)$ is defined by the parallel transport equation around a closed planar loop C parametrized by $\mathbf{x}(\tau)$ for $\tau \in [0,1]$, with $\mathbf{x}(0) = \mathbf{x}(1) \equiv \mathbf{x}_0$, which serves as a marked point on the loop, where we also specify the initial and final unit tangent vectors, $\hat{\mathbf{v}}_0 = \dot{\mathbf{x}}(0)/|\dot{\mathbf{x}}(0)|$ and $\hat{\mathbf{v}}_1 = \dot{\mathbf{x}}(1)/|\mathbf{x}(1)|$, respectively. The marked point is on a smooth segment of the loop if $\hat{\mathbf{v}}_0 = \hat{\mathbf{v}}_1$; otherwise it is at a cusp with opening angle $\pi \mp \cos^{-1} \hat{\mathbf{v}}_0 \cdot \hat{\mathbf{v}}_1$, where $\mp$ refer to opening angles $< \pi$ and $\geq \pi$, respectively, and $0 \leq \cos^{-1} \hat{\mathbf{v}}_0 \cdot \hat{\mathbf{v}}_1 < 2\pi$.

The equation for the holonomy

$$[i\partial_\tau + V(\tau)]\, \Phi(C) = 0 \quad , \quad V(\tau) \equiv \dot{x}^i(\tau) A_i(\mathbf{x}(\tau)) \tag{2.D.1}$$

is solved at the classical level by the Aharonov–Bohm phase.

$$\Phi^{\text{classical}} = \exp\left[i \int_0^1 d\tau\, V(\tau)\right] = \exp\left[i \int_C dx^i\, A_i\right] \tag{2.D.2}$$

The constraint (2.C.4) in the theory without sources forces A_i to be a pure gauge, and therefore $\Phi^{\text{classical}} = 1$, independent of the loop C.

To solve Eq. (2.D.1) at the quantum level we must recall that $[A_1(\mathbf{x})], A_2(\mathbf{x})] \neq 0$, therefore $[V(\tau), V(\tau')] \neq 0$, and the quantum holonomy operator is given by a path-ordered expression.

$$\Phi(C) = P \exp\left[i \int_0^1 d\tau\, V(\tau)\right] \tag{2.D.3}$$

To determine the action of $\Phi(C)$ on states we first need to undo the path ordering. Since the commutator $[V(\tau), V(\tau')]$ is a c-number, this yields

$$\begin{aligned}\Phi(C) &= \exp\left[-\frac{1}{2}\int_0^1 d\tau \int_0^\tau d\tau' \,[V(\tau), V(\tau')]\right] \exp\left[i\int_0^1 d\tau\, V(\tau)\right] \\ &= \exp\left[-\frac{i}{2\kappa}\int_0^1 d\tau \int_0^\tau d\tau'\, \dot{x}^i(\tau)\epsilon^{ij}\dot{x}^j(\tau')\delta^2\left(\mathbf{x}(\tau) - \mathbf{x}(\tau')\right)\right] \\ &\times \exp\left[i\int_0^1 d\tau\, \dot{x}^i(\tau)A_i\left(\mathbf{x}(\tau)\right)\right] \quad .\end{aligned} \tag{2.D.4}$$

In our chosen polarization, $\Phi(C)$ acts on functionals of the "coordinate" θ, and it is therefore convenient to reorganize (2.D.4) so that the "momentum" B stands on the right. To this end, we split the operator $V(\tau)$ into a self-commuting pair

$$\begin{aligned}V(\tau) &= V_1(\tau) + V_2(\tau) \\ [V_1(\tau), V_1(\tau')] &= [V_2(\tau), V_2(\tau')] = 0 \\ [V_1(\tau), V_2(\tau')] &= c\text{-number}\end{aligned} \tag{2.D.5}$$

Splitting the operator-valued exponential in (2.D.4) gives an additional phase, which combines with the first to yield

$$\begin{aligned}\Phi(C) &= \exp[i\gamma] \exp\left[i\int_0^1 d\tau\, V_1(\tau)\right] \exp\left[i\int_0^1 d\tau\, V_2(\tau)\right] \\ \gamma &= i\int_0^1 d\tau \int_0^\tau d\tau' \,[V_1(\tau), V_2(\tau')]\end{aligned} \tag{2.D.6}$$

With our polarization

$$\begin{aligned}V_1(\tau) &= \dot{x}^i(\tau)\partial_i\theta\left(\mathbf{x}(\tau)\right) = \frac{d}{d\tau}\theta\left(\mathbf{x}(\tau)\right) \quad , \\ V_2(\tau) &= \dot{x}^i(\tau)\epsilon^{ij}\partial_j^{-1}B\left(\mathbf{x}(\tau)\right) \quad ,\end{aligned} \tag{2.D.7}$$

γ becomes

$$\gamma = \frac{1}{\kappa}\int_0^1 d\tau \int_0^\tau d\tau' \frac{\partial}{\partial\tau}\dot{x}^i(\tau')\epsilon^{ij}\partial_j^{-1}\delta^2\left(\mathbf{x}(\tau) - \mathbf{x}(\tau')\right) \quad , \tag{2.D.8a}$$

where $\partial_j^{-1}\delta$ is the derivative of the two-dimensional Green's function.

$$\partial_j^{-1}\delta^2(\mathbf{r}) = \frac{1}{2\pi}\frac{r^j}{|\mathbf{r}|} = \frac{\epsilon^{jk}}{2\pi}\partial_k \tan^{-1}\frac{r^2}{r^1}$$

The formula relating $\partial_j^{-1}\delta$ to a derivative of $\tan^{-1}$ gives an alternative expression for γ.

$$\gamma = \frac{1}{2\pi\kappa}\int_0^1 d\tau \int_0^\tau d\tau' \frac{\partial}{\partial\tau}\frac{\partial}{\partial\tau'}\tan^{-1}\frac{x^2(\tau) - x^2(\tau')}{x^1(\tau) - x^1(\tau')} \tag{2.D.8b}$$

The derivation may be organized differently. We recognize that the phase arising from splitting $\exp\left[i\int_0^1 d\tau\, V(\tau)\right]$ into $\exp\left[i\int_0^1 d\tau\, V_1(\tau)\right]\exp\left[i\int_0^1 d\tau\, V_2(\tau)\right]$ vanishes because the loop is closed, leaving for the entire contribution to γ just the last quantity in (2.D.4), which arises from undoing the path-ordering, and which is determined by the polarization-independent $[A_i, A_j]$ commutator.

$$\gamma = -\frac{1}{2\kappa}\int_0^1 d\tau \int_0^\tau d\tau' \dot{x}^i(\tau)\epsilon^{ij}\dot{x}^j(\tau')\delta^2\left(\mathbf{x}(\tau)-\mathbf{x}(\tau')\right) \tag{2.D.8c}$$

Using the identity $2\pi\delta^2(\mathbf{r}) = \epsilon^{ij}\partial_i\partial_j \tan^{-1} r^2/r^1$ we can rewrite (2.D.8c).

$$\gamma = \frac{1}{4\pi\kappa}\int_0^1 d\tau \int_0^\tau d\tau' \left(\frac{\partial}{\partial\tau}\frac{\partial}{\partial\tau'} - \frac{\partial}{\partial\tau'}\frac{\partial}{\partial\tau}\right)\tan^{-1}\frac{x^2(\tau)-x^2(\tau')}{x^1(\tau)-x^1(\tau')} \tag{2.D.8d}$$

Equation (2.D.8c) exhibits the elusive nature of γ. Owing to the δ-function enforcing $\mathbf{x}(\tau) = \mathbf{x}(\tau')$, one might conclude that $\dot{x}^i(\tau)\epsilon^{ij}\dot{x}^j(\tau')$, and hence γ, vanishes. However, upon closer examination we recognize that the two-dimensional δ-function is a product of two one-dimensional δ-functions, each enforcing the same constraint on the one-dimensional variables τ, τ'. Thus the integrand in (2.D.8c) involves the ambiguous quantity $\dot{x}^i(\tau)\epsilon^{ij}\dot{x}^j(\tau)/\left|\dot{x}^1(\tau)\dot{x}^2(\tau)\right| \; \delta(\tau-\tau')\delta(\tau-\tau')$, and in the following a careful analysis is performed to obtain an unambiguous result. But it is already clear that γ is non-trivial owing to the continuum properties of space, which give rise to δ-functions. In a discretized world with Kronecker delta's, (2.D.8) does indeed vanish.

Our evaluation of γ is based on the following observation. In spite of the singular nature of (2.D.8c), one can begin with any of the other formulas (2.D.8a), (2.D.8b) and (2.D.8d), manipulate finite expressions and arrive at an unambiguous answer for γ. By this procedure, we shall derive below the result

$$\gamma = \frac{1}{2\pi\kappa}\Delta\Theta_{\mathbf{r}} - \frac{1}{4\pi\kappa}\Delta\Theta_{\mathbf{v}} \quad . \tag{2.D.9}$$

Here $\Delta\Theta_{\mathbf{r}}$ is the total angle accumulated at the marked point $\mathbf{x}_0$ when the loop is traversed by a vector based at $\mathbf{x}_0$,

$$\begin{aligned} \Theta_{\mathbf{r}}(\tau) &= \tan^{-1}\frac{r^2(\tau)}{r^1(\tau)} \quad , \\ \mathbf{r} &= \mathbf{x} - \mathbf{x}_0 \quad , \\ \Delta\Theta_{\mathbf{r}} &\equiv \Theta_{\mathbf{r}}(1) - \Theta_{\mathbf{r}}(0) \quad , \end{aligned} \tag{2.D.10a}$$

while the quantity $\Delta\Theta_{\mathbf{v}}$ is the accumulated angular change in the tangent to the curve.

$$\begin{aligned} \Theta_{\mathbf{v}}(\tau) &= \tan^{-1}\frac{\dot{r}^2(\tau)}{\dot{r}^1(\tau)} = \tan^{-1}\frac{\hat{v}^2(\tau)}{\hat{v}^1(\tau)} \\ \Delta\Theta_{\mathbf{v}} &\equiv \Theta_{\mathbf{v}}(1) - \Theta_{\mathbf{v}}(0) \end{aligned} \tag{2.D.10b}$$

The evaluation of (2.D.8) that gives (2.D.9) is performed without using regulators. However, to illustrate the subtlety of (2.D.8) we remark here that if regulators *are* introduced, for example by regulating the δ-function, and if the regularization preserves the fact that the Green's function derivative is odd under interchange of argument, $\partial_j^{-1}\delta^2(\mathbf{x}-\mathbf{y}) = -\partial_j^{-1}\delta^2(\mathbf{y}-\mathbf{x})$, so that $\partial_j^{-1}\delta^2(0)$ vanishes, then we again obtain a unique answer that does not depend on the details of the regularization.

However, this answer differs from (2.D.9): $\gamma^{\text{reg}} = \frac{1}{2\pi\kappa}\Delta\Theta_{\mathbf{r}}$, the contribution from the change in the tangent is missing.

We now present the derivation of (2.D.9). In (2.D.8a) or (2.D.8b) the τ derivative is interchanged with the τ' integral. Starting from (2.D.8a) we have

$$\begin{aligned}\gamma &= \frac{1}{\kappa}\int_0^1 d\tau \frac{\partial}{\partial\tau}\int_0^\tau d\tau'\, \dot{x}^i(\tau')\epsilon^{ij}\delta^2\left(\mathbf{x}(\tau)-\mathbf{x}(\tau')\right)\\ &\quad -\frac{1}{\kappa}\int_0^1 d\tau \left(\dot{x}^i(\tau')\epsilon^{ij}\partial_j^{-1}\delta^2\left(\mathbf{x}(\tau)-\mathbf{x}(\tau')\right)\right)\Big|_{\tau'=\tau}\\ &= -\frac{1}{\kappa}\int_0^1 d\tau'\, \dot{x}^i(\tau')\epsilon^{ij}\partial_j^{-1}\delta^2\left(\mathbf{x}(\tau')-\mathbf{x}_0\right)\\ &\quad +\frac{1}{\kappa}\int_0^1 d\tau \left(\dot{x}^i(\tau')\epsilon^{ij}\partial_j^{-1}\delta^2\left(\mathbf{x}(\tau')-\mathbf{x}(\tau)\right)\right)\Big|_{\tau'=\tau}\end{aligned} \tag{2.D.11a}$$

By expressing the derivative of the Green's functions in terms of the inverse tangent, or alternatively by beginning with (2.D.8b), we find

$$\begin{aligned}\gamma &= \frac{1}{2\pi\kappa}\int_0^1 d\tau' \frac{\partial}{\partial\tau'}\tan^{-1}\frac{x^2(\tau')-x_0^2}{x^1(\tau')-x_0^1}\\ &\quad -\frac{1}{2\pi\kappa}\int_0^1 d\tau\left(\frac{\partial}{\partial\tau'}\tan^{-1}\frac{x^2(\tau')-x^2(\tau)}{x^1(\tau')-x^1(\tau)}\right)\Big|_{\tau'=\tau}\end{aligned} \tag{2.D.11b}$$

[It is easy to see that the same formula emerges if one begins with (2.D.8d), treats the $\partial_\tau\partial_{\tau'}$ contribution as in (2.D.11b) and evaluates the $\partial_{\tau'}\partial_\tau$ contribution by first performing the τ' integral.] The integrand of the last term in (2.D.11b) is

$$\left(\frac{\partial}{\partial\tau'}\tan^{-1}\frac{x^2(\tau')-x^2(\tau)}{x^1(\tau')-x^1(\tau)}\right)\Big|_{\tau'=\tau} = \frac{1}{2}\frac{\partial}{\partial\tau}\tan^{-1}\frac{\dot{x}^2(\tau)}{\dot{x}^1(\tau)}$$

and so γ becomes

$$\gamma = \frac{1}{2\pi\kappa}\int_0^1 d\tau\frac{\partial}{\partial\tau}\tan^{-1}\frac{x^2(\tau)-x_0^2}{x^1(\tau)-x_0^1} - \frac{1}{4\pi\kappa}\int_0^1 d\tau\frac{\partial}{\partial\tau}\tan^{-1}\frac{\dot{x}^2(\tau)}{\dot{x}^1(\tau)} \tag{2.D.11c}$$

This establishes (2.D.9). [Note that if a regulator had been used for the δ function in (2.D.11a), then the last term of each equation in (2.D.11) would vanish, with natural regulators such that $\partial_j^{-1}\delta^2(0) = 0$, leaving, as commented above, for γ^{reg} just the first term on the right-hand side of (2.D.11c). A possible regularization would be to retain the kinetic Maxwell term, and then decouple it by passing with κ to infinity. It should be no surprise, in view of earlier discussions on this limit [see (2.C.18)], that different answers can be obtained. For more discussion, see the literature.[2e]

The value of γ depends on the curve. Consider first simple curves, traversed in the counterclockwise direction, without self-intersections. For smooth, simple curves, the angle swept out as the marked point is π, and

$$\frac{1}{2\pi\kappa}\Delta\Theta_{\mathbf{r}} = \frac{1}{2\kappa}\ . \tag{2.D.12a}$$

The angular change in the tangent is 2π, so

$$\frac{1}{4\pi\kappa}\Delta\Theta_{\mathbf{v}} = \frac{1}{2\kappa} \tag{2.D.12b}$$

and therefore

$$\gamma = 0 \tag{2.D.12c}$$

The same results hold if the curve has cusps, provided the marked point is not a cusp. However, when $\mathbf{x}_0$ does lie at a cusp where $\hat{\mathbf{v}}_0 \neq \hat{\mathbf{v}}_1$, there are shortfalls in the angular traversals and

$$\frac{1}{2\pi\kappa}\Delta\Theta_{\mathbf{r}} = \frac{1}{2\pi\kappa}\left(\pi \mp \cos^{-1}\hat{\mathbf{v}}_0 \cdot \hat{\mathbf{v}}_1\right) \tag{2.D.13a}$$

$$\frac{1}{3\pi\kappa}\Delta\Theta_{\mathbf{v}} = \frac{1}{4\pi\kappa}\left(2\pi \mp \cos^{-1}\hat{\mathbf{v}}_0 \cdot \hat{\mathbf{v}}_1\right) \tag{2.D.13b}$$

$$\gamma = \mp\frac{1}{4\pi\kappa}\cos^{-1}\hat{\mathbf{v}}_0 \cdot \hat{\mathbf{v}}_1 \tag{2.D.13c}$$

where $\mp$ refer to opening angles of the cusp $< \pi$ and $\leq \pi$, respectively.

We now discuss loops with intersections. We set the following conventions: the marked point lies on an outermost smooth segment of the loop, insuring that $|\Delta\Theta_{\mathbf{r}}| = \pi$, regardless of the number of intersections and cusps, since $\mathbf{x}_0$ does not lie at any of these exceptional points; the parametrization takes the contour through $\mathbf{x}_0$ in a counterclockwise direction thus fixing $\Delta\Theta_{\mathbf{r}} = \pi$; an intersection is defined by an actual crossing — touching contours are not intersections; the total number of intersections is ν and we do not consider loops with multiple intersections at the same point.

Only $\Delta\Theta_{\mathbf{v}}$ varies with the intersection number ν. For $\nu = 1$, there are two elementary intersections: the "figure eight" where the two sub-loops are traversed in opposite directions, so that $\Delta\Theta_{\mathbf{v}} = 0$, and the "nested loop" where the two sub-loops are traversed in the same direction, with $\Delta\Theta_{\mathbf{v}} = 4\pi$. Loops with higher ν can be constructed by superposing in various ways these two elementary blocks. For a given ν, the highest possible value for $\Delta\Theta_{\mathbf{v}}$, $\Delta\Theta_{\mathbf{v}}^{\max} = (\nu+1)2\pi$, is achieved by superposing like-oriented nested intersections. The lowest possible value, $\Delta\Theta_{\mathbf{v}}^{\min} = -(\nu-1)2\pi$, is obtained by building out of ν like-oriented intersections a loop whose direction has been reversed by a single figure eight intersection. The possible values of $\Delta\Theta_{\mathbf{v}}$, for fixed ν, interpolate between $\Delta\Theta_{\mathbf{v}}^{\min}$ and $\Delta\Theta_{\mathbf{v}}^{\max}$ in steps of 4π. The different allowed values for $\Delta\Theta_{\mathbf{v}}$, combined with the unique $\Delta\Theta_{\mathbf{r}} = \pi$, give a table of values for γ at fixed ν.

$$\gamma = \frac{m}{\kappa} \quad , \quad m \in \left[-\frac{\nu}{2}, -\frac{\nu}{2}+1, \ldots, \frac{\nu}{2}-1, \frac{\nu}{2}\right] \tag{2.D.14}$$

Finally, we return to the full holonomy operator, and determine its action on the state (2.C.8). In our polarization, we have

$$\Phi(C)\Psi(\theta) = \exp\left[i\gamma(C)\right]\exp\left[i\int_C dx^i\partial_i\theta\right]\exp\left[-\frac{i}{\kappa}\int_C dx^i\epsilon^{ij}\partial_j^{-1}\rho\right]\Psi(\theta) \quad . \tag{2.D.15}$$

Since the holonomy of the pure gauge $\partial_i\theta$ is trivial, $\exp\left[i\int_C dx^i\partial_i\theta\right] = 1$. The integral in the exponent of the last factor is evaluated by Stoke's law and gives

$$-\int_C dx^i\epsilon^{ij}\partial_j^{-1}\rho = \int_{S_C} d\mathbf{x}\,\rho = Q(C) \tag{2.D.16}$$

where $Q(C)$ is the total charge contained within the closed curve, with contributions appropriately signed if C is self-intersecting. [For a single point source of charge Q surrounded by a counterclockwise loop, $Q(C) = Q$.] Thus the physical state (2.C.8) is an eigenstate of the holonomy operator with eigenvalue $\phi(C)$.

$$\phi(C) = \exp\left(i\left[\gamma(C) + \frac{1}{\kappa}Q(C)\right]\right) \tag{2.D.17}$$

[Note that the holonomy operator commutes with the Gauss law operator because it is gauge invariant, and with the Hamiltonian (2.C.3) if $Q(C)$ is time-independent.]

Since holonomies around closed loops are the only gauge invariant and generally covariant observables, this shows that the $U(1)$ Chern–Simons theory in Minkowski space is characterized by the strengths of external charges, and by the vacuum holonomy $e^{i\gamma(C)}$, which is missed when the constraints are imposed before quantization. The vacuum holonomy, which can be evaluated without regularization, vanishes for simple loops provided the marked point is not at a cusp; otherwise, it is determined by the opening angle of the cusp. This rich structure is already present for the $U(1)$ Chern–Simons theory in Minkowski space.

The gauge structure in the Chern–Simons theory follows closely that of its topologically massive antecedent: both involve essentially the same cocycle in the action of the gauge group. However, the behavior of the quantum holonomy Φ is quite different. In the latter theory, the vacuum ground state — for the non-interacting Abelian model, the Gaussian (2.A.22) in $\mathbf{A}$ — is not an eigenstate of Φ, and the vacuum expectation value of the holonomy operator is infinite. In the Abelian Chern–Simons theory, the vacuum state — the only state for the source-free model in Minkowski space — is a Φ eigenstate with finite non-trivial eigenvalue, which is not seen when constraints are solved before quantization.

If expectation values of the holonomy operator in a topologically massive theory are to possess physical significance, Φ must be renormalized and its vacuum value is undefined. Neither regularization nor renormalization is needed in the Chern–Simons model; indeed since γ carries a rich loop dependence, a universal renormalization cannot remove the effect.

E. Anomalous Statistics and Spin of Charged Particles

We saw earlier that charged particles interacting through a Chern–Simons [Abelian] gauge field carry flux. This has the consequence that their spin and statistics is modified by the gauge-field interaction[10] — a result which can be established without reference to the detailed nature of the particle dynamics.[2d] Here we first show how the holonomy modifies statistics, and that spin adjusts so that the spin-statistics theorem is preserved. Later, we shall take a point-particle model for the matter and regain these results in an explicit manner.

Consider two identical particles, each with charge Q, and imagine a fixed-time test of statistics by carrying one particle around the other, corresponding to a double interchange of the particles. The wave function of the test particle will acquire, in addition to the conventional statistical factor, the phase

$$P\exp\left[iQ\int_C dx^i\,A_i\right] \quad ,$$

where C is a loop without self-intersections surrounding the particle. The state (2.C.8) is an eigenstate of this operator with eigenvalue $\exp\left[iQ^2/\kappa\right]$, apart from the vacuum contribution, which is absent provided the marked point is not at a cusp. The phase acquired by the wave function under a single interchange of the two particles is half the above, *i.e.* $Q^2/2\kappa$. Thus the statistics phase is an observable, whose value satisfies the spin-statistics theorem because particles in this theory carry anomalous spin S,

$$2\pi S = \frac{Q^2}{2\kappa} \tag{2.E.1}$$

as we now demonstrate.

Spin fractionization for the charged matter particles is due to the angular momentum of the gauge field associated with the magnetic flux created by the charged particle. The variation of the gauge field under an infinitesimal spatial rotation $\delta x^i = -\epsilon^{ij} x^j$,

$$\delta A_i = -x^j \epsilon^{jk} \partial_k A_i - \epsilon^{ij} A_j \quad , \tag{2.E.2}$$

leaves the Lagrangian corresponding to (2.C.1) invariant, provided the external current $\mathbf{j}$ is rotationally covariant.

$$x^j \epsilon^{jk} \partial_k j^i + \epsilon^{ij} j^j = 0 \tag{2.E.3}$$

For the conserved angular momentum operator we take

$$J = -\frac{\kappa}{2} \int_{\mathbf{x}} x^i \epsilon^{ij} \{A_j, B\} \quad . \tag{2.E.4}$$

J generates the transformation (2.E.2),

$$\delta A_i = i[J, A_i] \tag{2.E.5}$$

and commutes with the Hamiltonian (2.C.3) when the current is rotationally covariant, *i.e.*, (2.E.3) is satisfied. Also J is gauge invariant, as is seen by replacing A_j in (2.E.4) by $\partial_j \lambda$ and B by $-\rho/\kappa$ according to (2.C.4). When the charge density is spherically symmetric, an integration by parts yields a vanishing response to the gauge transformation. J is not obtained from Noether's energy-momentum tensor, rather from the symmetric tensor, which has no pure gauge field contribution owing to the topological nature of the Chern–Simons term; only the interaction contributes.

$$T^{\mu\nu} = g^{\mu\nu} j^\alpha A_\alpha - j^\mu A^\nu - j^\nu A^\mu \tag{2.E.6}$$

Thus we have

$$\begin{aligned} J = \int_{\mathbf{x}} \epsilon^{ij} x^i T^{0j} &= \int_{\mathbf{x}} \epsilon^{ij} x^i A_j \rho \\ &= -\kappa \int_{\mathbf{x}} \epsilon^{ij} x^i A_j B \quad , \end{aligned} \tag{2.E.7}$$

where (2.C.4) was used. In (2.E.4) the expression is symmetrized to insure Hermiticity; however, it differs from (2.E.7) by a commutator $i \int_{\mathbf{x}} \epsilon^{ij} \partial_j \delta^2(\mathbf{x}-\mathbf{x})$, which although involving $\delta^2(0)$ can be set to zero by an integration by parts. [Alternatively, $\partial_j \delta^2(\mathbf{x})$ is odd in its argument.]

Consider now the action of J on the state (2.C.8)

$$J\Psi(\theta) = \left[\int_{\mathbf{x}} \epsilon^{ij} x^i A_j \rho\right] \Psi(\theta) = \int \rho \epsilon^{ij} x^i \left[\partial_j \theta - \frac{1}{\kappa} \epsilon^{jk} \partial_k^{-1} \rho\right] \Psi(\theta) \tag{2.E.8}$$

The contribution proportional to $\partial_j \theta$ vanishes upon an integration by parts for spherically symmetric ρ. This shows that the physical state (2.C.8) is an angular momentum eigenstate with eigenvalue S.

$$\begin{aligned} S &= \frac{1}{\kappa} \int \rho x^i \partial_i^{-1} \rho \\ &= \frac{1}{2\pi\kappa} \int_{\mathbf{x}} \int_{\mathbf{y}} \rho(t,\mathbf{x}) \frac{\mathbf{x}\cdot(\mathbf{x}-\mathbf{y})}{|\mathbf{x}-\mathbf{y}|^2} \rho(t,\mathbf{y}) \\ &= \frac{1}{4\pi\kappa} \int_{\mathbf{x}} \int_{\mathbf{y}} \rho(t,\mathbf{x}) \left[1 + \frac{\mathbf{x}^2 - \mathbf{y}^2}{|\mathbf{x}-\mathbf{y}|^2}\right] \rho(t,\mathbf{y}) \\ &= \frac{Q^2}{4\pi\kappa} \quad , \end{aligned} \tag{2.E.9}$$

Equation (2.E.9) gives the fractional spin carried by the charged matter particles, and agrees with previous results. S is a sharp observable, which satisfies the spin statistics relation (2.E.1).

F. Point-Particles with Abelian Chern–Simons Gauge Fields

A possible model for charged matter consists of point particles.[11] The total Lagrangian is

$$L = L_{\text{matter}} + L_{\text{interaction}} + L_{\text{CS}} \tag{2.F.1}$$

where

$$L_{\text{matter}} = \frac{1}{2}\sum_{p=1}^{N} m_p v_p^2(t) \tag{2.F.2a}$$

$$\begin{aligned} L_{\text{interaction}} &= \sum_{p=1}^{N} e_p \Big(\mathbf{v}_p(t) \cdot \mathbf{A}\left(t, \mathbf{r}_p(t)\right) - A_0\left(t, \mathbf{r}_p(t)\right) \Big) \\ &= -\int_{\mathbf{x}} A_\mu(x) j^\mu(x) \\ j^\mu &= \sum_{p=1}^{N} e_p v_p^\mu(t) \delta^2\left(\mathbf{x} - \mathbf{r}_p(t)\right) = \left(\rho(x), \mathbf{j}(x)\right) \\ v_p^\mu &= (1, \mathbf{v}_p) \end{aligned} \tag{2.F.2b}$$

$$L_{\text{CS}} = \frac{\kappa}{2}\int_{\mathbf{x}} \epsilon^{ij} \dot{A}_i(x) A_j(x) - \kappa \int_{\mathbf{x}} A_0(x) B(x) \tag{2.F.2c}$$

We are considering N point-particles with coordinates $\mathbf{r}_p(t)$, $p = 1, \ldots, N$, which are the particle dynamical variables, and $\mathbf{v}_p(t) = \dot{\mathbf{r}}_p(t)$ are the velocities. The masses and charges are m_p and e_p, respectively. The second expression for the interaction Lagrangian makes use of the point-particle current, which is a δ-function. Thus the integral over all space evaluates $x = (t, \mathbf{x})$, the field point argument of $A_\mu(x)$, at $x = (t, \mathbf{r}_p(t))$. The time component A_0 has not been set to zero.

The [unordered] Euler–Lagrange equations of motion consist of the Lorentz force equation for the matter variables

$$m_p \dot{v}_p^i = e_p \left(E^i(\mathbf{r}_p) + \epsilon^{ij} v_p^j B(\mathbf{r}_p) \right) \tag{2.F.3a}$$

and a field-current identity that relates the electromagnetic fields to the matter currents.

$$E^i(x) = \frac{1}{\kappa}\epsilon^{ij} j^j(x) \tag{2.F.3b}$$

$$B(x) = -\frac{1}{\kappa}\rho(x) \tag{2.F.3c}$$

Point-particle electrodynamics suffers from well-known self-energy problems. Let us observe that these are absent here. Consider the equation of motion for a single particle, $N = 1$ and subscript p suppressed. The force in Eq. (2.F.3a) arises from the electromagnetic fields at the particle position $\mathbf{r}$; by (2.F.3b) and (2.F.3c) they are given by the charge and current densities evaluated at $\mathbf{x} = \mathbf{r}$. However from (2.F.2b) we see that at $\mathbf{x} = \mathbf{r}$ there appears the undefined quantity $\delta^2(\mathbf{r} - \mathbf{r})$ — the density function of a point-particle at its position. Fortunately this singular object is multiplied by a factor that vanishes, since according to the field-current identity, the quantity

$$E^i(\mathbf{x}) + \epsilon^{ij} v^j B(\mathbf{x}) = \frac{1}{\kappa}\epsilon^{ij}\left(j^j(x) - v^j \rho(x)\right) = \frac{1}{\kappa}\epsilon^{ij}\left(v^j - v^j\right)\delta^2(\mathbf{x} - \mathbf{r})$$

vanishes unambiguously; therefore we shall take it to be zero also at $\mathbf{x} = \mathbf{r}$. In other words the charge and current densities are regulated by non-singular expressions for the evaluation of the self-interaction, which is then shown to vanish.

With this prescription, particles interacting through Chern–Simons gauge fields do not experience self-interactions. Equations (2.F.3) combine to give for the particle coordinates a closed equation of motion, free from undefined quantities.

$$m_p \dot{v}_p^i = \epsilon^{ij} \frac{e_p}{\kappa} \sum_{q \neq p} e_q \left(v_q^j - v_p^j \right) \delta \left(\mathbf{r}_p - \mathbf{r}_q \right) \tag{2.F.4}$$

The Hamiltonian arising from the Lagrangian (2.F.1) and (2.F.2) is

$$H = \frac{1}{2} \sum_{p=1}^{N} m_p v_p^2 + \int_{\mathbf{x}} A_0(x) \left(\kappa B(x) + \rho(x) \right) \tag{2.F.5}$$

It is recognized that the Lagrange multiplier A_0 may be set to zero [by choosing the Weyl gauge] provided Eq. (2.F.3c) is imposed as a constraint. Thus the Hamiltonian is just the free particle one.

$$H = \frac{1}{2} \sum_{p=1}^{N} m_p v_p^2 \tag{2.F.6}$$

Although the gauge field is invisible in (2.F.6), its presence is felt in the commutator algebra.

The commutation relations between canonical variables follow in the usual way, except that vector potentials satisfy the Abelian version of (2.B.8). However, it is useful to present the algebra in terms of the gauge invariant velocity operator, which occurs in the Hamiltonian, rather than in terms of the canonical momentum.

$$\mathbf{p}_p = \frac{\partial L}{\partial \mathbf{v}_p} = m_p \mathbf{v}_p + e_p \mathbf{A}(\mathbf{r}_p) \tag{2.F.7}$$

Thus we have from (2.B.8),

$$\left[A^i(\mathbf{x}), A^j(\mathbf{y}) \right] = \frac{i}{\kappa} \epsilon^{ij} \delta^2(\mathbf{x} - \mathbf{y}) \tag{2.F.8}$$

The particle variables satisfy

$$\left[r_p^i, r_q^j \right] = 0 \tag{2.F.9a}$$

$$\left[r_p^i, m_q v_q^j \right] = i \delta^{ij} \delta_{pq} \tag{2.F.9b}$$

$$\left[m_p v_p^i, m_q v_q^j \right] = i \epsilon^{ij} \left(\delta_{pq} e_p B(\mathbf{r}_q) + \frac{1}{\kappa} e_p e_q \delta^2(\mathbf{r}_p - \mathbf{r}_q) \right) \tag{2.F.9c}$$

The velocity commutator does not vanish; rather it contains terms that arise from: (i) the fact that $\mathbf{v}$ differs from $\mathbf{p}$ by $\mathbf{A}(\mathbf{r})$, (2.F.7), and $\mathbf{p}$ does not commute with $\mathbf{A}(\mathbf{r})$ but produces the first term in the parenthesis of (2.F.9c); also (ii) the vector potentials do not commute, (2.F.8), giving rise to the second term in parenthesis of (2.F.9c). There is also the non-vanishing commutator between velocity and field.

$$\left[m_p v_p^i, A^j(\mathbf{x}) \right] = -\frac{i}{\kappa} \epsilon^{ij} e_p \delta^2(\mathbf{x} - \mathbf{r}_p) \tag{2.F.10}$$

Finally we remind that (2.F.3c) is imposed as a constraint.

Before proceeding with an analysis of the dynamical problem, it is interesting to record the symmetries of our theory.

First there are the spatial translation and rotation symmetries, under which the coordinates and fields change as

$$\begin{aligned} \delta r_p^i &= a^i \\ \delta A_\mu(t,\mathbf{x}) &= -a^i \partial_i A_\mu(t,\mathbf{x}) \qquad \text{(translations)} \end{aligned} \tag{2.F.11}$$

$$\begin{aligned} \delta r_p^i &= -\epsilon^{ij} r_p^j \\ \delta A_0(t,\mathbf{x}) &= -\epsilon^{jk} x^j \partial_\kappa A_0(t,\mathbf{x}) \\ \delta A_i(t,\mathbf{x}) &= -\epsilon^{jk} x^j \partial_\kappa A_i(t,\mathbf{x}) - \epsilon^{ij} A_j(t,\mathbf{x}) \quad \text{(rotations)} \end{aligned} \tag{2.F.12}$$

The Lagrangian (2.F.1), (2.F.2) is invariant, and the conserved constants of motion are the momentum $\mathbf{P}$ and angular momentum J, respectively.

$$\begin{aligned} \mathbf{P} &= \sum_{p=1}^N (m_p \mathbf{v}_p + e_p \mathbf{A}(\mathbf{r}_p)) + \kappa \int_{\mathbf{x}} \mathbf{A} B \\ &= \sum_{p=1}^N \mathbf{p}_p + \kappa \int_{\mathbf{x}} \mathbf{A} B \end{aligned} \tag{2.F.13}$$

$$\begin{aligned} J &= \sum_{p=1}^N (\mathbf{r}_p \times (m_p \mathbf{v}_p + e_p \mathbf{A}(\mathbf{r}_p))) + \kappa \int_{\mathbf{x}} (\mathbf{x} \times \mathbf{A}) B \\ &= \sum_{p=1}^N \mathbf{r}_p \times \mathbf{p}_p + \kappa \int_{\mathbf{x}} (\mathbf{x} \times \mathbf{A}) B \ . \end{aligned} \tag{2.F.14}$$

The second formula in both expressions makes use of the canonical momentum, (2.F.7). The vector potential, evaluated at $\mathbf{x} = \mathbf{r}_p$, which is combined with $m_p \mathbf{v}_p$ to form the canonical momentum, arises from the integration Lagrangian (2.F.2b); the last term in (2.F.13) and (2.F.14), proportional to κ, arises from the Chern–Simons kinetic term.

Let us observe that $\sum_{p=1}^N \mathbf{r}_p \times \mathbf{p}_p$ possesses integer eigenvalues when acting on single valued wave functions. Thus point-particles, interacting with gauge fields that are governed by Chern–Simons dynamics, possess in their angular momentum a contribution additional to the usual integer. The extra term is not quantized but is determined by the gauge field and the Chern–Simons coupling strength κ. This is consistent with the results of the external source analysis, presented earlier, and it will be shown later that in fact there is complete agreement.

Note that for point-particles moving in *prescribed external* gauge fields [rather than *dynamical* ones], the last term in (2.F.13) and (2.F.14) is missing, and $\mathbf{A}(\mathbf{r}_p)$ is a given function of $\mathbf{r}_p$. Therefore, in the external field problem the angular momentum spectrum comprises the conventional integers, even though it differs from the kinematical momentum.[12]

$$\sum_{p=1}^N \mathbf{r}_p \times \mathbf{p}_p = \sum_{p=1}^N \mathbf{r}_p \times m_p \mathbf{v}_p + \sum_{p=1}^N e_p \mathbf{r}_p \times \mathbf{A}(\mathbf{r}_p)$$

[This point is occasionally confused in the literature.]

In addition to the symmetries under the above spatial transformations, the theory is also invariant against transformations of time: obviously time translation $[t \to t + t_0]$ is a symmetry leading to energy [= Hamiltonian] conservation; but there are two further, unexpected time transformations that leave the action invariant: time dilation $[t \to \lambda t]$, and time special conformal transformation $[1/t \to 1/t + 1/t_0]$. Together, the three form a dynamical $SO(2,1)$ symmetry group of conformal transformations.[2f]

Infinitesimally we have for these time reparametrizations

$$\delta t = -f(t) \tag{2.F.15a}$$

$$f(t) = \begin{cases} 1 & \text{translation} \\ t & \text{dilation} \\ t^2 & \text{special conformal transformation} \end{cases} \tag{2.F.15b}$$

The dynamical variables transform as

$$\begin{aligned} \delta_f \mathbf{r}_p(t) &= f(t)\mathbf{v}_p(t) - \frac{1}{2}\dot{f}(t)\mathbf{r}_p(t) \\ \delta_f A_0(t,\mathbf{x}) &= \partial_t\big(f(t)A_0(t,\mathbf{x})\big) + \frac{1}{2}\dot{f}(t)\mathbf{x}\cdot\nabla A_0(t,\mathbf{x}) - \frac{1}{2}\ddot{f}(t)\mathbf{x}\cdot\mathbf{A}(t,\mathbf{x}) \\ \delta_f \mathbf{A}(t,\mathbf{x}) &= \partial_t\big(f(t)\mathbf{A}(t,\mathbf{x})\big) - \frac{1}{2}\dot{f}(t)\mathbf{A}(t,\mathbf{x}) + \dot{f}(t)\mathbf{x}\cdot\nabla\mathbf{A}(t,\mathbf{x}) \end{aligned} \tag{2.F.16}$$

and the Lagrangian changes by a total time derivative. The constants of motion arising from the three transformations are, respectively

$$H = \frac{1}{2}\sum_{p=1}^{N} m_p v_p^2 + \int_{\mathbf{x}} A_0(\kappa B + \rho) \tag{2.F.17a}$$

$$D = tH - \frac{1}{4}\sum_{p=1}^{N} m_p\left(\mathbf{r}_p\cdot\mathbf{v}_p + \mathbf{v}_p\cdot\mathbf{r}_p\right) - \frac{1}{2}\int_{\mathbf{x}} \mathbf{x}\cdot\mathbf{A}(\kappa B + \rho) \tag{2.F.17b}$$

$$K = -t^2 H + 2tD + \frac{1}{2}\sum_{p=1}^{N} m_p r_p^2 \tag{2.F.17c}$$

Of course (2.F.17a) coincides with the Hamiltonian of (2.F.5).

Returning now to the dynamical problem, we observe that when the constraint (2.F.3c) is imposed, all reference to gauge fields disappears from the equation of motion (2.F.4). Hence, we can set $\kappa B + \rho$ to zero throughout, and suppress the gauge degrees of freedom. Therefore, only the dynamical algebra (2.F.9) is relevant, and now it takes the form

$$\left[r_p^i, r_q^j\right] = 0 \tag{2.F.18a}$$

$$\left[r_p^i, m_q v_q^j\right] = i\delta^{ij}\delta_{pq} \tag{2.F.18b}$$

$$\begin{aligned} \left[m_p v_p^i, m_q v_q^j\right] &= i\frac{\epsilon^{ij}}{\kappa}\Big((1-\delta_{pq})e_p e_q \delta^2(\mathbf{r}_q - \mathbf{r}_q) \\ &\quad - \delta_{pq}\sum_{n\neq p} e_p e_n \delta^2(\mathbf{r}_p - \mathbf{r}_n)\Big) \end{aligned} \tag{2.F.18c}$$

The consistency of these commutation relations is established by realizing the operators through their action on functions of $\mathbf{r}_p$, with $\mathbf{r}_p$ acting by multiplication, while $m_p\mathbf{v}_p$ is redefined as

$$m_p v_p^i = p_p^i - \frac{e_p}{2\pi\kappa}\epsilon^{ij}\sum_{q\neq p} e_q \frac{(r_p^j - r_q^j)}{|\mathbf{r}_p - \mathbf{r}_q|^2} \tag{2.F.19}$$

with $\mathbf{p}_p$ acting as $-i\nabla_{\mathbf{r}_p}$.

The symmetry generators (2.F.13), (2.F.14) and (2.F.17) become

$$\mathbf{P} = \sum_{p=1}^{N} m_p \mathbf{v}_p \tag{2.F.20}$$

$$J = \sum_{p=1}^{N} \mathbf{r}_p \times m_p \mathbf{v}_p \tag{2.F.21}$$

$$H = \frac{1}{2}\sum_{p=1}^{N} m_p v_p^2 \tag{2.F.22a}$$

$$D = tH - \frac{1}{4}\sum_{p=1}^{N} m_p(\mathbf{r}_p\cdot\mathbf{v}_p + \mathbf{v}_p\cdot\mathbf{r}_p) \tag{2.F.22b}$$

$$K = -t^2 H + 2tD + \frac{1}{2}\sum_{p=1}^{N} m_p r_p^2 \tag{2.F.22c}$$

The presence of the interaction is hidden, but it is in evidence in the commutators (2.F.18) and in the relation (2.F.19) between canonical momentum and velocity, which implies that the effective particle Lagrangian is[11]

$$L_{\text{effective}} = \sum_{p=1}^{N}\left(\frac{1}{2}m_p v_p^2 + e_p \mathbf{v}_p\cdot\mathbf{a}_p\right) \tag{2.F.23a}$$

$$\begin{aligned} a_p^i(\mathbf{r}_1,\ldots,\mathbf{r}_N) &= \frac{1}{2\pi\kappa}\epsilon^{ij}\sum_{q\neq p} e_q\frac{(r_p^j - r_q^j)}{|\mathbf{r}_p - \mathbf{r}_q|^2} \\ &= \frac{1}{2\pi\kappa}\epsilon^{ij}\frac{\partial}{\partial r_p^j}\sum_{q\neq p} e_q \ln|\mathbf{r}_p - \mathbf{r}_q| \\ &= -\frac{1}{2\pi\kappa}\frac{\partial}{\partial r_p^i}\sum_{q\neq p} e_q\theta_{pq} \end{aligned}$$

$$\tan\theta_{pq} = \frac{y_p - y_q}{x_p - x_q} \tag{2.F.23b}$$

Explicitly, (2.F.23a) reads

$$L_{\text{effective}} = \frac{1}{2}\sum_{p=1}^{N} m_p v_p^2 + \frac{1}{2\pi\kappa}\sum_{\substack{p,q=1\\ p<q}}^{N} e_p e_q \frac{(\mathbf{v}_p - \mathbf{v}_q)\times(\mathbf{r}_p - \mathbf{r}_q)}{|\mathbf{r}_p - \mathbf{r}_q|^2} \tag{2.F.24}$$

Note that the angular momentum (2.F.21) is not constructed from the canonical momentum, hence as already remarked, its eigenvalues are non-integral. All the constants of motion $C = \mathbf{P}$, J, H, D and K generate the appropriate transformations (2.F.11), (2.F.12) and (2.F.16) upon commutation.

$$\delta \mathbf{r}_p = i\,[C, .\mathbf{r}_p] \tag{2.F.25}$$

In particular, the equation of motion (2.F.4), properly symmetrized, emerges upon commuting $\mathbf{v}_p$ with H. Also the $SO(2,1)$ generators satisfy the conformal Lie algebra,

$$[D,H] = -iH \quad , \quad [D,K] = iK \quad , \quad [H,K] = 2iD \tag{2.F.26}$$

which may be presented in the more familiar Cartan basis by forming linear combinations with the help of a fixed, positive parameter a of time dimensionality

$$\mathcal{R} = \frac{1}{2}\left(\frac{1}{a}K + aH\right) \tag{2.F.27a}$$

$$\mathcal{S} = \frac{1}{2}\left(\frac{1}{a}K - aH\right) \tag{2.F.27b}$$

$$L_{\pm} = (\mathcal{S} \pm iD) \tag{2.F.27c}$$

$\mathcal{S}$ and D act as non-compact two-dimensional boost generators, while $\mathcal{R}$ generates the rotations that form the compact $SO(2)$ subgroup of $SO(2,1)$.

$$[\mathcal{R}, L_{\pm}] = \pm L_{\pm} \tag{2.F.28a}$$

$$[L_+, L_-] = -2\mathcal{R} \tag{2.F.28b}$$

J commutes with the conformal generators; it rotates $\mathbf{P}$ in the proper manner.

$$\left[J, P^i\right] = i\epsilon^{ij}P^j \tag{2.F.29}$$

The momentum commutes with H; the commutators with the remaining conformal generators are

$$[D,\mathbf{P}] = -\frac{i}{2}\mathbf{P} \tag{2.F.30}$$

$$[K,\mathbf{P}] = -i\left(t\mathbf{P} - \sum_{p=1}^{N} m_p \mathbf{r}_p\right) \tag{2.F.31}$$

Equation (2.F.30) shows that the scale dimension of the momentum is 1/2, opposite to that of the coordinate $\mathbf{r}_p$. Since the commutator of two constants of motion is again a constant of motion, the right-hand side of (2.F.31) shows that center-of-mass motion is free. This is a consequence of the evident invariance of our theory against Galileo boosts, which are generated by $t\mathbf{P} - \sum_{p=1}^{N} m_p \mathbf{r}_p$.

That the generators are indeed constants of motion may be established by use of the formula

$$\frac{dC}{dt} = \frac{i}{\hbar}[H, C] + \frac{\partial C}{\partial t} \tag{2.F.32}$$

It follows from (2.F.32) that all three generators are time-dependent. Note however, D and K do not commute with the Hamiltonian; their total time derivative vanishes owing to the explicit time-dependence seen in (2.F.22b) and (2.F.22c).

We conclude this discussion of point-particle/Chern–Simons dynamics by recording the Casimir operator $\mathcal{J}^2$ of the SO(2,1) group.

$$\mathcal{J}^2 = \mathcal{R}^2 - \mathcal{S}^2 - D^2 = \frac{1}{2}(KH + HK) - D^2 \tag{2.F.33}$$

G. Quantum Dynamics

The Schrödinger equation governing dynamics of particles interacting with Chern–Simons gauge fields is inferred from (2.F.19), (2.F.22a) and (2.F.23).

$$i\frac{\partial}{\partial t}\Psi(t;\mathbf{r}_1,\ldots,\mathbf{r}_N)=H\Psi(t;\mathbf{r}_1,\ldots,\mathbf{r}_N) \tag{2.G.1a}$$

$$H=\sum_{p=1}^{N}\frac{1}{2m_p}\left(\frac{1}{i}\nabla_{\mathbf{r}_p}-e_p\mathbf{a}_p\right)^2 \tag{2.G.1b}$$

The wave function Ψ is single-valued. We may however make use of the formulas in (2.F.23b) to express $\mathbf{a}_p$ as a gradient, and remove the interaction in (2.G.1) by redefining the phase of the wave function.[13]

$$\Psi(\mathbf{r}_1,\ldots,\mathbf{r}_N)=e^{i\Theta}\Psi^0(\mathbf{r}_1,\ldots,\mathbf{r}_N) \tag{2.G.2}$$

$$\Theta=\sum_{\substack{p,q=1\\p<q}}^{N}\nu_{pq}\theta_{pq} \tag{2.G.3a}$$

$$\nu_{pq}=-\frac{e_pe_q}{2\pi\kappa} \tag{2.G.3b}$$

Then Ψ^0 satisfies the *free* Schrödinger equation.

$$i\frac{\partial}{\partial t}\Psi^0(t;\mathbf{r}_1,\ldots,\mathbf{r}_N)=H^0\Psi^0(t;\mathbf{r}_1,\ldots,\mathbf{r}_n) \tag{2.G.4a}$$

$$H^0=e^{-i\Theta}H^0e^{i\Theta}=\sum_{p=1}^{N}\left(-\frac{1}{2m_p}\nabla^2_{\mathbf{r}_p}\right) \tag{2.G.4b}$$

Even though H^0 is a sum of one-body Hamiltonians, Ψ^0 cannot be chosen as a simple product of one-body eigenstates [plane waves] because Ψ^0 satisfies complicated aperiodicity conditions, which must hold so that Ψ be single-valued. Of course Ψ^0 is a *superposition* of plane waves, however, determining the precise superposition, which when multiplied by $e^{i\Theta}$ gives a single-valued wave function, is a challenging, non-trivial problem that has been solved only for the two-body case.

The two-body problem is tractable because of the center-of-mass reduction, wherein only the relative coordinate, $\mathbf{r}=\mathbf{r}_1-\mathbf{r}_2$, experiences the interaction, while the center-of-mass coordinate $\mathbf{R}=\frac{m_1\mathbf{r}_1+m_2\mathbf{r}_2}{m_1+m_2}$ moves freely. By setting total momentum to zero, the two-body wave function depends only on $\mathbf{r}$ and satisfies

$$i\frac{\partial}{\partial t}\psi(t;\mathbf{r})=h\psi(t;\mathbf{r}) \tag{2.G.5}$$

where the Hamiltonian for relative motion is

$$h=\frac{1}{2M}\left(\mathbf{p}-\mathbf{a}(\mathbf{r})\right)^2 \tag{2.G.6}$$

M is the reduced mass, $M^{-1}=m_1^{-1}+m_2^{-1}$, and the vector potential $\mathbf{a}$ gives rise to a vortex with flux $\Phi=-\frac{e_1e_2}{\kappa}$

$$\begin{aligned}a^i(\mathbf{r})&=-\frac{\Phi}{2\pi}\epsilon^{ij}\frac{\hat{r}^j}{r}=-\frac{\Phi}{2\pi}\epsilon^{ij}\partial_j\ln r\\&=\frac{\Phi}{2\pi}\partial_i\theta=\nu\partial_i\theta\end{aligned} \tag{2.G.7}$$

$$\mathbf{r}=(r\cos\theta,r\sin\theta)$$

Hence (2.G.5) requires, after the usual separation of time,

$$\psi(t;\mathbf{r}) = e^{-iEt}\psi_E(\mathbf{r}) \tag{2.G.8}$$

solving the following eigenvalue problem.

$$-\frac{1}{2M}\left(\nabla - i\nu\nabla\theta\right)^2 \psi_E(\mathbf{r}) = E\psi_E(\mathbf{r}) \tag{2.G.9}$$

The eigenfunctions necessarily have theform

$$\psi_E(\mathbf{r}) = e^{i\nu\theta}\psi_k^0(\mathbf{r}) \tag{2.G.10}$$

where $\psi_k^0(\mathbf{r})$, though governed by the free Hamiltonian

$$h^0 = e^{-i\nu\theta} h\, e^{i\nu\theta} \tag{2.G.11}$$

$$h^0\psi_k^0 = \frac{k^2}{2M}\psi_k^0 \ , \qquad E = \frac{k^2}{2M} \tag{2.G.12}$$

is not a conventional plane wave, owing to a non-trivial boundary condition,

$$\psi_k^0(r,\theta=0) = e^{-i2\pi\nu}\psi_k^0(r,\theta=2\pi) \tag{2.G.13}$$

which must be met so that $\psi_E(\mathbf{r})$ is single-valued.

Since rotation by 2π corresponds to double exchange of particles, we see that ψ_k^0 acquires a statistics factor $-\pi\nu = e^2/2\kappa$, in agreement with (2.E.1) for $e_1 = e_2$. Moreover, the [relative] angular momentum

$$\begin{aligned} J &= \mathbf{r}\times M\mathbf{v} = \mathbf{r}\times\mathbf{p} - \mathbf{r}\times\mathbf{a} \\ &= \mathbf{r}\times\mathbf{p} + \frac{e^2}{2\pi\kappa} \end{aligned} \tag{2.G.14}$$

indicates that each particle possesses additional spin $e^2/4\pi\kappa$ again in agreement with (2.E.1).

[The angular momentum operator acting on the multi-valued wavefunction ψ_k^0 is

$$\begin{aligned} J^0 &= e^{-i\nu\theta} J\, e^{i\nu\theta} = e^{-i\nu\theta}\frac{1}{i}\frac{\partial}{\partial\theta}e^{i\nu\theta} + \frac{e^2}{2\pi\kappa} \\ &= \frac{1}{i}\frac{\partial}{\partial\theta} + \nu + \frac{e^2}{2\pi\kappa} = \frac{1}{i}\frac{\partial}{\partial\theta} \ , \end{aligned}$$

i.e. it is just the angular derivative. Nevertheless, its eigenvalues are non-integral, just as those of J in (2.G.14), since $J^0 = \frac{1}{i}\frac{\partial}{\partial\theta}$ acts on multi-valued functions which satisfy (2.G.13). It should further be emphasized that, as we have already stated, the reason that J is just $r\times M\mathbf{v}$ and not $\mathbf{r}\times\mathbf{p}$ is because our effective particle theory arises from a *dynamical* model for the gauge potential, where the dynamics is governed by the Chern–Simons term. If the problem (2.G.5) and (2.G.6) is viewed as describing single particle motion in an *externally* prescribed gauge potential $\mathbf{a}(\mathbf{r})$, then the correct angular momentum *is* $\mathbf{r}\times\mathbf{p}$, with integral eigenvalues.[12]]

The Schrödinger equation (2.G.5), (2.G.6) and (2.G.7) leads only to scattering, and the scattering amplitude has been obtained long ago by Aharonov and Bohm, and later by Ruijsenaars.[14] More recently the problem has re-emerged in the context of planar gravity.[3e]. As we shall see, there too one seeks free solutions with unconventional boundary conditions.

I shall now present the solution, using the gravitational techniques, which have the advantage of giving an explicit wave function, in the form of a contour integral.[2f]

$$\psi_k^0(\mathbf{r}) = \oint \frac{dz}{2\pi} e^{i\mathbf{k}(z)\cdot\mathbf{r}} \rho(z) \tag{2.G.15}$$

Here $\mathbf{k}(z) = (k\cos z, k\sin z)$ and it is obvious that $\psi_k^0(\mathbf{r})$ satisfies the free equation. That it also satisfies (2.G.13) requires a special contour C and weight function ρ, which we now derive, by considering the scattering problem, the radial equation and the phase shifts.

The *Ansatz* $u_E^j(r)\frac{e^{ij\theta}}{\sqrt{2\pi}}$ is made for $\psi_E(\mathbf{r})$ in (2.G.9); j is an arbitrary integer to insure single valuedness, $j = 0, \pm 1, \pm 2, \ldots$; $u_E^j(r)$ satisfies the radial equation,

$$\left(-\frac{1}{r}\frac{d}{dr}r\frac{d}{dr} + \frac{(j-\nu)^2}{r^2}\right) u_E^j(r) = k^2 u_E^j(r) \tag{2.G.16}$$

and is given by a Bessel function.

$$u_E^j(r) = \sqrt{M} J_{|j-\nu|}(kr) \tag{2.G.17}$$

The angular momentum of this partial wave is $j-\nu$; see (2.G.14). The normalization is fixed by

$$\int_0^\infty r dr\, u_E^j(r) u_E^j(r) = \delta(E - E') \tag{2.G.18a}$$

and insures

$$\int_0^\infty dE\, u_E^j(r) u^j E(r') = \frac{1}{r}\delta(r - r') \tag{2.G.18b}$$

When the plane wave identity

$$e^{ikr\cos\theta} = \sum_{j=-\infty}^{\infty} e^{ij(\theta+\frac{\pi}{2})} J_j(kr) \tag{2.G.19}$$

is recalled, we are led to form the scattering solution by

$$\psi_E(\mathbf{r}) = \sum_{j=-\infty}^{\infty} e^{i(\delta_j + \frac{\pi}{2}j)} u_E^j(r) \frac{e^{ij\theta}}{\sqrt{2\pi}} \tag{2.G.20}$$

where the phase shift δ_j of $u_E^j(r)$ relative to $J_j(kr)$ is identified from their large r asymptotes.

$$\delta_j = \begin{cases} \nu\frac{\pi}{2} & j > [\nu] \\ -\nu\frac{\pi}{2} - j\pi & j \le [\nu] \end{cases} \tag{2.G.21}$$

Here the brackets [ ] indicate integer part. The energy independence of the phase shifts is a consequence of scale invariance.[15]

From (2.G.20) and (2.G.21), the wave function is constructed as

$$\begin{aligned}\psi_E(\mathbf{r}) &= \sqrt{\frac{M}{2\pi}} \sum_{j=[\nu]+1}^{\infty} e^{i\frac{\pi}{2}(\nu+j)} J_{j-\nu}(kr) e^{ij\theta} \\ &+ \sqrt{\frac{M}{2\pi}} \sum_{j=-\infty}^{[\nu]} e^{-i\frac{\pi}{2}(\nu+j)} J_{\nu-j}(kr) e^{ij\theta}\end{aligned} \tag{2.G.22}$$

The sums are performed by using the Schläfli contour representation for the Bessel function.

$$J_\alpha(kr) = e^{i\frac{\pi}{2}\alpha} \oint_s \frac{dz}{2\pi} e^{-ikr\cos z} e^{iz\alpha} \tag{2.G.23}$$

The Schläfli contour C_s begins at $z = -3\pi/2 + i\infty$, descends to slightly above the real axis, passes from $z = -3\pi/2 + i(0^+)$ to $z = \pi/2 + i(0^+)$ and ascends to $z = \pi/2 + i\infty$. The sums are now geometric, and give for the two terms in (2.G.22), respectively,

$$\begin{aligned}\psi_E(\mathbf{r}) &= \left(\frac{M}{2\pi}\right)^{1/2} \oint_s \frac{dz}{2\pi} e^{-ikr\cos s} e^{i([\nu]\pi - \{\nu\}z + [\nu]\theta)} \frac{-1}{1+e^{-i(z+\theta)}} \\ &+ \left(\frac{M}{2\pi}\right)^{1/2} \oint_s \frac{dz}{2\pi} e^{-ikr\cos s} e^{i([\nu]\pi - \{\nu\}z - [\nu]\theta)} \frac{1}{1+e^{i(z-\theta)}} \\ &= \left(\frac{M}{2\pi}\right)^{1/2} \oint_{-s} \frac{dz}{2\pi} e^{-ikr\cos s} e^{i([\nu]\pi + \{\nu\}z + [\nu]\theta)} \frac{1}{1+e^{i(z-\theta)}} \\ &+ \left(\frac{M}{2\pi}\right)^{1/2} \oint_s \frac{dz}{2\pi} e^{ikr\cos s} e^{i([\nu]\pi + \{\nu\}z + [\nu]\theta)} \frac{1}{1+e^{i(z-\theta)}} \\ \{\nu\} &\equiv \nu - [\nu]\end{aligned} \tag{2.G.24a}$$

In passing from the first to the second equality, the change of variables $z \to -z$ is performed in the first integral. As a consequence, the integration contour for that integral, now called C_{-s}, becomes the mirror image of the Schläfli contour C_s. [C_{-s} starts from $3\pi/2 - i\infty$, ascends to $3\pi/2$ below the real axis, passes to $-\pi/2$ and descends to $-\pi/2 - i\infty$.] As a further consequence, the integrands of the two contour integrals become identical [we use $e^{i[\nu]\pi} = e^{-i[\nu]\pi}$]. To proceed, contours are shifted: C_{-s} is shifted by $\pi/2$ to the left, and C_s by $\pi/2$ to right, so that the vertical portions of both contours are at $z = \pm\pi$. The last step is to redefine the integration variable by $z = -z' + \theta - \pi$. The z' integral now runs over the contour C depicted in Fig. 1a, and $\psi_E(\mathbf{r})$ is represented by [z' is renamed z]

$$\begin{aligned}\psi_E(\mathbf{r}) &= \left(\frac{M}{2\pi}\right)^{1/2} e^{i\nu\theta} \oint \frac{dz}{2\pi} e^{ikr\cos(z-\theta)} \frac{e^{-i\{\nu\}z}}{1-e^{-iz}} \\ &= \left(\frac{M}{2\pi}\right)^{1/2} e^{i\nu\theta} \oint \frac{dz}{2\pi} e^{i\mathbf{k}(z)\cdot\mathbf{r}} \frac{e^{-i\{\nu\}z}}{1-e^{-iz}}\end{aligned} \tag{2.G.24b}$$

[A constant phase factor has been suppressed.] Thus we have derived the representation (2.G.15), with contour C as in Fig. 1a and $\rho(z)$ given by

$$\rho(z) = \left(\frac{M}{2\pi}\right)^{1/2} \frac{e^{-i\{\nu\}z}}{1-e^{-iz}} \tag{2.G.25}$$

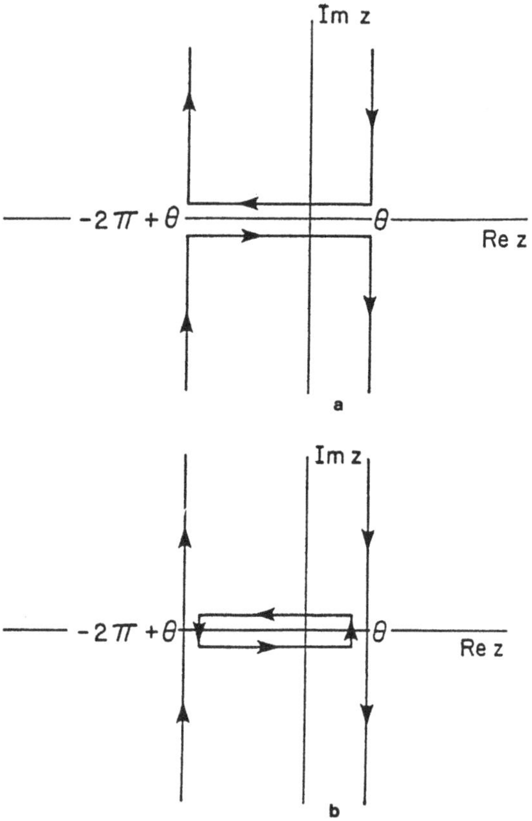

Fig. 1. (a) Integration contour C for the wave function (2.G.24b). The pole at the origin is avoided. (b) Contour $\tilde{C}$ equivalent to contour C. The pole at the origin is enclosed, giving rise to the incoming wave. The vertical contours produce the scattered wave.

That $\psi_E(\mathbf{r})$ as given by (2.G.24b) satisfies the free Schrödinger equation is obvious; that it is single-valued — periodic in θ with 2π period — is more easily seen in (2.G.24a).

The contour C avoids the pole in $\rho(z)$ at $z = 0$. However, we may alternatively enclose the pole and replace the contour C by the three-segmented contour $\tilde{C}$, depicted in Fig. 1b. The portion encircling the pole is evaluated by Cauchy's residue theorem, contributing $(M/2\pi)^{1/2} e^{i(kr\cos\theta+\nu\theta)}$; the portions arising from the vertical axes are presented in terms of real integrals by setting $z = \theta - 2\pi + iy$ and $z = \theta + iy$. Evidently, this separation decomposes the total scattering wave function $\psi_E(\mathbf{r})$ into an incoming wave [pole contribution] and the scattered wave [vertical contour contributions].

$$\psi_E(\mathbf{r}) = \left(\frac{M}{2\pi}\right)^{1/2} (\psi^{\text{in}}(\mathbf{r}) + \psi_E^{\text{sc}}(\mathbf{r}) \tag{2.G.26a}$$

$$\psi_E^{\text{in}}(\mathbf{r}) = e^{i(kr\cos\theta + \nu\theta)} \tag{2.G.26b}$$

$$\psi_E^{\text{sc}}(\mathbf{r}) = e^{i[\nu]\theta} e^{i\nu\pi} \sin\nu\pi \int_{-\infty}^{\infty} \frac{dy}{\pi} e^{ikr\cosh y} \frac{e^{\{\nu\}y}}{e^{y-i\theta} - 1} \tag{2.G.26c}$$

The large r asymptote of $\psi_E^{\text{sc}}(\mathbf{r})$ defines the scattering amplitude $f(\theta)$ through the formula

$$\psi_E^{\text{sc}}(\mathbf{r}) \underset{r\to\infty}{\longrightarrow} \sqrt{\frac{i}{r}} f(\theta) e^{ikr} \tag{2.G.27}$$

Although the integral (2.G.26c) for $\psi_E^{\text{sc}}(\mathbf{r})$ cannot be performed, its limit at large r can be evaluated. The circular wave formula (2.G.27) is found, with scattering amplitude

$$f(\theta) = \frac{1}{\sqrt{2\pi k}} e^{-i\{\nu\}\theta} e^{i(\nu+1/2)(\theta+\pi)} \frac{\sin\nu\pi}{\sin\theta/2} \tag{2.G.28}$$

Equations (2.G.26b) and (2.G.28) are essentially the results of Aharonov and Bohm and Ruijsenaars;[14] note especially that the incoming wave is not a plane wave, but is modulated by the additional phase $e^{i\nu\theta}$.

Since $\mathbf{r} = \mathbf{r}_1 - \mathbf{r}_2$, $e^{i\mathbf{k}(z)\cdot\mathbf{r}}$ is a product of two plane waves

$$\begin{aligned} e^{i\mathbf{k}(z)\cdot(\mathbf{r}_1-\mathbf{r}_2)} &= \psi_1^z(\mathbf{r}_1)\psi_2^z(\mathbf{r}_2) \\ \psi_1^z(\mathbf{r}) &= e^{i\mathbf{k}(z)\cdot\mathbf{r}} \\ \psi_2^z(\mathbf{r}) &= e^{-i\mathbf{k}(z)\cdot\mathbf{r}} \end{aligned} \tag{2.G.29}$$

Hence the representation (2.G.24b) shows explicitly how products of one-body plane waves are superposed to form our solution.[2f]

$$\psi_E^0(\mathbf{r}) = \left(\frac{M}{2\pi}\right)^{1/2} \oint \frac{dz}{2\pi} \psi_1^z(\mathbf{r}_1)\psi_2^z(\mathbf{r}_2) \frac{e^{-i\{\nu\}z}}{1 - e^{-iz}} \tag{2.G.30}$$

As yet we do not have a similar closed form for the N-body wave function. The problem is reminiscent of the δ-function interaction on a line. There too the many-body wave function is obtained by superposing one-body wave functions in a fashion prescribed by the Bethe *Ansatz*. Perhaps similar ideas will prove useful here.

Finally, we conclude that the two-body relative coordinate problem of course also possesses the $SO(2,1)$ symmetry, with generators given by the relative coordinate parts of the two-body generators (2.F.22)

$$H = (h) = \frac{1}{2} M v^2 \tag{2.G.31a}$$

$$D = tH - \frac{1}{4} M(\mathbf{r}\cdot\mathbf{v} + \mathbf{v}\cdot\mathbf{r}) \tag{2.G.31b}$$

$$K = -t^2 H + 2tD + \frac{1}{2} M r^2 \tag{2.G.31c}$$

The algebraic properties of these quantities hold as before; now they are based on the dynamical algebra

$$[r^i, r^j] = 0 \ , \quad [r^i, Mv^j] = i\delta^{ij} \ , \quad [Mv^i, Mv^j] = i\epsilon^{ij} 2\pi\nu\, \delta(\mathbf{r}) \tag{2.G.32}$$

The Casimir in (2.F.33) can be expressed in terms of the angular momentum (2.G.14)

$$\mathcal{J}^2 = \frac{1}{4}(J^2 - 1) \tag{2.G.33}$$

Since the eigenvalues of J are $j - \nu$, those of $\mathcal{J}^2$ are $\frac{1}{4}\left((j-\nu)^2 - 1\right)$, and the entire motion at fixed angular momentum is described by a single, irreducible, unitary and infinite-dimensional representation of $SO(2,1)$. We have already remarked that the energy independence of the phase shifts is a consequence of the symmetry. Because of the higher symmetry, the time-dependent Schrödinger equation (2.G.5) can be separated in coordinates other than the usual time and space. Indeed group theory may be used to give a complete, alternative analysis of the problem.[2f]

3. PLANAR GRAVITY

A. Introduction

The equations for Einstein's theory of gravity — general relativity — can be presented in any space-time with dimension d equal to or greater than three: The Einstein tensor

$$G_{\mu\nu} \equiv R_{\mu\nu} - \frac{1}{2} g_{\mu\nu} R \tag{3.A.1}$$

vanishes in the absence of matter sources,

$$G_{\mu\nu} = 0 \tag{3.A.2a}$$

while in their presence it is proportional to the energy-momentum tensor of matter, $T_{\mu\nu}$.

$$G_{\mu\nu} = 2\pi G\, T_{\mu\nu} \tag{3.A.2b}$$

Here $R_{\mu\nu}$ and R are traces of the four-index Riemann tensor $R_{\alpha\mu\beta\nu}$ in which all local geometrical information about the space-time is encoded. G is the gravitational coupling constant — the generalization to other dimensions of Newton's constant; in (3.A.2b) G enters with an unconventional normalization that is convenient for the subsequent analysis. The reason that (3.A.2) cannot be posited in two space-time dimensions is because there $G_{\mu\nu}$ vanishes identically. [However other geometrical equations have been proposed at $d = 2$.[16]]

It is obvious from (3.A.2b) that when space-time is flat, *i.e.* when the Riemann tensor vanishes, so also does the Einstein tensor and $T_{\mu\nu}$ must be zero. In general, the converse does not hold: absence of matter implies vanishing Einstein tensor, but the Riemann tensor need not be zero so that empty space-time need not be flat. However, in three dimensions the Riemann tensor is linearly related to the Einstein tensor,

$$R^{\alpha\mu}_{\beta\nu} = \epsilon^{\alpha\mu\gamma} \epsilon_{\beta\nu\delta} G^{\delta}_{\gamma} \tag{3.A.3}$$

so that the vanishing of the latter implies the vanishing of the former: empty space-time is necessarily locally flat.[17]

Several consequences follow immediately: since the vacuum state [empty space-time] is locally flat, there are no gravitational waves in the classical theory, and upon quantization there are no quantum gravitons. Sources produce curvature, but only locally at the location in space-time of the sources. Forces between sources are not mediated by graviton exchange, since there are no gravitons. Rather interactions arise because the locally flat space-time possesses in the large non-trivial geometrical and topological structure that gives rise to non-trivial motions. It also follows that the non-relativistic limit of Einstein's general relativity in three-dimensions is not three-dimensional Newtonian gravity, which involves a conventional force law that decreases with the inverse power of the distance.

It is the purpose of our research program to study in three-dimensional space-time the classical and quantum motions of matter that interacts gravitationally.[3] Since there are no propagating gravitational degrees of freedom, the problem is tractable, and we can learn much about the puzzles that are encountered when a geometrical theory is confronted by quantum mechanics. In four dimensions these puzzles exist as well, and it is my opinion that understanding them is important for understanding quantum gravity; a task quite independent of and perhaps more fundamental than the task of overcoming the unrenormalizable infinities that pollute four-dimensional gravity, but are absent in three dimensions since non-renormalizable graviton exchange does not occur. To conclude these introductory remarks, I note the following points.

(a) The theory can be elaborated by adding a cosmological constant to the field equation. The vacuum is then a space of constant curvature, whose sign depends on the sign of the cosmological constant. While some investigations of such models have been performed, I shall not further discussthem here.[3c]

(b) Another elaboration of the conventional theory involves adding a topological term, analogous to the gauge theoretic modification.[3a] This Chern–Simons addition will be discussed below.

B. Classical Space-Times

We record several interesting space-times that arise from classical sources.[3b] We begin with a single massive but spinless point-particle. Without loss of generality the particle is taken to be at rest at the origin of the coordinate system, *i.e.* it is described by an energy-momentum tensor all whose components, except the energy density, vanish,

$$\begin{aligned} \sqrt{\det g_{\mu\nu}} T^{00} &= M\delta(X)\delta(Y) \\ T^{0i} &= T^{ij} = 0 \end{aligned} \tag{3.B.1}$$

Here M is the particle mass.

The task is to find the metric or equivalently to give a formula for the line element. Clearly it is non-trivial only in its spatial components,

$$(ds)^2 = (dt)^2 - (d\ell)^2 \tag{3.B.2}$$

and we need to find expressions for $(d\ell)^2$.

We recognize that we seek a space which is everywhere flat [$T^{\mu\nu}$ vanishes] except at the origin where a δ-function singularity concentrates the curvature. It is clear that the desired space is a cone, with the source particle positioned at the apex.[3b,4d,17] It remains to give an analytic description of this obvious geometrical fact.

To solve the Einstein equation (3.A.2b) with sources given by (3.B.1), it is necessary to choose a coordinate system, and the conical solution looks different in different coordinates. Of course only the two-dimensional spatial section needs to be considered.

Particularly useful coordinates, which lend themselves to a many-body generalization, are the conformal ones where the metric tensor is a multiple of the flat metric tensor; this can always be locally achieved in two dimensions. The conformally flat spatial metric that solves Einstein's equation then leads to the following spatial interval.

$$(d\ell)^2 = \frac{1}{R^{2GM}}\Big((dR)^2 + R^2(d\Theta)^2\Big) \tag{3.B.3}$$

Here the variables range over the conventional circles.

$$\begin{aligned} 0 \leq R &\leq \infty \\ -\pi \leq \Theta &\leq \pi \end{aligned} \tag{3.B.4}$$

While (3.B.3) certainly provides the desired solution, it does not seem to produce the cone described earlier. Nor is it manifest that the space is flat except at the origin.

All this can be seen by passing to another coordinate system, attained from (3.B.3) and (3.B.4) by a change of variables.

$$\begin{aligned} r &= \frac{R^{1-GM}}{1-GM} \\ \theta &= (1-GM)\Theta \end{aligned} \tag{3.B.5}$$

In terms of r and θ the spatial metric is flat, and the line-element is trivial.

$$(d\ell)^2 = (dr)^2 + r^2(d\theta)^2 \tag{3.B.6}$$

However, the range of the new variables is unconventional — an angular region is excised, since according to (3.B.4) the range of (r, θ) is

$$\begin{aligned} 0 \leq r &\leq \infty \\ -\pi(1-GM) \leq \theta &\leq \pi(1-GM) \end{aligned} \tag{3.B.7}$$

This describes a cone, with apex determined by GM. [Henceforth we take $GM \leq 1$. For $GM > 1$, the space changes character and the description becomes more complicated.[3b] At $GM = 1$, it is seen from (3.B.3) that space becomes a cylinder in the variable $r = \ln R$.]

In summary, we say that a point particle of mass M at the origin gives rise to a locally flat space-time, but the global identification of coordinate variables is unconventional and reveals the presence of a massive point-particle: the point (t, r, θ) is identified with

$$(t, r, \theta) \approx (t, r, \theta + 2\pi(1-GM)) \tag{3.B.8a}$$

In terms of a complex variable description of the space, $z = x + iy$, we identify z with

$$z \approx e^{-2\pi i GM} z \tag{3.B.8b}$$

This is the analog in planar gravity of the Schwarzschild solution.

To find the planar analog of the Kerr solution, we endow our point-particle at the origin with spin S, *i.e.* now the energy-momentum tensor possess non-trivial energy density and momentum density, the latter giving rise to no momentum but to angular momentum S.

$$\begin{aligned} \sqrt{\det g_{\mu\nu}}\,T^{00} &= M\delta(X)\delta(Y) \\ \sqrt{\det g_{\mu\nu}}\,T^{0i} &= \sqrt{\det g_{\mu\nu}}\,T^{i0} = S\epsilon^{ij}\partial_j\delta(X)\delta(Y) \\ T^{ij} &= 0 \end{aligned} \tag{3.B.9}$$

In the spatially conformal coordinate system, the metric that solves the field equation leads to a space-time interval, which is non-trivial in time as well as space.

$$(ds)^2 = (dt + GSd\Theta)^2 - \frac{1}{R^{2GM}}\Big((dR)^2 + R^2(d\Theta)^2\Big) \tag{3.B.10}$$

Once again, by a change of variables one may pass to a locally flat space-time, where the presence of a massive, spinning source is encoded in a non-trivial identification of coordinate variables. Defining new spatial variables as in (3.B.5) and also a new time variable τ by

$$\tau = t + GS\Theta = t + \frac{GS}{1-GM}\theta \tag{3.B.11}$$

we see that (3.B.10) becomes flat,

$$(ds)^2 = (d\tau)^2 - (dr)^2 - r^2(d\theta)^2 \tag{3.B.12}$$

but the required identification is

$$(\tau, r, \theta) \approx \big(\tau + 2\pi GS, r, \theta + 2\pi(1-GM)\big) \tag{3.B.13}$$

Time is helical, space is conical and there are closed time-like contours.

Note that specifying a solution is equivalent to specifying an element of the 2 + 1-dimensional Poincaré group that effects the identification (3.B.13).

The static one-body solution can be generalized to describe N particles located at $\mathbf{R}_i$, with masses M_i and spins S_i, $i = 1, \ldots, N$.[18] One finds in spatially conformal coordinates

$$(ds)^2 = \left(dt + G\sum_{i=1}^{N} S_i \frac{(\mathbf{R}-\mathbf{R}_i)}{|\mathbf{R}-\mathbf{R}_i|^2} \times d\mathbf{R}\right)^2 - \prod_{i=1}^{N} \frac{1}{|\mathbf{R}-\mathbf{R}_i|^{2GM_i}}(d\mathbf{R})^2 \tag{3.B.14}$$

The passage to locally flat coordinates is effected by first defining a new time τ.

$$d\tau = dt + G\sum_{i=1}^{N} S_i \frac{(\mathbf{R}-\mathbf{R}_i)}{|\mathbf{R}-\mathbf{R}_i|^2} \times d\mathbf{R} \tag{3.B.15}$$

This hides the spins in complicated identifications on τ. To flatten the spatial interval, it is useful to express it in complex variables $Z = X + iY$, *etc.*

$$(d\ell)^2 = \left(\prod_{i=1}^{N} \frac{1}{(Z-Z_i)^{GM_i}}\right) dZ \left(\prod_{i=1}^{N} \frac{1}{(\bar{Z}-\bar{Z}_i)^{GM_I}}\right) d\bar{Z} \tag{3.B.16}$$

Thus the definition

$$dz = \left(\prod_{i=1}^{N} \frac{1}{(Z-Z_i)^{GM_i}}\right) dZ \tag{3.B.17}$$

gives the flat spatial interval

$$(d\ell)^2 = dz\, d\bar{z} \tag{3.B.18}$$

but complicated identifications on the complex plane, which generalize (3.B.8b), reveal the presence of N particles with masses M_i. Unlike in the one-body problem, we cannot express z as a closed form function of Z, but for most purposes the integral expression suffices,

$$z = \int^{Z} dZ' \prod_{i=1}^{N} \frac{1}{(Z' - Z_i)^{GM_i}} \tag{3.B.19}$$

and it can be explicitly evaluated in special cases.

It is easy to show that the above solution also satisfies self-consistently the geodesic equation.[3b] Thus a static N-body configuration exists and is stable in three-dimensional space-time, in contrast to higher dimensions where gravitational attraction would prevent this. This demonstrates vividly the absence of Newtonian attraction in our theory.

With point-particle sources, the two-dimensional space is flat, but curvature is concentrated on a lower-dimensional sub-space: the zero-dimensional collection of points where the particles are located. One may next consider flat space with curvature concentrated on one-dimensional lines; *i.e.* string sources in the plane, which presumably correspond to domain walls in four-dimensional space-time, just as points on the plane correspond to strings in four-dimensional space-time.

When considering strings, it is natural to allow for tension along the string; otherwise the source is an uninteresting pulvarization of the point-particle — a "dust" string.

In the spinless case the results are simple and startling.[3f, 19] There are no open strings, only closed ones. A circular source at $r = a$ is described by an energy-momentum tensor whose non-vanishing components are

$$\sqrt{\det g_{\mu\nu}}\, T^0_0 = \mu\delta(r - a) \tag{3.B.20a}$$

$$\sqrt{\det g_{\mu\nu}}\, T^\theta_\theta = \tau\delta(r - a) \tag{3.B.20b}$$

Here μ and τ are mass and stress density/per unit length; the total mass is $M = 2\pi a\mu$; for a relativistic string $\tau = \mu$. The momentum density and the other stress components vanish. With this source the space-time interval in conformally flat spatial coordinates is

$$(ds)^2 = \begin{cases} \left(1 - 2\pi G a\tau \ln \frac{r}{a}\right)(dt)^2 - \left(\frac{a}{r}\right)^2 \left((dr)^2 + r^2(d\theta)^2\right) & r \geq a \\ (dt)^2 - (dr)^2 - r^2(d\theta)^2 & r \leq a \end{cases} \tag{3.B.21}$$

The exterior spatial interval also reads $\left(d\, a \ln \frac{r}{a}\right)^2 + (ad\theta)^2$, which is a half-cylinder of radius a extending from infinity to $r = a$, where it is capped by the flat disk of the $r \leq a$ region. Moreover, the total mass $M = 2\pi a\mu$ is given by G^{-1}, so that

$$GM = 1 \tag{3.B.22}$$

We have seen earlier that for point-particles obeying (3.B.22) the space is a cylinder; here (3.B.22) is always obeyed for spinless strings under tension and the space is a capped cylinder.

Although for $\tau > 0$, g_{00} vanishes at a finite distance, this is not a conventional horizon because g_{00} does not change sign, but time does "stand still" there. Clearly there exist solutions with either sign of τ and unrelated to μ. However, for a relativistic string $\tau = \mu > 0$.

For more discussion on these extended objects and inclusion of spin, please consult the research papers.[3f, 19]

The simplest non-trivial dynamics arises when we consider the interaction of two point-particles with each other. As in other contexts, it is possible to pass to the center-of-mass frame where the relative coordinate moves in an effective potential that describes the interaction.[3d] The same problem arises without the center-of-mass reduction, but in the limit when one particle's mass becomes much larger than the other.[3e]

In view of this, it suffices to consider the problem of a test particle [mass m] moving in the field produced by the source particle [mass M] located at the origin.

The classical motion of a spinless test particle is easy to describe: in flat coordinates there is no deviation from straight-line motion. However, when the unconventional identification (3.B.13) is performed, we find a classical scattering angle,

$$\Delta\theta_{\text{classical}} = \pm\pi GM \tag{3.C.1}$$

and a classical time delay,

$$\Delta t_{\text{classical}} = \mp\frac{GS\pi}{1-GM} \tag{3.C.2}$$

where S is the source particle's spin, and the sign depends on which side the source is passed. The classical trajectories are depicted in Fig. 2 [ignore the dotted lines for the moment]. They depend only on the impact parameter, but not on the energy; the scattering angle does not vary with impact parameter, except in its sign.

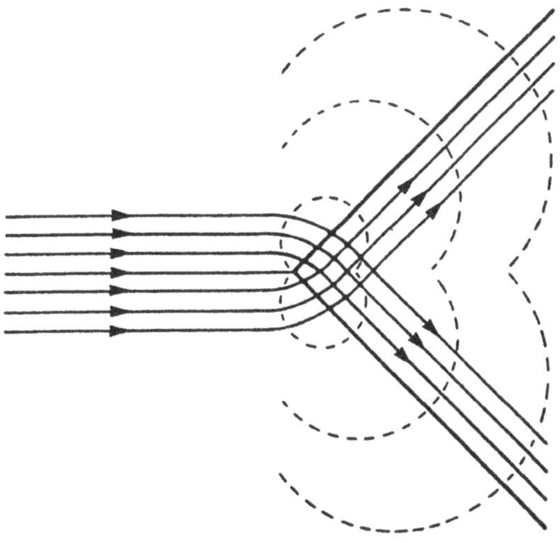

Fig. 2 Qualitatative pictorialization for scattering of waves on an obstacle at the origin. The two sharp lines are classical trajectories with scattering angle $\pm\pi GM$ in (3.C.1), the sign depending on which side the trajectory passes the source. The envelope to the right of the source, formed by heavy diagonal lines, is the sharp geometrical shadow. Broken lines represent diffraction on two sharp edges, even though no edge is actually present — the source [conical defect] produced the "edges."

Next we give a quantum mechanical description and to this end we solve a quantum mechanical equation appropriate to the test particle: Schrödinger or Klein–Gordon for spinless test particles; Dirac for spin 1/2 test particles, *etc.* [We do not second quantize the matter degrees of freedom.] The question that still must be considered is what interaction should we use to describe the influence of the source on the test particle.

The answer that we propose is that no interaction need be considered; rather we solve the free, non-interacting equation but impose on the solution a coordinate condition that reflects the identification (3.B.13).

For example, let us consider the simplest case first — a spinless test particle in a spinless source. The equation we propose to solve is the free [square-root] Klein–Gordon,

$$i\frac{\partial}{\partial t}\psi(t;r,\theta) = \sqrt{-\nabla^2+m^2}\,\psi(t;r,\theta) \tag{3.C.3}$$

with the requirement that

$$\begin{aligned} \psi(t;r,\theta) &= \psi(t;r,\theta+2\pi\alpha) \\ \alpha &= 1-GM \end{aligned} \tag{3.C.4}$$

[If non-relativistic motion is of interest, the non-local "square root" operator is replaced by $m - \nabla^2/2m$, which leads to the free Schrödinger equation, with boundary conditions (3.C.4). The mathematical analysis is identical.]

The solution of (3.C.3), which satisfies (3.C.4) is constructed along the same lines as the Aharonov–Bohm scattering solution discussed in detail earlier. I shall not repeat that presentation, beyond remarking that time, radial and angular variables are separated in the usual way, with partial waves carrying angular momentum, ℓ, which is not integer quantized, rather $\alpha\ell$ is an integer. This of course is a consequence of the fact that the angular range is $2\pi\alpha$, not 2π.

The scattering solution is given by a contour integral in which plane waves are superposed, with definite weight[3d,e]

$$\begin{aligned} \psi(t;r,\theta) &= e^{-iEt}\oint\frac{dz}{2\pi}e^{i\mathbf{k}(z)\cdot\mathbf{r}}\frac{1}{1-e^{iz/\alpha}} \\ &\equiv e^{-iEt}\psi(r,\theta) \end{aligned} \tag{3.C.5}$$

Here $E = \sqrt{k^2+m^2}$ and $\mathbf{k}$ is the vector of magnitude k, rotated by the contour integration variable z: $\mathbf{k} = (k\cos z,\, k\sin z)$. That (3.C.5) satisfies (3.C.3) is obvious, that also the boundary condition (3.C.4) is obeyed depends on the specific weight function in (3.C.5) and also on the contour, which is depicted in Fig. 3a.

The weight function has poles on the real axis at $z = z_n = 2\pi n\alpha$ and the contour C avoids them. However, the contour may be deformed as in the discussion of the Aharonov–Bohm problem. We can consider the equivalent, three segment contour $\tilde{C}$, depicted in Fig. 3b, where the poles are encircled and also there are integrals along the vertical lines. The contribution from the encircled poles is evaluated by Cauchy's theorem; it gives the incoming wave. The remaining integrals along the vertical lines give the scattered wave, but the integrations cannot be evaluated, so no closed form is available. Nevertheless, the large r asymptote is accessible, and the scattering amplitude is determined explicitly.[3d,e]

$$\psi(r,\theta) = \psi^{\text{in}}(r,\theta) + \psi^{\text{sc}}(r,\theta) \tag{3.C.6}$$

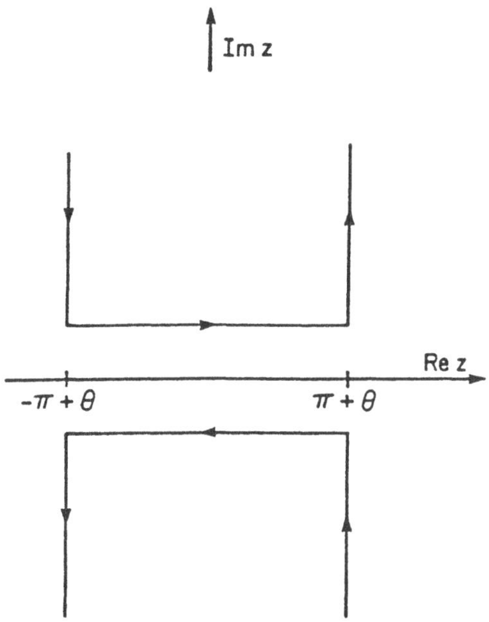

(a)

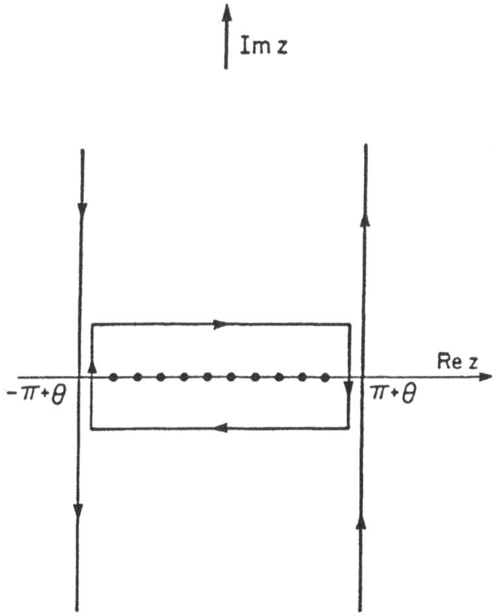

(b)

Fig. 3: (a) Integration contour C for the representation of $\psi(r,\theta)$ in (3.C.5). (b) Integration contour $\tilde{C}$ for the representation of $\psi(r,\theta)$ equivalent to that in (a) but giving rise to the decomposition $\psi = \psi^{\rm in} + \psi^{\rm sc}$. The incoming wave $\psi^{\rm in}$ is given by the [negative] Cauchy contour around the poles at $z = 2\pi n\alpha$, indicated by heavy dots. The integrals along the left and right verticle contours determine the scattered wave $\psi^{\rm sc}$, whose large distance asymptote defines the scattering amplitude f.

$$\psi^{\text{in}}(r,\theta) = \alpha \sum_n{}' e^{ik(z_n)\cdot r} \tag{3.C.7a}$$

$$\psi^{\text{sc}}(r,\theta) = i\int_{-\infty}^{\infty} \frac{dy}{2\pi} e^{ikr\cosh y} \left[\frac{1}{1-e^{i\frac{\pi}{\alpha}}e^{-\frac{1}{\alpha}(y+i\theta)}} - \frac{1}{1-e^{-i\frac{\pi}{\alpha}}e^{-\frac{1}{\alpha}(y+i\theta)}}\right]$$

$$\underset{r\to\infty}{\longrightarrow} \sqrt{\frac{i}{r}} f(\theta) e^{ikr} \tag{3.C.7b}$$

$$f(\theta) = \frac{1}{2\sqrt{2\pi k}}\left[\left(\operatorname{ctn}\frac{\theta-\pi}{2\alpha} - i\right) - \left(\operatorname{ctn}\frac{\theta+\pi}{2\alpha} - i\right)\right] \tag{3.C.8}$$

The prime on the sum in (3.C.7a) indicates that z_n must lie in the interval $[-\pi+\theta, \pi+\theta]$. Note that the incoming wave is not a plane wave, rather it is a superposition of variously rotated plane waves. This is analogous to the modulated plane wave found in the Aharonov–Bohm analysis.

We observe that the scattering amplitude $f(\theta)$ in (3.C.8) is real and vanishes when $1/\alpha$ is an integer. Also there are singularities at finite values of θ, where either of the two cotangents blows up. Finally, the optical theorem, which in two dimensions and with our normalization reads

$$\operatorname{Im} f(0) = \sqrt{\frac{k}{4\pi}} \int d\theta\, |f(\theta)|^2 \tag{3.C.9}$$

fails because the left-hand side vanishes and the right-hand side diverges. Nevertheless, there is no loss of unitarity: one can verify from the exact solution (3.C.6) – (3.C.7) that the probability current is conserved. The peculiarities of the scattering amplitude are presumably related to the long-range nature of the "interaction": no matter how far the scattered particle is from the source, it remains on a cone. An interesting problem that here remains is the study of how a wave packet evolves in time.

Going beyond the simplest case, we consider the situation that arises when both the source and the test particle are spinning. The source spin S is arbitrary; for the test particle we consider spins 0 and 1/2, solving the Klein–Gordon and Dirac equations, respectively, but now with the more elaborate identification (3.B.13). One may again give a contour integral representation for the wave function, obtain the incoming wave by performing a Cauchy contour integral, and deduce an explicit formula for the scattering amplitude. The result is an elegant generalization of (3.C.8), which can be presented in universal form, provided the following definitions are made.

$$\begin{aligned} S^s &= \text{spin of source [can be arbitrary, previously called } S]. \\ S^t &= \text{spin of test particle [actual calculations done only for } S^t = 0, 1/2]. \\ E^s &= \text{energy of source [taken to be } M]. \\ E^t &= \text{energy of test particle } \left(E^t = \sqrt{k^2+m^2}\right) \ . \end{aligned} \tag{3.C.10}$$

The scattering amplitude is[3d,e]

$$f(\theta) = \frac{e^{-i[\omega]\theta/\alpha}}{2\sqrt{2\pi k}}\left[e^{-i\{\omega\}\pi/\alpha}\left(\operatorname{ctn}\frac{\theta-\pi}{2\alpha} - i\right) - e^{i\{\omega\}\pi/\alpha}\left(\operatorname{ctn}\frac{\theta+\pi}{2\alpha} - i\right)\right] \tag{3.C.11}$$

Here ω is the symmetric cross product

$$\omega = E^s S^t + E^t S^w \tag{3.C.12}$$

while the square and curly brackets denote integer and fractional part, respectively.

$$\omega = [\omega] + \{\omega\} \tag{3.C.13}$$

For the spinless test particle, $S^t = 0$, one can determine from the phase shift $\delta(E)$ the time-delay by Wigner's formula. Agreement with the classical result (3.C.2) is found.

$$\Delta T = 2\frac{\partial \delta(E)}{\partial E} \tag{3.C.14}$$

We may understand the scattering amplitude as arising from diffraction effects [like in physical optics] which supplement the classical trajectories [whose analogy is geometrical optics]. These diffraction patterns are indicated by the dotted arcs in Fig. 2 and the two terms in (3.C.8) correspond to the two branches. We observe that scattering consists of a rotation through the angle $\pm\pi GM$, and we recall that in the presence of spin a rotation is accompanied by a phase change in the wave function. This explains the emergence of the additional phases in (3.C.11) as compared to (3.C.8).

The analysis of the Dirac equation is especially interesting owing to the fact that the Dirac Hamiltonian ceases to be self-adjoint on a conical, time-helical space time.[3e] [The same malady afflicts the Dirac equation in the presence of a vortex — the spinning Aharonov–Bohm effect.[20]] Of course the derivatives are formally Hermitian, but consideration of the boundary conditions indicates that a self-adjoint extension, depending on parameters, must be made and different physical results emerge with different values for the parameters. [In deriving Eq. (3.C.11) a definite choice is made to insure universality — but other choices are possible.]

In physical terms what is seen here is the failure of the point-particle description. Extended, smooth objects — described *e.g.* by fields — would lead to a self-adjoint Hamiltonian and in the point-particle limit various parameters, characterizing the extended object, survive as boundary terms on the particle surface and provide the missing information. The situation is similar to what is found for the Dirac equation with a [Dirac] point magnetic monopole. The Hamiltonian needs a one-parameter self-adjoint extension.[21] When a smooth 't Hooft–Polyakov monopole is considered, the parameter is identified as the QCD vacuum angle.[22] For the gravitational [and vortex] problems it remains an open question what model for the extended particle gives a physical origin to the mathematically necessary self-adjoint extension parameters.

The loss of self-adjointness appears to be related to the closed time-like curves that are present in a background metric arising from a spinning source.

We conclude this discussion of quantum motion by remarking that the true two-body problem — in contrast to its test particle source-particle equivalent description — is solved on a space with deficit angle given by the eigenvalues of the two-body Hamiltonian.[3d] This truly "Machian" behavior raises conceptual puzzles — for example it is impossible to superpose or compare energy eigensolutions. Moreover, the three- or more-body problem has thus far not been resolved [apart from a very easy special case[3e]] owing in part to difficulties in describing the multi-conical space on which the physical motion takes place.

D. Topological Elaborations

Up to now the discussion has been based on the three-dimensional version of the Einstein equation (3.A.2). However, in complete analogy to three-dimensional gauge theories, it is possible to modify (3.A.2) by an additional term, because in three dimensions there exists another second rank tensor that is symmetric and covariantly conserved. Sometimes called the *Cotton tensor*, its form is

$$C^{\mu\nu} = \frac{1}{2\sqrt{\det g_{\mu\nu}}} \epsilon^{\mu\alpha\beta} D_\alpha R^\nu_\beta + \mu \leftrightarrow \nu \tag{3.D.1}$$

Symmetry is manifest, covariant conservation follows from the Bianchi identities. $C^{\mu\nu}$ is traceless as follows from (3.D.1). also with the help of Bianchi identities.

$$C^\mu_\mu = 0 \tag{3.D.2}$$

Moreover, $C^{\mu\nu}$ may be viewed as the three-dimensional conformal tensor — an odd-parity analog of the Weyl tensor, the latter vanishing identically at $d = 3$. [That is why the Riemann tensor is determined by the Einstein tensor.] $C^{\mu\nu}$ is invariant against conformal redefinition of the metric tensor $g^{\mu\nu}(x) \to \lambda(x) g^{\mu\nu}(x)$ and vanishes if and only if space-time is conformally flat, $g_{\mu\nu}(x) = \lambda(x)\eta_{\mu\nu}$. We may supplement/replace the left-hand of (3.A.2) by the addition of a multiple of $C^{\mu\nu}$.[3a]

$$G^{\mu\nu} + \frac{1}{\kappa} C^{\mu\nu} = 0 \tag{3.D.3a}$$

$$G^{\mu\nu} + \frac{1}{\kappa} C^{\mu\nu} = 2\pi G T^{\mu\nu} \tag{3.D.3b}$$

[Also a cosmological constant can of course be added to the equation with or without sources, (3.D.3a) or (3.D.3b) respectively — we shall not do so.]

From its definition (3.D.1), we see that $C^{\mu\nu}$ is of one derivative order higher than $G^{\mu\nu}$, hence the dimension of κ is mass. Analysis of the linearized approximation yields dramatic results. While in the absence of the modification, there are no gravitational excitations, the addition "liberates" a previously "confined" graviton, which now becomes a single propagating mode; moreover, it is massive, while retaining general covariance. The spin is ± 2, the sign being correlated with the sign of κ. [The triple derivative nature of the differential equations (3.D.3) does not give rise to acausality; here, the conformal invariance comes into play, removing possibly dangerous terms from $C^{\mu\nu}$.]

$G^{\mu\nu}$ is obtained variationally from the Einstein–Hilbert action. Similarly, $C^{\mu\nu}$ may be obtained variationally from the Chern–Simons action, for the local Lorentz group in 2+1 dimensions — $SO(2,1)$. Constructing that quantity as in a gauge theory from the connection — either Christoffel or spin — but viewing the connection as a function of the fundamental dynamical variable — either the metric tensor or the *dreibein* respectively — and varying the dynamical variable gives $C^{\mu\nu}$.[3a]

Thus we see that the proposed modification is the complete analog of the situation in the gauge theory, and for that reason the model (3.D.3) is called *topologically massive gravity*.

However, no quantization condition need be imposed on κ.[23] Non-trivial homotopies in a non-compact group like $SO(2,1)$ coincide with those of its maximal compact subgroup, here $SO(2)$; but $SO(2)$ is trivial in this respect, so the gravitational Chern–Simons action is invariant, just as the field equations are covariant, and κ is unrestricted.

It is not known whether topologically massive gravity is renormalizable.

Of course a theory based solely on the Chern–Simons action/Cotton tensor field equation may also be considered.[3a] Here again, there are no propagating degrees of freedom, and due to the tracelessness of $C^{\mu\nu}$, only massless sources, with trace-free energy-momentum tensor can be coupled. However, owing to its triple derivative structure, the topological term is ***not*** natural for a low energy description, in contrast to the gauge theoretic Chern–Simons term. On the contrary. The Einstein/Hilbert theory is dominant at low energies, while the Chern–Simons/Cotton term dominates at high energy.

I conclude this discussion of topological elaborations on planar gravity by the following observations.

(a) Just like the gauge theoretic Chern–Simons term, the gravitational $SO(2,1)$ Chern–Simons term is induced by virtual fermions.[24] This raises a puzzle about our treatment of quantum scattering, when the matter degrees of freedom are *second* quantized fermions and the "bare" gravitational action is just the conventional Einstein–Hilbert action. On the one hand the bare gravitational action suggests that there are no propagating gravitational degrees of freedom. On the other, fermion loops induce a Chern–Simons action which when considered together with the bare action indicates the presence of massive, propagating gravitons. So which viewpoint is correct? Is the emergent "graviton" a fermion/anti-fermion bound state? How should perturbative calculations be organized?

(b) The fact that in planar Einstein gravity, the gravitational field is locally determined by matter sources is analogous to the situation in gauge theoretic Chern–Simons theory. Indeed the analogy exposes an identity: the Einstein–Hilbert action is also the Chern–Simons term for $ISO(2,1)$, the inhomogeneous $(2+1)$-dimensional Lorentz group, *i.e.* the Poincaré group.[25] There are six generators: J^μ rotations and P^μ translations. With these we associate respectively the "gauge" connections ω^μ and e^μ — the spin connection and *dreibein* — and use an off-diagonal "trace," $\langle P^\mu P^\nu\rangle = 0$, $\langle J^\mu J^\nu\rangle = 0$, $\langle J^\mu P^\mu\rangle = \delta^{\mu\nu}$, to construct the Chern–Simons term. The result is the Einstein–Hilbert action in first-order form.

(c) The Lagrangian for topologically massive gravity consists of $\mathcal{L}_{\mathrm{EH}} + \frac{1}{\kappa}\mathcal{L}_{\mathrm{CS}}$, the Einstein–Hilbert Lagrangian summed with κ^{-1} times the Chern–Simons term. Equivalently we may write it as $\mathcal{L}_{\mathrm{CS}} + \kappa\mathcal{L}_{\mathrm{EH}}$, and view the higher derivative $\mathcal{L}_{\mathrm{CS}}$ as the "kinetic" term and $\kappa\mathcal{L}_{\mathrm{EH}}$ as the "mass" term. The former possesses more symmetry than the latter — it is conformally invariant. In some sense that is "too much" symmetry, and no propagation is possible with just the kinetic term. Inclusion of the less symmetric mass [Einstein–Hilbert] term lowers the symmetry and "liberates" the previously confined graviton. One may even promote κ to a scalar field φ with its own [unspecified] dynamics. The combination $\mathcal{L}_{\mathrm{CS}} + \varphi\mathcal{L}_{\mathrm{EH}} + \mathcal{L}_\varphi$ can be conformally invariant for suitably chosen $\mathcal{L}_\varphi$. Then an expansion about $\langle\varphi\rangle = 0$ contains no propagating gravitons, while the symmetry breaking starting point $\langle\varphi\rangle = \kappa$ liberates the graviton.[26].

(d) Some classical solutions to topologically massive gravity have been found. They are planar analogs of Gödel universes.[27]

REFERENCES

1. G. 't Hooft, in *Proceedings of XIX Schladming School, Acta. Phys. Austr. Supple XXII* **531** (1980).

2. Papers from which the lectures on planar gauge theories are drawn:

   a. R. Jackiw and S. Templeton, *Phys. Rev. D* **23**, 2291 (1981); J. Schonfeld, *Nucl. Phys.* **B185**, 157 (1981).

   b. S. Deser, R. Jackiw and S. Templeton, *Phys. Rev. Lett.* **48**, 975 (1982), *Ann. Phys.* (NY) **140**, 372 (1982), (E) **185**, 406 (1988).

   c. S. Deser and R. Jackiw, *Phys. Lett.* **B139**, 371 (1984); L. Faddeev and R. Jackiw, *Phys. Rev. Lett.* **60**, 1692 (1988).

   d. G. Dunne, R. Jackiw and C. Trugenberger, *Ann. Phys.* (NY) **194**, 197 (1989).

   e. G. Dunne, R. Jackiw and C. Trugenberger, *Phys. Rev. D* **41**, xxx (1990).

   f. R. Jackiw, *Ann. Phys.* (NY) (in press).

   Much of this material is summarized in:

   g. R. Jackiw, in S.Treiman, R. Jackiw, B. Zumimo and E. Witten, *Current Algebra and Anomalies*, (Princeton University Press/World Scientific, Princeton, NJ/Singapore, 1985).

   h. R. Jackiw in Lectures presented at the *Fifth Jorge Swieca School* (Campos de Jordão, São Paulo, Brazil, 1989), to be published in the Proceedings.

   The above papers and reviews should be consulted for reference to other literature on ths subject.

3. Papers from which the lectures on planar gravity are drawn:

   a. S. Deser, R. Jackiw and S. Templeton, *Phys. Rev. Lett.* **48**, 975 (1982), *Ann. Phys.* (NY) **140**, 372 (1982), (E) **185**, 406 (1988).

   b. S. Deser, R. Jackiw and G. 't Hooft, *Ann. Phys.* (NY) **152**, 220 (1984).

   c. S. Deser and R. Jackiw, *Ann. Phys.* (NY) **153**, 405 (1984).

   d. G. 't Hooft, *Comm. Math. Phys.* **117**, 685 (1988).

   e. S. Deser and R. Jackiw, *Comm. Math. Phys.* **118**, 495 (1988); P. Gerbert and R. Jackiw, *Comm. Math. Phys.* **124**, 229 (1989).

   f. S. Deser and R. Jackiw, *Ann. Phys.* (NY) **192**, 352 (1989).

   Much of this material is summarized in:

   g. R. Jackiw, *Nucl. Phys.* **B252**, 343 (1985).

   h. R. Jackiw, in *Proceedings of the XVII International Colloquium on Group Theoretic Methods in Physics* (in press).

   The above papers and reviews should be consulted for reference to other literature on the subject.

4. a. L. Alvarez-Gaumé and E. Witten, *Nucl. Phys.* **B234**, 269 (1983).

   b. E. Witten, *Comm. Math. Phys.* **121**, 351 (1989); Y.-H. Chen, F. Wilczek, E. Witten and B. Halperin, *Intl. Jnl. Mod. Phys.* **B3**, 1001 (1989).

   c. A. Polyakov, *Mod. Phys. Lett.* **A3**, 325 (1988).

   d. J. Gott and M. Alpert, *Gen. Rel. Grav.* **16**, 243 (1984); J. Gott, *Ap. J.* **288**, 42 (1985).

   e. E. Witten, *Nucl. Phys.* **B311**, 46 (1988/89); J. Horne and E. Witten, *Phys. Rev. Lett.* **62**, 501 (1989); S. Carlip, *Nucl. Phys.* **B324**, 106 (1989).

5. B. Rosenstein, B. Warr and S.-H. Park, *Phys. Rev. Lett.* **62**, 1433 (1989), *Phys. Rev. D* **39**, 3088 (1989); B. Rosenstein and B. Warr, *Phys. Lett.* **B218**, 465 (1989), **B219**, 469 (1989), Texas preprint UTTG-18-89 (1989).

6. N. Redlich, *Phys. Rev. Lett.* **52**, 18 (1984), *Phys. Rev. D* **29**, 2366 (1984).

7. W. Siegel, *Nucl. Phys.* **B156**, 135 (1979).

8. C. Hagen, *Ann. Phys.* (NY) **157**, 342 (1984), *Phys. Rev. D* **31**, 2135 (1985).

9. D. Gonzales and N. Redlich, *Ann. Phys.* (NY) **169**, 104 (1984).

10. Polyakov, Ref. [4c]; G. Semenoff, *Phys. Rev. Lett.* **61**, 517 (1988); Dunne *et al.*, Ref. [2d]; T. Matsuyama, *Phys. Lett.* **B228**, 99 (1989).

11. S. Zhang, T. Hanson and S. Kivelson, *Phys. Rev. Lett.* **62**, 8 (1989); Chen *et al*; Ref. [4b]; Jackiw, Ref. [2h].

12. R. Jackiw and N. Redlich, *Phys. Rev. Lett.* **50**, 555 (1983).

13. D. Arovas, J. Schrieffer, F. Wilczek and A. Zee, *Nucl. Phys.* **B251**, 117 (1985).

14. Y. Aharonov and D. Bohm, *Phys. Rev.* **115**, 485 (1959); S. Ruijsenaars, *Ann. Phys.* (NY) **146**, 1 (1983).

15. R. Jackiw, *Phys. Today* **25** No. 1, 23 (1972).

16. R. Jackiw and C. Teitelboim in *Quantum Theory of Gravity*, S. Christensen, ed. (Adam Hilgar, Bristol, UK, 1984); A. Polyakov, *Mod. Phys. Lett.* **A2**, 893 (1987); K. Isler and C. Trugenberger, *Phys. Rev. Lett.* **63**, 834 (1989); A. Chamseddine and D. Wyler, *Phys. Lett.* **B228**, 75 (1989), Zürich University preprint (1989).

17. A. Staruszkiewicz, *Act. Phys. Pol.* **24**, 735 (1963).

18. G. Clément, *Int. Jnl. Theor. Phys.* **24**, 267 (1985).

19. G. Grignani and C.-K. Lee, *Ann. Phys.* (NY) (in press); G. Clément, *Ann. Phys.* (NY) (in press).

20. P. Gerbert, *Phys. Rev. D* **40**, 1346 (1989).

21. A. Goldhaber, *Phys. Rev. D* **16**, 1815 (1977); C. Callias, *Phys. Rev. D* **16**, 3068 (1977).

22. B. Grossman, *Phys. Rev. Lett.* **50**, 464 (1983); H. Yamagishi, *Phys. Rev. D* **27**, 2383 (1983); E. D'Hoker and E. Farhi, *Phys. Lett.* **127B**, 360 (1983).

23. R. Percacci, *Ann. Phys.* (NY) **177**, 27 (1987).

24. L. Alvarez-Gaumé, S. Della Pietra and G. Moore, *Ann. Phys.* (NY) **163** 288 (1985); M. Goni and M. Valle, *Phys. Rev. D* **34**, 648 (1986); I. Vuorio, *Phys. Lett.* **B175**, 176 (1986); J. van der Bij, R. Pisarski and S. Rao, *Phys. Lett.* **B179**, 87 (1986).

25. A. Achúcarro and P. Townsend, *Phys. Lett.* **B180**, 89 (1986); Witten, Ref. [4e].

26. S. Deser and Z. Yang, Brandeis University preprint BRZ TH-273 (1989).

27. I. Vuorio, *Phys. Lett.* **B163**, 91 (1985); R. Percacci, P. Sodano and I. Vuorio, *Ann. Phys.* (NY) **176**, 344 (1987); M. Ortiz, Cambridge University preprints, DAMTP R-89/13, 17 (1989).

INTRODUCTION TO CONFORMAL FIELD THEORY AND INFINITE DIMENSIONAL ALGEBRAS

David Olive
Imperial College
London SW7 2BZ, UK

LECTURE 1

In 1909, soon after Einstein's formulation of the theory of special relativity, based on the Lorentz invariance of Maxwell's equations, it was discovered by Bateman and by Cunningham[1] that these equations possessed an even larger symmetry, conformal symmetry. Conformal transformations are those space-time transformations preserving angles but not necessarily lengths. For a metric $diag(1,1,..1,-1,-1,..-1)$ with p entries $+1$ and q entries -1, and with $p+q$ larger than one, they constitute the group $so(p+1,q+1)$ which has finite dimension. But for two Euclidean dimensions, the conformal group possesses infinite dimension, as is familiar from electrostatics. If we consider holomorphic maps of the complex variable $z=x+iy, z\mapsto z'=\gamma(z)$ with $\partial\gamma/\partial\bar{z}=0$, angles are preserved. Near the identity map, $\gamma(z)=z+\sum_n \epsilon_n z^{n+1}$, with all coefficients ϵ_n small, and $f(\gamma(z))-f(z)=\sum_n \epsilon_n z^{n+1}\partial f/\partial z$. The differential operator $l_n=z^{n+1}\partial/\partial z$ satisfies the infinite dimensional algebra $[l_m,l_n]=(m-n)l_{m+n}$. This is the Virasoro algebra[2] without central extension.

The same algebra arises in Lorentzian two dimensions, periodic in space. Space-time transformations are generated by the energy momentum tensor, $\theta^{\mu\nu}=\theta^{\nu\mu}$. Scale symmetry implies that this is traceless, $\theta^{\mu}_{\mu}=\theta^{tt}-\theta^{xx}=0$, in order that the scale current $x_\mu\theta^{\mu\nu}$ be conserved. This in turn implies conservation of the conformal currents. Hence, in a local theory, conformal invariance can result from the absence of a fundamental scale. In two dimensions it is convenient to use light cone variables $x^{\pm}=(t\pm x)/\sqrt{2}$. Then θ^{+-} and θ^{-+} vanish, leaving only θ^{++} and θ^{--} which are right moving and left moving, respectively, and hence called "chiral". Appropriate Fourier components of these two quantities generate two commuting copies of the Virasoro algebra;

$$[L_m,L_n]=(m-n)L_{m+n}+(c/12)m(m^2-1)\delta_{m+n,0}, \tag{1.1a}$$

$$[L_m,c]=0, \tag{1.1b}$$

and similar equations for $\overline{L}_m$ and $\bar{c}$ which commute with L_m and c.

Note the "central extension" which satisfies all necessary Jacobi identies. Any consistent central extension can be put into the above form upon a redefinition of the L_n.

Physics, Geometry, and Topology
Edited by H. C. Lee
Plenum Press, New York, 1990

Note also that $L_0, L_{\pm 1}$ satisfy a three dimensional subalgebra, $sl(2,R)$, in which c does not appear (thereby explaining our choice of the form of the central extension).
In physical applications we require

$$L_0 + \overline{L}_0 \geq \text{const.},$$

as this quantity can be interpreted as a Hamiltonian, or a dilatation generator. As L_0 and $\overline{L}_0$ belong to commuting Virasoro algebras, it follows that

$$L_0 \geq \text{const}, \quad \overline{L}_0 \geq \text{const}. \tag{1.2}$$

We shall also require the representation space to carry a positive definite scalar product such that

$$L_n^\dagger = L_{-n}. \tag{1.3}$$

These properties we call "unitarity ". They are not well founded in all the physical applications but make a convenient working hypothesis which simplifies subsequent mathematical arguments.
In a given conformal field theory (CFT), c and $\bar{c}$ are definite numbers on the whole Hilbert space $\mathcal{H}$ of the theory. Because of the unitarity, $\mathcal{H}$ decomposes into irreducible representations (irreps) of $Vir_L \oplus Vir_R$, (the left-handed and right-handed Virasoro algebras). In order that the conformal symmetry have maximum predictive power we would like this decomposition to be finite in the sense that only a finite number of irreps occur. Such theories have now been classified, and it turns out that many other significant CFT's can be treated similarly by extending Vir to a yet larger symmetry with a finite decomposition. Examples work via affine Kac-Moody algebras or via supersymmetry, but there are many other possible means of extension, all of which it seems, can be approached via affine Kac-Moody algebras.

ANALYSIS OF UNITARY IRREPS OF *VIR*

We can choose as basis the eigenstates of L_0. Let $|h>$ denote such an eigenstate with eigenvalue h. Then, by (1.1),

$$L_0 L_n |h> = (h-n) L_n |h,>$$

i.e. L_n lowers L_0 by n. But the spectrum of L_0 is bounded below, (1.2), and so there must exist "ground states ", $|h>$, such that

$$L_n|h> = 0, \quad n \geq 1; \qquad L_0|h> = h|h>. \tag{1.4}$$

Consider the space spanned by states obtained by acting on ground states $|h>$ with a finite product of L's with negative suffices:

$$\{\prod L_{-n_1} L_{-n_2} .. \ |h>\} \tag{1.5}$$

This space is invariant under the action of Vir and forms an irrep of it, which can be labelled (h,c) as all the scalar products between states in (1.5) can be evaluated in terms of these two numbers. The states (1.5) need not be linearly independent. Linear combinations with zero norm must vanish by the unitarity assumption. No linear combination can have negative norm, and this places severe restrictions on c and h, as we illustrate by considering the state $L_{-n}|h>$ for $n \geq 1$.

$$0 \leq \| L_{-n}|h> \|^2 = < h|L_n L_{-n}|h> = < h|[L_n, L_{-n}]|h>$$

by (1.4). By (1.1)

$$= (2nh + cn(n^2-1)/12)|||h>||^2. \tag{1.6}$$

Taking $n = 1$ we deduce $h \geq 0$, i.e. $L_0 \geq 0$, and by taking n very large, we deduce $c \geq 0$.

In fact, if $|\psi> = L_{-2n}|h>$, $|\phi> = L_{-n}L_{-n}|h>$,

$$\det \begin{pmatrix} <\psi|\psi> & <\psi|\phi> \\ <\phi|\psi> & <\phi|\phi> \end{pmatrix} = 4n^3h^2(4h-5n),$$

if c vanishes. If $h > 0$, this expression is negative for large enough n. As this is impossible, we conclude that h vanishes whenever c does. In this situation we deduce by (1.6) that $L_{-n}|h> = 0$ for all n so that Vir acts trivially in a unitary representation with zero c. Discounting this possibility as an empty theory we reach the important conclusion that the central term c is essentially positive;

$$c > 0 \tag{1.7}$$

The basic idea of this result goes back to Schwinger[3] although the details of this version are due to Gomes[4].

THE VACUUM

For any valid value of c, valid values of h should always include zero as it is a physical requirement that we have a special ground state, called the vacuum, $|0>$, with h (and $\bar{h}$) zero. By (1.6), the vacuum is also annihilated by L_0 and $L_{\pm1}$, i.e. the generators of the aforementioned $sl(2, R)$ subalgebra, and is the only ground state with this property. It is assumed to be unique in any physical theory.

THE FQS THEOREM

Unitary, positive L_0 irreps of Vir must have either

$$c \geq 1, \quad \text{or, if} \quad 0 < c < 1$$

$$c = 1 - \frac{6}{(m+2)(m+3)} \tag{1.8}$$

and

$$h = h_{p,q} = \frac{[(m+3)p - (m+2)q]^2 - 1}{4(m+2)(m+3)}. \tag{1.9}$$

Note that the $h_{p,q}$ are positive rational numbers, and that for any m, $h_{1,1} = 0$, thereby automatically supplying the vacuum required by physical principles.

If, for example, $m = 1$ so $c = 1/2$, then by (1.9)

$$h_{1,1} = 0, \quad h_{2,1} = 1/2, \quad h_{2,2} = 1/16. \tag{1.10}$$

The proof of the above result, due to Friedan, Qiu and Shenker[5], based on earlier work of Belavin, Polyakov and Zamolodchikov[6] and of Kac[7], in turn based on the structure of the no ghost theorem of string theory[8], is rather technical, synthesising the requirement that the space (1.5) be positive definite. We shall be satisfied with the special cases presented above which illustrate the general principle of the method.

Note that the proof does not guarantee the existence of the corresponding unitary representations.

CONCEPT OF PRIMARY FIELD

A conformal field theory possesses an infinite number of fields, amongst which there are some (hopefully a finite number) obeying a particularly simple transformation law with respect to the conformal transformations $\gamma(z)$:

$$U(\gamma(z))\phi(z,\bar{z})U(\gamma(z))^{-1} = (\frac{d\gamma}{dz})^h \phi(\gamma(z),\bar{z}), \tag{1.11a}$$

and similarly for the antiholomorphic transformation $\bar{\gamma}(\bar{z})$. Such a field is called "primary "and is characterised by two "conformal weights ", the numbers h and $\bar{h}$. We shall show later that this usage of the symbol h agrees with our previous usage. It is readily checked that such a transformation law (1.11) satisfies the group property. The notion has been advocated by Belavin, Polyakov and Zamolodchikov[6]. If γ is close to the identity so that $\gamma(z) = z + \sum_n \epsilon_n z^{n+1}$ and $U(\gamma(z)) = 1 + \sum_n \epsilon_n L_n$ we have

$$[L_n, \phi(z,\bar{z})] = z^{n+1}\frac{d\phi}{dz} + hz^{n+1}\phi. \tag{1.11b}$$

This version of the condition has been familiar for twenty years in string theory[9].

CORRELATION FUNCTIONS OF PRIMARY FIELDS

These are vacuum expectation values of products of primary fields:

$$< 0|\phi_1(z_1)\phi_2(z_2).. \quad |0>, \tag{1.12}$$

where we have taken each field ϕ to be chiral (i.e. independent of $\bar{z}$) for simplicity of notation. The conformal transformations generated by $L_0, L_{\pm 1}$, the $sl(2,R)$ subalgebra, have the fractional linear form

$$\gamma(z) = \frac{(az+b)}{(cz+d)}, \quad ad - bc = 1. \tag{1.13}$$

As $U(\gamma)|0>= |0>$, for such γ, by the properties of the vacuum, the correlation function (1.12) equals, by (1.11a)

$$\prod_i (cz_i + d)^{-2h_i} < 0|\phi_1(\gamma(z_1))\phi_2(\gamma(z_2)).. \quad |0> .$$

This relation is sufficient to determine the one, two and three point correlation functions up to an overall constant. We find

$$< 0|\phi(z)|0 >= \text{const. } \delta_{h,0}, \tag{1.14}$$

and

$$< 0|\phi_1(z_1)\phi(z_2)|0 >= \text{const.}\delta_{h_1,h_2}/(z_1 - z_2)^{2h_1}, \tag{1.15}$$

using the identity

$$\gamma(z_1) - \gamma(z_2) = (z_1 - z_2)/(cz_1 + d)(cz_2 + d).$$

Thus correlation functions exhibit a power law decay with separation in space, the power being given by $2h$.

In statistical mechanics models of materials capable of making second order phase transitions, the correlation length specifies the scale of an exponential decay in separation at non-critical temperature. This correlation length diverges at the critical temperature, leaving a power law decay in separation, given by a "critical exponent ", which can be measured in the laboratory.

Nearly twenty years ago Polyakov[10] proposed that, because of the disappearance of a fundamental scale at the critical temperature, and because of the local interactions, conformal symmetry should apply at the critical temperature. These ideas have been vindicated by their fruition in recent years, and it is this which provides much of the justification for the renaissance of interest in conformal symmetry.

LECTURE 2

Recall that the Hilbert space $\mathcal{H}$ of a conformal field theory decomposes into irreps of $Vir_L \oplus Vir_R$, as

$$\mathcal{H} = \sum_{h,\bar{h}} N_{h\bar{h}} \quad (c,h) \otimes (\bar{c},\bar{h}), \tag{2.1}$$

where

$$N_{00} = 1, \tag{2.2}$$

in view of the assumed uniqueness of the vacuum. The other $N_{h\bar{h}}$ are the integers denoting the multiplicities of the representations indicated. We would like the sum of these integers to be finite as mentioned earlier, so that $\mathcal{H}$ decomposes finitely.

TWO POINT CORRELATION FUNCTION OF TWO NON-CHIRAL PRIMARY FIELDS

Repeating the previous analysis yields

$$< 0|\phi_1(z_1,\bar{z}_1)\phi_2(z_2,\bar{z}_2)|0 >= \delta_{h_1h_2}\delta_{\bar{h}_1\bar{h}_2}\text{const.}(z_1 - z_2)^{-2h_1}(\bar{z}_1 - \bar{z}_2)^{-2\bar{h}_1}. \tag{2.3}$$

Introducing polar co-ordinates $z_1 - z_2 = re^{i\theta}$, we find

$$\sim r^{-2(h+\bar{h})}e^{-2i\theta(h-\bar{h})}, \tag{2.4}$$

so that we can identify

$$h + \bar{h} = \text{critical exponent or anomalous dimension of } \phi, \tag{2.5a}$$

$$h - \bar{h} = \text{spin of } \phi. \tag{2.5b}$$

We see that the correlation function (2.3) is single valued in space (i.e. under $\theta \mapsto \theta + 2\pi$) if the spin, $h - \bar{h}$, is an integer or half integer. However, unlike the case in higher dimensions, there is no topological reason why this should be so.

If ϕ_1 and ϕ_2 are fields of the same species, we can effect an interchange by $\theta \mapsto \theta + \pi$. The result is a sign change $(-1)^{spin} = \pm 1$ according as the spin is an integer or half integer. Thus results the usual connection between spin and statistics.

ISING MODEL

Comparing the behaviour (2.4) with the known behaviour of the Ising model as explained, for example, in Baxter's book[11], we see that the "spin variable " has $(h, \bar{h}) = (1/16, 1/16)$. Comparing with (1.10) suggests that $c = 1/2$ at the critical temperature of the Ising model. This agrees with Onsager's result[12] of 1944 in which he used a Jordan-Wigner transformation[13] to convert the spin variables into a fermion field. It is known from the Ramond-Neveu-Schwarz fermionic string theory[14] that a single fermi field indeed yields $c = 1/2$. There the analogue of the spin variable is the "fermion emission vertex", much studied in the early 1970's[15].

FUNDAMENTAL PROPERTY OF PRIMARY FIELDS

This is the theorem: If $\phi(z, \bar{z})$ is a primary field with conformal weights $(h, \bar{h})$. Then

$$\phi(0,0)|0 >= |h, \bar{h} >, \tag{2.6}$$

a ground state (1.4).

Let us suppose ϕ is chiral, for simplicity. Then $\psi(z) = z^{-2h}\phi(1/z^*)^\dagger$ is the "conjugate "primary field to ϕ possessing the same weight. So

$$< 0|\psi(z)\phi(z')|0 >= \text{const. } (z - z')^{-2h} = \text{const.} z^{-2h} \sum_{n=0}^{\infty} (z'/z)^n \binom{2h + n - 1}{n}, \tag{2.7}$$

using the binomial theorem and arranging to exhibit that the coefficients are positive. Alternatively, inserting a complete set of eigenstates of L_0

$$= \sum_{h'} < 0|\psi(z)|h' >< h'|\phi(z')|0 > .$$

As L_0 is the dilatation operator, $\phi(z) = z^{L_0}\phi(1)z^{-L_0}$, with the same for ψ, we have

$$= z^{-2h} \sum_{h'} (z'/z)^{h'-h} | < h'|\phi(1)|0 > |^2. \tag{2.8}$$

Comparing powers of z'/z in (2.7) and (2.8), we see

$$h' = h + n, \qquad n = 0, 1, 2, 3, .. \ .$$

So $\phi(z)$ maps the vacuum irrep into that with ground state $|h >$. All the matrix elements $| < h + n|\phi(1)|0 > |^2$ are proportional to the constant in (2.7). We infer that $\phi(0)|0 >$ exists and that

$$L_n\phi(0)|0 >= [L_n, \phi(0)]|0 >, \qquad n \geq 0,$$

$$= [z^{n+1}\frac{d\phi}{dz} + h(n+1)z^{n+1}\phi(z)]_{z=0}|0 >$$

$$= \begin{cases} 0 & n \geq 0 \\ h\phi(0)|0 > & n = 0 \end{cases}$$

Hence $\phi(0)|0 >= |h >$ and, more generally, the result (2.6).

Thus the primary fields map the vacuum irrep into those irreps labelled by their conformal weights. We expect a one-to-one correspondence between the irreps in the sum (2.1) and the primary fields. Thus there are $N_{h,\bar{h}}$ primary fields of weight $(h, \bar{h})$. The primary field corresponding to the vacuum is simply the unit operator.

The physical importance of the result is that it establishes that the number h originally defined by the ground state (1.4) is the same as the conformal weight carried by the primary field (1.11) corresponding to that ground state, and hence the same as the value of the critical exponent governing the power law decay in separation of the correlation function of that primary field. The conclusion is that, in two dimensions at least, critical exponents are purely numbers determined by the representation theory of the algebra of conformal symmetry. In particular they do not depend on details of the inter-molecular forces. This explains why it was so useful to introduce the concept of "universality"[16].

ASSIGNMENT OF PRIMARY FIELDS

It follows that the determination of the representation content (2.1) is equivalent to the assignment of primary fields for the theory. This assignment has to satisfy certain consistency conditions.

(a) Operator Product Expansion

$$\phi_1(z_1)\phi_2(z_2) = \phi_3(z_1)/(z_1 - z_2)^{h_1+h_2-h_3} + \text{less singular terms.} \tag{2.9}$$

The coefficient of the leading term has to be primary if ϕ_1 and ϕ_2 are. The statement of the multiplicities occurring here constitute what is known as the fusion rule algebra. This algebra is abelian, associative and finite dimensional if there is a finite number of primary fields, that is, if the sum in (2.1) is finite.

(b) Cardy's postulate of modular invariance of the partition function

The partition function of the conformal field theory is the following quantity

$$Z(\tau) = Tr(q^{L_0-c/24} \quad \bar{q}^{\bar{L}_0-\bar{c}/24})_{\mathcal{H}}, \tag{2.10}$$

where

$$q = e^{2\pi i\tau}, \qquad q = e^{-2\pi i\bar{\tau}}. \tag{2.11}$$

As the trace in (2.10) is taken over the space (2.1)

$$Z(\tau) = \sum_{h,\bar{h}} N_{h,\bar{h}} \quad \chi_h(q)\chi_{\bar{h}}(\bar{q}), \tag{2.12}$$

where

$$\chi_h(q) = Tr(q^{L_0-c/24})_{(c,h)} \tag{2.13}$$

is a Virasoro character for the irrep (c, h).

Cardy[17] proposed convincing arguments which we shall not review here to the effect that $Z(\tau)$ should be invariant under the action of the modular group acting on τ:

$$\tau \mapsto \tau' = \frac{a\tau + b}{c\tau + d}, \qquad a, b, c, d \text{ integers}, \ ad - bc = 1. \tag{2.14}$$

The modular group is generated by

$$\left.\begin{aligned} T: \tau \mapsto \tau' &= \tau + 1 \\ S: \tau \mapsto \tau' &= -1/\tau \end{aligned}\right\}, \tag{2.15}$$

and so it is sufficient to check invariance with respect to these.

By the structure of the irrep,

$$\left(\chi_{(h,c)}\right)^T = e^{2\pi i(h-c/24)} \quad \chi_{(h,c)}, \tag{2.16}$$

and Z is invariant under the action of T if

$$\sum_{h,\bar{h}} N_{h,\bar{h}} \quad [1 - e^{2\pi i(h-\bar{h}-(c-\bar{c})/24}] \quad \chi_h \chi_{\bar{h}} = 0.$$

As the characters are linearly independent, each coefficient vanishes. Thus a primary field with conformal weights $(h, \bar{h})$ occurs in the theory only if $h - \bar{h} - (c - \bar{c})/24$ is an integer. But $N_{00} = 1$ by the uniqueness of the vacuum, so $c - \bar{c}$ must be an integer multiple of 24. Further we conclude that $h - \bar{h}$ is an integer. Thus, by (2.5), all primary fields have to have integer spins by Cardy's condition. In the Ising model, the aforementioned fermion fields ψ_L and ψ_R are not primary but their product $\psi_L \psi_R$ is. This carries the weights (1/2,1/2).

Notice that c and $\bar{c}$ need only be equal if they are both less than 24. Later on we shall mention interesting theories in which they are unequal.

The characters of the Virasoro irreps with $c < 1$ exhibit the following remarkable behaviour under S, (2.15)

$$\left(\chi_{(h,c)}\right)^S = \sum_{h'} S_{hh'} \chi_{(h',c)}. \tag{2.16}$$

A way of proving this will be mentioned later. The inclusion of the $c/24$ in the definition (2.13) is motivated by the simplicity of the result (2.16).

As the matrix S is unitary and independent of τ,

$$SS^\dagger = 1, \tag{2.17}$$

we see that

$$N_{h\bar{h}} = \delta_{h\bar{h}} \tag{2.18}$$

is always a solution to Cardy's condition as the corresponding partition function is then modular invariant (if $c < 1$). This partition function is called the diagonal invariant but it implies that all the primary fields are spinless. We should like to find more interesting solutions for the multiplicities so that the primary fields carry spin. The Ising model has only the diagonal invariant. Hence the primary fields have weights (0,0), (1/2,1/2) and (1/16,1/16) by (1.10).

E. Verlinde[18] found how to express the matrix S in (2.16) in terms of the structure constants of the fusion rule algebra. This was a surprising connection of ideas, and Moore and Seiberg[19] have carried these ideas further as Seiberg explained in his lectures[20].

Let us comment that one extremely important field already mentioned, the energy-momentum tensor $\theta^{++}(z) = \sum_n z^{-n-2} L_n$, is not primary as

$$\theta^{++}(0)|0> = L_{-2}|0>$$

which is not a ground state as it belongs to the vacuum irrep. $\theta^{++}(z)$ is said to be a descendant of the unit operator as this is the primary field corresponding to the vacuum irrep.

NEED FOR THE EXTENSION OF THE VIRASORO ALGEBRA

When $c > 1$ the states (1.5) are linearly independent. This is why there exist unitary representations of Vir for any such c. It follows that for any Virasoro irrep with $c > 1$

$$Tr(q^{L_0 - c/24})_{(c,h)} = q^{h-c/24} \prod_{n=1}^{\infty} (1 - q^n)^{-1}. \tag{2.19}$$

The modular transformation S, see (2.15), acting on this does not produce a finite linear combination of similar expressions with the same c but different h. It follows that unlike the situation for $c < 1$ there can never be a finite number of primary fields if $c > 1$. In fact a finite number of primary fields requires c and $\bar{c}$ to be equal and less than one.

However there are interesting conformal field theories with $c > 1$ and the most natural way to relax our assumption concerning the finite decomposability of $\mathcal{H}$ is to suppose that there exists an extension of the Virasoro algebra acting on the Hilbert space $\mathcal{H}$ of the theory in such a way that $\mathcal{H}$ decomposes finitely into irreps of this extended algebra. We can then appropriately modify the definition of ground state and primary field and investigate to what extent an analysis similar to that above can still apply. In fact there are many ways of achieving such extensions, via affine Kac-Moody algebras, via super conformal algebras, via parafermions and so on. It turns out that the first possibility is the most powerful, apparently including all the others known, in a sense to be explained.

LECTURE 3

EXTENSION OF THE VIRASORO ALGEBRA BY AN UNTWISTED AFFINE KAC-MOODY ALGEBRA

Both for the reasons just stated, and in order to understand better the theories with $c < 1$, we consider the extension of the Virasoro algebra (1.1) by

$$[T_m^a, T_n^b] = i f^{abc} T_{m+n}^c + km\delta^{ab}\delta_{m+n,0}, \tag{3.1}$$

$$[L_m, T_n^a] = -nT_{m+n}^a, \tag{3.2}$$

where k commutes with all the other quantities. Equation (3.1) specifies the affine untwisted Kac-Moody algebra[21] $\hat{g}$. The suffices are integers. Because they are conserved across the equations they constitute an integer grading. Equation (3.2) states that the "current " $\sum_n z^{-n-1} T_n^a$ has unit conformal weight. Notice that $\{T_0^a\} = g$, the compact Lie algebra with totally antisymmetric structure constants f^{abc}, whose affinisation is $\hat{g}$, and that g commutes with Vir.

When the central terms c and k vanish (1.1), (3.1) and (3.2) admit a nice geometrical interpretation. (3.1) is the Lie algebra of the group of maps of a circle into the Lie group G obtained by exponentiating g. This group is called the loop group. The Virasoro algebra is the algebra of diffeomorphisms of the circle.

In addition to the previous unitarity assumptions, (1.3) etc., we suppose

$$T_n^{a\dagger} = T_{-n}^a. \tag{3.3}$$

As T_n^a lowers the L_0 eigenvalue by n in the same manner as L_n, we must modify the notion of "ground state "

$$L_n|h,A>= T_n^a|h,A>= 0, \qquad n \geq 1, \tag{3.4a}$$

$$L_0|h,A>= h|h,A>, \qquad T_0^a|h,A>= \sum_B |h,B> t_{BA}^a. \tag{3.4b}$$

We see that now the ground states can form a unitary representation of g with generators t^a.

THE ROOT SYSTEM OF $\hat{g}$

Although the algebra $\hat{g}$ has infinite dimension, its root system spans only one more dimension than that of the finite dimensional Lie algebra g. To see its structure, introduce a Cartan Weyl basis for g, $\{H^i, E^\alpha\}$ and a corresponding basis for $\hat{g}$. In this notation, take as Cartan subalgebra of $\hat{g}$ augmented by L_0:

$$(H_0^i, k, -L_0)$$

We find from (3.1) and (3.2) that the following are roots with step operators E_n^α:

$$(\alpha^i, 0, n)$$

while the following roots display a rank g-fold degeneracy as they possess rank g linearly independent step operators H_n^i:

$$(0,0,n)$$

This accounts for all the generators of $\hat{g}$ augmented with L_0. The first type of root is called "real "or "space-like "and possesses the single step operator. The second type of root is called "imaginary "or "light-like "and is degenerate in the sense explained. This root system can be split into two disjoint parts, a set of positive roots, those with $n > 0$, or $n = 0$ and $\alpha > 0$ in the sense of g, and the remainder, the negative roots which are also the negatives of the positive roots. Just as for finite dimensional Lie algebras, we can choose "simple "roots so that any positive root can be expressed as a sum of them. They are:

$$a^i = (\alpha^i, 0, 0), \qquad i = 1, 2, .. \text{ rank } g, \quad \alpha^i \text{ a simple root of } g, \tag{3.5a}$$

$$a^0 = (-\psi, 0, 1), \qquad \psi \text{ the highest root of } g. \tag{3.5b}$$

It is possible to form a Dynkin diagram for $\hat{g}$ from these as Patera explained in his lectures[22]. It is obtained from that of g by adding one point in a special way.

EXTREME GROUND STATES

By analogy with the procedure for irreps of g we can single out of the degenerate ground states (3.4), "extreme ground states "by adding to (3.4) the conditions

$$E_0^\alpha|h,\lambda>= 0, \quad \alpha > 0, \qquad H_0^i|h,\lambda>= \lambda^i|h,\lambda>. \tag{3.6}$$

The significance of these extreme ground states is that they are annihilated by the step operators E_a for the positive roots a of $\hat{g}$. If a is, in addition, real, then

$$[E_a, E_{-a}] = 2a.H/a^2, \quad [a.H/a^2, E_{\pm a}] = \pm E_{\pm a}.$$

Thus we obtain an $su(2)$ subalgebra of $\hat{g}$ associated with each real positive root, a. As $2a.H/a^2$ is the analogue of $2T_3$ it should have an integer spectrum since the $su(2)$ is represented unitarily when $\hat{g}$ is. Further if $|h, \lambda>$ is an extreme ground state

$$0 \leq \| E_{-a}|h, \lambda> \|^2 = < h, \lambda|[E_a, E_{-a}]|h, \lambda> = < h, \lambda|2a.H/a^2|h, \lambda> . \qquad (3.7)$$

Choosing a to be one of the simple roots (3.4) in turn furnishes the most restrictive conditions on λ and k. If $a = (\alpha^i, 0, 0)$, we find

$$2\lambda.\alpha^i/(\alpha^i)^2 = 0, 1, 2, 3, ..$$

Hence λ is an integral dominant weight of g. If $a = a_0$, (3.5b), equation (3.7) yields

$$-2\lambda.\psi/\psi^2 + 2k/\psi^2 = 0, 1, 2, 3, 4.. \qquad (3.8)$$

As the first term is a negative integer we deduce that

$$2k/\psi^2 \equiv x = \text{level} = 0, 1, 2, 3, 4..$$

Thus the central term k is quantised in this simple way in terms of an integer x, called the "level "of the representation. Furthermore (3.8) also implies that

$$2\lambda.\psi/\psi^2 \leq x, \qquad (3.9a)$$

or

$$\lambda.\psi \leq k \qquad (3.9b)$$

There are only a finite number of dominant integral weights satisfying this inequality. This result was recovered by Seiberg during his lectures[20] starting from Chern-Simons theory, that is, a much more geometrical starting point. A further consequence of (3.9) is that any dominant integral weight which satisfies the condition for a given level automatically satisfies that condition for any higher level. In particular $\lambda = 0$ is a valid integral dominant weight for any level of any $\hat{g}$. Again the mathematical structure permits one and only one vacuum, just as we would wish on physical grounds.

Let us illustrate with $\hat{su}(2)_x$. Conventionally $\psi = 1$ in this case so that (3.9a) reads $\lambda \leq x/2$. As the spin λ can only take the values $0, 1/2, 1, 3/2, ..$ when it is integral and dominant, we see that λ can only take the $x + 1$ values

$$\lambda = 0, 1/2, 1, 3/2, .. \qquad x/2. \qquad ((3.10)$$

The review article by Goddard and Olive[23] or the book by Kac[24] presents the above results in greater detail but it can already be seen that the representation theory of $\hat{g}$ unifies ideas from finite dimensional Lie algebra theory and quantum field theory,

PRIMARY FIELDS FOR THE EXTENDED ALGEBRA

As we have seen, the extension of the algebra lead to a modification by restriction of the concept of ground state in equation (3.4). Correspondingly we must modify the notion of primary field by adding to (1.11b) the condition[25]

$$[T_n^a, \phi(z)] = z^n \phi(z) t^a. \tag{3.11}$$

Again $\phi(z)|0>$ is a ground state, but now in the modified sense (3.4) as

$$\begin{aligned} T_n^a \phi(0)|0> &= [T_n^a . \phi(0)]|0> \qquad \text{if } n \geq 0 \\ &= z^n \phi(z)|_{z=0}|0> t^a \\ &= \begin{cases} 0 & n \geq 1 \\ \phi(0)|0> t^a, & n = 0. \end{cases} \end{aligned}$$

SUGAWARA CONSTRUCTION

Currents which are analogues in four dimensions of the affine Kac-Moody currents are sources of gauge fields in the $g = u(2)$ electro-weak gauge theory. Before the importance of gauge theory was appreciated, much emphasis was placed on "current algebra " in the 1960's. Gellmann argued that it should determine the dynamics of the theory and hence the energy momentum tensor. Sugawara and Sommerfield independently proposed a formula[26] that achieved this, in retrospect, most naturally in two dimensions with conformal symmetry, $\theta^{++} \sim \sum_a j^{a+} j^{a+}$.

In the present notation, this construction reads, for g simple or abelian

$$\mathcal{L}_m^g = \frac{1}{2k + Q_\psi} \begin{smallmatrix}\times \\ \times\end{smallmatrix} \sum_{a=1}^{dim g} \sum_{n=-\infty}^{\infty} T_{m+n}^a T_{-n}^a \begin{smallmatrix}\times \\ \times\end{smallmatrix}. \tag{3.12}$$

The crosses denote the normal ordering whereby T_n^a is shifted to the right or left according as the suffix n is positive or negative. This normal ordering is necessary to ensure that this bilinear operator has finite matrix elements with respect to states in the irrep built on the ground states (3.4). The process upsets the algebraic properties which must be carefully recalculated in the manner of the classic string theory calculations. The prefactor quoted in (3.12) assures

$$[\mathcal{L}_m^g, T_n^a] = -n T_{m+n}^a. \tag{3.13}$$

The quantity Q_λ denotes the quadratic Casimir operator for an irrep of g with highest weight λ. As ψ, the highest root, is the highest weight of the adjoint representation, Q_ψ is given by

$$Q_\psi \delta^{ab} = \sum_{m,n} f^{amn} f^{bmn}. \tag{3.14}$$

Furthermore $\mathcal{L}_m^g$ satisfies the Virasoro algebra (1.1) with

$$c^g = \frac{2k \dim g}{2k + Q_\psi} = \frac{x \dim g}{x + \tilde{h}(g)}, \tag{3.15}$$

the result of another tedious calculation. $\tilde{h}(g) = Q_\psi / \psi^2$ is known as the dual Coxeter number of g, and can be shown to be an integer taking the following values for the simple lie algebras

$$\begin{pmatrix} g: & su(n) & so(n); n \geq 5 & sp(n) & E_8 & E_7 & E_6 & F_4 & G_2 \\ \tilde{h}(g): & n & n-2 & n+1 & 30 & 18 & 12 & 9 & 4 \end{pmatrix}. \tag{3.16}$$

These results may be derived from the known structure of the root system of g from which also follows

$$c^g = \text{rank } g + \frac{(x-1)n_L + (x - (S/L)^2)n_S}{x + \tilde{h}(g)}, \tag{3.17}$$

where n_L and n_S denote the number of long and short roots of g, and L/S is the ratio of their lengths. As the second term is positive or zero

$$\text{rank } g \leq c^g \leq \dim g. \tag{3.18}$$

In particular, c^g is rational and exceeds one and so lies outside the interesting range (1.8).

If g is abelian, its structure constants vanish, and hence so do Q_ψ and $\tilde{h}$. Hence

$$c^g = \text{rank } g = \dim g. \tag{3.19}$$

As $\hat{g}$ then reads

$$[T^i_m, T^j_n] = km\delta^{ij}\delta_{m+n,0}, \tag{3.20}$$

it can be realised in terms of the derivative of a set of rank g spinless scalar boson fields. In this case the Sugawara construction (3.11) reduces to the canonical energy momentum tensor for them and coincides with Virasoro's construction in bosonic string theory where the abelian group is the translation group in space time.

If g is not simple but compact, it can be decomposed

$$g = g_0 + g_1 + g_2 + .. \qquad + g_k,$$

where g_0 is abelian and the remaining pieces simple. Then

$$\mathcal{L}^g = \sum_{i=0}^{k} \mathcal{L}^{g_i}, \qquad c^g = \sum_{i=0}^{k} c^{g_i}. \tag{3.21}$$

We can simplify $\mathcal{L}^g_0$ as

$$\mathcal{L}^g_0 = \frac{\sum_a T^a_0 T^a_0 + 2\sum_a \sum_{n=1}^{\infty} T^a_{-n} T^a_n}{2k + Q_\psi}$$

and deduce the eigenvalue h^g on a ground state (3.4)

$$h^g_\lambda = \frac{Q_\lambda}{2k + Q_\psi}. \tag{3.22}$$

In particular h^g_λ vanishes if and only if λ does, a possibility that can occur for any $\hat{g}_x$ as already explained. This is the vacuum.

Thus, given a unitary representation of $\hat{g}_x$, we have constructed an automatic representation of the Virasoro algebra, which is also unitary. Thus $\hat{g}$ symmetry guarantees conformal symmetry as is illustrated by the Wess-Zumino-Witten model[27,28]. As each $\hat{g}_x$ irrep decomposes into an infinite number of Virasoro irreps the character

$$\chi^g(q)_{(x,\lambda)} = tr\left(q^{L_0 - c/24}\right)_{(x,\lambda)} \tag{3.23}$$

is an infinite sum of Virasoro characters of the type (2.19) for fixed c given by (3.15) and variable h. Yet it was discovered by Kac and Peterson[28] to display a beautiful behaviour in response to the action of an element $\mathcal{M}$ of the modular group (2.14) on $\tau = (\ln q)/2\pi i$:

$$\chi^g_{(x,\lambda)}\Big|^{\mathcal{M}} = \sum_{\lambda'} S_{\lambda,\lambda'}(\mathcal{M})\chi_{(x,\lambda')}, \tag{3.4}$$

where S is again a finite, unitary matrix independent of q. Kac and Peterson noted that the simple behaviour (3.24) required both the inclusion of the $c/24$ in the definition (3.23) and the Sugawara expression (3.22) for the ground state eigenvalue of L_0.

One can enquire whether the Sugawara construction provides the most general solution to equation (3.2). If one puts

$$L_m = \mathcal{L}^g_m + K_m$$

one finds that K_m commutes with T^a_m and hence $\mathcal{L}^g_n$ so that K_m satisfies the Virasoro algebra itself. Thus the ambiguity in L_m consists in the addition of a Virasoro algebra K_m corresponding to an extra independent degree of freedom. Thus the Sugawara construction gives, in a well defined sense, the minimum solution to (3.2).

LECTURE 4

COSET CONSTRUCTION

The argument given at the end of the preceding lecture provides a simple trick for obtaining unitary representations of Vir with $c < 1$ from unitary representations of affine Kac-Moody algebras (3.2), together with other information that we shall describe[30,31].

For simpleness of explanation, let us temporarily suppose that g is a simple Lie algebra containing a Lie subalgebra h. If h is also simple, it possesses an index of embedding I in g which is an integer $1, 2, 3, \ldots$. A level x representation of $\hat{g}$ then decomposes into representations of $\hat{h}$ with level xI. We write $\hat{g}_x \supset \hat{h}_{xI}$.The Sugawara construction (3.12) can be performed both for $\hat{g}$ and $\hat{h}$, and yields, respectively

$$[\mathcal{L}^g_m, T^a_n] = -nT^a_{m+n}, \qquad T^a \in g$$

$$[\mathcal{L}^h_m, T^a_n] = -nT^a_{m+n}. \qquad T^a \in h$$

So, subtracting,

$$[K_m, T^a_n] = 0 \qquad T^a \in h \tag{4.1}$$

since $h \subset g$ and, by definition

$$\mathcal{L}^g_m = \mathcal{L}^h_m + K_m. \tag{4.2}$$

Because of (4.1) and the Sugawara construction (3.12), K_m commutes with $\mathcal{L}^h_n$. It is then easy to check that K_m satisfies a Virasoro algebra (1.1) with c-number

$$c^K = c^g - c^h, \tag{4.3}$$

Unlike c^g and c^h, c^K need not be larger than unity, but, like them, it must be positive (or zero) as K_0 must have a spectrum bounded below so that the result (1.7) applies.

If the spectrum of K_0 were not bounded below, K_n for $n > 0$, which commutes with all the $\mathcal{L}^h_m$, could be used to lower the $\mathcal{L}^g_0$ eigenvalue indefinitely, contradicting its positivity. We conclude

$$c^K \geq 0. \tag{4.4}$$

COSETS WITH POSITIVE C

The result readily extends to the case in which g is semisimple, using (3.21). A particularly interesting coset is $su(2) \times su(2)$ divided by the diagonal $su(2)$ subalgebra with the levels indicated below

$$K = s\hat{u}(2)_m \times s\hat{u}(2)_1 / s\hat{u}(2)_{m+1} \tag{4.5}$$

has

$$c^K = 1 - 6/(m+2)(m+3),$$

that is, the FQS series (1.8), since by (3.15), $s\hat{u}(2)_m$ has $c = 3m/(m+2)$. This finally verifies that the FQS series representations can indeed be constructed in a unitary manner[30].

Using characters for the $c < 1$ Virasoro irreps due to Rocha-Caridi[32] it is possible to calculate the decomposition of the irrep $(m, \ell) \otimes (1, \epsilon)$ of the numerator of K in (4.5) into irreps of $s\hat{u}(2)_{m+1} + Vir_K$:

$$(m, \ell) \otimes (1, \epsilon) = \oplus \sum (m+1, \ell') \otimes (c^K, h_{2\ell+1, 2\ell'+1}), \tag{4.6}$$

where the sum is over ℓ' and is restricted so that $\ell - \ell' - \epsilon$ is an integer and (3.10) is satisfied, with λ and x replaced by ℓ' and $m+1$ respectively.

The relation between the characters of these irreps can be recovered from (4.6) by taking traces of appropriate quantities. From this it is possible to deduce (i) the modular transformation properties of the $c < 1$ characters, and in particular an explicit formula for the matrix S in (2.16), given the formulae of Kac and Peterson for the matrix S relevant to the $s\hat{u}(2)_m$ characters, and (ii) all possible modular invariant partition functions for $c < 1$, given those for $s\hat{u}(2)_m$.

All possible cosets with c^K less than one have been listed by Bowcock and Goddard[33] but there are many others of interest. For example,

$$s\hat{u}(2)_m \times s\hat{(2)}_2 / s\hat{u}(2)_{m+2}$$

yields the FQS superconformal series, while

$$\hat{g}_x \times \hat{g}_{\tilde{h}(g)} / \hat{g}_{x+\tilde{h}(g)}$$

yields more cosets on which the superconformal algebra can be realised. The "Z_k parafermions" can be realised on the cosets

$$s\hat{u}(2)_k / \hat{u}(1) \quad \sim \quad s\hat{u}(k)_1 \times s\hat{u}(k)_1 / s\hat{u}(k)_2.$$

Many, if not all, rational conformal field theories can be found in this manner or else in variations of it. This is the basis for the claim, made above, that the extension of the Virasoro algebra by affine Kac Moody algebras is, in this sense, universal. Recall that Seiberg, in his lectures, explained how to obtain these coset models out of Chern-Simons gauge theories in three dimensions.

THE SITUATION OF VANISHING C^K

If c^K vanishes, then so do all K_n, as K_0 is positive and the argument leading to (1.7) therefore applies. Thus the equality of c^g and c^h furnishes the necessary and sufficient condition for the equality of $\mathcal{L}^g$ and $\mathcal{L}^h$ when g contains h in the sense already explained. This can be thought of as a quantum equivalence between the $\hat{g}$ and $\hat{h}$ theories.

We have already seen a non-trivial coset for which c^K vanishes. This is g/t where g is a simply laced, simple Lie algebra, that is of A, D or E type, while t is the Cartan subalgebra and the level of g unity, as we saw from (3.17) and (3.19). The equality of the two Sugawara constructions for g and t relates the sum of the squares of the step operator currents of g to the sums of the squares of the t currents, which, as we remarked, can be thought of as derivatives of spinless boson fields. This relation suggests that the step operator currents can also be expressed in terms of these boson fields and, indeed, this is the substance of the vertex operator construction[34] of $\hat{g}_1$. This can be used to verify the identity between $\mathcal{L}^g$ and $\mathcal{L}^h$. This sort of argument has also been used to infer quite new constructions of affine Kac Moody algebra representations of, for example, $\hat{g}_1$ when g is not simply laced so that additional fermions of unconventional type are required[35].

If $c^g = c^h$, we say that h is "conformally embedded"in g. As c^K is a strictly increasing function of the level x of $\hat{g}$ when g is simple, the conformal embedding can only occur when x is unity, as in the above example.

MODULAR INVARIANT PARTITION FUNCTIONS FOR $\hat{g}_1$

Instead of (2.1), we now suppose that the structure of the Hilbert space of the theory is given by

$$\mathcal{H} = \oplus \sum N_{\lambda,\overline{\lambda}}(x, \lambda, \overline{x}, \overline{\lambda}), \tag{4.7}$$

where the sum is finite, in the sense that $\sum N_{\lambda,\overline{\lambda}} < \infty$. Since each term in (4.7) contains an infinite number of irreps of $Vir_L \otimes Vir_R$, the decomposition (2.1) would emphatically not be finite. The partition function is again given by (2.10) and required to be modular invariant. By the same argument, this is only possible if c and $\overline{c}$ differ by an integer multiple of 24. One possibility is that g is $E_8 + E_8 + E_8$ with x equal to unity and $\overline{x}$ vanishing. However we shall discard such possibilities in what follows and assume the equality of x and $\overline{x}$, thereby assuring the equality of c and $\overline{c}$. Then

$$Z(\tau) = \sum_{\lambda,\overline{\lambda}} \chi^g_{x,\lambda} N_{\lambda,\lambda'} \chi^g_{\overline{x}\lambda'}, \tag{4.8}$$

where the sum over the modified "conformal blocks "is finite. Again, because of (3.23) and the unitarity of S, $N_{\lambda,\lambda'} = \delta_{\lambda,\lambda'}$ is a solution, but again somewhat uninteresting. However there is now a simple way of using conformal embeddings to generate new modular invariant partition functions, first observed by Victor Kac[36], and relying on a result of Goddard and Olive[37] to the effect that: an irrep of $\hat{g}_1$ decomposes finitely into irreps of $\hat{h}_I$ if, and only if $h \subset g$ is a conformal embedding.

For definiteness let us take $g \supset h$ with both algebras simple. If the embedding is conformal, the levels are 1 and I, respectively. Let $D^g_{x,\lambda}$ denote the irrep of $\hat{g}$ at level x with highest g weight λ. Then the decomposition of this irrep can be written

$$D^g_{1,\lambda} = \oplus \sum_{\mu} A_{\lambda\mu} D^h_{I,\mu}. \tag{4.9}$$

The integers $A_{\lambda\mu} \geq 0$ are the multiplicities. The content of the theorem is that $\sum_\mu A_{\lambda\mu}$ is finite if, and only if the embedding is conformal.

If $h \subset g$ is not conformal , the coset construction tells us that there is a Virasoro algebra $Vir^{g/h}$, commuting with $\hat{h}$ and acting unitarily on the $\hat{g}$ irrep. As Vir only has infinite dimensional irreps when $c > 0$, the multiplicities are all either infinite or zero, that is the decomposition is not finite. Conversely, if the embedding is conformal, the values of μ on the right hand side of (4.9) run over a finite number of values determined by the inequality (3.9), (with x replaced by I). The corresponding values of h are given by (3.22) and apply equally as eigenvalues of $\mathcal{L}_0^g$ and $\mathcal{L}_0^h$ as these are equal in conformal embeddings. The largest of these values is therefore finite. But the subspace of the $\hat{g}$ irrep with smaller $\mathcal{L}_0^g$ eigenvalues is finite dimensional so that the sum in (4.9) is finite and the result established.
Taking traces of (4.9) yields the following relation between characters

$$\chi_{1,\lambda}^g = \sum_\mu A_{\lambda\mu} \chi_{I,\mu}^h. \tag{4.10}$$

This can be inserted into any modular invariant partition function (4.8) for $\hat{g}$ to give

$$Z = \chi_I^h A^T N A \chi_I^h, \tag{4.11}$$

that is, a modular invariant partition function of the form (4.8) but for $\hat{h}_I$ instead of $\hat{g}_1$ and with N replaced by

$$N' = A^T N A.$$

This matrix, like N is symmetric, with integer entries greater than or equal to zero. Further, as $A_{\lambda 0} = \delta_{\lambda 0}$, $N'_{00} = N_{00}$ and so the vacuum remains unique if originally so. The conclusion is that non-trivial modular invariant partition functions can be constructed for $\hat{h}_I$ whenever h is embedded conformally in g and I is the index of embedding.

MODULAR INVARIANT PARTITION FUNCTIONS FOR $s\hat{u}(2)_x$

Examples relevant to $s\hat{u}(2)$ and hence $c < 1$ partition functions are

$$(\hat{A}_1)_{10} \subset (\hat{B}_2)_1 \quad \text{and} \quad (\hat{A}_1)_{28} \subset (\hat{G}_2)_1.$$

These give modular invariant partition functions for $s\hat{u}(2)$ at levels 10 and 28 respectively, numbers which happen to be two less than the dual Coxeter numbers (3.16) of E_6 and E_8 respectively. However, a more systematic and complete construction of these partition functions, due to Nahm[38], uses the fact that a straightforward extension of the above argument yields modular invariant partition functions for h if $h \times h' \subset g$ is conformal by an elimination of h'. Therefore if $h = A_1$ we can use the conformal embeddings

$$(\hat{A}_1)_{r-1} \times \hat{A}_{r-2} \times U_1 \subset s\hat{o}(4(r-1)),$$

$$(\hat{A}_1)_{2(r-2)} \times \hat{D}_{r-2} \times \hat{A}_1 \subset s\hat{o}(8(r-2)),$$

$$(\hat{A}_1)_{10} \times \hat{A}_5 \subset \hat{so}(40), \qquad (4.12)$$
$$(\hat{A}_1)_{16} \times \hat{D}_6 \subset \hat{so}(64),$$
$$(\hat{A}_1)_{28} \times (\hat{E}_7)_{12} \subset \hat{so}(112).$$

These may seem rather recondite, but in fact, they can be read off standard mathematical tables listing irreducible symmetric spaces, making use of the following theorem by Goddard, Nahm and Olive[39]: *$c^{so(n)/h}$ vanishes if, and only if there exists a Lie algebra h' such that h'/h is a symmetric space of dimension n.*

The proof can only be sketched here. As $\hat{so}(n)$ must have have level 1 for the embedding to be conformal, it can be represented by bilinears in fermions, a construction familiar as the "quark model "in particle physics and one that can be regarded as the affinisation of the Clifford algebra construction of representations of the Lie algebra of the orthogonal groups. According to (3.15), and the fact that $so(n)$ has dimension $n(n-1)/2$ and dual Coxeter number $n-2$, the Virasoro c for $\hat{so}(n)_1$ is simply $n/2$, the same as that for the canonical energy momentum tensor of Dirac for n real, free, massless fermions, as we mentioned in connection with the Ising model and string theory. Hence by an argument of the type already used, there is a quantum equivalence equating the Sugawara construction for $\hat{so}(n)_1$ with the Dirac energy momentum tensor which, of course, is bilinear in the fermions, but involves a derivative of them. Since the currents are bilinear in fermions the Sugawara construction apparently contains terms quadrilinear in the fermions. This appearance must be deceptive in the case mentioned, and the quadrilinear terms must cancel out, and this can be checked explicitly. As $\hat{h}_I \subset \hat{so}(n)_1$, $\hat{h}_I$ can also be represented bilinearly in the n fermions which transform according to the n dimensional representation of $so(n)$ implied by the statement. If further this embedding is conformal, the Sugawara construction for $\hat{h}$ equals the Dirac construction, and so the terms quadrilinear in the fermions must yet again cancel out. These can be isolated by replacing the normal ordering with respect to currents by a new one with respect to the fermions and denoted by open dots. So if the current is

$$\sum_n z^{-n} T_n^a = i \sum_{A,B} \psi^A M_{AB}^a \psi^B / 2$$

where $\psi^A(z), A = 1,2,..n$, are the fermion fields, the term quadrilinear in fermions, which in fact must cancel to zero, is proportional to

$$\sum_{aABCD} {}^{o}_{o}\psi^A(z) M_{AB}^a \psi^B(z) \psi^C(z) M_{CD}^a \psi^D(z){}^{o}_{o}$$

$$= (1/3) \sum_{ABCD} {}^{o}_{o}\psi^A \psi^B \psi^C \psi^D{}^{o}_{o} \{\sum_a (M_{AB}^a M_{CD}^a + M_{BC}^a M_{AD}^a + M_{CA}^a M_{BD}^a\}$$

using the antisymmetry properties of the normal ordered product of the fermions. The condition that the embedding be conformal thus reduces to the vanishing of the expression inside the curly brackets. This expression resembles a Jacobi identity and, indeed, it is the condition that the space h'/h be symmetric as the quantities f^{abc} and M_{AB}^a actually constitute the structure constants of h'. The above argument can be reversed and extended to cover the situation when h and h' are not necessarily simple. The symmetric spaces which are responsible for the conformal embeddings (4.12) are, respectively

$$
\begin{aligned}
A_r/A_{r-2} \times U_1 \times A_1 \\
D_r/D_{r-2} \times A_1 \times A_1 \\
E_6/A_5 \times A_1 \\
E_7/D_6 \times A_1 \\
E_8/E_7 \times A_1 .
\end{aligned}
\tag{4.13}
$$

Notice that the algebra in the numerator is always simply laced and possesses a dual Coxeter number two more than the $\hat{A}_1$ level indicated in (4.12). The particular symmetric spaces in (4.13) have a very special structure because of the A_1 factor in the denominator. The representation carried by the tangent space, and hence the n fermions always has the form $(n/2, 2)$ where the $n/2$ denotes an irrep of the factor of the denominator besides $su(2)$. The only irrep of the $su(2)$ occurring has two dimensions, and so is pseudoreal. As the tangent space is real the $n/2$ irrep must also be pseudoreal, from which it follows that n is a multiple of four. Thus the $su(2)$ generators can be thought of as Pauli matrices and the symmetric spaces (4.13) have a special quaternionic structure. Actually, it is easy to prove that there exists one, and only one, such quaternionic symmetric space for each choice of simple Lie algebra in the numerator of the quotient. However only those which are simply laced need be considered in order to find the complete set of modular invariant partition functions for $s\hat{u}(2)_x$, previously established, using brute force methods, by Cappelli, Itzykson and Zuber[40], who recognised the various superficial connections to the numerator algebras, now explained by Nahm's construction.

These results posc many intriguing questions. For example, there is a well known connection between the simply laced Lie algebras A, D and E with the finite subgroups of $so(3)$ or $su(2)$. In his lectures[20], Seiberg mentioned an alternative and more recent construction of the modular invariant $su(2)$ partition functions related to this. There should be a direct connection between this and the work of Nahm. Pasquier[41] has reformulated the Q-state Potts model on a square lattice and found generalisations capable of second order phase transitions which also possess an A, D and E classification which corresponds almost, if not quite exactly, to the $c < 1$ theories consequent upon the results for $s\hat{u}(2)_x$, via the coset construction (4.5).

There have been many other exciting recent developments such as new and interesting super coset models which I regret I shall have no time to explain. Let me finish by reminding you that this rich area of development in two dimensional conformal quantum field theory has been achieved using techniques hitherto underused in this context, namely the very clear and precise methods of the theory of Lie algebras and their representations. It is, of course, a challenge to rederive and reinterpret these new results according to the more traditional, geometrical methods and progress already made in this direction was described by Seiberg, in his lectures.

REFERENCES

1) H.Bateman, *Proc. London Math. Soc.* **8** (1910) 223.
E.Cunningham, *Proc. London Math. Soc.* **8** (1909) 77.

2) M.A. Virasoro, *Phys. Rev.* **D1** (1970) 2933.

3) J. Schwinger, *Phys. Rev. Lett.* **3** (1959)296.

4) J.F.Gomes, *Phys. Lett.* **171B** (1986)75.

5) D.Friedan, Z.Qiu and S.Shenker, *Phys. Rev. Lett.* **52** (1984) 1575, *Commun. Math. Phys.* **107** (1986) 535.

6) A.A.Belavin, A.M.Polyakov and A.B.Zamolodchikov, *Nucl. Phys.* **B241** (1984) 333.

7) V.Kac, *Lecture notes in Physics* **94** (1979)441.

8) R.Brower, *Phys. Rev.* **D6** (1972)1655,

P.Goddard and C.Thorn, *Phys. Lett.* **40B** (1972)235.

9) S. Fubini and G. Veneziano, *Nuovo Cimento* **67A** (1970) 29.

10) A.M.Polyakov, *JETP Lett.* **12** (1970) 381.

11) R. Baxter, *Exactly Solved Models in Statistical Mechanics*, (Academic Press, 1982).

12) L. Onsager,*Phys. Rev.* **65** (1944) 117.

13) P Jordan and E.P. Wigner, *Zeit. Phys.* **47** (1928) 631.

14) P. Ramond, *Phys. Rev.* **D3** (1971) 2415.

A.Neveu and J.H. Schwarz, *Nucl. Phys.* **B31** (1971) 86, *Phys. Rev.* **D4** (1971) 1109.

15) C.B. Thorn, *Phys. Rev.* **D4** (1971) 1112,

E. Corrigan and D. Olive, *Nuovo Cimento* **11A** (1972) 749.

16) M. Fisher, *Phys. Rev. Lett.* **16** (1966) 11.

17) J.L. Cardy, *Nuclear Physics* **B270[FS16]** (1986) 186.

18) E. Verlinde, *Nucl. Phys.* **B300** (1988) 360.

19) G. Moore and N. Seiberg, *Phys. Lett.* **B212** (1988) 451.

20) N. Seiberg, *Lectures at Banff Summer School.*

21) V.G. Kac, *Funct. anal. Appl.* **1** (1967) 328,

R.V. Moody, *Bull. Amer. Math. Soc.* **73** (1967) 217.

22)J. Patera, *Lectures at Banff Summer School*.

23) P. Goddard and D. Olive, *Int. J. Mod. Phys.* **A1** (1986) 303, and in *Kac-Moody and Virasoro Algebras, a Reprint volume for Physicists*, (World Scientific, 1988), 8.

24) V.G.Kac, *Infinite Dimensional Lie Algebras*, (second edition, Cambridge University Press, 1985).

25) V.Knizhnik and A.B.Zamolodchikov, *Nucl. Phys.* **B247** (1984) 83.

26) H.Sugawara, *Phys. Rev.* **170** (1968) 1659,

C.Sommerfield, *Phys. Rev.* **176** (1968) 2019.

27) J.Wess and B. Zumino, *Phys. Lett.* **37B** (1971) 95.

28) E.Witten, *Commun. Math. Phys.* **92** (1984) 455.

29) V.G.Kac and D.H.Peterson, *Advances in Math.* **53** (1984) 125.

30) P.Goddard, A.Kent and D.Olive, *Phys. Lett.* **152B** (1985) 88, *Commun. Math. Phys.* **103** (1986) 105.

31) For earlier, special cases, see K.Bardakci and M.Halpern, *Phys. Rev.* **D3** (1971) 2493,

S.Mandelstam, *Phys, Rev.* **D7** (1973) 3777.

32) A.Rocha-Caridi, in *Vertex Operators in Mathematics and Physics, Proceedings of a Conference November 10-17, 1983,* eds. J.Lepowsky, S.Mandelstam and I.M.Singer (Springer-Verlag, 1984), 451.

33) P.Bowcock and P.Goddard, *Nucl. Phys.* **B285 FS[19]** (1987) 651.

34) G.Segal, *Commun. Math, Phys.* **80** (1981) 301.

I.B.Frenkel and V.G.Kac, *Invent. Math.* **62** (1980) 23.

35) P.Goddard, W.Nahm, D.Olive and A.Schwimmer, *Commun. Math. phys.* **107** (1986) 179,

D.Bernard and J. Thierry-Mieg, *Commun. Math. Phys.* **111** (1987) 181.

36) V.G. Kac, unpublished talk at Montreal.

37) P.Goddard and D.Olive, *Nucl. Phys.* **B257 [FS14]** (1985) 226.

38) W.Nahm, *Duke Math. J.* **54** (1987) 579.
39) P.Goddard, W.Nahm and D.Olive, *Phys. Lett.* **160B** (1985) 111.
40) A.Cappelli, C.Itzykson and J.-B. Zuber, *Commun. Math. Phys.* **113** (1987) 1.
41) V.Pasquier, *Nucl. Phys.* **B285 [FS19]** (1987) 162.

LECTURES ON RCFT*

Gregory Moore

Institute for Advanced Study, Princeton, New Jersey 08540, USA
and
Department of Physics, Yale University, New Haven
Connecticut 06511-8167, USA

and

Nathan Seiberg†

Institute for Advanced Study, Princeton, New Jersey 08540, USA
and
Department of Physics & Astronomy, Rutgers University
Piscataway, New Jersey 08855-0849, USA

We review some recent results in two dimensional Rational Conformal Field Theorey. We discuss these theories as a generalization of group theory. The relation to a three dimensional topological theory is explained and the particular example of the Chern-Simons-Witten theory is analyzed in detail. This study leads to a natural conjecture regarding the classification of all RCFT's.

* *Given by G. Moore in the Trieste spring school 1989 and by N. Seiberg in the Banff summer school 1989.*

† *On leave from the Department of Physics, Weizmann Institute of Science, Rehovot 76100, Israel.*

Physics, Geometry, and Topology
Edited by H. C. Lee
Plenum Press, New York, 1990

1. Introduction – a trip to the Zoo

The fundamental principles of string theory are not yet known. Since conformal field theory [1] plays a crucial role in string theory, many researches believe that a detailed study of conformal field theory will bring us closer to the concepts underlying string theory. It is hoped that a better understanding of the mathematical foundations of conformal field theory will lead to interesting and relevant generalizations of CFT, which might in turn lead to progress in string theory. There are other good reasons to study CFT, on the one hand, the study of CFT might eventually be useful in identifying 2D critical phenomena in nature and on the other it has lead to beautiful results and applications in pure mathematics, and promises to lead to more.

Motivated by the desire to understand better the mathematical structure of conformal field theories one turns to the problem of classifying theories. We are not so much interested in the final list of theories as we are in the techniques used to obtain such a list, and the mathematical structures characteristic of members on that list.

General conformal field theories have not yet been attacked in any meaningful way, but the study of an interesting subclass of theories has been very successful in the past two years. In order to motivate and define these theories let us recall that some theories have the beautiful properties that their correlation functions, partition functions etc. have very simple analyticity properties in the moduli. The prototype of such behavior is the holomorphic factorization of determinants on Riemann surfaces:

$$det\bar{\partial}\partial \sim |F(\tau)|^2$$

which plays a key role in the Belavin-Knizhnik theorem of string theory. Should we focus on this criterion? No: the theories which have this property are too simple – they are basically free theories (on the world sheet!). Holomorphic factorization admits a generalization which leads to a very rich class of conformal field theories, namely, the rational conformal field theories (RCFT). These may be characterized by saying that all correlation functions, partition functions, etc. can be expressed in terms of finite sums of analytic times anti-analytic functions:

$$\langle\phi\cdots\phi\rangle \sim \sum_{i=1}^{N<\infty} |\mathcal{F}_i|^2$$

More formally, RCFT's are distinguished amongst the set of all conformal field theories by the existence of a holomorphic (and anti-holomorphic) monodromy-free subalgebra $\mathcal{A}$ (and $\bar{\mathcal{A}}$) of the operator product algebra such that the physical Hilbert space can be decomposed into a finite sum of $\mathcal{A} \times \bar{\mathcal{A}}$ representations:

$$\mathcal{A} = \oplus_{i=1}^{N} \mathcal{H}_i \otimes \bar{\mathcal{H}}_i$$

In fact, known theories satisfying this criterion comprise a veritable zoo.

Let us collect some specimens from this zoo. The oldest and most venerable are surely the current algebras - also known as Wess-Zumino-Witten [2] theories. These current algebras have various extended algebras (a notion we explain below). So far, all known extended algebras are related to orbifolds [3] of WZW theories by a subgroup of the center. Another venerable example of rational theories are the minimal models of BPZ [1] and FQS [4] (and their N=1 and N=2 generalizations). These are based on the chiral algebra of the (N=1,or N=2 super-) Virasoro algebra itself, and these have rather nontrivial extensions known as W-algebras and their generalizations, W_n-algebras [5]. In addition there are various species of parafermions [6]. Between 1984 and 1986 it was realized [7] [8] [9] that parafermions and the various discrete series could be obtained by the GKO coset construction [10]. Indeed, any coset construction based on two rational chiral algebras will define a rational conformal field theory. Finally whenever the chiral algebra has a discrete symmetry we can form an orbifold [3] theory [11].

Clearly, this zoo should be organized. By trying to formulate all these theories in a unified way, we are led to conjecture: *all* RCFT's are related to certain deformations of groups, this deformation can be described axiomatically or in terms of 3D Chern-Simons-Witten (CSW) gauge theory and is closely related to certain quantum groups. A cynical version of this conjecture would state that nothing new has been found since 1986, so we must be done.

The purpose of these lectures is to make a case that the conjecture is not cynical but based on the insight that RCFT is closely related to group theory, and in fact must be defined by axioms closely related to those defining groups.

These lectures are not meant to be a review of the subject of RCFT. Many groups have contributed to this subject from various points of view. In particular, a completely independent line of development, beginning with the classic papers of Doplicher, Haag, and Roberts, and using the conceptual framework of algebraic quantum field theory has led to similar results [12]. For the most part we will review our own work on the subject [13] [14] [15] [16] [17] and will present it from the point of view developed in these references.

We will assume the reader has some familiarity with conformal field theory, e.g. we will assume familiarity with the material covered in standard review lectures [18]. We have included many exercises, hoping they will help the reader study the subject. It is a good idea to try to work out at least some of them in order to practice the formalism in the text. The answers to most of these exercises can be found in standard CFT reviews or in our papers [13]-[17].

In the next section we give several different definitions of chiral vertex operators. These allow us to have an operator formalism for calculations of conformal blocks and lead to the definition of the duality matrices. In the third section we examine the consistency conditions these matrices have to satisfy. The complete set of independent identities of these matrices is found in section 4. In the fifth section we describe the Tannaka-Krein approach to group theory which is similar to the structure we found in sections 2 - 4. This leads us to the conclusion that RCFT is a generalization of group theory. In section 6 we combine the left moving and right moving conformal blocks into a consistent conformal field theory. Section 7 is devoted to a general discussion about the relation between two dimensional duality (as described in the previous sections) and three dimensional general covariance. This general discussion is made more explicit in sections 8 - 10. In the eighth section we have some comments about quantum groups and the relation of quantum groups to knot invariants and RCFT. In sections 9 - 10 we consider an explicit example of a topological three dimensional field theory. This is the Chern-Simons-Witten (CSW) theory. We first discuss the canonical quantization of the theory (section 9) and explain the connection between the theory and two dimensional conformal field theory. We then consider different gauge groups in three dimensions (section 10) and show that all known RCFT can be obtained by an appropriate gauge group in three dimensions. Our conclusions are summarized in section 11 where we also present some conjectures about the classification of RCFT.

2. Chiral Vertex Operators and Duality Matrices

We need a formalism for manipulating holomorphic parts of vertex operators. Vertex operators will be replaced by objects known as chiral vertex operators (CVO's) [19] [20] [13][14][15] the distinction being that chiral vertex operators are purely holomorphic and keep track of the various internal states and couplings used to form a conformal block. Rather than give the definition immediately, let us build up to it.

Consider the minimal Virasoro models. For every triplet i, j, k of Virasoro representations and $\beta \in \mathcal{H}_j$ we define

$$\Phi^{j,\beta}_{i,k}(z) : \mathcal{H}_k \rightarrow \mathcal{H}_i$$

by its matrix elements.

First, consider β to be a highest weight vector $\beta = |j>$. For the primaries in $\mathcal{H}_i$ and $\mathcal{H}_k$ we have

$$< i|\Phi_{ik}^{j,\beta}(z)|k >=\| \Phi_{ik}^{j} \| z^{-(\Delta_j+\Delta_k-\Delta_i)}$$

where Δ is the conformal dimension of field. We can compute matrix elements between descendants using the Virasoro algebra and the rule

$$\left[L_n, \Phi_{ik}^{j,\beta}(z)\right] = \left(z^{n+1}\frac{d}{dz} + (n+1)z^n\Delta(\beta)\right)\Phi .$$

This only defines Φ on Verma modules. Demanding that Φ is defined on the irreducible quotients forces some of the constants $\| \Phi_{ik}^{j} \|$ to vanish.

• Exercise 2.1 *Null vectors at work*

a.) Suppose ϕ has a nonvanishing weight. Show that if $|0\rangle$ is the $sl(2)$ invariant vacuum then the null vector $L_{-1}|0\rangle$ implies that $||\Phi_{00}^{\phi}|| = 0$.

b.) Consider the Ising model with primary fields $1, \psi, \sigma$ of dimensions $0, 1/2, 1/16$. Use the null vector

$$(L_{-2} - \frac{3}{2}L_{-1}^2)|\psi\rangle = 0$$

to show that $||\Phi_{\psi\psi}^{\psi}|| = 0$.

We initially define the fusion rule $N_{jk}^{i} = 0, 1$ according to whether $\| \Phi \|$ must be zero or not. Having defined $\Phi_{ik}^{j,\beta}$ for highest weight states, we can define it for descendants $\beta = L_{-I}|j>$ (and their linear combinations) by contour integrals:

$$\Phi_{ik}^{j,\beta}(z) \equiv \oint d\xi_1(\xi_1 - z)^{n_1+1}T(\xi_1)\dots \oint d\xi_\ell(\xi_\ell - z)^{n_\ell+1}T(\xi_\ell)\Phi_{ik}^{j,|j>}(z) .$$

For simplicity we will often restrict ourselves to the minimal models. However, we will occasionally point out new elements that arise in more general RCFT's. For example, we define chiral vertex operators for affine Lie algebras $\hat{g}$. Each $\hat{g}$ representation $\mathcal{H}_i$ contains a ground state representation $W_i \subset \mathcal{H}_i$ for the finite dimensional algebra g. We first define $\Phi_{ik}^{j,\beta}(z)$ for $\beta \in W_j$. By commutation with the generators of $\hat{g}$ it suffices to define the matrix elements between $\alpha \in W^i$, $\gamma \in W_k$

$$\left\langle\alpha|\Phi_{ik}^{j\beta}(z)|\gamma\right\rangle = t_{\beta\gamma}^{\alpha} z^{-(\Delta_1+\Delta_j-\Delta_n)}$$

where $t_{\beta\gamma}^{\alpha} \in Inv\left(\bar{W}^i \otimes W^j \otimes W^k\right)$ is an invariant tensor. Other matrix elements and the definition for β a descendent can be carried out exactly as before. Again the null vectors will only allow one to define Φ consistently starting with a *subspace* of $Inv(\bar{W}^i \otimes W^j \otimes W^k)$. This subspace of good couplings

$$V^i_{jk} \subset Inv\left(\bar{W}^i \otimes W^j \otimes W^k\right)$$

is called the space of 3-point couplings and $N^i_{jk} = dim\ V^i_{jk}$ are the fusion rules. Notice that in this case, unlike the discrete series, the integers N^k_{ij} are not all zero and one – in some cases there exists more than one invariant coupling. Also, the representations are not all self conjugate. In other words, $N^j_{i0} = \delta^j_i$ but $N^0_{ij} = \delta_{\bar{i}j}$ where $\bar{i}$ is the conjugate of i. In more general theories there are CVO's which vanish for three primary fields but do not vanish for the descendants.

CVO's give an operator formalism for computing conformal blocks. For example, the conformal blocks of the 4-point function for 4 primaries in the minimal models are

$$\mathcal{F}^{ijk\ell}_p(z_2,z_3) = \left\langle i|\Phi^j_{ip}(z_2)\Phi^k_{p\ell}(z_3)|\ell\right\rangle \sim \qquad (2.1)$$

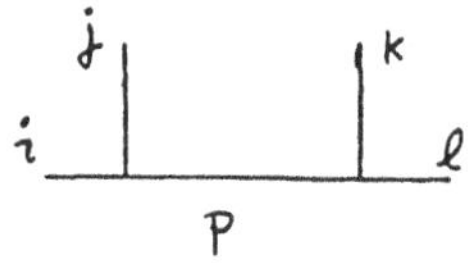

where the rhs of the above equation illustrates a useful pictorial notation for conformal blocks.

The physical correlation function is given (in the diagonal theory) by

$$\left\langle \phi^i|\phi^j(z_2)\phi^k(z_3)|\phi^\ell\right\rangle = \sum_p d_p|\mathcal{F}_p|^2$$

where d_p are constants independent of z and $\bar{z}$. This correlator looks like it depends on many choices. Duality states that many of those choices don't affect the above final result. More precisely, part of duality states that the physical correlators are independent of the choice of basis of conformal blocks. In particular, the order of $\phi^j\phi^k$ on the lhs is irrelevant so one could also have used the blocks

$$\mathcal{F}^{ikj\ell}_p(z_3,z_2) = \left\langle i|\Phi^k_{ip}(z_3)\Phi^j_{p\ell}(z_2)|\ell\right\rangle . \qquad (2.2)$$

But these blocks must give the same correlation function.

• Exercise 2.2 *Trivial fact of life.* Show that if $\{f_i\},\{g_i\},\{h_i\},\{k_i\}$ are four sets of linearly independent analytic functions such that

$$\sum_{i=1}^{N} f_i\bar{g}_i = \sum_{i=1}^{M} h_i\bar{k}_i$$

then $N = M$, and $\vec{f} = A\vec{h}\ \vec{g} = (A^{-1})^\dagger\vec{k}$ for some invertible matrix A.

From the above exercise it follows that the two sets of blocks (2.1) and (2.2) are linearly related, and in fact, by considering descendants we have an operator identity:

$$\Phi^j_{ip}(z_1)\Phi^k_{p\ell}(z_2) = \sum_q B_{pq}\begin{bmatrix} j & k \\ i & \ell \end{bmatrix} \Phi^k_{iq}(z_2)\Phi^j_{q\ell}(z_1) \tag{2.3}$$

where that the coefficients B are the same for the primaries and all the descendants.

If one thinks carefully about the above argument he will note that we must choose cuts since the $\mathcal{F}$'s are not globally defined and have monodromy. So we choose the cut: $z_1 - z_2 \in \mathbb{R}^+$. In order to compare (2.1) and (2.2) we must use analytic continuation, and we can only compare these functions on their common domain of definition. In the z_2 plane we find that (2.1) and (2.2) are defined on the following regions:

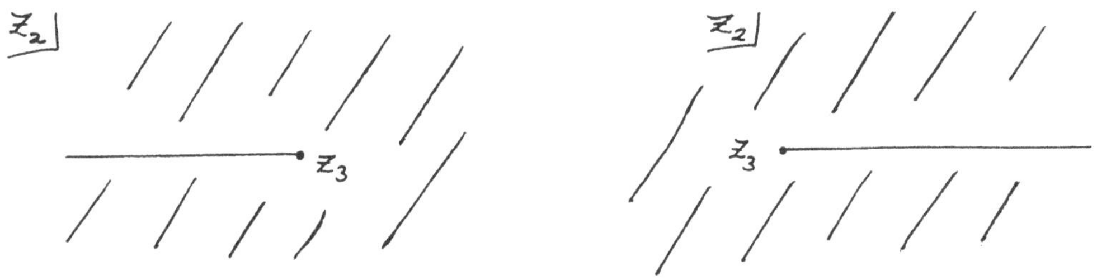

hence the overlap consists of two components and there are in principle two distinct B matrices. We define $B(+)$ by (2.3) for $\mathrm{Im}(z_1 - z_2) > 0$. For $\mathrm{Im}(z_1 - z_2) < 0$ we have, in general a different matrix $B(-)$. If the sign is omitted, we refer to $B(+)$.

• Exercise 2.3 *Relation to BPZ.* Compare the above discussion with section four of [1] and show that the definition of conformal blocks as matrix elements of Φ corresponds with that of BPZ.

All of this has been derived in the simplified notation appropriate for the minimal models, but these considerations apply to arbitrary RCFT's. In the general case, when the space of three-point couplings V^i_{jk} is a vector space of dimension larger than one we have linear transformations

$$B_{pq}\begin{bmatrix} j & k \\ i & \ell \end{bmatrix} \; : \; V^i_{jp} \otimes V^p_{k\ell} \longrightarrow V^i_{kp} \otimes V^p_{j\ell} \, .$$

The other part of the algebra of the Φ operators follows from the operator product expansion. We have

$$\Phi^j_{ip}(z_1)\Phi^k_{p\ell}(z_2) = \underbrace{\sum_q \mathcal{F}_{pq}\begin{bmatrix} j & k \\ i & \ell \end{bmatrix}}_{\substack{\text{Summarizes the} \\ \text{representation-theoretic} \\ \text{content of the operator}}} \; \underbrace{\sum_{Q \in \mathcal{H}_q} \Phi^{q,Q}_{i\ell}(z_2)\langle Q|\Phi^j_{qk}(z_{12})|k\rangle}_{\text{sum over descendants}} \tag{2.4}$$

• **Exercise 2.4** *Defining the F Matrix.* **Prove that the operator product expansion of two Φ operators has the form**

$$\Phi^i_{lp}(z_1)\Phi^j_{pr}(z_2) = \sum_k F_{pk}\begin{bmatrix} i & j \\ l & r \end{bmatrix} \sum_{K\in\mathcal{H}_k} \Phi^{k,K}_{lr}(z_2)\langle K|\Phi^i_{kj}(z_1 - z_2)|j\rangle$$

(Hint: Write out the operator product expansion with arbitrary coefficients. Use translation and scaling invariance to determine some of the structure of the coefficients. Now take the operator product expansion with a third operator Φ and demand consistency with braiding.)

Now going back to our blocks $\mathcal{F}^{ijkl}_p$ we see that we can insert the operator product expansion and define a new basis of conformal blocks, which we may denote pictorially:

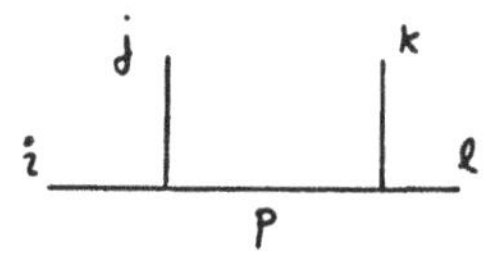

$$= \sum_q F_{pq}\begin{bmatrix} j & k \\ i & l \end{bmatrix}$$

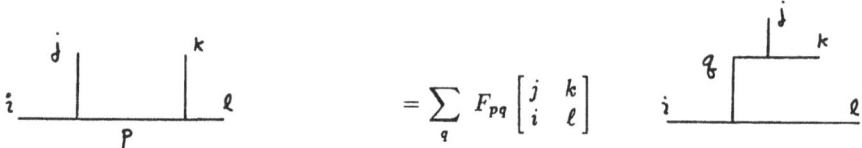

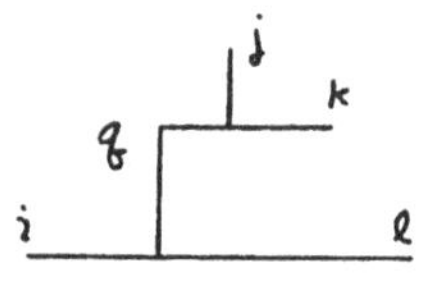

In the general case we have a linear transformation

$$F_{pq}\begin{bmatrix} j & k \\ i & l \end{bmatrix} : V^i_{jp} \otimes V^p_{kl} \rightarrow V^i_{ql} \otimes V^q_{jk}$$

The F, B transformations are the basic duality transformations. The reader may well ask why these objects are of interest. We may answer with two immediate consequences of these considerations.

First point: Already the *existence* of B, F have interesting consequences. Since they are defined by a change of basis, the transformation

$$B : \oplus_p V^i_{jp} \otimes V^p_{k\ell} \longrightarrow \oplus_p V^i_{kp} \otimes V^p_{j\ell}$$

is an isomorphism. Therefore, matching dimensions, we have

$$\sum_p N^i_{jp} N^p_{k\ell} = \sum_p N^i_{kp} N^p_{j\ell} \, .$$

This defines Verlinde's fusion rule algebra: [21]

• Exercise 2.5 *Fusion Rule Algebra.* Using the fact that B and F define isomorphisms show that the matrices

$$(\phi_k)^i_j \equiv N^i_{kj}$$

form a commutative associative algebra. This algebra is known as Verlinde's fusion rule algebra.

• Exercise 2.6. *Examples of Fusion Rule Algebras.*

a.) Show that the fusion rule algebra for the rational torus (see section 10) is $\mathbb{Z}/N\mathbb{Z}$.

b.)Write out the algebra for the Ising model. Try to determine all physically acceptable fusion rule algebras with three self-conjugate primaries.

c.) Show that the FRA for the WZW model $SU(2)_k$ (the subscript denotes the level) is generated by elements ϕ_ℓ, $\ell = 0, 1/2, \ldots, k/2$ with

$$\phi_{\ell_1}\phi_{\ell_2} = \sum_{|\ell_1-\ell_2|}^{min(\ell_1+\ell_2, k-(\ell_1+\ell_2))} \phi_\ell$$

by considering the null vector $J_{-1}^{k-2\ell+1}|\ell;\ell\rangle$. (Sec, e.g. [22].)

d.) Consider the WZW model $SU(3)_2$. Show that the fusion rule algebra for the six integrable representations $\mathbf{1}, \mathbf{3}, \mathbf{3}^*, \mathbf{6}, \mathbf{6}^*, \mathbf{8}$ can be determined purely from the known group theoretic decompositions and consistency conditions on the FRA. Note in particular that $N_{\mathbf{888}} = 1$ whereas in group theory it is equal to two.

Second point: Next, the matrix B^2 is not an identity matrix, precisely because of the cuts. In fact, B^2 is exactly the monodromy matrix for the analytic continuation of z_1 around z_2 for the vector of blocks $\mathcal{F}^{ijk\ell}_p(z_1, z_2)$. That is, if $\gamma(s)$ is the following curve:

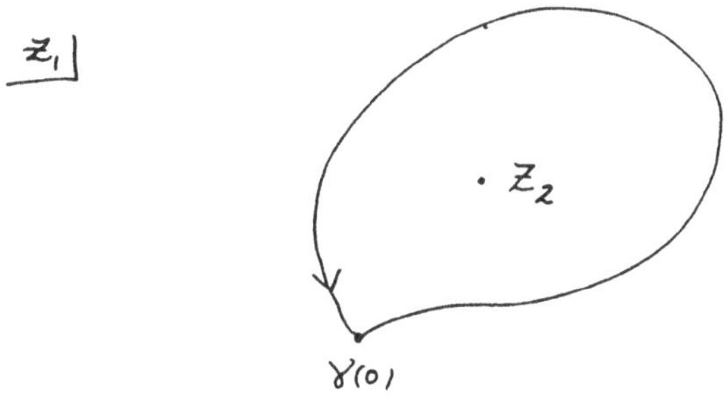

A curve in z_1 plane surrounding z_2

Then one can compute the monodromy as in

• Exercise 2.7 *Monodromy of the blocks.* Show *carefully* that upon analytic continuation we have:

$$\mathcal{F}_p^{ijk\ell}(\gamma(2\pi), z_2) = \sum_q \left(B\begin{bmatrix} k & j \\ i & \ell \end{bmatrix} B\begin{bmatrix} j & k \\ i & \ell \end{bmatrix} \right)_{pq} \mathcal{F}_q^{ijk\ell}(\gamma(0), z_2) \ .$$

Now, the monodromies of conformal field theory are related to the mutual locality factors and therefore to the conformal weights. Thus, the primitive hope is that nontrivial identities on B, F matrices are so restrictive that one can solve them and thus classify RCFT's. This is too naive, but it is on the right track. At any rate, with this hope in mind it is clearly wise to get better acquainted with B as in the following exercise:

• Exercise 2.8 *B and F with the unit operator.* By setting various external representations in the four-point function to be the identity we obtain a computable three-point function. Use this observation to evaluate the F and B matrices in the special cases that one of the fields is the identity. Notice that

$$B\begin{bmatrix} i & j \\ 0 & k \end{bmatrix}$$

defines a linear map

$$\Omega^i_{jk} : V^i_{jk} \to V^i_{kj}$$

which may be interpreted as the square-root of a mutual locality factor (compare the previous exercise). Show that

$$(\Omega^i_{jk})^2 = e^{2\pi i(\Delta_j + \Delta_k - \Delta_i)}$$

Therefore

$$\Omega^i_{jk} = \xi^i_{jk} e^{\pi i(\Delta_j + \Delta_k - \Delta_i)}$$

where $\xi = \pm 1$. In simple RCFT's like the discrete series ξ is always $+1$. In other theories ξ can be -1. For example, in $\hat{SU}(2)$ KM, the sign ξ corresponds to the symmetry or antisymmetry of the tensor coupling the representations. Show that in this example

$$\xi^i_{jk} = (-1)^{2(i+j+k)}$$

where the representations are labeled by their spin (which is integer or half integer). For simplicity, we will limit ourselves in some of the formulae below to the case $\xi = 1$.

From this discussion it is clear that we need to understand the identities on B, F. A number of questions arise: How can we obtain nontrivial identities? What is the full set? What is the minimal set of independent relations? To understand these identities we should understand better what a CVO is. Therefore, let us broaden our viewpoint on chiral vertex operators so that we see more clearly the S_3 symmetry of three-point couplings which is fundamental to duality. Instead of choosing the state β to define Φ we should consider a *single* linear operator

$$\begin{pmatrix} i \\ jk \end{pmatrix} : \mathcal{H}_j \otimes \mathcal{H}_k \longrightarrow \mathcal{H}_i \ .$$

that commutes with contour deformation of the chiral algebra. We would like to give this operator a geometrical interpretation. Namely, suppose we have representation spaces on three circles as in the following picture:

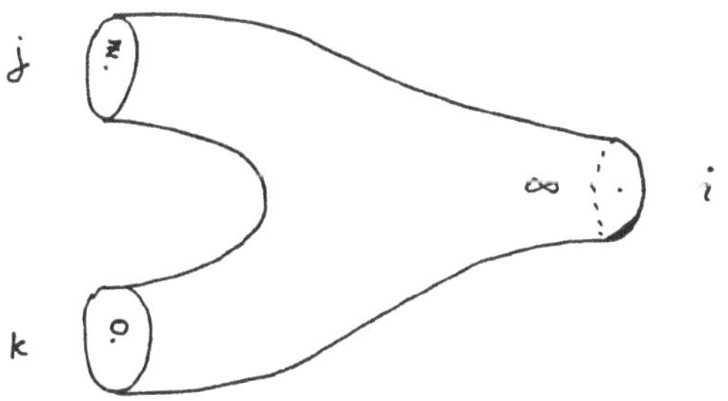

three-holed sphere with rep spaces on the three holes

Placing one of the holes about the point at ∞ we can define the Virasoro generators acting on the Hilbert space $\mathcal{H}_i$ at ∞ by:

$$L_n^{(\infty)} = \oint_{C_\infty} \zeta^{n+1} T(\zeta) d\zeta \ ,$$

But, since T is analytic, these can be deformed to generators around zero and z

$$\begin{aligned} L_n^{(\infty)} &= \oint_0 \zeta^{n+1} T(\zeta) + \oint_z \zeta^{n+1} T(\zeta) \\ &= L_n(0) + \sum_{k=0}^{\infty} \begin{pmatrix} n+1 \\ k \end{pmatrix} z^{n+1-k} L_{k-1}(z) \ . \end{aligned} \tag{2.5}$$

The chiral vertex operators commute with contour deformation, so we look for operators that satisfy:

$$L_n(\infty) \begin{pmatrix} i \\ jk \end{pmatrix}_z (\beta \otimes \gamma) = \begin{pmatrix} i \\ jk \end{pmatrix}_z \left[\left(\sum \begin{pmatrix} n+1 \\ k \end{pmatrix} z^{n+1-k} L_{k-1}^{(z)} \beta \right) \otimes \gamma + \beta \otimes L_n(0) \gamma \right] \tag{2.6}$$

for any states β, γ. This equation can be interpreted as follows. Think of $L_n(z)$ as a set of Virasoro operators acting on a Hilbert space at z, $\mathcal{H}_z$. Then $L_n(z) \otimes L_m(0)$ acts on the Hilbert space $\mathcal{H}_z \otimes \mathcal{H}_0$. The operators $L_n(\infty)$ act on the tensor product $\mathcal{H}_z \otimes \mathcal{H}_0$. They satisfy the Virasoro algebra with the *same* value of the central charge as $L_n(z)$. Therefore, equation (2.5) defines a map Δ_z from the Virasoro algebra, $\mathcal{A}$ to $\mathcal{A} \otimes \mathcal{A}$

$$\Delta_z(L_n) = 1 \otimes L_n + \sum_{k=0}^{\infty} \binom{n+1}{k} z^{n+1-k} L_{k-1} \otimes 1 .$$

This "comultiplication" allows us to take tensor products of Virasoro modules with given central charge. Then CVO's are "intertwining operators" for this notion of tensor product. (More on this below.) The above considerations generalize to arbitrary chiral algebras.

We must specify the z-dependence of these operators completely and this leads to the condition that

$$\frac{d}{dz} \binom{i}{jk}_z (\beta \otimes \gamma) = \binom{i}{jk}_z (L_{-1}\beta \otimes \gamma) \tag{2.7}$$

In RCFT's there is a finite-dimensional space, V^i_{jk} of operators satisfying (2.6) and (2.7) and we take these equations as our final definition of the CVO's. The connection to our previous description is that

$$\binom{i}{jk}_z (\beta \otimes \cdot) = \Phi^{j,\beta}_{ik}(z)$$

• Exercise 2.9 Prove the equivalence of these two definitions.

The superiority of our final definition is evident since we can now understand more clearly in terms of the formula (2.6) the statement that the CVO is an operator associated to a 3-holed sphere. Furthermore, it suggests a natural generalization, since we can consider more complicated situations - say, a 4-holed sphere. There will be a finite dimensional vector space $V^i_{jk\ell}$ of operators

$$\mathcal{H}_j \otimes \mathcal{H}_k \otimes \mathcal{H}_\ell \longrightarrow \mathcal{H}_i$$

which commute with contour deformation on the surface:

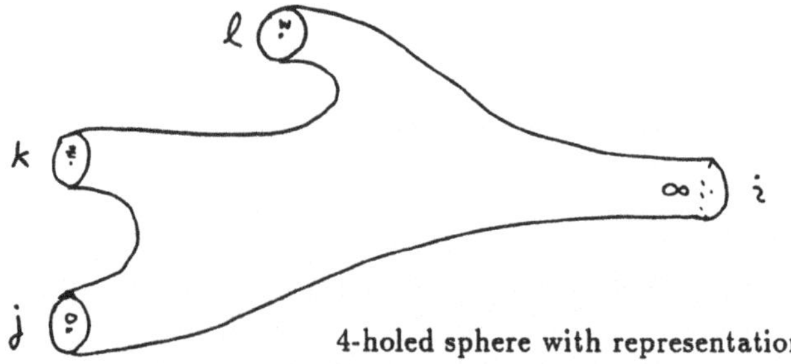

4-holed sphere with representations at each hole

The space of these operators is the same as the space of conformal blocks. This must be true since they are determined by the same equations (which follow from contour deformation arguments) used in more standard descriptions of conformal field theory [1][23]. The new spaces $V^i_{jk\ell}$ can be understood in terms of the simpler spaces of 3-point couplings. Geometrically, we can represent the 4-holed sphere as sewn 3-holed spheres. Analytically, we can use completeness of states to write operators in $V^i_{jk\ell}$ as compositions of CVO's.

Each sewing has a corresponding composition of CVO's and a corresponding decomposition of $V^i_{jk\ell}$ into simpler spaces:

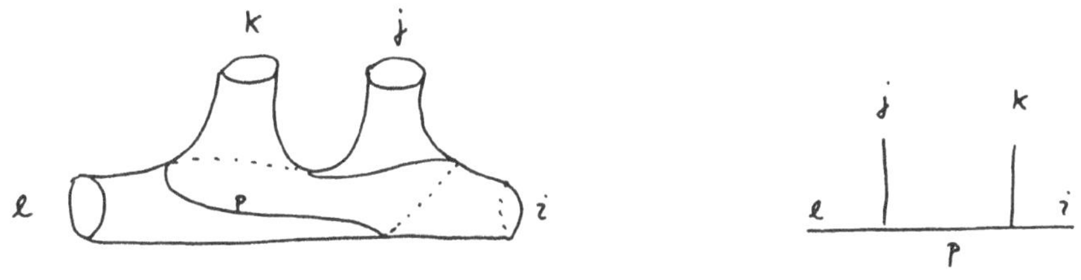

$$V^i_{jkl} \cong \oplus V^l_{jp} \otimes V^p_{ki}$$

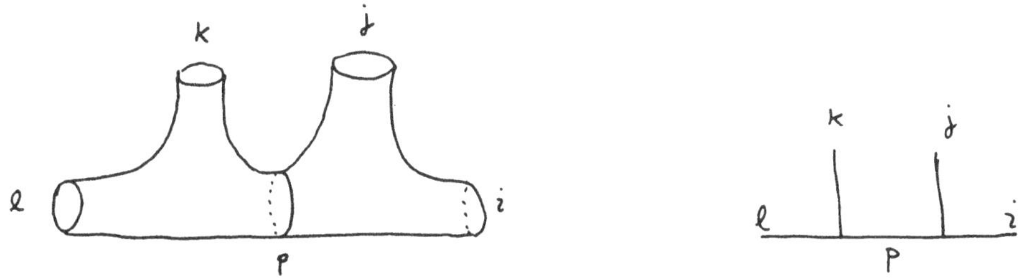

$$V^i_{jkl} \cong \oplus V^l_{kp} \otimes V^p_{ji}$$

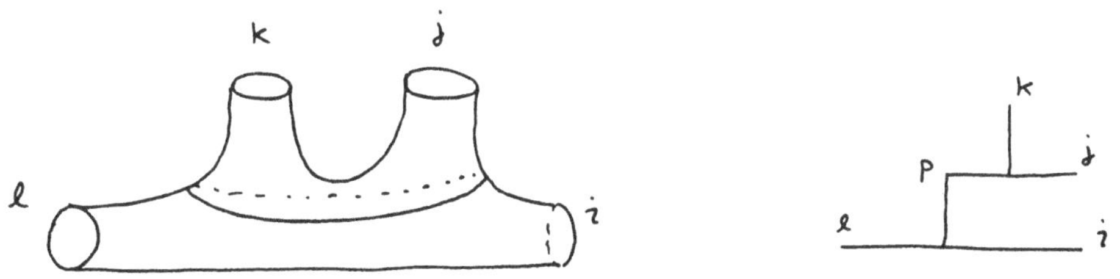

$$V^i_{jkl} \cong \oplus V^l_{pi} \otimes V^p_{kj}$$

Note that each of the sewings corresponds to a different asymptotic region of Teichmüller space. The general construction is the following - any ϕ^3 diagram can be thickened to give a surface - we can put FN length/twist coordinates on that surface and the region with small lengths corresponds to a region in Teichmuller space. In the asymptotic regions of Teichmuller space where the length coordinate goes to infinity the Riemann surface looks like a ϕ^3-diagram. In this limit the amplitudes of the conformal field theory and the conformal blocks have poles. The leading singularity corresponds to keeping only one intermediate state in the corresponding channel.

Thus different sewings simply correspond to different bases for $V^i_{jk\ell}$. The braiding/fusing isomorphisms express the relationships between these sewings. They are computed from the projectively flat connection on moduli space - according to the picture of the Friedan-Shenker modular geometry [24].

Finally we need the following remark - The compositions described so far only give us $g = 0$ surfaces. For CVO's of type $\underline{i \,|^j\, i}$ we can sew to get:

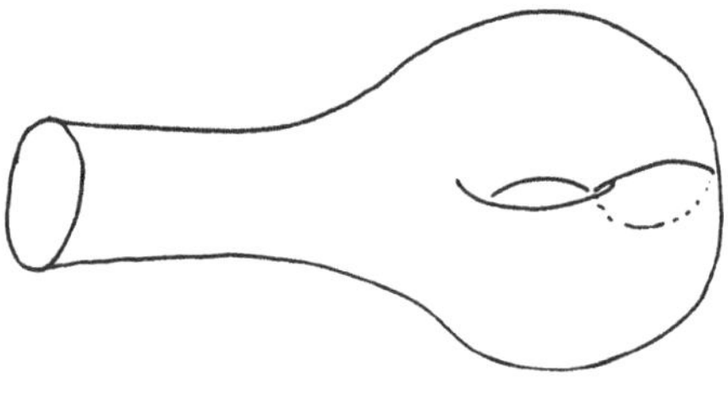

one holed torus obtained by sewing

The space of such conformal blocks with channel i will be the space V^i_{ji}, and the space of all one-point blocks is $\oplus_i V^i_{ji}$. In formulas, if we put a state β at a puncture on the torus we may form

$$\chi^{j,\beta}_i(z) = Tr q^{L_0 - c/24}\left(\Phi^{j,\beta}_{ii}(z)\right)(dz)^{\Delta(\beta)} \tag{2.8}$$

Here z is a point on the complex plane, but the trace essentially identifies $z \sim qz$ so that we actually compute a torus amplitude. If β is a Virasoro primary these blocks form a representation of the modular group with the matrix:

$$S(j) : \oplus_i V^i_{ji} \longrightarrow \oplus_i V^i_{ji} \tag{2.9}$$

In terms of sewings we are relating the following two diagrams

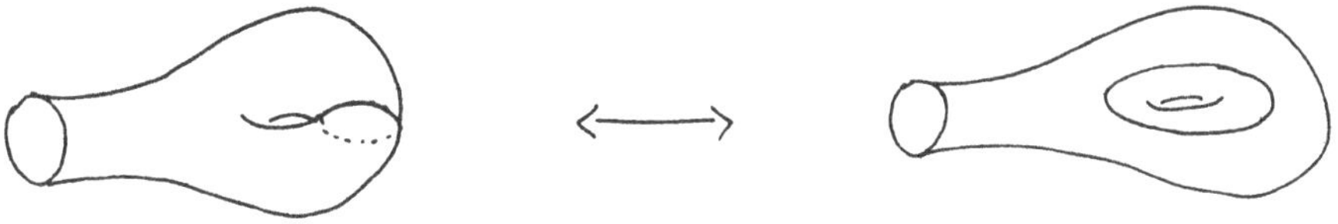

sewings for S

All these remarks generalize. As first emphasized by Friedan and Shenker [24], to every Riemann surface Σ we associate a vector space of conformal blocks $\mathcal{H}(\Sigma)$. This space is intrinsic but can be expressed in terms of the V^i_{jk} in many ways. Each such expression may be associated with a dual diagram. (Which, by its associated pants decomposition is correlated with an asymptotic region of Teichmüller space). Different decompositions of the *same* vector space must be related by isomorphisms. The specific isomorphisms follow from the existence of a projectively flat connection on moduli space. These isomorphisms are known as duality transformations.

An important point is that, in RCFT, all duality transformations can be expressed in terms of a finite number of basic duality transformations. Thus, we need only deal with a finite amount of data. This statement is intuitively obvious. It can be proved [25] that all sewings can be obtained from one another by the two basic moves

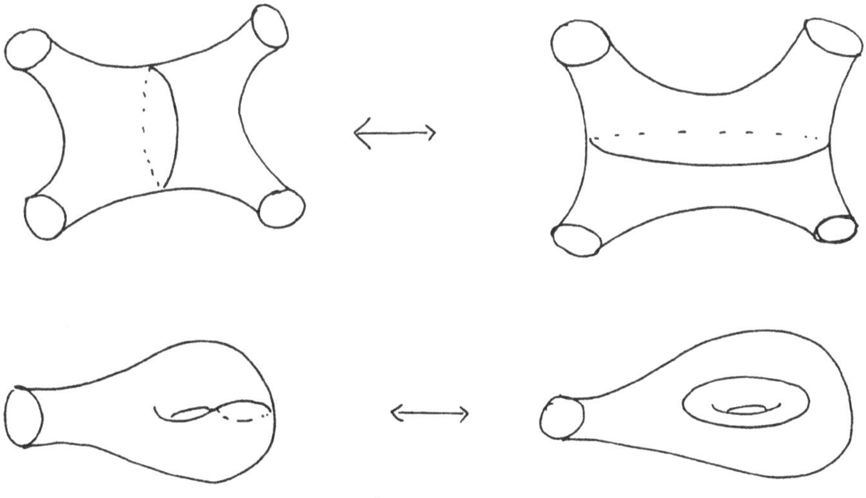

moves on four holed sphere and 1 holed torus

From this, taking into account twists around tubes one sees that all duality transformations can be written in terms of F,B,S and $e^{2\pi i c/24}$.

• Exercise 2.10 *Simple Moves.* Decompose the following move ("S for the two-point function") into steps of simple moves:

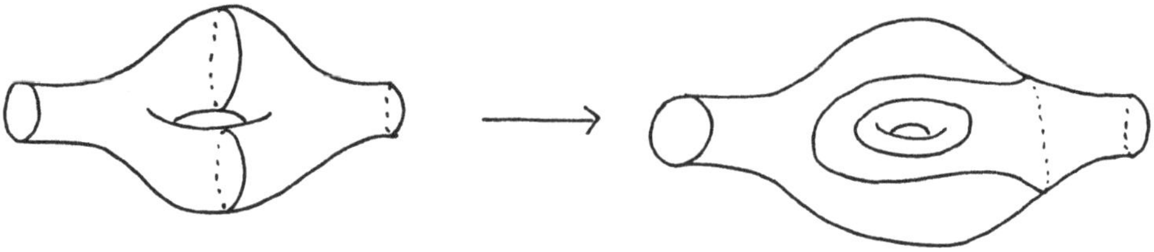

3. Duality Identities

The transformations F, B, S satisfy a large number of nontrivial identities. These identities can be understood in three ways:

a.) The algebra of operators Φ must be consistent.

b.) The monodromies of conformal blocks form representations of the modular group (duality groupoid).

c.) Different paths of the basic transformations F, B, S relate the same basis of blocks. Thus the identities are intimately connected with the geometry of moduli space.

The simplest example of an identity is the Yang-Baxter relation because it follows immediately from the exchange algebra of the Φ operators. Consider the following sewings for the 5-point function:

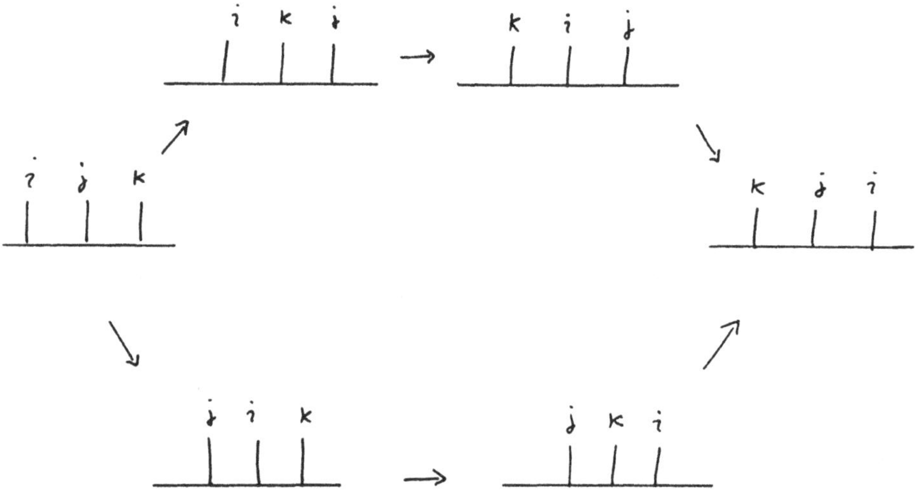

hexagon

implying an equation of the form $BBB = BBB$ (see below).

• Exercise 3.1 *Yang-Baxter Equation.* Derive the Yang-Baxter equation for the B-matrix by considering the product of three chiral vertex operators and demanding consistency of the braiding algebra.

It is very useful to introduce another pictorial formalism for deriving relations between braiding/fusing matrices. We imagine the the braiding and fusing matrices are like amplitudes between conformal blocks with "time" flowing upward as in the following picture:

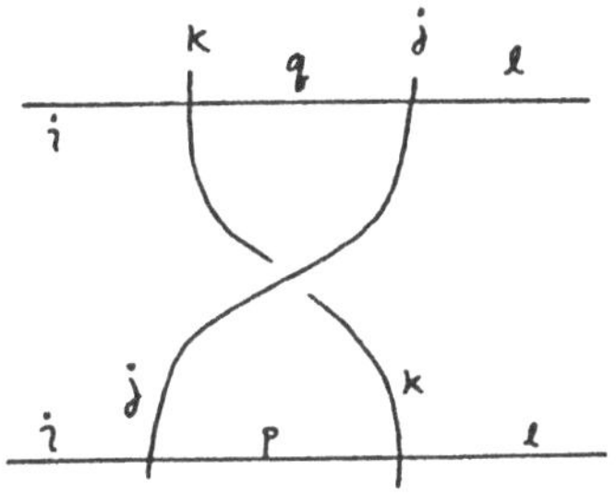

$$\sim \quad B_{pq}\begin{bmatrix} j & k \\ i & l \end{bmatrix}$$

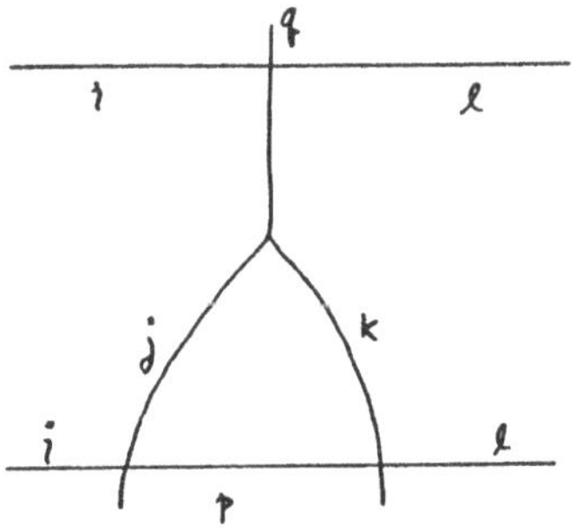

$$\sim \quad F_{pq}\begin{bmatrix} j & k \\ i & l \end{bmatrix}$$

Other pictures for B, F

(In the 3 dimensional point of view we will see that this interpretation of time can be taken quite literally.) Then we can picture the Yang-Baxter relations as follows:

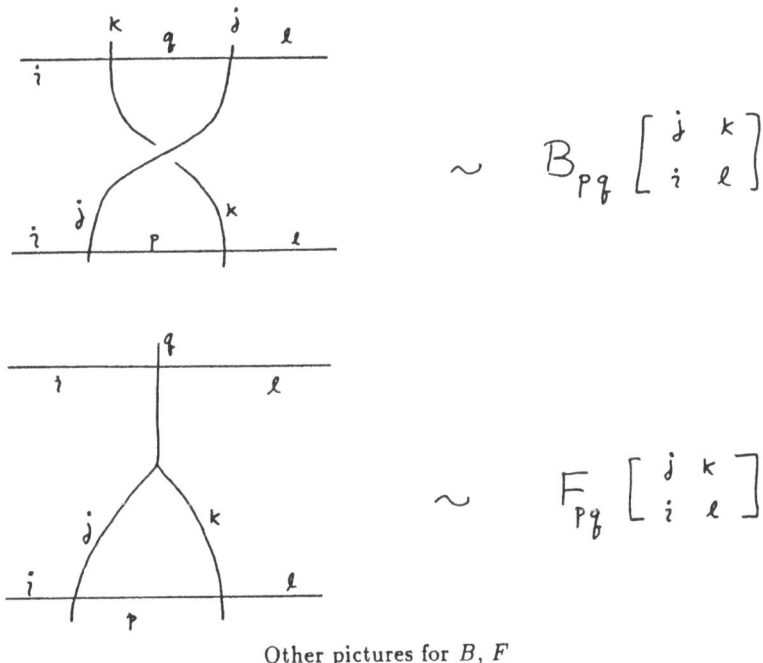

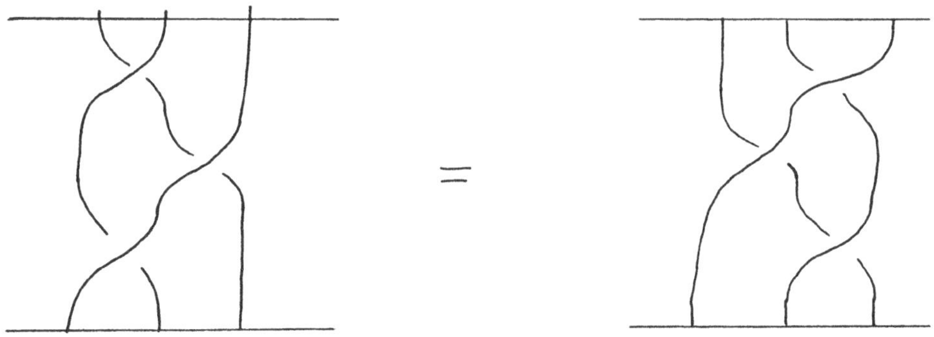

usual picture for Yang-Baxter

Another such identity is the braiding/fusing or pentagon identity which, in terms of duality diagrams may be represented as:

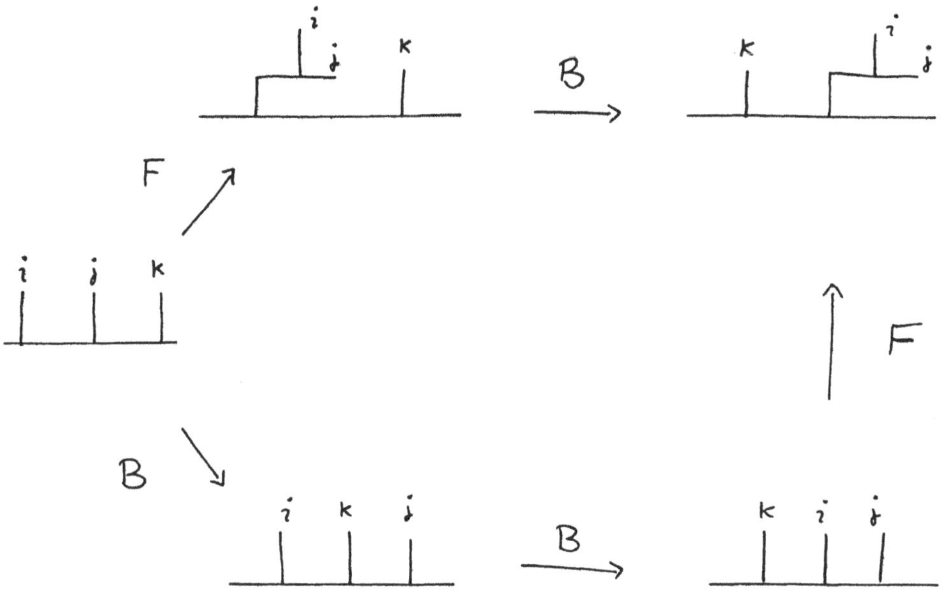

pentagon of dual diagrams

or in the other pictorial notation may be represented as:

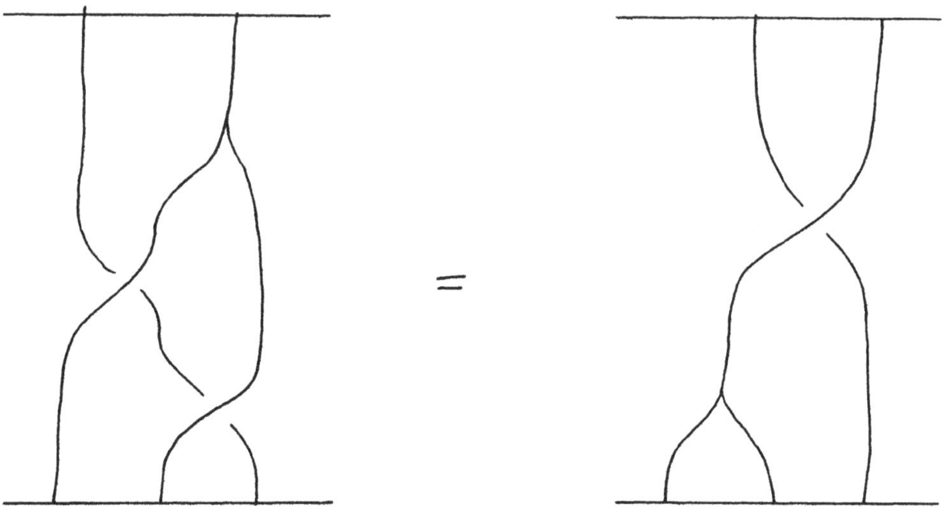

braiding/fusing

Clearly, by looking at more and more complicated graphs we will obtain more and more complicated identities. These identities can be neatly characterized as follows. Form a simplicial complex whose different vertices represent different ϕ^3 decompositions of conformal blocks. Join two vertices, if they are related by a simple move B or F. Every loop on the resulting one-complex gives a relation on duality matrices.

• Exercise 3.2 *The duality complex.* Write out the simplicial complex for the five-point function. Keep initial and final representations fixed in all moves. (Warning: This takes some time.)

These identities, and their graphical relations are a great deal of fun to play with - but there are a large number of indices and one can only understand them once he has worked them out for himself. Therefore we urge the reader to work through the following exercise.

• Exercise 3.3 *Systematic Derivation of Equations.* Consider the 4-point function complex. Show that the closed loop of moves:

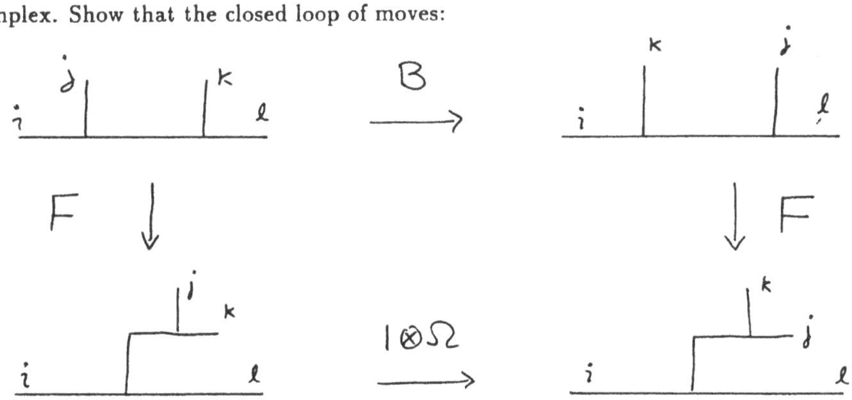

leads to the equation

$$\sum_{p'} B_{pp'}\begin{bmatrix} j & k \\ i & l \end{bmatrix}(\epsilon) F_{p'q}\begin{bmatrix} k & j \\ i & l \end{bmatrix} = F_{pq}\begin{bmatrix} j & k \\ i & l \end{bmatrix} e^{-i\pi\epsilon(\Delta_k+\Delta_j-\Delta_q)} \tag{3.1}$$

The ϵ denotes the sense of the braiding. Note that this identity shows that the eigenvalues of B are the square roots of mutual locality factors. Interpret (3.1) graphically:

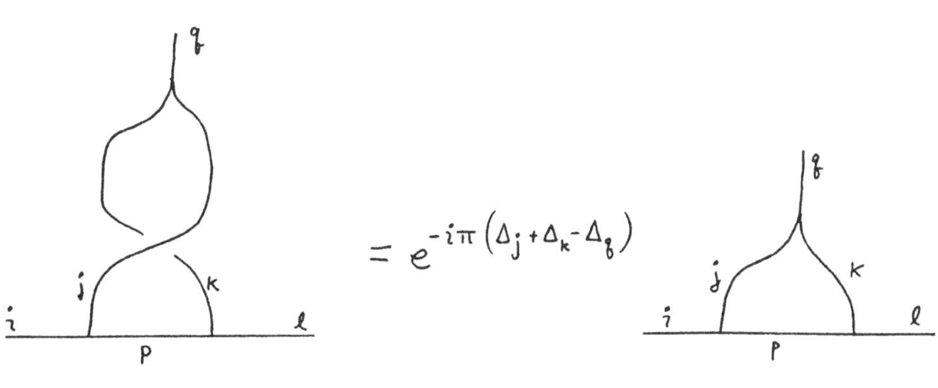

Write similar equations involving F^{-1}, B^{-1}. Now consider the braiding/fusing identity:

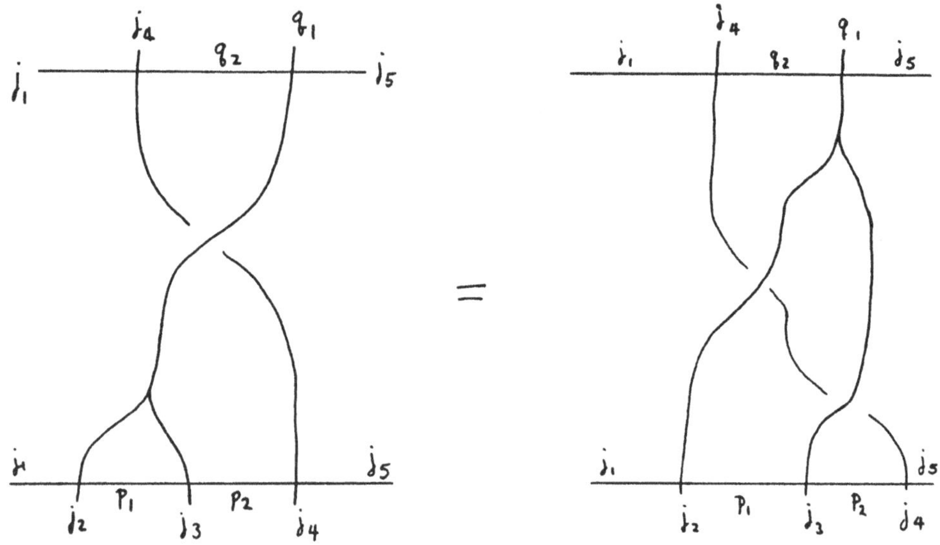

Write the corresponding equation:

$$F_{p_1q_1}\begin{bmatrix} j_2 & j_3 \\ j_1 & p_2 \end{bmatrix} B_{p_2q_2}\begin{bmatrix} q_1 & j_4 \\ j_1 & j_5 \end{bmatrix}(\epsilon) = \sum_s B_{p_2s}\begin{bmatrix} j_3 & j_4 \\ p_1 & j_5 \end{bmatrix}(\epsilon) B_{p_1q_2}\begin{bmatrix} j_2 & j_4 \\ j_1 & s \end{bmatrix}(\epsilon) F_{sq_1}\begin{bmatrix} j_2 & j_3 \\ q_2 & j_5 \end{bmatrix} \tag{3.2}$$

Now specialize this equation by putting $j_5 = 0$, the identity representation, and derive:

$$F_{pq}\begin{bmatrix} j & k \\ i & l \end{bmatrix} = e^{-i\pi\epsilon(\Delta_i+\Delta_k-\Delta_p-\Delta_q)} B_{pq}\begin{bmatrix} j & l \\ i & k \end{bmatrix}(\epsilon) \tag{3.3}$$

Now use the relation (draw the picture!):

$$\sum_{p'} B_{pp'}\begin{bmatrix} j & k \\ i & l \end{bmatrix}(\epsilon) B_{p'q}\begin{bmatrix} k & j \\ i & l \end{bmatrix}(-\epsilon) = \delta_{p,q} \tag{3.4}$$

to derive the following two consequences. First

$$\sum_{p'} F_{pp'}\begin{bmatrix} j & k \\ i & l \end{bmatrix} F_{p'q}\begin{bmatrix} l & k \\ i & j \end{bmatrix} = \delta_{p,q} \tag{3.5}$$

and next

$$\sum_{p'} e^{2\pi i\epsilon\Delta_p} B_{pp'}\begin{bmatrix} j & l \\ i & k \end{bmatrix}(\epsilon) e^{2\pi i\epsilon\Delta_{p'}} B_{p'q}\begin{bmatrix} l & j \\ i & k \end{bmatrix}(\epsilon) = \delta_{p,q} e^{2\pi i(\Delta_i+\Delta_k)} \tag{3.6}$$

Interpret (3.5) graphically as a relation following from a closed loop of dual diagrams on the duality complex:

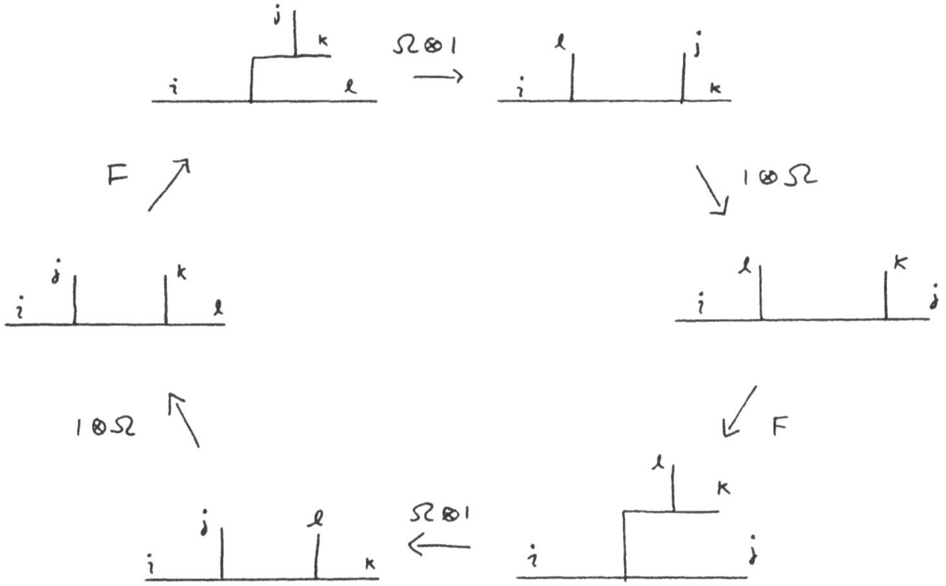

Note that the closed loop is a hexagon.

Note that the determinant of (3.6) gives an interesting constraint on the weights of a rational conformal field theory [26].

Now substitute (3.3) back into (3.1) to get

$$\sum B_{ps}\begin{bmatrix} j & k\\ i & l\end{bmatrix}(\epsilon)e^{i\pi\epsilon\Delta_s}B_{sq}\begin{bmatrix} i & j\\ k & l\end{bmatrix}(\epsilon) = e^{i\pi\epsilon\Delta_p}B_{pq}\begin{bmatrix} i & k\\ j & l\end{bmatrix}(\epsilon)e^{i\pi\epsilon\Delta_q}e^{-2\pi i\epsilon\Delta_j} \qquad (3.7)$$

Write (3.7) in terms of F and interpret in terms of dual diagrams via the hexagon identities as follows:

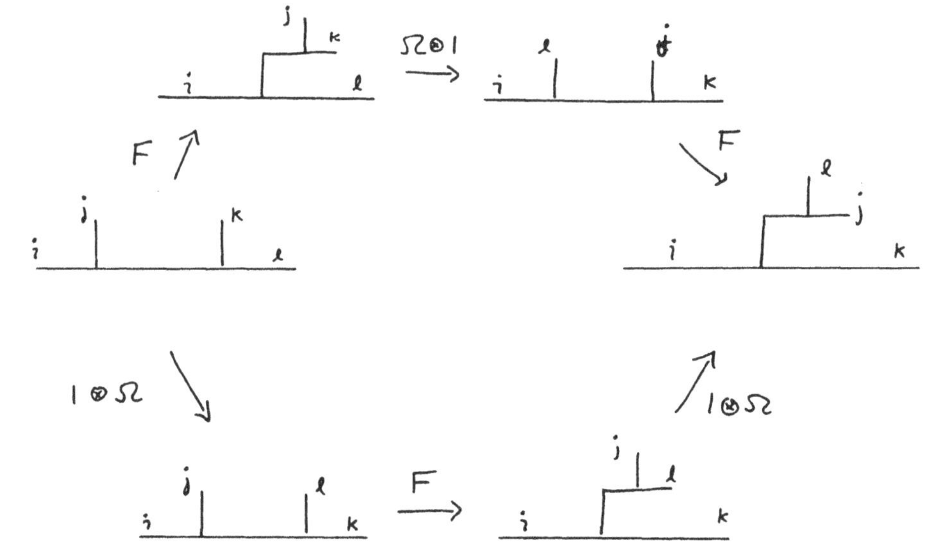

We have thus found three hexagon identities. Show that any one of these hexagons can be deduced from the other two, so there are only two independent hexagons. We will adopt the last two we have just derived.

Now use the equation for B in terms of F to rewrite the braiding/fusing identity:

$$\sum_s F_{p_2 s}\begin{bmatrix} j & k\\ p_1 & b\end{bmatrix} F_{p_1 l}\begin{bmatrix} i & s\\ a & b\end{bmatrix} F_{sr}\begin{bmatrix} i & j\\ l & k\end{bmatrix} = F_{p_1 r}\begin{bmatrix} i & j\\ a & p_2\end{bmatrix} F_{p_2 l}\begin{bmatrix} r & k\\ a & b\end{bmatrix} \tag{3.8}$$

Interpret this identity as a pentagonal loop of dual diagrams.

Finally, write out the equation corresponding to the figure:

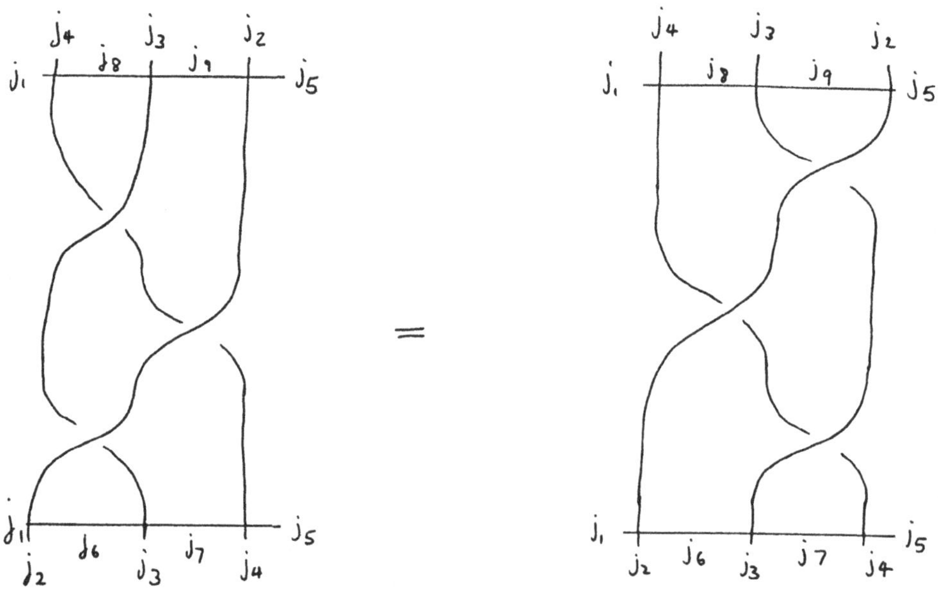

pictorial representation of the Yang-Baxter equation

giving the Yang-Baxter equation:

$$\sum_p B_{j_6 p}\begin{bmatrix} j_2 & j_3\\ j_1 & j_7\end{bmatrix}(\epsilon) B_{j_7 j_9}\begin{bmatrix} j_2 & j_4\\ p & j_5\end{bmatrix}(\epsilon) B_{p j_8}\begin{bmatrix} j_3 & j_4\\ j_1 & j_9\end{bmatrix}(\epsilon) =$$
$$\sum_p B_{j_7 p}\begin{bmatrix} j_3 & j_4\\ j_6 & j_5\end{bmatrix}(\epsilon) B_{j_6 j_8}\begin{bmatrix} j_2 & j_4\\ j_1 & p\end{bmatrix}(\epsilon) B_{p j_9}\begin{bmatrix} j_2 & j_3\\ j_8 & j_5\end{bmatrix}(\epsilon) \tag{3.9}$$

Show that by putting $j_1 = 0$ or $j_5 = 0$ we recover the two hexagon identities. Show moreover that the full Yang-Baxter equation may be deduced from the pentagon and hexagon identities. (Hint: Bring all the B matrices to one side of the equation. Insert $FF^{-1} = 1$ and use the braiding/fusing identity repeatedly.)

The two hexagons and the pentagon are the fundamental genus zero identities.

• Exercise 3.4 *Gauge Choices*. Note that we did not specify the normalizations $||\Phi^{i}_{jk}||$ in the definition of the chiral vertex operators. How do the F, B matrices change under a rescaling by λ^{i}_{jk}? We refer to such a change as a change of gauge. Show that the polynomial equations of the pervious exercise are gauge invariant.

• Exercise 3.5 *Symmetries of the F matrix*. Show, in the case of the discrete series that the matrices satisfy:

$$F_{pp'}\begin{bmatrix} j & k \\ i & l \end{bmatrix} = F_{pp'}\begin{bmatrix} i & l \\ j & k \end{bmatrix} = F_{pp'}\begin{bmatrix} l & i \\ k & j \end{bmatrix}$$

Show that these symmetries are gauge invariant. Interpret these symmetries pictorially. In theories other than the minimal models these symmetries typically hold only up to signs. (These signs are described precisely in [15].) When a special choice of gauge is made these matrices sometimes have much more symmetry, similar to the tetrahedral symmetries of Racah coefficients (see below).

If we move on to higher genus we get new identities on duality matrices. For example from the one-point block we obtain, as described above, $S_{ii'}(j)$. As is well-known, when the torus is represented as the quotient of the plane by a lattice the square of the transformation S is a 180 degree rotation around the puncture at z, so $log z \to -log z$ and we have (in the case where all the representations are self conjugate)

$$S^2(j) = \pm C e^{-i\pi\Delta(\beta)} \tag{3.10}$$

where C is the conjugation matrix on representations and the sign is very similar to the quantity ξ discussed above, and again arises from the symmetry or antisymmetry of a coupling. Similarly we have

$$(ST)^3 = S^2 \tag{3.11}$$

where $T_{jk} = e^{2\pi i(\Delta_j - \frac{c}{24})}\delta_{jk}$. Moving on to several punctures on the torus a new element appears. We may always fix one operator at the standard basepoint, but then there is nontrivial monodromy under the diffeomorphisms which move each of the points around the nontrivial homology cycles of the surface, and around each other.

For a famous example we have for the 2-point function.

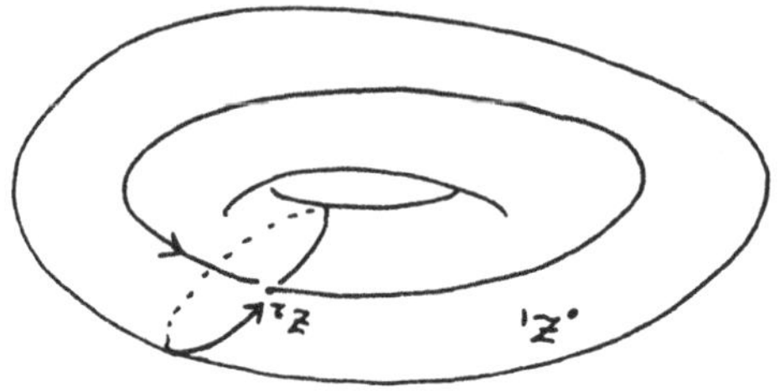

two points on a torus with a, b curves

As indicated before, each of these monodromies may be expressed in terms of F, B, S. Then the relations of the modular group of the n-holed torus imply identities on duality matrices. For example denoting the monodromies of conformal blocks obtained by dragging one operator around the a, b cycles by the same letters a, b we have

$$Sa\, S^{-1} = b\ .$$

The a, b monodromies can be expressed in terms of F, B matrices. Thus, the above equation implies a new identity relating F, B, S. Clearly, these considerations extend to any number of punctures at any genus.

Below we'll begin to bring some order to this chaos of identities. But first let us show that some of these identities can lead to very nice consequences indeed.

For example, the relation $SaS^{-1} = b$ leads to a proof of Verlinde's formula [21]:

$$N_{ijk} = \sum_{p} \frac{S_{ip}S_{jp}S_{kp}}{S_{0p}}$$

(Here $S = S(0)$, i.e. the transformation matrix on vacuum characters.) To prove this one looks at the blocks:

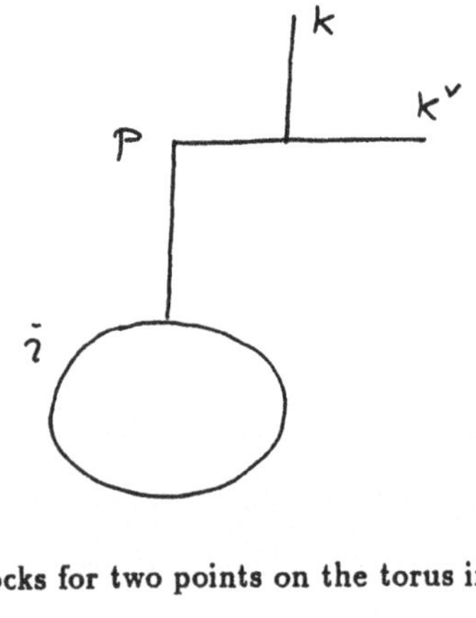

blocks for two points on the torus in one basis

and computes the a, b monodromies for

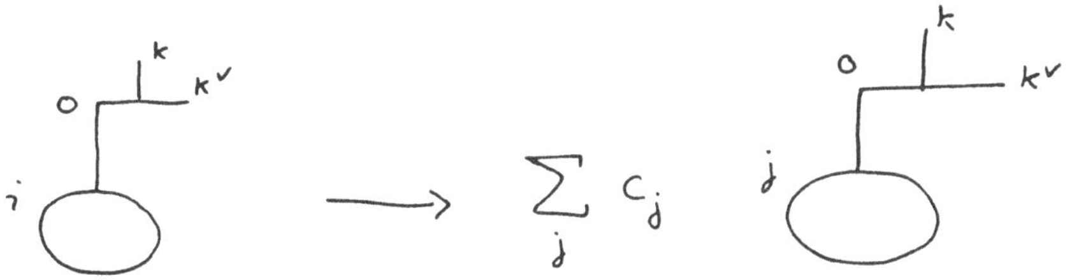

transformation of these blocks

Then using the fact that S converts a to b monodromies gives the result. Details are left to the following exercise:

• Exercise 3.6 *Proof of Verlinde's Formula*

Verlinde conjectured that the matrix $S = S(0)$ diagonalizes the fusion rule algebra in [21]. There are now, superficially, three different proofs of this statement [13], [27] [28] but all are really equivalent. We will return to a version of Witten's proof later. For now, we proceed with the least elegant, but most straightforward approach.

Consider the discrete series for simplicity. Show that Verlinde's formula

$$\frac{S_{ij}S_{jk}}{S_{0j}} = \sum_{l} N_{ikl} S_{lj}$$

follows from the modular relation $SaS^{-1} = b$ by considering the submatrix element illustrated below:

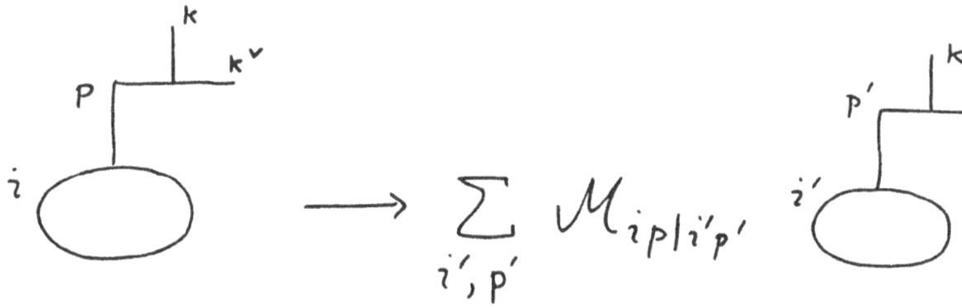

restrict to:

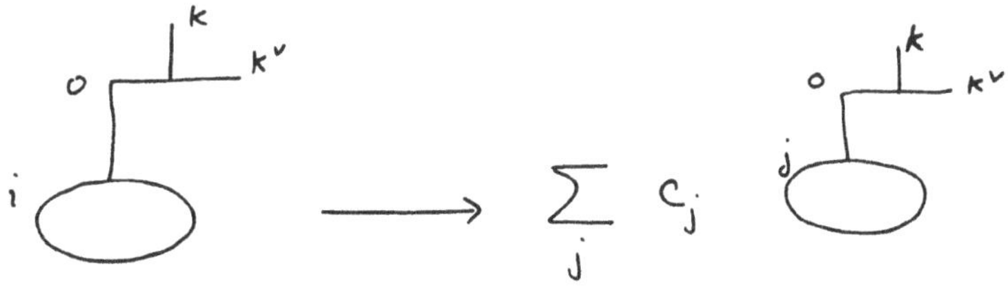

submatrix needed for a proof of Verlinde's formula

Relate the above basis of blocks to the basis

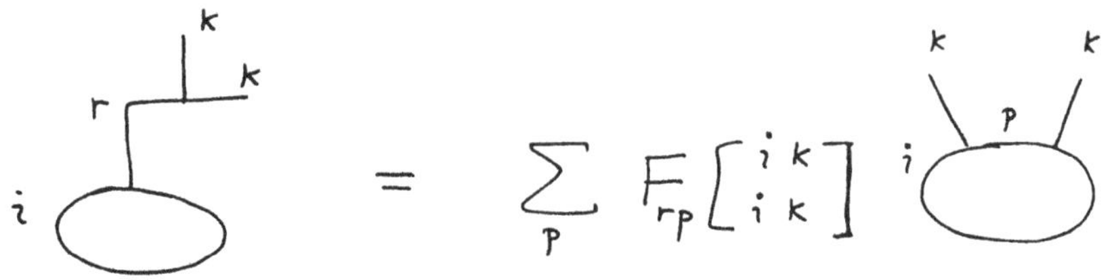

a different basis for the two point function on the torus

Show that the a monodromy in this basis is just $e^{2\pi i(\Delta_i - \Delta_p)}$. Use the identity

$$F_{k0}\begin{bmatrix} i & i \\ j & j \end{bmatrix} F_{0j}\begin{bmatrix} k & i \\ k & i \end{bmatrix} = F_{00}\begin{bmatrix} i & i \\ i & i \end{bmatrix} \equiv F_i$$

to simplify the b-monodromy in the original basis and obtain:

$$\sum_j S_{ij} \frac{\left(B\left[\begin{smallmatrix} j & k \\ j & k \end{smallmatrix}\right] B\left[\begin{smallmatrix} k & j \\ j & k \end{smallmatrix}\right] \right)_{00}}{F_k} S_{jl} = N_{ikl}$$

From this derive Verlinde's formula, and show also that

$$\frac{S_{jk}}{S_{00}} = \frac{\left(B\left[\begin{smallmatrix} j & k \\ j & k \end{smallmatrix}\right] B\left[\begin{smallmatrix} k & j \\ j & k \end{smallmatrix}\right] \right)_{00}}{F_k F_j}$$

Note especially the formula for $j = 0$. An argument analogous to the above holds for an arbitrary RCFT.

From Verlinde's formula we can deduce many interesting things. As a simple example we can describe the fusion rules for Kac-Moody algebras in a rather elegant way [21]:

• Exercise 3.7 *Geometry of the Kac-Moody Fusion Rules.* From Verlinde's formula and the formula for the matrix S of the Weyl-Kac characters show that the one-dimensional representations of the fusion rule algebra:

$$\phi_m \phi_l = \sum_i N^i_{ml} \phi_i$$

in the level k WZW theory are just given by

$$\lambda^{(j)}_m = ch_m\left(2\pi \frac{\mu_j + \rho}{k+h}\right)$$

Here ch_m is the character in the representation m, μ_j is the highest weight of the representation j, ρ is the Weyl vector, i.e., half the sum of the positive roots, and h is the dual Coxeter number.

Using this result characterize the fusion rule algebra for the level k WZW theory in terms of reflections in the hyperplane $x \cdot \psi = k+1$, where ψ is the highest root.

• Exercise 3.8 *Verlinde's Dimension Formula*

a.) Go to the dual basis for the vacuum characters of the form

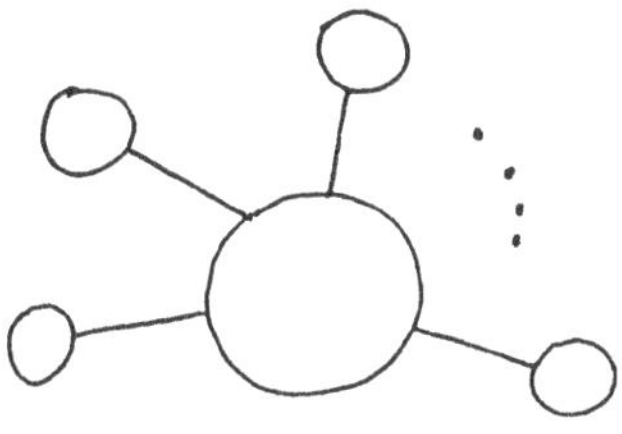

A circle with mirrors emanating from it.

and use Verlinde's formula to show that the dimension of the Friedan-Shenker vector bundle is [21]

$$dim\mathcal{H}(\Sigma_g) = \sum_p (\frac{1}{S_{0p}})^{2(g-1)}$$

b.) Go to the dual basis for the n-point functions of the form

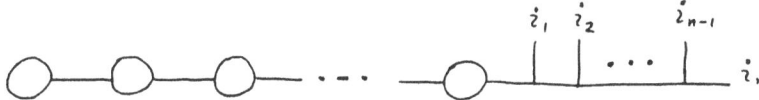

to show that the formula for the case with punctures in representations $i_1, \ldots, i_n$ is given by

$$dim\mathcal{H}(\Sigma_g;(P_1,i_1),\ldots(P_n,i_n)) = \sum_p (\frac{1}{S_{0p}})^{2(g-1)} \frac{S_{i_1 p}}{S_{0p}} \cdots \frac{S_{i_n p}}{S_{0p}}$$

c.) Verify that $S^2 = C$ guarantees the dimensions behave as expected under sewing.

d.) Substitute the Kac-Peterson formula for S_{ij} into the formula of part (a) to show that for level k WZW theory with simple and simply connected group G we have [29]:

$$dim(\mathcal{H}(\Sigma_g)) = (k+h)^{g-1}(|\Lambda_{rt}^*/\Lambda_{rt}|)^{g-1} \sum_\lambda \frac{1}{\prod_{\alpha\in\Delta}(1-e^{i(\alpha,\theta_\lambda)})^{g-1}}$$

Here h is the dual Coxeter number, Λ_{rt} is the root lattice, Δ is the set of roots, the sum runs over weight vectors λ defining level k integrable highest weight representations of the current algebra, and $\theta_\lambda = 2\pi\frac{\lambda+\rho}{k+h}$ is the conjugacy class canonically associated to the Kac-Moody integrable representation λ. Verlinde has conjectured that this formula can be derived as a fixed point theorem, but such an interpretation has not yet been given.

e.) Write the formula explicitly for $\hat{SU}(2)_k$ and show that, as $k \to \infty$, we have $dim\mathcal{H}(\Sigma) \sim k^{3g-3}$. This behavior is very natural from the Chern-Simons gauge theory viewpoint explained below.

---

4. Completeness

In the previous section we said that all duality transformations are expressed in terms of a finite amount of data: F, B, S. However, there seemed to be a proliferation of identities. The completeness theorem states that, in fact the number of independent identities is finite.

From the exercises you know that a special case of the B-matrix is

$$\Omega^i_{jk} : V^i_{jk} \to V^i_{kj}$$

a pictorial representation of Ω

Its eigenvalues are just the square roots of mutual locality factors.

The basic genus zero identities are

1) The pentagon

$$\sum_s F_{p_2 s}\begin{bmatrix} j & k\\ p_1 & b\end{bmatrix} F_{p_1 l}\begin{bmatrix} i & s\\ a & b\end{bmatrix} F_{sr}\begin{bmatrix} i & j\\ l & k\end{bmatrix} = F_{p_1 r}\begin{bmatrix} i & j\\ a & p_2\end{bmatrix} F_{p_2 l}\begin{bmatrix} r & k\\ a & b\end{bmatrix} \tag{4.1}$$

2) The two hexagons

$$\Omega^m_{lk}(\epsilon) F_{mn}\begin{bmatrix} j & k\\ i & l\end{bmatrix}\Omega^l_{jk}(\epsilon) = \sum_r F_{mr}\begin{bmatrix} j & l\\ i & k\end{bmatrix}\Omega^i_{kr}(\epsilon) F_{rn}\begin{bmatrix} k & j\\ i & l\end{bmatrix} \tag{4.2}$$

3) At $g = 1$ there are 3-more identities:

$$S^2(j) = \pm C e^{-i\pi\Delta_j}$$

$$(ST)^3 = S^2$$

$$SaS^{-1} = b$$

Using these identities we can check all the relations on F, B, S following from duality on all surfaces. One would like to present the equations in the most economical possible way. In fact, the last torus equation $SaS^{-1} = b$, which is rather complicated when written out with all its indices, contains a great deal of redundant information. Some of the equations implied by $SaS^{-1} = b$ can be used to solve for $S(p)$ in terms of the braiding and fusing matrices and the normalization term $S_{00}(0)$. In the case of the discrete series, the explicit formula one finds is

$$S_{ij}(p) = S_{00}(0)e^{-i\pi\Delta_p} \frac{F_{i0}\begin{bmatrix} i & i \\ p & p \end{bmatrix}}{F_p F_{p0}\begin{bmatrix} j & j \\ j & j \end{bmatrix} F_{p0}\begin{bmatrix} i & i \\ i & i \end{bmatrix}} \sum_r B_{pr}\begin{bmatrix} i & j \\ i & j \end{bmatrix}(-) B_{r0}\begin{bmatrix} j & i \\ i & j \end{bmatrix}(-) \tag{4.3}$$

and a similar formula holds for an arbitrary RCFT. (The only complication in the general case are some signs measuring the antisymmetry of certain couplings.) This expression is a generalization to arbitrary p of the expression in [13][14]. A nontrivial computation (outlined in section seven below) shows that once this expression is substituted into the remaining equations implied by $SaS^{-1} = b$ one finds no new conditions on F, B. Hence, in specifying the fundamental equations, the above three torus equations can be replaced by the definition (4.3) together with the constraint of the first two torus equations, determining that S define a representation of the modular group.

• Exercise 4.1 *Example of the Ising model.* Check (4.3) in the Ising model. In this case we have three representations $1, \psi, \sigma$ with the famous fusion rule algebra:

$$\begin{aligned} \psi \times \psi &= 1 \\ \psi \times \sigma &= \sigma \\ \sigma \times \sigma &= 1 + \psi \end{aligned} \tag{4.4}$$

choose a gauge by demanding that:

$$F\begin{bmatrix} \sigma & \psi \\ \sigma & \psi \end{bmatrix} = F\begin{bmatrix} \psi & \sigma \\ \psi & \sigma \end{bmatrix} = F\begin{bmatrix} \psi & \psi \\ \sigma & \sigma \end{bmatrix} = F\begin{bmatrix} \sigma & \sigma \\ \psi & \psi \end{bmatrix} = 1 \tag{4.5}$$

Then show, either by solving the polynomial equations, or by using explicit conformal blocks that we have

$$\begin{aligned} F\begin{bmatrix} \psi & \psi \\ \psi & \psi \end{bmatrix} &= 1 \\ F\begin{bmatrix} \sigma & \psi \\ \psi & \sigma \end{bmatrix} &= -1 \\ F\begin{bmatrix} \sigma & \sigma \\ \sigma & \sigma \end{bmatrix} &= \frac{1}{\sqrt{2}}\begin{pmatrix} 1 & 1 \\ 1 & -1 \end{pmatrix} \\ B\begin{bmatrix} \sigma & \sigma \\ \sigma & \sigma \end{bmatrix}(-) &= \frac{e^{-i\pi/8}}{\sqrt{2}}\begin{pmatrix} 1 & i \\ i & 1 \end{pmatrix} \end{aligned} \tag{4.6}$$

And substitute these into (4.3) . Note, in particular that for the one-point function of ψ the block $\eta(\tau)(dz)^{1/2}$ gives $S(\psi) = e^{-i\pi/4}$, as predicted by (4.3) .

Strictly speaking – only the following cases have been carefully checked in all details: ($g = 0$, n holes), ($g = 1$, n holes), (g, $n = 0$). We have no doubt that the remaining cases will also work (an argument is given in [15]), but what is needed is a better understanding of why the result should be true which will lead to a more conceptual proof, which should handle all cases simultaneously.

Here we will describe part of the $g = 0$ case in detail. To begin recall the generators and relations for the modular group of the sphere with n holes. The generators are: Firstly, R_i = a Dehn twist around the ith hole. Equivalently, this is a transformation on a local choice of coordinate $dz \rightarrow e^{2\pi i}dz$. Secondly, ω_i = interchange holes i and $i + 1$ The action of the generators ω_i may be pictured as follows:

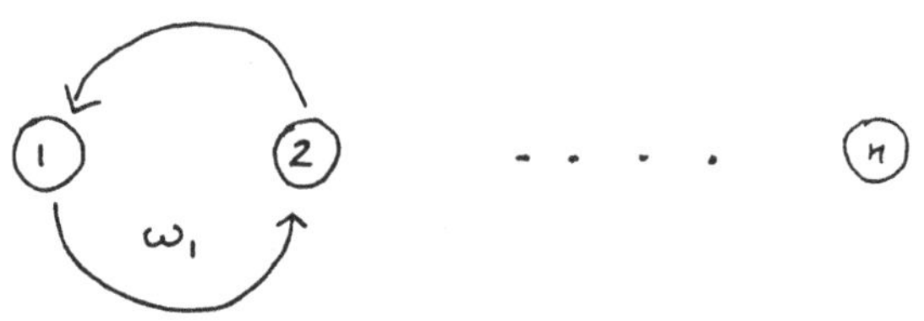

Illustrations of one of the generating modular transformations.

The idea of the proof is the following. Recall the simplicial complex from section 3 which is built by declaring that:

vertices $\longrightarrow$ dual diagrams

edges $\longrightarrow$ simple moves

Define a 2-complex by filling in all faces corresponding to pentagons/hexagons and - in the high genus case - the torus relations. There are no new relations, if the resulting complex is simply connected.

The question can be reduced - in a way which will be indicated below to checking the relations of the modular group. So let's worry about these. The relations we must check are:

A.

$$\omega_i\omega_j = \omega_j\omega_i \ |i - j| \geq 2$$

$$\omega_i R_j = R_j \omega_i$$

B.

$$\omega_i \omega_{i+1} \omega_i = \omega_{i+1} \omega_1 \omega_{i+1}$$

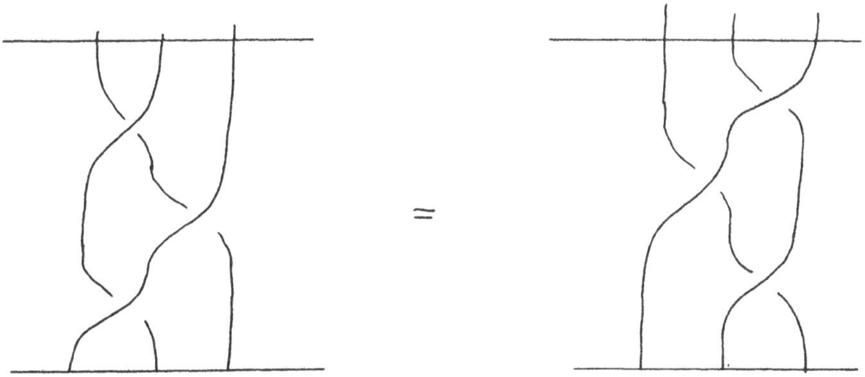

C.

$$\omega_1 \cdots \omega_{n-1}^2 \cdots \omega_1 = R_1^2$$

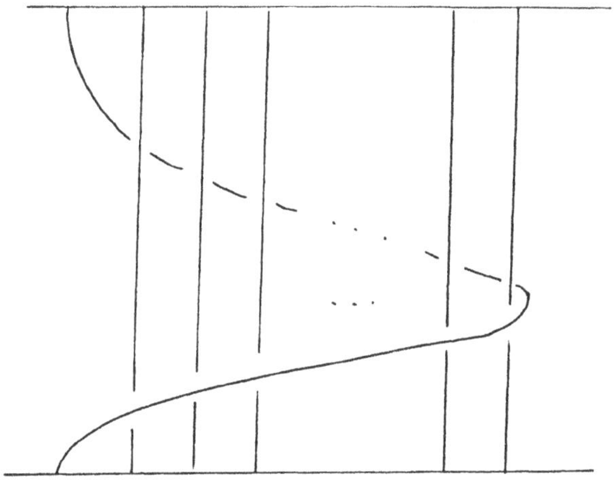

D.

$$(\omega_1 \cdots \omega_{n-1})^n = \Pi R_i$$

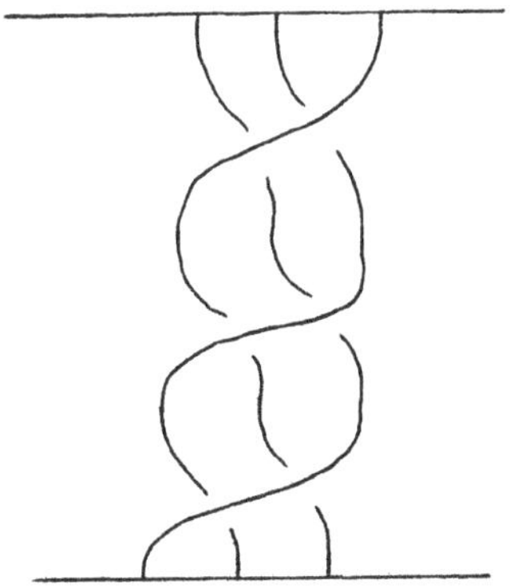

Now checking these relations is quite easy. We use the basis of blocks:

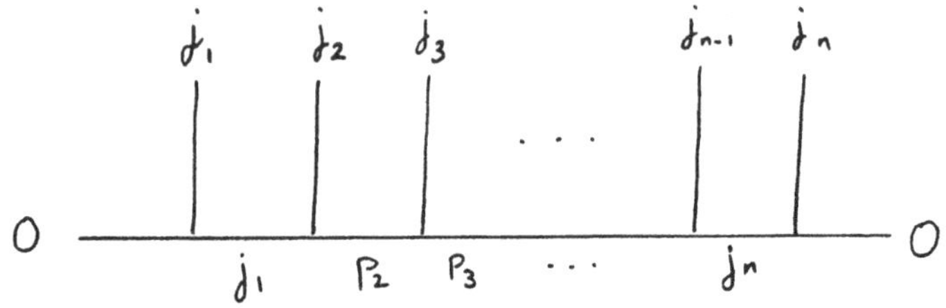

multiperipheral basis

So the representation is just:

$$\rho(R_k) = e^{2\pi i \Delta j_k}$$

$$\rho(\omega_k) = 1 \otimes \cdots \otimes B \begin{bmatrix} j_k & j_{k+1} \\ p_{k-1} & p_{k+2} \end{bmatrix} \otimes \cdots \otimes 1$$

Relation (A) is obviously satisfied. One easily checks that (B) follows from the Yang-Baxter relations. To check (C) we use braiding fusing:

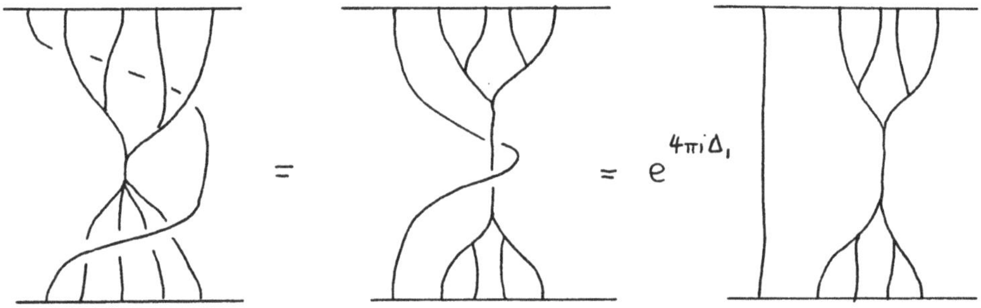

Finally we check (D) similarly.

• Exercise 4.2 *The barber pole.* Use the braiding/fusing identity and induction to verify the barber pole relation:

$$(\omega_1 \cdots \omega_{n-1})^n = \prod R_i$$

With considerably more work we can go on to check the modular relations at high genus. An example of a rather tractable one is:

• Exercise 4.3 *A Simple High-Genus Relation*

a.) Rewrite the equation $(ST)^3 = S^2$ as the equation $\alpha\beta\alpha = \beta\alpha\beta$ where α, β are Dehn-twists around the a, b cycles, respectively. (Hint: Show that $\alpha = T^{-1}$ and $\beta = TST$.)

b.) Verify geometrically the relation $\alpha\beta\alpha = \beta\alpha\beta$ in the modular group at any genus from the configuration of curves shown below:

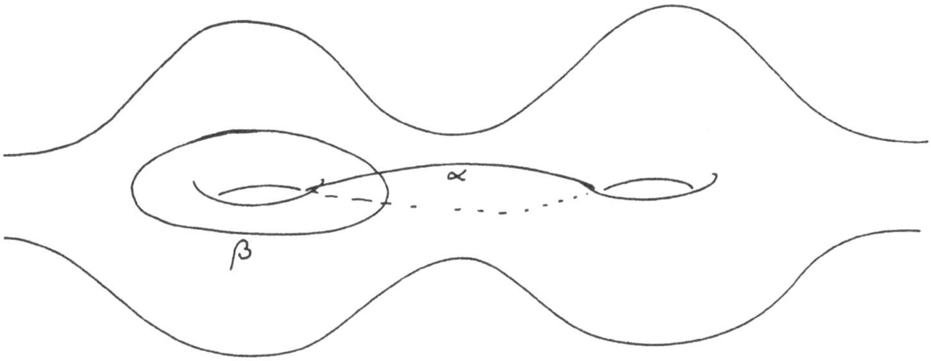

(Hint: Show that the product of Dehn twists $\alpha\beta\alpha^{-1}$ is a single Dehn twist around the image under α of the curve β.)

c.) Why is this not a new high genus relation on duality matrices?

One should still prove that it is enough to check the relations of the modular group. On the sphere, the argument is inductive in the number of external lines. The basic idea in the proof is to use the pentagon to show that there are no new identities from a set of duality transformations starting and ending in the multiperipheral basis. Then, it follows that every closed loop of transformations in the duality complex is homotopically equivalent to a closed loop of transformations in the multiperipheral basis. These transformations form the modular group. Since all the relations in this group are satisfied in this basis, there are no new identities.

The completeness theorem strongly suggests that the equations come close to defining RCFT. Specifically, what it does show is that a solution to the equations allows one to define transition functions for a compatible family of Friedan-Shenker vector bundles on all moduli spaces. This statement can be reformulated in a language currently much in vogue, which we now explain.

In Friedan-Shenker modular geometry the existence of a projectively flat vector bundle means that the data defining the bundle is essentially topological, involving (projective) representations of the Teichmüller modular group. Graeme Segal abstracted the concept, implicitly used from the earliest days of dual model theory and somewhat more precisely described in [1][24][21][13] to the notion of a *modular functor*. A modular functor may be specified by the following data and axioms:

Axioms for a Modular Functor

Data:

1. Representation labels: A finite set I of labels (i.e. the representations of the chiral algebra) with a distinguished element $0 \in I$ and an involution $i \to i^\vee$ such that $0^\vee = 0$.

2. Conformal blocks: A map

$$(\Sigma, (i_1, v_1, P_1), \ldots (i_n, v_n, P_n)) \to \mathcal{H}(\Sigma; (i_1, v_1, P_1), \ldots (i_n, v_n, P_n)))$$

from oriented surfaces with punctures, each puncture P_r being equipped with a direction v_r and a label i_r, to vector spaces.

3. Duality transformations: A linear transformation $\mathcal{H}(f) : \mathcal{H}(\Sigma_1) \to \mathcal{H}(\Sigma_2)$ associated to an automorphism $\Sigma_1 \to \Sigma_2$ (and similarly for punctures).

Conditions:

1. Functoriality: $\mathcal{H}(f)$ depends only on the isotopy class of f. Thus the mapping class group acts on $\mathcal{H}(\Sigma)$, (and similarly for punctures).

2. Involution: If bar denotes reversal of orientation and application of the involution to the representations then $\mathcal{H}(\bar{\Sigma}) \cong \mathcal{H}(\Sigma)^\vee$.

3. Multiplicativity: $\mathcal{H}(\Sigma_1 \coprod \Sigma_2) \cong \mathcal{H}(\Sigma_1) \otimes \mathcal{H}(\Sigma_2)$.

4. Gluing: Pinching $(\Sigma, (i_1, v_1, P_1), \ldots (i_n, v_n, P_n))$ along a cycle to obtain a surface (possibly connected or disconnected) $(\tilde{\Sigma}, (i_1, v_1, P_1), \ldots (i_n, v_n, P_n), (j, v, P), (j^\vee, v, \tilde{P}))$ with a pair of identified punctures $P, \tilde{P}$ defines vector spaces related by

$$\mathcal{H}(\Sigma; (i_1, v_1, P_1), \ldots (i_n, v_n, P_n)) \cong \oplus_{j \in I} \mathcal{H}(\tilde{\Sigma}; \ldots (i_n, v_n, P_n), (j, v, P), (j^\vee, v, \tilde{P})).$$

5. Normalization. $\mathcal{H}(S^2; (j, P)) \cong \delta_{j,0} \cdot \mathbb{C}$.

The all-important gluing axiom may be illustrated by the figure:

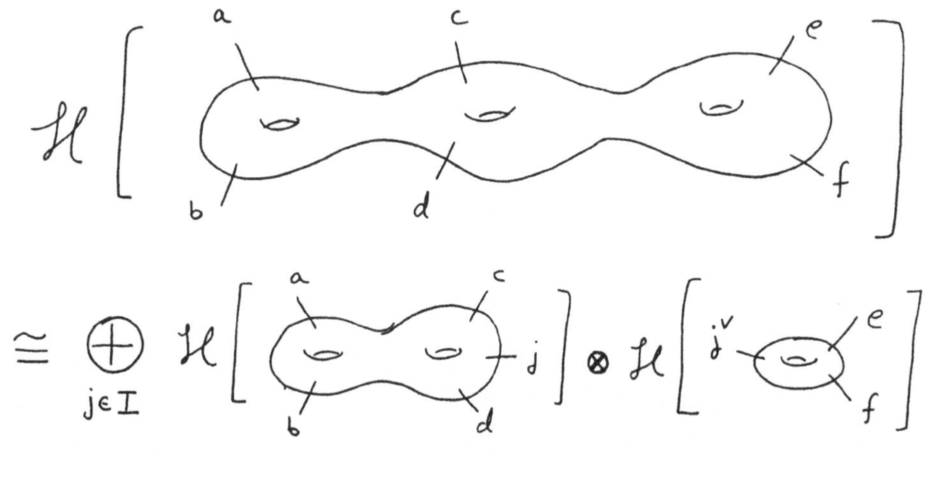

The directions v_r at the punctures are needed to keep track of the nontrivial effects of Dehn twists around the punctures. Geometrically they are needed since conformal blocks should be thought of as differentials on the surface Σ, i.e., $\mathcal{F} \sim f(z_1, \ldots z_n, \ldots)(dz_1)^{\Delta_1} \cdots (dz_n)^{\Delta_n}$. This subtlety, which shows up in the three-dimensional point of view in the need for framings of links, was first emphasized in [26].

In an obvious way one can change the definitions to define a modular functor which is projective, unitary, and so forth. In this language the completeness theorem states that from a finite amount of data F, B, S satisfying a finite number of conditions one can construct a projective modular functor.

The idea of a modular functor is truly beautiful and allows us to ask many interesting questions in a succinct way. For example we may ask to what extent a modular functor characterizes a rational conformal field theory. Since there are nontrivial theories with trivial modular functors this is a serious question. Or, we may ask if every modular functor arises in some conformal field theory. Simply defining the bundles is not enough for defining physical correlation functions. Whether these bundles have reasonable sections which correspond to blocks in a CFT is another matter which remains undecided. However, there is a closely related problem in mathematics where the answer is known to be in the affirmative in a suitably defined sense, namely the Tannaka-Krein approach to group theory – so we discuss this next.

5. Tannaka-Krein theory and Modular Tensor Categories

Let us switch our attention momentarily to an apparently different problem - we want to characterize the sets:

$\underline{\text{Rep}}(\text{G}) = \{V | V \text{ is finite dimensional representation of G}\}$

For example, let us consider G to be a compact Lie group, then there are the most important elements

$$R_i = \text{irreducible representations}$$

Moreover, we can decompose

$$R_i \otimes R_j = \cdots \underbrace{\otimes R_k \otimes \cdots \otimes R_k \otimes}_{n_{ij}^k \text{ times}} \cdots$$
$$= \oplus_k V_{ij}^k \otimes R_k$$

with

$$\dim V_{ij}^k = n_{ij}^k$$

The spaces V_{ij}^k are characterized as the space of a certain kind of intertwiner. Recall that if

W_{1,ρ_1} and W_{2,ρ_2} are two representations (that is, $\rho_1 : G \longrightarrow \mathrm{End}(W_1)$ is a homomorphism, etc.) Then an intertwiner $T : W_1 \longrightarrow W_2$ is a group - equivariant map, i.e.

$$\begin{array}{ccc} W_1 & \stackrel{T}{\rightarrow} & W_2 \\ \rho_1(g)\downarrow & & \downarrow \rho_2(g) \\ W_1 & \stackrel{T}{\rightarrow} & W_2 \end{array} \tag{5.1}$$

commutes for all $g \in G$. In this language $V_{ij}^{k} = \{intertwiners : R_k \longrightarrow R_i \otimes R_j\}$ e.g. in $SU(2)$ the space of intertwiners is always zero or one-dimensional and is spanned by

$$\begin{pmatrix} J \\ j_1 j_1 \end{pmatrix} = \sum_{M,M_1,M_2} |m_1 j_1 m_2 j_2 >< m_1 j_1 m_2 j_2 | MJ >< M, J|$$

where $< m_1 j_1 m_2 j_2 | MJ >$ are Clebsch-Gordon coefficients.

Now we will examine some nontrivial properties satisfied by these vector spaces, these follow from rather obvious isomorphisms of representations. First, we have the evident isomorphism $\Omega : R_i \otimes R_j \cong R_j \otimes R_i$ since the map $x \otimes y \longrightarrow y \otimes x$ is an intertwiner.

Therefore, if we decompose the above tensor products of representations we learn that:

$$\Omega : V_{ij}^{k} \cong V_{ji}^{k}$$

When manipulating these spaces of intertwiners it is good to develop a pictorial notation. Denote

$$V_{ij}^{k} = \underline{k \;|^{i}\; j}$$

Then

$$\Omega : \underline{k \;|^{i}\; j} \longrightarrow \underline{k \;|^{j}\; i}$$

Note well that an obvious consequence of the fact that the transformation Ω squares to one is that

$$\Omega^2 = 1$$

as a transformation on V_{ij}^{k}. Thus, when $i = j$ we can diagonalize Ω, the eigenvalues are ± 1 depending on the symmetry of the coupling.

Now consider the second evident isomorphism:

$$F : R_{j1} \otimes (R_{j2} \otimes R_{j3}) \cong (R_{j1} \otimes R_{j2}) \otimes R_{j3}$$

$$x \otimes (y \otimes z) \longrightarrow (x \otimes y) \otimes z$$

When decomposing in terms of irreducible representations we meet compositions of intertwiners, for example we find:

$$\begin{pmatrix} i \\ j_1 p \end{pmatrix} \begin{pmatrix} p \\ j_2 j_3 \end{pmatrix} : R_{j_1} \otimes R_{j_2} \otimes R_{j_3} \rightarrow R_i$$

Carrying our pictorial notation further we denote the tensor product of a spaces of intertwiners by

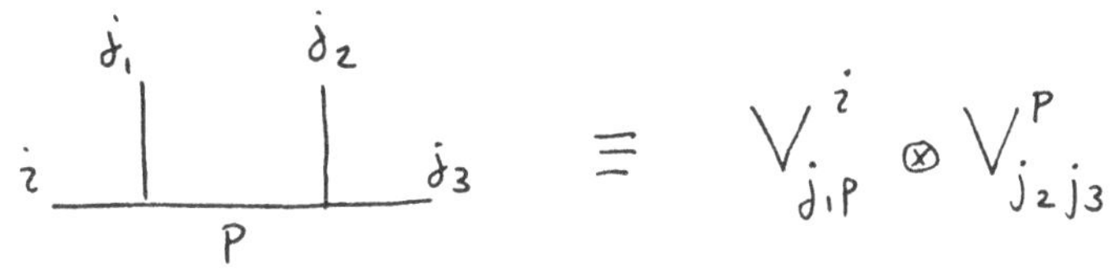

If we have a direct sum of these spaces over "intermediate" representations, then we denote the resulting vector space by

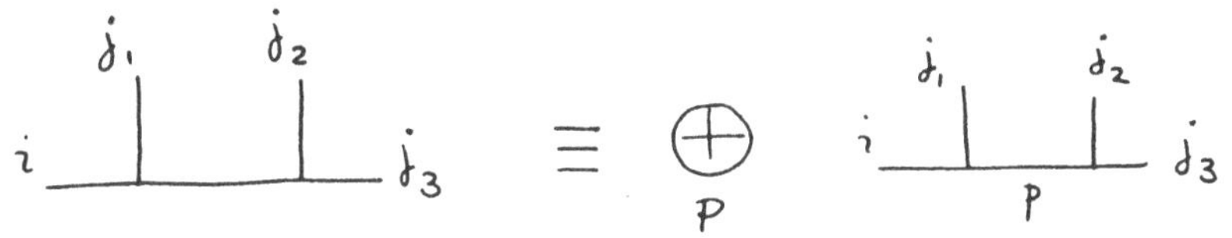

Thus, decomposing the second isomorphism in terms of irreducible representations we learn that there must be a transformation:

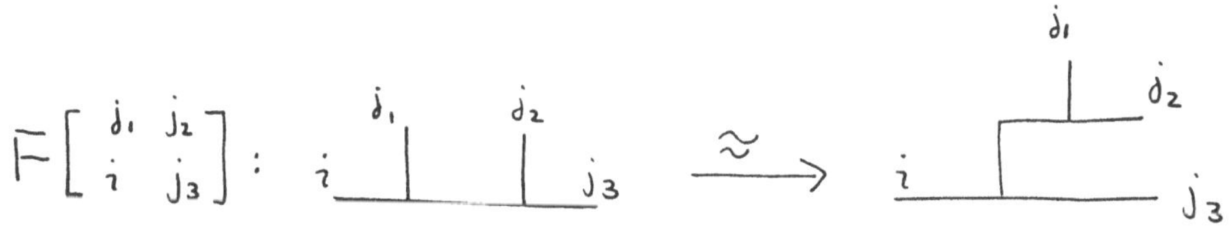

picture for F

Or, in formulas:

$$F\begin{bmatrix} j_1 & j_2 \\ i & j_3 \end{bmatrix} : \oplus_r V^i_{j1,r} \otimes V^r_{j2,j3} \longrightarrow \oplus_s V^i_{s,j_3} \otimes V^s_{j1,j2}$$

In the physics literature the intertwiners are known as Clebsch-Gordan coefficients (3j symbols) and the F's are known as $6j$ or Racah coefficients. Moreover, the fact that F is an isomorphism implies that n^k_{ij} defines a commutative associative algebra which is, in fact, the character ring of the group.

Now the two isomorphisms of representations Ω and F satisfy simple compatibility conditions. The first is the pentagon relation:

$$\begin{array}{ccccc} R_1 \otimes (R_2 \otimes (R_3 \otimes R_4)) & \xrightarrow{F} & (R_1 \otimes R_2) \otimes (R_3 \otimes R_4) & \xrightarrow{F} & (R_1 \otimes R_2) \otimes R_3) \otimes R_4 \\ \downarrow 1 \otimes F & & & & \downarrow F \otimes 1 \\ R_1 \otimes (R_2 \otimes R_3) \otimes R_4) & & \xrightarrow{F} & & (R_1 \otimes (R_2 \otimes R_3)) \otimes R_4 \end{array} \tag{5.2}$$

for representations $R_1, \ldots, R_4$. The second is the hexagon relation:

$$R_1 \otimes (R_2 \otimes R_3) \xrightarrow{F} (R_1 \otimes R_2) \otimes R_3 \xrightarrow{\Omega} R_3 \otimes (R_1 \otimes R_2)$$

$$\downarrow 1 \otimes \Omega \qquad\qquad\qquad\qquad\qquad\qquad \downarrow F \tag{5.3}$$

$$R_1 \otimes ((R_3 \otimes R_2) \xrightarrow{F} (R_1 \otimes R_3) \otimes R_2 \xrightarrow{\Omega \otimes 1} (R_3 \otimes R_1) \otimes R_2$$

Decomposing these relations in terms of irreducible representations we learn that F, Ω satisfy two corresponding compatibility conditions

$$\sum_s F_{p_2 s}\begin{bmatrix} j & k \\ p_1 & b \end{bmatrix} F_{p_1 l}\begin{bmatrix} i & s \\ a & b \end{bmatrix} F_{sr}\begin{bmatrix} i & j \\ l & k \end{bmatrix} = F_{p_1 r}\begin{bmatrix} i & j \\ a & p_2 \end{bmatrix} F_{p_2 l}\begin{bmatrix} r & k \\ a & b \end{bmatrix} \tag{5.4}$$

$$\Omega^m_{lk} F_{mn}\begin{bmatrix} j & k \\ i & l \end{bmatrix} \Omega^l_{jk} = \sum_r F_{mr}\begin{bmatrix} j & l \\ i & k \end{bmatrix} \Omega^i_{kr} F_{rn}\begin{bmatrix} k & j \\ i & l \end{bmatrix} \tag{5.5}$$

In the case of $SU(2)$, these relations are known in the physics literature as the Biedenharn sum-rule and Racah's sum-rule.

In category theory there is a theorem, called the MacLane coherence theorem that states that the above two identities are the full set of independent identities on F, Ω.

Let us describe the idea of the proof:

Define a simplicial complex where vertices correspond to dual diagrams and edges correspond to simple moves between diagrams. Label these edges by F, Ω etc. Fill in the pentagons and hexagons to get a two-complex, and show that the resulting two-complex is simply connected. There are two kinds of loops, those involving only the F move and those involving F, Ω. Define the following composite move:

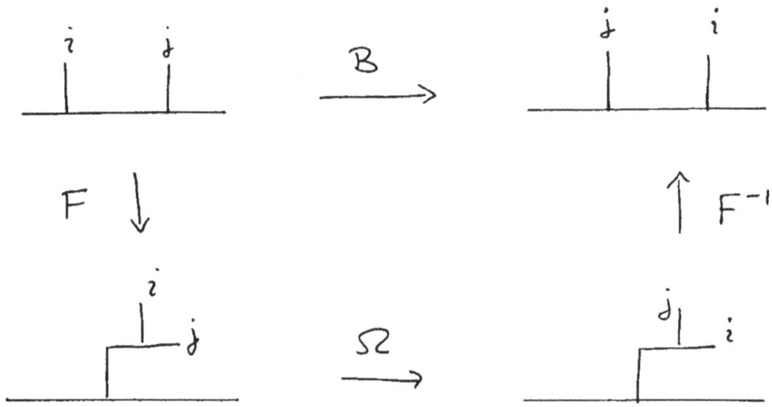

B from F and Ω

By the pentagon and hexagon we can deform all loops to those involving only multi-peripheral diagrams:

multiperipheral diagrams

Then we need only check that B satisfies the relations of the symmetric group.

There is a clear analogy here with rational conformal field theory, and we have now arrived at the point we were at with RCFT. In the case of group theory it turns out one can go further and state a partial converse to the above results. We would like to know if all solutions to the above axioms in fact come from group theory. It turns out there are solutions to the previous equations that do not come from groups, but we can eliminate these by adding two more axioms.

The first axiom corresponds to the existence of the trivial representation $R_{i=0} = \mathbb{C}$. Note that we have:

$$V_{i0}^{j} \cong V_{0i}^{j} \cong \delta_i^j \mathbb{C}$$

Every representation has a conjugate representation:

$$(R_i) = R_{\acute{i}}$$

and $R_i \otimes R_j$ contains the singlet only if $j = \acute{i}$, so

$$V_{ij}^{0} = \delta_{ij} \mathbb{C} \ .$$

The second axiom that we must add, which is due to Deligne, [30] involves the special fusion coefficient (for the case $\Omega_{ii}^{0} = 1$)

$$F_i = F_{00} \begin{bmatrix} i & i \\ i & i \end{bmatrix} :$$

Namely consider the composition

$$R_0 \longrightarrow (R_i) \otimes R_i \longrightarrow R_0$$

$$1 \longrightarrow \Sigma v_\alpha \otimes v_\alpha \longrightarrow dim\ R_i$$

We have a map of a one-dimensional vector space to itself, which is, canonically, a complex number. One can compute the value of this number by decomposing in terms of intertwiners, and one finds the answer $\frac{1}{F_i}$.

• Exercise 5.1 *Deligne's condition in terms of* F_i. By considering the sequence

$$R_i \cong R_0 \otimes R_i \to (R_i^{\vee} \otimes R_i) \otimes R_i \to (R_i \otimes R_i^{\vee}) \otimes R_i \to R_i \otimes (R_i^{\vee} \otimes R_i) \to R_i \otimes R_0 \cong R_i$$

and decomposing the tensor products into irreducible representations, show that

$$\frac{1}{F_i} = \dim\ R_i$$

• Exercise 5.2 *Another proof of Deligne's Condition in terms of* F_i. Consider the "group theoretic one-loop two-point function":

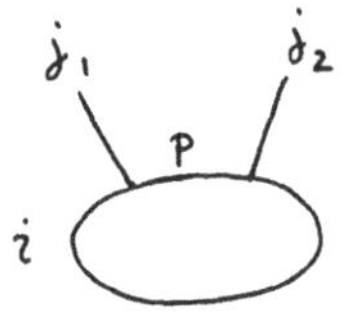

$$= \qquad tr_{R_i}\rho_i(g)\begin{pmatrix} i \\ j_1 p\end{pmatrix}(\beta_1 \otimes \cdot)\begin{pmatrix} p \\ j_2 i\end{pmatrix}(\beta_2 \otimes \cdot)$$

Group theoretic two point function

Consider the "monodromy" under $\beta_2 \to \rho_{j_2}(g)\beta_2$, Where $\binom{i}{jk}$ denote intertwiners. Using the basis of tensors:

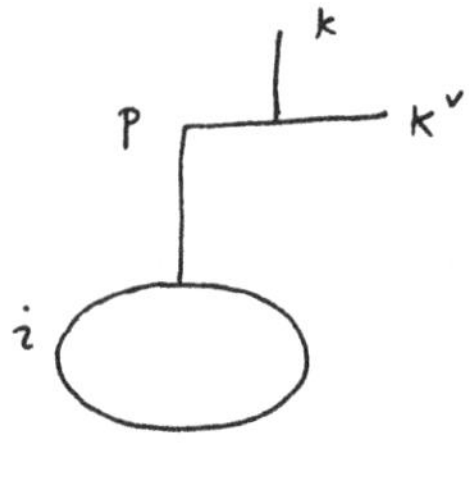

show that the monodromy is just:

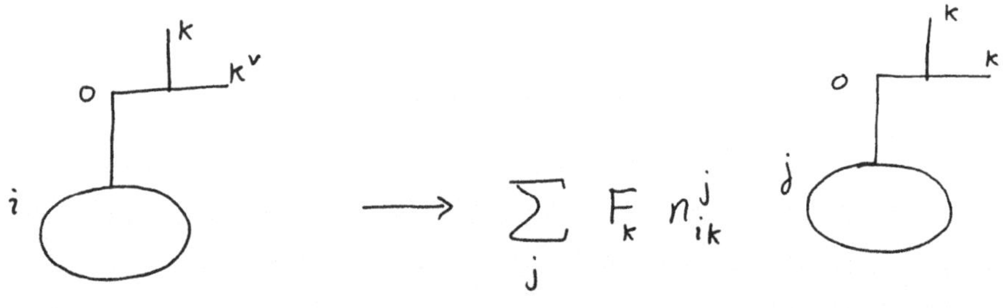

$$\longrightarrow \sum_j F_k\, n_{ik}^{j}$$

Take the limit $g \to 1$. Show that the other terms vanish and deduce that

$$\frac{1}{F_k} dim\ R_i = \sum_j n_{ik}^j dim\ R_j$$

and hence

$$\frac{1}{F_k} = dim\ R_k$$

The nontrivial result is that these axioms now characterize group theory. This is due to Deligne [30] and, in a slightly different form to Doplicher and Roberts [31] [32]. More precisely, suppose we are given the following:

Axioms for a Tannakian Category

Data:

1. An index set I with a distinguished element 0 and a bijection of I to itself written $i \mapsto i^{\vee}$.

2. Vector spaces: $V_{jk}^i\ i, j, k \in I$, with $dim V_{jk}^i = N_{jk}^i < \infty$

3. Isomorphisms:

$$\Omega_{jk}^i : V_{jk}^i \cong V_{kj}^i$$
$$F\begin{bmatrix} j_1 & j_2 \\ i_1 & k_2 \end{bmatrix} : \oplus_r V_{j_1 r}^{i_1} \otimes V_{j_2 k_2}^r \cong \oplus_s V_{s k_2}^{i_1} \otimes V_{j_1 j_2}^s \tag{5.6}$$

Conditions:

1. $(i^{\vee})^{\vee} = i$ and $0^{\vee} = 0$.

2. $V_{0j}^i \cong \delta_{ij} C \quad V_{ij}^0 \cong \delta_{ij} C \quad V_{jk}^i \cong V_{ji}^k \quad (V_{jk}^i)^{\vee} \cong V_{jk}^i$

3. $\Omega_{jk}^i \Omega_{kj}^i = 1$.

4. The identities:

$$F(\Omega \otimes 1)F = (1 \otimes \Omega)F(1 \otimes \Omega)$$
$$F_{23}F_{12}F_{23} = P_{23}F_{13}F_{12}$$

5. The normalization condition:

$$F_i^{-1} \in \mathbb{Z}_+$$

From such a set of axioms we can reconstruct a group for which V_{jk}^i are the intertwiners, F, the Racah coefficients, etc.

The proof of this result is rather involved, but it would probably be worthwhile to sketch the main ideas of reconstruction which proceeds as follows:

a) Define vector spaces $R_i = \mathbb{C}^{n_i}$, obviously. (In category theory these correspond to simple objects which we must realize with honest vector spaces.)

b) Define the space of intertwiners (morphisms) to be:

$$\begin{aligned} Hom(R_i \longrightarrow R_i) &= \mathbb{C} \\ Hom(R_i \longrightarrow R_j) &= O \ i \neq j \end{aligned}$$

and extend by linearity to $Hom(\oplus R \longrightarrow \oplus R)$.

c) Define tensor products: $R_i \otimes R_j \cong \oplus V_{ij}^k \otimes R_k$. That is, V_{ij}^k is a set of intertwiners. Now we define the set **Rep** = {all sums, products, quotients, duals of the R_i}.

d) Finally define the set of families of linear transformations:

$$\begin{aligned} \mathcal{G} = \{(\lambda_x)_{x \in \mathbf{Rep}} | \forall x, \lambda_x &: x \longrightarrow x \textit{ is an invertible linear transformation.} \\ \lambda_{x \otimes y} &= \lambda_x \otimes \lambda_y \\ T : x \rightarrow y & \text{ an intertwiner} \Rightarrow T\lambda_x = \lambda_y T\} \end{aligned}$$

$\mathcal{G}$ is a group: This is the group we want! One might naturally wonder whether, had one started with a group G, produced the objects F, Ω etc. and formed the group Aut, one would have recovered the same group G. This is settled in the following exercise.

• Exercise 5.3 *On Reconstruction* [30][33]. Suppose one begins with a compact group G and constructs the spaces V_{jk}^i as above. We will indicate why the reconstructed group $\mathcal{G}$ defined by the abstract procedure given here is exactly the original group G.

a.) Note first that $G \subset \mathcal{G}$. Note that every $g \in G$ defines a family $\{\lambda_X\}_{X \in \mathbf{Rep}}$ via $\lambda_X(g) = \rho_X(g)$, where ρ_X is the representation defined by X.

b.) Show that if $\vec{v} \in X$ is fixed by all of G, i.e., if

$$\forall g \in G : \rho_X(g)\vec{v} = \vec{v}$$

then it is fixed by all of $\mathcal{G}$, i.e.,

$$\lambda_X(\vec{v}) = \vec{v}$$

for any family satisfying the defining axioms of $\mathcal{G}$. (Hint: Show that $\lambda_{R_0} = 1$, and that $z \rightarrow z\vec{v}$ is an intertwiner $\mathbb{C} \rightarrow X$.) If G is a continuous Lie group we conclude that there are no "broken" generators in $\mathcal{G}/G$ and hence that $\mathcal{G} = G$.

c.) More generally suppose that $G \subset \mathcal{G}$ is a proper subgroup. Then there is some $\lambda_X \in End(X)$ which is not in the set $\{\rho_X(g) | g \in G\} \subset End(X)$. Use the fact that G is compact to show that there must exist a polynomial P on $End(X)$ which vanishes on $\{\rho_X(g) | g \in G\}$, but not at λ_X. Show that the space S of polynomials of degree $\leq deg(P)$ on $End(X)$ is a representation of G. Note that $P \in S$ violates (b), to conclude that $\mathcal{G} = G$.

In the above characterization of a Tannakian category we have worked directly with the data V_{jk}^i etc. Alternatively we could have defined the category more directly in terms of objects, with a tensor product of objects satisfying pentagon and hexagon conditions identical to (5.2) and (5.3) , and with some axioms relating to the unit object and dual objects. This is the definition one finds in the literature.

The situation arising in RCFT is more complicated than the one we have described for the Tannakian categories. In RCFT the index set I is finite. Moreover $\Omega^2 \neq 1$. This is crucial: it is the characteristic that leads to interesting monodromy and hence interesting braid representations. The pentagon relation remains but there are two hexagon relations involving Ω and Ω^{-1}. The category so defined (equivalently, the category defined by axioms on objects and morphisms of objects) is closely related to what is known as a "compact braided monoidal category" which was studied in [34]. Different definitions differ slightly on such details as whether $\check{}$ is involutive, or whether the set I should be finite or not. Thus, roughly speaking, the duality properties of RCFT's on the plane are characterized by "compact braided monoidal categories." Well defined RCFT's have more structure and must be defined on all Riemann surfaces. By the completeness theorem it suffices to define $S(p) : \oplus V_{pi}^i \to \oplus V_{pi}^i$ according to (4.3) and impose the relations of the modular group. We will call the category defined by these axioms a *modular tensor category*. More precisely we have

---

Axioms for a Modular Tensor Category

Data:

1. A finite index set I with a distinguished element 0 and a bijection of I to itself written $i \mapsto i\check{}$.

2. Vector spaces: V_{jk}^i $i, j, k \in I$, with $dim V_{jk}^i = N_{jk}^i < \infty$

3. Isomorphisms:

$$\Omega_{jk}^i : V_{jk}^i \cong V_{kj}^i$$

$$F\begin{bmatrix} j_1 & j_2 \\ i_1 & k_2 \end{bmatrix} : \oplus_r V_{j_1 r}^{i_1} \otimes V_{j_2 k_2}^r \cong \oplus_s V_{s k_2}^{i_1} \otimes V_{j_1 j_2}^s \tag{5.7}$$

4. A constant $S_{00}(0)$.

Conditions:

1. $(i\check{})\check{} = i$, $0\check{} = 0$.

2. $V_{0j}^i \cong \delta_{ij} C \quad V_{ij}^0 \cong \delta_{i\check{j}} C \quad V_{jk}^i \cong V_{j\check{i}}^{\check{k}} \quad (V_{jk}^i)\check{} \cong V_{\check{j}\check{k}}^{\check{i}}$

3. $\Omega_{jk}^i \Omega_{kj}^i \in End(V_{jk}^i)$ is multiplication by a phase.

4. The identities:

$$F(\Omega^\epsilon \otimes 1)F = (1 \otimes \Omega^\epsilon)F(1 \otimes \Omega^\epsilon)$$

$$F_{23}F_{12}F_{23} = P_{23}F_{13}F_{12}$$

for $\epsilon = \pm 1$.

5. The identities

$$S^2(p) = \pm e^{-i\pi\Delta_p} C$$
$$(ST)^3 = S^2$$

where $S(p) \in End(\oplus V_{pi}^{i})$ is defined by

$$S_{ij}(p) = S_{00}(0) e^{-i\pi\Delta_p} \frac{F_{i0}\left[\begin{smallmatrix} i & i \\ p & p \end{smallmatrix}\right]}{F_p F_{p0}\left[\begin{smallmatrix} j & j \\ j & j \end{smallmatrix}\right] F_{p0}\left[\begin{smallmatrix} i & i \\ i & i \end{smallmatrix}\right]} \sum_r B_{pr}\begin{bmatrix} i & j \\ i & j \end{bmatrix}(-) B_{r0}\begin{bmatrix} j & i \\ i & j \end{bmatrix}(-)$$

C represents the action of $\check{}$, the numbers $\pm e^{-i\pi\Delta_p}$ may be deduced from Ω, and $T : V_{ji}^{i} \to V_{ji}^{i}$ is scalar multiplication by $e^{2\pi i(\Delta_i - c/24)}$ for a constant c. (For more details see [15].)

Just as for Tannakian categories we could define modular tensor categories more directly in terms of objects and axioms on the tensor products of objects. In these terms one must define the analog of (4.3) . This may be done in terms of a generating set of simple objects R_i by defining a single morphism S of the object $\oplus_i R_i \otimes R_{\check{i}}$ to itself as follows:

$$\begin{aligned} \oplus_i R_i \otimes R_{\check{i}} &\longrightarrow \oplus_{i,j} R_i \otimes R_j \otimes R_j\check{} \otimes R_i\check{} \\ &\xrightarrow{\Omega^2 \otimes 1 \otimes 1} \oplus_{i,j} R_i \otimes R_j \otimes R_j\check{} \otimes R_i\check{} \otimes \\ &\xrightarrow{\Omega^{-1} \otimes \Omega} \oplus_{i,j} R_j \otimes R_i \otimes R_i\check{} \otimes R_j\check{} \otimes \\ &\longrightarrow \oplus_j R_j \otimes R_j\check{} \end{aligned} \tag{5.8}$$

Similarly one may use Ω to define the data $\pm e^{i\pi\Delta_j}$ as a morphism $R_j \to R_j$ and from this define T on $\oplus R_i \otimes R_{\check{i}}$ and impose a relation on S^2 (relating it to Ω) and the relation $(ST)^3 = S^2$.

The name *modular tensor category* was suggested by Igor Frenkel and we will adopt it. We thank him for discussions on this subject and for urging us to express the definition of S, (4.3) , in terms of simple objects, along the lines of (5.8).

As we have mentioned, the above axioms are sufficient for establishing the relation $Sa = bS$. Thus we may summarize the main result of [13][15] in the statement that a modular tensor category (henceforth MTC) is equivalent to a modular functor. As in section four we may ask whether all MTC's are associated to some RCFT, and to what extent an MTC characterizes the original RCFT.

From the analogy of Tannakian categories and MTC's one naturally wonders whether there is a reconstruction theorem for MTC's analogous to Deligne's theorem. This is not known at present, but there is some good evidence that such a statement exists. First, there is an analog of the integrality condition in RCFT. From the proof of Verlinde's formula one finds

$$\frac{1}{F_i} = \frac{S_{0i}}{S_{00}}$$

We have already noted that classically the quantity on the LHS is related to the dimension. The quantity on the RHS has been interpreted as the "relative dimension" of the representation spaces. Note that [35]

$$\frac{\text{"}dim H_i\text{"}}{\text{"}dim H_0\text{"}} = \lim_{q\to 1} \frac{tr_{H_i}\ q^{L_0-c/24}}{tr_{H_0}\ q^{L_0-c/24}} = \frac{S_{0i}}{S_{00}}$$

All this strongly suggests that some axioms additional to the above polynomial equations in fact characterize RCFT's - and that classifying solutions to these equations is the same as classifying RCFT's.

The relation between the axioms of RCFT as discussed above and the Tannaka-Krein approach to group theory becomes more complete in a certain limit of RCFT. Some RCFT's are labeled by a parameter k such that they simplify considerably in the $k \to \infty$ limit. In this limit the conformal dimensions of all the primary fields approach zero. More generally, there is a subset of the primary fields with a closed fusion rule algebra (namely, if i and j are in the subset then $N_{ij}^l \neq 0$ only for l in the set) whose conformal dimensions approach an integer in the $k \to \infty$ limit. We define this limit as the *classical limit of the RCFT.* Examining our axioms at genus zero in this limit we see that they simplify. In particular, since the relevant Δ's are integers,

$$\Omega^2 = 1\ . \tag{5.9}$$

Therefore, there are no monodromies in the classical theory and the two hexagons are the same equation. In this limit the axioms of a RCFT are identical to those of group theory in the Tannaka-Krein approach. Since classical RCFT is the same as group theory, it is natural to conjecture that *quantum RCFT is a generalization of group theory.* We'll return to this conjecture below. For the moment we note the following correspondences between group theory and conformal field theory:

Group	Chiral algebra
Representations	Representations
Clebsch-Gordan coefficients/Intertwiners	Chiral vertex operators
Invariant tensors	Conformal blocks
Symmetry of couplings	Ω
Racah coefficients (6j symbols)	Fusion matrix

It is also interesting to examine a larger class of CFT's. We refer to them as "quasirational CFT's." In these theories the chiral algebra has an infinite number of irreducible representations . However, the fusion rules are finite, i.e. for given i, j, N_{ij}^k is non zero only for a finite number of representations k. Because of this condition, the formalism of the CVO and the duality matrices on the plane is still applicable. Consequently, the polynomial equations on the plane (the pentagon and the two hexagons) are satisfied. One can still define $S(p)$ by (4.3) but since the number of irreducible representations is infinite, the torus polynomial equations are not obviously present. The category of representation spaces of the chiral algebra of a quasirational conformal field theory is also a generalization of a tensor category. A well known example of such a theory is the Gaussian model at an irrational value of the square of the radius.

Finally, we must not lose sight of the fact that many interesting irrational (non-quasirational) CFT's exist and that the challenge to understand their structure remains unanswered.

6. Combining leftmovers with rightmovers

CFT is not just the study of chiral algebras and their representations. In order to have a consistent conformal field theory, we need to put together left and right-movers to obtain correlation functions with no monodromy.

The left and right chiral algebras $\mathcal{A}$, and $\bar{\mathcal{A}}$ are the algebras of purely holomorphic and anti-holomorphic fields. We can decompose the total Hilbert space of the theory into irreducible representations: $H_r \otimes H_{\bar{r}}$, so t he partition function is:

$$Tr_H \; q^{L_0 - c/24} \; \bar{q}^{\bar{L}_0 - c/24} = \sum_{r,\bar{r}} h_{r\bar{r}} \chi_r(q) \chi_{\bar{r}}(\bar{q}).$$

The nonnegative integers $h_{r\bar{r}}$ characterize the field content of the theory.

We can write the physical conformal fields in terms of the chiral vertex operators as

$$\phi^{jm,\bar{j}\bar{m}}(z,\bar{z}) = \sum_{j,k} d^{(i\bar{i})}_{(j\bar{j})(k,\bar{k})} \Phi^{j,m}_{ik}(z) \bar{\Phi}^{\bar{j},\bar{m}}_{\bar{i}\bar{k}}(z) \tag{6.1}$$

We assume for simplicity that there is only one field with representation $(i,\bar{i})$ in the theory. Below we'll show that this assumption is always satisfied.

Now the physical correlation function must be independent of the choice of blocks, so there are certain conditions on the d-coefficients. For example, from invariance of the partition functions under $T : \tau \to \tau + 1$, we see that $h_{i\bar{i}} = 0$ unless $\Delta_i - \Delta_{\bar{i}} \in \mathbf{Z}$ Proceeding more systematically we could have deduced this from an analysis of 2 and 3 point functions.

Moving on to the four-point function, we must have the same correlator from either basis of blocks:

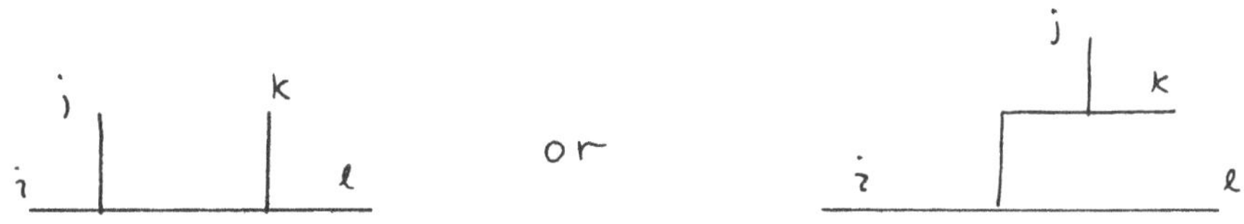

s and t channel blocks relevant for the four point function

this implies

$$\sum_{p,\bar{p}} d^{(i\bar{i})}_{(j\bar{j})(p\bar{p})}\, d^{(p\bar{p})}_{(k\bar{k})(l\bar{l})}\, F_{pq}\begin{bmatrix} j & k \\ i & l \end{bmatrix} \bar{F}_{\bar{p}\bar{q}}\begin{bmatrix} \bar{j} & \bar{k} \\ \bar{i} & \bar{l} \end{bmatrix} = d^{(i\bar{i})}_{(q\bar{q})(l\bar{l})}\, d^{(q\bar{q})}_{(j\bar{j})(k\bar{k})} \tag{6.2}$$

- Exercise 6.1 *Monodromy invariance*

a.) Write out the conditions on d following from locality of the three-point function.

b.) Show that the invariance of the physical correlator under B is guaranteed by the condition of part (a) together with the equation for F (6.2).

By using the operator product expansion for chiral vertex operators together with (6.2) we may deduce that

$$\begin{aligned}\phi^{j,m;\bar{j},\bar{m}}(z,\bar{z})\phi^{k,n;\bar{k},\bar{n}}(z,\bar{z}) = \sum_{p',\bar{p}'} d^{(p'\bar{p}')}_{(j\bar{j})(k\bar{k})} \sum_{P\in\mathcal{H}_p,\,P'\in\mathcal{H}_{p'}} & \phi^{p',P';\bar{p}',\bar{P}'}(w,\bar{w}) \\ & \langle P'|\Phi^{j,m}_{p'k}(z-w)|n\rangle\overline{\langle P'|\Phi^{j,m}_{p'k}(z-w)|n\rangle}\end{aligned} \tag{6.3}$$

Again there is a nice analog of this equation in group theory.

Recall that for a compact group the Hilbert space of L^2 functions on the group has an orthonormal basis given by the matrix elements $D^R_{\mu\nu}$ in the irreducible representations R. The operator $U(D^R_{\mu\nu})$ on $L^2(G)$ given by multiplication of functions may be represented in terms of intertwiners as

$$U(D^R_{\mu\nu}) = \sum_{R_1,R_2}\sum_{a\in V^{R_1}_{RR_2}} \begin{pmatrix} R_1 \\ RR_2 \end{pmatrix}(\mu\otimes\cdot)_a \begin{pmatrix} R_1^{\vee} \\ R^{\vee}R_2^{\vee} \end{pmatrix}(\nu\otimes\cdot)_{a^{\vee}} \tag{6.4}$$

Where we sum over a basis of intertwiners and $a^{\vee}$ is a basis dual to a. The algebra of functions on the group manifold is given by

$$D^{R_1}_{\mu_1\nu_1}D^{R_2}_{\mu_2\nu_2} = \sum_{R,\gamma_1,\gamma_2}\sum_{a\in V^R_{R_1R_2}} D^R_{\gamma_1\gamma_2}\langle R,\gamma_1|\begin{pmatrix} R \\ R_1R_2 \end{pmatrix}_a(\mu_1\otimes\mu_2)\rangle\langle R,\gamma_2|\begin{pmatrix} R^{\vee} \\ R_1^{\vee}R_2^{\vee} \end{pmatrix}_{a^{\vee}}(\nu_1\otimes\nu_2)\rangle$$

Thus we see that in the $k\to\infty$ limit of WZW models the operator product expansion of the fields with $\Delta\to 0$ becomes the algebra of functions on the group G, thus providing an explicit example of an old idea of Dan Friedan's for the reconstruction of manifolds from the operator product expansion of CFT. In fact, as described later, in the specific example of current algebra the above ope for finite k is closely related to the algebra of functions on a quantum group. For further discussion of these and related ideas see [36].

We now show how the above equations can be used to deduce some general theorems about the operator content of rational conformal field theories.

• Exercise 6.2 *No representation appears more than once.* Consider a RCFT where some representations occur more than once (either $h_{r\bar{r}} > 1$ or both $h_{r\bar{r}}$ and $h_{r\bar{r}'}$ are non zero for $\bar{r} \neq \bar{r}'$).

a. Add indices in equation (6.1) to describe this situation.

b. Rewrite equation (6.2) for this case.

c. Study the four point function of $\langle \phi\phi\phi'\phi' \rangle$ where ϕ and ϕ' transform the same under $\mathcal{A}$ (the representation r) but they are different conformal fields (they might or might not transform the same under $\overline{\mathcal{A}}$) and assume for simplicity that all the representations are self conjugate. Use (3.5) to bring $\overline{F}$ to the other side of the equation and study it for the case where the intermediate representation is 0 on both sides. Simplify the equation by using the fact that the $\mathcal{A}$ ($\overline{\mathcal{A}}$) includes *all* the holomorphic (antiholomorphic) fields i.e. the identity operator is the only primary field under $\mathcal{A} \otimes \overline{\mathcal{A}}$ which is holomorphic. The ope of $\phi\phi$ contains the identity operator and $\phi\phi'$ does not contain the identity operator. Use this fact to show that one side of the equation vanishes. The other side is proportional to F_r and does not vanish. Therefore, we are led to a contradiction and no representation can appear more than once.

Notice that in proving this result one uses only the equations on the plane and not the equations on the torus. Hence, this result applies not only in RCFT but also in quasirational theories. On the other hand, this result is not true in theories which are not quasirational [11]. A $\mathbb{Z}_2$ orbifold of the Gaussian model at an irrational value of the square of the radius is not quasirational – the ope of two twist fields includes all the untwisted representations. Since the previous proof does not apply, we are not surprised to see the same representation appearing more than once in the spectrum.

Similarly there is an equation for the d's following from the modular invariance of the $g = 1$, one-point functions.

• Exercise 6.3 *Equation for d from genus one.* Write the equation for invariance under $S(p)$ for every p. Remember that the characters of the one point function on the torus are defined as differential forms i.e. they have a z-dependence $\sim (dz/z)^{\Delta(p)}$ (otherwise they are not invariant). Therefore, there is a phase relating S of the left-movers to S of the right-movers.

At this point one may wonder whether there will be further constraints on the d coefficients from duality *invariance* of correlation functions on other Riemann surfaces. The answer is no. Since duality matrices defining an MTC allow us to define duality matrices on all surfaces we know that the conformal blocks are duality *covariant*. To check invariance of left-right combinations of blocks we merely have to check invariance under the generators of duality transformations. Since an MTC defines a modular functor, the generators can be taken to be those duality transformations represented by F, B, S. Thus the above duality invariance conditions suffice to guarantee invariance on all surfaces. A similar conclusion was reached independently in [37].

• Exercise 6.4 *Every representation of $\mathcal{A}$ occurs in the spectrum.* Show that $S(0)$ is unitary. Use this to show that one of the equations of the previous exercise can be written as

$$\sum_{\bar{j}} h_{i\bar{j}} \bar{S}_{\bar{j}\bar{k}} = \sum_{j} S_{ij} h_{j\bar{k}} \tag{6.5}$$

Use $h_{0\bar{i}} = h_{i\bar{0}} = \delta_{i0}$, i.e. $\mathcal{A}$ ($\overline{\mathcal{A}}$) includes *all* the holomorphic (antiholomorphic) fields, to show that there is no r such that $h_{r\bar{j}} = 0$ for every $\bar{j}$. Hence, no representation can be omitted.

From the last exercises we conclude: If the chiral algebras, $\mathcal{A}$ and $\bar{\mathcal{A}}$ are maximally extended, $h_{r,\bar{r}}$ must be a permutation matrix. We are now ready to tackle

---

• Exercise 6.5 *The left movers are paired with the right movers by an automorphism of the fusion rule algebra.* Use Verlinde's formula relating the fusion rules to S and (6.5) to prove this.

We conclude that $FRA(\mathcal{A}) = FRA(\overline{\mathcal{A}})$ and the pairing of the left movers and the right movers is an automorphism of the fusion rule algebra:

$$h_{r,\bar{r}} = \delta_{r,w(\bar{r})}$$

where

$$N_{\bar{i}\bar{j}\bar{k}} = N_{w(\bar{i})w(\bar{j})w(\bar{k})}$$

The main point here is that the classification of RCFT's is a two-step process. First we classify all chiral algebras and their representation theory, then we look for all automorphisms of the fusion rule algebras.

• Exercise 6.6 *No New Conditions on F.* For a unitary diagonal (i.e. $h_{i\bar{i}} = \delta_{i\bar{i}}$) theory, assuming F is real and the fusion rules are zero and one, show that the operator product coefficients may be written

$$d_{ijk}^2 = \frac{F_{0k}\left[\begin{smallmatrix} i & j \\ i & j \end{smallmatrix}\right]}{F_{k0}\left[\begin{smallmatrix} j & j \\ i & i \end{smallmatrix}\right]}$$

a.) Use the polynomial equations to show that d is totally symmetric.

b.) Substitute the above equation back into the full set of equations for d_{ijk} on the plane. Show that the resulting identities are guaranteed by the polynomial equations.

• Exercise 6.7 *Open Problem.* How general is the result of the previous exercise? Do the equations for the torus one-point function follow from the other identities? (Felder and Silvotti [38] have shown that for the discrete series the answer is yes, by direct calculation.) What about non-unitary theories? What about arbitrary fusion rules? Is this true for the non-diagonal theories – when a non-trivial automorphism is used to pair left and right movers?

• Exercise 6.8 *Modular Invariance of $A_1^{(1)}$ Characters*

a.) Find the automorphisms of the fusion rule algebra for the level k $SU(2)$ WZW model.

b.) Impose other necessary conditions, e.g. the monodromy invariance of the two-point function.

c.) Using the above point of view interpret the other modular invariants of $A_1^{(1)}$ characters.

• Exercise 6.9 *Automorphisms of Kac-Moody Fusion Rules.* Using Verlinde's formula for N_{ijk} and Kac's formula for S_{ij}, show how automorphisms of the extended Dynkin diagrams can define automorphisms of the fusion rule algebra. An application of this fact can be found in [39].

• Exercise 6.10 *the d-coefficients and gauge invariance.* How does d transform under the gauge transformations of rescaling the chiral vertex operators? Show that the equations for d are gauge invariant.

• Exercise 6.11 *Modular invariants for the rational torus.* As we will see in section 10 below, the Gaussian model at radius squared $R^2 = \frac{p}{2q}$ has a chiral algebra which depends only on the quantity pq. Compute the automorphisms of the fusion rule algebra of the rational torus and show that they define the different models for which $pq = p'q'$, but $p/q \neq p'/q'$.

The analogy between conformal field theory and group theory continues to hold for the combination of left movers with right movers. We can add to the table at the end of section 5 a few more rows:

Functions on the group	Physical fields
Product of functions on the group	Operator product expansion
Average over the group of a product of functions	Physical correlation function

---

• Exercise 6.12 *Analogy with group theory.* Explain the table. Show that it corresponds to the diagonal theory.

The equations for the ope coefficients d can be interpreted as defining a metric [24] on the vector space of the conformal blocks. Therefore, if all the d's are real and positive (and therefore we can pick the gauge $d = 1$), the vector space of the conformal blocks is a Hilbert space. This interpretation will play an important role in the following sections where this Hilbert space will appear in the quantization of a quantum mechanical system.

7. 2D Duality vs. 3D General Coordinate Invariance

Many people have noticed that RCFT's lead to knot invariants [20][40] [41] [27][42] [43]. One way of producing knot invariants is to view the B matrices as "transition amplitudes" of conformal blocks, then defining an appropriate trace (Markov trace) on these amplitudes the resulting polynomials are, in fact, knot invariants. There is an alternative formalism, used in [40] and elaborated upon in [42][43] which dispenses with the need for a trace at the cost of introducing some new moves. With these new moves the knot invariant becomes the transition amplitude for proceeding from the "null block" to itself with an intervening knot projection. We will present these results from our point of view using the formalism of the previous sections.

Consider the planar projection of a knot from S^3, e.g.

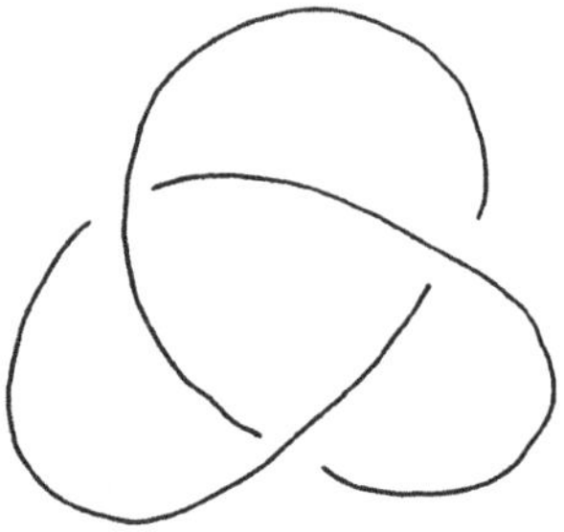

A projection of a knot on a plane

We assign a number to this figure by using the graphical formalism described above. For this, we label every line by a representation of a chiral algebra and also label the areas bounded by the lines by such representation. We assign factors of B to

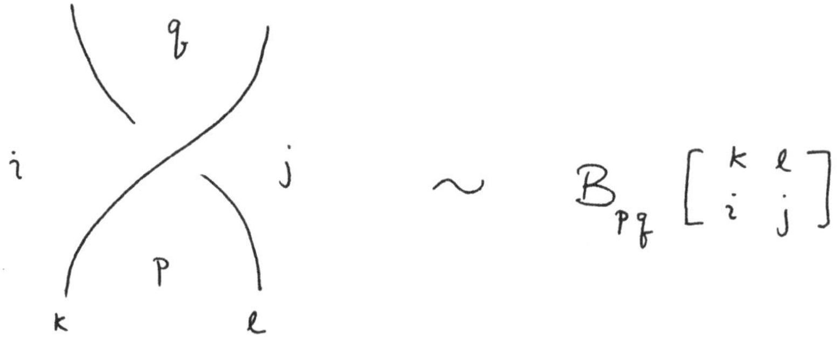

Graphical rules for computing a knot invariant

The knot that we consider is a "framed knot." It looks like a ribbon and hence

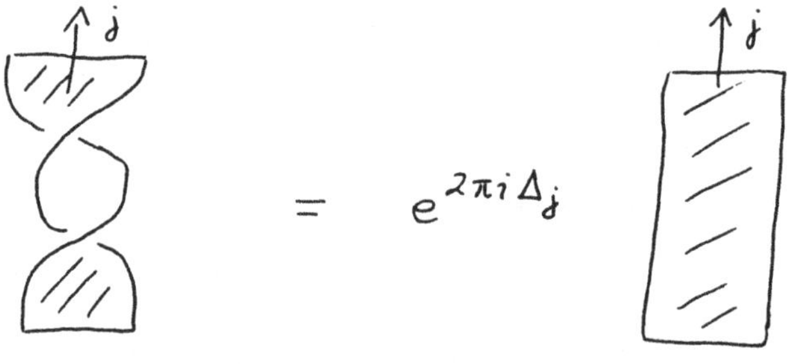

A non-trivial operation on a framed knot

The operation in the figure corresponds to a factor of $e^{2\pi i\Delta_i}$ in the knot invariant. We also need to introduce two new operations on lines for pair creation/annihilation:

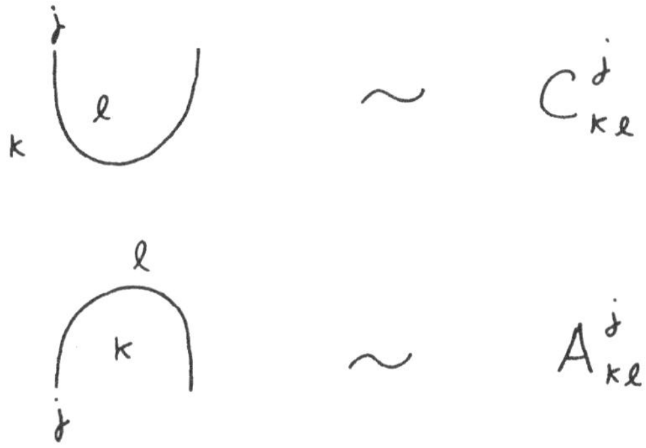

Pair creation and annihilation moves

The factors for these operations are determined by requiring that:

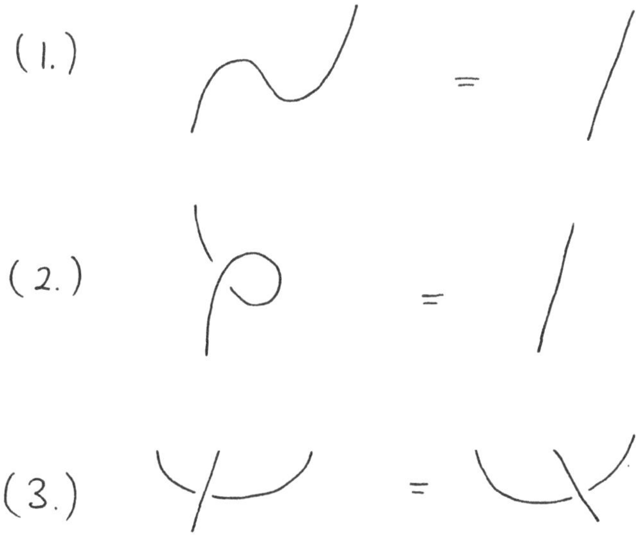

Consistency conditions on pair creation and annihilation

We make the *ansatz*

$$A^{j}_{ik} = \alpha_j F_{k0} \begin{bmatrix} i & i \\ j & j \end{bmatrix}$$

$$C^{j}_{ik} = \beta_j F_{0k} \begin{bmatrix} i & j \\ i & j \end{bmatrix}$$

and deduce from the first consistency condition that

$$\alpha_i \beta_i = \frac{1}{F_i}$$

Since for a closed graph there is always an equal number of α_i and β_i, we can set, without loss of generality, $\alpha_i = \beta_i = \frac{1}{\sqrt{F_i}}$.

This result leads to a new interpretation of Deligne's condition discussed earlier. It is simply the requirement that the value of a circle is a trace. Hence it should be an integer in group theory.

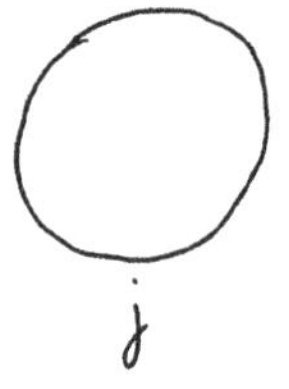

$$= n_j = \dim R_j \in \mathbb{Z}_+$$

Deligne's condition

In RCFT it is the relative dimension, as explained in the above. We will see below how this follows from the three-dimensional viewpoint.

• Exercise 7.1 *No more consistency conditions.* Show that consistency conditions (2) and (3) are automatically satisfied by using the polynomial equations discussed above and this value of A^i_{jk} and C^i_{jk}.

The non-trivial problem in knot theory is to prove that this procedure leads to a knot invariant. In other words, different projections of the same knot to two dimensions lead to the same result for the knot invariant. From the discussion in the previous sections and these exercises, it is clear that the polynomial equations guarantee this fact and we indeed find a knot invariant from every RCFT.

• Exercise 7.2 *Reidemeister Moves.* In the combinatorial approach to knot theory one must check the Reidemeister moves

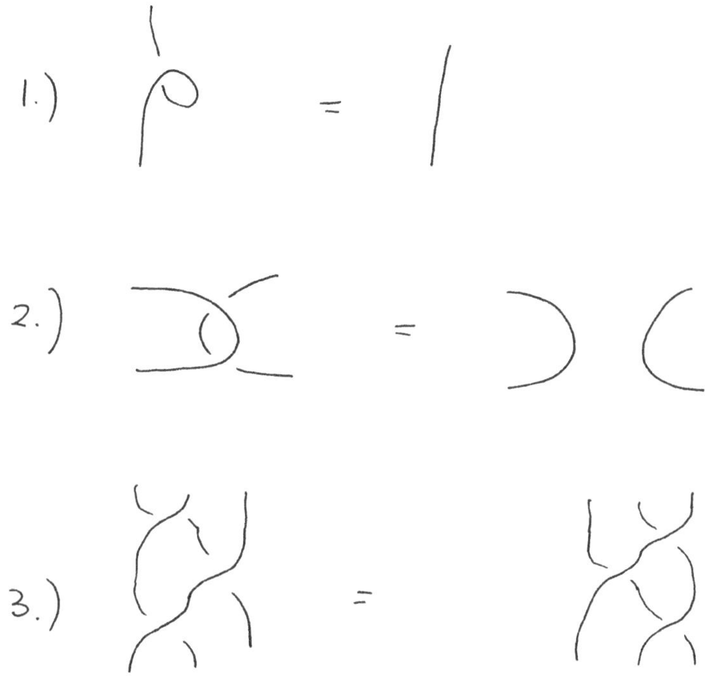

The three Reidemeister moves

Check these using the above formalism. Note that the first move is only satisfied up to phase. This may be fixed by discussing framed links or by introducing the writhe, following Kauffmann [44].

The analysis can easily be generalized to graphs with vertices, which are the analogs of the fusing move of conformal field theory. Define fusing and defusing moves

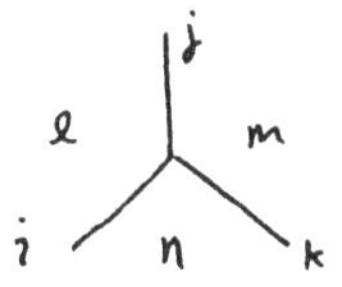

$$= f^{j}_{ik}\, F_{nj}\begin{bmatrix} i & k \\ l & m \end{bmatrix}$$

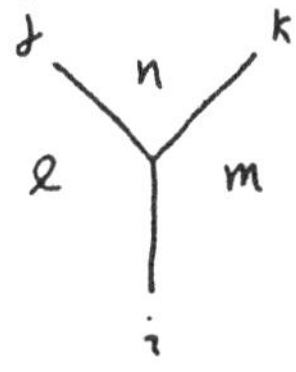

$$= f^{jk}_{i}\, F^{-1}_{in}\begin{bmatrix} j & k \\ l & m \end{bmatrix} = f^{jk}_{i}\, F_{in}\begin{bmatrix} m & k \\ l & j \end{bmatrix}$$

Fusing and defusing

• Exercise 7.3 *Consistency conditions on fusing and defusing.* Impose the relations

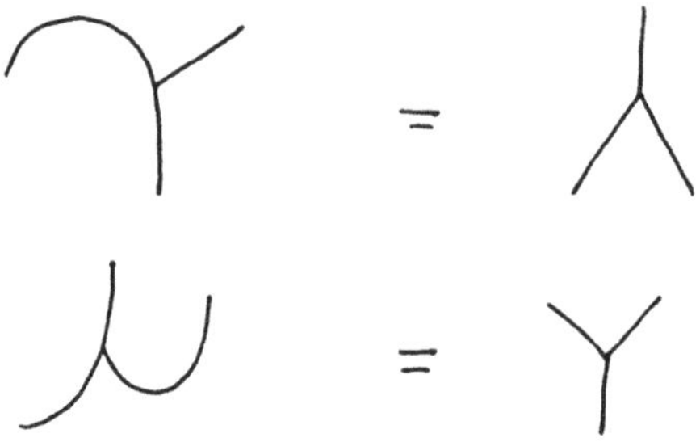

Consistency conditions on fusing and defusing

Derive $\frac{1}{\sqrt{F_i}} f_k^{ij} = \frac{1}{F_{k0}\left[\begin{smallmatrix} j & j \\ i & i \end{smallmatrix}\right]} f_{ik}^{j}$. Normalize the constants f such that if one of the lines corresponds to the identity representation, this line can be dropped from the graph and find the rules

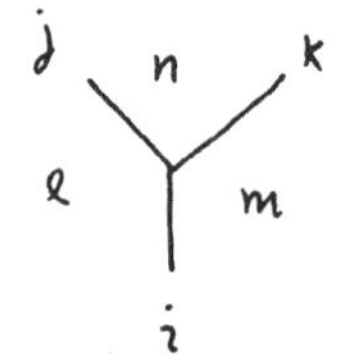

$$= \frac{F_{in}\left[\begin{smallmatrix} m & k \\ l & j \end{smallmatrix}\right]}{F_{i0}\left[\begin{smallmatrix} j & j \\ k & k \end{smallmatrix}\right]} \left(\frac{F_j F_k}{F_i}\right)^{1/4} \tag{7.1}$$

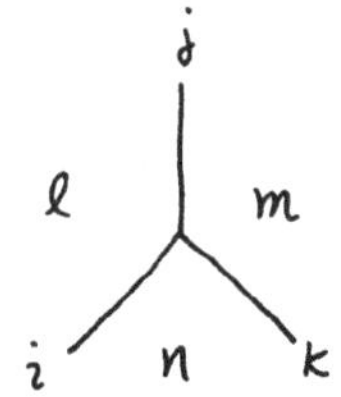

$$= F_{nj}\begin{bmatrix} i & k \\ l & m \end{bmatrix} \left(\frac{F_j}{F_i F_k}\right)^{1/4} \tag{7.2}$$

• Exercise 7.4 *Another consistency check.* Use the hexagon to show that

$$= e^{-i\pi(\Delta_j + \Delta_k - \Delta_i)}$$

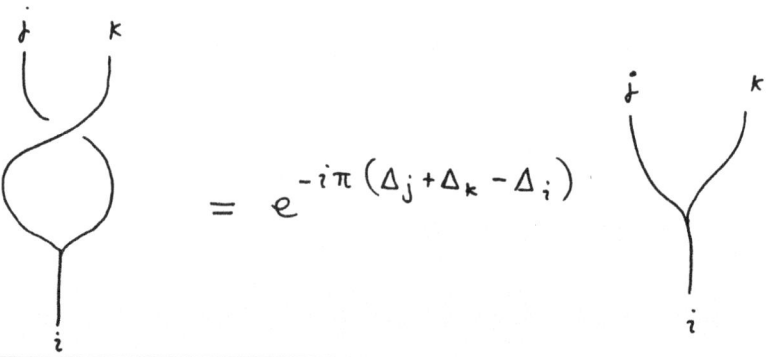

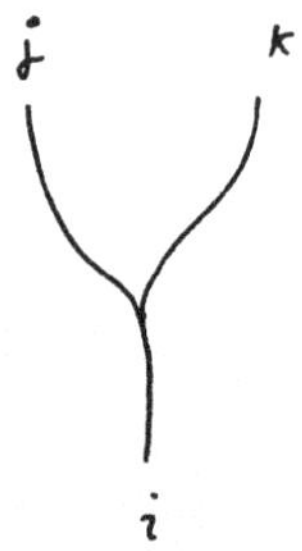

• Exercise 7.5 *Simple calculations.* Use the rules to compute the invariant of the graphs

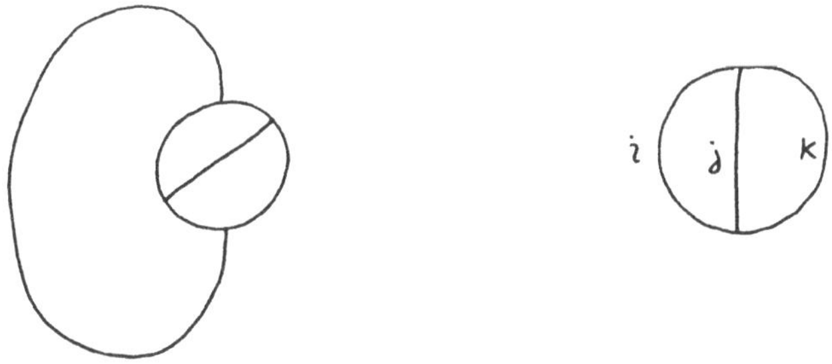

Two simple graphs

Use exercise 6.6 to write the second graph as $\frac{d_{ijk}}{\sqrt{F_i F_j F_k}}$ when the conditions of that exercise are fulfilled.

Using these rules one can compute invariants of knotted graphs. As in the case without the vertices, the polynomial equations guarantee the consistency.

• Exercise 7.6 *Gauge invariance.* Show that the invariant of knot without vertices is gauge invariant, i.e. it does not change if we rescale the CVO's and correspondingly the duality matrices. How do knots with vertices transform under such a rescaling? Interpret it.

It is convenient to pick the "good gauge"

$$F_{k0}\begin{bmatrix} i & i \\ j & j \end{bmatrix} = \sqrt{\frac{F_i F_j}{F_k}} \tag{7.3}$$

Write the fusing and the defusing rules in this gauge. Show that when the conditions of exercise 6.6 are fulfilled $d_{ijk} = 1$ in this gauge. Evaluate the two graphs in exercise 7.5 in this gauge. This gauge was used in [40][42].

• Exercise 7.7 *Symmetries of F.* Use the pentagon to show that

$$F_{n0}\begin{bmatrix} i & i \\ l & l \end{bmatrix} F_{pi}\begin{bmatrix} j & k \\ n & l \end{bmatrix} = F_{p0}\begin{bmatrix} k & k \\ l & l \end{bmatrix} F_{nk}\begin{bmatrix} i & j \\ l & p \end{bmatrix} \tag{7.4}$$

In the good gauge of exercise 7.6 this becomes

$$\sqrt{F_i F_p} F_{pi}\begin{bmatrix} j & k \\ n & l \end{bmatrix} = \sqrt{F_n F_k} F_{nk}\begin{bmatrix} i & j \\ l & p \end{bmatrix}$$

Define $W_{pi}\left[\begin{smallmatrix} j & k \\ n & l \end{smallmatrix}\right] = \sqrt{F_i F_p} F_{pi}\left[\begin{smallmatrix} j & k \\ n & l \end{smallmatrix}\right]$ and use the symmetries of exercise 3.5 to show that

$$W_{mn}\begin{bmatrix} i & j \\ k & l \end{bmatrix} = W_{kj}\begin{bmatrix} l & m \\ n & i \end{bmatrix} = W_{nm}\begin{bmatrix} j & l \\ i & k \end{bmatrix} = W_{nm}\begin{bmatrix} l & j \\ k & i \end{bmatrix} \tag{7.5}$$

These symmetries generate a tetrahedral symmetry generalizing the symmetry satisfied by $SU(2)$ Racah coefficients. Use the results of exercises 7.5 and 7.6 to explain the origin of this symmetry.

- Exercise 7.8 *Proof of the last equation on the torus*

The graphical formalism presented here is a very convenient tool in manipulating the duality matrices using the fundamental equations. We'll demonstrate this fact now by showing that the definition (4.3) of $S_{ij}(p)$ in terms of B and F satisfies the last equation on the torus $Sa = bS$. Consider the graph

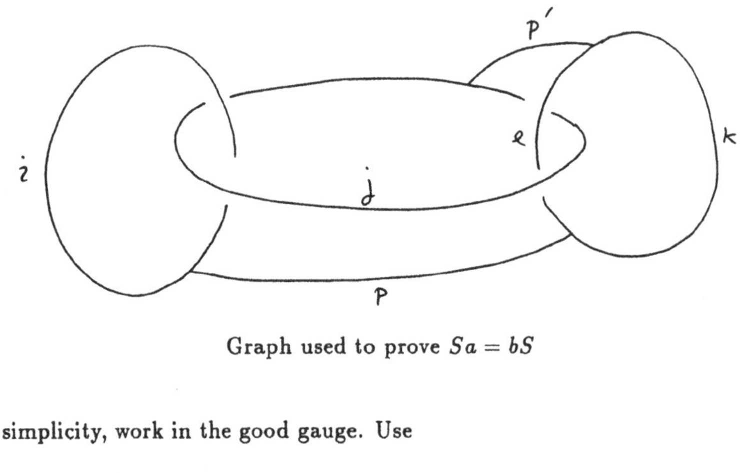

Graph used to prove $Sa = bS$

For simplicity, work in the good gauge. Use

$$S_{ij}(p) = S_{00}(0) e^{-i\pi\Delta_p} \frac{\sqrt{F_p}}{F_i F_j} \sum_r B_{pr}\begin{bmatrix} i & j \\ i & j \end{bmatrix}(-) B_{r0}\begin{bmatrix} j & i \\ i & j \end{bmatrix}(-) \tag{7.6}$$

to show that the graph has the value

$$\frac{1}{\sqrt{F_l F_k} S_{00}(0)} e^{i\pi\Delta_p} S_{ij}(p) \sum_r B_{pr}\begin{bmatrix} j & l \\ j & k \end{bmatrix}(+) B_{rp'}\begin{bmatrix} l & j \\ j & k \end{bmatrix}(+) \tag{7.7}$$

Now, deform the graph to

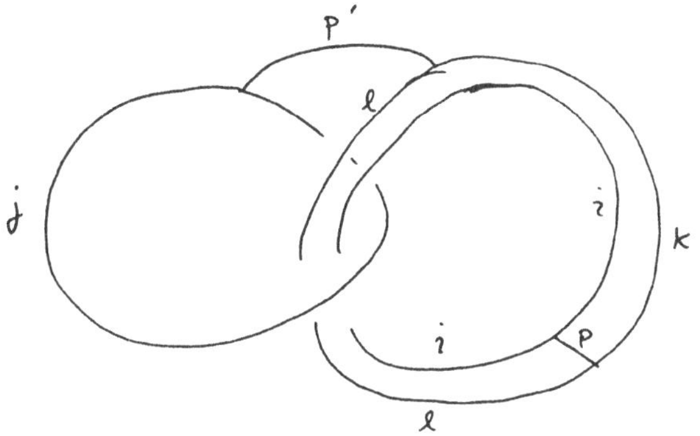

A deformation of the same graph

which differs from the original graph by a factor of $e^{i\pi(\Delta_k-\Delta_l)}$. Prove the identity

$\sum_s$ 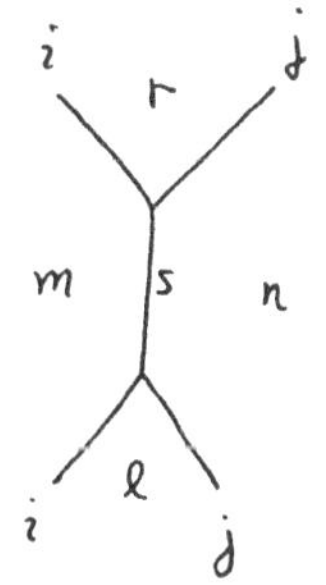 $\cdot \sqrt{\dfrac{F_i F_j}{F_s}} = \delta_{l,r}$

and use it to deform the graph to

$$\sum_s \sqrt{\frac{F_l F_i}{F_s}}\; e^{i\pi(\Delta_k-\Delta_l)}$$

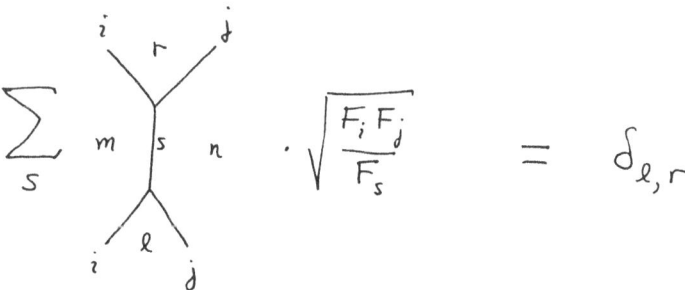

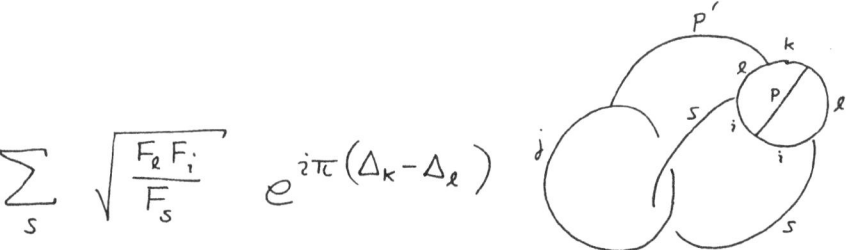

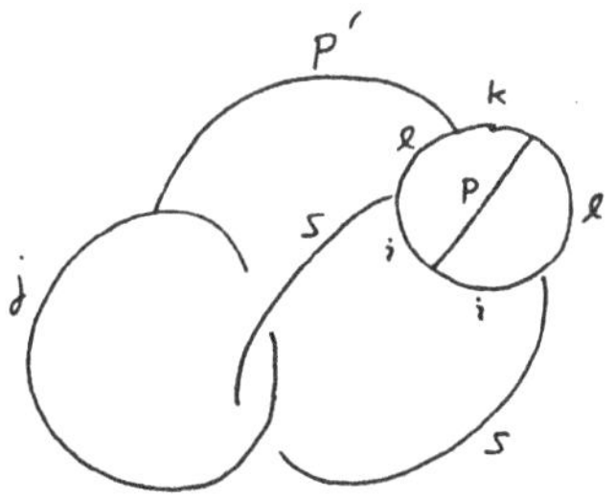

the original graph is equivalent to this graph

Turn this graph upside down and evaluate it. Use the symmetries of F and the expression for $S(p')$ to write it as

$$\frac{e^{i\pi(\Delta_k-\Delta_l)}}{S_{00}(0)\sqrt{F_k F_l}} \sum_s F_{ip'}\begin{bmatrix} s & s \\ k & l \end{bmatrix} F_{ps}\begin{bmatrix} i & l \\ i & k \end{bmatrix} S_{sj}(p') \tag{7.8}$$

Now express the a monodromy

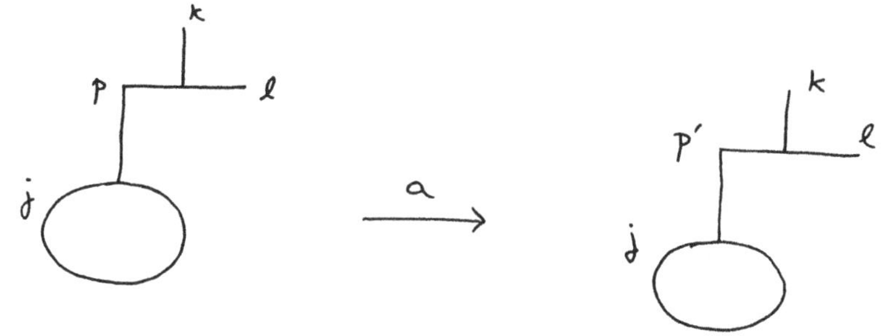

the general a monodromy

and the b monodromy

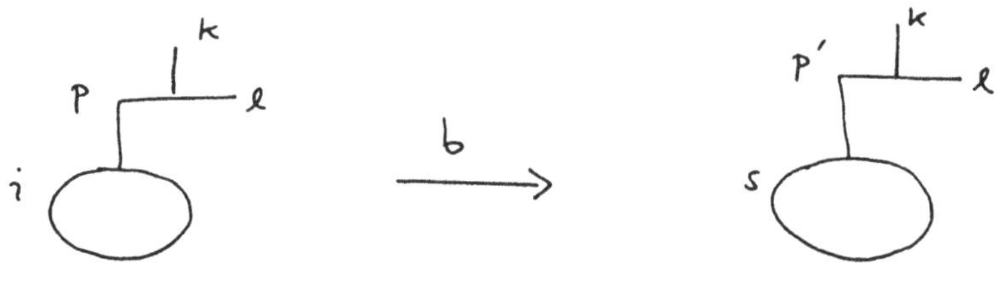

the general b monodromy

in terms of F and phases. Equate the two different expressions of the same graph (7.7) and (7.8) and use the expressions for these two monodromies to show that

$$Sa = bS$$

Therefore, this expression for S satisfies the last equation on the torus. Hence, this equation can be dropped from our list of axioms and be replaced by this definition of S.

• Exercise 7.9 *more identities for graphs.* Use the pentagon to show that

1.

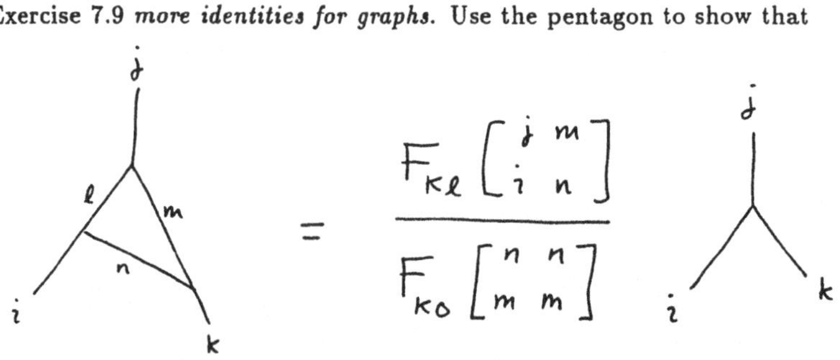

2. 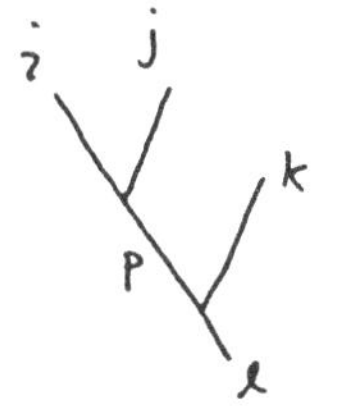 $= \sum_q F_{pq}\begin{bmatrix} j & k \\ i & l \end{bmatrix}$ 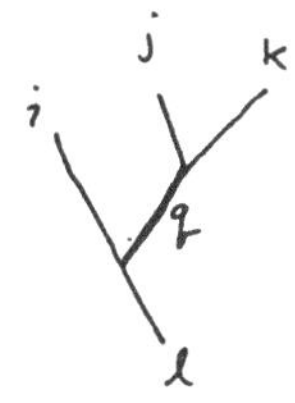

In all the manipulations with knots in S^3 we use only the polynomial equations on the plane. We do not need the torus equations. Therefore, quasirational as well as rational theories lead to knot invariants in S^3.

In the above discussion we have simply *defined* $S_{ij}(p)$ as a combination of certain duality matrices, exactly as in the axioms for a MTC. In order to see directly why, with this definition, S should be related to the modular group of the torus we must pause and discuss Witten's observation [27] that 2-dimensional duality (as axiomatized by the notion of a modular functor) is equivalent to 3 -dimensional general covariance.

One recent application of the knot invariants arising in RCFT has been to the construction of invariants of three manifolds [27][41][43] [45]. These applications are simply one facet of the current interest in studying the geometry and topology of manifolds via quantum field theory, through the general notion of topological QFT's. These were introduced by Witten and recently axiomatized by Atiyah. In 2 + 1 dimensions the Atiyah-Witten axioms, which summarize the formal properties of path integrals for topological field theories, are closely connected to the notion of a modular functor. To see this recall that the Atiyah-Witten axioms are [46] [47],

---

Axioms for a Topological Field Theory

Data:

1. A map from closed oriented d-manifolds to complex finite dimensional vector spaces $\Sigma \rightarrow \mathcal{H}(\Sigma)$.

2. A distinguished vector $Z(Y) \in \mathcal{H}(\Sigma)$ associated to $d+1$-manifolds such that $\Sigma = \partial Y$. (In particular if Y is closed $Z(Y)$ is a complex number.)

Conditions:

1. Naturality. If $f : \Sigma_1 \rightarrow \Sigma_2$ is an automorphism there is an isomorphism $\mathcal{H}(f) : \mathcal{H}(\Sigma_1) \rightarrow \mathcal{H}(\Sigma_2)$ satisfying $\mathcal{H}(f_1 f_2) = \mathcal{H}(f_1)\mathcal{H}(f_2)$. There is a similar naturality condition on the vectors $Z(Y)$.

2. Duality. $\mathcal{H}(\Sigma^*) \cong \mathcal{H}(\Sigma)^*$.

3. Multiplicativity. $\mathcal{H}(\Sigma_1 \coprod \Sigma_2) \cong \mathcal{H}(\Sigma_1) \otimes \mathcal{H}(\Sigma_2)$. Moreover $\mathcal{H}(\phi) \cong \mathbb{C}$.

4. Gluing. If Y and Y' are glued along a d-manifold Σ (with opposite orientations for Σ) to form $\tilde{Y}$ then

$$Z(\tilde{Y}) = \langle Z(Y), Z(Y') \rangle$$

The above makes sense since the opposite orientations of Σ allow us to pair a space with its dual.

5. Completeness. The states $Z(Y)$ for all Y with $\partial Y = \Sigma$ span $\mathcal{H}(\Sigma)$.

(Note: Atiyah adds a sixth axiom that $Z(Y^*) = Z(Y)^*$, but we will not need this.). Clearly for the case $d = 2$ the above notion is very close to that of a modular functor, in particular in any attempt to pass from one to the other the vector spaces $\mathcal{H}(\Sigma)$ are surely the same. Nevertheless, there are some things to prove. The precise connection was worked out in [48] [49]. To pass from a modular functor to a topological theory the main problem is to construct the vector $Z(Y)$ from the data of the modular functor. This was done in [48][49] by choosing a Morse function, using the data of the modular functor to define "transition amplitudes" between critical points of the Morse function and then checking that the choice of Morse function does not lead to ambiguities. To pass from the topological theory to the modular functor the main problem is to produce the finite set of labels (of "representations") and their fusion rule algebra, etc. An argument that this can be done is presented in [48]. The labels are a basis for the vector space $\mathcal{H}(torus)$.

The advantage of the point of view of modular functors and topological field theories is that for any system satisfying the axioms one can compute quantities for nontrivial graphs and nontrivial manifolds via the gluing axiom. In particular, one can compute various quantities using the notion of surgery.

If $\mathcal{H}(\Sigma)$ is an n dimensional vector space, any collection of $n+1$ vectors $Z_i \in \mathcal{H}(\Sigma)$ is linearly dependent; i.e. there are coefficients a_i such that $\sum_i a_i Z_i = 0$. This leads to a linear relation between the invariants of different manifolds. Let $Z_i = Z(Y_i)$ for $n+1$ different Y_i. Then,

$$\sum_i a_i Z(\tilde{Y}_i) = \sum_i a_i \langle Z(Y), Z(Y_i) \rangle = 0 \tag{7.9}$$

where $\tilde{Y}_i$ is obtained by gluing Y to Y_i along some d-fold Σ.

Rather than continuing in complete generality, we focus on the particular topological field theory corresponding to a RCFT. As explained above, the labels of the representations label a basis of $\mathcal{H}(T^2)$. The three manifold Y can have links carrying these labels (also links with vertices) and these links may terminate at the boundary of Y. For example, for Y a three ball with the link

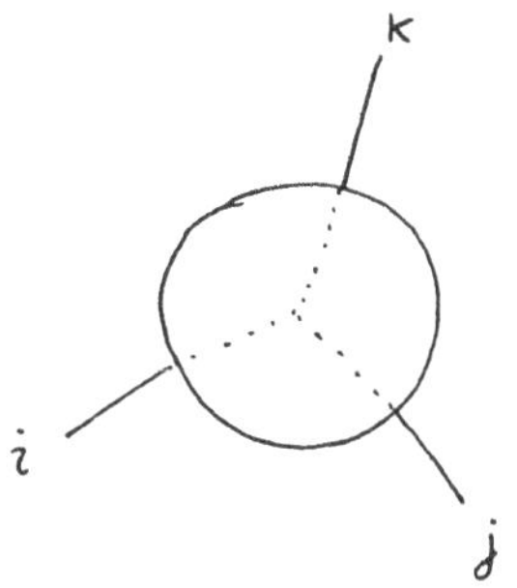

a link in a three ball

we find a vector $v \in \mathcal{H}(S^2_{ijk})$ where S^2_{ijk} is a sphere with three labeled points i, j, k. By the correspondence of a topological field theory and RCFT, $\mathcal{H}(S^2_{ijk}) \cong V_{ijk}$ and its dimension is N_{ijk} (if $N_{ijk} > 1$, we should specify the kind of coupling which is used in the vertex in the link). Continuing to assume for simplicity that $N_{ijk} = 0, 1$, the vector $\tilde{v} \in \mathcal{H}(S^2_{ijk})$ corresponding to

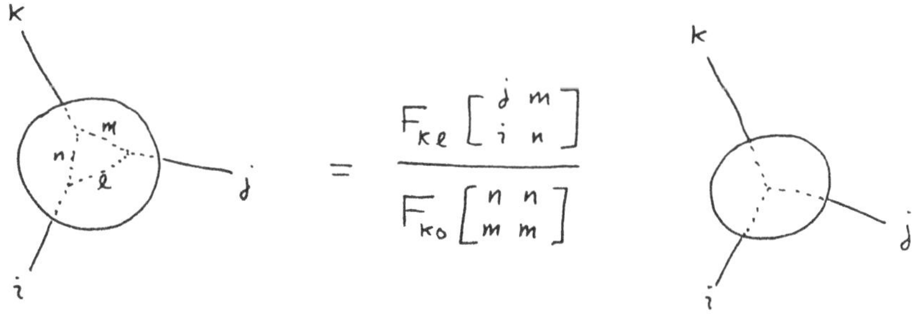

another link in a three ball

is proportional to the original one $\tilde{v} = xv$.

Now, consider a complicated three manifold Y with a link

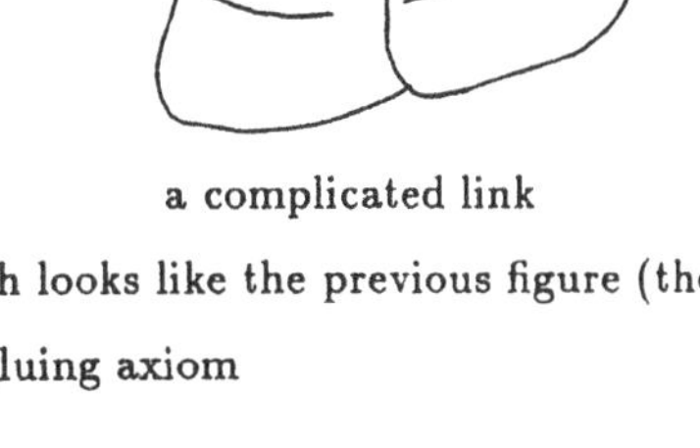

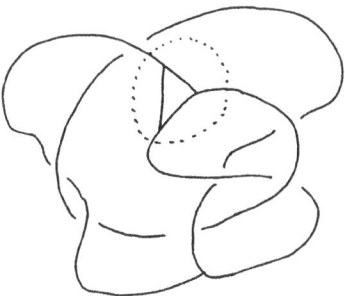

a complicated link

Remove the three ball which looks like the previous figure (the dashed line) to obtain the three manifold $\tilde{Y}$. By the gluing axiom

$$Z(Y) = \langle \tilde{v}, Z(\tilde{Y}) \rangle = x^* \langle v, Z(\tilde{Y}) \rangle = x^* Z(Y')$$

where Y' is the same as Y except that the ball is replaced by the simple link. This procedure simplifies the computation of $Z(Y)$ by relating it to a simpler object $Z(Y')$.

• Exercise 7.10 *Interpretation of previous results.*

a.) Use this understanding to interpret the first relation in exercise 7.9. Express x in terms of the duality matrices.

b.) Repeat this analysis for the sphere with four labels $ijkl$. Show that the vectors

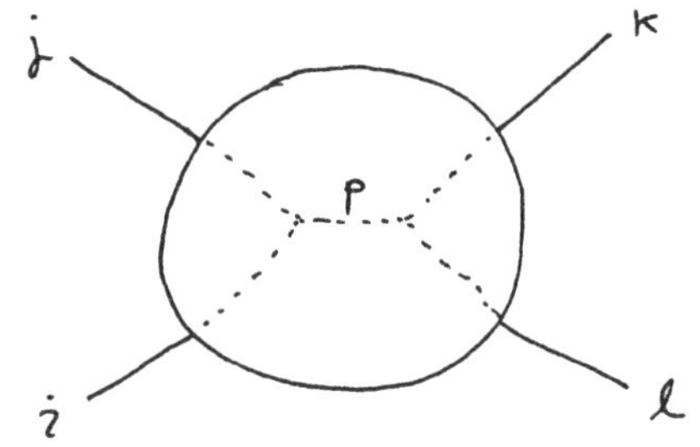

a basis for $\mathcal{H}(S^2_{ijkl})$

for all p span $\mathcal{H}(S^2_{ijkl})$. The vector of a given p corresponds in the RCFT to the conformal block with the representation p in the intermediate channel. The second relation in exercise 7.9 expresses duality in RCFT. Interpret it from three dimensions.

c.) Cut the tetrahedron graph (the first figure in exercise 7.5) along the lines i, j, l, n and express the invariant of the graph as an inner product of two vectors in $\mathcal{H}(S^2_{ijln})$. Use part b of this exercise to explain why the tetrahedron graph is proportional to F.

d.) Interpret the equations for the ope coefficients d as determining a metric on $\mathcal{H}$ as mentioned in the end of section 6. Use this fact to interpret the second graph in exercise 7.5 as $\frac{d_{ijk}}{\sqrt{F_i F_j F_k}}$.

e.) Interpret the gauge invariance as a freedom in the normalization of the vectors in $\mathcal{H}(\Sigma)$.

This interpretation is more powerful when combined with the notion of surgery [27]. First notice that $\mathcal{H}(T^2)$ is spanned by $v_i = Z(M_i)$ where M_i is a solid torus with a line with the label i around the non-contractible cycle. Consider a three manifold Y_i with a closed line with the label i. Removing a solid torus M_i surrounding the line from Y_i, we find the three manifold $\tilde{Y}$. By the gluing axiom, $Z(Y_i) = \langle Z(\tilde{Y}), v_i \rangle$. Now consider another three manifold X obtained by interchanging the a and b cycles [1] on the boundary of M_i and then gluing it back to $\tilde{Y}$. The relevant inner product is

---

[1] The a cycle is the contractible cycle inside M_i; however, there is an ambiguity in what we mean by the b cycle. We will return to this ambiguity shortly.

$$Z(X) = \sum_j S_{ij} \langle Z(\tilde{Y}), v_j \rangle = \sum_j S_{ij} Z(Y_j)$$

As before, we succeeded to express Z of some manifold in terms of Z's of other (simpler) manifolds. Using this procedure it is possible to compute Z for every manifold [27].

• Exercise 7.11 *Ambiguity in surgery.* Show that the ambiguity associated with the choice of the b cycle corresponds to the application of T in RCFT. Therefore, it is related to the fact that the lines have to be framed. How does the framing remove the ambiguity?

• Exercise 7.12 *Some calculations using surgery.*

a.) The invariant for two parallel nonbraiding (="cabled") lines W_i, W_j in $S^2 \times S^1$ is N_{0ij}. Why?

b.) Think of $S^2 \times S^1$ as two solid tori whose toroidal boundaries are identified via the identity map $(\sigma^1, \sigma^2) \to (\sigma^1, \sigma^2)$. Change the identification to the transformation: $S : (\sigma^1, \sigma^2) \to (-\sigma^2, \sigma^1)$. Show that the resulting three-manifold is just S^3.

c.) Suppose the two solid tori of part (b) contain lines W_i and W_j respectively. Each line wraps along the noncontractible direction. Show that the resulting configuration in S^3 is just:

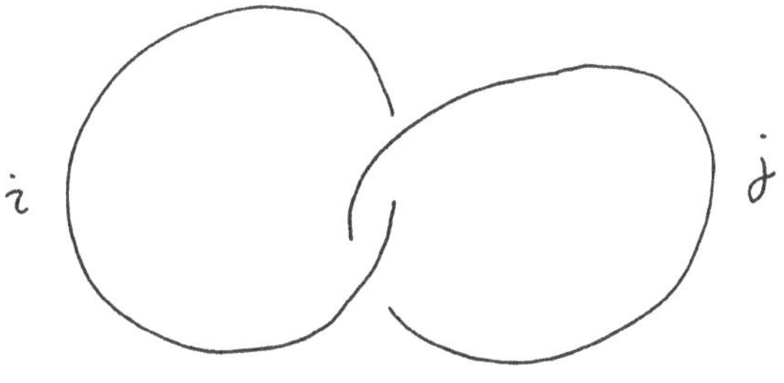

A configuration of lines in S^3

and therefore the invariant of this graph is S_{ij}.

d.) Using the graphical formalism described above, compute the figure in part (c) and rederive the formula

$$\frac{S_{ji}}{S_{00}} = \frac{\left(B\begin{bmatrix} j & i \\ j & i \end{bmatrix} B\begin{bmatrix} i & j \\ j & i \end{bmatrix} \right)_{00}}{F_i F_j}$$

we derived in a previous exercise. Notice that the graphical rules did not include an overall normalization factor of S_{00} for every graph in S^3. This factor is natural from the surgery point of view if the invariant in part a of this exercise is normalized to be N_{0ij}.

e.) Compute the invariant for two cabled lines W_i and W_j in $S^2 \times S^1$ as before but this time connected by a line with the label p:

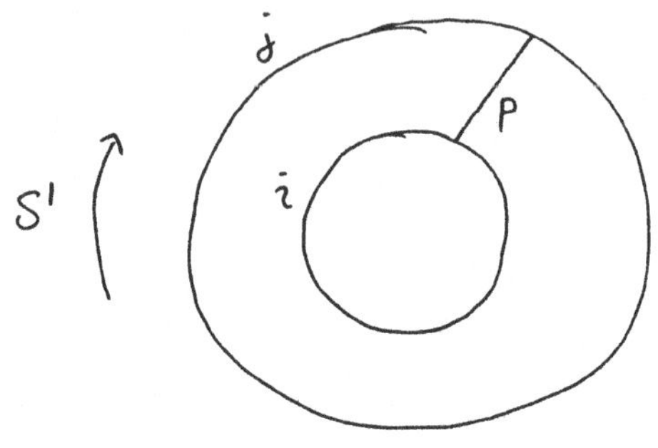

a configuration in $S^2 \times S^1$

f.) Perform surgery as above using $S(p)$ and turn this into

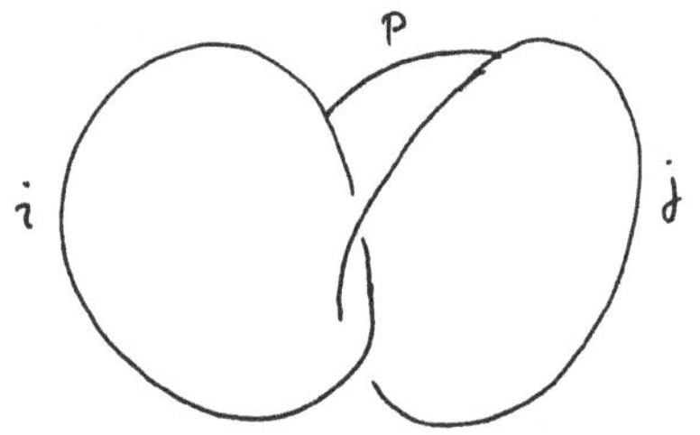

the previous graph after surgery

in S^3. Compute this graph using our rules and derive equation (4.3). (Because of the framing, there is a phase ambiguity. The phase $e^{-i\pi\Delta_p}$ is determined by consistency.)

• Exercise 7.13 *Verlinde's formula from Surgery.* We outline a slightly modified proof of E. Witten of Verlinde's formula.

a.) Consider the configuration:

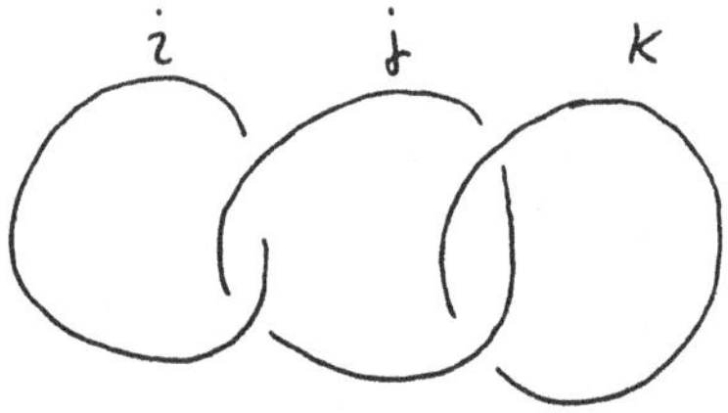

A configuration used in the proof of Verlinde's formula

Using the graphical rules and the above formula for S in terms of B show that this has the value:

$$\frac{S_{ij}S_{jk}}{S_{0j}}$$

b.) Rewrite the above as

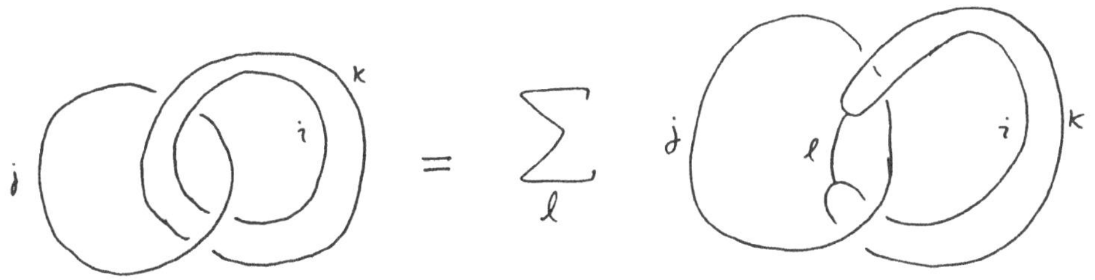

Use the identity $FF^{-1} = 1$ and the braiding/fusing identity to rewrite this as:

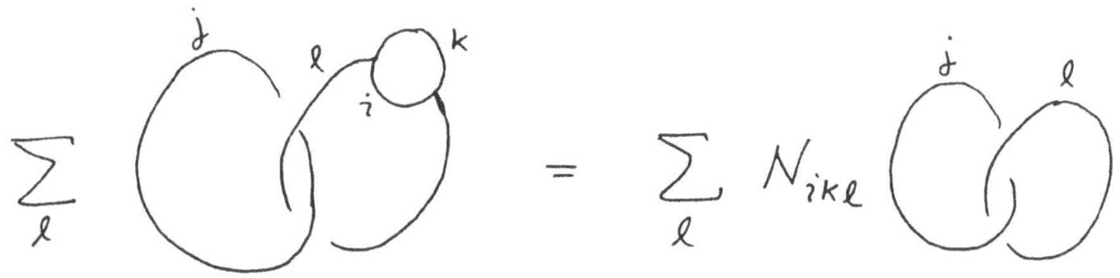

From this derive Verlinde's formula.

• Exercise 7.14 *a and b monodromies for the two point function on the torus.* Relate the graph

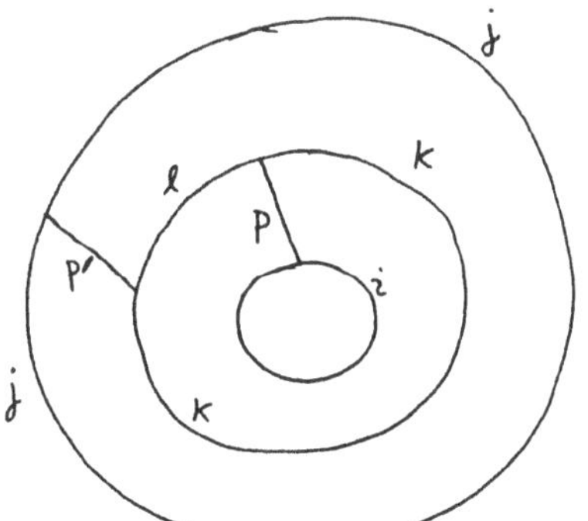

graphical formulas for the b monodromy

in $S^2 \times S^1$ to the b monodromy. Use surgery to relate it to the figure used in exercise 7.8. Find a graph in $S^2 \times S^1$ for the a monodromy and use surgery to relate it to the figure used in exercise 7.8. Thus making the previous proof of $Sa = bS$ somewhat intuitive.

We see that the information in surgery is equivalent to the information in the equation $Sa = bS$ which in turn is equivalent to the formula for $S(p)$ in terms of F and B.

We have seen that a RCFT defines a modular functor, which has been argued to give rise to a topological 2+1 dimensional theory. Recently L. Crane [45] has shown more directly that the data F, B, S can be used to construct invariants of framed 3-folds through the use of some theorems from combinatorial topology. For example, to identify the invariant associated to a closed 3-fold Y we use a "Heegaard splitting" whereby Y is represented as a glued pair of handlebodies Y_1, Y_2 which have as a common boundary the surface Σ. Y_1 is glued to Y_2 via a nontrivial diffeomorphism ϕ of Σ. Among the conformal blocks $\mathcal{H}(\Sigma)$ there is a distinguished (normalized) vector χ_0 defined by the condition that the trivial representation be present on all internal lines. Representing ϕ by the duality matrix $Z(\phi)$ we have the invariant $Z(Y) = \langle \chi_0, Z(\phi)\chi_0 \rangle$. Since the Heegaard decomposition is not unique it is nontrivial that $Z(Y)$ is an invariant. Using known facts about Heegaard splittings Crane shows that the axioms of an MTC guarantee that $Z(Y)$ is unambiguous up to a factor of $e^{2\pi i c/24}$. Yet another approach, due to Reshetikhin and Turaev [41] will be mentioned in the following section.

So far the discussion was very general and did not depend on a particular three dimensional theory. In [27] Witten considered the Chern-Simons-Witten gauge theory in three dimensions. This is a topological field theory and therefore the general analysis in this section applies there. Moreover, this theory can be solved exactly [27] and explicit expressions for the duality matrices can be obtained. The study of this theory is the subject of sections 9 and 10.

8. Quantum group solutions of the polynomial equations

This section contains some remarks intended for those already familiar with basic facts about quantum groups. Thus we assume some familiarity with [50] [40]. A nice review of the subject is [51].

If A is a Hopf algebra then the category of its finite dimensional representations $Rep(A)$ has a tensor product which may be defined by the comultiplication Δ. From the axioms satisfied by a comultiplication there will be an associativity constraint satisfying a pentagon consistency relation. In the previous terminology, the F matrix will exist and will satisfy the pentagon relation. In general there will be no commutativity constraint, i.e., there will be no analog of Ω. If A is a quasitriangular Hopf algebra (see [50], essentially it means that the comultiplication and opposite comultiplication are conjugate by a "universal" R matrix.) then there is a commutativity constraint, but in general $\Omega^2 \neq 1$. In this case there will be two hexagon conditions. These hexagon conditions are equivalent to Drinfeld's formulae $(\Delta \otimes 1)R = R_{13}R_{23}$ and $(1 \otimes \Delta)R = R_{12}R_{23}$. In this case $Rep(A)$ is a braided monoidal category. In [41] a central extension of a quasitriangular Hopf algebra is defined which these authors call a "ribboned Hopf algebra." The extra conditions specified for a ribboned Hopf algebra are such that in this case $Rep(A)$ is a "compact braided monoidal category," which in our terms means that when F, B matrices are suitably identified with quantum group Racah coefficients (in a way precisely analogous to the discussion of group theory above) then the genus zero axioms of a MTC are fulfilled. (Except, perhaps, for the finiteness of the index set I.) Correspondingly, in [41] $Rep(A)$ for a ribboned Hopf algebra is used to define invariants of knotted graphs [2] in $\mathbb{R}^3$.

An important special case of ribboned Hopf algebras is provided by the quantized universal enveloping algebras $U_q(\mathcal{G})$ for a Lie algebra $\mathcal{G}$. Applying the machine of [41] one may obtain invariants of knots in S^3 for any deformation parameter q. However when q is "rational," which means that $q^n = 1$ for some integer n, something more remarkable happens. In this case one may truncate the set of representations to a set of 'good' or 'type II' representations [52] [53], characterized as a minimal complete set of representations with nonvanishing quantum dimension, such that the truncated space of representations defines a modular tensor category.

The most famous and well-known example of this phenomenon is provided by $U_q(sl(2))$. In this case it has been shown that the braiding and Racah matrices for the case $q = e^{2\pi i/(k+2)}$ are *identical* to those of the conformal field theory $\widehat{su}(2)_k$ when we restrict the class of representations and invariant tensors to the "good" ones generated by irreducible representations of dimensions $\leq k+1$ and couplings satisfying the $\widehat{su}(2)_k$ fusion rules. The proof of this statement may be obtained as follows. One first computes the braiding matrices for spin 1/2 operators [20] and notices the exact correspondence with the corresponding quantum group objects. In conformal field theory the other braiding matrices may then be obtained by successive use of the braiding/fusing relation. Then one proves that it is valid to truncate the quantum group braiding/fusing relation so that it only includes the good representations. Another argument, using properties of Hecke and TLJ algebras has been advocated by Alvarez-Gaumé, Gomez, and Sierra [51]. With the coincidence of F, B matrices one may define S as in (4.3) and hence the restricted quantum group representation theory defines a MTC. Analogous statements exist for other $U_q(\mathcal{G})$ and full proofs for all cases have been published in [54]. The coincidence of F, B matrices has been widely noted and discussed. Just a few references include [20][54][15][55] [56] [57] [58] [51].

These observations allow one to give very explicit formulae for braiding/fusing matrices (which are more easily obtained by using quantum group technology). For example, very explicit formulae where written down in [40]. As a simple example we quote the well-

[2] More precisely, invariants of colored directed ribboned tangles.

known result for a braiding matrix of two spin 1/2 fields. The relevant space of conformal blocks is two-dimensional corresponding to intermediate spins $j \pm \frac{1}{2}$ and we have

$$B_{rs}\begin{bmatrix} \frac{1}{2} & \frac{1}{2} \\ j & j \end{bmatrix} = q^{1/4}\delta_{r,s} - q^{-1/4}\frac{\sqrt{S_r S_s}}{S_j}$$

where $S_j = sin\frac{\pi(2j+1)}{k+2}$. Alternatively this may be written

$$B\begin{bmatrix} \frac{1}{2} & \frac{1}{2} \\ j & j \end{bmatrix} = \frac{1}{[2j+1]}\begin{pmatrix} \frac{-q^{-(j+3/4)}}{\sqrt{q^{-1/2}[2j][2j+2]}} & \frac{\sqrt{q^{-1/2}[2j][2j+2]}}{q^{j+1/4}} \end{pmatrix}$$

where $[n] = (q^{n/2} - q^{-n/2})/(q^{1/2} - q^{-1/2})$.

In [41] Reshetikhin and Turaev represent 3-manifolds via surgery on links and use the surgery procedures of Witten to reduce the invariants of three-folds to those associated to links (or tangles). Their paper can be viewed as another construction of a three-dimensional topological field theory, starting from the MTC associated to the representation theory of $U_q(sl(2))$ for $q^{k+2} = 1$ (and, in principle, to other $U_q(\mathcal{G})$.) The link or tangle invariants are computed essentially as transition amplitudes of conformal blocks, along the lines described above.

The fact that the type II representation theory of $U_q(\mathcal{G})$ for rational deformation parameters coincides with the MTC of a canonically associated RCFT is still something of a mystery. The statement of this fact has been formulated in a number of conformal field theoretic constructions [51][57][59] [60] but these descriptions make use of the fact rather than explain it. Another connection of CFT to quantum groups has been noted in [61]. In [27] Witten proposed one approach to this problem, which, if successfully brought to conclusion would yield an adequate explanation. More recently Witten has proposed a different explanation in [62]. In the remainder of this section we present an alternative interpretation of Witten's idea.

We begin by noting that the quantum $3j$ symbols themselves may be seen to form an algebra. Namely, using the formalism of [40] we have

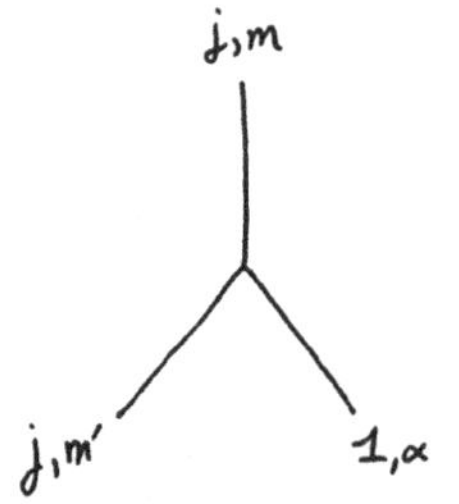

$$= \begin{bmatrix} j & 1 & j \\ m & \alpha & m' \end{bmatrix} \equiv (T_\alpha)_{mm'}$$

Graphical representation of a $3j$ symbol with one line carrying spin 1.

which we will take to define the matrix elements of three operators $T_{\alpha=-1,0,+1}$. By the very definition of Racah coefficients we may write

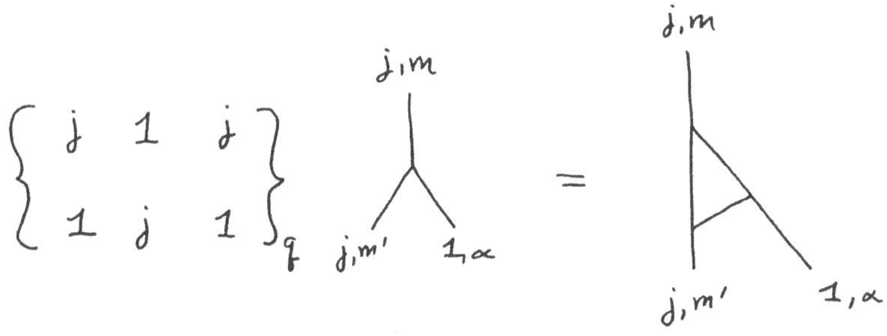

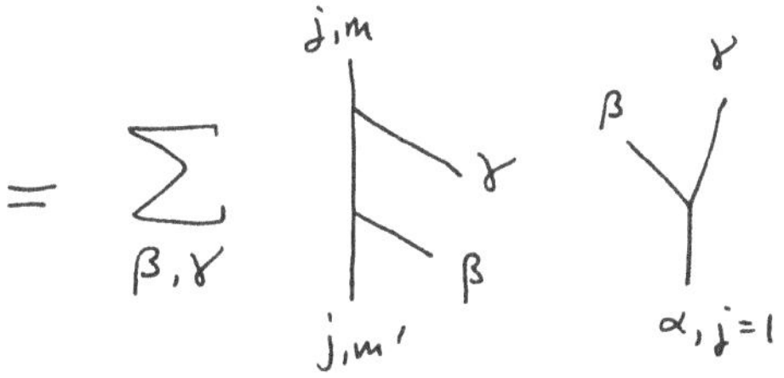

where

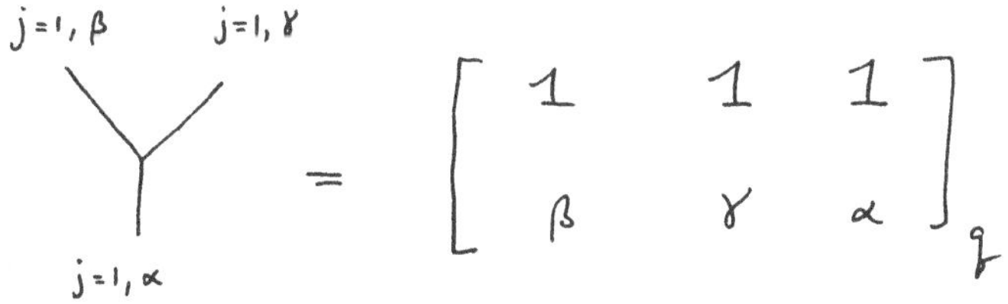

3j symbol for coupling three spin 1 representations

and we will denote the Racah coefficient by A_j.

Clearly the above formula may be regarded as defining an algebra for the T_α operators, the structure constants being defined by the 3j symbols for three spin 1 representations and the Racah coefficient A_j. That is, we may write:

$$\sum_{\beta,\gamma} \begin{bmatrix} 1 & 1 & 1 \\ \alpha & \beta & \gamma \end{bmatrix} T_\beta T_\gamma = A_j T_\alpha$$

For example, for $U_q(sl(2))$ one may easily compute:

$$q^{-1/2}T_+T_0 - q^{1/2}T_0T_+ = A_jT_+$$

$$T_+T_- - T_-T_+ = (q^{1/2} - q^{-1/2})T_0^2 + A_jT_0$$

$$q^{-1/2}T_0T_- - q^{1/2}T_-T_0 = A_jT_-$$

for any value of q. This is precisely the algebra derived in [62]. The reason for this is that graphs are computed with quantum Racah or 6j symbols. But, upon analytic continuation away from $|q| = 1$ the 6j symbols have large spin limits which are precisely 3j symbols. More precisely we have [40]

$\lim_{a\to\infty}$ 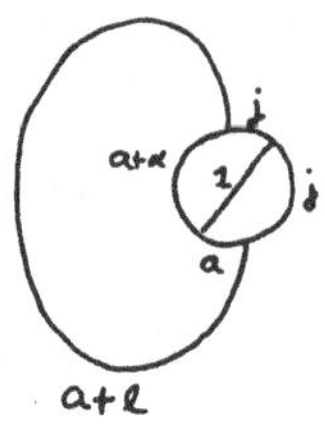

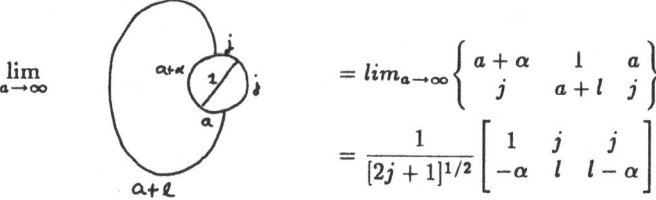

$$= lim_{a\to\infty}\left\{\begin{matrix} a+\alpha & 1 & a \\ j & a+l & j \end{matrix}\right\}$$

$$= \frac{1}{[2j+1]^{1/2}}\begin{bmatrix} 1 & j & j \\ -\alpha & l & l-\alpha \end{bmatrix}$$

Thus Witten's lassoing and limiting procedure produces the algebra of $3j$ symbols.

9. Chern-Simons-Witten gauge theory – Quantization

The discussion in section 7 was quite general. It can be made much more explicit in a particular field theory – the CSW theory[27]. This is a particular example (we will later mention a conjecture that this is essentially the only example) of a topological field theory. The theory is a gauge theory based on the gauge field $A = A^a_\mu \, T^a \, dx^\mu$ in some Lie algebra g with action

$$S = \frac{k}{4\pi}\int_Y Tr(AdA + \frac{2}{3}A^3)$$

for a three manifold Y. For simplicity we limit ourselves here to $SU(N)$ gauge theory with a trace in the fundamental representation ($TrT^aT^b = -\delta^{ab}$).

Clearly, the action is independent of the metric on Y. To prove that the theory is indeed topological, one needs to show that the measure of the functional integral is also independent of the metric. In what follows, we will assume that this is the case[3].

Perhaps the easiest way to understand the theory is by canonical quantization. Suppose we have a Riemann surface Σ and consider the theory on the 3-dimensional manifold $Y = \Sigma \times \mathbb{R}$.

If we canonically quantize the theory we obtain a space of physical states $H(\Sigma)$ associated to the surface Σ. Witten showed that these states have a natural interpretation in terms of the WZW model for g-current algebra at level k. Specifically:

$$\Sigma = \text{closed surface} \Rightarrow \quad H_\Sigma = \left\{\begin{array}{l} \text{vector space of} \\ \text{conformal block for} \\ \text{partition function} \\ \text{on } \Sigma \end{array}\right.$$

$$\Sigma = \left\{\begin{array}{l} \text{surface pierced by} \\ \text{Wilson line in} \\ \text{Representations } j_1, \cdots j_n \end{array}\right. \quad \Rightarrow \quad H_\Sigma = \left\{\begin{array}{l} \text{conformal blocks for n-point} \\ \text{function on } H_\Sigma \text{ for n fields} \\ \text{in the representations: } j_1 \cdots j_n \end{array}\right.$$

[3] In [63] Witten showed that the existence of the central charge in two dimensions is related to some dependence on the metric on Y – the theory depends on the "framing on Y".

Moreover for 3-manifolds interpolating between two surfaces Σ_1 and Σ_2 the path integral gives a transformation $H(\Sigma_1) \longrightarrow H(\Sigma_2)$. Witten shows that these transformations are just the duality transformations on the space of blocks. Why is it true? We will explain these matters in a simple physical way.

Choose $A_0 = 0$ gauge: If Σ has no boundary then

$$S = \frac{k}{4\pi} \int \epsilon^{ij} Tr A_i \frac{d}{dt} A_j$$

We then have a first order Lagrangian and therefore, the phase space is the space of gauge fields on Σ. The symplectic structure on this space leads to the commutation relations

$$\{A_i^a(x), A_j^b(y)\} = \epsilon_{ij}\delta^{(2)}(x-y)\frac{4\pi}{k}.$$

where $\int \delta^{(2)}(z-w)d^2z = 1$. It is convenient to pick a complex structure τ on Σ and to write

$$\{A_z^a(z), A_{\bar{z}}^b(w)\} = \delta^{(2)}(z-w)\frac{4\pi}{k}.$$

The wave functions in holomorphic quantization are holomorphic functions of A_z, $\psi = \psi(A_z)$. The Hilbert space is the space of all these functions. The physical space is the subspace of the Hilbert space which is invariant under the Gauss law.

• Exercise 9.1 *Gauss' law.* Show that

$$u(\epsilon) = \frac{ik}{4\pi} \int Tr(\epsilon F)$$

generates an infinitesimal gauge transformation by ϵ:

$$[u(\epsilon), A] = -D\epsilon$$

$$[u(\epsilon_1), u(\epsilon_2)] = u([\epsilon_1, \epsilon_2])$$

By integrating $u(\epsilon)$ the operator generating a finite transformation $g = e^{\epsilon}$ is

$$U(g) = e^{u(\epsilon)}$$

so

$$U(g)AU^{-1}(g) = gAg^{-1} - dgg^{-1}$$

Now how does it act on physical states? We certainly must have:

$$(U(g)\psi)(A_z) = e^{f(A_z;g)}\psi(A_z^g)$$

to find f, we impose the group law:

$$U(h)U(g) = U(gh)$$

and find:

$$f(A;gh) = f(A;h) + f(A^h;g) \bmod 2\pi i k$$

The solution is:

$$\begin{aligned} f(A_z;g) = & \frac{ik}{4\pi}\int Trg^{-1}\partial g g^{-1}\bar{\partial} g + k\Gamma_{WZ}(g) \\ & - \frac{ik}{2\pi}\int Tr(A_z g^{-1}\bar{\partial} g) \equiv ikS(g:A_z,0) \end{aligned}$$

So

$$(U(g)\psi)[A_z] = e^{ikS(g;A_z,0)}\psi[A_z^g]$$

This is the key equation. From it we may get the independent physical states as follows.

Physical states are invariant under the Gauss law - so we are looking for linearly independent solutions to the equation

$$\psi(A_z) = e^{ikS(g;A_z,0)}\psi(A_z^g)$$

Now, given *any* functional ψ_0 we can generate such a solution by

$$\psi_{phys} = \int DgU(g)\psi_0$$

i.e. we can write:

$$\psi_{phys}(A_z) = \int Dg\ e^{ikS(g;A_z,0)}\psi_0(A_z^g)$$

We will now carry this out for three examples: $\Sigma = T^2$, the torus; $\Sigma = S^2$ pierced by Wilson lines and $\Sigma =$ Disk.

From general principles we expect that H_Σ will be the space of characters of the affine Lie algebra. The easiest thing to do is choose a complex structure $z = \sigma^1 + \tau\sigma^2$ so we represent the torus by a parallelogram as usual. Define $A_z = \frac{\tau A_1 - A_2}{\tau - \bar{\tau}}$. So

$$[A_{\bar{z}}^a(x), A_z^b(y)] = \frac{-2\pi}{kIm\tau}\delta^{ab}\delta^{(2)}(x-y)$$

(In the equations above the factor $Im\tau$ was in the definition of the delta function.)

Now we use a basic fact: we can always gauge A_z to the constant Cartan:

$$A_z = hah^{-1} - \partial h h^{-1}$$

with h in the complexification of the gauge group where a is constant in the Cartan subalgebra. So - by the Gauss law it suffices to know the values $\psi[a_z]$ because $\psi[A_z] = e^{-ikS(h,a,0)}\psi[a]$. Now if we take the family of testfunctions for $\bar{J}_{\bar{z}}$, where $\bar{J}$ is a constant in the Cartan subalgebra,

$$\psi_0^J(A) = e^{\frac{ik}{2\pi}\int TrA_{\bar{z}}J}$$

then the corresponding physical states are

$$\psi^J_{phys}(a) = \int Dg\ e^{ikS(g,a_z,\bar{J})}e^{-\frac{ik}{2\pi}\int Tr(a\bar{J})}$$

where $S(g, A_z, A_{\bar{z}})$ is the gauged WZW action:

$$\begin{aligned} S(g, A_z, A_{\bar{z}}) =& \frac{ik}{4\pi}\int Trg^{-1}\partial gg^{-1}\bar{\partial}g + ik\Gamma \\ & -\frac{ik}{2\pi}\int Tr[Ag^{-1}\bar{\partial}g + \bar{A}\partial gg^{-1} + gAg^{-1}\bar{A} - A\bar{A}] \end{aligned}$$

The value of this path integral is well-known, it is just

$$= \sum_\lambda \psi_\lambda(\bar{J})\psi_\lambda(a)$$

where

$$\psi_\lambda(a) = e^{-\frac{kIm\tau}{2\pi}a^2}\ \chi_\lambda(\bar{\tau}, \frac{-iIm\tau}{\hbar}a)$$

where χ_λ are the Weyl-Kac characters. Thus - as we vary $\bar{J}$ we sweep out a space of states spanned by the characters.

• Exercise 9.2 *The Weyl Alcove.* Consider quantization of the Chern-Simons-Witten gauge theory on the torus with a real polarization, that is, $\psi = \psi[A_1(x)]$. Take the gauge group to be connected, simply connected and simply laced.

a.) Derive the Gauss law and show that ψ has support on those A_1 which are components of a flat connection. Thus the wavefunction is determined by its value for A_1 constant and in the Cartan subalgebra.

b.) Show that the Gauss law for the gauge transformations preserving the constant Cartan force ψ to be a periodic delta function whose support is at $\Lambda^{weight}/W \times k\Lambda^{root}$ where Λ^{weight} (Λ^{root}) is the weight (root) lattice and W is the Weyl group. The elements of this coset are in a natural one-to-one correspondence with the integrable highest weight representations of level k of the associated Kac-Moody algebra.

• Exercise 9.3 *Moduli Space of Flat Connections*

a.) In his original paper Witten first imposed the constraints and then quantized the resulting phase space. Show that this phase space is just the moduli space of flat connections on Σ.

b.) A flat connection is characterized by its holonomies, up to conjugation. Show that the real dimension of the resulting phase space is $(2g-2)dimG$ for the gauge group G on a surface of genus $g > 1$.

c.) Use the WKB approximation to show that the number of physical states grows as $k^{(g-1)dimG}$ and compare with exercise 3.8.

$\Sigma = S^2$ punctured by Wilson lines

The Wilson lines for finite transition amplitudes are

$$\langle m_f | P\ exp \int_C A\ | m_k \rangle$$

where the Wilson line carries some representation j and m_f, m_k are states in the representation j as in the following figure

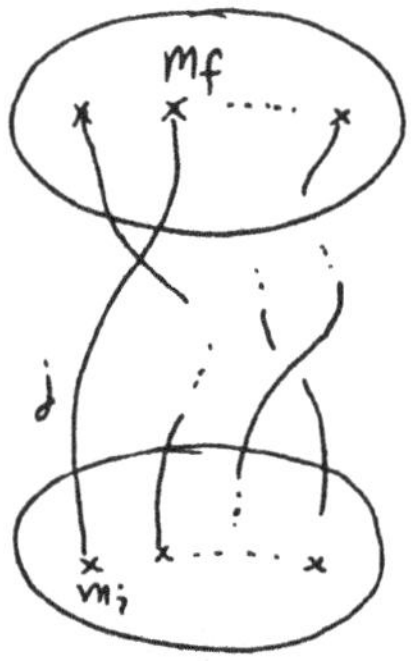

two sphere's with Wilson lines

Since the Hamiltonian of the theory is zero, finite time amplitudes are the same as overlaps of wavefunctions. So we see that the wavefunctions in the case with punctures are simply wavefunctionals valued in the tensor products of representations:

$$\vec{\psi}[A_z] = \sum_{m_i} \psi_{m_1 \cdots m_n}[A_z]\ |m_1\rangle \otimes \cdots \otimes |m_n\rangle$$

We know how Wilson lines transform under gauge transformation, so it is clear that the action of the Gauss law is just:

$$(U(g)\vec{\psi})[A_z] = e^{ikS(g;A_z,0)} \otimes_i \rho_i(g^{-1}(P_i))\vec{\psi}[A_z^g]$$

As before, we may use the basic fact that we can gauge away A_z i.e. $A_z = -\partial_z h h^{-1}$ Thus physical wavefunctions are completely determined by their value at $A_z = 0$:

$$\vec{\psi}^{phys}[A_z = 0] = \int Dg\ e^{ikS(g)} \otimes_i \rho(g^{-1}(P_i))\vec{\psi}_0(-\partial g g^{-1})$$

Now $\vec{\psi}_0$ is an arbitrary functional of the holomorphic current, so, by the holomorphic KM Ward identities we obtain a basis of physical states:

$$\vec{\psi}_{\vec{p}}(A_z) = e^{-ikS(h)} \otimes \rho_i(h(P_i))\vec{\mathcal{F}}_{\vec{p}}(\bar{z}_1, \dots \bar{z}_n)$$

From this example we see that the transition function given by the path integral for braided Wilson lines is indeed the appropriate duality matrix.

• Exercise 9.4 *Knizhnik-Zamolodchikov equations.*

a.) From the discussion of wavefunctions above write the Gauss law for the case of the sphere with sources as:

$$u(\epsilon) = \frac{k}{4\pi}\int Tr\epsilon F + \sum T_i^a \epsilon^a(P_i)$$

We would like to see how the wavefunctions change as the positions P_i of the sources change.

b.) Show that

$$[\mathcal{O}, u(\epsilon)] = \frac{\partial}{\partial \bar{z}_i} u(\epsilon)$$

for $\mathcal{O} = \rho_i(T^a)A_{\bar{z}}^a(P_i)$.

c.) Writing physical states as path integrals show

$$\bar{\partial}_i \vec{\psi}[A; P_i]|_{A=0} = \rho_i(T^a)A_{\bar{z}}^a(P_i)\int Dg e^{ik\left(S - \frac{1}{2\pi}\int TrAg^{-1}\bar{\partial}g\right)} \otimes_i \rho_i\Big(g^{-1}(z_i, \bar{z}_i)\Big)\vec{\psi}^0|_{A=0}$$

$$= \int Dg e^{ikS}\frac{1}{k}\bar{J}^a(z_i)\rho_i(T^a)\rho_i(g^{-1}(z_i, \bar{z}_i)) \otimes_{i\neq j} \rho_j(g^{-1}(P_j))\vec{\psi}_0$$

For simplicity (and WLOG) take $\vec{\psi}_0$ to be a constant tensor.

d.) We must define the singular product of operators at P_i. We do this by point splitting, then making an appropriate subtraction, which will be uniquely determined from self-consistency. Use the conformal field theory operator product relation (for a proof see [23].):

$$\bar{J}^a(\bar{\zeta})\rho_i(T^a)g^{-1}(z_i, \bar{z}_i) = \frac{C_i}{\bar{\zeta} - \bar{z}_i} + (k + h)\bar{\partial}_i g^{-1}(z_i, \bar{z}_i) + O(\zeta - z_i)$$

where h is the dual Coxeter number and $C_i = C_2(V^{j_i})$ is the Casimir of the representation V^{j_i}, to deduce that we must define the singular product of operators by

$$: \rho_i(T^a)\bar{J}^a(\bar{z}_i)\rho_i(g^{-1}(z_i, \bar{z}_i)) : \equiv \lim_{\zeta \to z}\Big[\rho_i(T^a)\bar{J}^a(\zeta)\rho_i\Big(g^{-1}(z_i, \bar{z}_i)\Big) - \frac{C_i}{\bar{\zeta} - \bar{z}_i} - h\bar{\partial}_i g^{-1}(z_i, \bar{z}_i)\Big]$$

e.) Plugging in this definition and using the Kac-Moody Ward identities for $\bar{J}$ show that physical states satisfy the Knizhnik-Zamalodchikov equations [23]

$$(k+h)\bar{\partial}_i\vec{\psi}[0;P_i] = \sum_{j\neq i} \frac{\rho_i(T^a)\rho_j(T^a)}{\bar{z}_i - \bar{z}_j}\vec{\psi}[0;P_i]$$

Σ=Disk=D

Finally, we consider the case of Σ with a boundary. In the case where Σ is a disk, H_Σ is the chiral algebra of the theory[27] .

We consider the path integral on $D \times \mathbb{R}$. Let us try to "evaluate" the path integral

$$\int \frac{DA}{vol\ G} e^{iS}$$

In order to do that we must decide on the appropriate boundary conditions. These are determined by demanding no boundary corrections to the equations of motion:

$$\delta S = \frac{k}{4\pi}\int_{\partial D\times\mathbf{R}} Tr(\delta A A) + \frac{k}{2\pi}\int_{D\times\mathbf{R}} Tr(\delta A F)$$

So we choose $A_0 = 0$ on the boundary. The gauge group appropriate for these boundary conditions is $\hat{G} = \{g : D \times \mathbb{R} \to G | g|_{\partial D\times\mathbf{R}} = 1\}$

Now let's decompose A into time and space components:

$$A = A_0 + \tilde{A}$$

so

$$d = dt\frac{\partial}{\partial t} + \tilde{d}$$

$$S = \frac{k}{4\pi}\int Tr\left(\tilde{A}\frac{\partial}{\partial t}\tilde{A}\ dt\right) + \frac{k}{2\pi}\int Tr A_0\left(\tilde{d}\tilde{A} + \tilde{A}^2\right).$$

Next, do the integral over A_0 giving

$$\int \frac{D\tilde{A}}{vol\ G}\delta(\tilde{F})e^{\frac{ik}{4\pi}\int_{D\times\mathbf{R}} Tr(\tilde{A}\frac{\partial}{\partial t}\tilde{A}\ dt)}.$$

We can solve this to get

$$\tilde{A} = \tilde{d}U\ U^{-1}$$

for $U : D \to G$, since D is simply connected.

Moreover, one can argue that there is no Jacobian

$$D\tilde{A}\delta(\tilde{F}) = DU$$

- Exercise 9.5 *No Jacobian.* Show that in the change of variables

$$\int DA\delta(F)\mathcal{O}(A) = \int DU\mathcal{O}(-U^{-1}dU)$$

for gauge invariant functionals $\mathcal{O}$.

Finally, we plug $\tilde{A} = -\tilde{d}UU^{-1}$ back into the Lagrangian to get:

$$S = \frac{k}{4\pi} \int_{\partial D \times \mathbf{R}} Tr\ U^{-1}\partial_\varphi U U^{-1}\partial_t U \ + \ k\Gamma(U)$$

where φ is the angular coordinate on the rim of the disk, and Γ stands for the Wess-Zumino functional. As is well known, this does not depend on the values of U on the interior - so we can divide out the volume of the gauge group to get the path integral

$$\int DU \ e^{ikS_{wzw}(U)}$$

where

$$U : \partial D \times \mathbb{R} \rightarrow G.$$

Quantization of this system is well-known to give the chiral algebra of the WZW model [2].

• Exercise 9.6 *A Disk with a source.* Work out the analogous change of variables for the case of a disk with a source in a representation λ. Represent the source by a quantum mechanics problem with the action [64]

$$\int dt Tr \lambda \omega^{-1}(\partial_0 + A_0)\omega(t).$$

Integrate over A_0 to find a constraint on $\tilde{A}$. Show that the holonomy of the flat connection around the source is determined by the representation of the source. Find the effective action on the boundary of $D \times \mathbb{R}$. Its quantization leads to the representation λ of Kac-Moody [65]. Use this Lagrangian to find the set of λ's which lead to inequivalent effective field theories and hence to the set of integrable representations of Kac-Moody.

• Exercise 9.7 *Two sources on* S^2. Repeat the analysis of the previous exercise for this case and prove that the Hilbert space is one dimensional if one source is in the conjugate representation to the other source and it is empty otherwise.

From these remarks we see that we can also learn about descendents from the $2+1$ dimensional viewpoint. Moreover, note that the quantization on the disk allows us to define a $2+1$ dimensional analog of a chiral vertex operator. Consider the following solid pants diagram threaded by three Wilson lines joined together with an invariant tensor a:

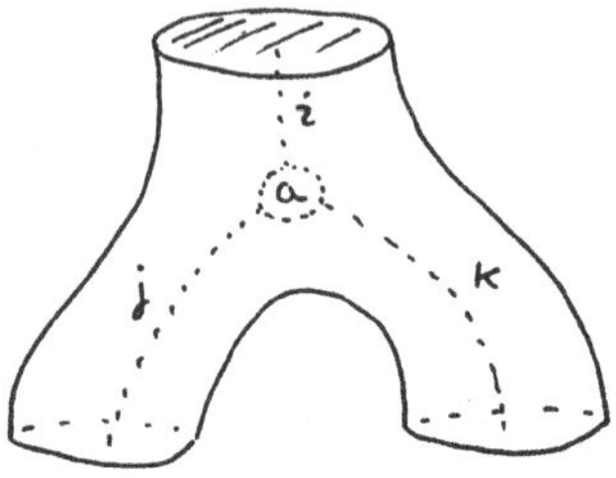

Solid pants diagram

The different boundaries are meant to reflect corresponding boundary conditions on the gauge field. From the above exercises we see that the path integral defines an operator from $\mathcal{H}_j \otimes \mathcal{H}_k$ to $\mathcal{H}_i$. Moreover, from general principles of CSW theory this operator has the braiding and fusing properties of a chiral vertex operator. Thus it is natural to suppose that it *is* a chiral vertex operator at some canonical value of z, but this has not yet been demonstrated.

Not all aspects of RCFT have been understood from the $2+1$ dimensional viewpoint. We end with the following exercise, part (c) of which is an open problem:

- Exercise 9.8 *Nontrivial Modular Invariants*

a.) Show that the natural inner product on quantum wavefunctions for CSGT with connected and simply connected gauge group defines a pairing of representations corresponding to the diagonal modular invariant.

b.) Give the $2+1$ dimensional interpretation of the unitarity of the matrix S.

c.) Find a natural interpretation of the nontrivial modular invariants especially exercise 6.5 from the $2+1$ dimensional viewpoint.

---

10. Chern-Simons-Witten gauge theory – Other RCFT's

In the previous section we saw how KM theories can be reconstructed from connected and simply connected gauge groups in three dimensions. It is therefore natural to ask if other RCFT's can be similarly related to CSW theory for different gauge groups. Here we will show that all known examples of RCFT arise from CSW theory for some gauge group.

Among the other known RCFT's there are three kinds:

1. Extended algebras. Examples include the rational torus, chiral algebras of D_n modular invariants(W-algebras), and other modular invariants obtained by orbifolds of WZW theories.

2. Coset models. Examples include various discrete series

3. Orbifolds of the above.

The holomorphic part of each of these theories can be given a CSGT interpretation:

1. Extended KM algebras

Most chiral algebras include high spin fields. Some of them can be obtained by adding extra holomorphic operators to a KM algebra. Theories not finitely decomposable in terms of KM or Virasoro representations might be finitely decomposable with respect to this larger algebra. For example, to form extended algebras one usually uses the "spectral flow" transformation associated to automorphisms of extended Dynkin diagrams. Thus, if we wish to extend level k $\hat{g}$-current algebra we begin with $\theta \in Center(G)$ and write $\theta = e^{2\pi\mu}$ for some weight vector μ. (For simplicity we take $G = SU(n)$, the discussion can be generalized.) The integrable level k representations are given by the points in the Weyl alcove

$$\Lambda_{weight}/W \times k\Lambda_{root}$$

The transformations $\lambda \to \lambda + k\mu$ is equivalent, via the affine Weyl group to a transformation $\lambda \to \mu(\lambda)$ of highest weight representations. For example for $SU(2)$ level k the spin j representation transforms by $j \to k/2 - j$.

Equivalently, we may consider the change in the currents obtained when the boundary conditions are twisted by the multiple-valued "gauge transformation"

$$\Omega(z) = z^{\theta} \tag{10.1}$$

which acts by

$$J(z) \to \Omega(z)J(z)\Omega^{-1}(z) - k\partial\Omega(z)\Omega^{-1}(z) \tag{10.2}$$

In modes we have:

$$\begin{aligned} H_n^i &\to H_n^i + k\theta^i\delta_{n,0} \\ E_n^\alpha &\to E_{n+\theta\cdot\alpha}^\alpha \\ L_n &\to L_n + \theta^i H_n^i + \tfrac{1}{2}k\theta^2\delta_{n,0} \end{aligned} \tag{10.3}$$

(E,H correspond to simple roots and Cartan elements, respectively) and in the special case of $SU(2)$ this becomes:

$$\begin{aligned} J_n^3 &\to J_n^3 + \frac{k}{2}\delta_{n0} \\ J_n^\pm &\to J_{n\pm 1}^\pm \\ L_n &\to L_n + \frac{1}{2}J_n^3 + \frac{k}{2}\delta_{n0} \end{aligned} \tag{10.4}$$

In general, for any subgroup $Z \subset Center(G)$ we can "mod out" by this action thus obtaining the extended chiral algebra

$$\mathcal{A} = \oplus_{\mu\in Z}\mathcal{H}_{\mu(0)}$$

A well known example is the rational torus. The toroidal $c = 1$ model with a boson $\phi \sim \phi + 2\pi R$ has a $U(1)$ KM symmetry generated by $J = \partial\phi$ when $R^2 = \frac{p}{2q}$ is rational there are extra holomorphic fields generated by

$$V = e^{\pm i\sqrt{2pq}\ \phi}$$

which generate a large algebra.

It can be shown that this process of extension of the algebra:

$$\{\partial\phi\} \rightarrow \left\{\partial\phi, e^{\pm\ i\sqrt{2N}\phi}\right\}$$

corresponds in CSW gauge theory to a change in the gauge group. Namely we can have an Abelian gauge field with action

$$S = \frac{ik}{8\pi}\int AdA$$

but it makes a big difference if the gauge group is $\mathbb{R}$ or $\mathbb{R}/\mathbb{Z} = U(1)$.

If the gauge group is $\mathbb{R}$, the allowed gauge transformations are $A \rightarrow A - d\epsilon(x)$ where $\epsilon : Y \rightarrow \mathbb{R}$ is a well-defined function. In that case:

1.) We can scale k out of the action

2.) The observables in the theory are the Wilson lines

$$e^{i\alpha \oint A}$$

Recall that the value of α defines a representation - this corresponds to a continuously infinite set of representations in CFT.

3.) No two Wilson lines are equivalent.

On the other hand, if the gauge group is $U(1)$ then around non contractible cycles ϵ is only well-defined modulo 2π, and this leads to some consequences:

1.) The theory only makes sense for $k = 0\ mod\ 4$

2.) The observables are

$$W_n(C) = e^{in\oint_C A} \quad n\epsilon Z$$

3.) Two Wilson lines can be equivalent

$$W_n(C) \cong W_{n+k/2}(C)$$

• Exercise 10.1 *Level k $U(1)$ Current Algebra*

a.) Compute explicitly the expectation values of Wilson lines in S^3 for the abelian case:

$$\int DAe^{\frac{ik}{8\pi}\int AdA}\prod_i e^{in_i\oint_{C_i} A} = exp\left[\frac{2\pi i}{k}\sum_{i,j} n_i n_j \Phi_{ij}\right]$$

where Φ_{ij} is the linking number. Φ_{ii} is ambiguous-but may be regularized and defined up to an integer.

b.) Show that the cross terms are invariant under the change $n \rightarrow n + \frac{k}{2}$. Show that the invariance of the self-linking number requires $k = 0 mod\ 4$.

c.) Perform a (singular) gauge transformation $A \to A + d\phi$ where ϕ is an angular variable around some Wilson line. Show that this changes $W_n \to W_{n+\frac{1}{2}k}$. This illustrates how changing the gauge group from $\mathbb{R}$ to $U(1) = \mathbb{R}/\mathbb{Z}$ brings about an identification of Wilson lines.

d.) If $k = 4N$ we refer to the corresponding CFT as $U(1)_N$, "level N $U(1)$ current algebra." Show that the conformal field theory is just the holomorphic part of the rational torus $R^2 = p/2q$ where $pq = N$.

e.) The Wilson line $W_{\frac{k}{2}}$ which is a non-trivial operator if the gauge group is $\mathbb{R}$ behaves like the identity operator when the gauge group is $U(1)$. The reason for this is the following. In the $U(1)$ theory one needs to sum over $U(1)$ bundles. The non-trivial bundles can be characterized by an insertion of an 'tHooft operator [66] in the functional integral of the $\mathbb{R}$ theory. Using part c of this exercise, show that the 'tHooft operator is equivalent to $W_{\frac{k}{2}}$. Since we have to sum over the insertions of such operators, the value of the functional integral is not modified if we add another one. Hence, this operator behaves like the identity operator. The two dimensional analog of this is the fact that the representation $\frac{k}{2}$ *extends* the $\mathbb{R}$ KM chiral algebra. This field becomes a descendent of the identity operator (under the larger chiral algebra) and its conformal blocks are the same as those of the identity.

f.) Show that the above considerations extend to any even integral lattice.

g.) Quantize the theory by canonical quantization on T^2 as in the previous section. Find the different states as the different representations of $U(1)_N$ and write their wave functions in terms of theta functions of higher level [67].

h.) Quantize the theory on a manifold with boundary. Find the extended chiral algebra by quantization on the disk (hint: because of the boundary conditions, there are non-trivial bundles corresponding to the insertion of $\frac{k}{2}$ in the $\mathbb{R}$ theory) and the different representations by quantization on a disk with a source.

i.) Show that the center of $\mathcal{A}(U(1)_N)$ is simply $\mathbb{Z}/2N\mathbb{Z}$. (Hint: We normally think of the gauge group of a $U(1)$ gauge theory, which is generated by

$$U(\epsilon) = e^{\frac{ik}{4\pi}\int \epsilon(x)F(x)}$$

for smooth functions ϵ as an abelian group. However we now allow functions like $\epsilon_P \sim \phi$ for ϕ an angular coordinate centered at any point P. Show that

$$U(\epsilon_P)U(\epsilon)U(\epsilon_P)^{-1} = e^{i\frac{k}{2}\epsilon(P)}U(\epsilon)$$

so that the group becomes nonabelian. Note that the elements of the center are in one-one correspondence with the representations of the rational torus chiral algebra.) Interpret the existence of this center from the two dimensional point of view (hint: the chiral algebra contains charged fields).

• Exercise 10.2 $G = SO(3) = SU(2)/\mathbb{Z}_2$

a. Show that the only representations which survive have odd dimension. Show moreover that to avoid global anomalies, or to have the extending Wilson line be invisible we must have $k = 0 mod 4$.

b. Show that by the singular gauge transformation we can prove equivalence of the Wilson lines

$$W_j(C) \cong W_{k/2-j}(C)$$

c. Show that the Wilson line $W_{k/4}$ is in fact not the simplest operator in the theory, rather we have $W_{k/4} = \mathcal{O}^+ + \mathcal{O}^-$ where the operators $\mathcal{O}^\pm$ cannot be simply expressed in terms of Wilson lines.

d. Find an expression for $\mathcal{O}^\pm$ in terms of $SU(2)$ theory. (Hint: consider a three point vertex of Wilson lines with one in the representation $k/2$.)

Quite generally one can show that all known extended algebras are obtained from 3 dimensional CSW gauge theories by changing the gauge group by

$$G \to G/Z$$

where Z is a subgroup of the center of G.

In going from G to $\tilde{G} = G/Z$, three changes in the possible representations take place:

a. Selection rule: of the representations of G current algebra only those which are invariant under Z should be kept.

b. Identification: different irreducible representations of G related by the s pectral fl ow operation are combined into one $\tilde{G}$ irreducible representation.

c. Fixed point: if the spectral flow has a fixed point, there are *different* $\tilde{G}$ representations which are the same as G representations.

These three rules generalize the three parts of the previous exercise.

• Exercise 10.3 *The three rules from canonical quantization.* Derive these three rules from canonical quantization on the torus. Hints:

1. Rule a follows from gauge transformations which wind around one cycle
2. Rule b from gauge transformations which wind around the other cycle.

3. Rule c is the most subtle. Twisted bundles on the torus are labeled by the subgroup Z used to divide the universal cover to obtain G. These bundles may be defined by cutting out a disc and using the transition function $g(\phi) = e^{i\phi\theta}$ where ϕ is an angular coordinate and θ is a weight vector. The flat gauge fields which are sections of the associated $ad(G)$ bundle are characterized [68] by conjugacy classes of solutions of

$$ABA^{-1}B^{-1} = e^{2\pi i\theta} \tag{10.5}$$

where $A, B \in \tilde{G}$ describe holonomies of the flat gauge field.

As a simple example, consider first the nontrivially twisted $SO(3)$ bundle on Σ_1. Without loss of generality we may rotate B into the maximal torus, taking $B = e^{2\pi i x T^3}$. Then A must be of the form wA_1 where w is in the Weyl group and A_1 is in the maximal torus. By conjugating with elements of the maximal torus we may set A_1 to one. Show that there is exactly one solution, $x = 1/4$ up to conjugacy. Thus the moduli space of twisted flat gauge fields consists of one point, and quantization gives one further state. Recall that in the conformal field theory there are two representations $\mathcal{H}^{\pm}_{k/4}$ of the $\mathcal{A}_k(SO(3))$ chiral algebra. Only one of these was accounted for from the quantization in the untwisted sector, the other comes from the twisted sector. Compare these two different irreducible representations with $\mathcal{O}^{\pm}$ in exercise 10.2. As an example, show that $\mathcal{A}(SO(3)_4) = \mathcal{A}(SU(3)_1)$ and recognize the two different representations of the $SO(3)$ theory as 3 and $\bar{3}$ of $SU(3)$.

These remarks generalize to arbitrary groups. Twisted bundles with transition function g_θ have one flat connection for the conjugacy class of each (discrete) solution x, w of $wxw^{-1} + \theta = x mod \Lambda_{rt}$ where x is in the Cartan subalgebra and w is in the Weyl group. Using the conjugacy freedom we can require that x is in the positive Weyl chamber, show that this equation then becomes exactly the condition for a weight $x = \lambda/k$ to be fixed by the spectral flow μ_θ. Thus, the states arising from quantization on the discrete set of points in the moduli space of twisted flat bundles exactly correspond to the different irreducible representations $\mathcal{H}^w_i$ arising from the representations fixed by subgroups of the spectral flow.

Using these considerations we can easily find new quantization conditions on k in the non-simply connected case (generalizing the $k = 0 mod 4$ in the $U(1)$ theory). The conformal dimension of the extending representation must be an integer. From the three dimensional point of view, this condition is the statement that there is no dependence on the framing of the 'tHooft operator which is used to described the twisted bundles – no global anomalies. The conformal dimension of the representation λ is $\Delta_\lambda = \frac{\lambda(\lambda+2\rho)}{2(k+h)}$. If the spectral flow is generated by the representation μ, the extending representation is $k\mu$ and its dimension is $\Delta_{k\mu} = \frac{k\mu(k\mu+2\rho)}{2(k+h)}$. The condition on k is that this number should be an integer. The same result has been obtained by other considerations in [69].

2. Coset models G/H

They may be obtained as follows: we take gauge fields

$$A^a,\ A^{\bar{a}} \epsilon Lie(G) \quad \bar{a} \text{ denote directions in } Lie(G)/Lie(H)$$

$$B^a \epsilon Lie(H)$$

and action

$$S = k_1 CS(A) - k_2 CS(B)$$

We must be careful to take the gauge group $(G \times H)/Z$ where Z is the common center of H embedded in G.

To see that this prescription is correct consider the quantization on the disk $D \times \mathbb{R}$, and let us reconsider the boundary conditions. Variation gives

$$\delta S = \frac{k_1}{4\pi} \int_{\partial D \times \mathbf{R}} Tr(\delta A A) - \frac{k_2}{4\pi} \int_{\partial D \times \mathbf{R}} Tr \delta B B + \text{ bulk terms}$$

One possibility is to choose $A_0 = B_0 = 0$ which leads to a $G \times H$ theory. However, when $H \subset G$ and $k_2 = \ell k_1$ (ℓ is the index of the embedding) we may choose instead the boundary conditions:

$$\begin{cases} A^a & = B^a \\ A_0^{\bar{a}} & = 0 \end{cases}$$

Performing the change of variables we had before we write (we have chosen $\ell = 1$ for simplicity)

$$A = -dUU^1$$

$$B = -dVV^{-1}$$

and get, as before

$$\int D\lambda DUDV \exp \Big[ikS_{wzw}(U) - ikS_{wzw}(V) + ik \int Tr\lambda(\partial_\varphi UU^{-1} - \partial_\varphi VV^{-1}) \Big]$$

where λ is a Lagrange multiplier enforcing the boundary condition $A^a = B^a$.

Making the change of variables $U \to gV$, $-\partial_\phi VV^{-1} \to a_\phi$, and $\lambda \to a_t$ we get the path integral

$$\int dadge^{ikS(g,a_\phi,a_t)}$$

which is the gauged WZW model, which is well-known [70] to be the path integral representation of the coset models. Actually, it is quite easy to see why this must be so. The phase spaces are the coadjoint orbits of the pair of $\hat{G}$ and $\hat{H}$ representations (Λ, λ):

$$(LG/T) \times (LH/T)^*$$

which, upon quantization give the space of states: $\mathcal{H}_\Lambda \otimes \mathcal{H}_\lambda^*$. Now we may impose the *first class* constraints: $\pi_H(\partial_\phi U U^{-1}) - \partial_\phi V V^{-1}$ (π_H is a projection from G to H) which is an H-current algebra with $k = 0$ to obtain the physical states:

$$(\mathcal{H}_\Lambda \otimes \mathcal{H}_\lambda^*)^{LH} \cong Hom_{LH}(\mathcal{H}_\lambda, \mathcal{H}_\Lambda) \cong \mathcal{H}_{\Lambda,\lambda}$$

where the final symbol is the space of states in the coset model, defined by the decomposition $\mathcal{H}_\Lambda = \oplus_\lambda \mathcal{H}_{\Lambda,\lambda} \otimes \mathcal{H}_\lambda$.

• Exercise 10.4 *Example of a Coset*

a.) Show, using CFT, that the coset model $U(1)_N \times U(1)_M / U(1)_{N+M}$ for the case that N, M have no common factors is equivalent to the rational torus $U(1)_L$ for $L = NM(N+M)$.

b.) Consider the expectation values of Wilson lines in S^3 for the action:

$$\frac{N}{2\pi} \int AdA + \frac{M}{2\pi} \int BdB - \frac{N+M}{2\pi} \int CdC$$

where A, B, C are three abelian gauge fields. Show that the expectation value is consistent with the result of part (a).

c.) Show that the quantization of this theory on T^2 leads to the correct answer only if the gauge group is $\frac{U(1)\times U(1)\times U(1)}{Z_2}$. In implementing our prescription, we have to view the chiral algebra $U(1)_N$ as non-abelian. See above, exercise 10.1.g.

• Exercise 10.5 *The $N = 0, 1$ discrete series and the role of the center.* Study the coset $\frac{SU(2)_k \times SU(2)_l}{SU(2)_{k+l}}$. For $l = 1$ this is the Virasoro discrete series and for $l = 2$ the super discrete series. The 3d gauge group is $\frac{SU(2)\times SU(2)\times SU(2)}{Z_2}$. The representations are labeled by three spins j_1, j_2, j_3 corresponding to the three $SU(2)$. Use rule a above to show that $j_1 + j_2 + j_3$ must be an integer. Use rule b above to show that the representation (j_1, j_2, j_3) is identified with the representation $(\frac{k}{2} - j_1, \frac{l}{2} - j_2, \frac{k+l}{2} - j_3)$. Use rule c to show that if both k and l are even, there are two different representations labeled by $(\frac{k}{2}, \frac{l}{2}, \frac{k+l}{2})$. Rule c applies in the superdiscrete series ($l = 2$) when k is even. What is the difference between the two representations in this case?

• Exercise 10.6 $c = 7/10$ This conformal field theory can be represented by a coset $\frac{SU(3)_2}{U(2)_2}$. Notice that in the coset we use $U(2)$ rather than $SU(2) \times U(1)$. Why? What are the irreducible representations of $U(2)_2$? What is the three dimensional gauge group? (Don't forget the common center.) What are the irreducible representations of the coset?

• Exercise 10.7 *Witten's Triple Cosets.* In [63] Witten proposed a generalization of the coset construction. Recall that in the coset construction the fields in the chiral algebra $\mathcal{A}(G/H)$ are all the fields in $\mathcal{A}(G)$ which commute with the fields in $\mathcal{A}(H)$. In particular, $\mathcal{A}(G/H)$ is a subalgebra of $\mathcal{A}(G)$. Thus, if we have a triple of inclusions $K \subset H \subset G$ then we may consider the fields in $\mathcal{A}(G)$ which commute with the fields in $\mathcal{A}(H/K)$. Witten defines this subalgebra of $\mathcal{A}(G)$ to be the triple coset algebra $\mathcal{A}(G/H/K)$. In this exercise we show that the construction of these algebras do not involve any new constructions other than those described above.

a.) As a warmup consider the explicit triple $\hat{SU}(N-1)_1 \subset \hat{SU}(N)_1 \subset \hat{SU}(N+1)_1$. Using the Frenkel-Kac construction of level one current algebra in terms of free scalar fields show that the triple coset is just $\hat{SU}(N-1) \times U(1)_{N(N+1)/2}$.

b.) More generally, show that $\mathcal{A}(G/H/K)$ always contains the subalgebra $\mathcal{A}(G/H) \times \mathcal{A}(K)$. Moreover these have the same central charge and are unitary theories. Thus $\mathcal{A}(G/H/K)$ may be expected to be at most an extended algebra of $\mathcal{A}(G/H) \times \mathcal{A}(K)$. Show that this is indeed the case by decomposing characters:

$$\begin{aligned}
\chi_0^G &= \sum_{\lambda} \chi_{0,\lambda}^{G/H} \chi_\lambda^H \\
&= \sum_{\lambda,\rho} \chi_{\lambda,\rho}^{H/K} \chi_\rho^K \chi_{0,\lambda}^{G/H} \\
&= \sum_{[\lambda,\rho]} \chi_{\lambda,\rho}^{H/K} \sum_{\mu \in C(\mathcal{A}(H)) \cap C(\mathcal{A}(K))} \chi_{\mu(\rho)}^K \chi_{0,\mu(\lambda)}^{G/H}
\end{aligned}$$

where $[\lambda,\rho]$ denotes equivalent pairs in the coset module. Thus, in particular, the character of the chiral algebra is just

$$\sum_{\mu \in C(\mathcal{A}(H)) \cap C(\mathcal{A}(K))} \chi_{\mu(0)}^K \chi_{0,\mu(0)}^{G/H}$$

which is a finite extension of $\mathcal{A}(G/H) \times \mathcal{A}(K)$.

c.) Show that this theory may be obtained from 2+1 dimensions using the (schematic) action $CS(G) - CS(H) + CS(K)$ with gauge group $(G \times H \times K)/Z$ and Z is generated by $(\theta,\theta,1)$ for $\theta \in C(G) \cap C(H)$ and by $(1,\theta,\theta)$ for $\theta \in C(H) \cap C(K)$.

---

3. Orbifolds

The MTC of rational orbifolds is fairly complicated in general. In the special case of a rational orbifold obtained from a theory with a trivial MTC, the rational orbifold MTC has a rather beautiful description given in [11][71]. If the finite group G is the orbifold group, the index set I consists of pairs $(\bar{g}, \alpha)$ where $\bar{g}$ is a conjugacy class in G and α is an irreducible representation of the centralizer subgroup of the conjugacy class. The basic data of the MTC can be described in terms of group cohomology [4]. In particular, the fusion rules are elegantly described as a multiplication law in the equivariant K-theory of G. Fortunately, one can demonstrate by rather general arguments that the holomorphic half of any rational orbifold model can be obtained from a 3D CSW gauge theory based on gauge groups which are not connected [5].

Let G be a connected group with a discrete automorphism group P. Then one can construct the semi-direct product group $P \ltimes G$. Quantizing the system on the disk and repeating the steps above, we find that the effective action is the WZW action for a field U on the boundary which takes values in G. The phase space is LG/G and leads to $\mathcal{A}(G)$, but because of P gauge invariance, the Hilbert space has to be truncated to the P invariant states (the states are in representations of P because P is an automorphism of G). This can be seen by considering the CSGT on $D \times S^1$. The functional integral in this case leads to the trace over the Hilbert space (since the Hamiltonian of the 3D theory vanishes, this trace is infinite). In the functional integral we need to sum over P bundles. This sum projects out the states which are not P invariant. Therefore, $\mathcal{A}(P \ltimes G) = \mathcal{A}(G)/P$. This is the chiral algebra of the orbifold constructed as G/P. By quantizing the system on other two surfaces with boundaries we obtain the other representations of the orbifold model.

Orbifolds and cosets are very similar in both two and three dimensions. In 3D we reduced the chiral algebra of the G theory by enlarging the gauge group. In 2D both theories are obtained by considering a G theory and gauging either a continuous subgroup, H/Z (to obtain G/H) or a discrete automorphism group, P (to obtain G/P). Finally note that the gauge group $(G \times H)/Z$ of the coset CSGT can also be written as $(H/Z) \ltimes G$ which is the same as the prescription for orbifolds. In the classical limit of these theories the integral weight fields have a closed ope. Therefore, there should be a one to one correspondence between these representations of the chiral algebra and representation spaces of some group. This group is the gauge group of the 3D theory.

[4] This is also true of theories with "abelian fusion rules" as explained in appendix E of [15].

[5] Initial work with E. Witten first suggested that $O(2)$ would reproduce the rational orbifold. This work motivated the general construction for orbifolds.

• Exercise 10.8 *The rational orbifold from* $O(2)$. Check that the $O(2)$ CSW gauge theory on T^2 leads to the correct number of representations. First use conformal field theory to find that for the rational orbifold of level N there are $N+7$ representations. When quantizing on T^2 the Hilbert space has several sectors. Show that from topologically trivial bundles (those which can be considered to be $SO(2)$ bundles) there are $N+1$ states. Find six twisted $O(2)$ bundles leading to six more states. Hence, the total number of states is $N+7$.

• Exercise 10.9 *A more complicated orbifold.* Study the orbifold $SU(2)_k/\mathbf{Z}_2 \times \mathbf{Z}_2$, where we take the quotient by 180° degree rotations around orthogonal axes. Unlike the previous exercise here two interesting subtleties arise. First, some of the twisted components of the Hilbert space have more than one state. Second, some of the twisted components in fact contribute no quantum states for some k's, because of a global anomaly in the appropriate sector. Show that the number of quantum states is $(11k+32)/2$ if k is even and $(11k+11)/2$ if k is odd. Derive the same result by the two dimensional considerations of [11].

The lesson that we learn from this is that all known RCFT's are equivalent to some CSW gauge theory for some compact gauge group. An arbitrary compact group may be disconnected (the quotient of G by its connected component being some finite group) and in turn the connected component may have a finite-sheeted cover consisting of a product of tori and simply connected simple factors. From the previous constructions we see that this full level of generality is needed to order the zoo of known rational conformal field theories.

When working with arbitrary compact groups a further subtlety arises which is analyzed in detail in [69][6]. In order to write the Chern-Simons action in the form

$$S = \frac{k}{4\pi} \int_Y Tr(AdA + \frac{2}{3}A^3)$$

one needs a trivial G-bundle over the three-manifold Y. By definition, the path integral for theories with G not connected and simply- connected include nontrivial G-bundles and one must find another definition of the action. This problem was solved in [69]. The upshot is that the appropriate data needed to specify the action is an element of the cohomology group $\lambda \in H^4(BG;\mathbb{Z})$. For a connected, simply-connected, simple group, $H^4(BG;\mathbb{Z}) = \mathbb{Z}$ and λ is simply the integer, usually called k, multiplying the Chern-Simons term. For arbitrary connected compact groups the data is equivalent to a nondegenerate symmetric

[6] We thank Dan Freed for very useful discussions on these matters.

invariant bilinear form on the Lie algebra, needed to define the notion of a trace. In the disconnected case there can be torsion and one must express the data as an element of $H^4(BG; \mathbb{Z})$.

In conclusion, the MTC's of all known RCFT's are organized by simply specifying the pair (G, λ) where G is a compact gauge group and λ is a cohomology class in $H^4(BG; \mathbb{Z})$.

11. Conclusions and Conjectures

In these lectures we tried to formulate RCFT in an axiomatic way. We were led to define certain axioms which have - rather remarkably - an analog in the TK approach to group theory. Even more remarkably, it is known in the group theory case that a single additional axiom: $F_i^{-1} \in \mathbb{Z}_+$ defines the representation theory of an algebraic group. (To obtain a compact group one has to say a bit more.) Moreover this crucial integrality condition has an analog in RCFT. Thus we conjectured that by adding some axioms to the polynomial equations on F, B, S we will define RCFT purely axiomatically. Making further progress from this point on is difficult: We know that reconstruction will be subtle because there exist nontrivial chiral algebras with one representation and no holonomy (e.g. those obtained from even self-dual lattices of dimension 0 mod 24). This raises a serious question as to how good the notion of a modular functor or a modular tensor category is at identifying a RCFT. Based on the absence of counterexamples we may hope that the only ambiguity comes from tensor products with $c = 24$ purely holomorphic CFT's.

Another difficulty is that it is not exactly obvious what we should say about F_i^{-1}. There should be some physical reason based solely on the defining axioms of conformal field theory for why these numbers should take on special values but no one has succeeded in elucidating such a reason [7]. Moreover, it is not obvious that there are not additional axioms with no group theoretic analog (just as there are additional polynomial equations with no group theoretic analog). Nevertheless it ought to be clear from our discussion that RCFT defines some mathematical structure generalizing group theory. Of course, reconstruction is much easier if you know what it is you are trying to reconstruct!

We saw in sections nine and ten that three-dimensional CSW gauge theories can be used to define the MTC of all known RCFT's by taking an appropriate compact gauge group (perhaps neither connected nor simply connected) and action (defined by an appropriate symmetric invariant nondegenerate bilinear form, or, more precisely, by an appropriate class in $H^4(BG; \mathbb{Z})$). Taking account of the general structure of compact groups we

---

[7] It has been pointed out by many authors that F_i^{-1} is an index for inclusions of finite von Neumann algebras. This is clearly the most fruitful interpretation from which to embark on an investigation of the analog of Deligne's condition.

saw that the full generality is needed to describe CFT's and that the extension from the case of simply-connected simple groups is not entirely trivial. Based on these observations one naturally guesses that the object we are trying to reconstruct is none other than compact CSW theory, and therefore that all RCFT's are equivalent to some compact CSW theory.

The equivalence between compact CSW theories and RCFT's is not one to one. First, there are CSW theories which do not correspond to any RCFT. For instance, if we repeat the G/H construction with H which is not a subgroup of G or with the two coupling constants, k_G for G and k_H for H which are not equal (or not proportional according to the index of the embedding) then the resulting theory does not describe the MTC of the holomorphic half of a RCFT. Also, different CSW theories might lead to the same RCFT. For instance, it is known that the same RCFT can sometimes be described as a coset in two different ways. Other identifications arise for low levels, for example $SU(2)_1$ is the same as $U(1)_1$. According to the philosophy of this paper, these isomorphisms should be viewed as the CFT version of the isomorphisms in the Cartan classification of Lie algebras for algebras of small rank, e.g. $su(2) \cong so(3)$, $su(4) \cong so(6)$, etc.

• Exercise 11.1 *A Sampling of Isomorphisms.* In the literature on CFT there are often several different realizations of the same theory. Identify the following isomorphisms:

a.) $SU(N)_1 \cong \mathbb{R}^{N-1}/\Lambda_{rt}$ where Λ_{rt} is the root lattice.

b.) $U(1)_2 \cong O(2)_1$.

c.) $\frac{SU(2)_N \times U(1)_{-N}}{\mathbb{Z}_2} \cong \frac{SU(N)_1 \times SU(N)_1 \times SU(N)_{-2}}{\mathbb{Z}_N}$

d.) $SO(3)_4 \cong SU(3)_1$

e.) $(SU(3)/\mathbb{Z}_3)_3 \cong SO(8)_1$

f.) $\frac{SU(2)_1 \times SU(2)_1 \times SU(2)_{-2}}{\mathbb{Z}_2} \cong (E_8)_1 \times (E_8)_1 \times (E_8)_{-2}$

g.) $\frac{SU(2)_3 \times SU(2)_1 \times SU(2)_{-4}}{\mathbb{Z}_2} \cong \frac{SU(3)_1 \times SU(3)_1 \times SU(3)_{-2}}{\mathbb{Z}_3}$

There is, at present, no general point of view on how to classify these isomorphisms.

The relation between the three dimensional and the two dimensional theories arises in two related ways. The Hilbert space of the theory on a manifold without a boundary is the space of conformal blocks. In this case one can study the dimensionality of the vector space and the action of the duality matrices. A more detailed connection between the theories arises upon quantization on a manifold with a boundary. Then all the states in the chiral algebra and in all its representations can be realized.

As we have mentioned, it is sometimes the case that two different theories have the same duality matrices. However, the structure of the representations is different. For instance, if we tensor a theory based on a $c = 24$ self dual lattice with any theory C the duality matrices are those of the theory C. The only difference is in the structure of the chiral algebra and its representations.

Correspondingly we may formulate a weak and a strong version of the conjecture alluded to throughout these lectures. The weak version states that the duality properties are reproduced by some CSW theory with compact group. More formally, we may state

Conjecture 1: The modular functor of any unitary RCFT is equivalent to the modular functor of some CSW theory defined by the pair (G, λ) with G a compact group and $\lambda \in H^4(BG; \mathbb{Z})$.

Let us make some remarks about this conjecture. First, as discussed at the end of section ten, if G is connected then λ may be thought of as the data needed to specify the normalizations of the traces in the Chern-Simons action. Alternatively, from the quantum group point of view, λ specifies the appropriate roots of unity required for various quantum deformations of relevant simple groups. Second, we expect that the gauge group must be compact for a simple reason. In the WKB approximation one obtains one quantum state for each unit of volume of phase space. The moduli spaces of noncompact groups are noncompact and hence quantization will lead to an infinite number of quantum states, that is, an infinite number of conformal blocks, so the corresponding two-dimensional theory cannot be rational. Recent work of H. Verlinde [72] suggests that this reasoning might be too naive at strong coupling, and that noncompact phase spaces might actually lead to finite dimensional spaces of states. Nevertheless, rational conformal field theories which do seem to be related to noncompact groups also have a description in terms of compact groups. Third, we limit our considerations to unitary theories because CSW theories, which are simply quantum mechanical systems with a finite number of degrees of freedom, are automatically unitary. Every known example of a unitary RCFT fits in with conjecture 1. The situation for nonunitary RCFT's is much less well understood, although there is some preliminary evidence that the correct organizing principle may be found in the theory of compact supergroups [73].

We have taken pains to state conjecture 1 precisely because it is the conjecture we understand best and in which we have the most confidence. Further conjectures in this section will be stated somewhat more loosely. We hope we have convinced the reader that there are substantial reasons for believing conjecture 1 is correct. As we have discussed, one might imagine a proof to proceed along lines very similar to the theorems of Deligne and Doplicher-Roberts. On the other hand, it would be fascinating if there were examples of "sporadic" modular tensor categories arising from conformal field theories. In the introduction we pointed out that an alternative statement of the conjecture says that all RCFT's have already been found. It was probably first stated by Emil Martinec [7] that the nontrivial RCFT's are essentially exhausted by the coset construction, and this was repeated in [9]. It has been reiterated many times in private by Bazhanov, Fröhlich, Gawedzki, Goddard, Reshetikhin, and perhaps others.

Conjecture 1 is a weak conjecture in the sense that it's truth would only classify modular functors of RCFT's. One may hope that a stronger version of the conjecture is true, namely

Conjecture 2: The chiral algebra of any unitary RCFT is the physical Hilbert space for canonical quantization of some CSW theory for an appropriate choice of compact gauge group, symmetric bilinear invariant nondegenerate form, and boundary conditions.

Obviously there is no counterexample to this conjecture, but there do exist some examples of chiral algebras which remain to be interpreted along the lines sketched above. Most notably, the chiral algebra of the Monster module remains uninterpreted [8].

• Exercise 11.2 *Open Problem.* Obtain the chiral algebra of all known $c = 24$ theories with trivial monodromy from quantization of some CSW theory on $D \times \mathbb{R}$.

• Exercise 11.3 *Dual of a RCFT.* Consider a RCFT with F, S, Ω, Δ, c. Show that since F, S, Ω satisfy the polynomial equations so do $F' = F^{-1}, S' = S^{-1}, \Omega' = \Omega^{-1}$. The conformal dimensions of these two solutions are related by $\Delta' = -\Delta mod 1$ and $c' = -c mod 8$. Sometimes there exists a RCFT with F', S', Ω' (remember, a solution of the polynomial equations does not guarantee that there exists a RCFT with these duality matrices). We define this theory as the dual of the original one.

a.) Show that a theory with one primary field is self dual.

b.) Show that the coset of a self dual theory by the chiral algebra $\mathcal{A}$ is a RCFT which is dual to the RCFT based on $\mathcal{A}$.

c.) Construct a self dual theory by appropriately coupling a theory and its dual.

d.) Use the self dual theory based on $E(8)_1 \times E(8)_1$ and part (b) of this exercise to show that the Ising model is dual to $E(8)_2$. A more sophisticated example of this phenomenon was studied in [74] where it was shown that a certain exceptional modular invariant of $F(4)$ KM is dual to $SU(3)_2$. Using part c a new self dual $c = 24$ theory can be constructed.

---

[8] We would like to thank W. Nahm for pointing this out to us.

e.) Show that the duality matrices of the dual theory can be obtained from three dimensions by reversing the sign of the action – reversing the orientation. Since exercise 10.2 is still an open problem, it is not clear if all the states in the chiral algebra and in its representations for every theory (in particular for the $F(4)$ theory of [74]) can be obtained from three dimensions.

Another conjecture, related to those above was posed by E. Witten [75]

Conjecture 3: All three dimensional topological field theories are CSW theories for some appropriate (super)-group.

As we have seen, any modular functor defines a three dimensional topological field theory so that the truth of conjecture 3 may be expected to imply that of conjecture 1, assuming there is no surprising need to resort to noncompact groups or supergroups.

Finally we should note that there has recently been much progress in abelianizing WZW theories [76] [77] [78] [79] [80] [81] [82] [83] and there has been related progress on abelianizing certain coset models. From this work one is naturally lead to wonder if Kadanoff's old idea that all CFT's are related to the gaussian model might in some sense be correct. More precisely, taking into account some of the recent bosonization results, reference [80] states

Conjecture 4: The chiral algebras and representations occuring in RCFT may always be expressed as cohomology spaces for sequences of Fock modules, and all CVO's of RCFT's may be expressed through free field constructions.

The ultimate reduction of RCFT to free field theory would not be in contradiction with the group-theoretic interpretation. Indeed, it is well-known that one can construct representations of groups with harmonic oscillators.

We hope that the truth or falsehood of these conjectures will be established in the near future. Looking beyond the subject of RCFT there are several horizons emerging involving various generalizations, extensions, and applications of the concepts we have used above, but which we have not even mentioned. It is not our intention to discuss these future directions here, should they bear fruit there will be no lack of opportunity for future discussion.

Acknowledgements

We thank L. Alvarez-Gaumé, T. Banks, V. Bazhanov, D. Bernard, M. Bershadsky, J. Birman, L. Crane, P. Deligne, R. Dijkgraaf, S. Elitzur, D. Freed, I. Frenkel, D. Friedan,

J. Frohlich, P. Ginsparg, P. Goddard, J. Harvey, V.F.R. Jones, D. Kazhdan, J. Lepowsky, E. Martinec, A. Morozov, H. Ooguri, V. Pasquier, M. Peskin, J. Polchinski, Z. Qiu, N. Reshetikhin, A. Schwimmer, G. Segal, S. Shatashvili, S. Shenker, E. Verlinde, H. Verlinde, E. Witten and S. Wolpert for useful discussions. We wish to thank the Institute for Advanced Study for the kind hospitalitly when most of this work was done. We would also like to thank the Aspen Center for Physics and C.C.N.Y. for hospitality. This work was supported by NSF grant PHY-86-20266, DOE contract DE-AC02-76ER02220, and DE-AC02-76ER03075.

References

[1] A. Belavin, A.M. Polyakov and A.B. Zamolodchikov, Nucl. Phys. **B241** (1984) 33.

[2] E. Witten, Comm. Math. Phys. **92** (1984) 455.

[3] L. Dixon, J.A. Harvey, C. Vafa and E. Witten, Nucl. Phys. **B261** (1985) 678; Nucl. Phys. **B274** (1986) 285.

[4] D. Friedan, Z. Qiu and S. Shenker, *in* Vertex operators in mathematics and physics, ed. J. Lepowsky et al. (Springer, Berlin 1984); Phys. Rev. Lett. **52** (1984) 1575.

[5] A.B. Zamolodchikov, Theo. Math. Phys. **65** (1986) 1205.

[6] A.B. Zamolodchikov and V.A. Fateev, Sov. Phys. JETP **62** (1985) 215.

[7] D. Kastor, E. Martinec and Z. Qiu, Phys. Lett. **200B** (1988) 434.

[8] J. Bagger, D. Nemeschansky and S. Yankielowicz, Phys. Rev. Lett. **60** (1988) 389; M.R. Douglas, "G/H conformal field theory", CALT-68-1453; F. Ravanini, Nordita-87/56-P; P. Goddard and A. Schwimmer, Phys. Lett. **206B** (1988) 62.

[9] F.A. Bais, P. Bouwknegt, M. Surridge and K. Schoutens, Nucl. Phys. **B304** (1988) 348; Nucl. Phys. **B304** (1988) 371.

[10] P. Goddard, A. Kent and D. Olive, Comm. Math. Phys. **103** (1986) 105.

[11] P. Ginsparg, Nucl. Phys. **B295** (1988) 153; R. Dijkgraaf, E. Verlinde and H. Verlinde, Comm. Math. Phys. **115** (1988) 649; R. Dijkgraaf, C. Vafa, E. Verlinde and H. Verlinde, Comm. Math. Phys. in press.

[12] See, e.g., J. Fröhlich, F. Gabbiani, and P.-A. Marchetti, "Braid Statistics in Three-Dimensional Local Quantum Theory," ETH-TH/89-36, and references therein.

[13] G. Moore and N. Seiberg, Phys. Lett. **212B**(1988)451.

[14] G. Moore and N. Seiberg, Nucl. Phys. **B313**(1989)16.

[15] G. Moore and N. Seiberg, Comm. Math. Phys. **123** (1989) 77.

[16] G. Moore and N. Seiberg, Phys. Lett. **220B**(1989)422.

[17] S. Elitzur, G. Moore, A. Schwimmer, and N. Seiberg, "Remarks on the Canonical Quantization of the Chern-Simons- Witten Theory," to be published in Nucl. Phys. B

[18] P. Goddard and D. Olive, Int. Journ. Mod. Phys. A **1** (1986) 303; M. Peskin, SLAC-PUB-4251; T. Banks, Lectures at the Theoretical Advanced Studies Institute, Santa Fe, 1987; P. Ginsparg, Les Houches lectures 1988, HUTP-88/A054.

[19] B. Schroer, Nucl. Phys. **B295** (1988) 4; "Algebraic Aspects of Non-Perturbative Quantum Field Theories," Como lectures; K.-H. Rehren, Comm. Math. Phys. **116** (1988) 675; J. Frohlich, "Statistics of Fields, the Yang-Baxter Equation, and the Theory of Knots and Links," lectures at Cargese 1987, to appear in *Nonperturbative Quantum*

Field Theory, Plenum; G. Felder and J. Frohlich, unpublished lecture notes; K.-H. Rehren and B. Schroer, "Einstein Causality and Artin braids," FU preprint 88-0439; G. Felder, J. Frohlich and G. Keller, "On the Structure of Unitary Conformal Field Theory I: Existence of Conformal Blocks" Comm. Math. Phys. in press and "On the Structure... II: Representation Theoretic Approach."

[20] A. Tsuchiya and Y. Kanie, in *Conformal Field Theory and Solvable Lattice Models*, Advanced Studies in Pure Mathematics, **16** (1988) 297; Lett. Math. Phys. **13** (1987) 303.

[21] E. Verlinde, Nucl. Phys. **B300** (1988) 360.

[22] D. Gepner and E. Witten, Nucl. Phys. **B278**(1986)493.

[23] V.G. Knizhnik and A.B. Zamolodchikov, "Current algebra and Wess-Zumino model in two dimensions", Nucl. Phys. **B247** (1984) 83.

[24] D. Friedan and S. Shenker, Nucl. Phys. **B281** (1987) 509; D. Friedan, Physica Scripta **T15** (1987) 72; D. Friedan and S. Shenker, Talks at Cargese and I.A.S. (1987) unpublished.

[25] A. Hatcher and W. Thurston, Topology **19** (1980) 221.

[26] C. Vafa, Phys. Lett. **206B** (1988) 421.

[27] E. Witten, Comm. Math. Phys. **121** (1989)351.

[28] J. Cardy, UCSBTH-89-06.

[29] E. Verlinde, in proceedings of the conference at Schloss Ringberg, April 1989

[30] P. Deligne and J.S. Milne, Lect. N. Math. **900**(Springer 1982); P. Deligne, *Catégories tannakiennes* IAS preprint.

[31] S. Doplicher and J.E. Roberts, in: Proc. of VIII Intl. Congress on Math. Phys., M.Mebkhout and R. Seneor (eds), Singapore: World Scientific 1989; see also: S. Doplicher, R. Haag and J.E. Roberts, Comm. Math. Phys. **23** (1971) 199; **35** (1974) 49.

[32] S. Doplicher and J. Roberts, "Monoidal C^*-Categories and a New Duality Theory for Compact Groups," Rome preprint URLS-DM/NS-88/006; "Why there is a field algebra with a compact gauge group describing the superselection structure in particle physics," Rome preprint I-00185.

[33] A.A. Kirillov, *Elements of the Theory of Representations*. Springer 1976.

[34] A. Joyal and R. Street, "Braided Monoidal Categories," Macquarie Math. Reports. Report no. 860081(1986)

[35] R. Dijkgraaf and E. Verlinde, "Modular invariance and the fusion algebra", presented at Annecy Conf. on conformal field theory.

[36] E. Martinec, "Criticality, Catastrophes, and Compactifications," to appear in the V.G. Knizhnik memorial volume.

[37] I. Frenkel, talk at meeting of Canadian Society of Mathematics, Vancouver, November, 1987; H. Sonoda, Nucl. Phys. **B311** (1988)401, 417.

[38] G. Felder and R. Silvotti, Comm. Math. Phys. **123**(1989)1.

[39] D. Bernard, Nucl. Phys. **B288**(1987)628

[40] N. Reshetikhin, LOMI preprints E-4-87,E-17-87; A.N. Kirillov and N. Reshetikhin, LOMI preprint E-9-88

[41] N. Reshetikhin and V.G. Turaev, "Ribbon Graphs and Their Invariants Derived from Quantum Groups," LOMI preprint; "Invariants of 3-Manifolds via link polynomials and quantum groups," preprint.

[42] E. Witten, "Gauge Theories and Integrable Lattice Models," IAS preprint IASSNS-HEP-89/11.
[43] J. Frohlich and C. King, "Two-Dimensional Conformal Field Theory and Three-Dimensional Topology," ETH-TH/89-9.
[44] L. Kauffmann, "Knot theory and statistical mechanics."
[45] L. Crane, "Topology of 3-Manifolds and Conformal Field Theories," Yale preprint YCTP-P8-89.
[46] M.F. Atiyah in the "Oxford Seminar on Jones-Witten Theory," 1988
[47] M.F. Atiyah, "Topological Quantum Field Theories," Publ. Math. IHES **68**(1988)175.
[48] M. Kontsevich, "Rational Conformal Field Theory and Invariants of 3-Dimensional Manifolds," Marseille preprint CPT-88/P.2189
[49] W. Thurston and E. Witten, unpublished.
[50] V. Drinfeld, "Quantum Groups," Proc. of the Intl. Conf. of Math. Berkeley, 1986.
[51] L. Alvarez-Gaumé, C. Gomez, G. Sierra, "Duality and Quantum Groups," CERN preprint CERN-TH.5369/89
[52] V. Pasquier and H. Saleur, SPhT/89-031.
[53] D. Kazhdan and I. Frenkel, Work in progress
[54] T. Kohno, in *Braids* Contemp. Math. vol 78; T. Kohno, "Quantized universal enveloping algebras and monodromy of braid groups," Nagoya preprint, 1988; V.G. Drinfeld, "Quasi-Hopf Algebras and Knizhnik-Zamolodchikov Equations," Kiev preprint, ITP-89-43E.
[55] A. Kirillov and N. Reshetikhin, in Proc. of the Conf.on Infinite Dimensional Lie Groups and Lie ALgebras, Marseille 1988, ed. by V.G. Kac.
[56] L. Alvarez-Gaumé, C. Gomez, and G. Sierra, Phys. Lett. **220B**(1989)142.
[57] G. Moore and N. Reshetikhin, "A Comment on Quantum Group Symmetry in Conformal Field Theory," IAS preprint, to appear in Nucl. Phys. B
[58] G. Felder, J. Fröhlich, and G. Keller, ETH preprint.
[59] J. Fröhlich et. al., talk at the Schloss Ringberg meeting.
[60] I. Frenkel, unpublished.
[61] J.-L. Gervais, "The quantum group structure of 2D gravity and minimal models," Ecole Normale preprint LPTENS 89/14.
[62] E. Witten, "Gauge Theories, Vertex Models, and Quantum Groups," IASSNS-HEP-89/32.
[63] E. Witten, "The Central Charge in Three Dimensions," IASSNS-HEP-89/38.
[64] A. Alekseev, L. Fadeev, and S. Shatashvili, J. Geom. Phys., in press.
[65] N. Pressley and G. Segal, *Loop Groups* (Oxford Univ. Press 1986).
[66] G. 't Hooft, Nucl. Phys. **B138** (1978) 1.
[67] D. Mumford, *Tata Lectures on Theta.* Prog. in Math. vol. 28. Birkhäuser.
[68] M.F. Atiyah and R. Bott, Phil. Trans. R. Soc. Lond. A **308** (1982) 523.
[69] R. Dijkgraaf and E. Witten, "Topological Gauge Theories and Group Cohomology," IASSNS-HEP-89/33, THU-89/9
[70] W. Nahm, Duke Math. J. **54** (1987) 579; K. Bardakci, E. Rabinovici and B. Saering, Nucl. Phys. **B299** (1988) 151; K. Gawedzki and A. Kupiainen, HU-TFT-88-29; 34; IHES/P/88/45; A.M. Polyakov, Lectures at Les Houches; D. Karabali, Q.-H. Park, H.J. Schnitzer and Z. Yang, BRX-TH-247.
[71] R. Dijkgraaf, "A Geometrical Approach to Two-Dimensional Conformal Field Theory," Utrecht thesis.

[72] H. Verlinde, "Conformal Field Theory, 2-D Quantum Gravity, and Quantization of Teichmüller Space," Princeton preprint, PUPT-89/1140.
[73] C. Crnkovic, G. Moore, N. Read, work in progress.
[74] A.N. Schellekens and S. Yankielowicz, CERN preprints CERN-TH-5344/89, 5377/89.
[75] E. Witten, "The search for higher symmetry in string theory," preprint IASSNS-HEP-88/55.
[76] M. Wakimoto, Comm. Math. Phys. **104**(1986)605.
[77] G. Felder, Nucl. Phys. **B317**(1989)215.
[78] N. Hitchin, in "Oxford Seminar on Jones-Witten Theory," Oxford preprint
[79] B. Feigin and E. Frenkel, Usp. Mat. Nauk. **43**(1988)227.
[80] A. Gerasimov, A. Marshakov, A. Morozov, M. Olshanetsky, S. Shatashvili, "WZW Model as a Theory of Free Fields," ITEP preprint 139-88 submitted to Nucl. Phys. ; A. Gerasimov, A. Marshakov, A. Morozov, ITEP 73-89; A. Morozov, ITEP 43-89.
[81] K. Gawedski, "Quadrature of Conformal Field Theories," IHES preprint.
[82] D. Bernard and G. Felder, "Fock Representations and BRST Cohomology in $SL(2)$ Current Algebra," ETH-TH/89-26.
[83] J. Distler and Z. Qiu, "BRS Cohomology and a Feigin-Fuks representations..." CLNS-89/911; D. Nemeschansky, "Feigin-Fuks representation of $su(2)$ Kac-Moody algebra,"; "Feigin-Fuks representation of string functions," USC-89/012.

CHERN-SIMONS GAUGE THEORY AND SPIN-STATISTICS CONNECTION IN 2 DIMENSIONAL QUANTUM MECHANICS

G.W. Semenoff*

Department of Physics, University of British Columbia
Vancouver, British Columbia, Canada V6T 2A6

I. Introduction

It has been known for some time that the possibilities for quantum statistics in 1 and 2 space dimensions are more general than in the familiar 3 dimensional world. There has recently been much interest in the 2 dimensional case where, besides the usual Bosons and Fermions there can be particles with intermediate statistics.

These are characterized by how their wavefunctions change when particle positions are exchanged. The wavefunction for indistinguishable particles must change by a phase. In the Bose case the phase is 1. In the Fermi case it is $e^{i\pi} = -1$. In 2 dimensions there can be particles for which this phase is $e^{i\alpha_C}$ where the real number α_C depends on the path C through which the particles are interchanged and needs not be an integral multiple of π. Both time reversal and 2 dimensional parity reverse the orientation of C and change the sign of α_C. Therefore the statistics can be unconventional only in quantum systems where both parity and time reversal invariance are broken or else where there are two kinds of particles with opposite statistics which transform into each other under parity and time reversal.

Exotic spin and statistics may play an important role in several condensed matter phenomena. There are speculations about exotic parity and time reversal violating states of matter confined to 2 dimensional spaces. These are particularly important to the phenomenology of the fractional quantum Hall effect where parity and time reversal symmetries are broken by an external magnetic field[1]. They have also recently conjectured to have something to do with high T_c superconductors[2].

In this Series of Lectures I shall discuss the fundamental reasons why exotic spin and statistics can occur in 2 dimensional quantum mechanics. The emphasis will be on possible physical realizations of this phenomenon and in particular the connection between spin and statistics. For the most part this will be a review. I believe that some of the discussion of the relationship between a many-body quantum theory with fractional statistics and a $U(1)$ gauge theory with a Chern-Simons term, some of the material on superconductivity and also the emphasis on the spin-statistics connection in a Chern-Simons gauge theory is original.

* This work is supported in part by the Natural Sciences and Engineering Research Council of Canada.

Physics, Geometry, and Topology
Edited by H. C. Lee
Plenum Press, New York, 1990

Historically this subject began with the work of Skyrme, Finkelstein and Rubenstein who recognized that topological effects in quantum field theory can lead to quantum states with unusual statistics and spin[3]. There, solitons in a nonlinear theory of scalar fields can behave like Fermions. Under certain circumstances the wavefunctionals of the field theory can be odd under exchange of soliton positions and the soliton states themselves can have the odd spin-parity characteristic of Fermionic states. Some of these ideas were generalized to quantum mechanical problems defined on multi-connected configuration spaces by Laidlaw and De-Witt[4]. There the essential ingredients of present day formulation of fractional statistics problem were given.

In 1 dimension the notion of statistics degenerates. Fermions have antisymmetric wavefunctions and Bosons have symmetric wavefunctions. However, a system of noninteracting Fermions on a line is equivalent to a system of Bosons with a short ranged hard core interaction. Also, spin can only be defined by helicity and there is no spin-statistics connection. In 1 space and 1 time dimension the Coleman-Mandelstam construction[5], and the vertex operator construction which is its natural generalization, make an explicit mapping between field theories with particles of differing statistics.

If the idea of exchange statistics is to be meaningful we require at least 2 space dimensions. There the spin of a particle is a rotation scalar and in a relativistic theory it is the time component of a 3-vector. The angular momentum algebra is abelian and does not constrain the eigenvalues of the angular momentum operator in the same way as in 3 and higher dimensions. This leaves open the possibility of fractional angular momentum and spin.

As we shall see later fractional statistics can arise from the rich homotopy of the configuration space of a gas of identical particles on a 2 dimensional space.

It was Liennaas and Myrlheim[6] who first noticed that particles in 2 space dimensions can have exotic statistics. They observed that if one considers bound states of charged particles and flux tubes the wavefunction for this system accumulates Bohm-Aharonov phases when the composite particles are transported around each other.

In a parallel development gauge field theories in three spacetime dimensions were found to have an interesting topological structure not found in their four dimensional counterpart. There the Chern-Simons three-form can be added to the action of the gauge theory[7,8]. In an interesting series of papers Deser, Jackiw and Templeton[8] showed that this term gives the photon, or gluon, a mass. They also showed that in non-Abelian theories, gauge invariance forces a quantization of the coefficient of the Chern-Simons term.

Some of this structure was also noted in the mathematics literature. Schwarz[9] showed that if one considers a gauge theory where the ordinary Maxwell or Yang-Mills kinetic term for the gauge fields is absent and the action and there appears only a Chern-Simons term the Fadeev-Popov determinants involved in fixing the gauge in this theory contained interesting mathematical quantities. The relationship between knot theory, intagrable models, conformal field theory and three dimensional Chern-Simons gauge theories is presently an active area of research in both mathematics and physics. This was first studied in a beautiful series of papers by Akutsu, Wadati and collaborators[10]. The work of Schwarz was later generalized by Witten[11] to get a very interesting relationship between this model with nondynamical matter and conformal field theory and integrable models in two spacetime dimensions. This complements earlier work on the relationship between integrable models and knot polynomials.

In the physics literature, Wilczek and Zee[12] identified fractional statistics as induced in a 2+1-dimensional $U(1)$ gauge theory where the gauge field has a Chern-Simons term. Also, its relation with the representation theory of the braid group was first studied by Wu[13].

Recently there has been much work in this field. The relationship between spin and statistics in a Chern-Simons theory was examined by Polyakov[14], in ref. 15 and 16 and by Dunne, Trugenberger and Jackiw[17]. A mathematical construction of a quantum field theory which exhibits particles with exotic statistics has been given by Frohlich and Marchetti[18]. They demonstrate that the vortices in the 3 spacetime dimensional Abelian Higgs model, which are topological solitons, correspond to asymptotic fields of the theory which have both fractional spin and exotic statistics. An important feature is the use of a gauge theory where the gauge field has a Chern-Simons term in the action.

II. Exotic Spin and Statistics in 2 Space Dimensions

The possibility that particles in a quantum mechanical system can have noncanonical spin and exotic exchange statistics is related with the connectivity of the rotation group and of the quantum mechanical configuration space respectively. For a system of N identical particles in a d dimensional space $\mathcal{R}^d$ the quantum mechanical probability measure

$$\rho(q_1, \cdots, q_N) = \psi^\dagger(q_1, \cdots, q_N)\,\psi(q_1, \cdots, q_N) \tag{2.1}$$

is a mapping from the space of positions of particles $\mathcal{R}^{dN}$ to the real numbers $\mathcal{R}$. It has a domain on a smaller set than the full position space. We shall call this set the quantum mechanical configuration space $\mathcal{C}$ such that

$$\mathcal{C} \xrightarrow{\rho} \mathcal{R}$$

Since identical particles are indistinguishable, ρ must be a symmetric function of its arguments.*

If the system also has a Pauli exclusion principle the probability measure should vanish whenever the positions of 2 or more particles coincide. This is a property of conventional Fermion wavefunctions which must be antisymmetric and therefore vanish when any 2 or more position arguements

* Particles are identical if the Hamiltonian has a permutation symmetry

$$H(q_1, p_1, q_2, p_2, \cdots, q_N, p_N) = H(q_{1'}, p_{1'}, q_{2'}, p_{2'}, \cdots, q_{N'}, p_{N'})$$

where $1', \ldots, N'$ is a permutation of $1, \ldots, N$. Energy eigenstates should carry a representation of the permutation group S_N and the probability measure should be symmetric.

are equal. It is also a characteristic of wavefunctions for particles with any exchange statistics which are not symmetric.

We construct $\mathcal{C}$ as follows: We first consider the position space of a gas of N particles moving on $\mathcal{R}^d$, $\mathcal{R}^{Nd}$ with points labelled by the vectors $(q_1, \ldots, q_N)$. Anticipating an exclusion principle [otherwise we could only get Bosons] we subtract from this the diagonal subspace

$$\mathcal{D} = \{(q_1, \ldots, q_N) : q_i = q_j \text{ for any i and j}\}$$

to get

$$\mathcal{R}^{Nd} - \mathcal{D}$$

Then we take into account the fact that particles are indistinguishable by factoring this space by the group of permutations S^N to get

$$\mathcal{C} = \frac{\mathcal{R}^{Nd} - \mathcal{D}}{S^N} \tag{2.2}$$

The permutation symmetry of the probability measure implies that the wavefunction must change by a phase when the positions of 2 or more particles are interchanged

$$\psi(\ldots, q_i, \ldots, q_j, \ldots) = e^{i\chi_{ij}} \psi(\ldots, q_j, \ldots, q_i, \ldots) \tag{2.3}$$

If this phase is nonzero, the wavefunction must vanish when $q_i = q_j$.

A given configuration of the system corresponds to a point on $\mathcal{C}$. To define the exchange of two particles we must specify a continuous path which is in fact a closed loop on $\mathcal{C}$ and represents an element of the fundamental group $\Pi_1(\mathcal{C})$. Compositions of successive interchanges of particles induces the multiplication of $\Pi_1(\mathcal{C})$ as the composition law for the phases $e^{i\chi_{ij}}$ which must therefore form a 1 dimensional unitary representation of $\Pi_1(\mathcal{C})$. which is the universal cover of $R^2 - q_1, \ldots, q_N$

This in turn implies that the wavefunction must carry this representation and that it lives on the universal cover of the configuration space, i.e. the smallest simply connected space $\hat{\mathcal{C}}$ such that

$$\mathcal{C} = \hat{\mathcal{C}}/\Pi_1(\mathcal{C}) \ .$$

If $d \geq 2$, $\mathcal{C}$ is connected, i,e,

$$\Pi_0(\mathcal{C}) = 0 \ .$$

Then $\hat{\mathcal{C}}$ can be constructed in the following way: We first choose a basepoint p_0 on $\mathcal{C}$ and take the set of all curves with one endpoint p_0 and other endpoints anywhere on $\mathcal{C}$. Then we identify all curves which have the same endpoints and which are equivalent under homotopy. [This set of representative curves of the equivalence classes is sometimes referred to as a standard grid.]

This construction produces $\hat{\mathcal{C}}$. We can define a projection from $\hat{\mathcal{C}}$ to $\mathcal{C}$ by associating each representive curve of a homotopy class with its endpoints. Thus $\hat{\mathcal{C}}$ together with this projection is a fiber bundle with base $\mathcal{C}$ and fibers $\Pi_1(\mathcal{C})$.

Wavefunctions are single-valued functions on $\hat{\mathcal{C}}$ which project to multivalued functions on $\mathcal{C}$. Since in $d \geq 3$,

$$\Pi_1\left(\mathcal{R}^{Nd} - \mathcal{C}\right) = 0 \quad ,$$

we have

$$\Pi_1\left(\mathcal{C}\right) \;=\; S^N \tag{2.4}$$

which has only totally symmetric or totally antisymmetric 1 dimensional unitary representations. Thus in greater than 2 dimensions wavefunctions for identical particles can be either symmetric or antisymmetric and we find conventional Bose and Fermi statistics, respectively.

On the other hand in 2 dimensions

$$\Pi_1\left(\mathcal{C}\right) \;=\; \mathcal{B}_N\left(\mathcal{R}^2\right) \tag{2.5}$$

which is an infinite discrete nonabelian group[19] , the Nth order braid group of the plane $\mathcal{R}^2$. It has an interesting 1 parameter family of 1 dimensional representations which are carried by the following function from $\hat{\mathcal{C}}$ to the circle $\mathcal{S}^1$

$$f_\alpha\left(q_1, \ldots, q_N\right) = \exp\left\{2i\alpha \sum_{i<j} \Theta\left(q_i - q_j\right)\right\} \tag{2.6}$$

Here $\Theta\left(q_i - q_j\right)$ is the multi-valued angle between the vector $q_i - q_j$ and a fixed reference direction. * f_α is a multivalued function on $\mathcal{C}$ and a single valued function on $\hat{\mathcal{C}}$. On the latter space it can be defined by integration along the representative curves which define $\hat{\mathcal{C}}$,

$$\left(\ln f_\alpha(q_1, \ldots, q_N)\right)_P = \int_{p_0}^{q_1,\ldots,q_n} dl \cdot \nabla \ln f_\alpha(l)$$

f_α carries a 1 dimensional unitary representation of the braid group. It changes by the phase $2\alpha i\left(\pi + 2\pi n\right)$ when the positions of 2 particles are interchanged. The additional $2\pi n$ depends on

---

* This angle is related to the Green function for the 2 dimensional Laplacian:

$$g(x) = \frac{1}{-\nabla^2}\delta(x) = \frac{1}{2\pi}\ln \mu|x|$$

where μ is an infrared cutoff. Θ has the property

$$\nabla_i \Theta(x) = -2\pi\epsilon_{ij}\nabla_j g(x)$$

and is independent of μ. It satisfies the Laplace equation and is not differentiable at the origin,

$$\nabla^2\Theta(x) = 0 \quad , \quad \nabla \times \nabla\Theta(x) = 2\pi\delta(x)$$

We remind the reader that in 2 dimensions the exterior product of 2 vectors is a scalar.

the number of other particles whose positions are linked by the exchange trajectory, with sign determined by the orientation of the trajectory. The function $f_\alpha(q_1, \ldots, q_N)$ can be used to characterize the wavefunction of a gas of particles with fractional statistics.

Not unrelated to the possibility of fractional statistics is the fact that the rotation group in 2 space dimensions is multiconnected

$$\Pi_1(SO(2)) = \mathcal{Z}$$

with universal covering group $\mathcal{R}^1$, the translation group of the real line. Wavefunctions carry representations of universal covering groups of symmetry groups. This implies that the eigenvalues of angular momentum need not be quantized as integers but can have values which are any real numbers. This is intuitive as the rotation group on 2 dimensions has only 1 generator so the algebraic constraints which restrict the spectrum of the generators in higher dimensions are absent there.

A way to characterize this possibility is by the spin parity which is defined by the change of phase of a wavefunction under a rotation through angle 2π. For example, for a single particle wavefunction the spin parity γ is defined by

$$e^{2\pi i L}\psi(q) = e^{2\pi i\gamma}\psi(q) \tag{2.7}$$

This change in phase leaves the probability density (2.1) invariant.

We can use the representation (2.6) to construct a wavefunction with exotic statistics from a wavefunction with Bose or Fermi statistics,

$$\psi(q_1, \ldots, q_N) = f_\alpha(q_1, \ldots, q_N)\,\psi_{symm}(q_1, \ldots, q_N) \tag{2.8}$$

where ψ_{symm} is a symmetric function of $q_1, \ldots, q_N$. If, for example, we consider a ideal gas with Hamiltonian

$$H = \sum_i \frac{1}{2m} p_i^2 \tag{2.9}$$

the transformation (2.8) can be used to map the free system onto a system of interacting Bosons with Hamiltonian

$$H = \sum_i \frac{1}{2m}(p_i + A(q_i))^2 \tag{2.10}$$

The statistical gauge potential is given by

$$A(q_i) = 2\alpha \sum_{j \neq i} \frac{\partial}{\partial q_i} \Theta(q_i - q_j) \tag{2.11}$$

This vector potential is single valued, * it obeys the Coulomb gauge condition

$$\frac{\partial}{\partial q_i} \cdot A(q_i) = 0$$

* Note that even though the angles $\Theta(q_i - q_j)$ are multivalued functions their derivatives are single valued.

and has the magnetic field

$$B(q_i) = 4\pi\alpha \sum_{j\neq i} \delta^2 (q_i - q_j) \tag{2.12}$$

which vanishes on the configuration space $\mathcal{C}$. Thus each particle sees the other particles as being attatched to a magnetic flux tube with flux 2α. Exotic statistics arise from the holonomy of the flat connections in (2.11). Physically it can be thought of as originating in the Bohm-Ahoronov effect of charged particles moving in the presence of each other's magnetic flux.

The Hamiltonian (2.10) follows from canonical quantization of the system with classical action

$$S = \int dt \sum_i \frac{m}{2}\dot{q}_i^2 - \sum_{i\neq j} 2\alpha \int dt \dot{q}_i \frac{\partial}{\partial q_i}\Theta(q_i - q_j) \tag{2.13}$$

$$= \int dt \sum_i \frac{m}{2}\dot{q}_i^2 - \alpha \sum_{i<j} \int dt \frac{d}{dt}\Theta(q_i - q_j) \tag{2.14}$$

For a periodic trajectory, the final term in (2.14) is the sum of the linking numbers of the trajectories of the particles.

An interesting generalization is to add to the action (2.14) the self-linking number of the particle trajectories. This quantity has various definitions. One sufficient for our purpose is to split the point particle into two points with infinitesimal separation, give some definition for the direction of the separation as the system evolves in time and to consider the linking number of the trajectories of the two points. The mathematical term for this proceedure for defining self-linking number is a *framing* of the trajectory.

Adding the self-linking number to the action has the obvious effect of altering the spin of particles. To see this, consider the Feynman path integral for a single particle and assume that the action contains the self-linking number of the particle's trajectory with coefficient $2\pi\gamma$. If we consider a trajectory where the frame of reference rotates adiabatically through angle 2π the self-linking number changes by 1 - the rotation puts 1 additional twist in the framing of the trajectory. Thus the wavefunction of the particle changes phase by $2\pi\gamma$ under this rotation, *i.e.* it now has spin parity γ. *

A similar argument for a gas of particles with action (2.14) [before the self-linking numbers are added] would produce and extra phase of $-2\pi\alpha N(N-1)$ under the adiabatic rotation where $N(N-1)$ are the number of pairs of particles. The self-linking numbers would contribute an extra $2\pi\gamma N$ so the total spin parity is $2\pi\,(N\gamma - N(N-1)\alpha)$. We will see this in a different way later.

* The point split definition of the angle $\Theta(q_i - (q_i + \epsilon))$ when included in the function f_α in (2.6) as

$$\exp\left\{ i\gamma \sum_i \Theta(q_i - (q_i + \epsilon_i)) \right\}$$

where ϵ_i are the infinitesimal point splitting vectors. This modifies the spin-parity of wavefunctions. For example a single-particle state obtains spin-parity γ.

If we adjust the coefficient of the self linking number to be $\gamma = -\alpha$ the action S can be obtained by introducing a statistical gauge field A_μ and

$$\exp\{iS\} = \int [dA_\mu(x,t)] \exp\left\{ i\int dt \sum_i \frac{m}{2}\dot{q}_i^2 + i\int d^3x \left(j^\mu A_\mu + \frac{1}{4\pi\alpha}\epsilon^{\mu\nu\lambda} A_\mu \partial_\nu A_\lambda \right) \right\} \tag{2.15}$$

where

$$j^\mu(x) = \sum_i \int \frac{d}{d(\tau)} q_i^\mu \tau \delta^3 (x - q_i(\tau)) \tag{2.16}$$

where $q_i^\mu(\tau)$ is the trajectory of the ith particle, and τ is the curve parameter. Performing the Gaussian integration in (2.15) obtains the action (2.14) [with $<$ replaced by $\leq$ in the final summation]. To see this we note that the solution of the field equation

$$\epsilon^{\mu\nu\lambda}\partial_\nu A_\lambda = -2\pi j^\mu$$

in the Coulomb gauge

$$\vec{\nabla} \cdot \vec{A} = 0$$

is

$$\vec{A}(x) = \alpha \sum_i \vec{\nabla}\Theta(x - q_i) \quad , \quad A_0(x) = \alpha \sum_i \dot{\vec{q}} \cdot \vec{\nabla}\Theta(x - q_i)$$

which upon substitution in (2.15) gives the action in equation (2.14) with the self-linking numbers included.

Note that this possibility of representing exotic statistics using a statistical quantum gauge field *requires* a spin-statistics connection: both the statistics parameter and the spin parity are given by the parameter α. The self-linking number of trajectories themselves are not unambiguously defined by the model but depend on the regularization. However, the spin parity, which is the change in the self-linking number under a 2π twist, is unambiguous. This fact has been seen before in theories with nondynamical matter and also in the context of topologically massive gauge theory with dynamical charged matter fields[16] and will be reviewed in Section IV. It is associated with the ultraviolet regularization which is needed for precise definition of the field theory in (2.15).

The shift of the spin of particles can be seen in the canonical formalism. Consider the gauge invariant angular momentum operator corresponding to the many particle theory (2.14),

$$L = \sum_i q_i \times p_i + \sum_i q_i \times A(q_i) \tag{2.17}$$

This operator generates a rotation and a gauge transformation which compensates the gauge variant rotation property of a wavefunction. The result is a gauge invariant rotation for the many-particle wavefunction. The final term can be written as

$$L = L_{canon.} + \sum_{i,j} 2\alpha q_i \cdot \frac{q_i - q_j}{(q_i - q_j)^2} \tag{2.18}$$

where the self-interaction terms have an implicit point splitting interaction and the canonical rotation generator is

$$L_{canon} = \sum_i q_i \times p_i$$

After symmetrizing the sum (2.18) becomes

$$= L_{canon} + \alpha N^2 \tag{2.19}$$

The resulting operator obviously commutes with the Hamiltonian. This shows that the spin parity of the N-particle wavefunction is αN^2.

If in $A(q_i)$ we did not include the self-linking contribution the N^2 in (2.19) would be replaced by $N(N-1)$, the number of pairs of particles. Thus we see that there is angular momentum α stored in the mutual interaction between each pair of particles. With the spin-statistics connection appropriate to the Chern-Simons theory there is also angular momentum α stored in the interaction of each particle with its own magnetic flux. We can think of this latter angular momentum as spin.

Finally, note that this large shift in the spin could be important in a realistic system where the quasiparticles have fractional statistics. If one couples an external magnetic field to the N-particle system by adding the term

$$S_{int} = -\frac{e}{c}\int dt \sum_i \dot{q}_i \cdot A_{ext}(q_i) \tag{2.20}$$

to the action, where

$$\vec{A}_{ext} = -\frac{1}{2}\vec{q} \times \vec{B}_{ext}$$

it is the gauge invariant angular momentum operator which couples to the magnetic field through the interaction

$$H_{int} = \frac{e}{mc} L \cdot B_{ext}. \tag{2.21}$$

and whose eigenvalues give the total magnetic moment of the system. It is intruguing that the spin component of the total angular momentum scales like the square of the number of paticles. This implies that the orbital part should also scale like the square of N so as to compensate. Otherwise the total magnetic moment would not scale correctly in the thermodynamic limit. This would also tend to minimize the kinetic energy which contains L^2. This in turn implies that the wavefunction is a rapidly varying function of angles.

In fact this is just the properties of Landau level wavefunctions which we would obtain by considering a set of electrons in an external constant background magnetic field with total flux αN. This sort of mean field theory has been suggested by Laughlin and collaborators[2]. It maps a perfect gas of particles with fractional statistics onto a quantum Hall effect problem. There is an interesting recent conjecture that such a system has a superconducting ground state[2].

In the following we shall discuss a generalization of the ideas presented in this Section to consider a quantum field theory of charged particles whose conserved $U(1)$ current couples to a statistical gauge field as in (2.15).

III. Many Particle Systems and Quantum Field Theory

In the last Section we saw that a quantum mechanical system of fractional spin and statistics particles could be described by a particular hybrid of a quantum field theory and point quantum mechanics where the field theory was a Chern-Simons gauge theory coupled to the particle currents. It is useful to consider a generalization of those arguments to a system where the charged particles are the elementary quanta of a quantum field theory.

We expect that, in the thermodynamic limit of N particle quantum mechanics, the large wavelength behavior can be described by a continuum Landau-Ginzburg theory and it is very reasonable that the statistical gauge field survives with a Chern-Simons term and couples to complex fields there. One would expect, by applying semiclassical arguments, that in the resulting quantum field theory the quasiparticles have exotic spin and statistics. It has been established by rather convincing arguments using the canonical Hamiltonian formalism of a field theory that this is indeed the case. In this Section, we shall review some of those arguments.

It is most straightforward to begin by constructing a second quantized version of the many particle quantum mechanics. For this we promote the Hamiltonian operator of the quantum mechanical system to a second quantized operator which has the same eigenvalues as the original operator in the N-particle sector. We begin with operators which create and annihilate particles at point q, $\psi(q)$ and $\psi^\dagger(q)$ respectively with the commutator algebra

$$[\psi(q), \psi^\dagger(q')] = \delta^2(q-q') \tag{3.1a}$$

$$[\psi(q), \psi(q')] = 0 = [\psi^\dagger(q), \psi^\dagger(q')] \tag{3.1b}$$

so that the state with N particles occupying positions $q_1, \ldots, q_N$ is

$$|q_1, \ldots, q_N> = \psi^\dagger(q_1) \ldots \psi^\dagger(q_N)|0> \tag{3.2}$$

where the vacuum $|0>$ is the state with no particles. The state is a symmetric function of $q_1, \ldots, q_N$. * The wavefunction of the N particle quantum mechanics is recovered from the wavefunction of the second quantized sytem $|\Psi>$ by

$$\psi(q_1, \ldots, q_N) = <q_1, \ldots, q_N|\Psi> \tag{3.3}$$

On the states (3.2) the charge density operator defined by

$$j^0(q) = \psi^\dagger(q)\psi(q) \tag{3.4}$$

has the eigenvalues

$$j^0(q)|q_1, \ldots, q_N> = \sum_i \delta^2(q-q_i)|q_1, \ldots, q_N> \tag{3.5}$$

* We could consider an analogous construction for Fermions using anticommuting operators and find a state which is a completely antisymmetric function of the position arguments.

which is the classical charge distribution of the N-particle system where the number density is concentrated at discrete points.

The wavefunction (3.3) is a symmetric function of the particle positions. To find wavefunctions which exhibit exotic exchange statistics we consider the following construction: Define the multivalued field operator

$$\hat{\psi}(q) = e^{i\alpha \int \Theta(q-q')j^0(q')}\psi(q)e^{i\alpha \int \Theta(q-q')j^0(q')} \tag{3.6a}$$

$$\hat{\psi}^\dagger(q) = e^{-i\alpha \int dq'\Theta(q-q')j^0(q')}\psi^\dagger(q)e^{-i\alpha \int dq'\Theta(q-q')j^0(q')} \tag{3.6b}$$

Here we have used the disorder operator

$$\exp\left\{2i\alpha \int d^2q'\Theta(q-q')j^0(q')\right\}$$

with the multivalued angle introduced in equation (2.6) and (2.8) to alter the statistics of the operator ψ.

For now, we shall be cavalier about the possible short distance singularity of this operator product. In order to make the particle spin well defined we shall have to split the points in the operator product in (3.6) - a proceedure analogous to framing the particle trajectories in the single particle quantum mechanics. There is another problem that the charge density in second quantization need not be concentrated at points but could be distributed.* Then the exponent in the disorder operator would be the multivalued angle function smeared over a continuous region - a quantity which does not exist as a differentiable function wherever the charge distribution has support. For the moment, we define the disorder operator by its action on states where the particles have fixed positions and the charge density is therefore concentrated at points and the exponent is well defined as a multivalued function:

$$\exp\left\{2i\alpha \int d^2q'\Theta(q-q')j^0(q')\right\}\psi^\dagger(q_1)\ldots\psi^\dagger(q_N)|0> =$$

$$= \exp\left\{2i\alpha \sum_i \Theta(q-q_i)\right\}\psi^\dagger(q_1)\ldots\psi^\dagger(q_N)|0> \tag{3.7}$$

This will be sufficient to render the operators in (3.6) well defined when we consider systems with finite numbers of particles since it defines their action on a complete set of basis vectors. Some additional care is needed in the thermodynamic limit. We will return to the issue of regularization later.

The state created from the vacuum by the operators (3.6) has the property

$$|\hat{q}_1,\ldots,\hat{q}_N>=\hat{\psi}^\dagger(q_1)\ldots\hat{\psi}^\dagger(q_N)|0> \tag{3.8}$$

* In fact, if the ground state is translation invariant its charge density must be uniformly distributed.

$$= e^{-2i\alpha\sum_{i\leq j}\Theta(q_i-q_j)}|q_1,\ldots,q_N> \tag{3.9}$$

so that

$$<\hat{q}_1,\ldots,\hat{q}_N|\Psi>= e^{i\alpha(2\sum_{i<j}+\sum_{i=j})\Theta(q_i-q_j)} <q_1,\ldots,q_N|\Psi> \tag{3.10}$$

which contains the mapping between symmetric and multi-valued wavefunctions which we found in equations (2.6) and (2.8). [As stated earlier, the $i = j$ terms in the exponent are defined by point-splitting regularization.]

We can use (3.8) and (3.9) to compute the commutation relation of the multivalued fields. The product $\hat{\psi}(q)\hat{\psi}(q')$ and $\hat{\psi}(q')\hat{\psi}(q)$ can be related only when we specify a trajectory on the 2 dimensional space which interchanges q and q'. This trajectory is a loop which is the composition of the curve $C_{qq'}$ along which q is taken to q' and the curve $C_{q'q}$ along which q' is taken to q. Then from (3.8) and (3.9) and using the identity

$$\Theta(q-q')-\Theta(q'-q)=\pi \bmod 2\pi$$

we deduce

$$\hat{\psi}(q)\hat{\psi}(q') = \hat{\psi}(q')\hat{\psi}(q)e^{2i\alpha(\pi+2\pi n)} \tag{3.11}$$

where n is the number of particle postions linked by the loop. It can be given by the integral of the charge operator over a disc with boundary the loop

$$n=\int_D d^2x j^0(x)$$

and the matrix elements of the charge operator are given by the eigenvalues in (3.5).

To construct a Hamiltonian which describes a perfect gas of particles with fractional statistics it is necessary to find a second quantized operator which for a fixed number of particles has the same eigenvalues as (2.10). For this we use the gauge field derived from the disorder operator

$$\hat{A}(q)=\exp\left\{-2i\alpha\int dq'\Theta(q-q')j^0(q')\right\}\frac{1}{i}\frac{\partial}{\partial q}\exp\left\{2i\alpha\int d^2q'\Theta(q-q')j^0(q')\right\}$$
$$=2\alpha\int d^2q'\frac{\partial}{\partial q}\Theta(q-q')j^0(q') \tag{3.12}$$

[Here the gradient operator commutes with the integration when the number of particles is finite and the charges are concentrated at points.] This operator, when operating on the states $|q_1,\ldots,q_N>$ defined in (3.2) has eigenvalues

$$\hat{A}(q)|q_1,\ldots,q_N>=2\alpha\sum_i\frac{\partial}{\partial q}\Theta(q-q_i)|q_1,\ldots,q_N> \tag{3.13}$$

This eigenvalue is just the classical gauge field which was a solution of the field equations for the Chern-Simons gauge theory in (2.15). Using this fact we construct the second quantized Hamiltonian

$$\hat{H}=\int d^2q\frac{1}{2m}\left(\frac{1}{i}\vec{\nabla}_q-\hat{A}(q)\right)\psi^\dagger(q)\cdot\left(\frac{1}{i}\vec{\nabla}_q+\hat{A}(q)\right)\psi(q) \tag{3.14}$$

where the ordering of the gauge field and the matter field in the covariant derivative is defined symmetrically,

$$\left(\vec{\nabla}-i\vec{A}\right)\psi \equiv \vec{\nabla}\psi - \frac{i}{2}\left(\vec{A}\psi+\psi\vec{A}\right)$$

The Hamiltonian (3.14) has the same eigenvalues as the first quantized Hamiltonian in (2.12).

$$\hat{H}|q_1,\ldots,q_N> = H|q_1,\ldots,q_N> \tag{3.15}$$

Note that the ordering of operators in (3.14) has to be chosen appropriately. Also, to define the self-interaction term in the classical gauge field in h, there is an implicit point splitting regularization of the product of the fields ψ and $\hat{A}$.

The gauge field $\hat{A}$ has the property

$$\vec{\nabla}\cdot\vec{\hat{A}}(q) \;=\; 0 \tag{3.16}$$

and has the operator valued magnetic flux

$$B(q) = \vec{\nabla}\times\vec{\hat{A}}(q) = 4\pi\alpha j^0(q) \tag{3.17}$$

To make the connection with a gauge theory we could treat $\hat{A}(q)$ as an independent field which is to be determined from the constraint (3.17) and the gauge fixing condition (3.16). The condition that the constraint (3.17) commutes with the Hamiltonian fixes the algebra of the gauge fields,

$$\left[A_i(q),A_j(q')\right] = 2\alpha i\epsilon_{ij}\delta^2(q-q') \tag{3.18}$$

In fact these are the Coulomb gauge condition, the Gauss' law constraint and the Dirac bracket commutator arising in the canonical formalism for the quantum field theory with action

$$S=\int d^3x\left\{i\psi^\dagger D_0\psi+\frac{1}{2m}\psi^\dagger\vec{D}^2\psi+\frac{1}{8\pi\alpha}\epsilon^{\mu\nu\lambda}A_\mu\partial_\nu A_\lambda\right\} \tag{3.19}$$

When restricted to states with fixed numbers of particles this quantum field theory therefore describes a many-particle quantum system where the particles have fractional spin and statistics. The essential feature is the coupling of the statistical gauge field to the conserved $U(1)$ charge of the system and the Chern-Simons kinetic term for the gauge field.

It is very reasonable to generalize this idea to realtivistic quantum field theory. Possible difficulties lie in the thermodynamic limit and in ultraviolet divergences introduced by the gauge interactions. When the kinetic term for the gauge field is only a Chern-Simons term, the gauge interactions introduce no additional dimensional parameters in the theory, and themselves have a dimensionless coupling constant. The gauge interaction is therefore strictly renormalizable and the field theory should be defined with an ultraviolet cutoff.

Notice that for the quantum field theory the constrained canonical quantization proceedure is not complete without an additional operator ordering prescription which fixes the way the operators $\hat{A}(q)$ and $\psi(q)$ are ordered in the Hamiltonian after the constraint equations are solved. This operator ordering affects the self-interactions of the particles and therefore the spin parity of particles in the theory. In (3.14) we have chosen a particular ordering which reproduces the canonical spin-statistics relation. However, it is possible to choose other orderings which give different spin parity than that expected from the spin-statistics connection[15]. It would be desirable to find an unambiguous way of finding the spin-statistics connection which does not depend on an ad-hoc choice of operator ordering.

It has been shown that one way of doing this is to regulate the ultraviolet structure of the field theory by adding an Maxwell term to the action,

$$-\frac{1}{4e^2}F_{\mu\nu}F^{\mu\nu}$$

and later taking the limit $e^2 \to \infty$ of the dimensional parameter e^2 which acts as a cutoff here. [16] In the Hamiltonian formalism of the resulting topologically massive gauge field theory the operator ordering is unambiguous and the correct spin-statistics connection is obtained. We review this developement in the following Section.

IV. Canonical Quantum Field Theory with Exotic Spin and Statistics

In the last Section we learned that a system with a finite number of particles with fractional spin and statistics is described by a $U(1)$ gauge theory where the kinetic term is a Chern-Simons term. Once we have constructed the Lagrangian, it is natural to consider the thermodynamic limit of the system. There it is very likely that the structure survives. One would still expect that the Chern-Simons term endows the charged quasiparticles with fractional spin and statistics. In this Section we shall argue that this is indeed the case.

There are several problems to be addressed in taking the thermodynamic limit. First, the charge of the ground state goes to infinity, with finite charge density. [In a relativistic field theory even the charge density goes to infinity.] Furthermore, the set of states of the system where the charge is concentrated only at points is no longer a good basis for the eigenfunctions of the Hamiltonian. It is therefore not clear that considerations based on the assumption that states with point-charge distributions were complete still apply there. These important points must be addressed in the following.

We shall consider a quantum field theory with a conserved $U(1)$ charge which we couple to a gauge field which has a Chern-Simons kinetic term

$$S = S_{\text{matter}} + \frac{1}{8\pi\alpha}\int d^3x \left(\epsilon^{\mu\nu\lambda}A_\mu\partial_\nu A_\lambda + A_\mu j^\mu\right) \tag{4.1}$$

The Maxwell equations which follow from this action

$$\epsilon^{\mu\nu\lambda}F_{\nu\lambda} = -4\pi\alpha j^\mu \tag{4.2}$$

contain the constraint

$$B(x) = \nabla \times A(x) = 4\pi\alpha j^0(x) \tag{4.3}$$

where

$$B(x) = \epsilon^{0ij}\nabla_i A_j \tag{4.4}$$

is the magnetic flux density which in 2 space and 1 time dimensions is the charge density corresponding to the topological current

$$\mathcal{J}^\mu = \epsilon^{\mu\nu\lambda}F_{\nu\lambda} \tag{4.5}$$

[The Bianchi identity is $\partial_\mu\epsilon^{\mu\nu\lambda}F_{\nu\lambda} = \epsilon^{\nu\mu\lambda}\partial_\mu\partial_\nu A_\lambda = 0$.]

The total magnetic flux is

$$\Phi = \frac{1}{2\pi}\int d^2x B(x) \tag{4.6}$$

The Maxwell equation here is not an equation of motion but rather an equation of constraint which connects the particle charge current with the topological current.

To see this more clearly we can solve the Maxwell equation in the Coulomb gauge

$$\nabla_i A_i = 0 \tag{4.7}$$

There the Maxwell equations take the form

$$\epsilon^{0ij}\nabla_i A_j = -4\pi\alpha j^0 \tag{4.8}$$

$$\epsilon^{i0j}(\partial_0 A_j - \nabla_i A_0) = -4\pi\alpha j^i \tag{4.9}$$

The first equation can be solved using the gauge condition

$$A_i = \epsilon_{0ij}\nabla^j\frac{1}{\nabla^2}j^0 \tag{4.10}$$

Then the second equation reads

$$\epsilon^{0ij}\nabla_j A_0 = \left(\delta_{ij} - \frac{\nabla_i\nabla_j}{\nabla^2}\right)j^j \tag{4.11}$$

and has the solution

$$A_0 = 4\pi\alpha\frac{1}{\nabla^2}\epsilon_{0ij}\nabla^i j^j \tag{4.12}$$

Here we see that the field equations are solved by relations between the gauge fields and the charged current which are stictly local in time. This is a typical feature of equations of constraint. The equation of motion simply impose relations on the field configurations at a fixed time rather than determining their time evolution. Furthermore, the solution of Maxwell equations exhausts the gauge field degrees of freedom.

If the charges were classical point charges with worldlines the curves $\vec{q}_i(t)$ so that

$$j^0(q,t) = \sum_i \delta(q - q_i(t)) \quad , \quad \vec{j}(q,t) = \sum_i \frac{d}{dt}\vec{q}_i(t)\delta(q - q_i(t)) \tag{4.13}$$

when the solutions (4.10) and (4.12) are substituted in the action (4.1) the result is the topological term which appears in (2.14). In the following we shall generalize this result to the case where, instead of classical point particles the charges are due to quantized fields.

To begin, we consider the example of a charged scalar field:

$$S = \int d^3x \left\{ (\partial_\mu + iA_\mu)\phi^* (\partial^\mu - iA^\mu)\phi - m^2\phi^*\phi + \frac{1}{4\theta}\epsilon^{\mu\nu\lambda}A_\mu\partial_\nu A_\lambda - \frac{1}{4e^2}F_{\mu\nu}F^{\mu\nu} \right\} \tag{4.14}$$

The coupling constant θ in the coefficient of the Chern-Simons term is dimensionless and e^2 has the dimension of mass. The Maxwell term is irrelevant in the low energy, $e^2 \to \infty$, limit. We are nominally interested in this theory where the Chern-Simons term is absent. Here, we have introduced it for the purpose of ultraviolet regularization. Furthermore, there is no symmetry to prevent its generation by quantum corrections. Even if it is absent at the tree level it is generated by radiative corrections[20] and it is therefore unnatural to set it to zero.

With a Maxwell term the gauge degrees of freedom are massive with tree level screening length $\frac{2\theta}{e^2}$. In the infrared limit $e^2 \to \infty$ and the propagating degrees of freedom of the gauge field are irrelevant. Only the statistics altering Chern-Simons term remains. The Maxwell term dominates the large momentum behavior of the theory where it cuts off the linear and logarithmic ultraviolet divergences. Thus we expect that at long wavelengths the fields which solve the model have altered statistics due to the Chern Simons term[22] whereas the fields which solve the ultraviolet limit of the theory have canonical statistics. The distance scale where the statistics changes is given by $\frac{2\theta}{e^2}$. It has been shown that in the absence of Fermions this parameter recieves no quantum corrections[20,21].

For canonical quantization we identify the canonical momenta,

$$\pi^* = \frac{\delta S}{\delta \dot{\phi}} = (D_0\phi)^* \quad , \quad \pi = D_0\phi \tag{4.15a}$$

$$\pi_0 \approx 0 \tag{4.15b}$$

$$\pi_i - \frac{1}{4\theta}\epsilon_{0ij}A_j = \frac{1}{e^2}F_{oi} \tag{4.15c}$$

According to Dirac's classification, since they have vanishing bracket with each other equation (4.15b) and (4.15c) are first class constraints[22]. The Hamiltonian is

$$H = \int d^2x \left\{ \pi^*\pi + \phi^* \left(\overleftarrow{\nabla} + i\vec{A} \right) \cdot \left(\vec{\nabla} - i\vec{A} \right) \phi + m^2\phi^*\phi - \right.$$
$$\left. -A_0 \left(j^0 + \frac{1}{2\theta}B + \frac{1}{e^2}\nabla_i F_{0i} \right) + \frac{1}{2e^2}F_{0i}^2 + \frac{1}{2e^2}B^2 \right\} \tag{4.16}$$

where

$$j^0 = i\left(\phi^*\pi - \pi^*\phi\right) \tag{4.17}$$

is the electric charge density and

$$B = \epsilon_{0ij}\nabla_i A_j \tag{4.18}$$

is the magnetic field. Conserving the first class constraint (4.15b) requires $\{\pi_0, H\} \approx 0$ which leads to

$$j^0 + \frac{1}{2\theta}B + \frac{1}{e^2}\nabla_i F_{0i} \approx 0 \tag{4.19}$$

Equation (4.19) is Gauss' law. It is this constraint which is ultimately responsible for the exotic statistics of charged particles in this model. For gauge invariant states it requires that charge density is accompanied by magnetic flux. The distributions of charge and magnetic flux differ only within the screening length of the gauge field. As a consequence, the motion of charged particles with spatial separation greater than $\frac{\theta}{e^2}$ in each other's magnetic fields produces Bohm-Aharonov phase factors in multi-particle wavefunctions. These phases can be considered either as a consequence of the dynamics where they appear in the time evolution of wavefunctions or, alternatively, they can be taken into account by a singular gauge transformation leading to multi-valued wavefunctions. This induced multi-valuedness on the confuguration space results in exotic statistics.

Equations (4.15b) and (4.19) contain two first class constraints. Thus, we are required to impose two additional gauge fixing conditions[22], which we choose as

$$A_0 \approx 0 \quad , \quad \vec{\nabla}\cdot\vec{A} \approx 0 \tag{4.20}$$

The quantum mechanical commutator for the gauge field and its canonical momentum is obtained from the Dirac bracket as

$$\left[A_i(\vec{x}), \pi_j(\vec{y})\right] = i\left(\delta_{ij} - \frac{\nabla_i\nabla_j}{\vec{\nabla}^2}\right)\delta(\vec{x}-\vec{y}) \tag{4.21}$$

and the Hamiltonian is

$$\begin{aligned} H = \int d^2x \Big\{ & \pi^*\pi + \phi^*\left(\overleftarrow{\nabla} + i\vec{A}\right)\cdot\left(\vec{\nabla} - i\vec{A}\right)\phi + m^2\phi^*\phi \\ & + \frac{e^2}{2}\left(j^0 + \frac{1}{2\theta}B\right)\frac{1}{-\vec{\nabla}^2}\left(j^0 + \frac{1}{2\theta}B\right) + \frac{e^2}{2}\pi_i^2 + \frac{1}{2e^2}B^2\Big\} \end{aligned} \tag{4.22}$$

Here both the gauge field and its canonical momentum are transverse, $\vec{\nabla}\cdot\vec{A} = 0$ $\vec{\nabla}\cdot\vec{\pi} = 0$ and the longitudinal part of $\vec{\pi}$ is explicitly taken account in the Coulomb interaction term.

The Coulomb interaction term is finite only if $B = -2\theta j^0$ over large distance scales. To see this note that the Coulomb potential is given by

$$(\vec{x}|\frac{1}{-\vec{\nabla}^2}|\vec{y}) = -\frac{1}{2\pi}\ln\mu|\vec{x}-\vec{y}| \tag{4.23}$$

where μ is an infrared cutoff. The cutoff appears in the Hamiltonian (4.22) in the term

$$\lim_{\mu\to 0+} -\frac{e^2}{4\pi}(\ln\mu)\left\{\int d^2x\left(j^0 + \frac{1}{2\theta}B\right)\right\}^2$$

The energy is independent of this cutoff only when

$$\int d^2x\left(j^0 + \frac{1}{2\theta}B\right) = 0 \tag{4.24}$$

[Otherwise the energy is logarithmically infrared divergent as we put $\mu \to 0$.] This together with Gauss' law (4.19) implies that the electric charge is screened

$$\int d^2x \nabla_i F_{0i} = 0 \tag{4.25}$$

This also implies that the in the presence of matter field charge $Q = \int d^2x j^0$ there is magnetic flux

$$\Phi = -\frac{\theta}{\pi} Q \tag{4.26}$$

where $\Phi = \frac{1}{2\pi}\int d^2xB$ and the spatial components of the gauge field therefore have a long range vortex-like component which gives the statistical interaction. With the gauge condition $\vec{\nabla}\cdot\vec{A}=0$ the asymptotic form of the gauge field is

$$\lim_{|\vec{r}|\to\infty} A_i(\vec{r}) = -\Phi\ \epsilon_{0ij}\frac{r^j}{\vec{r}^2} = \frac{\theta}{\pi} Q\ \epsilon_{0ij}\frac{r^j}{\vec{r}^2} \tag{4.27}$$

Thus the asymptotic gauge field contains magnetic flux proportional to the total charge of the system. The scalar fields couple to this long-range gauge field by minimal coupling in (4.22). * This long-ranged component of the interaction can be removed by a redefinition of the scalar fields in terms of the multi-valued operators

$$\hat{\phi}(\vec{x}) = \exp\left\{i\frac{\theta}{2\pi}\int d^2y\Theta(\vec{x},\vec{y})j^0(\vec{y})\right\}\phi(\vec{x})\exp\left\{i\frac{\theta}{2\pi}\int d^2y\Theta(\vec{x},\vec{y})j^0(\vec{y})\right\} \tag{4.28a}$$

$$\hat{\pi}(\vec{x}) = \exp\left\{i\frac{\theta}{2\pi}\int d^2y\Theta(\vec{x},\vec{y})j^0(\vec{y})\right\}\pi(\vec{x})\exp\left\{i\frac{\theta}{2\pi}\int d^2y\Theta(\vec{x},\vec{y})j^0(\vec{y})\right\} \tag{4.28b}$$

$$\hat{\phi}^*(\vec{x}) = \exp\left\{-i\frac{\theta}{2\pi}\int d^2y\Theta(\vec{x},\vec{y})j^0(\vec{y})\right\}\phi^*(\vec{x})\exp\left\{-i\frac{\theta}{2\pi}\int d^2y\Theta(\vec{x},\vec{y})j^0(\vec{y})\right\} \tag{4.28c}$$

$$\hat{\pi}^*(\vec{x}) = \exp\left\{-i\frac{\theta}{2\pi}\int d^2y\Theta(\vec{x},\vec{y})j^0(\vec{y})\right\}\pi^*(\vec{x})\exp\left\{-i\frac{\theta}{2\pi}\int d^2y\Theta(\vec{x},\vec{y})j^0(\vec{y})\right\} \tag{4.28d}$$

where

$$\Theta(\vec{x},\vec{y}) = \arctan\left(\frac{x_2-y_2}{x_1-y_1}\right) \tag{4.29}$$

is a multi-valued function giving the angle between the vector $\vec{x}-\vec{y}$ and the x_1-axis. It has the property that $\vec{\nabla}\Theta$ is single-valued and

$$\vec{\nabla}^2\Theta(\vec{x}) = 0\ ,\ \vec{\nabla}\times\vec{\nabla}\Theta(\vec{x}) = 2\pi\delta(\vec{x}) \tag{4.30}$$

Also,

$$\nabla_i\Theta(\vec{x},\vec{y}) = -\epsilon_{0ij}\nabla_j \ln\mu|\vec{x}-\vec{y}| \tag{4.31}$$

---

* Also, we observe that in the limit $e^2 \to \infty$ finiteness of the energy would require screening to arbitrarily short distance scales, $j^0 = -\frac{1}{2\theta}B$ and magnetic flux is rigidly tied to the matter charge density. This is similar to the effect of the statistical gauge field in the quantum mechanical case (2.12).

When j^0 represents an assembly of point charges the phase transformation in (4.28a-d) is

$$\int d^2y\Theta(\vec{x},\vec{y})j^0(\vec{y}) = \int d^2y\Theta(\vec{x},\vec{y})\sum_i \delta(\vec{y}-\vec{y}_i)e_i$$

$$= \sum_i \Theta(\vec{x},\vec{y}_i)e_i \tag{4.32}$$

However, when the charge has a continuous distribution some care must be taken to define the integral. Here it should have the properties

$$\vec{\nabla}^2 \int d^2y\Theta(\vec{x},\vec{y})j^0(\vec{y}) = 0 \tag{4.33a}$$

and

$$\vec{\nabla}\times\vec{\nabla}\int d^2y\Theta(\vec{x},\vec{y})j^0(\vec{y}) = 2\pi j^0(\vec{x}) \tag{4.33b}$$

The property (4.33b) can also be expressed in integral form

$$\oint_C d\vec{l}\cdot\vec{\nabla}\int d^2y\Theta(\vec{x},\vec{y})j^0(\vec{y}) = 2\pi\int\int_{\mathcal{D}} d^2xj^0(\vec{x}) \tag{4.34}$$

where the closed curve $\mathcal{C}$ is the boundary of the disc $\mathcal{D}$. This is most easily achieved by defining the integral as the continuous limit of a Riemann sum

$$\int d^2y\Theta(\vec{x},\vec{y})j^0(\vec{y}) = \lim_{\epsilon\to 0}\sum_i \epsilon^2\Theta(\vec{x},\vec{y}_i)j^0(\vec{y}_i) \tag{4.35}$$

where $\vec{y}_i$ form a lattice embedded in the plane R^2 and with spacing ϵ and $j^0(\vec{y}_i)$ is the value of the charge density on the lattice sites. If $j^0(\vec{y})$ has compact support then the sum (4.35) has a finite number of terms. It also has the requisite properties (4.33a-b) and (4.34). It can be shown that the continuum limit of (4.35) retains the property (4.34) when j^0 has compact support and is continuous and once differentiable.

Using the identity

$$\exp\left\{-i\frac{\theta}{\pi}\int d^2y\Theta(\vec{x},\vec{y})j^0(\vec{y})\right\}\phi(\vec{z})\exp\left\{i\frac{\theta}{\pi}\int d^2y\Theta(\vec{x},\vec{y})j^0(\vec{y})\right\} =$$

$$= \exp\left\{i\frac{\theta}{\pi}\Theta(\vec{x}-\vec{z})\right\}\phi(\vec{z}) \tag{4.36}$$

we see that the operators in (4.28a-d) obey the graded commutation relations

$$\hat{\phi}(\vec{x})\hat{\phi}(\vec{y}) - \exp\left\{i\frac{\theta}{\pi}\Delta\right\}\hat{\phi}(\vec{y})\hat{\phi}(\vec{x}) = 0 \tag{4.37a}$$

$$\hat{\phi}(\vec{x})\hat{\phi}^*(\vec{y}) - \exp\left\{i\frac{\theta}{\pi}\Delta\right\}\hat{\phi}^*(\vec{y})\hat{\phi}(\vec{x}) = 0 \tag{4.37b}$$

$$\hat{\phi}(\vec{x})\hat{\pi}(\vec{y}) - \exp\left\{i\frac{\theta}{\pi}\Delta\right\}\hat{\pi}(\vec{y})\hat{\phi}(\vec{x}) = 0 \tag{4.37c}$$

$$\hat{\phi}(\vec{x})\,\hat{\pi}^*(\vec{y}) - \exp\left\{i\frac{\theta}{\pi}\Delta\right\}\hat{\pi}^*(\vec{y})\,\hat{\phi}(\vec{x}) = i\delta(\vec{x}-\vec{y}) \tag{4.37d}$$

with the multi-valued phase

$$\Delta = \Theta(\vec{x},\vec{y}) - \Theta(\vec{y},\vec{x}) = \pi \bmod 2\pi n \tag{4.38}$$

The multi-valued nature of this phase factor is essential to the consistency of (4.37a-d). The graded commutation relations are indicative of exotic statistics. For example, when $\theta = \pi$ the variables anticommute and therefore behave as Fermions.

The operators $\hat{\phi}$, $\hat{\pi}$, $\hat{\phi}^*$ and $\hat{\pi}^*$ are not yet candidates for interpolating fields of asymptotic states. This is because they upset the condition that the states have finite energy (4.26). This can be remedied by augmenting them with an operator which creates magnetic flux,

$$u(\vec{x}) = \exp\left\{-i\frac{\theta}{\pi}\int d^2y\left(\vec{\pi}(y)\cdot\vec{\nabla} + \frac{1}{4\theta}\vec{A}(y)\times\vec{\nabla}\right)\Theta(x,y)\right\}$$

Then, the operators

$$\tilde{\phi}(\vec{x}) = u(\vec{x})\hat{\phi}(x) \tag{4.39a}$$

$$\tilde{\pi}(\vec{x}) = u(\vec{x})\hat{\pi}(x) \tag{4.39b}$$

$$\tilde{\phi}^*(\vec{x}) = u^\dagger(\vec{x})\hat{\phi}^*(\vec{x}) \tag{4.39c}$$

$$\tilde{\pi}^*(\vec{x}) = u^\dagger(\vec{x})\hat{\phi}^*(\vec{x}) \tag{4.39d}$$

commute with the constraint (4.26) and create states with finite energy and fractional spin and statistics when operating on other states with finite energy. We thus identify them as candidates for interpolating operators for the asymptotic fields of the theory. They have the algebra (4.37). As yet, little is known about concrete field theory representations of this algebra.

V. Fractional Spin

In order to examine the spin of particles in this model we must analyze its spacetime symmetries. Consider the gauge invariant, symmetric energy-momentum tensor

$$\begin{aligned} T_{\mu\nu} = {} & \phi^*\left(\overleftarrow{\partial}_\mu + iA_\mu\right)\left(\vec{\partial}_\nu - iA_\nu\right)\phi + \phi^*\left(\overleftarrow{\partial}_\nu + iA_\nu\right)\left(\vec{\partial}_\mu - iA_\mu\right)\phi - \\ & -g_{\mu\nu}\left(\phi^*\left(\overleftarrow{\partial}^\theta + iA^\theta\right)\left(\vec{\partial}_\theta - iA_\theta\right)\phi - m^2\phi^*\phi\right) - \frac{1}{e^2}F_{\mu\lambda}F_\nu{}^\lambda + \frac{1}{4e^2}g_{\mu\nu}F_{\lambda\rho}F^{\lambda\rho} \end{aligned} \tag{5.1}$$

This energy-momentum tensor is obtained by introducing a background 3-metric, covariantizing the action, taking a functional derivative with respect to the metric and then setting the metric flat and orthonormal. Note that the Chern-Simons term is convariant without reference to the metric and therefore does not contribute to the energy-momentum tensor.

From the energy and momentum densities

$$T_{00} = \pi^* \pi + \phi^* \left(\overleftarrow{\nabla} + i\vec{A}\right) \cdot \left(\vec{\nabla} - i\vec{A}\right) \phi + m^2 \phi^* \phi + \frac{1}{2e^2} F_{0i}^2 + \frac{1}{2e^2} B^2 \tag{5.2}$$

$$T_{0i} = \pi^* \left(\vec{\nabla}_i - iA_i\right) \phi + \phi^* \left(\overleftarrow{\nabla}_i + iA_i\right) \pi + \frac{1}{e^2} \epsilon_{0ij} F_{0j} B \tag{5.3}$$

where we have used the temporal gauge condition $A_0 = 0$, we construct the generators of the Poincare group

$$H = \int d^2x T_{00} \ , \ P_i = \int d^2x T_{0i} \ , \ L = \int d^2x \epsilon_{0ij} x_i T_{0j} \ , \ K_i = \int d^2x x_i T_{00} \tag{5.4}$$

Using the Dirac bracket (4.21) the commutator of electric fields is

$$\left[F_{0i}(\vec{x}), F_{0j}(\vec{y})\right] = -i\frac{e^4}{4\theta} \epsilon_{0ij} \delta(\vec{x} - \vec{y}) \tag{5.5}$$

Using this commutator it is straightforward to verify the relation

$$\left[K_i, K_j\right] = i\epsilon_{ij} L \tag{5.6}$$

of the Poincare algebra. The angular momentum operator is

$$L = \int d^2x \left\{ \pi^* \vec{r} \times \vec{\nabla}\phi + \vec{r} \times \vec{\nabla}\phi^* \pi - i\pi^* \vec{r} \times \vec{A}\phi + i\phi^* \vec{r} \times \vec{A}\pi - \frac{1}{e^2} x_i F_{0i} B \right\} \tag{5.7}$$

It can be written as a combination of the canonical generator of rotations, an operator containing the Gauss' law constraint (4.19) and a surface term[31]

$$\begin{aligned} L = & \int d^2x \left\{ \pi^* \vec{r} \times \vec{\nabla}\phi + \vec{r} \times \nabla\phi^* \pi + \pi_i \left(\vec{r} \times \nabla A_i + \epsilon_{0ij} A_j\right) \right\} + \\ & + \int d^2x \vec{r} \times \vec{A} \left(\frac{1}{e^2} \nabla_i F_{0i} + j^0 + \frac{1}{2\theta} B \right) + \int d^2x \nabla_i \cdot \left(\pi_i \vec{r} \times \vec{A} \right) \end{aligned} \tag{5.8}$$

When we impose the constraint (4.19), use the asymptotic form of the gauge field (4.27) and the fact that the electric field is screened (4.25) we obtain

$$L = L_{\text{canonical}} - \Phi \left(Q + \frac{\pi}{2\theta} \Phi \right) \tag{5.9}$$

The full angular momentum operator differs from the canonical angular momentum operator by a term proprotional which can be regarded as an operator-valued induced spin. However, because it is operator-valued the rotation of the scalar field $\phi(\vec{x})$ does not close,

$$e^{i\omega L} \phi(\vec{x}) e^{-i\omega L} = \exp(i\omega\Phi) \phi(\Lambda(\omega)\vec{x}) \tag{5.10}$$

In this sense the single-valued operators $\phi(\vec{x})$ do not represent the rotation group. The magnetic flux should be considered an operator on the same level as the electric charge. (5.10) indicates that the angular momentun does not generate a pure rotation of ϕ. On the other hand the multi-valued operators $\tilde{\phi}(\vec{x})$ defined in (5.28) represent the rotations with a phase

$$e^{i\omega L}\tilde{\phi}(\vec{x})\,e^{-i\omega L} = e^{i\frac{\theta}{2\pi}\omega}\,\tilde{\phi}(\Lambda(\omega)\,\vec{x}) \tag{5.11}$$

This is a result of the anomalous rotation property of the phase operator,

$$e^{i\omega L}\exp\left\{i\frac{\theta}{\pi}\int d^2y\Theta(\vec{x},\vec{y})\,j^0(\vec{y})\right\}e^{-i\omega L} = \exp\left\{i\frac{\theta}{\pi}\int d^2y\Theta(\Lambda(\omega)\vec{x},\vec{y})\,j^0(\vec{y})\right\} =$$

$$= e^{-i\frac{\theta}{\pi}Q\omega}\exp\left\{i\frac{\theta}{\pi}\int d^2y\Theta(\Lambda(\omega)\vec{x},\vec{y})\,j^0(\vec{y})\right\} \tag{5.12}$$

Note that when the statistics of $\tilde{\phi}$ are Fermionic, *i.e.* when $\theta = \pi$, (4.11) implies that $\tilde{\phi}$ also has the odd spin parity of Fermions. * Thus, the multi-valued fields have exotic statistics and spin-parity both characterized by the phase $\exp\{i\theta\}$. **

VI. Discussion

We have reviewed the arguments which show that topologically massive electrodynamcis is solved at long wavelengths by multi-valued quantum fields which create quantum states with exotic spin and statistics. The essential feature which arises is the modification of the angular momentum by an induced spin containing the magnetic flux and charge operators. As a consequence the canonical fields do not represent the rotation group. Furthermore, the anomalous transformation of the multi-valued phase makes the multi-valued operators represent the rotation group with anomalous spin parity. The multi-valued fields have the graded anomalous commutators which are expected of fields which create multi-valued states. Both the anomalous commutators and the induced spin are characterized by the phase $e^{\frac{i\theta}{\pi}}$.

Finally, there is an interesting lattice regularization of the models considered here which incorporates fractional statistics in a natural way and evades problems with ultraviolet regularization[23]. Although it is not completely clear that it corresponds to the latticization of a gauge theory with Chern-Simons term it is a promising new approach.

---

* This differs from the result of ref. 15 where the field $\tilde{\phi}$ had spin zero in the canonical theory with no Maxwell term. In that case, even though the statistics were independent of operator ordering and therefore regularization, the spin did depend on the prescription chosen. Here the ultraviolet divergences of the theory are regulated by the Maxwell term and the spin is unambiguous. It is intriguing that this regularization gives this generalized spin-statistics connection.

** Note that we could form a field which has interpolating statistics by the transformation

$$\hat{\phi}(\vec{x}) = \exp\left\{i\frac{\theta}{\pi}\int d^2yF(\vec{x},\vec{y})\,j^0(\vec{y})\right\}\phi(\vec{x})$$

where $F(\vec{x},\vec{y}) \rightarrow \Theta(\vec{x},\vec{y})$ when $|\vec{x}-\vec{y}| > \frac{2\theta}{e^2}$ and $F(\vec{x},\vec{y}) = 0$ for $|\vec{x}-\vec{y}| << \frac{2\theta}{e^2}$. However it would not give a representation of the rotations.

References

1. B. Halperin, Phys. Rev. Lett. 52 (1984), 1583; R. Laughlin, Phys. Rev. Lett. 50 (1983), 1395; T.H.Hansson, S.Kivelson and Zhang, Phys. Rev.Lett. 62 (1989), 82; S.Girvin in *The Quantum Hall Effect*, ed. by R. Prange and S. Girvin, Springer-Verlag 1987; G. W. Semenoff and P. Sodano. Phys. Rev. Lett. 57 (1986).
2. V.Kalmeyer and R.B.Laughlin, Phys.Rev.Lett. 59 (1987), 2095; R. B. Laughlin, Phys. Rev. Lett. 60 (1988), 2677; A. L. Fetter, C. B. Hanna and R. B. Laughlin, Phys.Rev.B39 (1989),9679; T. Banks and J. Lykken, Santa Cruz preprint, 1989.
3. T.H.R. Skyrme, Proc. Roy. Soc. Lond. A260 (1961), 127; D. Finkelstein, Jour. Math. Phys. 7 (1966), 1218; D. Finkelstein and J. Rubenstein, Jour. Math. Phys. 9 (1968).
4. C.M. De Witt and Laidlaw, Phys. Rev. D3 (1971), 1375.
5. S. Coleman, Phys. Rev. D11 (1975), 2088; S. Mandelstam, Phys. Rev. D11 (1975), 3026.
6. J. Leinaas and J. Myrlheim, Nuovo Cimento B37 (1977), 1.
7. W.Siegel, Nucl.Phys.B156, 135 (1979); J. Shonfeld, Nucl. Phys. D185 (1981), 157; R. Jackiw and S. Templeton, Phys. Rev. D24 (1981), 2291.
8. S. Deser, R. Jackiw and S. Templeton, Phys. Rev. Lett. 48(1982), 475; Ann. Phys. (N. Y.)140 (1982), 372.
9. A.S. Schwarz, Lett. Math. Phys. 2 (1978), 247.
10. Y. Akutsu and M. Wadati, J. Phys. Soc. Japan 56 (1987), 3039; J. Phys. Soc. Japan 59 (1987), 3034; Y. Akutsu, T. Deguchi and M. Wadati, J. Phys. Soc. Japan 56 (1987), 3464.
11. E. Witten, Comm. Math. Phys. 121 (1989), 351.
12. F. Wilczek and A. Zee, Phys. Rev. Lett. 51 (1984), 2250.
13. Yong-Shi Wu, Phys. Rev. Lett. 52 (1984), 2103; Yong-Shi Wu, Phys. Rev. Lett. 53 (1984), 111.
14. A. M. Polyakov, Mod. Phys. Lett. A3 (1988), 325;
15. G. W. Semenoff, Phys. Rev. Lett. 61 (1988), 517.
16. G. W. Semenoff and P. Sodano, Nucl. Phys. B328 (1989), 753.
17. G. Dunne, C. Trugenberger and R. Jackiw, Ann. Phys. (N.Y.) 1989, in press.
18. For a review of their work see J. Frohlich and P. Marchetti, this preceedings.
19. R. Fox and L. Neuwirth, Math. Scand. 10 (1961), 119; E. Fadell and J. Van Buskirk, Duke Math. 29 (1962), 243; E. Artin, Ann. Math. 48 (1947), 101.
20. G.Semenoff, P.Sodano and Y.-S. Wu, Phys.Rev.Lett. 62 (1989).
21. S. Coleman and B. Hill, Phys. Lett. 159B(1985) 184; Y. Kao and M. Suzuki, Phys. Rev. D31(1985), 2137; M. Bernstein and T. Lee, Phys. Rev. D32(1985), 1020.
22. P. A. M. Dirac, *Lectures on Quantum Mechanics*, Yeshiva Press, New York, 1969.
23. P. W. Anderson, S. John, G. Baskaran, B. Doucot and S.-D. Liang, Princeton preprint 1988; E.J. Mele and D. Morse, Philadelphia Preprint, 1989; J. Ambjorn and G.W. Semenoff, Phys.Lett. B226 (1989), 107; E. Fradkin, Urbana preprint, 1989.

YANG-BAXTER ALGEBRAS, INTEGRABLE THEORIES AND QUANTUM GROUPS

H.J. DE VEGA

Laboratoire de Physique Théorique et Hautes Energies
Laboratoire associé au CNRS UA 280
Université Pierre et Marie Curie
Tour 16 - 1er étage
4, Place Jussieu
75252 Paris Cedex 05 - France

Abstract

The Yang-Baxter algebras (YBA) are introduced in a general framework stressing their power to exactly solve the lattice models associated to them. The algebraic Bethe Ansatz is developed as an eigenvector construction based on the YBA. The six-vertex model solution is given explicitely.

It is explained how these lattice models yield both solvable massive QFT and conformal models in appropiated scaling (continuous) limits within the lattice light-cone approach. This approach permit to define and solve rigorously massive QFT as an appropiate continuum limit of gapless vertex models.

The deep links between the YBA and Lie algebras are analyzed including the quantum groups that underly the trigonometric/hyperbolic YBA. Braid and quantum groups are derived from trigonometric/hyperbolic YBA in the limit of infinite spectral parameter.

To conclude, some recent developments in the domain of integrable theories are summarized.

I. YANG-BAXTER ALGEBRAS

A Yang-Baxter (YB) algebra consists of a set of operators $T(\theta)$. They depend on the complex variable θ (the spectral parameter). Each operator $T(\theta)$ acts on two vector spaces, say $\mathcal{A}$ and $\mathcal{V}$. This means that they have two couples of indices of different kind, in general. The best way to work with Yang-Baxter algebras is to use graphical notation. It is defined as follows :

a) a line of different type is associated to each vector space

——— $= \mathcal{A}$, ~ ~ ~ ~ ~ ~ $= \mathcal{V}$, etc.

Fig.1 To each type of line is associated a vector space.

Physics, Geometry, and Topology
Edited by H. C. Lee
Plenum Press, New York, 1990

b) The intersection of two lines is associated to an operator $T(\theta)$ where θ is the angle between the two lines.

$$[\, T_{ab}{}^{(\mathcal{A},\mathcal{V})}(\theta)\,]_{\alpha\beta} \;=$$

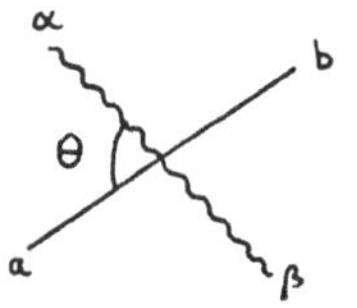

Fig.2 A YB generator is associated to the intersection of two lines.

c) There is summation over all states in the vector spaces associated to the lines between two vertices ["internal lines"].

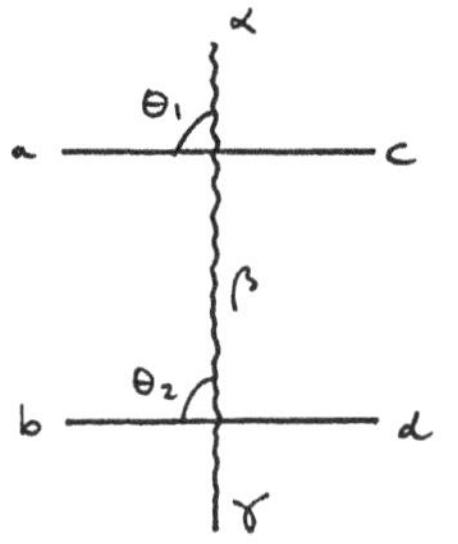

$$= \sum_{\beta} [\, T_{ac}(\theta_1)\,]_{\alpha\beta}\, [\, T_{bd}(\theta_2)\,]_{\beta\gamma}$$

Fig.3 There is a summation over the states of internal lines.

d) Left to right order in the formulas correspond to up to down in the pictures (see fig. 3).

This definition originates in two dimensional vertex models where the links can be in different states spanning a vector space, say $\mathcal{V}$ Here we consider the general case when the lattice is formed by different types of bonds (they are associated to vector spaces $\mathcal{A}$, $\mathcal{V}$, ...).

Let us call $\mathcal{I}$ the set of all vector spaces where the YB algebra (YBA) generators act

$$\mathcal{I} = \{\, V^I \,\}$$

$\mathcal{I}$ is also the set of different types of lines. The basic equation that characterizes the YBA is

$$T^{(K,I)}(\theta - \theta')\, T^{(K,J)}(\theta)\, T^{(I,J)}(\theta') \;=\; T^{(I,J)}(\theta')\;\; T^{(K,J)}(\theta)\, T^{(K,I)}(\theta - \theta') \qquad (1.1)$$

for all spaces V^I, V^J, $V^K \in \mathcal{I}$. Eq. (1.1) is called Yang-Baxter equation (YBE) or triangular relation or factorization equation. It can be represented graphically as

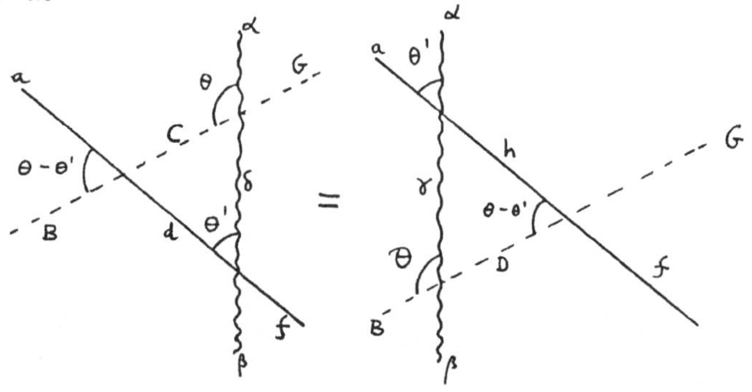

Fig.4 The Yang-Baxter equation (YBE) in its general form.

Here ——— $= V^I$, $\sim\sim\sim\sim\sim\sim$ $= V^J$, - - - - - $= V^K$. We used repeatedly the rules of Fig. 2 and 3 to write down Fig. 4. Eq. (1.1) writes putting all indices explicitly

$$\sum_{C,d,\delta} [\, T_{BC}{}^{(K,I)}(\theta-\theta')\,]_{ad}\,[\, T_{CG}{}^{(K,J)}(\theta)\,]_{\alpha\delta}\;[\, T_{df}{}^{(I,J)}(\theta')\,]_{\delta\beta} =$$

$$= \sum_{D,h,\gamma} [\, T_{ah}{}^{(I,J)}(\theta')\,]_{\alpha\gamma}\;[\, T_{BD}{}^{(K,J)}(\theta)\,]_{\gamma\beta}\;[\, T_{DG}{}^{(K,I)}(\theta-\theta')\,]_{hf} \quad . \qquad (1.2)$$

As one sees in fig. 4 the YBE says that one can displace any line through the intersection of other two provided the angles are kept fixed. This is called sometimes Z-invariance since it leaves the partition function unchanged [see below][1]. That is, we have invariance under parallel displacements in the lattice. Eqs.(1.1)-(1.2) [or fig. 4] shows the general YBE.

Eqs.(1.1) or (1.2) hold for all values of the external indices (a, B, β, f, G and α) and all values of the spectral parameters θ and θ'.

The YB generator associated to the intersection of two lines of the same type is called a R-matrix : $R^I(\theta) \equiv T^{(I,I)}(\theta)$,

$$R^{ab}{}_{cd}(\theta) = [\, T_{bc}{}^{(I,I)}(\theta)\,]_{ad} =$$

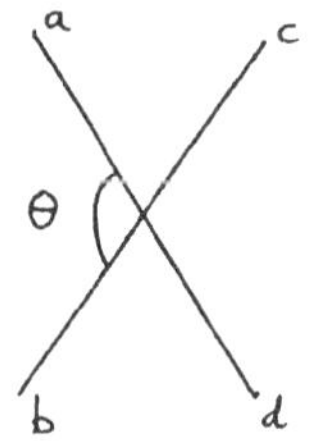

Fig.5 The R-matrix.

In the particular case when two of the vector spaces are identical, say $V^I = V^K = \mathcal{A}$ and $V^J = \mathcal{V}$, eqs. (1.1)-(1.2) become

$$\sum_{c,d,\delta} R^{ab}{}_{cd}(\theta-\theta')\;[\, T_{cg}(\theta)\,]_{\alpha\delta}\;[\, T_{df}(\theta')\,]_{\delta\beta} =$$

$$= \sum_{c,d,\gamma} [\, T_{ac}(\theta')\,]_{\alpha\gamma}\;[\, T_{bd}(\theta)\,]_{\gamma\beta}\;\; R^{cd}{}_{gf}(\theta-\theta') \,.$$

This can be rewritten in a more compact way as[3]

$$R(\theta-\theta')\,[\,T(\theta)\otimes T(\theta')\,] = [\,T(\theta')\otimes T(\theta)\,]\,R(\theta-\theta') \qquad (1.3)$$

where we use tensor product notation

$$(\,A \otimes B\,)_{ab,cd} \equiv A_{ac}\;B_{bd} \quad ,$$

and

$$R(\theta) = R^I(\theta) \qquad , \qquad T(\theta) = T^{(I,J)}(\theta)\,.$$

In eq. (1.3) an operator product in the space $\mathcal{V}$ is understood. The $\otimes$ means tensor product of the space $\mathcal{A}$ multiplied by itself. R acts in $\mathcal{A} \otimes \mathcal{A}$ as a matrix. The R-matrix associated to the space V_0 of lowest dimensionality in $\mathcal{G}$ as called the fundamental R-matrix. The fundamental R-matrix characterizes the YB algebra.

A YB algebra is then, generally speaking , a set of operators $T^{(K,I)}(\theta)$

fulfilling eq.(1.1) identically. Usually, this is an infinite set. Non-trivial examples of such algebras exist and simple cases will be considered below.

Let us see why YB algebras are connected deeply with integrable theories. Eq. (1.1) can be written as

$$T^{(K,J)}(\theta)\, T^{(I,J)}(\theta') = \{T^{(K,I)}(\theta-\theta')\}^{-1}\, T^{(I,J)}(\theta')\; T^{(K,J)}(\theta)\, T^{(K,I)}(\theta-\theta') \tag{1.4}$$

Taking the trace of eq. (1.4) in the space $V^K \otimes V^I$ yields

$$\tau_K(\theta)\ \tau_I(\theta') = \tau_I(\theta')\ \tau_K(\theta) \tag{1.5}$$

where we use the cyclic property of the trace and

$$\mathrm{Tr}_{V^I\otimes V^K}\left(T^{(I,J)} \otimes T^{(K,J)}\right) = \mathrm{Tr}_{V^I}\left(T^{(I,J)}\right)\ \mathrm{Tr}_{V^K}\left(T^{(K,J)}\right) \tag{1.6}$$

Here an operatorial product in the space V^J is understood. We denote by $\tau_I(\theta)$ and $\tau_K(\theta)$ the transfer matrices

$$\tau_I(\theta) = \mathrm{Tr}_{V^I}\left(T^{(I,J)}\right) = \sum_a T_{aa}(\theta)^{(I,J)} \tag{1.7}$$

$$\tau_K(\theta) = \mathrm{Tr}_{V^K}\left(T^{(K,J)}\right) = \sum_A T_{AA}(\theta)^{(K,J)} \tag{1.8}$$

$\tau_I(\theta)$ and $\tau_K(\theta)$ are operators acting on V^J. They form a **set** of families of commuting transfer matrices

$$[\ \tau_I(\theta)\ ,\ \tau_K(\theta')\ ] = 0\ ,\quad \forall\, \theta, \theta' \in \mathbb{C}\ ,\ \forall\, I, K \in \mathcal{J} \tag{1.9}$$

Moreover, series expanding in θ yields an infinite number of commuting operators acting on V^J.

$$[\ c_n{}^I\ ,\ c_m{}^K\ ] = 0\quad ,\ \forall\, n, m \geq 0\ ,\quad \forall\, I, K \in \mathcal{J} \tag{1.10}$$

Here $c_n{}^I$ are the expansion coefficients of $\tau_I(\theta)$ or $\log \tau_I(\theta)$ in powers of θ. The existence of an infinite number of commuting operators is the necessary condition to have a quantum integrable system with an infinite number of degrees of freedom. Actually, only in the thermodynamic limit this number of degrees of freedom is attained. In addition, the hamiltonian of the system (H) must commute with $\tau_I(\theta)$. Actually, H often expresses itself in terms of $\tau_I(\theta)$ either as the logarithmic derivative at $\theta = 0$ (spin chains, see eq.(2.13)) or in terms of $\log \tau_I(\theta)$ at some special value of θ (field theories in the light cone approach, see sec.IV).

Since the operators $\tau_I(\theta)$ are mutually commuting for all θ and V^I, one can expect to be able to diagonalize all of them simultaneously. This is actually possible through the algebraic Bethe Ansatz (BA). Moreover, in the BA, the eigenvectors and eigenvalues can be constructed using the YB algebra itself. This is probably the main application of YBA. They permit to built eigenvectors and eigenvalues of all $\tau_I(\theta)$ and operators $c_n{}^I$ derived from them in a purely algebraic framework.

A specially important YB equation follows when the three vector spaces in eq. (1.1) are equal : $V^I = V^J = V^K = \mathcal{A}$. In this case eqs.(1.1)-(1.2) can be written as

$$[\,1 \otimes R(\theta-\theta')\,][\,R(\theta)\otimes 1\,][\,1\otimes R(\theta')\,] =$$
$$= [\,R(\theta')\otimes 1\,][\,1\otimes R(\theta)\,][\,R(\theta-\theta')\otimes 1\,] \qquad (1.11)$$

In explicit notation this reads

$$\sum_{1\le c,d,e\le q} R^{cd}{}_{a_2a_1}(\theta-\theta')\; R^{b_1e}{}_{a_3c}(\theta)\; R^{b_2b_3}{}_{ed}(\theta') =$$
$$= \sum_{1\le m,n,p\le q} R^{mn}{}_{a_3a_2}(\theta')\; R^{pb_3}{}_{na_1}(\theta)\; R^{b_1b_2}{}_{mp}(\theta-\theta') \quad , \qquad (1.12)$$

where $q \equiv \dim\mathcal{A}$. This equation can be depicted as

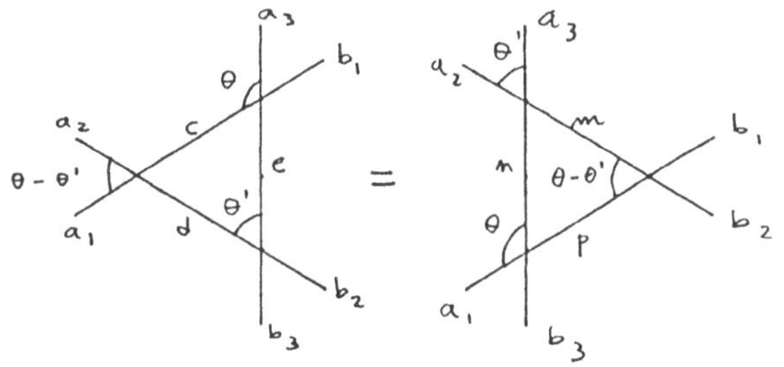

Fig.6 The YBE for the R-matrix.

We see that eq. (1.11) or (1.12) is a system of q^6 equations ($q = \dim \mathcal{A}$) with q^4 unknowns (the functions $R^{ab}{}_{cd}(\theta)$, $1 \le a, b, c, d \le q$). That is, one finds a heavily over-determined set of equations. The existence of a solution is clearly a necessary condition to have a YBA. Actually it is also a sufficient condition since one can define a YB generator acting on $\mathcal{A} \otimes \mathcal{A}$ as (see fig. 5)

$$[\,t_{ab}{}^{(\mathcal{A},\mathcal{A})}(\theta)\,]_{cd} = R^{ca}{}_{bd}(\theta) \qquad (1.13)$$

It obeys

$$R(\theta-\theta')\,[\,t^{(\mathcal{A},\mathcal{A})}(\theta)\otimes t^{(\mathcal{A},\mathcal{A})}(\theta')\,] = [\,t^{(\mathcal{A},\mathcal{A})}(\theta')\otimes t^{(\mathcal{A},\mathcal{A})}(\theta)\,]\,R(\theta-\theta')$$

which just follows by rewriting eq. (1.12) with the help of eq. (1.13).

The most remarkable fact in integrable theories is that eqs. (1.11) or (1.12) do admit a rich set of non-trivial solutions. Actually each solution exhibits some invariance which probably explains its very existence. That is, thanks to the presence of an invariance the number of actual independent equations is largely reduce from q^6.

A YB algebra is invariant [see eq. (1.3)] under a transformation $g \in \mathcal{G}$ in $\mathcal{A}$

$$T_{ab}(\theta) \;\rightarrow\; \sum_c g_{ac}\, T_{cb}(\theta) \qquad (1.15)$$

provided [4,5]

$$[\,g\otimes g\,,\,R(\theta)\,] = 0 \;,\; \forall\theta\in\mathbb{C}\;,\;\forall g\in\mathcal{G} \qquad (1.16)$$

This can be proven as follows from eq.(1.3). Let us multiply eq.(1.3) by $g \otimes g$ from the left. We find

$$(g \otimes g) R(\theta - \theta') (g^{-1} \otimes g^{-1}) \ [gT(\theta) \otimes gT(\theta')] = $$
$$= [gT(\theta') \otimes gT(\theta)] R(\theta - \theta')$$

Therefore $gT(\theta)$ obey a YB algebra with the same $R(\theta)$ as $T(\theta)$ provided

$$(g \otimes g) R(\theta - \theta') (g^{-1} \otimes g^{-1}) = R(\theta - \theta') ,$$

which is just eq.(1.16). This is clearly a sufficient condition of invariance. More generally, we find

$$[g_I \otimes g_J , T^{(I,J)}(\theta)] = 0 , \forall \theta \in \mathbb{C} , \forall g \in \mathfrak{G} \qquad (1.17)$$

where g_I and g_J are the representation of $g \in \mathfrak{G}$ acting on the vector spaces V^I and V^J respectively. For an infinitesimal transformation

$$g_I = 1 + i \varepsilon S_I \quad , \quad g_J = 1 + i \varepsilon S_J$$

where $\varepsilon \ll 1$ and S_I and S_J are the generators representation in V^I and V^J respectively. Hence eq. (1.17) yields

$$[S_I , T^{(I,J)}(\theta)] + [S_J , T^{(I,J)}(\theta)] = 0 \qquad (1.18)$$

The invariance of a YB algebra under a group $\mathfrak{G}$ can be formalized as follows :

Let us define the transformed YB operator $T_g^{(I,J)}(\theta)$ as follows :

$$T_g^{(I,J)}(\theta) \equiv g_I \ T^{(I,J)}(\theta) \ g_J^{-1} \qquad (1.19)$$

This definition includes an additional transformation g^{-1} in the vertical space compared with eq.(1.15). In this way $T_g^{(I,J)}(\theta)$ also obeys the YBE (1.1). This is easy to check , taking into account that

$$g_I \ T^{(I,J)}(\theta) \ g_J^{-1} = g_J^{-1} \ T^{(I,J)}(\theta) \ g_I \qquad (1.20)$$

There exists a direct connection between the kind of symmetry group $\mathfrak{G}$ of the YBA and the functional dependence on θ. This connection is displayed in table I. This is a first hint about the deep connection between YB algebras and Lie algebras and their deformations (quantum' groups).

TABLE I

Correspondence between the symmetry group of the Yang-Baxter algebras and the functional dependence of the YB operators on the spectral parameter θ.

$\mathfrak{G}$: symmetry group	θ-dependence in $R_{ab}^{cd}(\theta)$
discrete : Z_q	elliptic
continuos abelian : $U(1)^q$	trigonometric or hyperbolic
continuous non-abelian : U(q), O(q),....	rational

Another important invariance of YB algebras is the shift invariance. That is, if $T(\theta)$ is a YB generator, so it is

$$T(\theta - \alpha)$$

with fixed α. A look to eq. (1.3) shows that this is true since R depends on the difference $\theta-\theta'$, α must be the same so it drops.

Let us now discuss the most important property of YBA : the reproduction property. It can be stated as follows : if $t(\theta)$ obeys the YBA

$$R(\theta - \theta')\,[\, t(\theta) \otimes t(\theta')\,] = [\, t(\theta') \otimes t(\theta)\,]\, R(\theta - \theta')$$

with horizontal space $\mathcal{A}$ and vertical $\mathcal{V}$, so does

$$T_{ab}^{[N]}(\theta) = \sum_{a_1,\ldots a_{N-1}=1}^{N} t_{aa_1}(\theta) \otimes t_{a_1a_2}(\theta) \otimes \ldots\ldots \otimes t_{a_{N-1}b}(\theta) \quad , \tag{1.21}$$

with the same R-matrix. The auxiliary space for $T_{ab}^{[N]}(\theta)$ is also $\mathcal{A}$, the vertical one being

$$\mathcal{V}^{(N)} = \bigotimes_{1\leq j\leq N} \mathcal{V}_j \tag{1.22}$$

In order to show that $T_{ab}^{[N]}(0)$ [eq.(1.21)] fulfils the YBE (1.3) , the best is to use graphical methods.

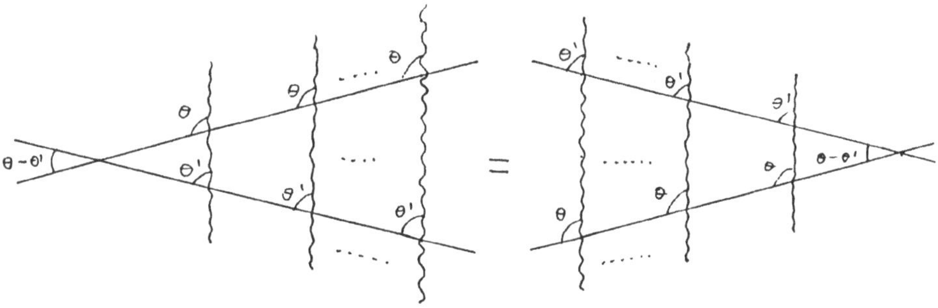

Fig.7 The YBE for the YB generator (1.19).

We must show that the equal sign holds in fig. 7. Remembering fig. 4, we can push one by one to the left the wavy lines in the lhs of fig. 7 through the R-matrix vertices leaving the expression invariant. After displacing all vertical wavy lines, we precisely get the rhs. This ends the proof. Of course, it can be also done analytically inserting eq.(1.20) in the lhs of eq.(1.3) and using eq.(1.19) repeatedly.

For N = 2 eq. (1.19) can be considered as a way of multiplying YB generators yielding new YB generators. This can be called a coproduct and shows that we have a Hopf algebra structure. More generally, if the YB generators are invariant under a group $\mathcal{G}$ [eq. (1.15)-(1.17)] we have as generator in $\mathcal{A} \otimes \mathcal{V}^N$,

$$T_{ab}{}^{[N]}(\theta,\bar\mu,\mathbf{g}) = \sum_{a_1,\ldots a_{N-1}=1}^{N} [\, g_1\, t(\theta-\mu_1)]_{aa_1} \otimes [\, g_2\, t(\theta-\mu_2)]_{a_1a_2} \otimes \ldots\ldots \otimes [\, g_N\, t(\theta-\mu_N)]_{a_{N-1}b} \tag{1.23}$$

It obeys the YB eq. (1.3) for any **fixed** transformations $\mathbf{g} = (g_1, \ldots, g_N)$ and $\bar\mu = (\mu_1,\ldots, \mu_N)$, with $g_i \varepsilon \mathcal{G}$, $\mu_i \varepsilon \mathbb{C}$, $1 \leq i \leq N$. The introduction of the parameters μ_i leads to integrable inhomogeneous vertex models[6].

There exists in addition, another coproduct multiplying the generators from right to left

$$\tilde T_{ab}{}^{[N]}(\theta,\bar\mu,\mathbf{h}) = \sum_{a_1,\ldots a_{N-1}=1}^{N} [\, h_1\, t(\theta-\mu_1)]_{a_1 b} \otimes [\, h_2\, t(\theta-\mu_2)]_{a_2a_1} \otimes \ldots\ldots \otimes [\, h_N\, t(\theta-\mu_N)]_{aa_{N-1}} \tag{1.24}$$

That is, $\tilde T_{ab}{}^{[N]}(\theta, \bar\mu, \mathbf{h})$ obeys the same YBA [eq. (1.3)] as $t_{ab}(\theta)$ does.

As we see, YBA are not Lie algebras since the sum of two generators $T(\theta)$ is **not** a YB generator. However, one finds for the YBA the analogous for most of the features of Lie algebras. The YBE (1.12) plays the role of the Jacobi identity in Lie algebras. The fundamental R-matrix being the analogue of the structure constants. There exists for YBA an "adjoint representation" [eq. (1.13)] provided by the R-matrix. We also have a "Cartan algebra" formed by the commuting transfer matrices $\tau_l(\theta)$ [eq. (1.7)]. A representation theory for YBA has been developped. That is, the construction of $T(\theta)$ for different spaces $(\mathcal{A}, \mathcal{V})$ given a fundamental R-matrix[3].

Actually there exist more general commuting transfer matrices than (1.7). It follows from eqs. (1.1) and (1.17) that the following operators on V^J :

$$\begin{aligned} \tau_{g_I}(\theta) &= Tr_{V^I}(\, g_I T^{(I,J)}\,) = \sum_{a,b} (g_I)_{ab}\, T_{ba}(\theta)^{(I,J)} \\ \tau_{g_K}(\theta) &= Tr_{V^K}(\, g_K\, T^{(K,J)}) = \sum_{A,B} (g_K)_{AB}\, T_{BA}(\theta)^{(K,J)} \end{aligned} \tag{1.25}$$

commute

$$[\ \tau_{g_I}(\theta)\ ,\ \tau_{g_K}(\theta')\ ] = 0\ ,\quad \forall\, \theta, \theta'\ \varepsilon\, \mathbb{C}\, , \forall\, g\ \varepsilon\, \mathcal{G} \tag{1.26}$$

Notice in eq. (1.24) that the transformation $g \in \mathcal{G}$ is the **same** in both transfer matrices.

It is legitime to call $T(\theta, \mathbf{g}, \bar\mu)$ [eq. (1.23)] a gauge transformation of $T(\theta)$ [eq. (1.22)]. We apply in eq.(1.23) a group symmetry transformation $(\, g_i\, , \mu_i\, , i = 1,\ldots.,N\,)$ that depends upon the site. This is a one-dimensional **local** gauge transformation on the lattice. Since the gauge transformed operator (1.22) or (1.23) obeys a YB algebra, these gauge transformations respect integrability[7].

Actually, a YB gauge transformed generator under $\mathcal{G}$ can be related with the untransformed one as follows [4]. Let us call $g_{\mathcal{A}}$ and $g_{\mathcal{V}}$ the group element representations acting on the horizontal and vertical spaces respectively and we set $\alpha_i = 0$, $i = 1, \ldots ,N$. Then eq.(1.20) implies

$$g_{\mathcal{A}} \quad t^{(\mathcal{A},\mathcal{V})}(\theta) \quad = \quad g_{\mathcal{V}}^{-1} \; t^{(\mathcal{A},\mathcal{V})}(\theta) \; [g_{\mathcal{A}}^{-1} \otimes g_{\mathcal{A}}^{-1}] \tag{1.27}$$

Now, we recognize in the l.h.s. of eq.(1.27) each of the operator factors in the r.h.s. of eq.(1.23). Inserting eq.(1.27) in eq.(1.23) yields after a little calculation:

$$T_{ab}^{[N]}(\theta, g) = \sum_{c} \bigotimes_{1\leq i\leq N} G_i^{-1} \; T_{ac}^{[N]}(\theta) \bigotimes_{1\leq i\leq N} G_i \; J_{cb} \tag{1.28}$$

where G_i^{-1} and G_i here act on the i-th vertical space with

$$G_i = \prod_{j=1}^{i} g_j \qquad \text{and} \qquad J = \prod_{j=1}^{N} g_j$$

That is, the gauge transformed $T_{ab}^{[N]}(\theta, \mathbf{g})$ can be obtained from the untransformed $T_{ab}^{[N]}(\theta)$ by a similarity transformation $\otimes_{1\leq i\leq N} G_i$ on the vertical space plus a right transformation J on the horizontal space. Hence, the gauge transformed transfer matrix,

$$\tau(\theta, \mathbf{g}) = \sum_{a} T_{aa}(\theta, \mathbf{g}) \tag{1.29}$$

can be written as

$$\tau(\theta,\mathbf{g}) = K \sum_{a,c} T_{ac}(\theta) \, J_{ca} \, K^{-1} = K \, \tau_J(\theta) \, K^{-1} \tag{1.30}$$

where $K \equiv \otimes_{1\leq i\leq N} G_i$.

$\tau_J(\theta)$ is precisely the generalized transfer matrix introduced by eq.(1.25). As we shall see in sec. II, it corresponds to twisted boundary conditions. The matrix J defines the twist in the vertical space [see eq.(2.2)]. Eq.(1.29) has a deep implication : local gauge transformations only affect the physical operators like the transfer matrix and the spin hamiltonian (generated by it) through a twist J on the boundary conditions and a similarity transformation K .

When $\theta = \theta'$ eq. (1.3) naturally suggests that R(0) is a multiple of the unit matrix in $\mathcal{A} \otimes \mathcal{A}$. When this happens the corresponding R-matrix is called regular. That is

$$R(0) = c \, 1 \tag{1.35}$$

where c is a numerical constant. This property can be represented graphically as follows (cf. fig. 5)

$$R^{ab}{}_{cd}(0) = c \, \delta_{ac} \, \delta_{bd} = \tag{1.36}$$

This property plays a key role in the theory of integrable models. First, it implies that the transfer matrices $\tau(\theta)$ built from R-matrices are generating functionals of local lattice operators. That is, those $\tau(\theta)$ following from eqs. (1.7) and (1.21) when $t(\theta)$ is given by eq. (1.13) (see eqs. (2.12)-(2.13)).

Secondly, the unitarity properties of $T(\theta)$ follows from eq. (1.35). Let us consider the YB equation (1.1) when a) $V^I = \mathcal{A}$, $V^J = V^K = \mathcal{V}$ and b) $V^I = \mathcal{V}$, $V^J = V^K = \mathcal{A}$. This gives respectively

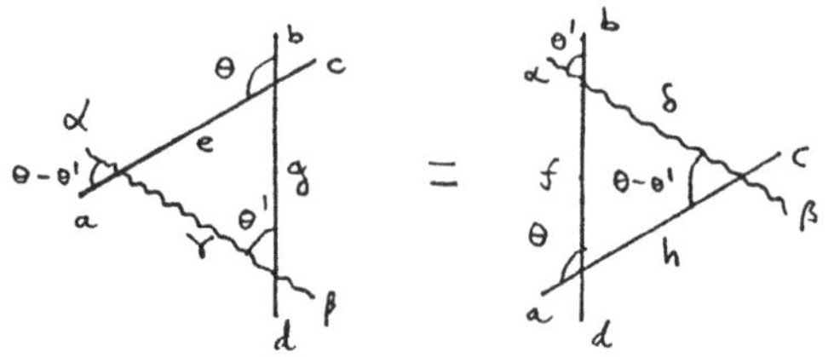

(1.37)

and

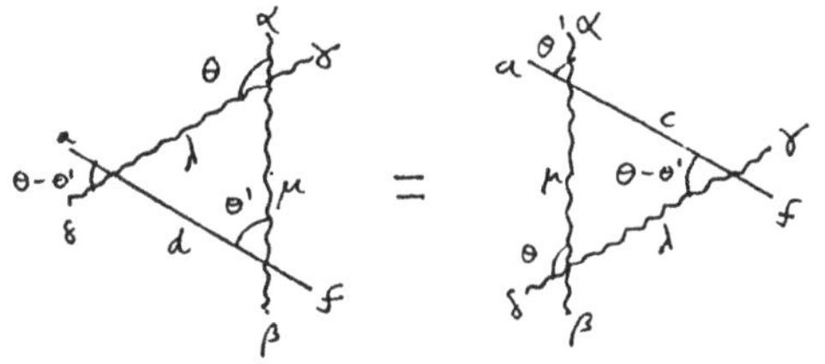

(1.38)

Now, if we set $\theta = 0$ in (1.37) and (1.38), we find using eq. (1.35) for $R^{\mathcal{A}}(\theta)$ and $R^{\mathcal{V}}(\theta)$,

$$M^{\alpha\beta}{}_{af}(\theta)\ \delta_{\delta\gamma} = \delta_{\alpha\beta}\ M^{\delta\gamma}{}_{af}(-\theta)$$
$$M^{\alpha\beta}{}_{ad}(\theta)\ \delta_{bc} = \delta_{ad}\ M^{\alpha\beta}{}_{bc}(-\theta) \tag{1.39}$$

Here,

$$M^{\alpha\beta}{}_{ab}(\theta) = [\, T_{ad}(-\theta)\,]_{\alpha\lambda}\, [\, T_{db}(\theta)\,]_{\lambda\beta} =$$

$=$ 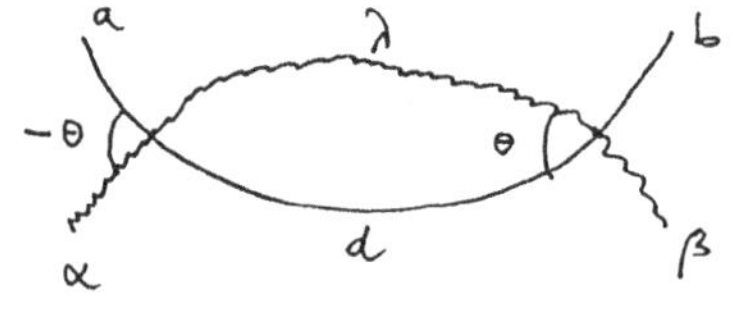

(1.40)

Eq. (1.39) shows that

$$M^{\alpha\beta}{}_{ab}(\theta) = \delta_{ab}\ \delta_{\alpha\beta}\ \rho(\theta) \tag{1.41}$$

where $\rho(-\theta) = \rho(\theta)$ is a c-number function. Eq. (1.37) is actually an operator product on **two** vector spaces $\mathcal{A}$ and $\mathcal{V}$ Keeping in mind this double matrix product, we find

$$T(\theta)\ T(-\theta) = \rho(\theta)\ \mathbf{1} \tag{1.42}$$

where $\mathbf{1}$ stands for the unit operator in $\mathcal{V} \otimes \mathcal{A}$. We have found that all YB generators possess an inverse provided their R-matrix is regular in the sense of eq. (1.35). That is

$$T^{-1}(\theta) = [\, 1 / \rho(\theta)\,]\ T(-\theta)$$

fullfils

$$T(\theta)\ T^{-1}(\theta) = T^{-1}(\theta)\ T(\theta) = \mathbf{1} \tag{1.43}$$

The antipode generator is defined by

$$T^{A}(\theta) \equiv T^{-1}(\theta)^{t} \tag{1.44}$$

where t means transpose in $\mathcal{A}$. That is

$$[\, T_{ab}(\theta)^t \,]_{\alpha\beta} = [\, T_{ba}(\theta)\,]_{\alpha\beta} \tag{1.45}$$

The antipode is an automorphism of the YB algebra. It follows from eqs. (1.3) and (1.43) that

$$R(\theta-\theta')\,[\, T^{A}(\theta) \otimes T^{A}(\theta')\,] = [\, T^{A}(\theta') \otimes T^{A}(\theta)\,]\, R(\theta-\theta') \tag{1.46}$$

The YB algebra possess therefore a Hopf algebra structure with antipode. Since the coproduct [eq. (1.21) for N = 2] is non-commutative as well as the usual product of $T(\theta)$, we have a non-commutative and non-cocommutative Hopf algebra. Actually there are many choices for the coproduct [eqs.(1.23)-(1.24)].

We shall consider in these lectures YB algebras and their applications to statistical models and quantum field theory. YB algebras also describe classically integrable field theories. That is integrable non-linear PDE[2,8]. Let us just derive the classical version of the YB algebras from the quantum one [eq.(1.3)]. Let us consider an infinite dimensional vertical space $\mathcal{V}$ (a Hilbert space) where a classical limit can be defined when some parameter h goes to zero. The horizontal space $\mathcal{A}$ is taken to be finite dimensional. Then, in the classical limit, we assume the Bohr correspondence principle to hold:

$$[\, T_{ab}(\theta)\,,\, T_{cd}(\theta')\,] \underset{h\to 0}{=} i h\, \{\, T_{ab}(\theta)\,,\, T_{cd}(\theta')\,\} \tag{1.47}$$

where $\{ A\,,\, B \}$ stands for the classical Poisson bracket.

When $h \equiv 0$, this conmutator vanishes and we find that we can set $R(\theta) = P$ [as defined by eq.(1.30)]. For small, but non-zero h, one can assume

$$R(\theta) = P\,[\, 1 + h\, r(\theta) + O(h^2)\,] \tag{1.48}$$

where the matrix $r(\theta)$ is called classical r-matrix. Inserting eq.(1.44) and (1.48) in eq.(1.3) yields to first order in h

$$\{\, T_{ab}(\theta)\,,\, T_{cd}(\theta')\,\} = r(\theta-\theta')_{ac,ef}\, T_{eb}(\theta)\, T_{fd}(\theta') - T_{ae}(\theta)\, T_{cf}(\theta')\, r(\theta-\theta')_{ef,bd} \tag{1.49}$$

or, in a more compact notation[2,8]:

$$\{\, T(\theta) \otimes, T(\theta')\,\} = [\, r(\theta-\theta')\,,\, T(\theta) \otimes T(\theta')\,] \tag{1.50}$$

where,

$$\{ A \otimes, B \}_{ac,bd} \equiv \{ A_{ab}\,,\, B_{cd} \}$$

Inserting eq.(1.48) in the YB equation (1.11) yields to order h^2 the so called classical YB equation for the r-matrix :

$$[\, r_{12}(\theta-\theta')\,,\, r_{13}(\theta)\,] + [\, r_{12}(\theta-\theta')\,,\, r_{23}(\theta')\,] + [\, r_{13}(\theta)\,,\, r_{23}(\theta')\,] = 0 \tag{1.51}$$

where $r_{ij}(\theta)$ is $r(\theta)$ on $\mathcal{V}_i \otimes \mathcal{V}_j$ and the unit matrix in $\mathcal{V}_k$ ($j \neq k \neq i$; i,j,k =

1, 2, 3). Notice that $R(\theta)$ as given by eq.(1.25) identically verifies the YBE (1.11) to order $\hbar^0$ and $\hbar^1$.

II. PHYSICAL REALIZATIONS OF YANG-BAXTER ALGEBRAS

In this section we shall describe YB algebras in two-dimensional statistical models, field theories and S-matrix theory.

We associate in sec. I a vector space V^I to each type of lines and a YB generator $T^{(I,J)}(\theta)$ to a pair of lines (I,J) intersecting with an angle θ. This can be immediately applied to a two-dimensional lattice of lines [fig. 8] intersecting at the sites. The vector spaces describe the possible local states of the bonds and the $t(\theta)$ describe the statistical weights of the different link configurations.

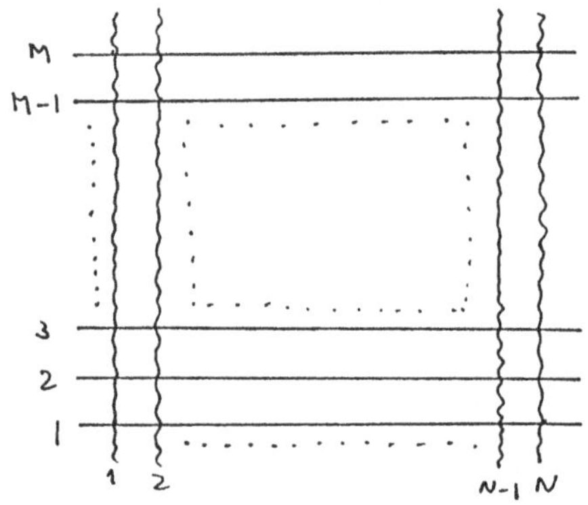

Fig.8 A N x M two dimensional lattice. The local states of horizontal (vertical) bonds belong to the vector space $\mathcal{A}$ ($\mathcal{V}$).

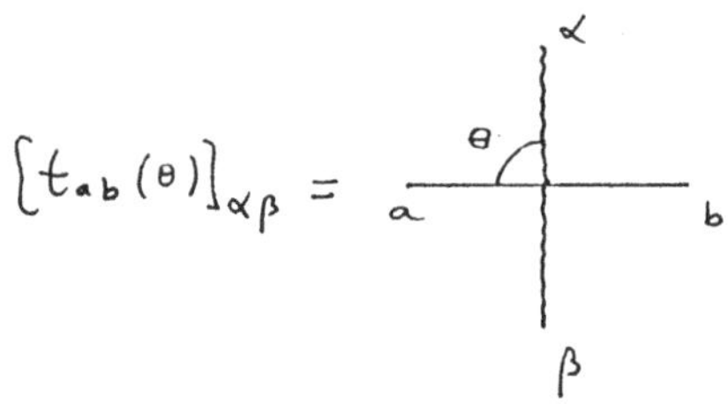

Fig.9 The local statistical weights w ($\alpha\ \beta$ | a b) depend on the states of the four bonds joining at a vertex.

That is the matrix element $[t_{ab}(\theta)]_{\alpha\beta}$ defines the probability for the local configuration depicted in fig. 9. The product of the local weights over all sites in the lattice yields the probability for such configuration of the whole system. Finally summing over all possible configurations gives the partition function Z. When periodic boundary conditions are used in both horizontal and vertical directions, Z expresses as

$$Z = \mathrm{Tr}_{\mathcal{V}} [\ \tau(\theta)^M\ ] \tag{2.1}$$

Actually eq. (2.1) holds irrespective of the YB equations.

The transfer matrices $\tau_g(\theta)$ [5] [eq. (1.22)] correspond to twisted boundary conditions. That is, when the operators at sites N+1 and 1 are related by the transformation g :

$$S_{N+1} = g\, S_1\, g^{-1} \ , \quad \forall S \ . \tag{2.2}$$

Here g acts in the appropriate representation of $\mathcal{G}$. Then, $\tau_{g_A}(\theta)$ is the transfer matrix. If we also impose twisted b.c in the vertical direction with a twist $h_{\mathcal{V}}$, Z writes

$$Z_{g,h} = \mathrm{Tr}_{\mathcal{V}} [\ \tau_{g_A}(\theta)^M\, h_V\ ] \tag{2.3}$$

Eq. (2.1) and (2.3) show how important is the knowledge of the eigenvalues of $\tau(\theta)$. Actually, just the largest eigenvalue $\Lambda^{[N]}{}_{MAX}(\theta)$ gives the free energy in the thermodynamic limit

$$f = -\lim_{N,M\to\infty} (\ 1\ /\ NM\ ) \log Z = -\lim_{N\to\infty} \{(1/N\ ) \log \Lambda^{[N]}{}_{MAX}(\theta)\} \tag{2.4}$$

(The dependence on the b.c. drops in the N = M = ∞ limit).

The lattice model here described is called a vertex model. It is homogeneous but not isotropic since horizontal and vertical lines are of different nature. Horizontal bonds live in states of the space $\mathcal{A}$ and vertical bonds in states of the space $\mathcal{V}$. One can even generalize these integrable vertex models taking lines at arbitrary intersection angles[9]. Also taking inhomogeneous weights $g^{\mathcal{A}}(x)\ h^{\mathcal{V}}(y)\ t(\ \theta - \alpha(x) - \beta(y)\ )$ that depend upon the horizontal (x) and the vertical (y) coordinates. Moreover, one could take the lattice lines from all possible vector spaces $V^I \in \mathcal{I}$ at will. All these models are integrable and solvable although inhomogeneous and anisotropic.

Let us now study the transfer matrices $\tau(\theta)$ as generating functionals of commuting local operators on the lattice. This is the case for R-matrix models (where $\mathcal{A} = \mathcal{V}$) when R is a regular R-matrix [eq. (1.30)]. We find from eqs. (1.13) and (1.30)-(1.31)

$$[\ t_{ab}(\ 0\ )\ ]_{cd} = c\ \delta_{bc}\,\delta_{ad} \tag{2.5}$$

Then for a N-site transfer matrix as defined by (1.7) and (1.19)

$$\tau(0)^{[N]}{}_{\mathbf{a}|\mathbf{b}} = c^N \prod_{i=1}^{N} \delta_{a_i, b_{i+1}} \tag{2.6}$$

$$\mathbf{a} \equiv (a_1, a_2, \ldots, a_N) \quad ; \quad \mathbf{b} \equiv (b_1, b_2, \ldots, b_N)$$

where $b_{N+1} \equiv b_1$. Here we assume periodic boundary conditions (PBC). The operator in the rhs of (2.6) is just the lattice unit shift operator to the right. Therefore, we can define the momentum operator as

$$\mathcal{P} \;=\; i \;\mathrm{Log}[\; c^{-N}\, \tau(0)^{[N]} \;] \tag{2.7}$$

Let us now show that the logarithmic derivative of $\tau^{[N]}(\theta)$ at $\theta = 0$ gives an operator coupling nearest neighbors.

Using eq. (1.31), $T_{ab}{}^{[N]}(0)_{c|d}$ and $\tau^{[N]}(0)_{c|d}$ can be drawn as follows

$T_{ab}{}^{[N]}(0)_{c|d} \;=$ 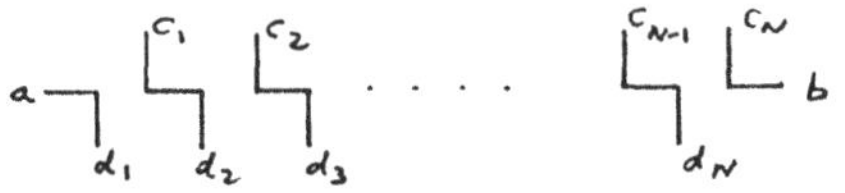 (2.8)

$\tau^{[N]}(0)_{c|d} \;=$ 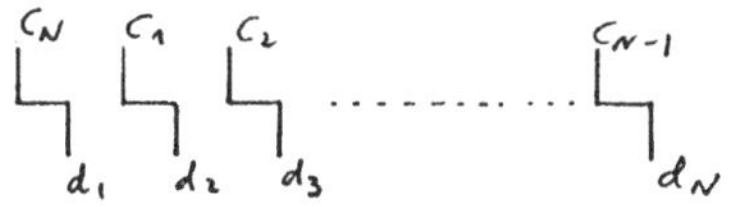 (2.9)

Similarly,

$\{\tau^{[N]}(0)\}^{-1}{}_{c|d} \;=$ 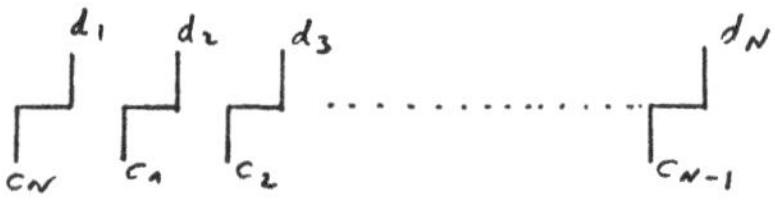 (2.10)

Now, if we compute $d/d\theta\; \tau(\theta)$ from eq. (1.19) we obtain N terms, each one containing $d/d\theta\; t^{(h)}(\theta)$, $1 \leqslant h \leqslant N$ and the others $t^{(\ell)}$ ($\ell \neq h$)) not derived. Hence, setting $\theta = 0$ yields

$\dot{\tau}^{[N]}(0) \;=$ 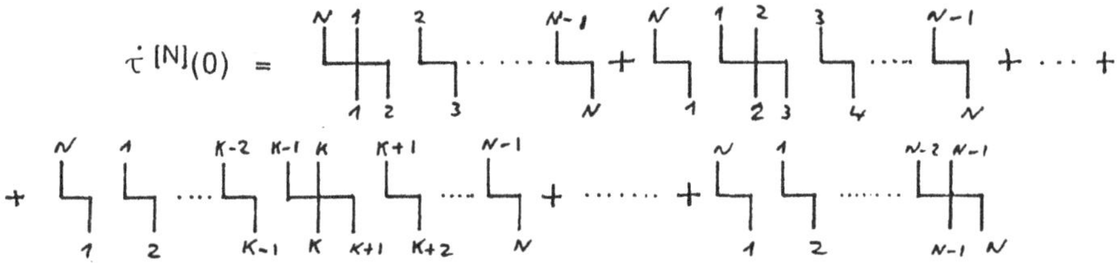 (2.11)

Here  stands for $\dot{R}(0)$

It is now very simply to perform the product $\tau^{[N]}(0)^{-1}\; \dot{\tau}^{[N]}(0)$ just combining eqs. (2.10) and (2.11) with the result

$\{\tau^{[N]}(0)\}^{-1}\, \dot{\tau}^{[N]}(0) \;=$  (2.12)

Therefore $\tau^{[N]}(0)^{-1}\; \dot{\tau}^{[N]}(0)$ is a sum of terms each one acting as an operator on two neighboring sites. Now, putting all factors

$$H \;=\; \partial/\partial\theta \;\log \tau^{[N]}(\theta)\; \Big|_{\theta=0} \;=\; \sum_{n=1}^{N} h_{n,n+1} \tag{2.13}$$

where the matrix elements of $h_{n,n+1}$ read

$$< c_n\, c_{n+1} \mid h_{n,n+1} \mid d_n\, d_{n+1} > \;=\; (1/c)\;\; \dot{R}(\,0\,)^{c_n\, c_{n+1}}{}_{d_n\, d_{n+1}} \tag{2.14}$$

Notice that $[\tau(\theta), \dot{\tau}(\theta)] = 0$ since $[\tau(\theta), \tau(\theta')] = 0, \; \forall \theta, \theta'$. Therefore,

$$\partial/\partial\theta \log \tau^{[N]}(\theta) = \dot{\tau}(\theta)\, \tau^{-1}(\theta) = \tau^{-1}(\theta)\, \dot{\tau}(\theta).$$

More generally the nth derivative of $\log \tau(\theta)$ at $\theta = 0$ is an operator that couples $(n+1)$ neighboring sites[10].

The previous derivation generalizes easily for the twisted transfer matrices (1.25) with $\mathcal{V}^I = \mathcal{V}^{(N)}$ [eq.(1.22)] and $\mathcal{V}^J = \mathcal{A}$. We find in this case

$$\tau_g(0)^{[N]}{}_{a|b} = c^N \prod_{i=1}^{N-1} \delta_{a_i,b_{i+1}}\, g_{a_N b_1} \qquad (2.15)$$

and

$\{\tau_g^{[N]}(0)\}^{-1}\, \dot{\tau}_g^{[N]}(0) =$ 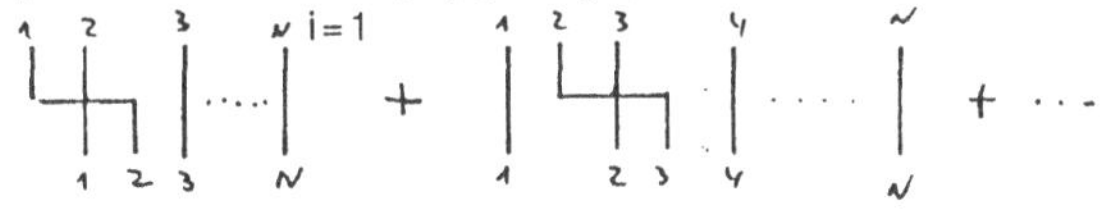

(2.16)

Only the last term differs from the PBC case (2.12). This last term is actually $h_{N,N+1}$ describing the interaction of the site N with the next one. In the PBC case (2.12) we set $N+1 \equiv N$. In eq.(2.15) we have indeed the $h_{N,N+1}$ as given by eq.(2.13) provided we make the identification (2.2) for the operators acting on $\mathcal{V}_1$ and $\mathcal{V}_{N+1}$. The factor $g_{a_N b_1}$ in eq.(2.15) has the same explanation. It is equivalent through eq.(2.2) to $\delta_{a_N b_{N+1}}$.

The operator H can be interpreted as a one-dimensional quantum hamiltonian. It is an operator coupling neighboring q-component "spins". The word spins only applies, rigorously speaking, when the fundamental R-matrix corresponds to the six or eight vertex model. That is, the underlying Lie algebra in A_1 and we have true SU(2) spins. Otherwise one finds SU(q) spins, O(q) spins, etc. Eq. (2.12) suggest that θ may be the imaginary time variale. This possibility has not been fully explored yet. Anyway it must be noticed that $\tau(\theta) \neq e^{\theta H}$.

Let us expand the gauge transformed transfer matrix $\tau(\theta,\mathbf{g})$ [eq.(1.29)] around $\theta = 0$. We find :

$$\tau(0,\mathbf{g})_{a|b} = c^N \prod_{i=1}^{N} (g_i)_{a_{i-1},b_i} =$$ 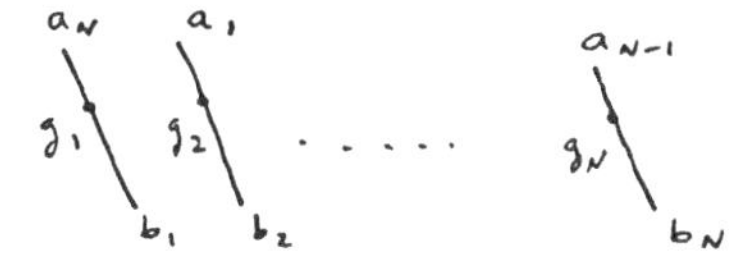

(2.17)

where 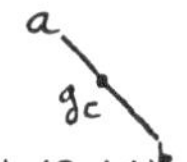 $= g_c : \mathcal{V}_b \to \mathcal{V}_a$. A calculation analogous to that from eq.(2.10)-(2.11) yields for $\tau(\theta,\mathbf{g})$ [7]

$$H[\mathbf{g}] = \partial/\partial\theta \log \tau^{[N]}(\theta,\mathbf{g})\,\big|_{\theta=0} = \sum_{n=1}^{N} (g^{-1})_{n+1}\, h_{n,n+1}\, g_{n+1} \qquad (2.18)$$

We see from eq.(2.18) that the way $\mathbf{g} = (g_1, \ldots, g_N)$ acts on the spin hamiltonian can be considered as a gauge transformation on the lattice. From the analysis in sec. I for $\tau(\theta,\mathbf{g})$ we conclude that $H[\mathbf{g}]$ is, up to a similarity transformation K, identical to $H[\mathbf{1}]$ with boundary conditions twisted by $J = \prod_{1\leq i\leq N} g_i$ [4].

One can decide to expand $\tau(\theta)$ around a point $\theta = \theta_0 \neq 0$ to generate commuting operators. The trouble is that these operators are in general non-local. That is, they couple all sites in the line. Only when $\tau(\theta_0)$ is the shift operator (2.6) the logarithmic derivative of $\tau(\theta)$ at θ_0 is local and the higher order derivatives multilocal. It is nevertheless interesting to expand or around $\theta = \infty$. Besides the quantum group generators [see sec.V], this expansion yields non-local integrable hamiltonians.
To conclude this section, let us sketch the classification of the known YB algebras. That is, solutions of eq.(1.1). All known solutions posses symmetries in the sense of eq.(1.15) [see table I]. Probably, the presence of such symmetries is the basic reason why a so heavily overdetermined set of equations [see eq.(1.12)] has non-trivial solutions.

Elliptic solutions depend on three continuous parameters θ (spectral parameter), γ (anisotropy parameter), and k (elliptic modulus). The trigonometric and hyperbolic solutions depend upon two parameters : θ and γ . They can be obtained from each other by "Wick rotation" : $\gamma \to i\gamma$, $\theta \to i\theta$. The degenerate limit $k \to 0$ ($k \to 1$) of the elliptic solutions yield trigonometric (hyperbolic) solutions. Trigonometric/hyperbolic solutions associated to all simple Lie algebras are known in vertex language[4,46,63,64].

The trigonometric/hyperbolic solutions are not invariant under the full Lie group $\mathfrak{G}$ [in the sense of eq.(1.17)] but only under its Cartan subalgebra. In addition they are invariant under the corresponding quantum group as we discuss in sec.V. The rational YB algebras follow from the trigonometric/hyperbolic ones in the isotropic limit $\gamma \to 0$. Besides these continuous parameters (k, γ, θ) , the YB operators are labeled by the representation spaces $\mathcal{A}$ and $\mathcal{V}$ where they act. As we have seen, each $T^{(\mathcal{A},\mathcal{V})}(\theta)$ defines an inequivalent physical vertex model. Actually there is a set of physical models attached to each $T^{(\mathcal{A},\mathcal{V})}(\theta)$: a vertex model, a spin hamiltonian, a conformal theory and a massive quantum field theory. The last two cases are exposed in ref.[4] and sec. IV.

Algebraic Bethe Ansatz solutions for a significant number of vertex models are known[17-22] and we believe that such constructions should exist for **any** solution of the YB equations. In sec. 3 we describe the simplest case : the six vertex model solution. Further algebraic Bethe Ansatz constructions (for richer models) can be found in refs.[4,17-22]. Let us recall some important features of the BA solutions.

The vërtex (and spin hamiltonians) turn to have a non-zero gap in the elliptic and hyperbolic regimes. This gap is identically zero in the trigonometric and rational regimes. Therefore, interesting scaling limits exist in these gapless regimes. Two inequivalent continuous limits exist : a) a massless limit yielding conformal invariant models b) a massive limit leading to integrable massive quantum field theories. These scaling behaviors are reviewed in refs.[4]. Let us just recall that all (known) conformal field theories derive from some integrable model in the continuous limit. The anisotropy parameter appears in the conformal dimensions (which vary continuously in general) but not in the central charge. Both depend on the YB algebra considered as well as on $\mathscr{A}$ and $\mathscr{V}$. The spectral parameter turns to be irrelevant (in the sense of crical phenomena) for this massless continuous limit. For massive scaling limits (yielding QFTs) the spectral parameter undergoes a dimensional transmutation and generates the mass scale together with the lattice spacing. The anisotropy parameter provides the (single) continuos coupling constant. The elliptic modulus vanishes and leaves no trace in the scaling limits[22].

We consider here solutions of the YB algebra where the operators $T^{(\mathscr{A},\mathscr{V})}(\theta)$ depend on a single argument θ. More generally, they can be functions of a pair of variables θ and θ' associated to each of the lines intersecting at the vertex, respectively. In refs.[23] there has been found solutions of such type associated to higher genus curves. These solutions are actually related to the YB algebra associated to the six vertex model[24]. For YB solutions of genus zero or one depending on (θ,θ') there always exist parametrizations where the dependence becomes on $\theta - \theta'$ and hence reduces to the one variable case treated here[1].

Besides the symmetry transformations (1.19) leaving invariant the YB algebras, one can also look for transformations not involving linear combinations of YB operators like

$$[T_{ab}(\theta)]_{\alpha\beta} \quad \rightarrow \quad C_{ab\alpha\beta}(\theta)\, [T_{ab}(\theta)]_{\alpha\beta} \tag{2.19}$$

Let us analyze these abelian transformations for the R-matrix,

$$R^{ab}{}_{cd}(\theta) \rightarrow \mathring{R}^{ab}{}_{cd}(\theta) \;=\; \exp[F_{abcd}(\theta)]\; R^{ab}{}_{cd}(\theta) \tag{2.20}$$

in the trigonometric/hyperbolic cases where

$$a + b = c + d \tag{2.21}$$

due to the U(1) symmetry(ies). One can seek for $F_{abcd}(\theta)$ an expansion of the type

$$F_{a_1a_2a_3a_4}(\theta) \;=\; \sum_{1\le i\le 4} a_i\, f_i(\theta) \;+\; \sum_{1\le i,j\le 4} a_i\, a_j\, f_{ij}(\theta) \;+\; \dots\dots\dots \tag{2.22}$$

Inserting eqs.(2.34)-(2.36) in eq.(1.12) and requiring $\mathring{R}^{ab}{}_{cd}(\theta)$ to be also a solution of the YBE (1.12) leads to a set of constraints on $f_i(\theta)$, $f_{ij}(\theta)$, ...We find after a long but straightforward algebra

$$F_{abcd}(\theta) = \mu\theta\,(a + c - b - d) + \nu\,(a + d - b - c) + \omega\,(ab - cd) + \dots \tag{2.23}$$

where μ, ν and ω are arbitrary parameters and we used also eq.(2.35). These transformations has been given in refs.(25) where they are called 'symmetry breaking transformations'. The reason of the name is that these transformations map R-matrices with P and T invariances [eqs.(2.30)] to R-matrices ($\mathring{R}$) where P and T do not hold.

III. THE SIX VERTEX MODEL AND ITS DESCENDANTS

The six vertex model corresponds to the trigonometric and hyperbolic solutions of the YBE (1.11) for $q = 2$. That is,

$$R(\theta) = \begin{pmatrix} a(\theta,\gamma) & 0 & 0 & 0 \\ 0 & c(\gamma) & b(\theta) & 0 \\ 0 & b(\theta) & c(\gamma) & 0 \\ 0 & 0 & 0 & a(\theta,\gamma) \end{pmatrix} \tag{3.1}$$

We have here three different regimes,

I) $a(\theta,\gamma) = sh(\gamma-\theta)$, $b(\theta) = sh\theta$, $c(\gamma) = sh\gamma$, $\gamma > \theta > 0$, $\gamma > 0$ in the antiferroelectric regime.

II) $a(\theta,\gamma) = \sin(\gamma-\theta)$, $b(\theta) = \sin\theta$, $c(\gamma) = \sin\gamma$, $0 < \gamma < \pi$, $\gamma > \theta > 0$ in the trigonometric regime. This regime is critical (gapless) and antiferroelectric.

III) $a(\theta,\gamma) = sh(\theta+\gamma)$, $b(\theta) = sh\theta$, $c(\gamma) = sh\gamma$, $\theta > 0$, $\gamma > 0$ in the ferroelectric regime.

The parameter γ describes the anisotropy of the model. The character of regimes I, II and III will be clear from the ground state and excitations obtained below.

This model enjoys the following symmetry group $\mathfrak{G}$ (in the sense of eq. (1.15)]

$$\mathfrak{G} = \{ \exp(i\, \alpha\, \sigma_z) \ , \ 0 \leq \alpha < 2\pi \ ; \ \sigma_x \ \} \tag{3.2}$$

That is $\mathfrak{G} = U(1) \otimes Z_2$. When $\gamma = 0$ this group enlarges to $SU(2)$. This point corresponds to a Kosterlitz-Thouless type transition as we will see below from the explicit solution.

It is called six vertex model, since the non-zero elements of the R-matrix, eq.(3.1) define six allowed configurations. The integrable eight-vertex model will not be considered here[1]. The state of a bond in the six-vertex (and eight-vertex) models is usually characterized by the sense of an arrow. This corresponds here to the values 1 or 2 of the vertical and horizontal indices. In fig. 10 the allowed configurations and their respective statistical weights are depicted.

It must be recalled that the trigonometric regime of the six-vertex model (II) describes the critical (zero gap) limit of the eight-vertex model[1]. As it will be clear from the solution one describes a critical line when γ varies from 0 to π. As a S-matrix eq.(3.1) for regime II describes the scattering of a particle and its antiparticle with a conserved U(1) charge[12]. The crossing symmetry (2.23) writes here

$$[P\, R(\theta)\,]^{t_1} = (1 \otimes \sigma)\, P\, R(-\theta - \gamma)\, (1 \otimes \sigma) \tag{3.3}$$

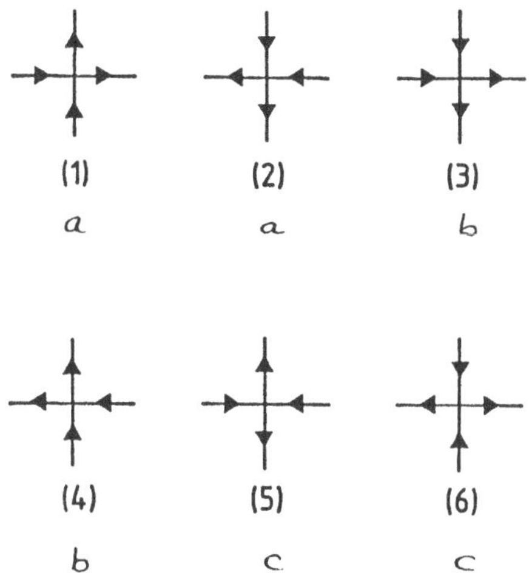

Fig.10 Allowed configurations in the six-vertex model and their statistical weights (see eq.(3.1)).

where
$$\sigma \equiv \begin{pmatrix} 0 & -1 \\ 1 & 0 \end{pmatrix} = -i\sigma_y$$

The YB generators read here (for one site)

$$t_{11}(\theta) = \begin{pmatrix} a(\theta,\gamma) & 0 \\ 0 & b(\theta) \end{pmatrix}, \qquad t_{22}(\theta) = \begin{pmatrix} b(\theta) & 0 \\ 0 & a(\theta,\gamma) \end{pmatrix} \tag{3.4}$$

$$t_{12}(\theta) = c(\gamma)\,\sigma_- \qquad , \qquad t_{21}(\theta) = c(\gamma)\,\sigma_+$$

The YB generator $T_{ab}^{[N]}(\theta)$ for a N-sites line follows from eq. (1.19) where one inserts the $t_{ab}(\theta)$ given by eq. (3.4). One can then set

$$T_{ab}^{[N]}(\theta) = \begin{pmatrix} A(\theta) & , & B(\theta) \\ C(\theta) & , & D(\theta) \end{pmatrix} \tag{3.5}$$

The one-dimensional quantum hamiltonian associated to the six-vertex model is the XXZ Heisenberg hamiltonian. One finds from eqs. (2.12)-(2.13) and (3.1)

$$h = [\ \cos\gamma + \sigma_x \otimes \sigma_x + \sigma_y \otimes \sigma_y + \Delta\ \sigma_z \otimes \sigma_z\ ]/(\ 2\sin\gamma\ )$$

where $\Delta = -\cos\gamma$ .Then

$$H_{XXZ} = -\tfrac{1}{2} \sum_{a=1}^{N} (\ \sigma_x^a \otimes \sigma_x^{a+1} + \sigma_y^a \otimes \sigma_y^{a+1} + \Delta\ \sigma_z^a \otimes \sigma_z^{a+1}\ ) \tag{3.6}$$

where we droped the terms proportional to the unit operator and redefined H by a factor $-\sin\gamma$.

The six vertex model enjoys gauge invariance (see sec.I) under the symmetry group (3.2). That is we can perform U(1) transformations $g_a = \exp(\ i\ \mu_a\ \sigma_z^a\ )$ at each site a of the line and $T(\theta,g)$ will continue to obey the YB algebra. In particular, the gauge transformed XXZ Heisenberg hamiltonian reads from eq.(2.18) and (3.6)[7]

$$H_{XXZ}[\bar{\mu}] = -\frac{1}{2}\sum_{1\le a\le N} \exp(-i\mu_{a+1}\sigma_z^{a+1})\,(\sigma_x^a \otimes \sigma_x^{a+1} + \sigma_y^a \otimes \sigma_y^{a+1} +$$

$$+ \Delta\, \sigma_z^a \otimes \sigma_z^{a+1})\exp(i\mu_{a+1}\sigma_z^{a+1}) \tag{3.7}$$

where $\bar{\mu} = (\mu_1, \ldots, \mu_r)$.

As discussed in sec.II , the expansion of $\tau(\theta)$ around any point $\theta = \theta_0 \neq 0$ yields non-local operators. Nevertheless, expanding around $\theta_0 = \infty$ leads to interesting objects : generators of the $SU(2)_q$ quantum group as we shall see in sec.VI ($q = \exp(i\gamma)$ or $\exp(-\gamma)$). Let us start to expand $A(\theta,\bar{\mu})$ and $D(\theta,\bar{\mu})$ around $\theta = \infty$. We find in the regime I[5]

$$A(\theta,\bar{\mu}) \underset{\theta\to\infty}{=} y^N \exp(\gamma S_z - \alpha)\,[\,1 + Q_+(\bar{\mu})/y^2 + O(1/y^4)\,]$$

$$D(\theta,\bar{\mu}) \underset{\theta\to\infty}{=} y^N \exp(-\gamma S_z - \alpha)\,[\,1 + Q_-(\bar{\mu})/y^2 + O(1/y^4)\,] \tag{3.8}$$

where $y \equiv \frac{1}{2}\exp(\theta + \gamma/2)$, $\alpha \equiv \sum_{a=1}^{N} \mu_a$, $S_z = \frac{1}{2}\sum_{a=1}^{N} \sigma_z^a$.

and

$$Q_\pm(\bar{\mu}) = \mathrm{sh}2\gamma \sum_{1\le a<b\le N} \sigma_-^a \exp(\mu_a + \mu_b) \prod_{c=a+1}^{b-1} \exp(\pm\gamma\,\sigma_z^c)\,\sigma_\pm^b \tag{3.9}$$

$$-\frac{1}{4}\sum_{a=1}^{N}\exp(2\mu_a)\,[\,\cosh\gamma - \mathrm{sh}\gamma\,\sigma_z^a\,]$$

For simplicity we have set **g = 1** in eq.(1.23).

The operator

$$Q_\mu(\bar{\mu}) \equiv \exp(i\mu)\,Q_+(\bar{\mu}) + \exp(-i\mu)\,Q_-(\bar{\mu}) \tag{3.10}$$

belong to the infinite sequence of operators generated by the transfer matrix

$$\tau_\mu(\theta) = \exp(-\gamma S_z + i\mu)\,A(\theta) + \exp(\gamma S_z - i\mu)\,D(\theta) \tag{3.11}$$

since

$$\tau_\mu(\theta) = y^N \exp(-\alpha)\,[\,1 + Q_\mu(\bar{\mu})/y^2 + O(1/y^4)\,] ,$$

and

$$[\,\tau_\mu(\theta), \tau_\mu(\theta')\,] = 0 \quad , \quad [\,\tau_\mu(\theta), Q_\mu(\bar{\mu})\,] = 0 .$$

These operators simplify in the isotropic limit $\gamma \to 0$, $\mu_a \to \kappa_a \gamma$, $\theta \to \gamma\lambda$. We find from eqs.(3.9)-(3.10)

$$Q_\mu(\bar{\mu}) \underset{\gamma\to 0}{=} -\frac{1}{2}N\cos\mu - \gamma(\nu\cos\mu + 2i\sin\mu\, S_z) - \frac{1}{4}\gamma^2\,[\,3N + \sum_{a=1}^{N}\kappa_a^2 -$$

$$4(S_x^2 + S_y^2)\,]\cos\mu + \gamma^2 \sin\mu\; h(\bar{\kappa}) + O(\gamma^3)$$

where $\nu = \alpha/\gamma$ for $\gamma \to 0$ and $h(\bar{\kappa})$ is the hermitean operator

$$h(\bar{\kappa}) \equiv \tfrac{1}{4}\sum_{1\le a\neq b\le N} \mathrm{sign}(a-b)\,(\varepsilon_{zij}\,\sigma_i^a\,\sigma_j^b) - \sum_{a=1}^{N} \kappa_a\,\sigma_z^a \tag{3.12}$$

Here $i,j = x, y$, $\varepsilon_{123} = +1$ and $\bar{\kappa} = (\kappa_1, \ldots, \kappa_N)$. $h(\bar{\kappa})$ can be interpreted as a non-local Dzialozhinski-Moriya interaction hamiltonian plus a coupling with an external magnetic field v_a pointing in the z direction. That is a magnetic field varying from site to site on the line. The properties of $h(\bar{\kappa})$ are analyzed in ref.[5]. This exactly solvable hamiltonian can describe disordered systems since the κ_a are completely arbitrary. For example they can follow some probability distribution.

The YB algebra defined by the R-matrix (3.1) yields some number of bilinear algebraic relations between the $T_{ab}^{[N]}(\theta)$. Let us just write down the more useful ones for the subsequent derivations

$$A(\theta)\, B(\theta') = g(\theta' - \theta)\, B(\theta')\, A(\theta) - h(\theta' - \theta)\, B(\theta)\, A(\theta') \tag{3.13}$$

$$D(\theta)\, B(\theta') = g(\theta - \theta')\, B(\theta')\, D(\theta) - h(\theta - \theta')\, B(\theta)\, D(\theta')$$

$$B(\theta)\, B(\theta') = B(\theta)\, B(\theta') \tag{3.14}$$

$$[\, C(\theta), B(\theta')\,] = h(\theta - \theta')\,\{ A(\theta')\, D(\theta) - A(\theta)\, D(\theta') \}$$

where $g(\theta) = a(\theta,\gamma) / b(\theta)$ and $h(\theta) = c(\gamma) / b(\theta)$.

Let us now proceed to construct the exact eigenvectors and eigenvalues of

$$\tau^{[N]}(\theta) = \mathrm{Tr}_{\mathcal{A}}\, T^{[N]}(\theta) = A(\theta) + D(\theta)\ , \tag{3.15}$$

using the algebraic Bethe Ansatz[17]. We shall assume N to be even. One notices that the ferromagnetic state

$$|\Omega\rangle = \begin{pmatrix} 1 \\ 0 \end{pmatrix} \otimes \begin{pmatrix} 1 \\ 0 \end{pmatrix} \otimes \cdots\cdots \otimes \begin{pmatrix} 1 \\ 0 \end{pmatrix} \tag{3.16}$$

is an eigenvector of $A(\theta)$ and $D(\theta)$

$$A(\theta)\,|\,\Omega\rangle = a(\theta,\gamma)^N\,|\,\Omega\rangle\ , \qquad D(\theta)\,|\,\Omega\rangle = b(\theta)^N\,|\,\Omega\rangle \tag{3.17}$$

In addition

$$C(\theta)\,|\,\Omega\rangle = 0 \tag{3.18}$$

whereas $B(\theta)|\Omega\rangle$ is non-zero and not proportional to $|\Omega\rangle$. The algebraic Bethe ansatz consist in looking for eigenvectors of $\tau(\theta)$ with the form

$$\Psi(\theta_1,\ldots,\theta_r) = B(\theta_1)\, B(\theta_2)\cdots\cdots B(\theta_r)\,|\,\Omega\rangle \tag{3.19}$$

Here, the complex number $\theta_1, \ldots, \theta_r$ will be determined by requiring that $\Psi(\theta_1,\ldots,\theta_r)$ is an eigenvector of $\tau(\theta)$.

In order to do that one applies $A(\theta) + D(\theta)$ to the r.h.s. of eq. (3.20) and pushes $A(\theta) + D(\theta)$ through the $B(\theta_j)$ with the help of eqs. (3.13). After using eqs. (3.13) r times, $A(\theta)$ and $D(\theta)$ reach $|\,\Omega>$ where their action is known from eqs. (3.17). These operations produced a lot of terms. Let us first write down explicitely those generated by the first term in eqs. (3.13) :

$$A(\theta)\,\Psi(\theta_1,\ldots,\theta_r) = a(\theta,\gamma)^N \prod_{j=1}^{r} g(\,\theta_j-\theta\,)\,B(\theta_1)\,B(\theta_2)\ldots\ldots\,B(\theta_r)\,|\,\Omega> +$$

$$+ \text{ unwanted terms } = \Lambda_+(\theta)\ \Psi(\theta_1,\ldots,\theta_r) + \text{ unwanted terms} \tag{3.20}$$

and an analogous formula for $D(\theta)\,\Psi(\theta_1,\ldots,\theta_r)$. The remaining terms are called "unwanted" since they are not proportional to $\Psi(\theta_1,\ldots,\theta_r)$ and hence they must finally cancel in order that $\Psi(\theta_1,\ldots,\theta_r)$ be an eigenvector of $\tau(\theta)$.

Now, let us concentrate in terms containing the vector

$$B(\theta)\,B(\theta_2)\ldots\ldots\,B(\theta_r)\,|\,\Omega>$$

They originate when the second term in eq. (3.13) is used to express $A(\theta)\,B(\theta_1)$ and the first term for the rest when $A(\theta_1)$ is pushed through $B(\theta_j)$ $(2 \le j \le r)$. Hence, one finds

$$-\,h(\theta_1-\theta)\,B(\theta)\,A(\theta_1)\,B(\theta_2)\ldots\ldots\,B(\theta_r)\,|\,\Omega> =$$

$$= -\,h(\theta_1-\theta)\,a(\theta_1,\gamma)^N \prod_{j=2}^{r} g(\theta_j-\theta_1)\,B(\theta)\,B(\theta_2)\ldots\ldots\,B(\theta_r)\,|\,\Omega> + \text{ other types of terms.} \tag{3.21}$$

It is now very easy to determine the remaining coefficients since $\Psi(\theta_1,\ldots,\theta_r)$ is a symmetric function of $\theta_1,\ldots,\theta_r$ due to eq. (3.14). Therefore one can permute θ_1 by θ_j in eq. (3.21) with the result

$$A(\theta)\,\Psi(\theta_1,\ldots,\theta_r) = \Lambda_+(\theta,\theta_1,\ldots,\theta_r)\,\Psi(\theta_1,\ldots,\theta_r) + \sum_{k=1}^{r} \Lambda_k{}^+(\theta,\theta_1,\ldots,\theta_r)\,\Psi_k(\theta,\theta_1,\ldots,\theta_r) \tag{3.22}$$

where

$$\Psi_k(\theta,\theta_1,\ldots,\theta_r) \equiv B(\theta) \prod_{1\le j\le r,\,j\ne k} B(\theta_j)\,|\,\Omega> \tag{3.23}$$

and

$$\begin{aligned} \Lambda_+(\theta,\theta_1,\ldots,\theta_r) &= a(\theta,\gamma)^N \prod_{j=1}^{r} g(\,\theta_j-\theta\,) \\ \Lambda_k{}^+(\theta,\theta_1,\ldots,\theta_r) &= h(\theta_k-\theta)\ \ a(\theta_k,\gamma)^N \prod_{1\le \ell \le r,\,\ell\ne k} g(\theta_\ell-\theta_k) \end{aligned} \tag{3.24}$$

One analogously finds

$$D(\theta)\,\Psi(\theta_1,\ldots,\theta_r) = \Lambda_-(\theta,\theta_1,\ldots,\theta_r)\,\Psi(\theta_1,\ldots,\theta_r) + \sum_{k=1}^{r} \Lambda_k{}^-(\theta,\theta_1,\ldots,\theta_r)\,\Psi_k(\theta,\theta_1,\ldots,\theta_r) \tag{3.25}$$

where

$$\Lambda_-(\theta,\theta_1,\ldots,\theta_r) = b(\theta)^N \prod_{j=1}^{r} g(\theta-\theta_j)$$

$$\Lambda_k^-(\theta,\theta_1,\ldots,\theta_r) = h(\theta-\theta_k)\ b(\theta_k)^N \prod_{1\le \ell \le r,\ell\neq k} g(\theta_k-\theta_\ell) \qquad (3.26)$$

Now, in order to get an eigenvector of $\tau(\theta)$ we must require

$$\Lambda_k^+(\theta,\theta_1,\ldots,\theta_r) + \Lambda_k^-(\theta,\theta_1,\ldots,\theta_r) = 0 \qquad (3.27)$$

This yields a set of r algebraic equation in θ_k $(1 \le k \le r)$ usually called Bethe Ansatz equations (BAE). Notice that the dependence on θ drops in eq.(3.27) thanks to the fact that $h(\theta)$ is an odd function. This could be expected since the commutativity of $\tau(\theta)$ for different θ suggest that its eigenvectors can be chosen θ-independent. More explicitely these BAE read,

$$[\,\mathrm{sh}(\lambda_j + i\gamma/2)\,/\,\mathrm{sh}(\lambda_j - i\gamma/2)\,]^N =$$

$$= -\prod_{k=1}^{r} [\mathrm{sh}(\lambda_j-\lambda_k+i\gamma)\,/\,\mathrm{sh}(\lambda_j-\lambda_k-i\gamma)] \qquad \text{regime II} \qquad (3.28)$$

$$[\,\sin(\lambda_j + i\gamma/2)\,/\,\sin(\lambda_j - i\gamma/2)\,]^N =$$

$$= -\prod_{k=1}^{r} [\sin(\lambda_j-\lambda_k+i\gamma)\,/\,\sin(\lambda_j-\lambda_k-i\gamma)] \qquad \text{regime I and III} \qquad (3.29)$$

Here we have introduced $\lambda_j \equiv i(\theta_j + \gamma/2)$, for regime III and $\lambda_j \equiv -i(\theta_j - \gamma/2)$ for regimes I and II, $1 \le j \le r$. Once the λ_j are found by solving eqs. (3.28) or (3.29) the eigenvalues $\Lambda(\theta)$ of $\tau(\theta)$ follow from eqs.(3.24) and (3.26) as

$$\Lambda(\theta,\theta_1,\ldots,\theta_r) = \Lambda_+(\theta,\theta_1,\ldots,\theta_r) + \Lambda_-(\theta,\theta_1,\ldots,\theta_r) \qquad (3.30)$$

We can assume $|\mathrm{Re}\,\lambda_j| \leqslant \pi/2$ in regimes I and III, whereas $-\infty < \mathrm{Re}\,\lambda_j < +\infty$ for regime II. The r.h.s. of eq. (3.30) would seem to have poles at $\theta = +i\lambda_j + \gamma/2$. However the corresponding residues identically vanish due to eqs. (3.28)-(3.29). Actually one can use this property as a short-cut to derive the BAE when the construction of the explicit eigenvectors is more involved.

Eq. (1.18) gives for the six-vertex model symmetry (3.2) (rotations around z)

$$[\,A(\theta)\,,S_z\,] = [\,D(\theta)\,,S_z\,] = 0$$

$$[\,S_z\,,B(\theta)\,] = -B(\theta) \quad , \quad [\,S_z\,,C(\theta)\,] = C(\theta) \qquad (3.31)$$

where $S_z = \frac{1}{2}\sum_{a=1}^{N} \sigma_z^a$ acts in the vertical space. Therefore $B(\theta)$ $[C(\theta)]$

lowers [raises] the z-component of the spin in one unit.

In particular we find that the state (3.19) is an eigenvectors of S_z

$$S_z \Psi(\theta_1,\ldots,\theta_r) = (N/2 - r) \Psi(\theta_1,\ldots,\theta_r) \tag{3.32}$$

This algebraic BA construction of the eigenvectors of $\tau(\theta)$ has some analogies with the angular momentum states obtained from the maximal weight state applying lowering operators S_-. Here $|\Omega>$ is also a maximal weight vector for the total spin and $B(\theta)$ lowers the spin by one unit [eq.(3.31)]. Nevertheless, to obtain exact eigenvectors of $\tau(\theta)$ the arguments $\theta_1,\ldots,\theta_r$ must fulfil the BAE (3.28)-(3.29). This notion is absent in Lie algebra constructions of eigenvectors.

The resolution of the BAE (3.28)-(3.29) is possible analytically for small r and N. For large r and N this is a formidable task. Only in the $N = \infty$, $r = \infty$ limit, things become easier since the roots become closer and closer for large N. The separation is O(1/N) between neighboring λ_j in the real axis. The density of roots is defined as

$$\rho(\lambda_j) = \lim_{N\to\infty} 1/[N (\lambda_{j+1} - \lambda_j)] \tag{3.33}$$

This function obeys a linear integral equation that follows from eqs.(3.28)-(3.29). Since it has a difference kernel, it can be solved in closed form by Fourier integrals or Fourier series. Once $\rho(\lambda)$ is known, the calculation of physical quantities like $\Lambda(\theta)$ (eigenvalue of $\tau(\theta)$) reduce to quadratures. That is

$$\lim_{N\to\infty} (1/N)\Lambda(\theta) = \int d\lambda \; \phi(\lambda+i\theta,\gamma/2) \, \rho(\lambda) \tag{3.34}$$

where

$$\phi(\lambda,\alpha) = i \operatorname{Log} \{ \sin(\lambda + i\alpha) / \sin(\lambda - i \alpha) \} \quad \text{, regime I}$$
$$\phi(\lambda,\alpha) = i \operatorname{Log} \{ \sinh(\lambda + i\alpha) / \sinh(\lambda - i \alpha) \} \quad \text{, regime II} \tag{3.35}$$

Besides real roots the BAE posses complex roots (see, for example refs.[26]). There exist an important literature on numerical solutions of the BAE[27]. Moreover, it is possible to solve the BAE asymptotically for N large but finite using the method proposed in ref.[28] , extended and generalized in refs.[29-33]. The finite size corrections turn out to be exponiantially small in N for regime I and power like in the gapless regime II .

In the QFT associated to vertex models, the vacuum (ground state) corresponds precisely to the antiferroelectric ground state. Let us concentrate on this state and excitations around it from now on. The operators $B(\theta_j)$ play here the rôle of creation operators of excitations over the bare vacuum $|\Omega\rangle$. That is pseudo-particles or "bare" particles. The antiferroelectric ground state is the analog of the filled Dirac sea for free fermions. However, the pseudoparticles are here not free, they interact through two-body interactions. The functions $\phi(\lambda_i - \lambda_j, \gamma)$ describe the two-body phase-shift associated to such interactions. These functions are equal to π when $\gamma = \pi/2$ in regime II : the free fermion (or Ising) case.

The BAE (3.28)-(3.29) can be rewritten as

$$\exp i[\, N\, \phi(\lambda_j, \gamma/2) - \sum_{k\neq j} \phi(\lambda_j - \lambda_k, \gamma\,)\,] = 1 \qquad (3.36)$$

The first term in the exponent, $N\phi(\lambda_j, \gamma/2)$, is just the momentum of the jth pseudoparticle times the number of sites. That is the phase for a free particle moving around this ring of length N. The second term can be interpreted as the phase shifts induced in the wave function of the jth pseudoparticle by the (pair) interaction with the rest of them. In other words eq. (3.29) ensures the periodicity of the "wave function" when turning aroung the ring. This interpretation of the BAE extends for more general models[4].

This concludes our exposition of the Bethe Ansatz solution of the six-vertex model. A more general Bethe Ansatz construction provides the eigenvectors of the eight-vertex model[17,34]. It has been shown recently that these eight-vertex eigenvectors become the six-vertex eigenvectors in the limit where the eight-vertex weights become those of the six-vertex model[35].

The Bethe Ansatz has been also generalized for multi-state vertex models. That is when dim$\mathcal{A}$ and/or dim $\mathcal{V}$ is larger than two. The resulting construction is a set of nested Bethe Ansatz. It is reviewed in ref.[4] .

IV. THE LIGHT-CONE LATTICE APPROACH

This approach starts by discretizing the two-dimensional Minkowski space-time in light-cone coordinates $x_\pm = x \pm t$. Space time is thus approximated by a diagonal lattice. This discretization scheme turns to be an useful regularization method for quantum field theories (integrable or not) since they become naturally connected with vertex models in their scaling limit[43-44].

The sites in the light-cone lattice (fig. 11) are considered as world

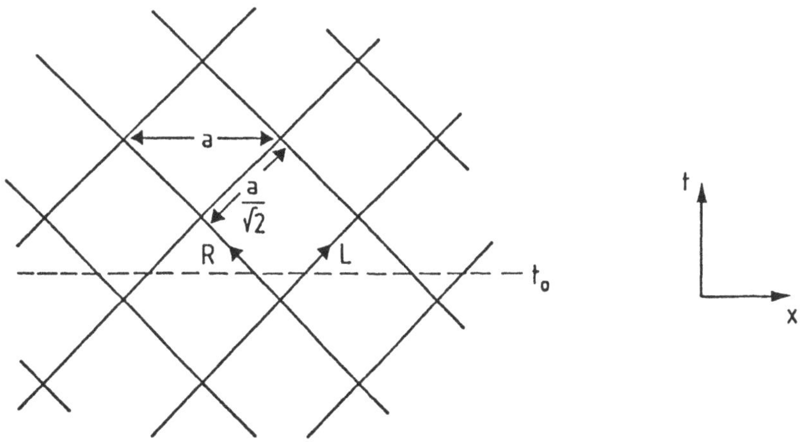

Fig.11 Discretized Minkowski space-time. Sites are world events joined by world lines of the bare particle propagation.

events. Each site (event) is joined by light-like links to its four nearest neighbours along x_+ and x_- . There diagonal links are possible world lines for the propagation upwards in time of "bare" massless particles. Particles on right-oriented (R) and left-oriented (L) links are called respectively right and left-movers.

One then associates microscopic amplitudes to each site (world event) where two oppositely oriented world lines cross. These amplitudes describe the different processes that can take place, and must verify general invariance properties like unitarity.

Let us start for the simplest case where each link describes only two different configurations. The general case of any number of states per link is considerezd later on. We assume that these two cases correspond to the presence or absence of a bare fermion without internal degrees of freedom. In general, there can be 16 different amplitudes per site corresponding to the 16 configurations (occupied/empty) of the four links joining there. Only U(1) invariant microscopic amplitudes will be considered here such that the number of particles is conserved at each site. U(1) transformations act on the link states by

$$\begin{aligned} |0\rangle &\rightarrow e^{i\lambda}\,|0\rangle \\ |1\rangle &\rightarrow e^{-i\lambda}\,|1\rangle \end{aligned} \qquad (4.1)$$

where $|0\rangle$ = (empty) and $|1\rangle$ = (occupied). Therefore, there are only six non-zero amplitudes as depicted in fig. 12. The correspondence with the general (non-symmetric) six-vertex model is evident.

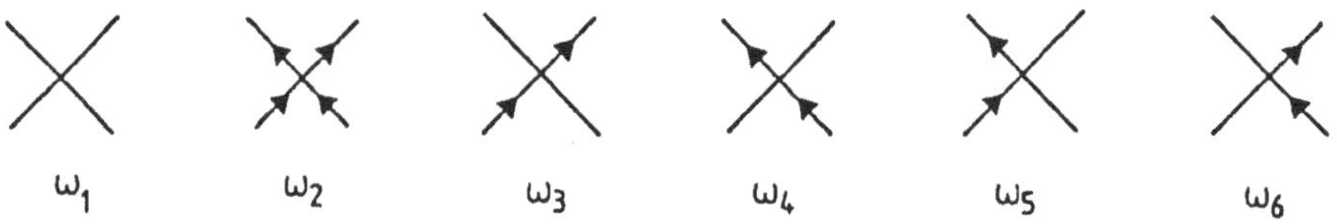

$| \equiv |0\rangle \equiv |\text{empty}\rangle$; $\quad$ ↑ $\equiv |1\rangle \equiv |\text{occupied}\rangle$

Fig.12 The six non-zero microscopic transition amplitudes. They coincide with the weights of fig.10.

Of course, space-time translational invariance implies that the amplitudes are the same in all sites of the lattice. It is natural (and causes no loss of generality) to set the nothing-to-nothing amplitude to be 1. Unitarity then requires

$$\Omega\,\Omega^\dagger = 1\ , \qquad \Omega = \begin{pmatrix} \omega_3 & \omega_5 \\ \omega_6 & \omega_4 \end{pmatrix}\ , \quad |\omega_2|^2 = 1 \qquad (4.2)$$

While ω_3 and ω_4 are naturally interpreted as amplitudes for free propagation (being therefore related to kinetic energies in the continuum limit), ω_5 and ω_6 play the role of mass terms since they couple right and left movers.

Symmetry under parity transformation holds if

$$\omega_3 = \omega_4 = b \quad , \quad \omega_5 = \omega_6 = c \tag{4.3}$$

This corresponds now to an integrable six vertex model. Unitarity now reads

$$|b|^2 + |c|^2 = 1 \quad , \quad b\bar{c} + \bar{b}c = 0 \tag{4.4}$$

One can organize these microscopic amplitudes at a site into a 4x4 unitarity "bare" S-matrix

$R^{ab}{}_{cd}(\theta) =$ 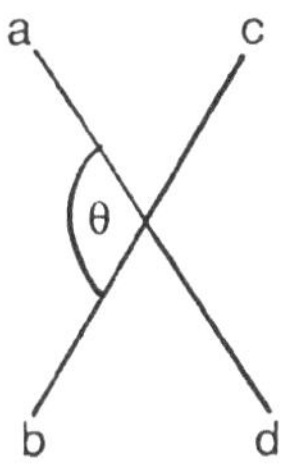

$$= \begin{pmatrix} 1 & 0 & 0 & 0 \\ 0 & c & b & 0 \\ 0 & b & c & 0 \\ 0 & 0 & 0 & \omega \end{pmatrix} \tag{4.5}$$

where $\omega \equiv \omega_2$ and $\alpha, \beta, \alpha', \beta'$ take the values 0 or 1 for empty or occupied links like in eq. (1.1).

The amplitude for a global process, from a given state at $t = t_0$ to another given state at a later time, is obtained by summing over the amplitudes of all allowed vertex configurations compatible with initial and final conditions and with boundary conditions. Each of these is given by a product of microscopic amplitudes ω_i. It clearly corresponds to the sum over all possible paths of an arbitrary, but constant in time number of particles. At any instant, a particle can move to the left or to the right at the speed of light. We are thus dealing with a discretization of Feynman path integral for fermions.

It is convenient to parametrize b and c following the constraints (4.4) as

$$b = b(\theta,\gamma) = \mathrm{sh}\theta / \mathrm{sh}(\theta - i\gamma) \quad , \quad c = c(\theta,\gamma) = \mathrm{sh}\gamma / \mathrm{sh}(\theta - i\gamma)$$
$$0 < \theta < \infty \quad , \quad 0 < \gamma < \pi \tag{4.6}$$

This makes (4.5) identical the six-vertex model R matrix (3.1) up to an overall factors $\mathrm{sh}(\theta - i\gamma)$ and a redefinition of $\theta \rightarrow i\theta$ when $\omega = 1$. Actually $\omega \neq 1$ corresponds to a six-vertex model in an external field.

Let us now describe the operator formalism for the light-cone approach[43]. The unit evolution operators in the light-cone direction (R or L) are given by simply juxtaposition of the microscopic S-matrices (4.5) at the same horizontal level. That is

```
      1 2   3 4   5 6   7 8               2N-1 2N
U_L =  X     X     X     X  ..................  X        (4.7)
     2N 1   2 3   4 5   6 7               2N-2 2N-1
```

$$U_R = \overset{1\;\;2}{\underset{2\;\;3}{X}}\;\overset{3\;\;4}{\underset{4\;\;5}{X}}\;\overset{5\;\;6}{\underset{6\;\;7}{X}}\;\overset{7\;\;8}{\underset{8\;\;9}{X}}\;\ldots\ldots\ldots\ldots\ldots\;\overset{2N\text{-}1\;\;2N}{\underset{2N\;\;1}{X}} \tag{4.8}$$

where the numbers 1, 2, 3,...,2N label the sites. Here time evolves upward, according to rule d) of sec. I. Here N is assumed to be even and $\alpha_{j+N} \equiv \alpha_j$. Notice that there are no summations in eq. (4.7)-(4.8). One can now define the two light-cone lattice evolution generators as

$$H \pm P = (2i/a) \operatorname{Log} U_{R_L}(\theta) \quad . \tag{4.9}$$

where H and P stand for lattice hamiltonian and momentum and a is the latttice spacing.

Eq. (4.9) is extremely suggestive since it provides a lattice version of field-theoretic H and P in terms of lattice vertex transfer matrices U_R and U_L. The natural question is now to find the eigenvectors of them. It will be shown now that this is possible using the techniques of sec. III and IV (and their generalizations) provided $R(\theta)$ verifies the YB algebra (1.12)[43].

Let us consider the row-to-row transfer matrix $\tau^{[N]}(\theta,\alpha)$ of eq. (1.23) [with $g_1 = \ldots = g_N = 1$] with the particular choice of inhomogeneities (recall N = even)

$$\mu_j = (-1)^{j+1}\,\theta$$
$$\bar{\theta} = (+\theta, -\theta, + \ldots\ldots\ldots, +\theta, -\theta) \tag{4.10}$$

It then follow from eqs.(1.23) and (1.35) using eq. (1.13) that

$$\tau(\theta,\bar{\theta}) = U_L(\theta) \tag{4.11}$$
$$\tau(-\theta,\bar{\theta}) = U_R(\theta)^\dagger \tag{4.12}$$

Let us check (4.12). Setting (4.10) in (1.13) and (1.23) yields

$$\begin{aligned}\tau(\theta,\bar{\theta})_{\bar{\alpha}|\bar{\sigma}} &= \sum_{\gamma_1,\ldots,\gamma_N} \delta^{\alpha_1}{}_{\gamma_1}\,\delta^{\gamma_1}{}_{\sigma_1}\,R(\theta)^{\alpha_2\gamma_3}{}_{\gamma_2\sigma_2}\ldots\ldots\delta^{\alpha_{N-1}}{}_{\gamma_{N-1}}\,\delta^{\gamma_N}{}_{\sigma_{N-1}} \\ R(\theta)^{\alpha_N\gamma_1}{}_{\gamma_N\sigma_N} &= \prod_{j=1}^{N/2} R(\theta)^{\alpha_{2j},\alpha_{2j+1}}{}_{\sigma_{2j-1},\sigma_{2j}} = U_L(\theta)_{\bar{\alpha}|\bar{\sigma}}\end{aligned} \tag{4.13}$$

after using eq.(1.35) repeatedly.

The key relations (4.9), (4.11)-(4.12) connect the lattice H and P with the row-to-row transfer matrices whose eigenvectors and eigenvalues can be constructed by the algebraic Bethe Ansatz developed in sec. III and IV. The light-cone or diagonal-to-diagonal transfer matrices result to be particular cases of the inhomogeneous row-to-row transfer matrices. The commutativity property (1.9) gives in addition :

$$[\,\tau(\lambda,\bar{\theta})\,,\,U_L(\theta)\,] = 0$$
$$[\,\tau(\lambda,\bar{\theta})\,,\,U_R(\theta)^\dagger\,] = 0\;,\;[\,U_L(\theta)\,,\,U_R(\theta)^\dagger\,] = 0 \tag{4.14}$$

One can consider the infinite sequence of commuting operators $(0 \leqslant K < \infty)$

$$c_K = \partial^K/\partial\lambda^K \; \mathrm{Log}\; \tau(\lambda,\bar{\theta})\big|_{\lambda=\theta} \tag{4.15}$$

They all commute with $U_L(\theta)$, $U_R^+(\theta)$ and with each other.

Let us now consider the continuum limit $(a \to 0)$ of the lattice models through eq. (4.9). The antiferroelectric ground state (regimes I and II) of $\tau(\lambda, \theta)$ corresponds just to the physical vacuum (filled Dirac sea) of the QFT defined by H and P. The bare vaccuum is the ferromagnetic state $|\Omega>$. The physical vaccuum follows here by BA by filling the bare one [eq.(3.19)] with **interacting** pseudoparticles (their phase shift is given by $\phi(\lambda_i-\lambda_j,\gamma)$) and not free particles as for the free Dirac field. Only when $\gamma = \pi/2$ the phase shift is a constant (π).

The particle states follow from the lowest excitations. Since a factor a^{-1} appears in $H \pm P$ [see eq. (4.9)] only gapless models yield finite energy states in the scaling limit. Moreover, in order to compute the energy and momentum in the scaling limit it is enough to know the eigenvalues of $\tau(\pm\theta,\bar{\theta})$ close to the bottom of the spectrum. The low-lying excitations are associated to holes and complex solutions with large (real) rapidity. Moreover, their eigenvalues normalized to the vacuum ones [as in eq. (3.55)] are independent of the inhomogeneities

Let us start by the fermion model (with $\omega = 1$) associated to the six-vertex models (eqs. (4.1)-(4.6)). We need the eigenvalues $\Lambda(\theta)$ of the transfer matrix $\tau(\theta)$. The solution of the six vertex BAE derived in sec.III yields for the excited states in the thermodynamic limit

$$\lim_{N\to\infty} \Lambda(\theta,\phi) / \Lambda_0(\theta) = \exp[-ig(\theta,\phi)] \tag{4.16}$$

where $g(\theta,\phi)$ reads for hole excitations

$$g(\theta,\phi) = 2\,\mathrm{arctg}[\exp\pi(\phi + i\theta)/\gamma] \tag{4.17}$$

and ϕ is the hole position. Combining eqs.(5.16)-(5.17) with eqs.(4.9)-(4.12) yields

$$\epsilon \pm p = -(1/a)\; g(-iv) \qquad , \; v \in \mathcal{R} \tag{4.18}$$

Let us start by a hole excitation. We find for large v from eq. (3.55) and (3.61) of ref.[4]

$$\underset{v\to+\infty}{g(\pm iv)} = \pm 2\exp[\pm\pi\theta_h/\gamma - \pi v/\gamma] + O(\exp[-2\pi v/\gamma]) \tag{4.19}$$

after discarding an irrelevant π (It does not contribute to the eigenvalue of $\tau(\theta)$ since the holes appear always by pairs). We find a relativistic spectrum provided $v \to +\infty$ when $a \to 0$ keeping fixed the renormalized mass

$$m = (4/a)\exp[-\pi v/\gamma] \tag{4.20}$$

The dispersion law results

$$\epsilon = m\cosh(\pi\theta_h/\gamma)$$
$$p = m\sinh(\pi\theta_h/\gamma) \tag{4.21}$$

So, $\pi\theta_h/\gamma$ is the physical rapidity of the particle [see eq. (2.14)]. The particles in eq.(4.21) correspond to the fermion or antifermion in the massive Thirring model or the sine-Gordon solitons. Besides these holes one finds complex solutions of the BAE disposed as strings (that is, with imaginary parts equally spaced by γ or $\pi-\gamma$). They provide relativistic particles in the **same** scaling limit (4.20) with masses

$$m_n = 2m \sin\{ \tfrac{1}{2} n\pi(\pi/\gamma - 1) \} \quad , \quad 1 \le n < [\![\pi/(\pi - \gamma)]\!] \qquad (4.22)$$

where $[\![x]\!]$ stands for integer part of x. Eq.(4.22) follows from eqs. (4.18) and (3.61) of ref.[4]. This set of particles are fermion-antifermion bound states. They relate semiclassically to the breathers of sine-Gordon as the fermions or holes relate to the sine-Gordon solitons. The S-matrix describing the two-body interaction of all these particles follows by direct calculation from the BAE using standard methods[49].

The preceding exposition of the light-cone lattice method applies to all gapless vertex models. In ref.[43] the models with rational R-matrices associated to simple Lie algebras are analysed. The $q(2q-1)$ vertex model associated to the deformed A_{q-1} algebra is also considered in its gapless regime[4,43].

Within this light-cone approach it is possible to construct explicitly the canonical bare fields on the lattice and to show that in the scaling limit (4.20) the massive Thirring model emerges[43].

Eq.(4.20) defines the renormalized scaling limit yielding rigorously all physical quantities. There is in addition the **bare** scaling limit giving the bare quantum fields in the continuum[43].

We use the word "rigorous" since we solve in this approach a lattice model exactly, then we take the infinite volume limit and finally the $a \to 0$ (scaling) limit. In other words, here one solves (exactly) a model with both UV and volume cutoffs and then lets the cutoffs to infinity in a precise way. This is clearly much better than the coordinate Bethe-Ansatz (CBA) where the UV cutoff is introduced after the obtention of the solution. For the MTM and the chiral Gross-Neveu model the results of the CBA coincide with the light-cone approach for on-shell magnitudes. Hence the CBA works well in these cases. This is not the case for the multiflavor Chiral fermion model treated in ref.[45] by CBA. As it is shown in ref.[43] the results of ref.[45] are not correct.

Starting from richer vertex models than the six-vertex a large set of QFTs arises[43]. Let us first summarize the integrable vertex models classification in terms of simple Lie Algebras. There exist an inequivalent vertex model for each representation space of each simple Lie algebra. In the rational case these models are invariant unde the corresponding Lie group in the sense of eq.(1.17). In the trigonometric/hyperbolic case the invariance under the Cartan algebra survives and for the rest we have the invariance under the quantum group in the sense of eq.(5.28).

In these models, the links can be in any of q different local states where q is the dimension of the respective representation space.As an illustration let us take the model associated to the A_{q-1} Lie algebra in the fundamental representation. The R-matrix reads for this case[4,46]:

$$R^{ab}_{ij}(\theta) = \delta_{ia}\,\delta_{jb}\,\mathrm{sh}\gamma\,\exp[\,\theta\,\mathrm{sign}(a-b)] + \mathrm{sh}\theta\,\delta_{ib}\,\delta_{ja} \qquad \text{for} \quad i\neq j,$$
$$R^{aa}_{ii}(\theta) = \delta_{ia}\,\mathrm{sh}(\,\gamma \pm \theta\,) \quad , \quad \pm \text{ for regime I / III.} \qquad (4.23)$$

Here $1 \le i, j, a, b, \le q$ where q can take any value ≥ 2 . In the trigonometric regime (regime II), some weights are complex. The R-matrix reads:

$$R^{ab}_{ij}(\theta) = \delta_{ia}\,\delta_{jb}\,\sin\gamma\,\exp[\,i\theta\,\mathrm{sign}(a-b)] + \sin\theta\,\delta_{ib}\,\delta_{ja} \qquad \text{for} \quad i\neq j,$$
$$R^{aa}_{ii}(\theta) = \delta_{ia}\,\sin(\,\gamma + \theta\,) \quad , \qquad (4.24)$$

This is a regular R-matrix, since

$$R^{ab}_{ij}(0) = \delta_{ia}\,\delta_{jb}\,\mathrm{sh}\gamma \quad (\text{or} \quad \delta_{ia}\,\delta_{jb}\,\sin\gamma\,) \qquad (4.25)$$

The YB algebra defined by the R-matrix (4.34) is invariant under the group

$$\mathcal{G} = [U(1)]^{q-1} \otimes Z_q \qquad (4.26)$$

The cyclic group Z_q is generated by powers of the matrix h defined as

$$h_{ab} = \delta_{a+1,b} \quad \text{with} \quad \delta_{a+q,b} = \delta_{a,b+q} = \delta_{a,b} \qquad (4.27)$$

We have $h^q = 1$.

The R-matrix (4.23) enjoys PT invariance but not P or T separately :

$$R^{ab}_{ij}(\theta) = R^{ij}_{ab}(\theta) \quad , \quad R^{ab}_{ij}(\theta) \neq R^{ba}_{ji}(\theta) \quad , \quad R^{ab}_{ij}(\theta) \neq R^{ji}_{ba}(\theta) \qquad (4.28)$$

There are q(2q-1) non-vanishing weights in the vertex model defined by this R-matrix. Eq. (4.28) tells that the model is invariant under a 180° rotation of the whole lattice. For $q = 2$ we recover the six-vertex model.

In this model the links can be in q different states. In regime III, for $\gamma > 1$, $2\theta + \gamma$ fixed, the model has a long-range generalised ferroelectric order and the dominant configurations are formed mostly by vertices of type R^{a1}_{a1} (for $\theta > 0$) and some $R^{a,q-1}_{a,q-1}$. There are q different predominant patterns following one from each other by shifting by one the state of all links. This generalises the six vertex ($q = 2$) situation [1]. Regimes I and III map into each other through $\gamma \to -\gamma + i\pi$.

The exact solution of this q(2q-1) vertex model can be found in ref.[4,18]. That is the eigenvectors and eigenvalues of the transfer matrix $\tau^{[N]}(\theta, \alpha)$ associated to the R-matrix (4.23)-(4.24).

The structure of the nested Bethe Ansatz (NBA) that solves each of these vertex models looks like the one of their respective Dynkin diagram. It must be noticed that a proof that these BAE lead to the eigenvectors and eigenvalues of the transfer matrix has been explicited only for a subset of models : those associated to U(N)[4,18], Sp(2N)[19], and SO(2N) [20] and some others. However, these statements are extremely likely to hold for all semisimple Lie algebras.

Let us describe the BAE for the trigonometric models. The derivation of these equations (for a subset of Lie algebras) is in refs.[19,20]. The eigenvalues of the transfer matrix can be written as a sum of terms. The dominant one in the infinite volume limit ($N \to \infty$) is

$$\lambda_\omega(\theta,\underline{\lambda}^{(j)}) =$$

$$= \prod_{a=1}^{N}\prod_{k=1}^{r}\prod_{j_k=1}^{p_k} \text{sh}[\, i(\theta-\theta_a) + \lambda^{(k)}_{j_k} - i\,\gamma\,(\omega_a,\alpha_k)\,] / \text{sh}[\, i(\theta-\theta_a) + \lambda^{(k)}_{j_k} + i\,\gamma\,(\omega_a,\alpha_k)\,] \tag{4.24}$$

for θ in the vicinity of $\theta = 0$, $|\theta| < \theta_0$. Here $\theta_1, \theta_2, \ldots\ldots, \theta_N$ are given numbers describing inhomogeneities of the lattice as discussed in sec.II [1,4,5] and $\underline{\lambda}^{(j)} = \{\lambda_{j_k}{}^{(k)}, 1 \leqslant j_k \leqslant p_j,\ 1 \leqslant k \leqslant r\}$. The ω_a are fundamental weights and α_k are the simple roots of G whose rank is r. (α, β) stands for the usual inner product in root space. The $\lambda_{j_k}{}^{(k)}$ $(1 \leqslant j_k \leqslant p_j,\ 1 \leqslant k \leqslant r)$ are solutions of the nested BAE (NBAE) :

$$\prod_{a=1}^{N} \text{sh}[\, \lambda^{(k)}_{j_k} - i\theta_a - i\,\gamma\,(\omega_a,\alpha_k)] / \text{sh}[\, \lambda^{(k)}_{j_k} - i\theta_a + i\,\gamma\,(\omega_a,\alpha_k)] =$$

$$= -\prod_{i=1}^{r}\prod_{j_i=1}^{p_i} \text{sh}[\, \lambda^{(k)}_{j_k} - \lambda^{(i)}_{j_i} - i\,\gamma\,(\alpha_k,\alpha_i)\,] / \text{sh}[\, \lambda^{(k)}_{j_k} - \lambda^{(i)}_{j_i} + i\,\gamma\,(\alpha_k,\alpha_i)] \tag{4.25}$$

$$1 \le j_k \le p_k \quad , \quad 1 \le k \le r \ .$$

Here the upper indices $^{(i)}$ label the steps in the NBA. Each step is associated to a simple root α_i. The structure of eq.(4.25) coincides with the respective Dynkin diagram: when two roots, say α_ℓ and α_i, are orthogonal, their associated parameters $\lambda_{j_l}{}^{(\ell)}$ and $\lambda_{j_i}{}^{(i)}$ $(1 \leqslant j_\ell \leqslant p_\ell,\ 1 \leqslant j_i \leqslant p_i)$ are not directly coupled through (4.25) since $(\alpha_i, \alpha_\ell) = 0$. It must be noticed that due to the orthogonality of fundamental weights and simple roots [47]

$$(\omega_a,\alpha_k) = \tfrac{1}{2}\,\delta_{ak}\ (\alpha_k,\alpha_k) \tag{4.26}$$

The normalization of the simple roots can be absorbed as a multiplicative factor on the $\lambda_{j_i}{}^{(i)}$

We refer to the original references and to the reviews [4] for the resolution of the NBAE. Let us just say that there is a ferroelectric regime (regime III) where the ground state (eigenvalue of $\tau(\theta)$ with maximal modulus) is a maximal weight vector. In regimes I and II the ground state is antiferroelectric with non-zero and zero gap respectively. There exists r branches of excitations formed by holes in the NBAE. In addition complex roots that can be interpreted as hole bound states appear in regime II for $0 < \gamma < \pi/2$ yielding further excitations. When non-fundamental representations or non-simply laced algebras (even in their fundamental representation) are considered the ground state turns to be formed by complex roots (strings for $N = \infty$). For example, this happens in the A_1 case for spin $S \geq 1$[48]. In order to derive a QFT in the continuum limit from these lattice models in the light-cone approach it is enough to know the excitation eigenvalues $\Lambda_k(\theta, \phi)$ of $\tau(\theta)$. We find

$$\lim_{N\to\infty} \Lambda_k(\theta,\phi) / \Lambda_0(\theta) = \exp[-i\, g_k(\theta,\phi)\,] \quad , \ 1 \le k \le r \tag{4.27}$$

where $\Lambda_0(\theta)$ is the ground state eigenvalue of $\tau(\theta)$, k labels the branch of excitation, ϕ the hole position in rapidity and

$$g_k(\theta, \phi) \underset{i\theta \to -\infty}{=} (m_k / \pi\gamma) \exp[\pm \kappa(\phi + i\theta)/\gamma] \{ 1 + o(\exp(-|\theta\delta|)) \} \tag{4.28}$$

where $\delta > 0$. The parameters κ and m_ℓ are given in Table II. κ is just 2π times the length squared of the shortest simple root in the normalization where [47]

TABLE II

Integrable QFT associated to trigonometric Yang-Baxter algebras in the light-cone approach. We indicate the underlying Lie Algebra $\mathfrak{G}$, the respective scale parameter κ (it coincides with the one-loop beta function) and the corresponding mass spectrum.

Lie Algebra	κ	m_k
A_n	$2\pi/(n+1)$	$\sin(\pi k/[n+1])$, $1 \le k \le n$
B_n	$\pi/(2n-1)$	$\sin(\pi k/[2n-1])$, $1 \le k \le n-1$, $m_n = 1/2$
C_n	$\pi/(n+1)$	$\sin(\pi k/2(n+1))$, $1 \le k \le n$
D_n	$\pi/(n-1)$	$\sin(\pi k/2(n-1)$, $1 \le k \le n-2$ $m_\pm = 1/2$
E_6	$\pi/6$	$m_1 = m_5 = m_6/2 = \sqrt{3}/2$ $m_2 = m_4 = (3+\sqrt{3})/2$ $m_3 = (3+\sqrt{3})/\sqrt{2}$
E_7	$\pi/9$	$m_2 = 2m_1 \cos\pi/18$, $m_3 = m_1/(2\sin\pi/18)$ $m_4 = 2m_2 \cos\pi/9$, $m_5 = 2m_2 \sin 2\pi/9$, $m_6 = 2m_1 \sin 2\pi/9$, $m_7 = 2m_1 \cos\pi/9$.
E_8	$\pi/15$	$m_2 = 2m_1 \cos\pi/5$, $m_3 = 2m_1 \cos\pi/30$, $m_4 = 2m_2 \cos 7\pi/30$, $m_5 = 2m_2 \cos 2\pi/15$, $m_6 = m_2 m_3/m_1$, $m_7 = m_2 m_4/m_1$, $m_8 = m_2 m_5/m_1$.
F_4	$\pi/9$	$m_4 = 2m_1 \cos 2\pi/9$, $m_3 = m_1/(2\sin\pi/18)$, $m_3 = 2m_1 \cos\pi/18$.
G_2	$\pi/6$	$m_2 = m_1 [(\sqrt{6}+\sqrt{2})/2]$.

$$B(E_\alpha , E_{-\alpha}) = 1 ,$$

and $B(x , y)$ is the Killing form.

Light-cone evolution operators can be defined through eqs.(4.7)-(4.9) for any R-matrix. Let us see that a relativistic dispersion law arises from any excitation spectrum as given by eq.(4.27). Let us call $E_\ell (\varphi)$ and $p_\ell (\varphi)$ the eigenvalues of H and P , respectively. Eqs. (4.9) and (4.27) yield

$$
\begin{aligned}
E_\ell(\theta_h) &\underset{i\theta \to \infty}{=} [\exp(-i\kappa\theta/\gamma)/(\pi a\gamma)]\, m_\ell \cosh(\kappa\theta_h/\gamma) + O(\exp(-2i\kappa\theta/\gamma)) \\
p_\ell(\theta_h) &\underset{i\theta \to \infty}{=} [\exp(-i\kappa\theta/\gamma)/(\pi a\gamma)]\, m_\ell \sinh(\kappa\theta_h/\gamma) + O(\exp(-2i\kappa\theta/\gamma))
\end{aligned}
\tag{4.29}
$$

It is then natural to define the scaling limit according to

$$a \to 0 , \quad i\theta \to \infty , \mu = \exp(-i\kappa\theta/\gamma)/(\pi a\gamma) = \text{fixed} \tag{4.30}$$

is the renormalised or physical mass scale and the particle mass spectrum of these integrable QFTs is given by

$$M_\ell = \mu \, m_\ell \tag{4.31}$$

We recognize in eq.(4.29) $\kappa\,\theta_h/\gamma$ as the physical particle rapidity.

This is a very general way of constructing integrable QFTs. The operators H and P given by eq. (4.9) are well defined on the lattice as well as all the higher conserved charges. In the continuum limit $a \to 0$, they provide the energy and momentum of a relativistic invariant QFT, as long as the spectrum of the initial vertex model is gapless. This is usually the case for rational or trigonometric weights. In addition to the particle spectrum , the S-matrix is exactly calculable from the BAE by standard methods[49,50].

As it was the case for the MTM, the evolution operators U_R and U_L are much simpler than H and P on the lattice. This was exploited before in ref.[43] to obtain the lattice field equations for the fermionic fields of the MTM regularized by the lattice. An analogous local construction would be very interesting to obtain in the general case of a Lie algebra G. We present in ref.[43] a lattice construction for the current operators for all rational models. The H and P are always given by eqs. (4.7) and (4.9). The renormalized scaling limit (4.30) yields the mass spectrum (4.31) [see Table II] .

In refs.[43] it is shown that we obtain in this scaling limit of rational vertex models associated to $\mathfrak{G}$ the non-abelian Thirring model associated to the group $\mathfrak{G}$. This theory, also called Chiral Gross-Neveu model, has as Lagrangian,

$$\mathcal{L} = i \bar\Psi \partial\!\!\!/ \Psi - g (\bar\Psi \gamma_\mu t^\alpha \Psi) (\bar\Psi \gamma^\mu t^\beta \Psi) K_{\alpha\beta} \tag{4.32}$$

Here Ψ transforms under the irreducible representation ρ of $\mathfrak{G}$, t^α are the $\mathfrak{G}$-generators in that representation and $K_{\alpha\beta}$ is proportional to the inverse of the Killing form. Actually the H and P constructed from eqs. (4.7)-(4.9) describe the zero-chirality sector of the model (4.32).

The field theoretic models discussed up to here correspond to finite dimensional $\mathcal{V}$ and $\mathcal{A}$. Namely, a finite dimensional vector space at each link in the light-cone lattice. This is clearly appropiate for fermion or parafermion fields. Since there exists infinite dimensional representations of YB algebras, also bosonic QFTs may be described in this framework.The SU(2) principal chiral model (PCM) is defined by the lagrangian

$$\mathcal{L} = \mathrm{Tr}[(g^{-1}\partial_\mu g)^2] \qquad (4.33)$$

where $g \in SU(2)$. The lagrangian (4.33) exhibits a $SU(2) \otimes SU(2)$ invariance.

The infinite spin representation of the SU(2) invariant R-matrix (rational limit of the six-vertex model) relates to the PCM as it is investigated in ref. [51]. For arbitrary spin S, this R-matrix writes[52]

$$R_{12}(\theta) = \Gamma(2S+1+i\,\theta)\,\Gamma(\mathbf{J}+1-i\,\theta) \,/\, [\Gamma(2S+1-i\,\theta)\,\Gamma(\mathbf{J}+1+i\,\theta)] \qquad (4.34)$$

where the operator $\mathbf{J}$ is defined by

$$\mathbf{J}(\mathbf{J}+1) = 2S(S+1) + 2\,\vec{S}_1 \otimes \vec{S}_2$$

$\vec{S}_1$ and $\vec{S}_2$ are spin S operators acting on the spaces $\mathcal{A}$ and $\mathcal{V}$ respectively $[(\vec{S}_1)^2 = (\vec{S}_2)^2 = S(S+1)]$. The light-cone hamiltonian (4.7)-(4.9) provides particle states that yield all particle masses and S-matrix amplitudes for the PCM letting $S = \infty$.

The conserved currents in the PCM read

$$J_\mu{}^L = g\,\partial_\mu(g^{-1}) \qquad \text{and} \qquad J_\mu{}^R = g^{-1}\partial_\mu g$$

They transform under the left and right SU(2) groups, respectively.

One obtains left or right transforming states depending upon we identify $J_\pm{}^L$ or $J_\pm{}^R$ with the spin operators in the lattice:

$$\begin{aligned} &(1/ga\theta) \sum_{1\leq i\leq 3} \sigma^i S^i{}_{2n} \;\to\; J_+{}^L(x) \text{ or } J_+{}^R(x) \\ &(1/ga\theta) \sum_{1\leq i\leq 3} \sigma^i S^i{}_{2n-1} \;\to\; J_-{}^R(x) \text{ or } J_-{}^L(x) \end{aligned} \qquad (4.35)$$

with $x = na$ for $a \to 0$ .That is, the H obtained through eq.(4.9) is not the full hamiltonian of the PCM as it is proven in refs.(51) and (53). There is a very simple explanation for this, the physical particle states for this model transform under the group $SU(2)_L \otimes SU(2)_R$ and from the present construction only left or right operators can be obtained. Therefore all states obtained in this way are left (or right) singlets. The detailed counting of states in ref.(51) is confirmed by the simple proof of ref. (53).

This whole construction generalizes to the SU(N) PCM. It also applies for Chiral fermion models and PCM with one anisotropy axis (trigonometric YB algebras)[54].

V. BRAID GROUPS AND QUANTUM GROUPS FROM YANG-BAXTER ALGEBRAS

Let us see first how braid groups follow from trigonometric/hyperbolic YBA. In the limit $\theta \to \pm\infty$ $[\pm i\infty]$ the hyperbolic/trigonometric generators behave as $e^{\pm K\theta}(e^{+K\theta})$ times a well defined operator (K being a constant). Since such exponential factor can be absorbed in $T(\theta)$ respecting the YBE, we can in general assume that the limit

$$\lim_{\theta \to \pm\infty} T(\theta) = T_{\pm} \tag{5.1}$$

is finite and non-trivial for hyperbolic (trigonometric) YB generators. Actually a symmetry breaking transformation (2.23) may be necessary in order to find a non-trivial limit. These limiting operators can be graphically represented as follows

T_+ T_-

Fig. 13 A braid from B_n.

Letting $\theta \to \pm\infty$, $\theta' \to \pm\infty$ with $\theta - \theta' \to \pm\infty$ in the hyperbolic regime of eq. (1.1) yields

$$T_{\pm}^{(K,I)}\, T_{\pm}^{(K,J)}\, T_{\pm}^{(I,J)} = T_{\pm}^{(I,J)}\, T_{\pm}^{(K,J)}\, T_{\pm}^{(K,I)} \tag{5.2}$$

In addition eq.(1.42) tells us that T_+ and T_- are inverses of each other

$$T_+ T_- = T_- T_+ = 1 \quad , T_{\pm} = T_{\pm}^{(I,J)} \tag{5.3}$$

A factor $\sqrt{\rho(\theta)}$ has been absorbed in the definition of $T(\theta)$ as in eq. (5.1).

If one consider R matrices instead of general YB operators $T^{(I,J)}(\theta)$, eqs. (5.2)-(5.3) read

$$R_{\pm}^{23} R_{\pm}^{12} R_{\pm}^{23} = R_{\pm}^{12} R_{\pm}^{23} R_{\pm}^{12} \tag{5.4}$$

$$R_+ R_- = R_- R_+ = 1 \tag{5.5}$$

where now $V^I = V^J = V^K = \mathcal{A}$, $R_{12} = R \otimes 1$, $R_{23} = 1 \otimes R$ and $R_{\pm} = T_{\pm}^{(\mathcal{A},\mathcal{A})}$.

Lines intersect in just two ways in the ultrarelativistic limit as depicted in fig.13. (For finite θ we have a continuous family of possibilities parametrized by the angle θ). We can interpret geometrically this two possibilities as one line being over the other or viceversa. This fact naturally connects YBA in the $\theta = \pm\infty$ limit with braids, knots and links. The YB property allowing to push lines through intersections (Z-invariance, see sec. I) means now simply continuous deformation of lines without tearing them. More technically this connects with the Reidemeister moves. An important problem is to compute link and knot invariants which are intrinsic and therefore the same for topologically invariant objects. This can be achieved by computing the partition function Z for $\theta = \pm\infty$ of an integrable model taking as lattice a given knot, braid or link. Since Z is precisely Z-invariant it will be a topological invariant.

Let us now study the connection of the ultrarelativistic limit of hyperbolic/trigonometric YB algebras with braids, knots, links and quantum groups. The matrices R_+ and R_- will give a representation of a braid group in the following way. Let us consider the operators $X_i(\theta)$ acting in the tensor produict of n auxiliary spaces $\mathcal{A}$[55]

$$X_i(\theta) = \underset{1}{1} \otimes \ldots\ldots\ldots \otimes \underset{(i,\,i+1)}{R(\theta)} \otimes \ldots\ldots\ldots \otimes \underset{n}{1} \tag{5.6}$$

That is,

$$(a_1 a_2 \ldots a_n \mid X_i(\theta) \mid b_1 b_2 \ldots b_n) = R^{b_i b_{i+1}}{}_{a_i a_{i+1}}(\theta) \prod_{k \neq i, i+1}^{n} \delta_{a_k, b_k} \tag{5.7}$$

They fulfil the relations

$$[X_i(\theta), X_j(\theta')] = 0 \qquad \text{if} \quad |i-j| \geq 2, \quad \forall\, \theta, \theta'$$

$$X_i(\theta)\, X_{i+1}(\theta+\theta')\, X_i(\theta') = X_{i+1}(\theta')\, X_i(\theta+\theta')\, X_{i+1}(\theta) \tag{5.8}$$

$$X_i(\theta)\, X_i(-\theta) = 1$$

The last two equations being a consequence of the YB equation (1.11) and the unitarity eq.(1.42), respectively.

These operators are clearly of "light-cone" type. They are closely related to the light-cone evolution operators discussed in sec. IV. We find

$$U_+ = X_1\, X_3 \ldots X_{N-1}$$
$$U_- = X_2\, X_4 \ldots X_N$$

where[43] (see eqs.(4.7)-(4.8))

$$U_R = V\, U_-$$
$$U_L = U_+\, V^\dagger$$

Here $V\ (V^\dagger)$ is the shift operator affecting one-half translation to the right (to the left). There are two different limits for hyperbolic (trigonometric) YB algebras : $\theta \to \pm\infty$ ($\pm\, i\,\infty$) . Upon adequately normalizing $R(\theta)$ and eventually applying a symmetry breaking transformation (2.23) in order to have a finite and non-trivial limit, we find

$$b_i = \lim_{\theta\to\infty} X_i(\theta) \qquad , \quad b_i^{-1} = \lim_{\theta\to-\infty} X_i(\theta) \tag{5.9}$$

and analogous relations for trigonometric YBA. The operators b_i $(1 \leq i \leq n)$ fulfil:

$$b_i\, b_j = b_j\, b_i \qquad \text{when} \quad |i-j| \geq 2\ ,$$
$$b_i\, b_{i+1}\, b_i = b_{i+1}\, b_i\, b_{i+1} \tag{5.10}$$

Let us briefly recall the notion of a braid group[56]. Braids are formed when n points in a straight line are connected by n lines with othe n points on a paralell line as shown in fig.14. When the lines connecting the points have no intersections, the braid is called trivial . A general n-braid is obtained from the trivial one applying succesively the operations b_i and / or the inverses b_i^{-1} ($1 \leq i \leq n-1$). The operations b_i and b_{i-1} are depicted in fig.15.

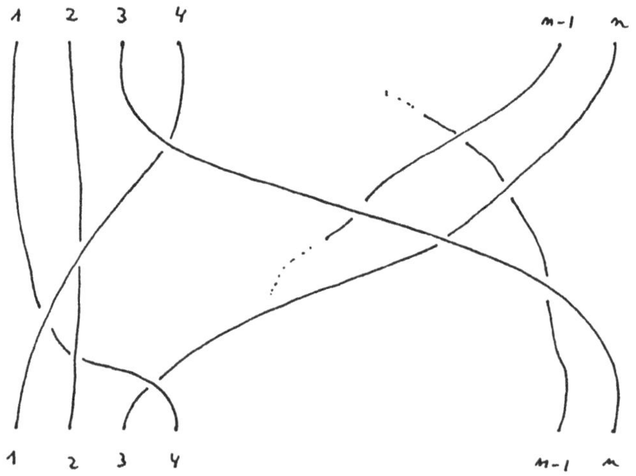

Fig.14 The elementary operations b_i and b_i^{-1} from the braid group B_n.

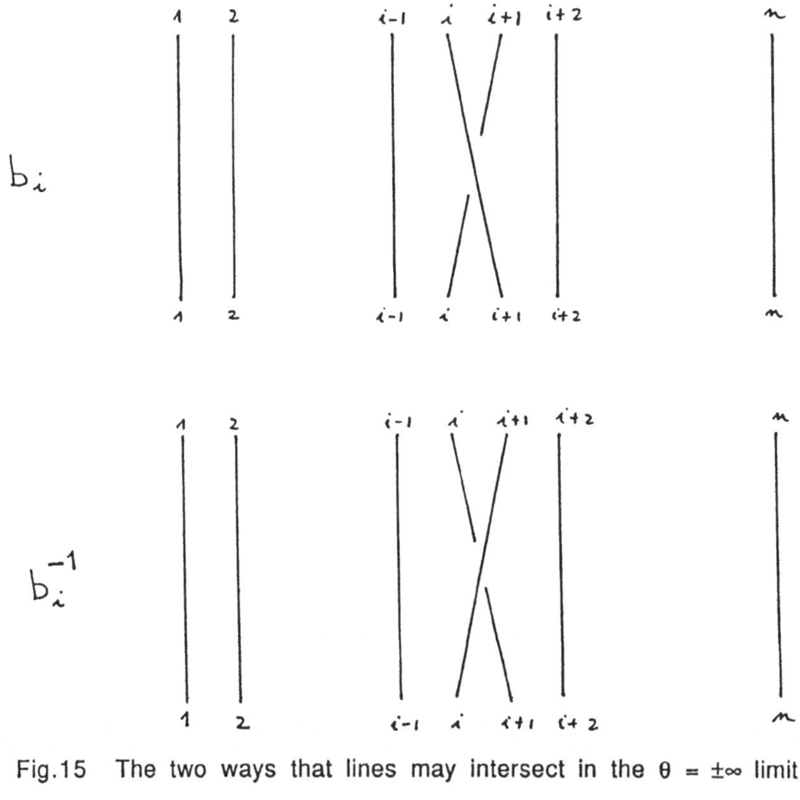

Fig.15 The two ways that lines may intersect in the $\theta = \pm\infty$ limit.

Then each topologically equivalent class of braids is identified with an element in B_n. Eq. (5.10) shows that the $\theta = \pm\infty$ limit of hyperbolic R-matrices provide a representation of B_n. This connection between YB algebras and braid groups revealed recently very fruitful to obtain knot invariants and link polynomials [57,58].

We want to remark that sometimes eq.(5.4) is called YB equation in the literature. This is somehow misleading since eq.(5.4) is only a particular case of the the YBE with spectral parameter eq.(1.1) or (1.11). It would be more appropiate to call eq.(5.4) 'braid equation'.

A braid invariant is directly obtained by taking the trace of the product of b_i's and b_i^{-1}'s that represent the given braid. Moreover, a knot or a link can be associated to many different braids (Alexander's theorem). The so-called Markov moves relate equivalent braids associated to the same link. Remarkably enough, it is possible to define a trace which is invariant under Markov moves and therefore yields a link invariant. This trace is analogous to Z with twisted boundary conditions.

The exchange of points in the n-point conformal blocks forming the conformal invariant correlation functions yields a representation of a braid group (5.10) [59]. We want to remark that the R-matrix associated to such braid groups defines a lattice statistical model whose critical behavior is described precisely by the conformal theory yielding this braid group.

Let us now discuss the quantum groups. They are related to trigonometric/hyperbolic YBA in the $\theta = \pm\infty$ limit (as the braid groups).
Let us start by the six-vertex case, where [eq.(3.4)]. In the $\theta = \pm\infty$ limit, the YB operators $T_{ab}(\theta)$ $(1\leq a,b\leq 2)$ yield for regime I [5]

$$T_{11}(\theta) \underset{\theta\to\pm\infty}{=} (y_\pm)^N \exp(\pm\gamma S_z)\,[1 + O(y_\pm^{-2})]\,, \tag{5.11}$$

$$T_{22}(\theta) \underset{\theta\to\pm\infty}{=} (y_\pm)^N \exp(\mp\gamma S_z)\,[1 + O(y_\pm^{-2})]\,, \tag{5.12}$$

$$T_{12}(\theta) \underset{\theta\to\pm\infty}{=} (y_\pm)^{N-1}\,\mathrm{sh}\,\gamma\, J_-(\mp\gamma)\,[1 + O(y_\pm^{-2})]\,, \tag{5.13}$$

$$T_{21}(\theta) \underset{\theta\to\pm\infty}{=} (y_\pm)^{N-1}\,\mathrm{sh}\,\gamma\, J_+(\pm\gamma)\,[1 + O(y_\pm^{-2})]\,, \tag{5.14}$$

where

$$y_\pm = \pm\tfrac{1}{2}\exp[\pm(\theta + \gamma/2)]\,, \qquad S_z = \tfrac{1}{2}\sum_{a=1}^{N}(\sigma_a)_z\,,$$

and

$$J_\pm(\gamma) = \sum_{k=1}^{N}\prod_{j=1}^{k-1}\exp\{\tfrac{1}{2}\gamma(\sigma_j)_z\}\,(\sigma_\pm)_k\prod_{l=k+1}^{N}\exp\{-\tfrac{1}{2}\gamma(\sigma_l)_z\} \tag{5.15}$$

(Compare with eqs.(3.8)-(3.9)).

The ultrarelativistic limit of the YB algebra relations (3.13) that followed from the R-matrix (3.1) yields the algebra of the operators $J_\pm(\gamma)$ and S_z. We find [1]

$$[\, J_+(\gamma)\, ,\, J_-(\gamma)\,] \;=\; \mathrm{sh}(2\gamma S_z)\, /\, \mathrm{sh}\,\gamma \tag{5.16}$$

$$[\, S_z\, ,\, J_\pm(\gamma)\,] \;=\; \pm J_\pm(\gamma) \tag{5.17}$$

Analogous relations hold with $\gamma \to i\gamma$ in the trigonometric regime. Eqs.(5.13)-(5.14) define the so-called $SU_\gamma(2)$ quantum group. That is, a one-parameter deformation of the SU(2) Lie algebra. Note that eqs.(5.13)-(5.14) for $\gamma \to 0$ reduces to the usual angular momentum algebra. The operators $J_+(-\gamma)$, $J_-(-\gamma)$ and S_Z obey the same algebra than $J_+(\gamma)$, $J_-(\gamma)$ and S_Z. In regime II these representations are complex conjugate of each other.

We want to notice that $J_\pm(\gamma)$ and S_Z have a coproduct structure heritated from the YB algebra [eq.(1.19)]. Let us call $J_\pm{}^{(N-1)}(\gamma)$, $S^{(N-1)}{}_Z$ ($J_\pm{}^{(1)}(\gamma)$, $S^{(1)}{}_Z$) the operators acting on (N-1) sites (the Nth site) of the line. Then, we find from eq.(5.15)

$$\begin{aligned} J_\pm{}^{(N)}(\gamma) &= \Delta^{(N-1,1)}(\, J_\pm\,) \equiv \\ \equiv\; J_\pm{}^{(N-1)}(\gamma) &\otimes \exp[\, -\gamma S^{(1)}{}_Z] \;+\; \exp[\, \gamma S^{(N-1)}{}_Z] \otimes J_\pm{}^{(1)}(\gamma), \end{aligned} \tag{5.18}$$

$$\begin{aligned} S^{(N)}{}_Z &= \Delta^{(N-1,1)}(\, S_Z\,) \equiv \\ \equiv\; S^{(N-1)}{}_Z &\otimes 1^{(1)} \;+\; 1^{(N-1)} \otimes S^{(1)}{}_Z\, , \end{aligned} \tag{5.19}$$

where $S^{(1)}{}_Z \equiv \frac{1}{2}\sigma^{(1)}{}_Z$ and $J_\pm{}^{(1)} \equiv \sigma^{(1)}{}_\pm$. Eqs.(5.18)-(5.19) define a coproduct endowing (5.16)-(5.17) with a Hopf algebra structure. That is, if $J_\pm{}^{(\mathcal{A})}(\gamma)$, $S^{(\mathcal{A})}{}_Z$ and $J_\pm{}^{(\mathcal{V})}(\gamma)$, $S^{(\mathcal{V})}{}_Z$ separately obey eqs.(5.16)-(5.17) as operators on $\mathcal{A}$ and $\mathcal{V}$, respectively, then $\Delta^{(\mathcal{A},\mathcal{V})}(\, J_\pm\,)$ and $\Delta^{(\mathcal{A},\mathcal{V})}(\, S_Z\,)$ as given by eqs.(5.18)-(5.19) **also** obey eqs.(5.16)-(5.17). Notice that $\Delta^{(\mathcal{A},\mathcal{V})}(\, J_\pm\,) \neq \Delta^{(\mathcal{V},\mathcal{A})}(\, J_\pm\,)$. That is this coproduct is not cocommutative (although $\Delta^{(\mathcal{A},\mathcal{V})}(\, J_Z\,) = \Delta^{(\mathcal{V},\mathcal{A})}(\, J_Z\,)$).

The physical interpretation of the coproduct definition for YB [eq.(1.19)] is transparent : it is the way to combine the operators associated to two independent sites in order to have the same mathematical structure (YB, there) for the combined operator as for each one separately. In most physical situations the coproduct is a simple direct sum. For example, the spin $\vec{S}$ of a two site lattice reads,

$$\vec{S} \;=\; \vec{S}^{(1)} \otimes 1^{(2)} \;+\; 1^{(1)} \otimes \vec{S}^{(2)}$$

where $\vec{S}^{(1)}$ ($\vec{S}^{(2)}$) is the spin of site 1 (2). For YBA the coproduct is depicted in fig.3 for two sites and in eq.(1.19) for N sites. This coproduct (1.19) ensures that $T_{ab}{}^{(N)}$ fulfils YB when each $t_{ab}{}^{(k)}$ ($1 \leq k \leq N$) obeys YB . The coproduct (5.19) is the consequence of the YB coproduct (1.19) for infinite θ.

One parameter deformations of all simple Lie algebras are known. They appear as ultrarelativistic limits of hyperbolic YB algebras. They can also be defined directly as deformations of classical Lie algebras using the Weyl basis.

It is possible to relate the operators $J_{\pm}(\gamma)$ with the usual spin operators $S_{\pm}$, S_z obeying

$$[S_+ , S_-] = 2S_z \quad , \qquad [S_z , S_{\pm}] = \pm S_{\pm} \tag{5.20}$$

Inserting the ansatz

$$J_+(\gamma) = S_+ f(\gamma, S_z, S) \ , \quad J_-(\gamma) = f(\gamma, S_z, S)\, S_- \tag{5.21}$$

in eqs. (5.16)-(5.17) yields the recursion relation

$$[f(\gamma, S_z-1, S)]^2 (S_z -1)[S(S+1) - S_z(S_z - 1)] - [f(\gamma, S_z, S)]^2 [S(S+1) - S_z(S_z + 1)] = \sin(2\gamma S_z) / \sin\gamma \tag{5.22}$$

where $S (S + 1) = (S_+ S_- + S_- S_+)/2 + (S_z)^2$, as usual. This has as solution[60]

$$f(\gamma, S_z, S) = \frac{1}{\sin\gamma} \sqrt{\frac{\sin[\gamma(S-S_z)] \sin[\gamma(S+S_z+1)]}{(S-S_z)(S+S_z+1)}} \tag{5.23}$$

The quadratic ("Casimir") operator commuting with $J_{\pm}(\gamma)$ and S_z writes here

$$\mathcal{C} = \tfrac{1}{2}[J_-(\gamma) J_+(\gamma) + J_+(\gamma) J_-(\gamma)] + \cos\gamma \sin^2(\gamma S_z) / \sin^2\gamma \tag{5.24}$$

This deformation of $\vec{S}^2$ has the value

$$\mathcal{C} = \sin[(S+1)\gamma] \sin[S\gamma] / \sin^2\gamma \tag{5.25}$$

In summary eqs. (5.21)-(5.23) explicitely display the $SU(2)_\gamma$ quantum group generators in terms of the usual SU(2) generators $S_{\pm}$, S_z.

The six-vertex model is invariant only under z-rotations and not under the full SU(2) group as long as $\gamma \neq 0, \pi$. Let us see how this a deformed SU(2) invariance appears using $J_+(\gamma)$, $J_-(\gamma)$ and S_z .This analysis is easier upon a symmetry breaking transformation (2.23) with $\mu = -\frac{1}{2}$, $\nu = \omega = 0$. Only the off-diagonal elements of $R^{ab}{}_{cd}(\theta)$ are affected by this transformation :

$$\mathring{R}^{12}{}_{12}(\theta) = \exp(\theta)\, c \ , \quad \mathring{R}^{21}{}_{21}(\theta) = \exp(-\theta)\, c$$

(It must be noticed that this form $\mathring{R}$ naturally emerges from the Toda field theory[46]). Then, the generalization of the invariance under $S_{\pm}$ for $\gamma \neq 0$ is[61]

$$[R(\theta) , \sigma_{\pm} \otimes \exp(-\tfrac{1}{2} \gamma \sigma_z) + \exp(\tfrac{1}{2} \gamma \sigma_z) \otimes \sigma_{\pm}] = 0 \tag{5.26}$$

This implies for the N-site YB operator $T_{ab}{}^{(\mathcal{A}.\mathcal{V})}(\theta)$ using eqs. (1.13) and (1.21) :

$$T^{(\mathcal{A}.\mathcal{V})}(\theta) [J_{\pm}(\gamma) \otimes \exp(-\tfrac{1}{2} \gamma \sigma_z) + \exp(\gamma S_z) \otimes \sigma_{\pm}] =$$

$$= [\sigma_{\pm} \otimes \exp(- \gamma S_z) + \exp(\tfrac{1}{2} \gamma \sigma_z) \otimes J_{\pm}(\gamma)] T^{(\mathcal{A}.\mathcal{V})}(\theta) \tag{5.27}$$

where $\otimes$ stands for tensor product between the auxiliary space $\mathcal{A} = \mathbb{C}^2$ (spin $\frac{1}{2}$) and the vertical space

$$\mathcal{V} = \bigotimes_{1\leq i\leq N} (\mathbb{C}^2)_i$$

Eq.(5.27) can also be derived from the asymptotic behavior of the YBE (1.3) using the eqs.(5.11)-(5.13) transformed for the $\mathring{T}(\theta)$.

In eq.(5.27) we recognize the coproduct (5.18)-(5.19). Therefore, we can express the 'quantum group invariance' in the compact form

$$T^{(\mathcal{A},\mathcal{V})}(\theta)\ \Delta^{(\mathcal{V},\mathcal{A})}(J_\pm) = \Delta^{(\mathcal{A},\mathcal{V})}(J_\pm)\ T^{(\mathcal{A},\mathcal{V})}(\theta) \tag{5.28}$$

where $\Delta^{(\mathcal{A},\mathcal{V})}(J_\pm)$ is given by eq.(5.18). The invariance under z-rotations (Cartan algebra here) writes in the same fashion :

$$T^{(\mathcal{A},\mathcal{V})}(\theta)\ \Delta^{(\mathcal{V},\mathcal{A})}(J_z) = \Delta^{(\mathcal{A},\mathcal{V})}(J_z)\ T^{(\mathcal{A},\mathcal{V})}(\theta) \tag{5.29}$$

where $\Delta^{(\mathcal{A},\mathcal{V})}(J_z)$ is given by eq.(5.19).

The z-rotations invariance yield for the transfer matrix

$$[\, S_z, \tau_\alpha(\theta)\,] = 0$$

where $\tau_\alpha(\theta) = \exp(i\alpha/2)\ A(\theta) + \exp(-i\alpha/2)\ D(\theta)$. Now, taking trace on $\mathcal{A}$ in eq.(5.27) yields

$$\tau_{-\gamma}(\theta)\ J_+(\gamma) - J_+(\gamma)\ \tau_\gamma(\theta) + 2\exp(-\gamma/2)\ \sinh\{\gamma(S_z - \tfrac{1}{2})\}\ C(\theta) = 0$$
$$\tau_{-\gamma}(\theta)\ J_-(\gamma) - J_-(\gamma)\ \tau_\gamma(\theta) + 2\exp(\gamma/2)\ \sinh\{\gamma(S_z + \tfrac{1}{2})\}\ B(\theta) = 0$$

Here we restricted ourselves to the "γ-deformation" of the SU(2) algebra. The γ-deformations of all simple Lie algebra are known[61]. They are also connected with the $\theta = \infty$ limit of YBA[4]. Moreover the quantum group invariance (5.28)-(5.29) generalizes to all trigonometric/hyperbolic YB algebras.

To conclude, I want to stress that YB algebras are more general and powerful tools than the quantum groups. Moreover, the elliptic YB algebras provide additional structures beyond the quantum groups.

VI. SURVEY OF RECENT PROGRESS

Let us very briefly summarize some recent progress in the domain of integrable theories. The full account can be find in the original references.

In ref.[65] we obtained the dominant finite size corrections to the free energy $f_L - f_\infty$ for a U(1)-invariant conformal 2D theory on a cilynder with arbitrarily twisted boundary conditions. We found

$$f_L - f_\infty = \pi(c - 24\ \Delta_\theta)\ /\ (6\ L) \tag{6.1}$$

where c is the central charge and Δ_θ is the conformal weight of the operator which realizes the twist by the angle θ. As examples of the general formula, the free chiral massless field and the massless Thirring model are worked out in detail. As an application, we identify the proper twist operator, in the scaling limit, of the XXZ quantum spin chain.

Eq.(6.1) was first derived by exact finite size calculations from the BAE (see ref.[4] for a review). Later we realize that these universal results could be derived independently of the integrability of the model only using the conformal invariance. The arguments in ref.[65] can be considered as the generalization to twisted boundary conditions of refs.[66].

In ref.[67] properties of the Thirring model solution obtained from the XXZ Heisenberg chain in the continuous limit are computed exactly using the Bethe Ansatz.

The fermion fields, vector current, axial anomaly and equal-time current commutators are constructed explicitly on the lattice and then its continuous limit is obtained. An explicit coupling constant dependence is found for the Schwinger term. This result, together with the current correlation behavior and the conformal weights of the fields here, shows that this Thirring model solution is new. That is, it is a spin 1/2 solution not contained in previous families of continuous solutions. More precisely, the continuous limit of the XXZ chain does not fit into the two-parameter solution of the Thirring model described by Klaiber[68]. Moreover, field theoretic anomalies are found using the Bethe Ansatz for the first time.

In ref.[42] exact relations on the lattice are found between SOS and vertex partition functions. We find that both partition functions are identical up to boundary conditions on the four sites at the corners. Therefore, for large size their difference will be much smaller than N^{-2}. This shows, for example, that both models have identical conformal properties (central charge and conformal weights).

In ref.[35] the critical limit of the eight-vertex model eigenvectors obtained by means of the generalised Bethe Ansatz is shown to give the six vertex eigenvectors as constructed in ref.[42]. Furthermore an explicit mapping is established between these eigenvectors and the usual Bethe Ansatz eigenvectors of the six-vertex model. This allowed us to show that the index ν labelling the eight-vertex eigenstates in refs.[17,34] becomes exactly the third component of the total spin in the critical limit. It turned out that this endomorphism between six-vertex eigenvectors involves the quantum group generators (5.12). Refs.[35] together with ref.[42] show the remarkable mathematical richness of the Bethe Ansatz constructions of eigenvectors already in the trigonometric case.

In ref.[22] the continuous limit of the eight-vertex model is shown to give solely the Massive Thirring Model (MTM). More precisely, using the light-cone lattice approach we find a whole class of lattice spacing (a) dependences in the eight vertex parameters (θ, γ, k) yielding a relativistic field theory in the $a = 0$ limit. This turns to be the MTM with no dependence on k. The exact excitation spectrum calculation presented in ref.[22], lead us to these conclusions. Analogous results are reached from a perturbative renormalization group study of the anisotropic current-current continuum fermion field theory.

This paper[22] gives a negative answer to the question : would the light-cone eight-vertex model give a new, more general, local quantum field theory in the continum limit ? At first, one might expect a positive answer on the basis of the relativistic S-matrix interpretation put forward in ref.[69] for the eight-vertex elliptic R-matrix. Actually the fact that this answer is negative is directly connected with the c-theorem.

The finite size resolution methods for the BAE proposed in ref.[28], generalized and extended in refs.[29-32] concerns models where the ground state is formed with real roots. When the ground state is formed by complex roots (strings in the infinite volume limit) only numerical results were available[27]. This applies to models like the spin S integrable spin chains and spin S vertex models for $S \geq 1$. In these models the roots are conveniently parametrized as

Spin 1 : $\lambda^{\pm}_{J} = \eta_{j} \pm i(1/2 + \delta_{j})$

Spin S : $\lambda^{m}_{J} = \eta^{m}_{J} + i (S - 1/2 - m) + \delta^{m}_{J}$ (6.2)

where $m = 0, 1, 2, ..., 2S-1$ and η_{J}, δ_{J}, η^{m}_{J} and δ^{m}_{J} are real quantities. We have singled out in eq.(6.2) the imaginary parts $\pm 1/2$ and $(S - 1/2 - m)$ which are expected by the string hypothesis. δ_{J} and δ^{m}_{J} vanish for $N \to \infty$ and fixed j .

In ref.[33] the Bethe Ansatz equations for spin S ($S \geq 1$) integrable vertex models (and magnetic chains) are investigated for finite size N. It is shown, that the finite size corrections to the imaginary parts of the roots (Bethe strings) for $N >> 1$ are given by

$$\delta^{m}_{J} = \alpha_m / [2N \, \sigma (\eta^{m}_{J})] \quad (6.3)$$

where m is the index of the root within the string [cfr. eq.(7.2)], η^{m}_{J} is the real part of the roots and

$$\sigma(\eta) = 1 / [2 \cosh(\pi\eta)] \quad (6.4)$$

is the density of the real parts. The constants α_m are determined by a set of algebraic equations, and are given explicitely by

$$\alpha_m = (1/\pi) \ln\{ \cos[\tfrac{1}{2}\pi(S - m - 1)/(S + 1)] / \cos[\tfrac{1}{2}\pi(S - m)/(S + 1)] \} \quad (6.5)$$

For the best known $S = 1$ case $\alpha_0 = \ln 2 / (2 \pi)$. Eqs.(6.2)-(6.3) hold in the isotropic case and generalizes to the anisotropic case through a rescaling of the BAE roots by a factor γ^{-1} for $0 \leq \gamma \leq \pi/(2S)$.

These results are found through a generalisation of the Euler-Maclaurin formula including non-analytic contributions in N^{-1} which turn out to be essential in the solution of the present problem. We want to notice that the explicit result for the α_m was obtained through an inexpected relation of the algebraic equations fulfilled by the α_m with the Chebyshev polynomials and their zeroes.

Besides the vertex models discussed in the previous sections, there exist a large class of interesting integrable lattice models formulated in face language[37-40]. It is therefore an important issue to generalize the vertex language methods to face language. In ref.[41] an analog of the Yang-Baxter Algebra (YBA) is defined in face language. The operators $t_{\alpha\alpha',\beta\beta'}(\theta)$ introduced in ref.[41] enjoy all essential properties of the vertex language YBA. Using this face YBA an algebraic Bethe Ansatz (BA) is constructed for SOS models (unrestricted IRF models). The face dual of the six-vertex model and the critical ABF model are worked out explicitely. Eigenvectors and eigenvalues of the transfer matrix are found and the corresponding BA equations derived and compared with the six vertex BAE.

The connection between Conformal Field Theories (CFT) and Integrable Field Theories (IFT) is obtained working solely in the continuum in refs.[70]. It appears that for any (rational) CFT there exist, directly on the continuum, one or more relevant perturbations leading to a massive IFT[70]. This led to the costruction by bootstrap of factorizable S-matrices describing the scattering of massive particles in the perturbed CFT . In ref.[71] the exact S-matrix of the affine E_8 Toda field theory is found by bootstrap and checked against tree level standard perturbation theory. At a certain purely imaginary value g_0 of the coupling constant g, this S-matrix coincides with that associated in ref.[70] to the critical Ising model in a magnetic field. This supports the idea that the affine E_8 TFT at $g = g_0$ describes the scaling limit of the Ising model in a field. The calculation in ref.[71] lead to a value for g_0 at which the E_8 TFT becomes strictly renormalizable. This means that the ultraviolet fixed point of the theory at $g = g_0$ is no longer the 8-componrnt massless Bose field (central charge $c = 8$). This is consistent with the expectation $c = 1/2$, which would follow from the identification of the $g = g_0$ E_8 TFT with the critical Ising model in a field. This latter has obviously the $c = 1/2$ Ising field theory as UV fixed point.

REFERENCES

[1]R.J. Baxter, Exactly solved models in Statistical Mechanics, Academic Press (1982).

[2]L.D. Faddeev, Soviet Sci. Review **C1**, 107 (1980) and Les Houches lectures, North Holland (1982) and references contained therein.

[3]P.P. Kulish and E.K. Sklyanin, in Tvärmine lectures, Springer Lectures in Physics, Vol. 151 (1981).

[4]H.J. de Vega, Int. J. Mod. Phys., **4**, 2371-2463 (1989) and in Advanced Studies in Pure Mathematics, vol.19, Kinokuniya-Academic,1989.

[5]H.J. de Vega, Nucl. Phys. **B240**, 495 (1984).

[6]R. J. Baxter, Studies in Appl. Math. **L51**, (1971).

[7]H. J. de Vega and E. Lopes, Phys. Lett. **B 186**, 180 (1987).

[8]see for example : L. D. Faddeev and L. A. Takhtadzhyan, 'Hamiltonian approach to Soliton Theory', Nauka 1986.

[9]R.J. Baxter, Phil. Trans. Roy. Soc. **289**, 315 (1978).

[10]M. Lüscher, Nucl. Phys. **B117**, 475 (1976).

[11]J.B. Mc Guire, J.M.P. **5**, 622 (1964).

[12]A.B. Zamolodchikov and Al.B. Zamolodchikov,Ann. Phys. **120**, 253 (1979).

[13]H.B. Thacker, Revs. Mods. Phys. **53**, 253 (1981).

[14]M. Lüscher, Nucl. Phys. **B135**, 1 (1978). See also: H.J. de Vega, H. Eichenherr and J.M. Maillet, Nucl. Phys. **B240**, 377 (1984) , Phys. Lett. **B132**, 337 (1983) and Comm. Math. Phys. **92**, 507 (1984).

[15]The Analytic S-Matrix. R. J. Eden, P. Landshoff, D. Olive and J.C.Polkinghorne, Cambridge Univ. Press (1966).

[16]A.B. Zamolodchikov, Comm. Math. Phys. **69**, 165 (1979). R. Shankar, Phys. Rev. Lett. **47**, 1177 (1981).

[17]L.D. Faddeev and L.A. Takhtadzhyan, Russ. Math. Surveys **34**, 11 (1979).

[18]O. Babelon, H. J. de Vega and C. M. Viallet, Nucl. Phys. **B200**, 266 (1982).

[19]L.A. Takhtadzhyan, Journal of Soviet Math. 2470 (1983).
P.P. Kulish and N. Yu. Reshetikhin, J. Phys. **A16**, L591 (1983).
N.Y. Reshetikhin, Theor. Math. Phys. **63**, 555 (1986).

[20]H.J. de Vega and M. Karowski, Nucl. Phys. **B280**, 225 (1987).

[21] Zhou Yu-Kui, Yan Mu-Lin and Hou Bo-Yu, J. Phys. **A 21**, L929 (1988).

[22]C. Destri and H. J. de Vega, LPTHE Paris preprint 89-20, to appear in Mod. Phys. Lett.

[23]H. Au-Yang, B. M. McCoy, J.H.H. Perk, S. Tang and M. L. Yan, Phys. Lett.**A 123**, 219 (1987).
B. M. McCoy, J.H.H. Perk, S. Tang and C. H. Sah, Phys. Lett. **A 125**, 9 (1987).
R. J. Baxter, J. H. H. Perk and H. Au-Yang, Phys. Lett. **A 128**, 138 (1988).

[24]V. V. Bazhanov and Yu. G. Stroganov, Canberra preprint, 1989.

[25]Y. Akutsu, T. Deguchi and M. Wadati, J. Phys. Soc. Japan, **56**, 3039 (1987).

[26]F. Woynarovich, J. Phys. **A15**, 2985 (1982).
C. Destri and J. H. Lowenstein, Nucl. Phys. **B205**, 369 (1982).
L. D. Faddeev and L.A. Takhtadzhyan, Zap. Nauch. Sem. LOMI, **109**, 134 (1982).
O. Babelon, H.J. de Vega and C.M. Viallet, Nucl. Phys. **B220**, 13 (1983) and 282 (1983).

[27]L. V. Avdeev and B. D. Dörfel, Theor. Math. Phys. **71**, 272 (1987).
F. C. Alcaraz, M. N. Barber and M.T. Batchelor,
Phys. Rev. Lett. **58**, 771 (1987) and Ann. Phys. **182**, 280 (1988).
F. C. Alcaraz and M. J. Martins, J. Phys. **A 22**, 1829 (1989) and **A 21**, 4397 (1988).

[28]H.J. de Vega and F. Woynarovich, Nucl. Phys. **B251**, 439 (1985).

[29]H.O. Martin and H.J. de Vega, Phys. Rev. **B32**, 5959 (1985).

[30]M.T. Batchelor et al.J. Phys.**A 20**,5677 (1987)
C. J. Hamer and M. T. Batchelor J. Phys. **A 21**, L173(1988).
F. Woynarovich and H.P. Eckle, J. Phys. **A 20**, L443 (1987).

[31]H.J. de Vega, J. Phys. **A 20**, 6023 (1987) and J. Phys. **A 21**, L1089 (1988).

[32]a) H.P. Eckle and F. Woynarovich, J. Phys. **A 20**, L97 (1987).
b) F. Woynarovich, Phys. Rev. Lett. **59**, 259 (1987).

[33]H. J. de Vega and F. Woynarovich, LPTHE Paris preprint 89-32.

[34]R. J. Baxter, Ann. Phys. **76**, 1, 25 and 48 (1973).

[35]C. Destri, H.J. de Vega and H.J. Giacomini, J. Stat. Phys. **56**, 291 (1989).

[36]J.H.H. Perk and F. Y. Wu, J. Stat. Phys. **42**, 727 (1986).

[37]G.E. Andrews, R. J. Baxter and P. J. Forrester,J. Stat. Phys. **35**, 193 (1984).

[38]M. Jimbo, T. Miwa and M. Okado, Comm. Math. Phys. **116**, 507 (1988)
and preprint RIMS-579 (1987).
E. Date, M. Jimbo, A. Kuniba T. Miwa and M. Okado,Nucl. Phys. **B 290**, 231 (1987).
A. L. Owczarek and R. J. Baxter, J. Stat. Phys. **49**, 1093 (1987).

[39]V. Pasquier, J. Phys. **A 20**, L217 and L221 (1987) and Nucl. Phys. **B 285**, 162 (1987).

[40] Y. Akutsu, T. Deguchi and M. Wadati, J. Phys. Soc. of Japan, **57**, 1173 (1988).

[41]H. J. de Vega, LPTHE preprint 89/23,to be published in Int. J. Mod. Phys.

[42] H. J. de Vega and H. J. Giacomini, J. Phys. **A 22**, 2759 (1989).

[43]C. Destri and H.J. de Vega, Nucl. Phys. **B290**, 363 (1987);
Phys. Lett. **B201**, 261 (1988) and J. Phys. **A22**, 1329 (1989).

[44]T.T. Truong and M.D. Schotte,Nucl. Phys. **B220**, 77 (1983) and **B230**, 1 (1984).
M.F. Weiss and M.D. Schotte, Nucl. Phys. **B225**, 247 (1983).
T.T. Truong in Non-linear Eqs. in Classical and QFT Springer
Lectures in Physics, Vol. **226** (1985), Ed. N. Sánchez.

[45]L.V. Avdeev and M.V. Chizov, Phys. Lett. **B184**,363 (1987).

[46]O. Babelon, H.J. de Vega and C.M. Viallet, Nucl. Phys. **B190**, 542 (1981).

[47]H. Freudenthal and H. de Vries, Linear Lie Groups, New-York, Academic Press.

[48]H. M. Babujan, Nucl. Phys. **B215**, 317 (1983).
P. P. Kulish and N. Yu. Reshetikhin, Zap. Nauk Sem. LOMI **101**, 101 (1981).
L. A. Takhtadzhyan, Phys. Lett. **A 87** , 479 (1982).

[49]V. E. Korepin, Theor. Math. Phys. **41**, N°2 (1979).
N. Andrei and J. H. Lowenstein, Phys. Lett.B **91**, 401 (1980).

[50]E. Ogievetski, N. Y. Reshetikhin and P. B. Wiegmann, Nucl. Phys. **B280**, 45 (1987).

[51]L. D. Faddeev and N. Y. Reshetikhin, Ann. Phys. **167**,227 (1986).

[52]P. P. Kulish, N. Y. Reshetikhin and E. K. Sklianin, L.M.P. **5**, 393 (1981).

[53]C. Destri and H. J. de Vega, Phys. Lett. **B201**, 245 (1988).

[54]A. N. Kirillov and N. Y. Reshetikhin in Proceedings of the Paris-Meudon
Colloquium. H. J. de Vega and N. Sanchez , Editors. World Scientific 1987.

[55] R. J. Baxter, J. Stat. Phys. **28**, 1 (1982).

[56]E. Artin, Ann. of Math. **48**, 101 (1947).
J. S. Birman, Braids et al., Princeton Univ. Press, 1974.

[57]T. Deguchi, M. Wadati and Y. Akutsu, Tokyo Univ. preprint, 1988.
M. Wadati and Y. Akutsu, C. M. P. **117**, 243 (1988).

[58]V. G. Turaev, LOMI preprint E-3-87, Leningrad.
N. Y. Reshetikhin, LOMI preprints E-4-87 and E-17-87, Leningrad.
V. Jones (unpublished).

[59]A. Tsuchiya and Y. Kanie, LMP **13**, 303 (1987) and
in Adv. Studies Pure Math., **16**, 297 (1988), Kinokuniya, Tokyo, 1988.
J. Fröhlich, Cargèse Lectures, 1987, Plenum New York.
K.H. Rehren, CMP **116**, 675 (1988).
K. H. Rehren and B. Schroer, Nucl.Phys. **B312**, 715 (1989).

[60]H. M. Babujian and A. M. Tsevelik, Nucl. Phys. **B265**, 24 (1986).

[61]M. Jimbo, L.M.P.**11**,247 (1986) and **10**, 63 (1985).
V. G. Drinfeld Dokl. Akad. Nauk, **283**, 1060(1985) and **296**, 13 (1987).

[62]E. Date, M. Jimbo, T. Miwa and M. Okado, RIMS preprint 590,(1987)
and references therein.

[63]M. Jimbo in Springer Notes in Physics, vol. 246, H. J. de Vega and N. Sanchez, editors
and Comm. Math. Phys. **102**, 537 (1986).
N. Yu. Reshetikhin, Theor. Math. Phys. **63**, 197 and 347 (1985).

[64]N. Yu. Reshetikhin and P. B. Wiegman, Phys. Lett. B **168**, 360 (1986).
E. I. Ogievetski and P. B. Wiegman, Phys. Lett. B **189**, 125 (1987).

[65]C. Destri and H. J. de Vega, Phys. Lett. **B223**, 365 (1989).

[66]H. W. Blöte, J. L. Cardy and M.P. Nightingale, Phys. Rev. Lett. **56**, 742 (1986).
I. Affleck, Phys. Rev. Lett. **56**, 746 (1986).

[67]H. J. de Vega and T. M. J. Simoes, Phys. Lett. B **217**, 142 (1989).

[68]B. Klaiber, Boulder Lectures, Vol. **XA**, 141 (1968).
J. A. Swieca, Fortschr. Phys. **25**, 303 (1977).

[69]A. B. Zamolodchikov, Comm. Math. Phys. **69**, 165 (1979).

[70] A. B. Zamolodchikov in Advanced Studies in Pure Mathematics, vol.19,Kinokuniya-Academic,1989 and Int. J. Mod. Phys. **16**, 4235 (1989).

[71] C. Destri and H. J. de Vega, LPTHE Paris preprint 89-33 (to appear in Phys. Lett. B).
H. W. Braden, E. Corrigan, P. E. Dorey and R. Sasaki, Durham preprint.

SYMMETRY AND FUNCTIONAL INTEGRATION

C.-M. Viallet

Laboratoire de Physique Théorique L.P.T.H.E.
Université Paris 6 / Tour 16 1er étage
4 Place Jussieu
F-752252 Paris Cedex 05
France

1 Introduction

The purpose of these lectures is to present the salient features of systems with symmetries, in view of their quantization. Among the different types of symmetries, an important class arises in the study of singular lagrangians (or equivalently constrained hamiltonians), of which the paradigm is the Yang-mills gauge lagrangian. We shall take this example as an illustration, although one should keep in mind that the features we will describe always appear, mutatis mutandis, for finite dimensional systems as well as for other field theories (e.g. gravitation or string theory).
One of the important points about the geometrical aspects of gauge theories is that they appear both in the hamiltonian and the lagrangian approach. I will thus first present this geometry, and only then turn to the hamiltonian and lagrangian descriptions. It shall become clear as we proceed that there is a nice interplay between differential geometrical aspects and purely algebraic aspects, more appropriate for the treatment of the quantum theory.
We start with the presentation of the basic geometry of a space of fields on which a group acts in the definite example of Yang-Mills theory, and introduce the space of all gauge potentials, the group of gauge transformations, and the resulting quotient space (orbit space) with its Riemannian geometry(paragraphs 2 to 5).
We then give the hamiltonian analysis of the constraints. We will not dwell here on purely homological and algebraic methods, which may take the place of differential geometric methods, and we refer to M. Henneaux's lectures in the same volume.
We next explain what are the Faddeev-Popov determinant, Gribov ambiguity, BRS operator, and ghost field in geometrical terms.
In the rest of the lectures we concern ourselves with the functional integral approach to the quantization of gauge theories, with a special emphasis on the case of anomalous theories, in the light of geometry.

Work supported by CNRS

2 Notations and basic objects

It has been realized for a long time that any proper description of gauge fields [1] requires some basic notions of differential geometry (like principal fibre bundles, connections, etc ...), and that these concepts flourished both in Mathematics and Physics [2].
However it will not be our purpose to explain why a gauge potential $A_\mu(x)$ is a component of a connection in some finite dimensional principal bundle, why the notion of a bundle is just the right one to accommodate fancy boundary conditions on the fields, or things of the sort. There are a number of references on the subject [3,4,5,6,7,8,9,10,11,12,13,14,15,16,17,18] see also [19]. We shall concentrate on the *geometry of the space of all fields.*

We deal with gauge theory over space-time M. Space-time will be of any dimension, especially 4-dimensional euclidean space-time in the covariant case, and $M = R \times V$ with V = 3-dimensional euclidean space in the hamiltonian formalism (resp. M or V are supposed to be compact and without boundary, which is a way of introducing a volume cut-off into the theory). It is important to notice that the geometry of the space of fields is in essence not sensitive to the dimension of M, except in $1+1$ dimensions where it somewhat degenerates in the hamiltonian formalism.

The structure group G will be a compact Lie group. Its Lie algebra is denoted by $\mathbf{g}$.

Gauge potentials are connections on a principal fibre bundle $P(M,G)$.
We use two interesting associated bundles, constructed from P:

$$E = P \otimes_{Ad} \mathbf{g}$$

the associated vector bundle with fibre $\mathbf{g}$, with the adjoint action of G on $\mathbf{g}$, and

$$F = P \otimes_{ad} G$$

the associated bundle with fibre the group G, with the adjoint action.
We also introduce spaces of forms on M with values in E:

$$\mathbf{A}^p = \Gamma(E \otimes \Lambda^p(M)) \qquad \mathbf{A} = \oplus \mathbf{A}^p.$$

If ω is a connection on P, we have a corresponding covariant derivative ∇ acting on $\mathbf{A}$

$$\nabla : \mathbf{A}^p \longrightarrow \mathbf{A}^{p+1}.$$

With the metric on M and a bi-invariant metric on G (denoted tr), we may define a scalar product in $\mathbf{A}^p$, using the Hodge $*$ operator. Recall that locally in a coordinate system, if φ is a p-form then $*\varphi$ is an $(n-p)$-form of components

$$(*\varphi)_{i_1 i_2 \ldots i_{n-p}} = \frac{1}{(n-p)!} g_{i_1\alpha_1} \cdots g_{i_{n-p}\alpha_{n-p}} \frac{\epsilon^{\alpha_1 \ldots \alpha_n}}{\sqrt{g}} \varphi_{\alpha_{n-p+1} \ldots \alpha_n}$$

where ϵ is the completely antisymmetric tensor, and g_{ij} are the components of the metric on M. We define the scalar product $(\ ,\ )$ in $\mathbf{A}^p$ by:

$$\forall \alpha \in \mathbf{A}^p, \forall \beta \in \mathbf{A}^p \quad (\alpha, \beta) = \int_M tr(\alpha \wedge *\beta).$$

The covariant derivative ∇ has an adjoint with respect to the scalar product $(\, , \,)$, the covariant divergence $\nabla^* : \mathbf{A}^{p+1} \longrightarrow \mathbf{A}^p$, such that

$$\forall \tau \in \mathbf{A}^p, \forall \xi \in \mathbf{A}^{p-1} \qquad (\tau, \nabla \xi) = (\nabla^* \tau, \xi).$$

We will use the covariant laplacian on $\mathbf{A}^0$:

$$\Box_\omega = \nabla_\omega^* \nabla_\omega : \mathbf{A}^0 \longrightarrow \mathbf{A}^0.$$

When the laplacian is invertible, we denote its inverse by G_ω:

$$G_\omega \cdot \Box_\omega = \Box_\omega \cdot G_\omega = 1.$$

3 The space $\mathcal{C}$ of connections and the group $\mathcal{G}$ of gauge transformations

The *local* expression of gauge transformations is very well known: the transformation is given by a G-valued function g on M, and the action on the components $A_\mu(x)$ is

$$A_\mu(x) \longrightarrow g^{-1}(x) A_\mu(x) g(x) + g^{-1}(x) \partial_\mu g(x).$$

The correct way of describing such a transformation on a connection ω on P comes as follows: a gauge transformation is an automorphism of P, which induces the identity mapping on the base space. Phrased differently, it is a mapping f of P into itself, which moves the points of P along fibres, and commutes with the group action on P: $\forall p \in P$, $f(p)$ belongs to the same fibre as p, and

$$\forall a \in G, \forall p \in P, f(p \cdot a) = f(p) \cdot a$$

where $p \cdot a$ denotes the right action of $a \in G$ on $p \in P$. A gauge transformation may equivalently be described by a G-valued function φ on P, since we can always write $f(p) = p \cdot \varphi(p)$. The equivariance property of f reads

$$\varphi(p \cdot a) = a^{-1} \varphi(p) a = ad_{a^{-1}}(\varphi(p)).$$

This last relation shows that we may consider gauge transformations as defined on M, provided their values are taken not in G, but in the bundle F introduced above. The product of gauge transformations is just the composition of mappings on P, and gives the pointwise product in G.

We denote by $\mathcal{G}$ the group of gauge transformations. $\mathcal{G}$ acts naturally on any connection on P by pull-back. Clearly, we recover that an element of $\mathcal{G}$ is, *locally*, a G-valued function on M, and that the usual gauge transformation formula is a change of coordinates under a change of sections of P. It is possible to show that $\mathcal{G}$ = space of sections of $F = \Gamma(F)$ is a true Lie group (although infinite dimensional). Its Lie algebra is the space of sections of E i.e. $\Gamma(E) = \mathbf{A}^0$.
The action of $\mathcal{G}$ on $\mathcal{C}$ is:

$$\omega \longrightarrow \omega \cdot g = \omega + g^{-1} \nabla g.$$

What is noticeable in this transformation law is that $\mathcal{C}$ *is not a vector space*. It is an *affine space*, since the difference τ of any two connections transforms covariantly.

Actually $\tau \in \mathbf{A}^1$, and thus the tangent space to $\mathcal{C}$ is canonically $\mathbf{A}^1$. We shall denote by $T_\omega(\mathcal{C})$ the tangent space to $\mathcal{C}$ at ω.

The gauge transformation formula reduces, for an infinitesimal gauge transformation $\xi \in \mathbf{A}^0$, to

$$\omega \longrightarrow \omega + \nabla\xi.$$

This gives the form of the elements of $T_\omega(\mathcal{C})$ which are tangent to the fibre through ω. These are the vertical vectors at ω. We denote by $V_\omega(\mathcal{C})$ the vector space of vertical vectors at ω.

From the expression of vertical vectors, it easy to see that the action of $\mathcal{G}$ on $\mathcal{C}$ has no fixed point, if for example we impose some normalization to the gauge transformations. It is sufficient to suppose that gauge transformations are normalized to unity at some point, or equivalently that infinitesimal transformations vanish at this point. For ω to be a fixed point of the infinitesimal transformation ξ, we have to have $\omega = \omega + \nabla\xi$, or $\Box_\omega\xi = 0$, which implies, with our hypothesis, $\xi = 0$.

4 A metric and a connection on $\mathcal{C}$

The scalar product (,) on $\mathbf{A}^1 \approx T_\omega(\mathcal{C})$ is a metric on $\mathcal{C}$. With that metric, $\mathcal{C}$ is flat, since (,) does not depend on ω. Moreover the metric on $\mathcal{C}$ is it gauge invariant. This is a basic point for what we will say later on.
It is a fundamental principle of the theory that two gauge potentials related by a gauge transformation are equivalent and describe the same physical reality [1]. This will also appear in the analysis of the lagrangian (see later).
The gauge fixing is just the choice of one representative in each equivalence class (orbit). We want to draw a surface in $\mathcal{C}$ which cuts all orbits once (define a section of the $\mathcal{G}$-bundle $\mathcal{C}$). We may do this locally around an origin ω_0 (reference connection) as follows: define the affine subspace $\mathcal{S}_0$ of $\mathcal{C}$

$$\mathcal{S}_0 = \{\text{all } \omega \in \mathcal{C} \text{ s.t. } \tau = \omega - \omega_0 \text{ is orthogonal to the orbit through } \omega_0\}$$

$\mathcal{S}_0$ is made out of points which depart from ω_0 perpendicularly to the orbit. These

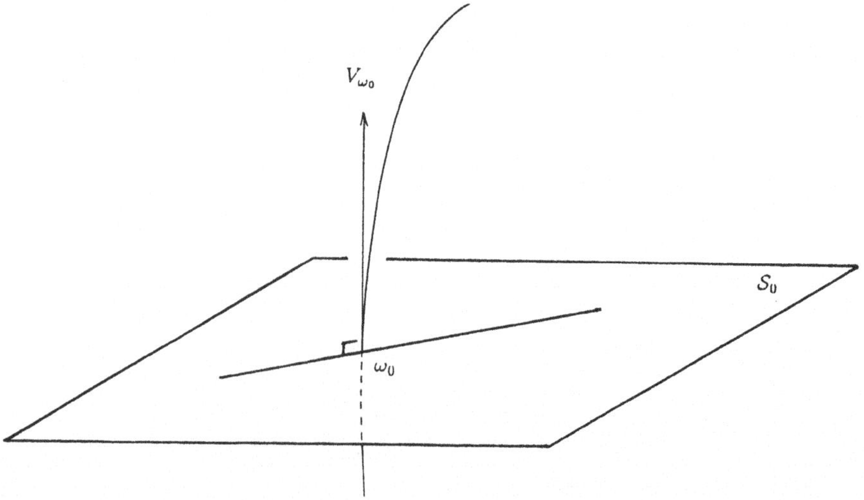

points verify

$$(\tau, \nabla_0 \xi) = 0 \quad \forall \xi \in \mathbf{A}^0,$$

or

$$(\xi, \nabla_0^* \tau) = 0 \quad \forall \xi \in \mathbf{A}^0,$$

or equivalently

$$\nabla_0^* \tau = 0 \tag{1}$$

This is a linear condition on τ, and defines what we call ***horizontal*** vectors at ω_0. We denote by H_{ω_0} the space of solutions of equation(1). Clearly $\mathcal{S}_0$ is the affine space generated by H_{ω_0} when ω_0 is taken as origin.

Claim: $\mathcal{S}_0$ is a good gauge section around ω_0 This is the covariant background gauge condition around ω_0 [20,11,22,23].
It was shown by topological methods [23,24] that there is no global section: if one goes sufficiently far away from ω_0 (within $\mathcal{S}_0$) one has to meet a point gauge related to ω_0. Our claim is that there is a region of finite radius around ω_0 in $\mathcal{S}_0$, where no two gauge related points exist, and that all orbits in the vicinity of the orbit through ω_0 cut $\mathcal{S}_0$ inside that region. We will return to the problem of gauge fixing later.

The previous result is the property of local triviality, basis of the stucture of fibre bundle of $\mathcal{C}$.

Actually, with some care taken of the spaces of functions we work with (Sobolev spaces), one can show that the action of $\mathcal{G}$ on $\mathcal{C}$ does define a nice fibre bundle, and that the orbit space is modelled on $\mathcal{S}_0$. [see [23,25,26,27], especially for the more delicate points of the definition of normalized group and of the restriction to irreducible connections].

Notice that a similar structure exists on the space of metrics on a riemannian manifold [28,29]: gauge transformations are replaced by diffeomorphisms, irreducible connections are replaced by metrics without isometries, and the same kind of objects on the spaces of metrics have exactly the same kind of structure. This is used in gravity theory and in string theory.

We denote by p the projection: $\mathcal{C} \longrightarrow \mathcal{C}/\mathcal{G}$.

It is easy to see that we have a *connection on* $\mathcal{C}$, with our horizontality condition: indeed if we define the 1-form χ on $\mathcal{C}$ with values in the Lie algebra of $\mathcal{G}$ (i.e. $\mathbf{A}^0$) by

$$\chi = G_\omega \nabla_\omega^*,$$

then:
—the kernel H_ω of χ at each point ω in $\mathcal{C}$ defines a distribution of horizontal spaces invariant by $\mathcal{G}$.
—the value of χ on a fundamental vector field ξ^* (vertical vector field on $\mathcal{C}$ generated by the infinitesimal action of $\xi \in \mathbf{A}^0$) is ξ itself.
—χ transforms with the adjoint representation of $\mathcal{G}$.
—the necessary regularity properties are satisfied.

Define the horizontal projection operator $\Pi_\omega : T_\omega(\mathcal{C}) \longrightarrow H_\omega$ by:

$$\Pi_\omega = 1 - \nabla_\omega G_\omega \nabla^*_\omega = 1 - \nabla_\omega \chi_\omega.$$

and the vertical projection operator

$$\nabla_\omega \chi_\omega : T_\omega(\mathcal{C}) \longrightarrow V_\omega.$$

The operator Π_ω verifies:

$$\Pi^2_\omega = \Pi^*_\omega = \Pi_\omega, \quad \text{and} \quad \chi_\omega \Pi_\omega = 0.$$

5 The metric on the orbit space $\eta = \mathcal{C}/\mathcal{G}$

We denote by η the quotient space $\mathcal{C}/\mathcal{G}$ [40,36]. We define a scalar product in the tangent space $T_a(\eta)$ at any point $a \in \eta$ as the one induced by (,):
If $X, Y \in T_a(\eta)$, choose any point ω in the fibre $p^{-1}(a)$ above a. The vectors X and Y have horizontal lifts τ_X and τ_Y at ω. By definition, the scalar product (metric on η) is:

$$g(X, Y) = (\tau_X, \tau_Y).$$

The gauge invariance of (,) ensures the independence of g on the choice of ω in $p^{-1}(a)$.
We can now compute the metric g in the local coordinate system centered at ω_0 and defined by $\mathcal{S}_0$.

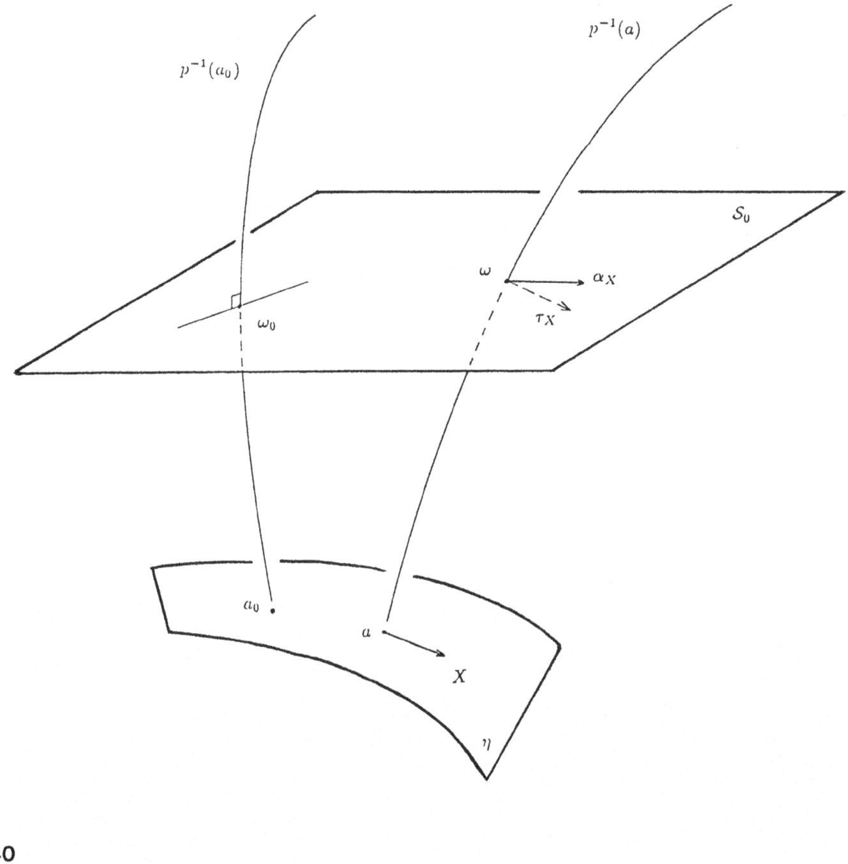

The vectors $X, Y \in T_a(\eta)$ have coordinates α_X and α_Y such that:

$$\nabla_0^* \alpha_X = \nabla_0^* \alpha_Y = 0,$$

or

$$\Pi_0 \alpha_X = \alpha_X, \qquad \Pi_0 \alpha_Y = \alpha_Y.$$

Clearly α_X is not the horizontal lift τ_X of X at $\omega \in \mathcal{S}_0$. The horizontal lifts of X and Y are:

$$\tau_X = \Pi_\omega(\alpha_X)$$

$$\tau_Y = \Pi_\omega(\alpha_Y).$$

Thus

$$g(X,Y) = (\Pi_\omega \alpha_X, \Pi_\omega \alpha_Y),$$

or

$$g(X,Y) = (\alpha_X, \Pi_0 \Pi_\omega \Pi_0 \alpha_Y).$$

This gives us the metric g at the point ω, in the coordinate system provided by the covariant background gauge at ω_0. Notice that it is ω–dependent. *The orbit space is not flat*, as we will see.

6 Dirac analysis of the lagrangian

This analysis of the lagrangian leads to the construction of the hamiltonian of the theory. We thus use the canonical formalism (non covariant) where time is separated from space [30,31,32,33,34,35,36]. Gauge potentials are time dependent connections on a bundle over 3–dimensional space V.

The action is:

$$S = \tfrac{1}{4} \int dt \int_V tr(F_{\mu\nu} F^{\mu\nu}),$$

with

$$F_{\mu\nu} = \partial_\mu A_\nu - \partial_\nu A_\mu + [A_\mu, A_\nu].$$

With our notations, the lagrangian is:

$$L = \tfrac{1}{2}(\dot{A} - \nabla A_0, \dot{A} - \nabla A_0) - \mathbf{V}(A)$$

where

$$\dot{A} = \frac{\partial A}{\partial t},$$

and

$$\mathbf{V}(A) = \tfrac{1}{4}(\Omega, \Omega),$$

with Ω the curvature 2–form of A. ($\Omega \in \mathbf{A}^2$).
The conjugate momenta are:

$$\mathbf{p} = \dot{A} - \nabla A_0$$

$$\mathbf{p}_0 = 0.$$

The last equation is the primary constraint and leads to the hamiltonian

$$H_0 = \tfrac{1}{2}(\mathbf{p}, \mathbf{p}) + \mathbf{V} + (\mathbf{p}, \nabla A_0) + (\lambda, \mathbf{p}),$$

where λ is a Lagrange multiplier ($\lambda \in \mathbf{A}^0$).
We get as a secondary constraint:

$$\{H_0, \mathbf{p}_0\} = \nabla^* \mathbf{p} = 0 \qquad \text{Gauss condition}$$

The hamiltonian becomes by incorporating Gauss condition:

$$H_T = H_0 + (\mu, \nabla^* \mathbf{p}),$$

yielding as equations of motion:

$$\begin{aligned} \dot{\mathbf{p}}_0 &= 0, \\ \dot{A}_0 &= \lambda. \end{aligned}$$

A_0 appears as an unphysical degree of freedom, which we have to discard. The true hamiltonian is thus:

$$H = \tfrac{1}{2}(\mathbf{p}, \mathbf{p}) + \mathbf{V} + (\xi, \nabla^* \mathbf{p}),$$

with ξ a lagrange multiplier ($\xi \in \mathbf{A}^0$). The equation of motion is

$$\dot{A} = \{H, A\} = \mathbf{p} + \nabla \xi.$$

The time evolution of A contains an horizontal part $\mathbf{p}$ ($\mathbf{p}$ is horizontal from Gauss condition) and a vertical part $\nabla \xi$ (pure gauge variation induced by the Lagrange multiplier). From Gauss condition we may express $\mathbf{p}$ in terms of A:

$$\mathbf{p} = (1 - \nabla G \nabla^*)\dot{A} = \Pi_A \dot{A},$$

and the true lagrangian is:

$$\mathcal{L} = \tfrac{1}{2}(\Pi_A \dot{A}, \Pi_A \dot{A}) - \mathbf{V}.$$

The lagrangian $\mathcal{L}$ is naturally defined on the orbit space. Both parts of $\mathcal{L}$ are gauge invariant, and the true configuration space appears to be the orbit space. The first term is a kinetic energy term constructed with the metric g on η. The second term is a potential part (magnetic part).

On the true configuration space, the lagrangian is of course not singular and of the typical form:

$$\mathcal{L}(q, \dot{q}) = \tfrac{1}{2} g(\dot{q}, \dot{q}) - \mathbf{V}(q),$$

where $q(t)$ is a trajectory on η, and $\dot{q}$ is the velocity.

7 The riemannian geometry of η

The previous paragraph shows that the classical evolution of the Yang-Mills fields is a motion on a non flat configuration space with a potential term $\mathbf{V}$. This motivates a detailed study [36] of the riemannian geometry of η. We will perform our computations in the local coordinate system given by the covariant background gauge condition around a reference connection ω_0.
Define the following operators, associated to a generic point $\omega \in \mathcal{S}_0$

$$\gamma : \mathbf{A}^0 \longrightarrow \mathbf{A}^0 \qquad \gamma = \nabla_0^* \nabla_\omega = \nabla_\omega^* \nabla_0.$$

γ is the Faddeev-Popov operator in the coordinate system we consider. γ is invertible if ω is sufficiently close to ω_0. Define also

$$P : \mathbf{A}^1 \longrightarrow \mathbf{A}^1 \qquad P = 1 - \nabla_0 \gamma^{-1} \nabla_\omega^*.$$

P is the projection on H_ω along V_0, and reduces to Π_0 if $\omega = \omega_0$.
We have a number of relations between Π and P, especially:

$$\Pi_0 \Pi_\omega \Pi_0 \cdot P^* P = P^* P \cdot \Pi_0 \Pi_\omega \Pi_0 = \Pi_0,$$

meaning that P^*P is the inverse of the metric in our coordinate system.
Finally define, for any $\tau \in \mathbf{A}^1$:

$$\begin{array}{llll} K_\tau & : \mathbf{A}^p \longrightarrow \mathbf{A}^{p+1} & & K_\tau(\xi) = [\tau, \xi] \\ K_\tau^* & : \mathbf{A}^{p+1} \longrightarrow \mathbf{A}^p & & \text{its adjoint} \end{array}$$

The riemannian connection D on η may easily be written for vector fields having constant coordinates X, Z (and thus commuting).

$$\begin{aligned} D_X Z = \tfrac{1}{2} P^* P \Big(& -\chi_\omega^* K_X^* \Pi_\omega Z - \Pi_\omega K_X \chi_\omega Z - \chi_\omega^* K_Z^* \Pi_\omega X - \Pi_\omega K_Z \chi_\omega X \\ & + [\chi_\omega X, \Pi_\omega Z] + [\chi_\omega Z, \Pi_\omega X] \Big). \end{aligned} \qquad (2)$$

The riemannian curvature tensor is

$$R(X,Y)Z = \Pi_0 \big(-2 K_Z G K_X^*(Z) - K_Y G K_X^*(Z) + K_X G K_Y^*(Z) \big).$$

(nb: this expression is valid at the center of coordinates). The sectional curvature in the 2–plane generated by the two orthogonal vectors X and Y is:

$$\mathcal{K}(X,Y) = 3 \big(K_X^*(Y), G K_X^*(Y) \big).$$

We see that η is of positive sectional curvature [36], see also [49]. However there is no strictly positive lower bound for $\mathcal{K}$.

8 Geodesics on the orbit space

It is clear that the geodesics of $\mathcal{C}$ are all straight lines in $\mathcal{C}$.
Moreover it is a general property that, for any group action on a riemannian manifold, and provided the metric is invariant by the group action, if one geodesic cuts one orbit perpendicularly at some point, then it cuts all orbits it meets perpendicularly [50]. It so happens that some straight lines in $\mathcal{C}$ have this property, as we may see directly: Suppose we consider the line through ω of unit vector τ:

$$\omega = \omega_0 + \lambda \tau \qquad (\lambda \in R)$$

Such a line is horizontal at ω_0 if $\nabla_0^* \tau = 0$. It is then horizontal at all its points since

$$\nabla_\omega^* \tau = \nabla_0^* \tau + \lambda K_\tau^*(\tau) = \nabla_0^* \tau = 0;$$

Therefore we have a notion of horizontal line in $\mathcal{C}$.

Claim: Geodesics on η are just the projection of horizontal lines. The proof is immediate from the geodesics equation [36].

Remark 1. If a_1 and a_2 are two points in η, we may evaluate the distance between a_1 and a_2. Take a generic point ω_1 (resp $\omega_2 = \omega_1 + \tau$) in $p^{-1}(a_1)$ (resp $p^{-1}(a_2)$). The distance in $\mathcal{C}$ between ω_1 and ω_2 is $\alpha = \sqrt{(\tau, \tau)}$. It is invariant by a simultaneous gauge transformation of ω_1 and ω_2. To define a distance d_η on η, we may take the minimum of α when ω_2 runs along its fibre a_2. When α is minimized, we have $\nabla_2^* \tau = 0$ and thus, at least locally, $d_\eta =$ *geodesic distance*.

Remark 2. Suppose we start from a point ω in $\mathcal{C}$, along some horizontal straight line; then the orbits we meet are all perpendicular to the line we follow, but they do not remain perpendicular to $\mathcal{S}_0$.

Remark 3. Since the metric g is defined via the connection χ (itself issued from the metric on $\mathcal{C}$), the projection $p : \mathcal{C} \longrightarrow \eta$ of horizontal lines preserves length. Thus η is geodesically complete, for all straight lines are of infinite length.

Remark 4. The property of the geodesics shows that the covariant background gauge around ω_0 yields a normal coordinate system at ω_0.

Remark 5. Similar properties hold true for the space of moduli of metrics.

9 The Gribov ambiguity in gauge fixing

Suppose we use the covariant background gauge around ω_0. The Faddeev-Popov operator γ is invertible as long as ω is in a neighbourhood of ω_0 (there exists such a neighbourhood); however if we go far enough from ω, then at the point $\omega = \omega_0 + \lambda\tau$, the operator $\gamma(\lambda) = \Box_0 + \lambda_0^* K_\tau$ may become non invertible. *This is where the Gribov ambiguity appears* [51,24,36,52,53].
It is the point where the coordinate system becomes singular. At this point there exist vectors v, which are vertical, but verify the gauge condition $\nabla_0^* v = 0$ (equivalent to saying that γ has a kernel: if $\gamma(\xi) = 0$, then $v = \nabla_\omega \xi$ is such a vector).
The point ω is the first focal point of ω_0 in the direction τ.
The picture is the following:

(see next page)

The vector v projects to zero on η. At the point ω, the projection p from $\mathcal{S}_0$ to η is singular. The region of $\mathcal{S}_0$ where $det\gamma \geq O$ is convex and is precisely the region where the coordinate system is non singular (the riemannian exponential is non singular). To know if and how that region covers the whole orbit space is an open question [54,55]. However, the best conjecture is that the orbit space is entirely covered by a region strictly smaller than the one delimited by the Gribov horizon.

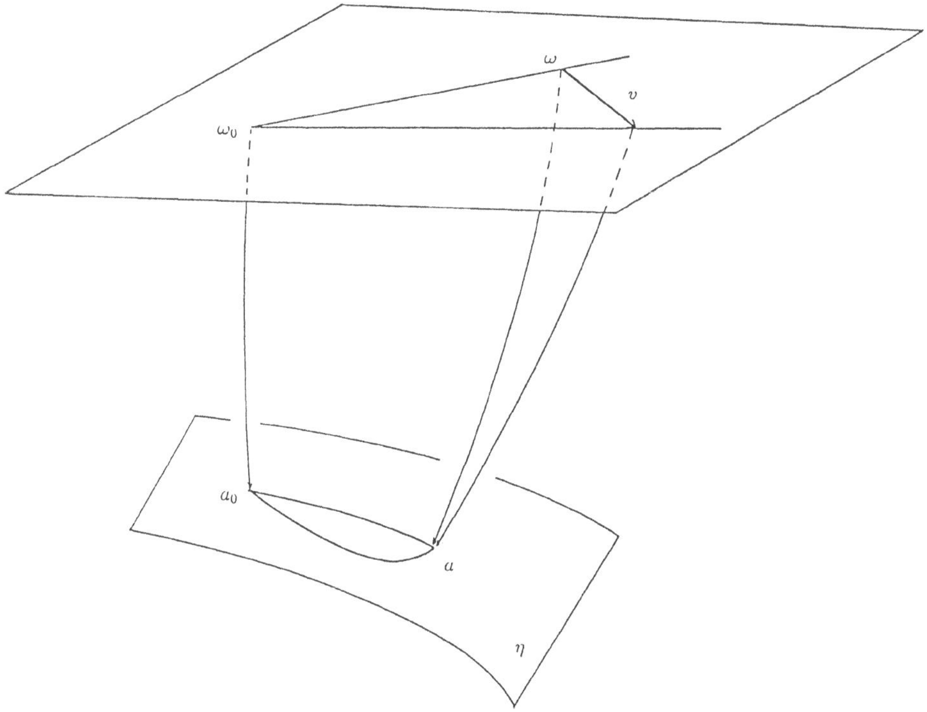

10 The Becchi-Rouet-Stora operator and the ghost field

The behaviour under gauge transformations of any function of the connections is easy to test: we just have to compute the derivatives of the function along the fibres. Notice that this will test infinitesimal gauge transformations. If $\mathcal{G}$ has more than one connected component, we stay within the component of the identity [56]. For infinitesimal gauge transformations, we may do the following [58,57,59]:
Let $d_{\mathcal{C}}$ be the exterior derivative on $\mathcal{C}$. Define the vertical part δ of $d_{\mathcal{C}}$ as follows. If φ is a q–form on $\mathcal{C}$, then:

$$\delta\varphi(X_1, X_2, \ldots, X_{q+1}) = d_{\mathcal{C}}\varphi(V_1, V_2, \ldots, V_{q+1}),$$

where V_i is the vertical part of X_i ($V_i = \nabla_\chi X_i$).

For a function on $\mathcal{C}$ (zero-form), we measure the variation along fibres.

Notice that this definition is similar to the definition of the covariant derivative (one would take horizontal parts and not vertical parts). However, contrarily to what happens for the covariant derivative, we have:

$$\delta^2 = 0,$$

by integrability of the distribution of vertical spaces.

δ is the Becchi-Rouet -Stora operator.

Let $\Omega^p(P)$ be the space of p-forms on P with values in the Lie algebra $\mathbf{g}$, which transform by ad under $\mathcal{G}$.
Let $S^{p,q}$ be the space q-forms on $\mathcal{C}$ with values in $\Omega^p(P)$, and which are invariant by $\mathcal{G}$. (and $S = \bigoplus_{p,q} S^{p,q}$).
The exterior differential d_P of P acts on S, by acting on the values.
But the exterior derivative $d_\mathcal{C}$ of $\mathcal{C}$ also acts on S. So does δ. We shall take into account the degree of the value by using $(-)^p d_\mathcal{C}$ on $S^{p,q}$ rather than $d_\mathcal{C}$ (resp. $(-)^p\delta$ rather than δ).

The function ω, defined on $\mathcal{C}$, and which to any connection on P associates its connection 1–form belongs to $S^{1,0}$.
$\delta\omega$ is a 1–form on $\mathcal{C}$ with values Ω^1.

$$\delta\omega(\tau) = -vertical\ part\ of\ \tau = -\nabla\chi(\tau).$$

Thus

$$\delta\omega = -\nabla\chi,$$

which is the B.R.S. transformation of the gauge potential.

The connection 1-form χ on $\mathcal{C}$ belongs to $S^{0,1}$.
Since the curvature 2–form $\mathcal{R} = d_\mathcal{C}\chi + \frac{1}{2}[\chi,\chi]$ of the connection χ is horizontal in $\mathcal{C}$, we have

$$\delta\chi = -\tfrac{1}{2}[\chi,\chi].$$

χ is the ghost field, and we have recovered its B.R.S. transformation.

11 The anomaly problem as a cohomological problem on $\mathcal{C}$

Quantum anomalies are the breaking, at the quantum level of the classical gauge symmetry: some quantum diagrams, involving fermion loops, generate after renormalisation, non invariant interactions [60,62,63,64,65,66,67] . We shall be more specific in the last two paragraphs.
For example, if we denote by $\Gamma(A)$ the quantum effective action of a background gauge potential in the presence of quantized Weyl fermions, $\Gamma(A)$ may not be gauge invariant . Equivalently

$$\Delta = \delta\Gamma \neq 0.$$

Δ is the anomaly.
From $\delta^2 = 0$, we see immediately that:

$$\delta\Delta = 0 \tag{3}$$

This is the Wess-Zumino consistency condition [68].
From the way the non invariance of Γ appears at the level of Feynman graphs, it is known that Δ is an integral over space-time of some polynomial in the fields and their derivatives. It is always possible to redefine Γ by such a polynomial: the anomaly Δ is spurious if it is of the form

$$\Delta = \delta(polynomial).$$

The problem of finding the true anomalies is thus a cohomological problem: we have to find $\Delta(A)$ (a vertical 1-form on $\mathcal{C}$) verifying $\delta\Delta = 0$ modulo the trivial solutions of the form $\delta M(A)$ with 'local' functions. [see [59]].

There is a simple way of producing solutions of equation(3), from the cohomology of the orbit space:
Suppose $[\varphi]$ is in $H^2(\eta)$, i.e. φ is a 2-form on η in the cohomology of η (e.g. de Rham cohomology although the precise definition of this cohomology needs some detail) [69,70,84].
The pull-back $\psi = p^*\varphi$ is a 2-form on $\mathcal{C}$ such that:

$$a) \qquad d_{\mathcal{C}}\psi = 0$$

$$b) \qquad \psi \text{ vanishes on vertical vectors.}$$

Since $d_{\mathcal{C}}$ has no cohomology on $\mathcal{C}$, there exists a 1-form θ on $\mathcal{C}$ such that

$$\psi = d_{\mathcal{C}}\theta.$$

Restricting θ to vertical vectors produces a solution of equation(3) of ghost degree one. What is remarkable is that on S^4, we get the usual chiral anomaly [71,72,73], although the condition of locality is absent in this approach.

It is an open problem to define the part of $H^*(\eta)$ which will give the correct (local) cohomology of δ.
Two different paths have been followed:
—take in $H^*(\eta)$ only the Chern character of the appropriate bundle. This is the index theorem approach, and in fact it links directly to the original problem of definition of functional determinant, at least when space-time is compactified to a sphere. This approach also applies to gravity [74].
—use a purely algebraic approach and limit oneself to some polynomials in the fields and their derivatives. This line was taken in [75,76,77,78,79].
The importance of the consistency equation is revealed not only in the problem of the usual chiral anomaly (first cohomology group of δ), but also, and with possible drastic consequences on our understanding of quantum gauge theories, in the study of Schwinger terms in the commutation of quantum currents (second cohomology group of d) [80,81].
It is worth noticing that the covariant anomaly also has an interpretation on $\mathcal{C}$[82].

12 Functional measures on the orbit space

Before describing the main features of the functional integral, it is interesting to give a remarkable operatorial relation between the metric on the orbit space and the Faddeev-Popov determinant.
Indeed we know that when using a functional integral formalism to write down rules of quantization for the Yang-Mills theory, one is lead to a functional measure which depends on the gauge condition [31,37,38,39,40].
It is very important to distinguish the hamiltonian formalism and the covariant formalism; fortunately enough the geometrical concepts we have introduced are pertinent to both cases.

As a first step we will compare (formally) the spectra of the operators γ (on $\mathbf{A}^0$) and the metric operator g (in the tangent space to η). The difficulty comes from the fact that γ essentially acts on vertical vectors, while g acts on horizontal vectors, and thus the two operators act on spaces of different dimensions.
Let us introduce the operator $Q : \mathbf{A}^1 \longrightarrow \mathbf{A}^1$ defined by

$$Q = \Pi_0 \nabla_\omega G_\omega \nabla_\omega^* = \Pi_0(1 - \Pi_\omega).$$

Q sends V_ω in H_0. Its adjoint is

$$Q^* = \nabla_\omega G_\omega \nabla_\omega^* \Pi_0 = (1 - \Pi_\omega)\Pi_0.$$

Q^* sends H_0 in V_ω.

On $\mathcal{S}_0$, the metric can be written

$$g = 1 - QQ^*.$$

Let $h : \mathbf{A}^0 \longrightarrow \mathbf{A}^0$ be the operator

$$h = G_\omega \nabla_\omega^* \nabla_0 G_0 \nabla_0^* \nabla_\omega = \chi_\omega(1 - \Pi_0)\nabla_\omega.$$

There exists an isomorphism between $\mathbf{A}^0$ and V_ω given by:

$$\nabla_\omega : \mathbf{A}^0 \longrightarrow V_\omega.$$

Its 'inverse' is given by $\chi_\omega : V_\omega \longrightarrow \mathbf{A}^0$.
Thus h is similar to $h' : V_\omega \longrightarrow V_\omega$

$$h' = \nabla_\omega \cdot h \cdot \chi_\omega.$$

It is easy to check that

$$h' = 1 - Q^*Q.$$

From the fact that QQ^* and Q^*Q have the same non zero spectrum, and that $det_{\mathbf{A}^1} g = det_{\mathcal{S}_0} g$, we get

$$det_{\mathcal{S}_0} g = det_{V_\omega} h' = det_{\mathbf{A}^0} h,$$

or, formally, by assuming that the determinant of a product is the product of determinants, we get the basic identity [40]:

$$det g \cdot det \Box_0 \cdot det \Box_\omega = (det \gamma)^2 \tag{4}$$

Denote by g_3 the metric on the true configuration space, and by g_4 the metric on the orbit space of 4–dimensional potentials.

In the canonical formalism, we see that the measure (up to constant factors) is

$$\prod_{time} \sqrt{det g_3}$$

a naive natural volume element for paths over the orbit space.

In the covariant formalism however we have:

$$\text{Faddeev-Popov determinant} = \sqrt{det g_4} \cdot \sqrt{det \Box_\omega}.$$

The factor $\sqrt{det\Box_\omega}$ being the scale of the fibre through ω [41], the covariant functional integral *is an integral over the whole space of connections rather than over the orbit space.*

Notice that the same phenomenon happens when one wants to integrate over the space of metrics an action which is invariant by diffeomorphisms, in string theory [47,48], as well as in gravity theory [42,43,44,46].

13 The problem of anomalous theories

We want to describe how the functional integration in the lagrangian (Lorentz invariant) formalism leads to quantization rules for field theories in the presence of anomalies, taking the definite example of Yang-Mills theory with fermions of a definite chirality (Weyl fermions).

The Lagrangian of the theory is

$$L = -\tfrac{1}{4} tr F_{\mu\nu}^2 + \overline{\psi}_L D(A) \psi_L$$

where

$$\psi_L = \text{ left fermion } = \frac{1-\gamma_5}{2}\psi_L$$

and $D(A)$ is the Dirac operator constructed with A, that is to say

$$D(A) = i\partial\!\!\!/ + A\!\!\!/.$$

The fermions belong to some unitary representation of the group G. The action is classically invariant by the gauge transformation

$$\begin{array}{ccc} A & \longrightarrow & A \cdot g \\ \psi & \longrightarrow & g^{-1} \cdot \psi \end{array}$$

where g acts on the fermion through the representation of group to which ψ belongs.

The appearance of gauge anomalies in the quantization of the theory has been up to now considered as dirimant in 4 space-time dimensions. This has lead to very stringent conditions on the construction of models [86,87] . For instance in the standard model for weak and electromagnetic interactions, where chiral fermions are used on purpose -to obtain parity non invariant interactions- the fermion content is not arbitrary, and one insists on cancellation of possible anomalies between the various fermions.

The belief that an anomalous theory cannot be quantized consistently (i.e. so as to obtain a Lorentz invariant unitary renormalizable theory), has been questioned recently [80,89,88], and proposals have been made. Such a proposal we will explain here. One should insist that this proposal is intended to apply in all dimensions (2, 4, and others). We will use the functional integral approach and the lagrangian formalism, to ensure Lorentz invariance, but at the risk of irremediably loosing unitarity. This is the approach of [90,91,92], and it is deeply motivated by the geometry.

Our purpose is to produce calculation rules for a perturbation theory (formal perturbation) from the usual functional integral

$$\int dA\, d\bar{\psi} d\psi \exp(\frac{i}{\hbar} S),$$

keeping in mind that the presence of the anomaly will force us to revise the usual rules. We should insist that we take the *same starting point* as in the non–anomalous case, but *expecting new perturbation rules.*

One of the strong motivations to do so is the work [47,48] about string theory, where the presence of the Weyl anomaly produces additional terms in the action which one has to take into account when constructing the quantum theory. Phrased differently it already appeared in [47] that the presence of an anomaly yields, at the quantum level to a field content which is different from the naive classical one. It is natural to believe that the features which appeared there will exist *for all theories with anomalies.*

In the next paragraph, we analyze the consequences of the results obtained above on the geometrical content of the Faddeev–Popov procedure, and study of the integration measure dA for the integration over gauge potentials. This integration is *not sensitive* to the presence of anomalies.

In the following paragraph, we examine the effect of the presence of anomalies. Indeed performing the fermionic integration goes through the definition of a fermionic measure

$$d\mu_A(\psi) = d\bar{\psi} d\psi \exp(\frac{i}{\hbar}\bar{\psi} D(A)\psi).$$

One should notice that if such a measure is ever defined, it will certainly be this 'gaussian' measure, and not the 'Lebesgue' measure $d\bar{\psi} d\psi$. The fermionic measure will then be dependent on the gauge potential. The anomaly is a non invariance of $d\mu_A(\psi)$ under gauge transformation. If γ is a gauge transformation one has:

$$d\mu_{A\cdot\gamma}(\gamma^{-1} \cdot \psi) = d\mu_A(\psi) \cdot J(A, \gamma),$$

where $J(A, \gamma)$ is a jacobian coming from the admissible non invariance of the renormalization procedure used to define $d\mu(\psi)$, and from the anomaly (the two add!). The invariance of the fermionic measure would read $J = 1$.

In the last paragraph, we consider the result of the integration over all fields, i.e. gauge potentials and fermions.

14 A measure without anomaly: dA

The Faddeev–Popov determinant was introduced in [37] in the process of eliminating an infinite integration and producing perturbation rules for Yang-Mills theory. However, one has to realize that this determinant appears as soon as one defines a theory with a symmetry. Indeed the volume element used to define the functional integral is constructed (formally) from a metric on the space of fields, and may just be thought of as the square root of the determinant of that metric. In order to preserve the symmetry at the quantum level, one uses an invariant volume element, and to that end , one uses an invariant metric on the space of fields.

Integration of an invariant function (exponential of the action) with an invariant measure evidently leads to an integral over the space of fields quotiented by the symmetry (orbit space). On the orbit space, the measure to use is straightforwardly the product of the of two pieces: the first is the measure induced on the orbit space by the original one, and the second is the volume of the orbit above each point.

This is best pictured in a naive example of finite dimension (the plane). Suppose $f(x,y)$ is a function of two variables (x and y), independent of y –i.e. invariant by the symmetry operation which translates y– and is integrated with the invariant measure $dx\ dy$, over the triangle

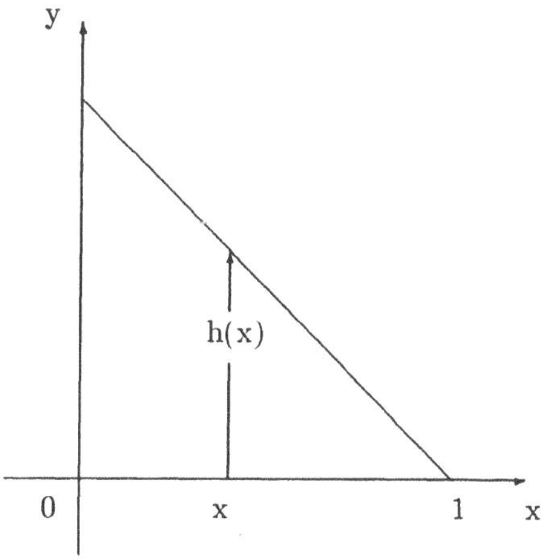

The orbit space is the real segment $(0,1)$ and the induced measure is just dx. The integral

$$I = \int f(x,y)dxdy$$

may be written

$$I = \int f(x)h(x)dx,$$

where $h(x)$ is the lengtht of the orbit above x. We see how the presence of the factor $h(x)$ weights differently different points $x \in (0,1)$.

The same phenomenon happens for any theory with symmetry, and appears explicitly when one chooses adequate coordinates in the space of fields. This choice of coordinates is dictated by the existence of the action of the symmetry group on this space. One chooses a surface $\mathcal{S}$ which cuts, at least locally, the orbits (local gauge section).

A generic point A in the space of fields has as coordinates two objects: first a specific field a which sits at the intersection os $\mathcal{S}$ and the orbit through A, second the group element g which takes a into A. This way of coordinatizing the space of fields implies only that we have a nice group action on the space of fields, just as is the case for Yang–Mills [23,25,24]. In these coordinates, the volume element dA can be rewritten

$$da = d\nu(a) \cdot \rho(a) \cdot dg,$$

where

$\cdot d\nu(a)$ is the induced measure over the orbit space, in the local coordinate system given by the section $\mathcal{S}$. This volume element was shown above to be the square root of the determinant of the induced metric, thus invariant by changes of coordinates, i.e. invariant by changes of local section. This invariance shows that the problem of patching different coordinate charts, *is automatically taken care of.*

$\cdot\rho(a)dg$ is the integration measure over the orbit, explicitely built from a volume element over the symmetry group, and the scale with which the group is pictured in the space of fields.

One should make two remarks at this point:

1. A choice of gauge is merely a choice of coordinates if one does not drop the coordinate g, and is then meaningful, even if one is to integrate a function which is not gauge invariant, as will happen later.
2. The picture is *essentially* different in the hamiltonian formalism where the group coordinate is eliminated a priori.

15 A measure with an anomaly

The anomaly was understood in [94,95] to be a non invariance of the fermionic measure. Rather than talking about a measure $d\overline{\psi}d\psi$, we prefer (see above) to use

$$d\mu_A(\psi) = d\overline{\psi}d\psi \exp(\frac{i}{\hbar}\overline{\psi}D(A)\psi).$$

The existence of the anomaly, seen as a non invariance of $d\mu_A$, creates -through the explicit dependence on A– a gauge dependence of the fermionic measure.

The transformation law

$$d\mu_{A\cdot\gamma}(\gamma^{-1}\cdot\psi) = d\mu_A(\psi)\cdot J(A,\gamma) \tag{5}$$

reflects this dependence. The jacobian factor arising in (5) is a calculable function of the gauge field –considered here as an external field– and does not depend on ψ.

The value of the jacobian J may be obtained by a direct integration of (5), yielding

$$J(A,\gamma) = \frac{detD(A\cdot\gamma)}{detD(A)} \tag{6}$$

where the determinants are calculated on the adequate spaces of fermions. This relation leads to the value of the jacobian, as the integral along a path in the group $\mathcal{G}$, of the anomaly of $\Gamma(A)$, and directly to a cocycle relation for J:

$$J(a,g) = J(a,\gamma)\cdot J(a\cdot\gamma,\gamma^{-1}g) \tag{7}$$

It is very important at this stage to emphasize the ambiguity in the value of J, which comes from the ambiguity in the expression of the anomaly. One has the possibility of redefining, in the spirit of renormalization theory [93], the jacobian by a factor of the form $\exp(P(a\cdot g)-P(a))$

$$J(a,g)\sim J(a,g)\exp(P(a\cdot g)-P(a)) \tag{8}$$

where P is some *polynomial* in the fields (i.e. a local counterterm to the action). To say there is an anomaly is precisely saying that

$$J(a,g)=\exp\Big(\ln detD(a\cdot g)-\ln detD(a)\Big)$$

canot be put to 1 by a change of the type (8). This is actually the origin of the *cohomological* problem underlying the calculation of possible anomalies. Here it is the cohomology of the group $\mathcal{G}$ (see above where we had a problem of cohomology of its algebra).

Finally, once a form of the anomaly is chosen, the jacobian is nothing but the exponential of the Wess-Zumino action [68]

$$J(a,g)=\exp(WZ(a,g)).$$

Remark 1. One has to notice that the cocycle relation (7) ensures that the transformation law (5) is meaningful, since it allows the composition of transformations.

Remark 2. If the number of dimensions of space-time is larger than two, we do not know a closed expression for the effective action $\Gamma(A)=\ln detD(A)$. However we know its variation along an orbit, given by $J(A,g)$. It is then very natural to prefer as coordinate system for the gauge potentials, the one we have chosen, since the dependence along the orbit will be calculable.

16 The integral over all fields

We start the integral

$$Z\int dA\; d\mu_A(\psi)\;\exp(\frac{i}{\hbar}S_{YM}).$$

We choose one section S as above and reexpress the same integral in the coordinates (a,g,ψ), getting

$$Z=\int d\nu(a)\;\rho(a)\;dg\;d\mu_A(\psi)\;\exp(\frac{i}{\hbar}S_{YM}).$$

The dependence in g remains only in

$$d\mu_A(\psi)=d\mu_{a\cdot g}(\psi)=d\mu_a(g^{-1}\psi)\cdot J(a,g).$$

We may redefine the fermion variable, since it appears only in the measure, and get

$$Z=\int d\nu(a)\;\rho(a)\;dg\;J(a,g)\;d\mu_a(\psi)\exp(\frac{i}{\hbar}S_{YM}(a)) \tag{9}$$

An equivalent rewriting is obtained using Faddeev-Popov ghost c an dinatighost $\bar{c}$, since *the measure over the gauge potentials is the same as in the non anomalous case*:

$$Z=\int d\bar{c}dcdadgd\bar{\psi}d\psi\exp\Big(\frac{i}{\hbar}(S_{original}+S_{gauge\;fixing}+S_{FP}+WZ(a,g)\Big) \tag{10}$$

The dependence in g is only through the Wess-Zumino action and could be eliminated if we were dealing with Dirac fermions (non anomalous case). In this case the integral of dg would just be factored out.
The explicit appearance of the field g in the action we want to take for constructing perturbation theory promotes this field to the rank of true degreee of freedom of the quantum theory, while it disappeared from the classical theory. However the exact status of g as a physical field is the main unsolved problem in this approach.
We can easily prove the independence of Z on the choice of gauge. This shall produce a B.R.S. invariance of the quantum action of the theory [58,67].
Two choices of gauge are related by a family of gauge transformations describing how, on each orbit, one goes from one to the other.

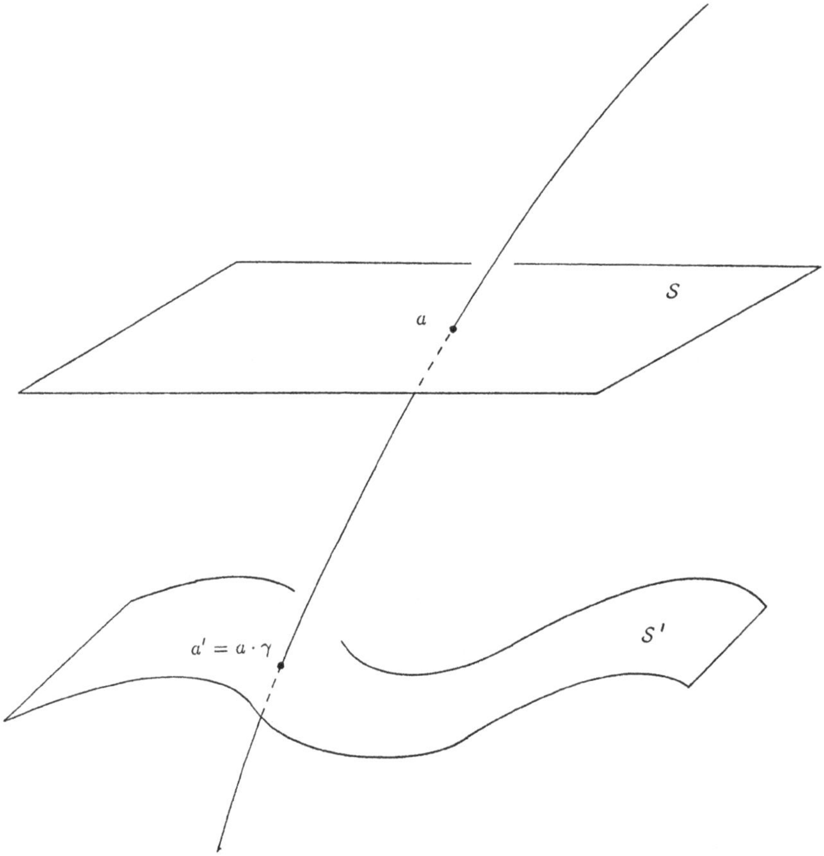

For each point a of S, one has a gauge transformation γ such that $a' = a \cdot \gamma$. Clearly γ *varies from point to point.* The change of coordinates is then

$$\begin{aligned} a &\longrightarrow a' = a \cdot \gamma \\ g &\longrightarrow g' = \gamma^{-1} g \\ \psi &\longrightarrow \psi' = \gamma^{-1} \cdot \psi \end{aligned}$$

The measure $d\nu(a)$ is independent of the choice of gauge.
The measure dg is independent of the choice of gauge, and so is $\rho(a)$.
Finally the measure

$$J(a,g)\cdot d\mu_a(\psi)$$

is independent of the choice of gauge, since

$$J(a\cdot\gamma,\gamma^{-1}g)d\mu_{a\cdot\gamma}(\gamma^{-1}\cdot\psi)=J(a\cdot\gamma,\gamma^{-1}g)\cdot J(a,\gamma)\cdot d\mu_a(\psi)=J(a,g)\cdot d\mu_a(\psi)$$

by the cocycle condition (7).

One should notice here that *B.R.S. invariance is the independence on the choice of gauge.*

Relation (10) should be taken as a starting point for a perturbation theory. A number of results have been obtained for the two dimensional theories, abelian or not. They conclude to the existence of a consistent unitary theory, with one more degree of freedom than in the non anomalous case. See for example [96,81].

The challenging problem is the definition of the theory in 4 dimensions, where the main difficulty is to concile renormalizability and unitarity. Preliminary results show that the two may contradict.

References

[1] C. N Yang, R. L. Mills. "Conservation of isotopic spin and isotopic gauge invariance". Phys. Rev. 96 (1954), p. 191.

[2] C. N. Yang. in "Chern Symposium". June 1979. Berkeley.

[3] E. Lubkin. "Geometric definition of gauge invariance". Ann. Phys. (NY) 23 (1963), p. 233.

[4] R. Kerner. "Generalization of the Kaluza-Klein theory for an arbitrary non abelian gauge group".JAnn. Inst. H. Poincaré A9 (1968), p. 143.

[5] A Trautman. "Fibre bundles associated with space-time". Rep. Math. Phys. 1 (1970), p. 29.

[6] T. T. Wu, C. N. Yang. "Concept of non integrable phase factors and global formulation of gauge fields". Phys. Rev. D12 (1975), p. 3845.

[7] Y. M. Cho. "Higher dimensional unifications of gravitation and gauge theories". J.M.P. 16 (1975), p. 2029.

[8] W. Dreschsler, M.E. Mayer. "Fiber Bundle Techniques in Gauge Theories". Lect. Notes in Physics 67. Springer (1977).

[9] S.J. Avis, C. Isham. Cargèse Lectures 1977.

[10] M.F. Atiyah. "Geometry of Yang-Mills fields". Acc. Naz. dei Lincei. Pisa (1979).

[11] M. Daniel, C.M. Viallet. "The geometrical setting of gauge theories of the Yang-Mills type". Rev. Mod. Phys. 52 (1980), p. 175.

[12] T. Eguchi, P.B. Gilkey, A.J. Hanson. " Gravitation, gauge theory, and differential geometry". Phys. Reports 66 (1980), p. 213.

[13] Y. Choquet-Bruhat, C. de Witt-Morette, M. Dillard-Bleick. "Analysis, Manifolds and Physics", 2nd. ed. (1982).

[14] J. Madore. "Geometric methods in classical field theory". Phys. Rep. 75 (1981), p. 125.

[15] A. Trautman. "Differential geometry for physicists". (1984) Bibliopolis.

[16] N. Steenrod. "The topology of fibre bundles". Princeton Math. Series 14 (P.U.P. 1951).

[17] A. Lichnerowicz. "Théorie globale des connexions et des groupes d'holonomie". Ed. Cremonese. Roma (1955).

[18] S. Kobayashi, N. Nomizu. "Foundations of differential geometry". Vol1(1963), Vol2(1969). Wiley.

[19] R. Utiyama. "Invariant theoretical interpretation of interaction". Phys. Rev. 101 (1956), p. 1597.

[20] M. F. Atiyah, N. Hitchin, I. M. Singer. "Self duality in four dimensional riemannian geometry" Proc. R. Soc. A362 (1978), p. 425.

[21] M. Daniel, C. M. Viallet. "The gauge fixing problem around classical solutions of the Yang-Mills theory". Phys. Lett. 76B (1978), p. 458.

[22] V. Moncrief. "Gribov degeneracies: Coulomb gauge conditions and initial value constraint". Journ. Math. Phys. 20 (1979), p. 579.

[23] M. S. Narasimhan, T. R. Ramadas. "The geometry of SU(2) gauge fields". Comm. Math. Phys. 67 (1979), p. 21.

[24] I. M. Singer. "Some remarks on the Gribov ambiguity". Comm. Math. Phys. 60 (1978), p. 7.

[25] P. K. Mitter, C.M. Viallet. "On the bundle of connections and the gauge orbit manifold in Yang-Mills theory". Comm. Math. Phys. 79 (1981), p 43.

[26] W. Kondracki.. J. S. Rogulski. "On the stratification of the orbit space for the action of automorphisms of connections".Preprint P. A. N. 1983.

[27] W. Kondracki. P. Sadowski. "Geometric structure of the orbit space of gauge connections". Inst. Math. Polish Ac. of Sciences (1984).

[28] D. Ebin. "The manifold of riemannian metrics". Proc. AMS Symp. Pure Math. XV (1968), p. 11.

[29] J. P. Bourguignon. "Une stratification de l'espace des structures riemanniennes". Comp. Math. 30 (1975), p. 1.

[30] P. A. M. Dirac. "Lectures on quantum mechanics" (1964) Belfer Series.

[31] L. D. Faddeev. "The Feynman integral for singular lagrangians". Theor. Math. Phys. 1 (1969), p. 3.

[32] A. Hanson, T. Regge, C. Teitelboim. "Constrained hamiltonian systems". Acc. Naz. dei Lincei. (1969).

[33] E. S. Fradkin, G. S. Vilkovisky. "Quantization of relativistic systems with constraints. Equivalence of canonical and covariant formalims in quantum theory of gravitational field". CERN Preprint TH 2332 (1977) (unpublished).

[34] K. Sundermeyer. "Constrained dynamics". Lect. Notes in Phys. Springer (1982).

[35] M. Henneaux. "Hamiltonian form of the path integral for theories with a gauge freedom" Physics Reports 126 1985.

[36] O. Babelon, C.M. Viallet. "The riemannian geometry of the configuration space of gauge theories". Comm. Math. Phys. 81 (1981), p. 515.

[37] L. D. Faddeev, V.N. Popov. "Feynman diagrams for the Yang-Mills field". Phys. Lett. 25B (1967), p. 29.

[38] R. P. Feynman. "Quantum theory of gravitation". Acta Phys. Pol. 24 (1963), p. 697.

[39] B. S. de Witt in "Relativity, groups and topology" (1964) Blackie and son.

[40] O. Babelon, C.M. Viallet. "The geometrical interpretation of the Faddev-Popov determinant". Phys. Lett. 85B (1979), p. 246.

[41] A. S. Schwartz. "Instantons and fermions in the field of instantons". Comm. Math. Phys. 64 (1979), p. 233.

[42] B. S. de Witt. "Quantum theory of gravity I. The canonical theory". Phys. Rev. 160 (1967), p. 1113.

[43] B. S. de Witt. "Quantum theory of gravity II. The manifestly covariant theory". Phys. Rev. 162 (1967), p. 1195.

[44] B. S. de Witt. "Quantum theory of gravity III. Applications of the covariant theory". Phys. Rev. 162 (1967), p. 1239.

[45] P. Ellicot, G. Kunstatter, D. J. Toms. "Geometrical derivation of the Faddeev-Popov Ansatz" To appear in International Journal of Modern Physics.

[46] R. Catenacci, M. Martellini. "On a geometrical interpretation of the Faddeev-Popov determinant for pure quantum gravity". Phys. Lett. 138B (1984), p. 263.

[47] A. Polyakov. "Quantum geometry of bosonic strings".Phys. Lett. 103B (1981), p. 207.

[48] A. Polyakov. "Quantum geometry of fermionic strings". Phys. Lett. (1981), p. 211.

[49] D. Groisser, T. H. Parker. "The riemannian geometry of the Yang-Mills moduli space". Comm. Math. Phys. 112 (1987), p. 663.

[50] R. Bott in "Representation theory of Lie groups". London Math. Soc. Lect. Notes 34 Cambridge Univ. Press (1979).

[51] V. N. Gribov. "Quantization of non abelian gauge theories". Nucl. Phys. B139 (1978), p. 1.

[52] T. P. Killingback. "The Gribov ambiguity in gauge theories on the four-torus". Phys. Lett. 138B (1984), p. 87.

[53] D. Zwanziger. "Non perturbative modification of the Faddeev-Popov formula and banishment of the naive vacuum". Nucl. Phys. B209 (1982),p. 336.

[54] M. A. Semenov-Tian-Shansky, V. A. Franke. "A variational principle for the Lorentz condition and restriction of the domain of path integration in non abelian gauge theory" Zapiski Nauch. Sem. Leningrad (1982), english transl. Jour. Sov. Math. (1986), p. 1999.

[55] G. Jona-Lasinio, C. Parinello. "On the stochastic quantization of gauge theories" Phys. Lett. B213 (1988), p. 466

[56] P. Nelson, L. Alvarez-Gaumé. "Hamiltonian interpretation of anomalies". Comm. Math. Phys. 99 (1985), p. 103.

[57] C. Becchi, A. Rouet, R. Stora."Renormalization of gauge theories". Ann. Phys.98 (1976), p. 287.

[58] C. Becchi, A. Rouet, R. Stora. "Renormalization of the abelian Higgs-Kibble model". Comm. Math. Phys.42 (1975),127.

[59] L. Bonora, P. Cotta-Ramusino. "Some remarks on BRS transformations, anomalies, and the cohomology of the Lie algebra of the group of gauge transformations". Comm. Math. Phys. 87 (1983), p. 589.

[60] S. Adler, W. Bardeen. "Absence of higher order corrections in the anomalous axial-vector divergence equation". Phys. Rev. 182 (1969), p. 1517.

[61] W. Bardeen. "Anomalous Ward identities in spinor field theory". Phys. Rev. 184 (1969), p. 1848.

[62] R. Jackiw, K. Johnson. "Anomalies of the axial-vector current". Phys. Rev. 182 (1969), p. 1459.

[63] S. Adler. "Perturbation theory anomalies". Brandeis Summer School 1970. Deser, Grisaru, Pendleton eds. M.I.T. Press.

[64] R. Stora. Erice Lectures 1975 and Cargèse Lectures 1976.

[65] B. Zumino. Les Houches Lectures 1983.

[66] R. Stora. Cargèse Lectures 1983 and Gift Lectures 1985.

[67] C. Becchi. Les Houches Lectures 1983.

[68] J. Wess, B. Zumino. "Consequences of anomalous Ward identities". Phys. Lett. 37B (1971), p. 95.

[69] M. F. Atiyah, R. Bott. "The Yang-Mills equations over Riemann surfaces". Phil. Trans. R. Soc. A308 (1982), p. 523.

[70] M. Asorey, P. K. Mitter. "On geometry, topology and q sectors in a regularized quantum Yang-Mills theory". Preprint CERN TH3424 (1982).

[71] L. Alvarez-Gaumé, P. Ginsparg. "The topological meaning of non-abelian anomalies". Nucl. phys. B243 (1984), p. 449.

[72] M. F. Atiyah, I. M. Singer. "Dirac operators coulped to vector potentials". Proc. Nat. Ac. Sc. U.S.A. 81 (1984), p. 2597.

[73] I. M. Singer. "Families of Dirac operators with applications to physics". Conference en l'Honneur d'E. Cartan. Astérisque(1984).

[74] O. Alvarez, I. M. Singer, B. Zumino. "Gravitational anomalies and the family's index theorem". Comm. Math. Phys. 96 (1984), p. 409.

[75] J. A. Dixon. "Cohomology and renormalisation of gauge theories" (1976), unpublished.

[76] J. Thierry-Mieg. "Classification of the Yang-Mills anomalies in even and odd dimension". Phys. Lett. 147B (1984), p. 430.

[77] M. Dubois-Violette, M. Talon, C.M. Viallet. "Results on BRS cohomologies in gauge theory". Phys. Lett. 158B (1985), p. 231.

[78] M. Dubois-Violette, M. Talon, C.M. Viallet. "BRS algebras. Analysis of the consistency equations in gauge theory". Comm. Math. Phys. 102 (1985), p. 105.

[79] M. Dubois-Violette, M. Talon, C.M. Viallet. "Anomalous terms in gauge theory: relevance of the structure group". Ann. Inst. Henri Poincaré 44 (1986), p. 103.

[80] L. D. Faddeev. "Operator anomaly for the Gauss law". Phys. Lett.145B (1984), p. 81.

[81] R. Jackiw. "Chern-Simons terms and cocycles in physics and mathematics". to appear in E. S. Fradkin Festschrift. Adam Hilger. Bristol (1986).

[82] D. Bao. V. P. Nair. "A note on the covariant anomaly as an equivariant momentum mapping". Comm. Math. Phys. 101 (1985), p. 437.

[83] S. Deser, R. Jackiw, S. Templeton. "Topologically massive gauge theories". Ann. Phys. 140 (1982), p. 372.

[84] M. Asorey, P. K. Mitter. "Cohomology of the Yang-Mills gauge orbit space and dimensional reduction" .Phys. Lett. 153B (1985), p. 147.

[85] M. Asorey. P. K. Mitter. "Regularized continuum Yang-Mills process and Feynman-Kac functional integral". Comm. Math. Phys. 80 (1981), p. 43.

[86] C. Bouchiat, J. Iliopoulos, Ph. Meyer. "An anomaly version of Weinberg's model" Phys. Lett. 38B (1972), p. 519.

[87] D. Gross, R. Jackiw. "Effects of anomalies on quasi-renormalizable theories". Phys. Rev. D6 (1972), p. 477.

[88] R. Jackiw, R. Rajaraman. "Vector-Meson mass generation by chiral anomalies" Phys. Rev. Lett. 54 (1985), p. 1219

[89] L. D. Faddeev, S. L. Shatashvili. "Realization of the Schwinger term in the Gauss law and the possibility of correct quantization of a theory with anomalies". Phys. Lett. 167B (1986), p. 225.

[90] O. Babelon, F. A. Schaposnik, C.-M. Viallet. "Quantization of gauge theories with Weyl fermions". Phys. Lett. B177 (1986), p. 385.

[91] K. Harada, I. Tsutsui. "On the path integral quantization of anomalous gauge theories". Phys. Lett. B183 (1987), p. 311.

[92] A. V. Kulikov. "Dynamic conservation of anomalous current in gauge theories" Serpukhov preprint ihep 86-33 (1986).

[93] O. Piguet, A. Rouet. Physics Reports 76C (1981).

[94] K. Fujikawa. "Path integral measure for gauge invariant fermions theories" Phys. Rev. Lett. 42 (1979), p. 1195.

[95] K. Fujikawa. "Path integral for gauge theories with fermions". Phys. Rev. D21 (1980), p. 2848

[96] R. Rajaraman. "Hamiltonian formulation of the anomalous chiral Schwinger model". Phys. Lett. B154. (1985), p. 305.

TOPOLOGICAL ASPECTS OF THE QUANTUM HALL EFFECT

Yong-Shi Wu*

School of Natural Sciences
Institute for Advanced Study
Olden Lane, Princeton, New Jersey 08540, U.S.A.

Abstract: Topological aspects of the quantum Hall effect, including both the integral and fractional cases, are discussed. These include the quantized Hall conductance (for both IQHE and FQHE) as a topological invariant, the role of fractional statistics in the FQHE and the ground state degeneracy of the FQH states on a compactified space. To make the lecture notes self-contained, the background material is introduced in great detail.

1. Introduction

When I was asked to give a lecture on the Quantum Hall Effect (QHE) at the Banff Summer School where most students have particle physics background, I completely agreed with the organizers that the QHE should become part of the training for all graduate students in physics, since the subject is so deeply related to fundamental principles of physics and so strongly connected to the frontier of topological investigations in physics. In the last forty years, the vital interactions between quantum field theory and condensed matter physics have proved beneficial to both sides. The QHE has exhibited an interesting interplay between gauge fields, two dimensionality and topology, all of which are at the center of attention in contemporary theoretical physics. In particular, the QHE is related to Chern-Simons gauge theory, which recently is a popular topic in mathematical physics, and probably to high-T_c superconductivity (at least to one school of the high-T_c theory), which is currently under intensive study in condensed matter physics. Many problems in the QHE (particularly in the FQHE) remain a challenge and recently there are revived interests in the QHE on the off-diagonal long range order, on the topological order, on the transition regions between plateaus and on the edge excitations, and so forth. The progress in understanding anyon superconductivity, which originated from analogy with the FQHE, may eventually feedback to promote our understanding of the latter.

The most remarkable and fascinating features of the QHE are, of course, the appearance of an integral or fractional quantum number, the quantized Hall conductance, as a measurable many-body quantity and the great precision with which the quantization is observed in dirty condensed

* *On sabbatical leave from Department of Physics, University of Utah, Salt Lake City, Utah 84112, U.S.A.*

matter samples. The two features combined together make the quantized Hall conductance distinct from other known examples of similar quantum numbers in physics. For example, the quantization of circulation in superfluid He has not been tested with great precision, and on the other hand the quantized flux in superconducting ring is not directly a many-body quantity. It is well-known that the integer or rational quantum numbers occurring in quantum theory are of two different types. Spin is an example of first type, which is directly related to a symmetry, rotational symmetry in this case; though there are some topological aspects in the quantization of spin, spin itself is not a quantity of topological origin. The quantized Hall conductance is a good example of the other type, which has no symmetry origin but is identifiable directly as a topological invariant. Topological quantization, such as the magnetic charge for a monopole, has emerged in particle physics for many years, but up to now there is no convincing experimental evidence. The QHE really provides us a rare opportunity to enjoy the interplay between topology and physics and to put our understanding under test. Now we believe that there is some non-trivial topological order (or structure) in the QH states and, in addition to the quantized Hall conductance, there should be more features or characterizations which are of topological origin. Though topological aspects do not exhaust all essential features, the study of them will continually promote our understanding of the QH states as a new macroscopic quantum phenomenon.

The lecture notes are meant to be self-contained. So we first introduce the experimental observations in Sec. 2 and then discuss in Sec. 3 the classical and quantum dynamics of a single electron in a magnetic field. Sec. 4 starts with a critical review of Laughlin's argument for the IQHE and proceeds to show the topological basis for the quantization of Hall conductance, which Laughlin's argument seems to fail to provide. Both the case of non-interacting electrons in a periodic potential and the more general case with various complications allowed are discussed, and the Hall conductance is shown to be a topological invariant (the first Chern class on an appropriate bundle). Sec. 5 is devoted to Laughlin's wave functions for the FQHE, providing a basis to the next two sections. In Sec. 6 we present theoretical evidence for our belief that quasiparticles in an FQH system obey fractional statistics and explain how it helps to understand the occurrence of stable fractions for the filling factor of the system. Finally in Sec. 7 we discuss why the ground state degeneracy on a torus is responsible to the fractional quantization, how to show the existence of this degeneracy and how it is related to edge excitations, which is recently a focus of attention in this field.

2. Brief Summary of Experimental Facts

Experimentally the QHE is observed[1-3] for 2-d electron systems at low temperatures and in strong magnetic fields. The electrons are trapped in a thin layer (~ 100Å) at the interface between semiconductor and insulator or between semiconductors. The mostly used devices in observing the QHE are the Si MOSFET *(Metal-Oxide-Semiconductor-Field-Effect-Transistor)* and the GaAs/Ga_xAl_{1-x}As ($0<x<1$) hetero-structure. Typically for the IQHE, the temperature range is $T\sim 1$-4°K and the magnetic field is about $B\sim 3$-15 Tesla (1 Tesla $= 10^4$ Gauss). For the FQHE, the temperature is even lower: $T\sim 20$-100 mK and the magnetic field needs to be stronger: $B\sim 15$-30 Tesla. The low temperature is needed to quantum mechanically freeze the degree of freedom for motion in the perpendicular direction in the ground state so that the system can really be treated as a 2-d system. The strong magnetic field makes the system to be in the quantum limit so that electrons fill, from the bottom up, the Landau levels of the cyclotron motion *(see below, Sec. 3).*

An important physical parameter of the 2-d electron gas is the so-called filling factor defined by

$$\upsilon = \frac{\#\ \text{of electrons}}{\#\ \text{of Landausites}} = \frac{N_e}{N_\phi} = \frac{N_e}{\phi / \phi_0} \qquad (2.1)$$

where N_e is the number of electrons, N_ϕ is the degeneracy of each Landau level which is just equal to the total magnetic flux though the planar system in units of flux quantum $\phi_0 = hc/e$ (-e being the electronic charge). Experimentally, υ can be adjusted either by source or drain control or by changing the magnetic field. In physics υ represents the number of filled Landau levels at T=0.

Now we are ready to describe the experimental observations of the QHE. The data is normally presented as two curves: ρ_{xy} vis B and ρ_{xx} vis B. *See Figs. 1-4.* Here we suppose an electric field is applied in the y-direction in the sample plane, and the Hall current is in the x-direction. $\rho_{xx} = V_x/I_x$ and $\rho_{xy} = V_y/I_x$ are, respectively, the longitudinal and transverse resistance. With fixed number of electrons, change in B means actually change in υ.

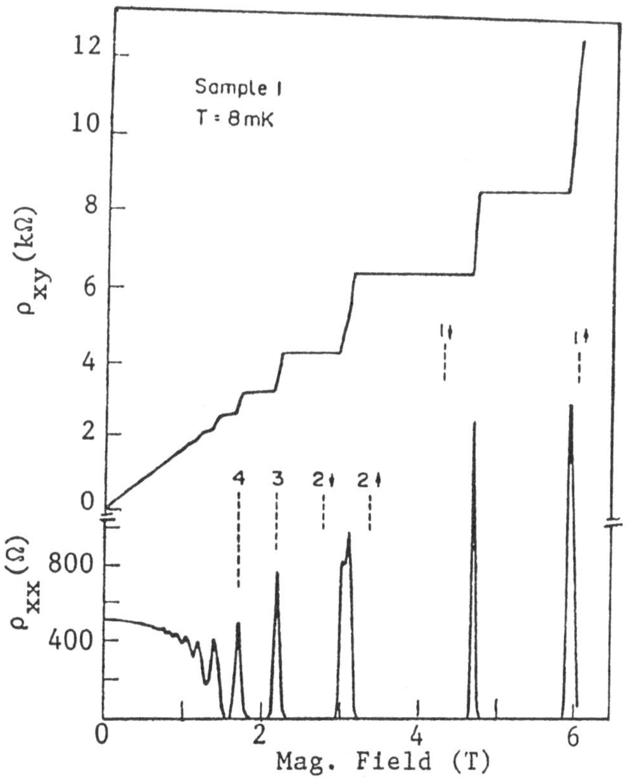

Fig. 1. A sample of the IQHE. *(Ref. 4)*

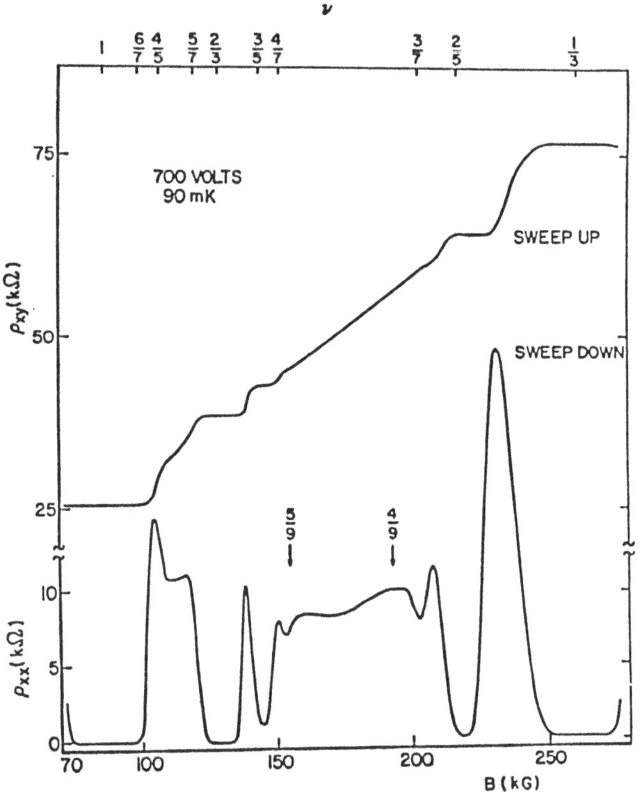

Fig. 2. A sample of the FQHE. *(Ref. 5)*

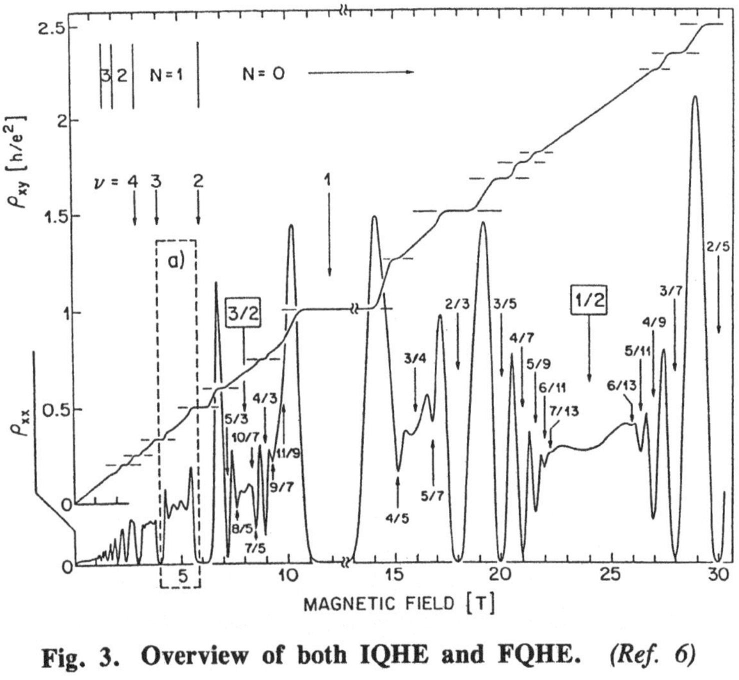

Fig. 3. Overview of both IQHE and FQHE. *(Ref. 6)*

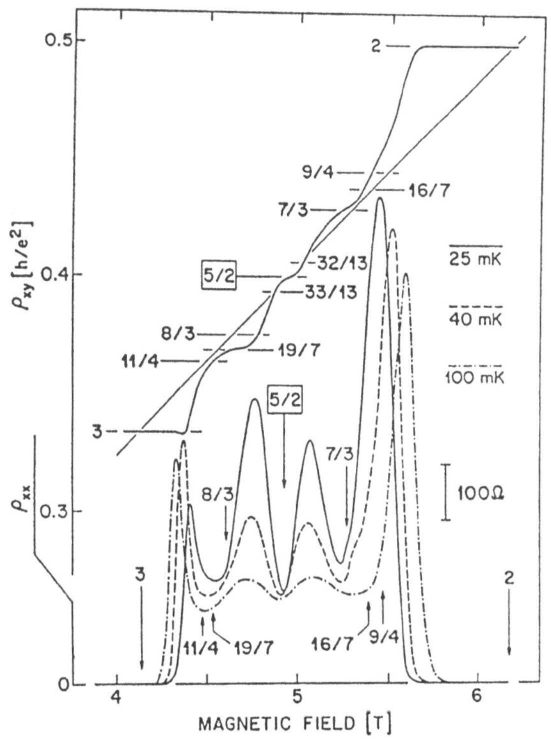

Fig. 4. Overview of both IQHE and FQHE *(cont.)*
Region (a) in Fig. 3 enlarged. *(Ref. 6)*

From Figs. 1-4, the basic feature of the experimental curves is the development of the Hall resistance plateaus at filling factors near all low-lying integers $\upsilon=1,2,3,\cdots$ and near some special fractions such as $\upsilon=1/3$, 2/3, 2/5, 3/5, ••• and 4/3, 5/3, •••. Corresponding to the plateaus in ρ_{xy}, there are valleys in ρ_{xx} with $\rho_{xx}\approx 0$ or at least sharp dips in ρ_{xx}. When ρ_{xx} is zero, the inverse of ρ_{xy} gives us the Hall conductance σ_{xy} *(see Sec. 3 below).* Another remarkable feature of the QHE is that at the plateaus the Hall conductance is quantized to be the corresponding integer[1] or fractional[2] filling factor in units of e^2/h:

$$\sigma_H \equiv \rho_{xy}^{-1} = \upsilon \frac{e^2}{h} \tag{2.2}$$

Note that the unit here, e^2/h, is expressed only in terms of fundamental constants. When υ = integers we have the IQHE and when υ = fractions for the FQHE. On the look, the curves have similar features in both cases. But the FQHE corresponds to partially filled Landau levels, so the theoretical explanation for it is more complicated.

The third feature, which makes the QHE practically useful and scientifically fascinating, is the high accuracy with which the quantization (2.2) of σ_H is observed. For the IQHE, $\Delta\upsilon\leq 10^{-8}$ and for the FQHE, $\Delta\upsilon\leq 10^{-5}$. In particular, the IQHE can be used to make the standard for resistance and to improve the measurements of the fundamental physical constants. Also it is notable that almost all filling factors corresponding to the FQHE have an odd denominator with only one exception ($\upsilon=5/2$). This is the so-called "odd-denominator rule".

Recently there are experimental studies on the transition region between plateaus and on the edge currents in the sample, but we do not have time to discuss about them. Their explanation is still an open question, so hereafter we will restrict ourselves to discuss topological aspects of the QHE which are directly related to the observed features we have described above.

3. Motion of Single 2-d Electron in Magnetic Field

First let us consider the motion of an electron confined on a plane in the presence of a perpendicular magnetic field. The discussion of a 2-d non-interacting electron gas is directly reduced to this case. Also the knowledge of the motion of a single electron is the starting point for dealing with the complicated interacting system.

1. Classical Mechanics

Set $z=x+iy$, $v=\dot{z}\equiv dz/dt$, $E=E_x+iE_y$. The Newton-Lorentz equation is

$$\dot{v} = -\frac{eE}{m} + i\omega_c v \tag{3.1}$$

where $\omega_c \equiv eB/m_c$ is the cyclotron frequency, with B the perpendicular magnetic field. m is the electron mass, (-e) the electron charge. We also have assumed that there is an electric field (E_x,E_y) in the plane. If E and B are constant both in space and time, then the solution to eq. (3.1) is given by

$$v(t) = \frac{e_0 E}{im\omega_c} + v_0 e^{i\omega_c t} \tag{3.2}$$

where v_0 is the initial complex velocity: $v_0=v_{ox}+iv_{oy}$. The second term represents the cyclotron motion, while the first term the drift motion with the velocity $v_d=-icE/B$. Note that

$$(v_d)_x = cE_y/B, \qquad (v_d)_y = -cE_x/B \tag{3.3}$$

so the drift velocity is in the direction of $\vec{E}\times\vec{B}$ and is independent of the sign of the charge.

For a non-interacting electron gas, the drift current or the Hall current is given by $j^H=-nev_d$, where n is the density of electrons. Thus,

$$(j^H)_x = -nce\,E_y/B, \qquad (j^H)_y = nce\,E_x/B \tag{3.4}$$

The Hall current (density) $\vec{j}^H$ is perpendicular to the applied electric field $\vec{E}$. In general if one introduces the conductivity matrix (σ_{ij}) by

$$\begin{pmatrix} j_x \\ j_y \end{pmatrix} = \begin{pmatrix} \sigma_{xx} & \sigma_{yy} \\ \sigma_{yx} & \sigma_{yy} \end{pmatrix} \begin{pmatrix} E_x \\ E_y \end{pmatrix} \tag{3.5}$$

then eq. (3.4) gives the Hall conductivity

$$\sigma^H_{xy} = -\sigma^H_{yx} = -nce/B, \qquad \sigma_{xx} = \sigma_{yy} = 0 \tag{3.6}$$

We emphasize that this result is derived in classical mechanics for a non-interacting 2-d electron gas.

Remark. The resistance matrix (ρ_{ij}) is the inverse of the conductivity matrix: $(\rho_{ij}) = (\sigma_{ij})^{-1}$. So eq. (3.6) leads to

$$\rho_{xx} = \rho_{yy} = 0, \qquad \rho_{xy} = \sigma_{yx}^{-1} = B/nce_0 \tag{3.6'}$$

To incorporate a non-vanishing longitudinal resistance, one has to take into account the collisions between electrons and to consider the Langevin equation

$$\frac{d}{dt} <v> = -\frac{eE}{m} + i\omega_c <v> - \frac{<v>}{\tau} \tag{3.7}$$

where we have used the average velocity $<v>$ to replace v in eq. (3.1) and have added the third term to represent the effect of collisions. τ is the relaxation time. In a steady state (in equilibrium), the left-hand side of eq. (3.7) vanishes and one has for the current density

$$j = -ne <v> = \frac{\sigma_o E}{1 - i\omega_c\tau} \tag{3.8}$$

with $\sigma_o \equiv ne^2\tau/m$. For resistances, it follows that

$$\rho_{xx} = \rho_{yy} = 1/\sigma_o, \quad \rho_{xy} = -\rho_{yx} = \omega_c\tau/\sigma_o \tag{3.9}$$

but for conductivities one has

$$\begin{cases} \sigma_{xx} = \sigma_{yy} = \sigma_o/[1+(\omega_c\tau)^2] \\ \sigma_{xy} = -\sigma_{yx} = -\sigma_o\omega_c\tau/[1+(\omega_c\tau)^2] \end{cases} \tag{3.10}$$

In particular,

$$-\sigma_{xy} = \frac{nce}{B} - \frac{\sigma_{xx}}{\omega_c\tau} \tag{3.11}$$

which reduces to eq. (3.6) only when $\tau \to \infty$ or $\sigma_{xx} \to 0$.

Note. In obtaining eqs. (3.10) we have used

$$\sigma_{xx} = \frac{\rho_{xx}}{\rho_{xx}^2 + \rho_{yy}^2}, \qquad \sigma_{xy} = \frac{-\rho_{xy}}{\rho_{xx}^2 + \rho_{yy}^2}$$

2. *Quantum Mechanics*

One needs to solve the Schrödinger equation

$$H\psi \equiv \frac{1}{2m}[(p_x + \frac{e}{c}A_x)^2 + (p_y + \frac{e}{c}A_y)^2]\psi = \varepsilon\psi \tag{3.12}$$

For a uniform magnetic field B we take the Landau gauge $A_x = -B_y$, $A_y = 0$. In this gauge $[H, p_x] = 0$, so we set

$$\psi = \exp\{ik_x x\}\phi(y) \tag{3.13}$$

with

$$\frac{d^2\phi}{dy^2} + \frac{2m}{\hbar^2}\{\varepsilon - \frac{m}{2}\omega_c^2(y-y_o)^2\}\phi=0 \tag{3.14}$$

where

$$y_o \equiv \frac{cp_x}{eB} = \frac{c\hbar k_x}{eB} \tag{3.15}$$

Eq. (3.14) is the Schrödinger equation with a harmonic well with the center at $y=y_o$, so one obtains

$$\phi(y) = \phi_n(y-y_o) \equiv \exp\left\{-\frac{m\omega_c}{2\hbar}(y-y_o)^2\right\} H_n\left[\sqrt{\frac{m\omega_c}{\hbar}}(y-y_o)\right] \tag{3.16}$$

which corresponds to

$$\varepsilon_n = (n+\tfrac{1}{2})\hbar\omega_c \tag{3.17}$$

The energy levels, the so-called Landau levels, are equally spaced like those of a harmonic oscillator with spacing $\hbar\omega_c$. The Landau orbital (3.16) peaks at $y=y_o$; in other words, the parameter y_o defined by eq. (3.15) represents the peak position, or the site, of the Landau orbital (3.16).

One key feature of the solutions is that eq. (3.17) is independent of y_o, and the Landau site y_o depends only on k_x. So each Landau level is highly degenerate. If the system is of finite size (with area L_xL_y) then the allowed values of k_x are discrete: $\Delta k_x=2\pi/L_x$. Moreover the condition $0<y_o<L_y$ implies that $0<|k_x|<(e_oB/\hbar c)L_y$; so the number of allowed values of k_x is finite and given by

$$\frac{e_oBL_y}{\hbar c}\bullet\frac{1}{\Delta k_x} = \frac{e_oBL_xL_y}{hc} = \frac{\phi}{\phi_o}\,. \tag{3.18}$$

Thus the degeneracy, or the number of Landau sites, for each Landau level is just the total flux ϕ ($\equiv BL_xL_y$) threading the planar system in units of the flux quantum ϕ_o ($\equiv hc/e$).

For a 2-d electron gas, if the filling factor is υ, then in total there are $\upsilon\phi/\phi_o$ electrons, which correspond to a surface density $n=(\upsilon\phi/\phi_o)/L_xL_y = \upsilon eB/hc$. If the gas is non-interacting and the classical result (3.6) applies, then one would expect

$$\sigma^H_{xy} = -\frac{nce}{B} = -\upsilon\frac{e^2}{h} \tag{3.19}$$

Now let us show that indeed this classical result for the drift current is still correct in quantum mechanics, although the center of the quantized Landau orbit cannot have definite values for both the guiding center coordinates x_o and y_o simultaneously. (In the gauge we are using, y_o is definite but x_o is not; *see eqs. (3.15) and (3.13)).*

To see this, let us suppose $E_x=0$ and $E_y=E$. In the presence of the electric field, eq. (3.12) is now changed to

$$\tilde{H}\tilde{\psi} = \frac{1}{2m}[(p_x+\frac{e}{c}A_x)^2 + (p_y+\frac{e}{c}A_y)^2]\tilde{\psi}-eE_y\tilde{\psi} = \tilde{\varepsilon}\tilde{\psi} \qquad (3.20)$$

Again, $[\tilde{H},p_x]=0$ and one sets

$$\tilde{\psi} = e^{ik_x x}\tilde{\phi}(y) \qquad (3.21)$$

$$\frac{d^2\tilde{\phi}}{dy^2} + \frac{2m}{\hbar^2}\{\tilde{\varepsilon} - \frac{m}{2}\omega_c^2(y-y_0)^2 + eEy\}\tilde{\phi}=0 \qquad (3.22)$$

The effect of adding the electric field is to shift

$$y_0 \to \tilde{y}_0 = y_0 + \frac{eE}{m\omega_0^2} \qquad (3.23)$$

and

$$\varepsilon_n \to \tilde{\varepsilon}_n = (n+\tfrac{1}{2})\hbar\omega_c + eE\tilde{y}_0 - \frac{m}{2}(\frac{cE}{B})^2 \qquad (3.24)$$

The wave functions are $\tilde{\phi}(y) = \phi_n(y-\tilde{y}_0)$ with ϕ_n given by eq. (3.16). It is easy to see that

$$\begin{cases} \langle v_y\rangle = \frac{1}{m}\langle p_y\rangle = \frac{1}{m}\int dx\,dy\;\tilde{\psi}_n^*(-i\hbar\frac{\partial}{\partial y})\tilde{\psi}_n = 0 \\ \langle v_x\rangle = \frac{1}{m}\langle(p_x - \frac{eB}{c}y)\rangle = \frac{hk_x}{m} - \frac{eB}{mc}\tilde{y}_0 = \frac{cE}{B} \end{cases} \qquad (3.25)$$

From $j_x = -ne\langle v_x\rangle$ and $\sigma_{xy} = j_x/E_y$ one obtains eq. (3.19).

If the system is circular (rather than rectangular), one may use the symmetric gauge $A_x=-B_y/2$, $A_y=B_x/2$. The eigenvalues of eq. (3.12) are still given by eq. (3.17), but the wave functions become

$$\phi_{m,n} = \exp\left\{\frac{1}{4}\frac{x^2y^2}{l^2}(\frac{\partial}{\partial x}+i\frac{\partial}{\partial y})^m(\frac{\partial}{\partial x}-i\frac{\partial}{\partial y})^n\exp\left\{-\frac{x^2+y^2}{4l^2}\right\}\right. \qquad (3.20')$$

where $l\equiv(\hbar c/eB)^{1/2}$, m is the angular momentum, on which the energy does not depend. For the lowest Landau level n=0,

$$\phi_{m,o} = \frac{1}{\sqrt{m!}}z^m\exp\left\{-\frac{|z|^2}{4l^2}\right\} \qquad (3.21')$$

If there are random impurities, the Landau levels are expected to broaden into energy bands. Theoretically we are not quite sure, but the experimental existence of a non-zero quantized Hall conductance is sufficient evidence for the existence of extended states in the presence of magnetic fields *(see below, Sec. 4.1).* The fact that at Hall plateaus $\rho_{xx}=0$ indicates that the Fermi level is in the region of localized states.

4. Quantized Hall Conductance as Topological Invariant (the IQHE Case)

1. Laughlin's Argument for the IQHE

The high precision (up to 10^{-8}) of the observed integral quantization of Hall conductance naturally motivates the idea that this exact quantization must be deeply related to fundamental principles of physics. Soon after the discovery of the IQHE, Laughlin[7] has given a very elegant and general argument for the exact quantization, making clever use of gauge invariance and a non-trivial geometry: a cylinder threaded by a flux-tube. Laughlin's argument has played a very important role in later developments of our theoretical understanding of not only the IQHE but also of the FQHE, including the edge excitations in both cases. Also this argument has been later refined and elaborated into a rigorous mathematical formalism, which shows that the quantized Hall conductance is actually a topological invariant. The topological approach[8,9] was first proposed by Thouless, Kohmoto, Nightingale and den Nijs for the non-interacting case with a periodic potential present. It has been further generalized by Niu, Thouless and Wu[10], and by Avron and Seiler[11], to general many-body cases allowing electron-electron interactions, weak random impurities and crystal imperfections. These topological formalisms have refined and elaborated Laughlin's original argument in two aspects, *i.e.* in uncovering the topological origin of exact quantization and in clarifying the role of the (non-) degeneracy of the ground state. Despite these, the essence of the topological approaches is still that of Laughlin's argument, namely the interplay of gauge invariance and non-trivial topology. Here we give a critical review of this fundamental argument.

Following Laughlin, let us consider a cylindrical system, pierced everywhere by a magnetic field normal to its surface. In the presence of disorder, the Landau levels have been broadened into bands of extended states separated by tails of localized states. The Fermi level is assumed to be in a mobility gap where the localized states occupy. *(See Fig. 5.)*

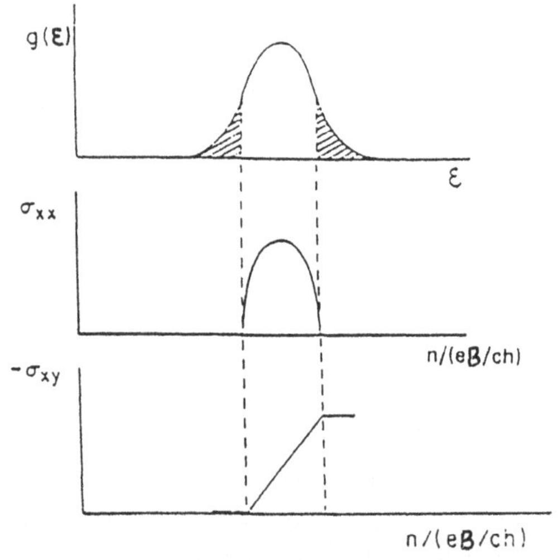

Fig. 5. The structure of broadened Landau levels and the density of states The shaded region represents localized states; unshaded extended states

Now let a flux-tube thread through the hole formed by the cylindrical surface with flux adiabatically turned on from 0 to ϕ_0 (unit flux). The gauge invariance of the electromagnetic field[12] tells us that a unit flux ϕ_0 in the thin tube is equivalent to no flux at all, since it can be gauged away by a gauge transformation. (Note the latter is not legal unless the flux is an integral multiple of ϕ_0.) Therefore after adding a unit flux ϕ_0, the Hamiltonian is back to that with no flux, and the system is back to itself with possibly excitation or de-excitation of the original one, by the quantum adiabatic theorem[13]. On the other hand, there is a gap in the spectrum of the system which is supposed to remain unclosed during this process, so an adiabatic change of the many-body Hamiltonian cannot excite quasiparticles across this gap. [For this to be true, the flux can not pierce the system; when the flux pierces the system a quasiparticle excitation is possible *(see Sec. 5.2).*] Therefore adding a unit flux can only produce an excitation of charge transfer, *i.e.* the energy increases due to the net transfer of N (integer) electrons from one edge to the other. (For N to be integer, the ground state has to be non-degenerate otherwise because of the wave nature of electrons in quantum mechanics, the transfer of a fraction of electron is possible and indeed this occurs in the FQHE. The non-degeneracy condition was not stated in Laughlin's original paper. The role of this condition can be clearly seen in the topological approach, *see below Sec. 5.2 and 5.3.*) The energy change for such a charge-transfer is obviously

$$\Delta U = Ne\, E_y L_y = Ne\, V_y \tag{4.1}$$

and the current around the cylinder is given by

$$I_x = c\,\frac{\Delta U}{\Delta \phi} = \frac{cNeV_y}{hc/e} = N\,\frac{e^2}{h}\,V_y \tag{4.2}$$

Therefore, the Hall conductance is always quantized:

$$\sigma_H = \frac{I_x}{V_y} = N\,\frac{e^2}{h} \qquad \text{(N: integer)} \tag{4.3}$$

To see more clearly how there can be a charge transfer between edges, let us examine the non-interacting case. From eqs. (3.15) and (3.23) we see that the uniform increment $\Delta A = \Delta\phi/L$ in A_x, due to a small change $\Delta\phi$ in ϕ, leads formally $\Delta p_x = \frac{e}{c}\Delta A$ and therefore $\Delta y_0 = \Delta A/B$. Thus the centre of the Landau site is shifted by $\Delta A/B$. After adding a unit flux $\Delta\phi = hc/e$, the Landau site is just shifted by one:

$$\Delta y_0 = \frac{\Delta A}{B} = \frac{hc}{eBL_x} = \frac{L_y}{\phi/\phi_0} = \text{site spacing} \tag{4.4}$$

Therefore, one electron is transferred from one edge to the other for each fully filled Landau level. The integer N is directly related to the number of fully occupied Landau levels in the non-interacting case:

$$N = \nu \Rightarrow \sigma_H = \nu\,\frac{e^2}{h} \tag{4.5}$$

If the system is dirty, the above argument can not tell us what the integer N should be. It can be zero, as is in most systems with gaps. For N to be non-zero, there should be a long-range phase coherence around the cylinder which needs the existence of extended electronic states in the sample. For if all states are localized, the only effect of the flux in the hole is to multiply each localized wave function by a gauge factor $\exp\{ieAx/\hbar c\}$ (x is the coordinate of the point where the state is localized) and the energy change and current are both zero. For extended states such a gauge transformation is illegal unless $\phi = n\phi_o$, so that a non-vanishing Hall current is possible.

Experimentally, the samples are always dirty and interacting systems, and the observed quantization N on a plateau is always directly related to a nearby integral filling factor υ by eq. (4.5). To establish this relation, one needs the topological invariance of the Hall conductance σ_H and applies it to the process, in which various interactions in the sample are turning off until the system becomes non-interacting, assuming the gap remains open the whole way. This is the subject of the next two subsections.

2. *Non-interacting Electrons in a Periodic Potential*

In this case we have the Hamiltonian for a single electron

$$\hat{H} = \frac{1}{2m} (\vec{p} + \frac{e}{c} \vec{A})^2 + V(x,y) \tag{4.6}$$

where $V(x+a,y) = V(x,y+b) = V(x,y)$, (a,b) being lattice spacing. Again we take the Landau gauge: $A_x=0$, $A_y= Bx$.

Let us introduce the generators of magnetic translation:

$$\begin{cases} \Pi_x = p_x + \frac{e}{c} A_x + \frac{e}{c} By = - i\hbar \frac{\partial}{\partial x} + \frac{e}{c} By \\ \Pi_y = p_y + \frac{e}{c} A_y - \frac{e}{c} Bx = - i\hbar \frac{\partial}{\partial y} \end{cases} \tag{4.7}$$

They commute with the kinematic momentum operators

$$[\Pi_i, p_j + \frac{e}{c} A_j] = 0 \qquad \text{(for i,j = x,y)} \tag{4.8}$$

Therefore, the magnetic translation by a lattice spacing in either x or y direction is a symmetry: if we define

$$T_a = \exp\{\frac{i}{\hbar} a\Pi_x\} , \qquad T_b = \exp\{\frac{i}{\hbar} b\Pi_y\} \tag{4.9}$$

then

$$[T_a,H] = [T_b,H] = 0 \tag{4.10}$$

However, T_a and T_b do not commute with each other; they rather satisfy

$$T_aT_b = T_bT_b\, e^{-i2\pi\phi} \tag{4.11}$$

where $\phi \equiv (eB/hc)ab$ is the flux per unit cell in units of flux quantum. Now we assume $\phi=p/q$ where p and q are two integers which are mutually prime to each other. Then $T_{qa} \equiv (T_a)^q$ commutes with T_b, since qa and b form a bigger magnetic unit cell, which is the smallest area which contains an integral multiple of flux quanta. Therefore, we can simultaneously diagonalize H, T_{qa} and T_b.

Applying the Bloch theorem, one has

$$T_{qa}\psi = e^{ik_1qa}\psi , \qquad T_b\psi = e^{ik_2b}\psi \tag{4.12}$$

then

$$0 \le k_1 \le 2\pi/qa , \qquad 0 \le k_2 \le 2\pi/b \tag{4.13}$$

Note that the magnetic Brillouin zone is q times smaller than the original one in the absence of a magnetic field. The wavefunction satisfying eq. (4.12) is of the form

$$\psi^{(\alpha)}_{k_1k_2}(x,y) = e^{i(k_1x+k_2y)}\, U^{(\alpha)}_{k_1k_2}(x,y) \tag{4.14}$$

with $U^{(\alpha)}$ satisfying the twisted periodic conditions

$$\begin{cases} U^{(\alpha)}_{k_1k_2}(x+qa,y) = e^{-i2\pi py/b}\, U^{(\alpha)}_{k_1k_2}(x,y) \\ U^{(\alpha)}_{k_1k_2}(x,y+b) = U^{(\alpha)}_{k_1k_2}(x,y) \end{cases} \tag{4.15}$$

where α is the band index. Substituting eq. (4.14) into eq. (4.6), we see that $U^{(\alpha)}_{k_1k_2}$ satisfies the Schrodinger equation

$$\tilde{H}(k_1,k_2)U^{(\alpha)}_{k_1k_2}(x,y) = \varepsilon^{(\alpha)}_{k_1k_2}\, U^{(\alpha)}_{k_1k_2}(x,y) \tag{4.16}$$

with

$$\tilde{H}(k_1,k_2) = \frac{1}{2m}\left(-i\hbar\frac{\partial}{\partial\vec{x}} + \hbar\vec{k} + \frac{e}{c}\vec{A}\right)^2 + V(x,y) \tag{4.17}$$

From the band theory, this implies that a Landau level in the presence of a magnetic field is split into a number of bands by the lattice potential, or that a band in the lattice potential is split into a number of sub-bands by the magnetic field. The details of the splitting pattern depend on the lattice potential V(x,y).

The non-interacting electrons fill the single-electron energy levels from the bottom up. Now we are ready to show that if the Fermi level lies in a gap between two sub-bands, then the Hall conductance is an integer in units of e^2/h. Let us start with the Kubo formula in the linear response theory for σ_{xy}:

$$\sigma_{xy} = ie^2h \sum_{\varepsilon_\alpha<E_F<\varepsilon_\beta} \frac{(V_x)_{\alpha\beta}(V_y)_{\beta\alpha}-(V_y)_{\alpha\beta}(V_x)_{\beta\alpha}}{(\varepsilon_\alpha-\varepsilon_\beta)^2} \tag{4.18}$$

where we sum over all states α below E_F and states β above E_F and $(\vec{V})_{\alpha\beta}$ is the matrix element

$$(\vec{V})_{\alpha\beta} = \frac{1}{m} \delta_{k_1 k_1'} \delta_{k_2 k_2'} \int_0^{qa} dx \int_0^{b} dy\, U^{(\alpha)*}_{k_1 k_2} (-i\hbar \frac{\partial}{\partial \vec{x}} + \frac{e}{c} \vec{A}) U^{(\beta)}_{k_1' k_1'}$$

Note that by definition $\alpha \neq \beta$ ($\because$ $\varepsilon_\alpha < E_F < \varepsilon_\beta$), so that $<\alpha|\vec{k}|\beta> = \vec{k} <\alpha|\beta> = 0$. This leads to (for i=1,2)

$$\hbar (V_i)_{\alpha\beta} = (\varepsilon_\beta - \varepsilon_\alpha) < \alpha|\frac{\partial \beta}{\partial k_i}> = (\varepsilon_\alpha - \varepsilon_\beta) < \frac{\partial \alpha}{\partial k_i}|\beta> \tag{4.19}$$

In fact,

$$\hbar (V_i)_{\alpha\beta} = <\alpha|\frac{\partial \hat{H}}{\partial k_i}|\beta> = <\alpha|\frac{\partial}{\partial k} (\hat{H}|\beta>) - <\alpha|\hat{H}|\frac{\partial \beta}{\partial k_i}>$$

$$= <\alpha|\frac{\partial}{\partial k_i} (\varepsilon_\beta|\beta>) - <\alpha|\hat{H}|\frac{\partial \beta}{\partial k_i}> = (\varepsilon_\beta - \varepsilon_\alpha) < \alpha|\frac{\partial \beta}{\partial k_i}>$$

where we have used $(\partial \varepsilon_\beta / \partial k_i) <\alpha|\beta> = 0$. Substituting eq. (4.19) into (4.18) one has

$$\sigma_{xy} = \frac{ie^2}{\hbar} \sum_{\varepsilon_\alpha < E_F < \varepsilon_\beta} \left\{ <\frac{\partial U^\alpha}{\partial k_1}|\beta> <\beta|\frac{\partial U^\alpha}{\partial k_2}> - <\frac{\partial U^\alpha}{\partial k_2}|\beta> <\beta|\frac{\delta U^\alpha}{\partial k_1}> \right\}$$

Using

$$\sum_{\varepsilon_\alpha < E_F < \varepsilon_\beta} \{ |\alpha> <\alpha| + |\beta> <\beta| \} = 1$$

We finally rewrite the Hall conductance into the form

$$\sigma_{xy} = \sum_{\varepsilon_\alpha < E_F} \sigma^{(\alpha)}_{xy} \tag{4.20}$$

$$\sigma^{(\alpha)}_{xy} = \frac{e^2}{h} \frac{1}{2\pi i} \int d^2k \left\{ <\frac{\partial U^\alpha}{\partial k_2}|\frac{\partial U^\alpha}{\partial k_1}> - <\frac{\partial U^\alpha}{\partial k_1}|\frac{\partial U^\alpha}{\partial k_2}> \right\} \tag{4.21}$$

where the integral $\int d^2k$ is taken over the magnetic Brillouin zone $(0 \leq k_x \leq 2\pi/q_a,\ 0 \leq k_y \leq 2\pi/b)$.

Eq. (4.20) tells us that the Hall conductance σ_{xy} is the sum of the contributions $\sigma^{(\alpha)}_{xy}$ from each filled Landau sub-bands α below the Fermi level. More important, eq. (4.21) implies that $\sigma^{(\alpha)}_{xy}$ in units of e^2/h is a topological invariant, the so-called first Chern class, which can only be integers. To see this, we note that if we take the phases of wavefunctions $U^{(\alpha)}_{k_1 k_2}(x,y)$ to be smoothly varying inside the magnetic Brillouin zone, then by using Stokes' theorem in eq. (4.21) we get a line integral along the boundary of the zone

$$U^{(\alpha)}_{xy} = \frac{e^2}{h} \bullet \frac{-1}{2\pi i} \oint_C dk_i <U^\alpha|\frac{\partial}{\partial k_i}|U^\alpha> = \frac{e^2}{h} \bullet n_\alpha \tag{4.22}$$

where n_α is an integer, since the line integral just represents the total phase change of the wave function $U^{(\alpha)}_{k_1k_2}$ along the boundary which, by the single-valuedness of $U^{(\alpha)}_{k_1k_2}$, must be an integral multiple of 2π. Alternatively the set of the wavefunctions $\{U^{(\alpha)}_{k_1k_2}(x,y);\ 0\le k_1\le 2\pi/qa,\ 0\le k_2\le 2\pi/b\}$ forms a cross-section of a (twisted) complex line bundle over the magnetic Brillouin zone, whose topology is actually a torus. A gauge potential (or Berry's connection) is given by

$$A^{(\alpha)}_i(k_1,k_2) = -i < U^{(\alpha)} | \frac{\partial}{\partial k_i} | U^{(\alpha)} > \tag{4.23}$$

and the integrand in eq. (4.21) is nothing but the field strength of this potential. So $\sigma^{(\alpha)}_{xy}$ is the first Chern number of the bundle, which represents the total "magnetic flux" through the torus and is well-known to be quantized. Eqs. (4.20) and (4.21), together with the above topological-invariant interpretation, was first obtained by Thouless, Kohmoto, Nightingale and den Nijs.[8]

3. The General Case

The realistic samples exhibiting IQHE, of course, are not the idealistic non-interacting electron gas in a perfect crystal. Rather there are impurities, imperfections and electron-electron interactions. Can we generalize the derivation of the integrally quantized Hall conductance as a topological invariant to the general case with various complications allowed? An indication that this is possible is Prange's explicit proof[13] that if there is an isolated δ-function impurity, even though it binds a localized state, the remaining delocalized states carry exactly enough extra current to compensate for its loss. The discussion of the general case allowing all possible complications was given first independently by Niu, Thouless and Wu[10] and by Avron and Seiler[11].

For a realistic sample, certainly one needs to consider a many-body Hamiltonian to incorporate various interactions:

$$H = \sum_j \frac{1}{2m}\left[-i\hbar \frac{\partial}{\partial \vec{r}_j} + \frac{e}{c} \vec{A}(\vec{r}_j)\right]^2 + \sum_j U(\vec{r}_j) + \sum_{i<j} V(|\vec{r}_i-\vec{r}_j|) \tag{4.24}$$

where $\vec{r}_j$ are 2-d coordinates of the jth electron; the second term $U(\vec{r}_j)$ stands for the interactions of the electron with the positively charged background and with impurities; the third term the interactions among electrons. In the following derivation we do not need the detailed knowledge of these interactions. To generalize the derivation in the last sub-section, we note that the key point there is that the Hamiltonian (4.17) contains two parameters (k_1,k_2), which essentially form a torus. How to introduce similar parameters in the many-body Hamiltonian? Our key observation is that the twisted boundary conditions for the many-body wave function will do the job:

$$\begin{cases} \psi(\cdots,x_i+L_1,\cdots) = e^{i\theta_1} e^{-i(eB/\hbar)y_i L_1} \psi(\cdots,x_i,\cdots) \\ \psi(\cdots,y_i+L_2,\cdots) = e^{i\theta_2} \psi(\cdots,y_i,\cdots) \end{cases} \tag{4.25}$$

Here (θ_1,θ_2) are parameters characterizing the boundary conditions on the sample, which is assumed to be a rectangle with sides L_1 and L_2. Obviously (θ_1,θ_2) live on a torus: $0\le\theta_1,\theta_2\le 2\pi$. Because of the identity of electrons, (θ_1,θ_2) are the same for all electrons. A more symmetric and covariant form of the boundary condition is

$$t_i(L_1)\psi = e^{i\theta_1}\psi\,, \qquad t_i(L_2)\psi = e^{i\theta_2}\psi \tag{4.25'}$$

where $t_i(L_1)$ and $t_i(L_2)$ are magnetic translations for the ith electron, similar to T_a, T_b in eq. (4.9). In the particular gauge $\vec{A}=(0,Bx)$, eqs. (4.25′) reduce to eqs. (4.25). Here we have assumed that the flux through the sample, $L_1L_2(eB/hc) \equiv \phi/\phi_0 = N_\phi$, is an integer in units of ϕ_o, so that $t_i(L_1)$ commutes with $t_i(L_2)$ and, therefore, the two conditions in eqs. (4.25′) are compatible to each other. For the many-body system, the Kubo formula takes the form

$$\sigma_H = \frac{ie^2\hbar}{L_1L_2}\sum_{n>0}\frac{(V_x)_{on}(V_y)_{no} - (V_y)_{on}(V_x)_{no}}{(E_o-E_n)^2} \tag{4.26}$$

where 0 refers to the ground state and n(>0) excited states;

$$V_x = \sum_i \frac{1}{m}\left(-i\hbar\frac{\partial}{\partial x_i}\right),\quad V_y = \sum_i \frac{1}{m}\left(-i\hbar\frac{\partial}{\partial y_i} + \frac{e}{c}Bx_i\right) \tag{4.27}$$

In order to absorb the parameters (θ_1,θ_2) characterizing the boundary conditions into the Hamiltonian, we make the following gauge transformation:

$$\psi_n = \exp\left\{i\frac{\theta_1}{L_1}(x_1+\cdots+x_N) + i\frac{\theta_2}{L_2}(y_1+\cdots+y_N)\right\}\phi_n \tag{4.28}$$

Then, the Hamiltonian for the new wave function ϕ_n is

$$\left\{\begin{array}{l} H\to\tilde{H} = H\left[-i\hbar\dfrac{\partial}{\partial x_i}\to -i\hbar\dfrac{\partial}{\partial x_i} + \dfrac{\hbar\theta_1}{L_1}\,,\ -i\hbar\dfrac{\partial}{\partial y_i}\to i\hbar\dfrac{\partial}{\partial y_i}+\dfrac{\hbar\theta_2}{L_2}\right] \\[2ex] V_x\to\tilde{V}_x = \dfrac{L_1}{\hbar}\dfrac{\partial\tilde{H}}{\partial\theta_1}\,,\ V_y\to\tilde{V}_y = \dfrac{L_2}{\hbar}\dfrac{\partial\tilde{H}}{\partial\theta_2}\end{array}\right. \tag{4.29}$$

Substituting in eq. (4.26) and with the same manipulations similar to those between eqs. (4.18) and (4.20), one obtains

$$\sigma_H = \frac{ie^2}{\hbar}\sum_{n>0}\frac{1}{(E_o-E_n)^2}\left\{<\phi_0|\frac{\partial\tilde{H}}{\partial\theta_1}|\phi_n><\phi_n|\frac{\partial\tilde{H}}{\partial\theta_2}|\phi_0> - (\theta_1\Leftrightarrow\theta_2)\right\}$$

$$= \frac{ie^2}{\hbar}\left\{<\frac{\partial\phi_o}{\partial\theta_1}|\frac{\partial\phi_o}{\partial\theta_2}> - <\frac{\partial\phi_o}{\partial\theta_2}|\frac{\partial\phi_o}{\partial\theta_1}>\right\} \tag{4.30}$$

By this expression, σ_H depends on (θ_1,θ_2), the parameters in the boundary conditions, which seem to be fixed. However, when we measure σ_H in a sample, there must be an external electric field $\vec{E}$ which drives the boundary phases varying in time. In fact, since $\vec{E}=(-1/c)\partial\vec{A}/\partial t$, imposing $\vec{E}$ leads to adding a term $-c\vec{E}t$ to $\vec{A}$. If $\vec{E}$ is small, one may apply the adiabatic approximation, *i.e.* use the above result with the time-varying parameters $\theta_i(t) \equiv \theta_i-(e/\hbar)E_iL_it$. (In the thermodynamic limit, the bulk properties of the sample is isotropic, so we do not need to assume $\vec{E}$ is only in x- or y-direction.) So the measured Hall conductance $\bar{\sigma}_H$ should be obtained by averaging eq. (4.30) over (θ_1,θ_2):

$$\bar{\sigma}_H = \frac{1}{(2\pi)^2}\int_0^{2\pi} d\theta_1 \int_0^{2\pi} d\theta_2\ \sigma_H(\theta_1,\theta_2)$$

$$= \frac{e^2}{h} \cdot \frac{1}{2\pi i}\iint d\theta_1 d\theta_2 \left\{ \langle\frac{\partial\phi_o}{\partial\theta_2}|\frac{\partial\phi_o}{\partial\theta_1}\rangle - \langle\frac{\partial\phi_o}{\partial\theta_1}|\frac{\partial\phi_o}{\partial\theta_2}\rangle \right\} \tag{4.31}$$

The form of this expression is exactly the same eq. (4.21); by the same reasoning below eq. (4.21), $\bar{\sigma}_H$ is quantized to be an integral multiple of e^2/h. Also it acquires the interpretation as a topological invariant.

Before discussing the consequences of eq. (4.31) I would like to first emphasize the implicit assumptions we have made in the derivation of integral quantization of the Hall conductance. First we have assumed that the ground state of the system has to be separated from other states by energy gaps which do not become zero either in the thermodynamic limit or in the whole range of $0 \le \theta_1,\ \theta_2 \le 2\pi$. Otherwise linear response theory would not apply because the Zener tunneling would become important for infinitesimal gaps. The arguments of topological invariance also require the opening of the gap plus the assumption that the Fermi level lies always in this gap. Finally the integral quantization is true only when the ground state is non-degenerate; For this guarantees that the ground state comes back to itself (up to a phase) as, say, θ_1 changes from 0 to 2π, so that the integral (4.31) is over a compact torus. Under these assumptions our above derivation is valid, no matter what complications (impurities, imperfections or electron-electron interaction) may exist in the sample.

The topological-invariant interpretation of σ_H provides us a rationale for why its integral quantization has been so precise as up to 10^{-8} in some experiments. (Remember that the real samples are always very "dirty" and vary case to case. Generally one does not expect high precision in measuring a quantity in condensed matter physics.) Also it explains why the IQHE is stable against weak perturbations and various complications that may happen in real samples. Especially the topological invariance explains the existence of the plateaus for σ_H provided that the gap persists and the Fermi level remains in the gap. Finally, if the ground state of the sample can be obtained by continuously turning on the impurities and various interactions from non-interacting electrons at an integral filling with a persisting gap and with the Fermi level remaining in that gap, then the topological invariance asserts that the Hall conductance must be locked at the integer corresponding to that integral filling. This is what we believe happens for states on the plateaus of σ_H in the IQHE, which establishes the relationship between quantized σ_H and the nearest integral filling factor as observations show.

5. Laughlin Wave Functions for the FQHE

1. Similarities and Distinctions between FQH and IQH States

Now we proceed to discuss the FQHE, which looks to the eye like the IQHE, except that the Hall conductance is fractional in units of e^2/h at certain fractional fillings *(see Sec. 2).* The most notable case is the $\upsilon=1/3$ FQHE with $\sigma_H=(1/3)(e^2/h)$. Despite the similarity with the IQHE, a FQH ground state must differ from the IQH states at least in that the former cannot be obtained from non-interacting electrons by continuously turning on interactions without undergoing a sort of "phase transition"; for otherwise if none of the conditions for the topological arguments is violated, one would have integral rather fractional Hall conductance. This suggests that a FQH state must be a new type of many-body condensate, which has certain unusual properties that normal IQH states do not have, albeit sharing some other properties with the latter.

The most important common feature of the FQH and IQH states turns out to be that both of them are incompressible quantum fluid states (in the absence of impurities), because there are downward cusps in E_g (the ground state energy *vs.* B (the magnetic field) for fixed electron density n at particular $B_0=\upsilon^{-1}(hc/e)n$, or in E_g *vs.* n for fixed B at particular $n_0=\upsilon(eB/hc)$, with υ being integers or simple rationals. A downward cusp implies a gap in density of states (spectral function of Green's function) at T=0 with the chemical potential in that gap. So microscopically the incompressibility is equivalent to the property that all quasiparticle excitations above the ground state have a gap (or cost a finite amount of energy). For free electrons, integral $\upsilon=1,2,3,..$ correspond to completely filled Landau levels, and it is easy to understand the incompressibility of the IQH states in view of Fermi statistics. However at fractional values of υ, which correspond to only partially filled Landau levels, the incompressibility is a non-trivial property that originates from the interactions between electrons.

In addition to explaining the incompressibility, the theory of FQHE must be capable of explaining the distinct features of FQHE which are not shared by IQH states. Intuitively the IQHE may be viewed as a spectroscopy of the electron charge; by analogy, one expects that the FQHE should be interpreted in terms of spectroscopy of a quasiparticle charge that is fractional. Namely if fractionally charged quasiparticles exist and behave more or less like electrons or holes in the IQHE, then the Hall plateaus observed in the FQHE can be understood as due to localization of these quasi-particles. Furthermore, the formation of other stable, incompressible FQH states at values of υ other than 1/3, including the odd-denominator rule for $\upsilon<1$, should be explicable by condensation of the fractionally charged quasiparticles (to form new interpenetrating fluids), which turn out to obey exotic fractional statistics. Another feature of the FQH states is the ground state degeneracy on a torus. The existence of this degeneracy can be inferred from a generalized argument[14] of gauge invariance a lá Laughlin. It is well-known that a FQH ground state is non-degenerate on a disc or a sphere geometry. Thus for a FQH system the ground-state degeneracy must depend on the topology of the surface that the system lives in. In this and the following sections we will concentrate on these distinct features of the FQHE which can be viewed as manifestations of a non-trivial topological structure, or topological order, contained in the FQH states.

2. *Laughlin Wave Function for υ=1/m Ground States*

Let us start with Laughlin's microscopic theory[16] of the FQH system with $\upsilon=1/m$ (m odd). We will ignore the spin degrees of freedom which are frozen out by a strong magnetic field. In a study of the three-electron case, Laughlin made the observation that electrons in a strong magnetic field, lying in the lowest Landau level, like to avoid each other. Motivated by this observation, he has proposed a many-body wave function for the ground state on a circular disc geometry as follows:

$$\Psi_0^{(m)} = \prod_{i<j}^{N} (z_i - z_j)^m \exp\left\{-\frac{1}{4}\sum_{i=1}^{N} z_i \bar{z}_i\right\} \tag{5.1}$$

Here $z_i = x_i + y_i$ in units of the magnetic length $l = \sqrt{\hbar c/eB}$, which we have set equal to one, is the comoplex coordinate for the i-th electron. The integer m has to be odd, in order for the wave function to be totally anti-symmetric. The mathematical features and physical meanings of the Laughlin wave function (5.1) are the following[17]:

1) The prefactor $f(z_1,..,z_N) \equiv \prod_{i<j}^{N} (z_i - z_j)^m$ is an analytic function of z_i, so the wave function (5.1) is comprised solely of single-electron wave functions lying in the lowest Landau level (in the symmetric gauge $(A_x, A_y) = (\frac{B}{2}y, -\frac{B}{2}x)$; *see eq. (3.21)*). This should be a good approximation, since in a strong magnetic field, the Coulomb potential is "small" compared to the cyclotron energy $\hbar\omega_c$.

2) The prefactor is a homogeneous polynomial in z_i of degree $M = mN(N-1)/2$. So (5.1) is an eigenstate of total angular momentum with M.

3) The prefactor is of the Jastrow form with each factor $(z_i - z_j)^m$ having a m-th order zero at $z_i = z_j$. So the electrons want badly to keep far apart from each other, and each electron sees m zeros bound to the positions of the other electrons.

4) If z_i goes around z_j counter-clockwise by an angle ϕ, then the factor $(z_i - z_j)^m$ gives rise to a phase $e^{im\phi}$. Thus, the phase correlation of electrons is as if each electron carries m flux quanta.

5) The square of the modulus of (5.1) can be written as a classical Boltzmann distribution

$$|\Psi_0^{(m)}(z_1,\bullet\bullet\bullet,z_N)|^2 = \exp\{-\beta V_{eff}(z_1,\bullet\bullet\bullet,z_N)\} \tag{5.2}$$

With the fictitious temperature is set to $1/\beta = m$ such that

$$V_{eff}(z_1,\bullet\bullet\bullet,z_N) = -2m^2 \sum_{i<j} ln|z_i - z_j| + \frac{m}{2}\sum_k |z_k|^2 \tag{5.3}$$

coincides with the potential energy of a 2-d one-component plasma: the first term in (5.3) represents the repulsion between particles of charge m via the Coulomb interaction which is logarithmic in two dimensions; the second term is the attraction of these charges to the origin due to a uniform neutralizing background on the same circular disc geometry of charge density $\rho_0=1/2\pi l^2$. This analog reduces the calculation of expectation values with the wavefunction (5.1) to a corresponding one in the equivalent plasma. In particular, the neutrality of the plasma tells us that the electron density in the state (5.1) is uniform and equal to

$$\rho_e^{(m)} = \frac{\rho_o}{m} = \frac{1}{2\pi m l^2} = \frac{1}{m} \bullet \frac{eB}{hc} \tag{5.4}$$

In other words, the state (5.1) has the filling factor $\upsilon=1/m$, so that on the average there are m flux quanta for one electron.

6) In eqs. (5.2) and (5.3) the ratio of potential energy to temperature is $\Gamma=2m$, which means the fictitious temperature of the equivalent 2-d plasma is sufficiently high for m=1,3,4,•••, the cases of interest for FQHE. Monte Carlo studies of 2-d plasma have shown that the equivalent plasma at such temperatures is a liquid rather than a Wigner crystal and so is the FQH states (5.1) with $\upsilon=1$, 1/3, 1/5, •••. The liquid nature of the FQH states guarantees the uniformity on length scales smaller than the inter-particle spacing.

In summary, the Laughlin wave function (5.1) describes a uniform circular liquid droplet for a FQH state with $\upsilon=1/m$ (m odd), in which electrons lie in the lowest Landau orbit and keep far apart from each other, and have phase correlations as if each electron carries m flux quanta. The justification that such a state gives the ground state for the corresponding FQH system comes from two arguments. The first is a variational one, *i.e.* the ground state energy evaluated with (5.1) is substantially below the competing Hartree-Fock-Wigner-crystal charge-density-wave states as $\upsilon>0.1$ or $m<10$. The second argument, given by Haldane,[18] goes as follows: first in a pseudo-potential approach[19] it is shown that the Laughlin wave function is the exact and unique ground state of a truncated Hamiltonian representing short-range components of the electron-electron interactions. Then one may argue that when the long-range part of the interactions is continuously turning back, no phase transition in the ground state occurs; at least this can be verified by numerical studies of finite-size systems.[19] It is clear from this approach that the Laughlin wave function (5.1) is exceedingly rigid and unresponsive to variations in the form of e-e repulsive potential and particularly to variations in its short-range part. This is exactly what one expects from the property that in the state (5.1) electrons keep far apart from each other as a consequence of the Jastrow form.

3. *Laughlin Wave Function for Quasiparticles*

To generate a fractionally charged quasi-particle one may follow a thought experiment[16]: pierce the ground state (5.1) with an infinitesimally thin flux-tube at the point Z_0 and adiabatically turn on the flux from zero to a flux quantum $\phi_0\equiv hc/e$. During the process the state of the quantum liquid droplet remains to be always an eigenstate of the changing Hamiltonian. Since the changing flux induces a circular electric field, due to the Hall transport the surrounding electrons will flow inward to or outward from the point Z_0, depending on the direction of the flux. So as $\phi\to\phi_0$ we end up with some positive or negative charge accumulated in an area of size l around Z_0. By gauge invariance (a flux quantum is equivalent to no flux), this final state is an excited state of the original Hamiltonian. This excitation is a quasiparticle: either a quasihole or a quasi-electron, depending on the sign of the accumulated charge. From the known FQHE: $\sigma_H=(1/m)(e^2/h)$, it is easy to calculate the accumulated charge to be $\pm(1/m)e$. Thus, the quasi-particles are fractionally charged.

For definiteness, let us consider quasi-holes with $+(1/m)e$. They are formed by depletion of (1/m) electrons in an area of size l. So it is natural to suggest the many-body wavefunction for a state with a quasi-hole centered at Z_0 to be of the form

$$\Psi^{(m)}_{+1}(z_1,\cdots,z_N) = \prod_{i=1}^{N} (z_i-Z_0) \prod_{i<1}^{N} (z_i-z_j)^m \exp\{-\frac{1}{4}\sum_l |z_l|^2\}$$

$$= \prod_{i=1}^{N} (z_i-Z_0)\, \Psi^{(m)}_0 (z_1,\cdots,z_N) \qquad (5.5)$$

Numerical calculations indicate this wavefunction for the quasi-hole is rather accurate: it really accumulates an excess charge $+e/m$. Also it is shown numerically that the creation energy of a quasi-hole is about 0.026 e^2/l which represents a non-vanishing energy gap. This tells us that the FQH fluid is incompressible. The incompressibility comes from the rigidity of the Laughlin wavefunction (5.1) and the Coulomb interactions which give rise to finite energy cost for fractionally charged quasi-particles. Laughlin has also shown[17] that quasi-particles act like electrons or holes in the lowest Landau level. Not only their size, charge and energy do not depend on Z_0, but also they execute cyclotron motion like electrons, except that the orbit radius is $m^{1/2}$ times large (due to $e^*=e/m$).

From the argument given above one can see that if we increase or decrease the total flux through the system so that υ deviates from the exact fraction 1/m, then we will generate fractionally charged quasi-particles in the system. Like in the IQHE, the impurities cause the localization of these quasi-particles which leads to the observed Hall plateau with σ_H locked at $(1/m)(e^2/h$. The existence of Hall plateaus at other values of υ is explained by condensation of quasi-particles to form new penetrating incompressible quantum fluids. At which fraction υ the new stable "daughter" states can be formed depends on the statistics of the quasi-particles, in a way similar to the fact that the original stable states occur at $\upsilon=1/m$ with m odd in view of the Fermi statistics of electrons. In the following we will present a direct and elegant derivation of fractional statistics starting from Laughlin wavefunctions.

6. Fractional Statistics and the FQHE

1. Fractional Statistics of the Quasiparticles

According to Arovas, Schrieffer and Wilczek,[20] the charge and statistics of the quasiparticles can be determined by a direct method based on the concept of Berry's phase.[21] To determine the charge, one calculates the change of phase γ of the quasi-hole wavefunction (5.5) as the quasi-hole location Z_0 adiabatically moves around a closed loop of radius R enclosing flux ϕ. We note that Z_0 appears as a parameter in the many-body wavefunction (5.5), so that the phase change consists of usual "dynamical" phase $\int^t E(\tau)d\tau$, where $E(\tau)$ is the energy of the state, plus Berry's phase which is independent of how slowly the loop is transversed. (For details about Berry's phase, see Vinet's lectures in these proceedings.) Berry's phase $\gamma(t)$ satisfies

$$\frac{d\gamma}{dt} = i < \Psi^{(m)}_{+1}(Z_0)|\frac{d}{dt}|\Psi^{(m)}_{+1}(Z_0)>$$
$$= i < \Psi^{(m)}_{+1}(Z_0)|\frac{d}{dt}\sum_i \ln\,[z_i - Z_0(t)]|\Psi^{(m)}_{+1}(Z_0)> \tag{6.1}$$

To make it manageable, we note that the electron density in the state is given by

$$\rho_{+1}(z) = < \Psi^{(m)}_{+1}(Z_0)|\sum_i \delta(z_i - z)|\Psi^{(m)}_{+1}(Z_0)> \tag{6.2}$$

Thus

$$\frac{d\gamma}{dt} = i \int d^2z\, \rho_{+1}(z)\, \frac{d}{dt}\, \ln[z - Z_0(t)] \tag{6.3}$$

Now integrate $Z_0(t)$ along the loop C in a counter-clockwise sense. The time-derivative term has non-vanishing contribution only from z inside C which is $-2\pi i$:

$$\gamma = \int \frac{d\gamma}{dt}\, dt = 2\pi \int_D d^2z\, \rho_{+1}(z) \tag{6.4}$$

where D is the region enclosed by C. If we ignore the correction to the background density ρ_0 arising from the quasi-particle, which is of the order of $(l/R)^2$ (l being the magnetic length), we obtain Berry's phase

$$\gamma = 2\pi \int_D d^2z\, \rho_0 = 2\pi <n>_D = 2\pi\upsilon\phi/\phi_0 \tag{6.5}$$

where $<n>_D$ is the number of electrons in D. This extra phase should be interpreted as the Aharanov-Bohm phase that a charge e* would gain in moving around this loop:

$$(e^*/\hbar c) \oint_C \vec{A}\cdot\vec{dl} = 2\pi(e^*/e)\phi/\phi_0 \tag{6.6}$$

Comparing (6.6) with (6.5), this determines the charge e* of the quasi-hole to be $e^* = \upsilon e$. This derivation shows that the charge of the quasiparticle and the charge density of the ground state are essentially the same thing.

To determine the statistics of the quasi-particles, one needs to consider the state with two quasi-holes at Z_1 and Z_2: by generalizing (5.5) we have

$$\Psi^{(m)}_{+2}(Z_1,Z_2) = \prod_i (z_i - Z_1)(z_i - Z_2)\; \Psi^{(m)}_0(z_1,\cdots,z_N) \tag{6.7}$$

Now let Z_1 move slowly around a closed loop C, then a formula similar to eq. (6.4) obtains for Berry's phase. If Z_2 is enclosed by C we have

$$\gamma = 2\pi <n>_D = 2\pi\upsilon(\phi/\phi_0 - 1) \tag{6.8}$$

The first term is the same as before and the second term is due to the depletion of electrons in the configuration of the second quasi-hole at Z_2. This extra phase is independent of small deformations of C and is interpreted as the exchange (or statistics) phase $e^{-2i\theta}$. (The factor 2 is because moving Z_1 around Z_2 once is equivalent to exchanging them twice.) Therefore the statistics parameter of a quasi-hole is determined as

$$\theta = -\Delta\gamma/2 = \upsilon\pi \tag{6.9}$$

For example, in the FQH state with $\upsilon=1/3$, one has $\theta=\pi/3$. Since generally υ is fractional, so we have fractional statistics for the quasi-particles in an FQH state.

From the derivation it is clear that the statistics parameter is directly related to the depletion number in the quasi-particle configuration. The latter is a definite number, because the FQH state is incompressible. If it was not so, the depleton number might not be well-defined and fractional statistics would no longer make sense. For example, the above derivation can be applied mathematically to the discussion of statistics for vortices in superfluid He^4 film. However due to the compressibility of superfluid He^4, the depleton number for a vortex depends on the loop C and logarithmically diverges for very large loops and, therefore, it does not make sense to consider assigning a fractional statistics to the vortices in superfluid He^4 film.[22]

An astute reader may point out that Berry's phase for a closed loop is intrinsically ambiguous up to $2\pi N$ and so is the statistics parameter determined from it up to π. However this ambiguity can be solved by considering the $\upsilon=1$ case, in which the quasi-particle is just usual electrons or holes obeying Fermi statistics ($\theta=\pi$).

2. *The Odd-denominator Rule and Fundamental Selection Rule*

A remarkable empirical observation is that all stable FQH states for $\upsilon<1$ occur at fractional fillings with an odd denominator. Since this rule is valid without exception, there must be a deep reason for it. Tao & Wu[23] first pointed out that the odd-denominator rule originates from the Fermi statistics of electrons, which are constituent particles of real FQH systems.

Their argument goes as follows. It is well-accepted that a quasi-hole is formed in the incompressible fluid with $\upsilon=1/m$ by a 2-d bubble of a size such that 1/m of an electron is removed. So it is very natural to expect that if m quasi-holes are present at the same location Z_0, then the state should be the same as if one electron is removed. In other words, a small cluster of m quasi-holes is equivalent to a usual hole. Though this property is not generally true in more complex fluid systems, it is correct in the spinless or spin-polarized FQH fluids with constituent particles having "hard-cores" or short-ranged repulsive interactions. Indeed this is a well-known property of the Laughlin wave functions: making m "quasiparticles" at Z_0 is represented by $\prod_{i=1}^{N} (Z_0-z_i)^m \; (z_1,\cdots,z_N)$; from the form (5.5) of $\Psi_0^{(m)}$, the latter wavefunction is nothing but the ground state $\Psi_0^{(m)}(Z_0,z_1,\cdots,z_N)$ of N+1 electrons with one electron removed from the zero-angular-momentum orbital centered at Z_0, $\exp\{-\frac{1}{4}|Z_0|^2\}$. Thus the statistics of m quasi-holes should be the same as that of a usual hole.

Given the statistics $\theta/\pi=1/m$ from eq. (6.10), the statistics of the cluster of m quasi-holes is given by

$$\frac{\Theta_m}{\pi} = (\frac{\theta}{\pi})\cdot m^2 = m \qquad (6.10)$$

Here we have used the fact that the exchange of two clusters contains m^2 pair-exchanges.[24] Thus we have

$$e^{i\Theta_m} = e^{im\pi} = (-1) \qquad (6.10')$$

Here -1 represents the Fermi statistics of the usual hole. It is obvious that m must be odd.

For a general filling fraction $\upsilon=p/q$, the argument can be formulated in a way independent of the Laughlin wavefunctions. Let a thin flux-tube pierces the fluid with unit flux slowly turned on, then p/q of the particles must be removed around it by the radial Hall current (with conductance $\sigma_H=(p/q)(e^2/h)$) induced by the transient electric field. Such a combination (unit flux-tube with p/q particle charge) is the so-called anyon.[25] From the Aharanov-Bohm phase produced by exchanging two anyon quasiparticles, one infers that the statistics of the anyon is $\theta/\pi=p/q$. A small cluster of q such anyon quasiparticles should be equivalent to removal of p particles, or simply to p holes. The statistics of q anyons is $\Theta/\pi=(p/q)q^2=pq$ and the statistics of p particles is $(\mp 1)^p$. Equating the two leads to the relation

$$(-1)^{pq} = (\mp 1)^p \begin{cases} (-) \text{ - sign for fermions} \\ (+) \text{ - sign for bosons} \end{cases} \qquad (6.11)$$

This is the fundamental selection rule for the stable filling fractions at which the FQHE occurs. For electronic FQH systems, if p and q are mutually prime, then q must be odd, giving the "odd-denominator rule". For bosonic cases, the fractions with both p and q odd are excluded. Bosonic FQH systems, though not realistic, can be created in computer simulations or be considered in some theoretical approaches. In all cases the above selection rule is verified without exception. (The 5/2 FQH state is beyond the scope of the above discussion, since it involves electrons with both spin orientations.)

Generalizing the argument a bit, it is easy to realize that the composite object consisting of q quasiparticles and p particles in the FQH state with $\upsilon=p/q$ should be a "charge-zero boson", if the selection rule (6.11) is satisfied. Read[26] and Haldane[27] has recently suggested that there is a new off-diagonal long range order in the QHE exhibited through the correlation function of such composite objects and, therefore, the incompressible QH fluids may contain Bose condensation of these composite objects.

3. *Fractional Statistics and the Hierarchy Scheme*

The Laughlin many-electron wave functions are easy to construct only for simple filling fractions like $\upsilon=1/m$ (m odd). More general incompressible FQH fluid states may be constructed by the hierarchical scheme.[18,19]. The basic idea is that the dominant interaction between the quasi-particles is the short-range repulsive part of the pair interaction and the quasiparticles act like electrons in the lowest Landau level.

Therefore the Laughlin wavefunction can be applied to the interacting quasiparticle gas, which is present at filling factors near the stable value of the parent incompressible state and will condensate to form a new interpenetrating incompressible quasiparticle fluid at appropriate filling factors. These new stable filling fractions must be related to the fractional statistics of the quasi-particle the way the original fractions $\upsilon_0=1/m$ (m odd) to the Fermi statistics of electrons. The derivation of the hierarchy of FQH states along this line of thought was given by Halperin,[28] who first suggested the relevance of fractional statistics to the FQHE.

In analog to the Laughlin wavefunction (5.5), the wavefunction for the quasi-holes in a parent fluid with $\upsilon_0=1/m$ is of the form

$$\Psi_0'(Z_1,\bullet\bullet\bullet,Z_{N'}) = \prod_{j<k}^{N'} (Z_j-Z_k)^{m_1} \exp\left\{-\frac{1}{4}\left|\frac{e}{e*}\right|\sum_k^{N'} |Z_k|^2\right\} \tag{6.12}$$

Since the quasi-holes obey statistics with $\theta/\pi=1/m$, the exponent m_1 in the prefactor can only be

$$m_1=2p_1+\alpha_1/m \tag{6.13}$$

where $\alpha_1=+1$ for quasiholes (and -1 for quasi-electrons), and p_1 is a positive integer. Again the probability distribution $|\Psi_0'|^2$ can be interpreted as that of a classical one-component plasma of charges m_1 in a uniform background with charge e*/e. So the density of the plasma is fixed by the charge neutrality condition, which gives the number of quasi-holes in an area $2\pi l^2$ to be just

$$n_1 = |e^*/e|/m_1 = 1/mm_1 = 1/(2p_1m+\alpha_1) \tag{6.14}$$

Since each quasi-hole has charge α_1/m, the filling factor for electrons in the new stable state is

$$\upsilon_1 = \upsilon_0-\alpha_1 n_1/m = 2p_1/(2p_1m+\alpha_1) \tag{6.15}$$

For m=3, $p_1=1$ we have $n_1=1/7$ (or 1/5), $\upsilon_1=2/7$ (or 2/5) for the quasi-hole (or quasi-electron) fluid. Higher order states can be constructed in a similar way, reproducing observed fractions.

Obviously this discussion can be extended to a hierarchical scheme; namely one may consider quasi-particles in the fluid of quasi-particles in the fluid of and so on and so forth.. The iterative equation for the charge q_{s+1} of a quasi-particle at level s+1 is

$$q_{s+1} = -\alpha_{s+1}q_s/m_{s+1} \qquad (\alpha_{s+1}=\pm 1) \tag{6.16}$$

where

$$m_{s+1} = 2p_{s+1}+\alpha_{s+1}/m_s \qquad (p_{s+1}>0,\text{integer}) \tag{6.17}$$

The stable filling factor at level s+1 satisfies

$$\upsilon_{s+1} = \upsilon_s + \frac{|q_s|q_{s+1}}{m_{s+1}} \tag{6.18}$$

It can be shown that υ_{s+1} can be expressed as a continued fraction in terms of the sequences $\{p_i,\alpha_i\}$:

$$\upsilon_{s+1} = \cfrac{1}{2p_1 + \cfrac{\alpha_1}{2p_2 + \cfrac{\alpha_2}{2p_3 + \cdots}}} \tag{6.19}$$

This is the hierarchy that Haldane obtained[18] in a different way.

7. Ground State Degeneracy of FQH Systems

1. The Extended Laughlin Argument

The IQHE has been explained by Laughlin's argument, making clever use of a non-trivial topology: a cylinder threaded by a flux tube, combined with the gauge invariance argument. *(See Sec. 4.1.)* Tao & Wu[14] have extended it to incorporate the FQHE without violating gauge invariance, and pointed out that the condition which spoils the integral but favors the fractional quantization is the ground state degeneracy of FQH systems on certain compactified space geometry. (In the present case it is the torus geometry with periodic boundary condition on the two edges of the cylinder.)

The ground-state non-degeneracy condition for the IQHE was not explicitly stated in Laughlin's original argument, but emphasized by Tao & Wu[14]. They pointed out that this condition, in addition to the torus geometry and the gauge invariance argument (one pure flux quantum is equivalent to no flux†), is necessary to guarantee an integral Hall conductance. So to obtain a fractional $\sigma_H=(p/q)(e^2/h)$, the system needs to be q-fold degenerate on a torus. More rigorously one can see this from the derivation of the Hall conductance as a topological invariant by Niu, Thouless & Wu[10], presented in Sec. 4.3. There we have seen that whenever the ground state is nondegenerate and is separated from the excited states by a finite energy gap for all values of the boundary phases (θ_1,θ_2), the ground state ϕ_0 will map back to itself when either θ_1 or θ_2 changes from 0 to 2π. Then the Hall conductance, when averaged over θ_1 and θ_2 on the torus $(0\le\theta_1<2\pi,\ 0\le\theta_2<2\pi)$, is the integral of Berry's curvature over the torus, which must be quantized as integers in e^2/h. To obtain fractional quantized σ_H, the ground state must be degenerate. Now let us start from one of the ground states, now written as ϕ_1. Changing θ_1 or θ_2 from 0 to 2π may not map ϕ_1 back to itself. Suppose one needs to change θ_1 from 0 to $2\pi q$ (q is an integer) to make ϕ_1 back to itself. Then the averaged Hall conductance can be written as an integral over the torus $0\le\theta_1<2\pi q$, $0\le\theta_2<2\pi$:

$$\sigma_H = \bar{\sigma} = \frac{e^2}{hq}\int_0^{2\pi q} d\theta_1 \int_0^{2\pi} d\theta_2 \frac{1}{2\pi i}\left\{\left\langle\frac{\partial\phi_1}{\partial\theta_2}\middle|\frac{\partial\phi_1}{\partial\theta_1}\right\rangle - (\theta_1\Leftrightarrow\theta_2)\right\} \tag{7.1}$$

† *For anyons (charges threaded by flux), the condition is a bit different. For example, it is two flux quanta which are equivalent to no flux for the flux threading an electron.*

where the factor q comes from the averaging procedure. Now the integral must be an integer, as an integral of Berry curvature over a compact surface always is, so in general σ_H becomes a fraction with denominator q. Note that the above integral over $0 \leq \theta_1 < 2\pi$, $0 \leq \theta_2 < 2\pi$ is no longer an integral over a torus, since the integrand are not the same at the edges $\theta_1 = 0$ and $\theta_1 = 2\pi$.

Experimentally σ_H is p/q times e^2/h when the filling factor $\upsilon \approx p/q$. To make the above argument capable of explaining this, one needs to further prove that

(1) The integer q defined above to make a ground state back to itself is the same for either of the ground states we start with and is identified to the denominator of the filling factor.

(2) The integral in eq. (7.1) is also independent of the choice of the ground state and is actually the numerator integer p of υ.

We leave the proof of *(1)* to the next subsection. For the proof *(2)*, one may invoke the topological invariance, assuming the gap is unclosed by turning off various interactions in the Hamiltonian. Then the integral is unchanged. For a non-interacting gas, σ_H is known to be p/q at the filling factor $\upsilon = p/q$, and so is the σ_H calculated from eq. (7.1). The existence of a plateau near $\upsilon = p/q$ is explained by localization of quasiparticles by impurities, similar to the IQHE cases.

2. *The Ground State Degeneracy on a Torus*

A general method to prove the existence of degeneracy in quantum mechanics is to find two non-commuting symmetry operators each of which commutes with the Hamiltonian. In the present situation such operators are provided by magnetic translations. The infinitesimal magnetic translations of a single electron are defined by eqs. (4.7). In a many-body system one can define the magnetic translation

$$t_i(\vec{a}) = \exp\{\tfrac{i}{\hbar}\, \vec{a} \cdot \vec{\Pi}_i\} \tag{7.2}$$

for each electron i. In the absence of impurities but in the presence of electron-electron interactions the total magnetic translation

$$T(\vec{a}) = \prod_{i=1}^{N} t_i(\vec{a}) \qquad (\vec{a}\text{: 2-d vector}) \tag{7.3}$$

leaves the many-body Hamiltonian invariant.

For definiteness, to impose the torus geometry let us consider the periodic boundary condition imposed on the many body wavefunction

$$t_j(\vec{L}_1)\psi = t_j(\vec{L}_2)\psi = \psi \tag{7.4}$$

with $\vec{L}_1 = L_1\vec{e}_x$, $\vec{L}_2 = L_2\vec{e}_y$. So the wavefunction is unchanged when one of the electrons is magnetically translated by $\vec{L}_1$ or $\vec{L}_2$ across the plane. This corresponds to the case with $\theta_1 = \theta_2 = 2n\pi$ in eqs. (4.25′). Suppose there are

$$N_\phi = \frac{L_1 L_2}{2\pi l^2} = \text{integer} \tag{7.5}$$

flux quanta through the surface, and there are

$$N_e = \upsilon N_\phi = (p/q)N_\phi \tag{7.6}$$

electrons, with p,q mutually prime integers. The translations which also leave the boundary condition (7.4) invariant are

$$T_1 \equiv T(\vec{L}_1/N_\phi) \ , \qquad T_2 \equiv T(\vec{L}_2/N_\phi) \tag{7.7}$$

and their powers. These are the symmetry operators we are looking for, since T_1 and T_2 do not commute:

$$T_1T_2 = T_2T_1 \exp\{-iN_e|\vec{L}_1\times\vec{L}_2|/N_\phi^2 l^2\} = T_2T_1\, e^{-i2\pi p/q} \tag{7.8}$$

One may choose a ground state ϕ_0 to be an eigenstate of T_2, then the q states defined by

$$\phi_n = T_1^n\, \phi_0 \qquad (n=0,1,\cdots,q\text{-}1) \tag{7.9}$$

are degenerate, *i.e.* have the same energy with ϕ_0. Furthermore, they are all eigenstates of T_2 but with different eigenvalues:

$$T_2\phi_n = e^{-i2\pi np/q}\, T_2\phi_0 \tag{7.10}$$

and therefore they are orthogonal to each other. In other words, the ground states must form representations of the Heisenberg algebra (7.8), whose irreducible representation is q dimensional. So the ground state must be at least q-fold degenerate.[10] If there is no accidental degeneracy, there are exactly q ground states.

In general, the presence of disorder or random impurities breaks the invariance under many-body translations (7.7). This means that the ground state degeneracy is lifted by weak impurity potentials. If the impurity potentials are weaker than the energy gap, one may use the first-order degenerate perturbation theory in the ground state subspace. This approach has been taken by Tao & Haldane,[29] and Wen & Niu.[30] In general, there is a unique ground state for each set of boundary phases (θ_1,θ_2). Does this imply that the Hall conductance should be an integer according to the theory of topological invariant? The answer is negative, because the temperature in realistic samples is believed to be higher than the energy splitting caused by impurities. The lowest q states split from the ground states are equally populated below the Fermi level, so the discussion in the last subsection still applies.

Wen & Niu[30] has recently proved, by using the effective Ginsburg-Landau theory for the FQH systems, that the ground state degeneracy of a FQH system with $\upsilon=p/q$ on a Riemann surface with genus g is q^g-fold, even though the translation invariance no longer holds on a Riemann surface with g>1. This degeneracy is shown to be invariant against weak but otherwise arbitrary perturbations. This result explicitly shows that the ground state degeneracy of a FQH system depends on the global topology of the surface. Originally the ground state degeneracy was controversial for a while: Laughlin's wavefunction for a disc geometry[16] and Haldane's generalization for a spherical geometry[18] were both found to be non-degenerate. On the other hand, the degeneracy argued by Tao & Wu for the torus geometry[14] was seen in computer simulations[31] and in explicit construction of Laughlin wavefunction on a torus.[32] Later the speculations about symmetry breaking from this degeneracy turned out to be incorrect, since the degeneracy on a torus can be interpreted as the degeneracy of center-of-mass motion.[33] Wen & Niu resolved the puzzle. They showed that the degeneracy is dependent on spatial topology and preserved in the thermodynamic limit even though the translational or rotational symmetries may be lost on a high-genus Riemann surface. This implies that the degeneracy of FQH states is a reflection of some topological order in the system and that the degeneracy should not be interpreted as a symmetry breaking of usual type, nor should it be disregarded as merely the center-of-mass degeneracy. Very recently Wen[34] has developed the concept of topological order and has attempted to use the non-abelian Berry connection in the ground state subspace to fully characterize the topological order of the FQH states.

3. Gapless Current-carrying Edge States and Topological Considerations

This is a subject currently attracting a lot of interest. Almost immediately after Laughlin's argument, Halperin[35] has noticed an important consequence of the gauge-invariance argument. Namely there should exist current-carrying states in a QH (both IQH and also FQH) system with boundary; these states are localized on the boundary but extended around the perimeter. Because of recent experimental observations of the edge states in FQH systems,[36] they have attracted more and more attention.[37] We do not have time to enter into this subject, but wish to make the following comment.

There should be a close relationship between the edge states on the boundary and the degeneracy of the FQH states on a compact surface, because both are required by gauge invariance. This can be seen more clearly from the cylinder geometry in Laughlin's original argument. In the $\upsilon=1/3$ FQHE case, as the flux threading the hole enclosed by the cylinder increases from zero to ϕ_0 (unit flux), a charge-1/3 quasi-particle is transferred from one edge to the other. Only as the flux increases to $3\phi_0$, an electron or usual hole is transferred across the system between the two edges. Since the basic constituents for the edge states are still electrons, the Hilbert space of edge states should be understood from the point of view of electrons. If so, with unit flux turned on, the edge state should not be in the same sector in the edge-state Hilbert space as before, because of the transfer of a charge-1/3 quasiparticle between edges. Thus, the edge Hilbert space should contain three sectors and different sectors are related by adiabatically turning on unit flux. Only turning on three unit flux does not change the sector. Intuitively one may expect the correspondence of the three sectors in the edge Hilbert space on a cylinder and the 3-fold ground state degeneracy on a torus.

Actually such a correspondence can be realized by bending the cylinder and bringing the edges together in contact to each other in physical space. With the edge potential barrier present and tunneling between edges ignored, the system is like the cylinder one we discussed above. If we imagine to reduce the height of the edge barrier and to increase the tunneling of electrons between edges in a thought experiment, finally we can achieve the toroidal system with the disappearance of the edge barrier and with electrons going freely across the edges. In the latter situation, a gap is expected to be open in the edge Hilbert space and the state on the "would-be cut" where the edges are glued together to be in the ground state which is 3-fold degenerate. If one moves a charge-1/3 quasi-particle from the "would-be" cut around a non-shrinkable loop on the torus, this will bring a ground state to another degenerate one, since it corresponds to, in the cylinder geometry, what happens when a unit flux is turned on. It would be interesting to see how the physics discussed here is exhibited in an appropriate formalism for the Hilbert space of edge states in the cylinder geometry.

Acknowledgement – The author thanks his previous collaborators, Dr. R. Tao, Dr. Q. Niu and Prof. D. Thouless for many useful discussions and pleasant collaboration. He is particularly grateful to Prof. C.N. Yang who drew his attention to the quantum Hall effect and lower dimensional physics. The work was supported in part by U.S. NSF grant PHY-8706501 and the Monell Foundation through Institute for Advanced Study.

References

1. K. von Klitzing, G. Dorda & M. Pepper, Phys. Rev. Lett. **45**(1980)494
2. D.C. Tsui, H.L. Stormer & A.C. Gossard, Phys. Rev. Lett. **48**(1982)1559
3. For an excellent review book, see *The Quantum Hall Effect*, ed. by R.E. Prange & S.M. Girvin (Springer, New York, 1987)
4. G. Ebert, K. von Klitzing *et al.*, Solid State Commun. **58**(1982)95
5. A.M. Chang, P. Berglund, D.C. Tsui, H.L. Stormer & J.C.M. Huang, Phys. Rev. Lett. **53**(1984)997
6. R. Willet, J.P. Eisenstein, H.L. Stormer, D.C. Tsui, A. Gosssard & J.H. English, Phys. Rev. Lett. **59**(1987)1776
7. R.B. Laughlin, Phys. Rev. **B23**(1981)5632
8. D.J. Thouless, M. Kohmoto, M.P. Nightingale & M. den Nijs, Phys. Rev. Lett. **49**(1982)405
9. M. Kohmoto, Ann. Phys. (N.Y.) **160**(1985)355. See also J.E. Avron, R. Seiler & B. Simon, Phys. Rev. Lett. **51**(1983)51
10. Q. Niu, D.J. Thouless & Y.S. Wu, Phys. Rev. **B31**(1985)3372; see also Q. Niu & D.J. Thouless, J. Phys. **A17**(1984)2453
11. J.E. Avron & R. Seiler, Phys. Rev. Lett. **54**(1985)259
12. T.T. Wu & C.N. Yang, Phys. Rev. **D35**(1975)3849
13. L. Schiff, *Quantum Mechanics* (McGraw-Hill, N.Y., 1955), p.290
14. R. Tao & Y.S. Wu, Phys. Rev. **B30**(1984)1097
15. R.E. Prange, Phys. Rev. **B23**(1981)5632
16. R.B. Laughlin, Phys. Rev. Lett. **50**(1983)1395
17. See R.B. Laughlin in ref. 3
18. F.D.M. Haldane, Phys. Rev. Lett. **51**(1983)605
19. F.D.M. Haldane in ref. 3
20. D. Arovas, J.R. Schrieffer & F. Wilczek, Phys. Rev. Letts. **53**(1984)722
21. M.V. Berry, Proc. Roy. Soc. (London) Ser. A **392**(1984)45
22. F.D.M. Haldane & Y.S. Wu, Phys. Rev. Lett. **55**(1985)2887
23. R. Tao & Y.S. Wu, Phys. Rev. **B31**(1985)6859
24. D.J. Thouless & Y.S. Wu, Phys. Rev. **B31**(1985)1191
25. F. Wilczek, Phys. Rev. Lett. **49**(1982)957; Y.S. Wu, Phys. Rev. Lett. **53**(1984)111
26. N. Read, Phys. Rev. Lett. **62**(1989)86
27. F.D.M. Haldane (unpublished).
28. B.I. Halperin, Phys. Rev. Lett. **52**(1984)1583
29. R. Tao and F.D.M. Haldane, Phys. Rev. **B33**(1986)3844
30. X.G. Wen & Q. Niu, Santa Barbara preprint NSF-ITP-89-151
31. W.P. Su, Phys. Rev. **B30**(1984)1069
32. F.D.M. Haldane & D. Rezaya, Phys. Rev. **B31**(1985)2529

33. F.D.M. Haldane, Phys. Rev. Lett. **55**(1985)2095
34. X.G. Wen, Santa Barbara preprints NSF-ITP-89-107;
see also X.G. Wen, Phys. Rev. **B40**(1989)7387
35. B.I. Halperin, Phys. Rev. **B25**(1982)2185
36. R.J. Haug, A.H. MacDonald, P. Streda and K. von Klitzing, Phys. Rev. Lett. **61**(1988)2797;
S. Washsburn, A.B. Fowler, H. Schmid & D. Kern, Phys. Rev. Lett. **61**(1988)2801
37. X.G. Wen, Santa Barbara preprint NSF-ITP-89-157;
C.W.J. Beenakker, Phys. Rev. Lett. **64**(1990)216;
A.H. MacDonald, Phys. Rev. Lett. **64**(1990)220;
F.D.M. Haldane *(private communication).*

PART II

SEMINARS
on
PHYSICS, BRAIDS & LINKS

THE NONABELIAN CHERN-SIMONS TERM WITH SOURCES AND BRAID SOURCE STATISTICS

M. Bourdeau*

Department of Physics
Syracuse University
Syracuse, N.Y. 13244-1130

INTRODUCTION

The statistical properties of N identical particles are related to the fundamental group $\pi_1(Q)$ of Q, where Q, the configuration space is given by $\frac{(\mathbb{R}^d)^N \backslash \mathbb{R}^d_{\text{diagonal}}}{S_N}$, where d is the space dimension. In three or more dimensions, the group $\pi_1(Q)$ is the permutation group S_N. Each unitary irreducible representation (UIR) of $\pi_1(Q)$ corresponds to a distinct quantum theory, these quantum theories describing bosons, fermions and paraparticles in three or more dimensions.[1] In two space dimensions the situation is different: $\pi_1(Q)$ is the infinite braid group B_N[11], which has $N-1$ generators σ_α, these generators describing the exchange of particles α with $\alpha+1$. This group governs the statistics and has many UIR's which are not also UIR's of S_N, S_N being a factor group of B_N. We expect therefore a novel type of statistics, specific to two space dimensions. In 2+1 dimensions, it has been known for some time[2] that a Lagrangian consisting of an abelian Chern-Simons term for charged particles leads to 'fractional statistics'. This statistics seems to be the one obeyed by the Laughlin[3] quasiparticles of the fractional quantum Hall effect. This work

* This work is done in collaboration with A.P. Balachandran and S. Jo.

Physics, Geometry, and Topology
Edited by H. C. Lee
Plenum Press, New York, 1990

concentrates on the nonabelian generalization of the $U(1)$ sources and the abelian Chern-Simons term and whether this generalization leads to novel statistics. This inquiry is of relevance for example for the strongly coupled Hubbard model which has an $SU(2)$ gauge invariance[4] and where it is possible that the effective action of the connection fields contains a Chern-Simons term.

In two dimensions the solution for the gauge connection for the Lagrangian of N identical charged sources interacting with an abelian Chern-Simons term is unique up to gauge equivalence. The generalized statistics is determined by the strength θ of the Chern-Simons term. We will see that the situation for the nonabelian Chern-Simons term with sources is quite different: there exist many gauge inequivalent solutions for the gauge connection and each class of gauge potentials leads to a different (braid) statistics. We find also that there is no obvious correlation between the spin and statistics of the particles.

The outline is as follows: we shall first describe the model of N identical particles in interaction with a nonabelian Chern-Simons term, then find all solutions of the field equations with sources and describe how they fall in equivalence classes; the quantization of the system will be discussed and finally, the relation between spin and statistics of these particles.

DESCRIPTION OF THE MODEL

Consider G to be a simple compact connected Lie group with Lie algebra $\underline{G}$, for example, $G = SU(M)$. The sources for the $SU(M)$ gauge fields are point particles. Then, the Lagrangian for N identical point particles carrying nonabelian charges I^{α} with a Chern-Simons term can be written down as:

$$
\begin{aligned}
L = \sum_{\alpha} \frac{1}{2} m[\dot{z}^{a(\alpha)}(t)]^2 + i\,\mathrm{Tr}[K^{(\alpha)} s^{(\alpha)}(t)^{-1} D_t s^{(\alpha)}(t)] \\
+ \int d^3x \frac{k}{4\pi} \epsilon^{\mu\nu\lambda}\, \mathrm{Tr}[A_\mu \partial_\nu A_\lambda + \frac{2}{3} A_\mu A_\nu A_\lambda]
\end{aligned}
\qquad (1)
$$

where $z^{\alpha}(t)$ are the space-time coordinates of the particle α:

$$
\begin{aligned}
z^{a(\alpha)}(t) &= x^{a(\alpha)}(t), \quad \text{for} \quad a = 1, 2 \\
z^{0(\alpha)}(t) &= t
\end{aligned}
$$

and where we define an isospin-like variable:

$$I^{(\alpha)}(t) = js^{(\alpha)}(t)K^{(\alpha)}s^{(\alpha)}(t)^{-1} = I_j^{(\alpha)}(t)\tau_j \quad \text{with} \quad s^{(\alpha)}(t) \in G.$$

$K^{(\alpha)}$ determines the UIR of G to which the particle belongs. On quantization, $I_j^{(\alpha)}(t)$ become the generators of the representation of $\underline{G}$ which characterizes the source.

The coupling

$$D_t s^{(\alpha)}(t) \equiv \dot{s}^{(\alpha)}(t) + \dot{z}^{\mu(\alpha)}(t)A_\mu(z^{(\alpha)}(t))s^{(\alpha)}(t).$$

The last term in L is $\int \mathcal{L}_{CS}$ with $A = A_\mu dx^\mu$, the Yang-Mills potential for 2+1 dimensional space-time and the group G. The coefficient in $\mathcal{L}_{CS}$ is quantized, the quantization condition being $k \in \mathbf{Z}$.

We neglect the kinetic energy term in this Lagrangian. It is probable that the physical effects predicted by this model and others where the kinetic energy is also neglected survive for sufficiently large length scales because the long range fields to leading order appear to be caused solely by the contribution of the Chern-Simons term to the field equations.

The variations in A , $s^{(\alpha)}(t)$ and $x^{(\alpha)}(t)$ give the following analogues of Wong's equations[5]:

$$\begin{aligned} \frac{k}{2\pi}F_{ab}(x) &= -\epsilon_{ab}\sum_\alpha \delta^2[x - x^{(\alpha)}(t)]I^{(\alpha)}(t), \\ \frac{k}{2\pi}F_{a0}(x) &= \epsilon_{ab}\sum_\alpha \delta^2[x - x^{(\alpha)}(t)]I^{(\alpha)}(t)\dot{z}^{b(\alpha)}(t), \end{aligned} \tag{2}$$

$$\text{where} \quad \epsilon_{ab} = \epsilon_{0ab}$$

$$\text{and} \quad \text{where} \quad F_{\mu\nu} = \partial_\mu A_\nu - \partial_\nu A_\mu + [A_\mu, A_\nu]$$

$$\frac{dI^{(\alpha)}(t)}{dt} + \dot{z}^{\mu(\alpha)}(t)[A_\mu(z^{(\alpha)})(t), I^{(\alpha)}(t)] \equiv D_t I^{(\alpha)}(t) = 0, \tag{3}$$

$$m\ddot{x}^{a(\alpha)}(t) = i\,\mathrm{Tr}[I^{(\alpha)}_{\cdot}(t)F^{a}_{\mu}(z^{(\alpha)}(t))]\dot{z}^{\lambda(\alpha)}(t). \tag{4}$$

where the last equation is the nonabelian version of the Lorentz force equation.

Classically, the motion in the spatial coordinates is free, as seen by substituting Eqs. (2) in (4):

$$\ddot{z}^{a(\alpha)}(t) = 0.$$

Quantum mechanically, however, the sources will in general scatter.

SOLUTIONS TO THE FIELD EQUATIONS

We can now solve for the preceding equations in a particular gauge. It can be shown that any solution is gauge equivalent to such a solution[6]. We will first assume that no two sources have the same first coordinate at a given time: $[x^{1(\alpha)}(t) - x^{1(\beta)}(t) \neq 0 \quad \text{if} \quad \alpha \neq \beta]$. The general solution for A can then be obtained as follows: choose $K^{(\alpha)}$ to be any fixed point in the orbit of $I^{(\alpha)}$ under the adjoint action:

$$K^{(\alpha)} = g^{(\alpha)}(t)I^{(\alpha)}(t)g^{(\alpha)}(t)^{-1}.$$

Then at time t,

$$\begin{aligned} A = A_\mu dx^\mu &= 0 && \text{in} \quad \mathbb{R}^2\backslash\cup_\alpha\Delta^{(\alpha)}(t), \\ &= -\frac{i}{k}\,K^{(\alpha)}\,\frac{\partial\psi^{(\alpha)}}{\partial x^\mu}\,dx^\mu && \text{in} \quad \Delta^{(\alpha)}(t). \end{aligned} \tag{5}$$

As shown in Figure 1. below, $\Delta^{(\alpha)}$ is a thin strip in $\mathbb{R}^2$ going from $x^{(\alpha)}(t)$ along the negative 2 axis. The tails are assumed to be thin enough as not to overlap $[\Delta^{(\alpha)} \cap \Delta^{(\beta)} = \emptyset \quad \text{if} \quad \alpha \neq \beta]$. $\psi^{(\alpha)}$ is an angle like function which (at time t) is constant in $\mathbb{R}^2\backslash\Delta^{(\alpha)}(t)$ and increases by 2π as $\Delta^{(\alpha)}(t)$ is crossed from left to right below the source position $x^{(\alpha)}(t)$.

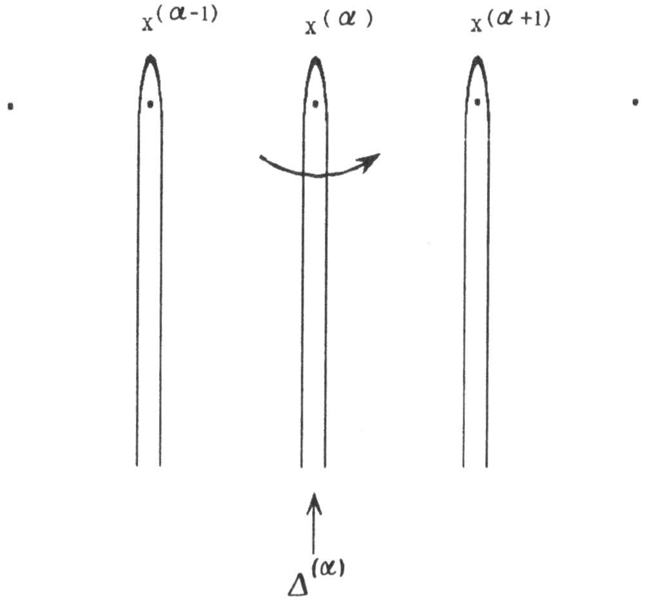

Fig.1. $\psi^{(\alpha)}$ increases by 2π when $\triangle^{(\alpha)}$ is crossed along the arrow.

By Stokes' theorem, and using the relation:

$$\int_{C^{(\alpha)}} \frac{\partial \psi^{(\alpha)}}{\partial x^a}\, dx^a = 2\pi,$$

for a counterclockwise loop enclosing only the source α at time t, we get:

$$(\partial_a \partial_b - \partial_b \partial_a)\psi^{(\alpha)}(x) = \epsilon_{ab}\, 2\pi\, \delta^2[x - x^{(\alpha)}(t)].$$

Using this, and the fact that $\psi^{(\alpha)}$ can be regarded as a function of $(x^a - x^{a(\alpha)}(t))$, one can check that A satisfies (2) in the gauge where $I^{(\alpha)}(t) = K^{(\alpha)}$, where $K^{(\alpha)}$ are independent of time.

The holonomies associated with the gauge connection are:

$$W^{(\alpha)} = P\exp\Big[\int_{C^{(\alpha)}} A\Big] = \exp\Big[-\frac{2\pi i}{K^{(\alpha)}}\Big] \quad . \tag{6}$$

Altough the connection looks abelian, the holonomies $W^{(\alpha)}$ don't necessarily commute for specific choices of $K^{(\alpha)}$ and k.

There are two important remarks about these solutions which come about as follows:

If two solutions $A^{(L)}$ and $A^{(M)}$ are gauge related by a gauge transformation g_{LM}, then their holonomies are in the same conjugacy class. Conversely, if their holonomies are in the same conjugacy class, that is if there exists an element $h_{LM} \in G$ independent of α which fulfills $W^{(L)(\alpha)} = h_{LM}\, W^{(M)(\alpha)}\, h_{LM}^{-1}$, then there exists g_{LM} which gauge transforms $A^{(M)}$ to $A^{(L)}$. However there is no guarantee that g_{LM} is well defined at the source positions. A requirement on g_{LM} to partially avoid such singularities is:

$$d^2 g_{LM} = 0. \tag{7}$$

This condition has the following consequences:

-All solutions of (2) are gauge related to a solution of the type (5).
-Two connections which are solutions of (2) (with sources characterized by internal vectors $K^{(\alpha)}$ and $\overline{K}^{(\alpha)}$) are gauge related if and only if there exists a $h \in G$ (independent of α) such that

$$hK^{(\alpha)}h^{-1} = \overline{K}^{(\alpha)}.$$

A detailed proof of the above can be found in Ref. 6.

Now, in the abelian Chern-Simons theory, the solution for A is unique up to a gauge transformation. In the nonabelian problem, as we have just seen, the situation is different: there exist many gauge inequivalent solutions and as we shall see later, the statistics depends on the equivalence class of solutions one is considering.

LAGRANGIAN FORMALISM FOR THE SOURCES AFTER ELIMINATION OF THE GAUGE FIELD

We will now eliminate the field A and obtain an effective Lagrangian for the sources. The Lagrangian formalism for a point particle with internal symmetry will be discussed. The requirement that L_{EFF} give classically free motion in the particle coordinates then leads to an interesting quantization condition (similar to the Dirac quantization condition for the charge-monopole system[6]), as we shall see in the next section.

The Lagrangian before elimination of the gauge potential is:

$$\begin{aligned} L = &\sum_{\alpha} \frac{1}{2} m [\dot{z}^{a(\alpha)}(t)]^2 + ij \operatorname{Tr}[K^{(\alpha)} s^{(\alpha)}(t)^{-1} D_t s^{(\alpha)}(t)] \\ &+ \int d^3x \frac{k}{4\pi} \epsilon^{\mu\nu\lambda} \operatorname{Tr}[A_\mu \partial_\nu A_\lambda + \frac{2}{3} A_\mu A_\nu A_\lambda] \end{aligned} \tag{8}$$

Note that if we redefine $s^{(\alpha)}(t) = s'^{(\alpha)}(t) f^{(\alpha)}$, $K^{(\alpha)}(t) = f^{(\alpha)^{-1}} K'^{(\alpha)}(t) f^{(\alpha)}$ in (8), we can arrange to have the same $K^{(\alpha)}$'s in (8) as in the gauge potential equation (5). Then the $K^{(\alpha)}$'s coming in the definition of $I^{(\alpha)}$'s are the same as the ones coming in the formula for the gauge potential. This is done for simplicity. We can now substitute in the source part of the Lagrangian the potential due to all the particles $\beta \neq \alpha$. We are neglecting self interactions here.

The potential seen by particle α due to $N-1$ particles is given in a general gauge by

$$gA^{(\alpha)}g^{-1} + gdg^{-1} = g\Big[-\frac{i}{k} \sum_{\beta \neq \alpha} K^{(\beta)}\, d\psi^{(\beta)}\Big] g^{-1} + gdg^{-1} , \tag{9}$$

where $A^{(\alpha)}$ is the solution given in (5):

$$A^{(\alpha)} = -\frac{i}{k} \sum_{\beta \neq \alpha} K^{(\beta)}\, d\psi^{(\beta)} \tag{10}$$

We can infer now from (9) and (8) that

$$L_{EFF} = L_{EFF}^{(\alpha)}$$

$$= \sum_{\alpha} \frac{1}{2} m [\dot{z}^{a(\alpha)}(t)]^2 + j \operatorname{Tr}[K^{(\alpha)} S^{(\alpha)}(t)^{-1} D_t S^{(\alpha)}(t)], \tag{11}$$

$$S^{(\alpha)}(t) = g(z^{(\alpha)})^{-1} s^{(\alpha)}(t), \tag{12}$$

$$D_t S^{(\alpha)}(t) \equiv \dot{S}^{(\alpha)}(t) + \frac{1}{2} \dot{z}^{\mu(\alpha)}(t) [A_{\mu}^{(\alpha)}(z^{(\alpha)})] S^{(\alpha)}(t). \tag{13}$$

Note that $S^{(\alpha)}$ is gauge invariant.

The factor of $\frac{1}{2}$ in (13) comes about by eliminating the Chern-Simons term from the action[2]:

The terms in the action involving A (upto terms coming from the gauge transformation which are not of interest here) are:

$$S' = \int d^3x [\mathcal{L}_{CS} + Tr J^{\mu}(x) A_{\mu}(x)], \tag{14}$$

with

$$J^{\mu} = i \sum_{\alpha} \int d\tau \delta^3 [x - z^{(\alpha)}(\tau)] S^{(\alpha)}(t) K^{(\alpha)} S^{(\alpha)}(t)^{-1}.$$

The variation of this expression under an infinitesimal variation of $z^{(\alpha)}$ must coincide with the variation of the interaction term in the effective Lagrangian for particles. One then finds[6] that the term in the effective Lagrangian arising from S' is:

$$\frac{1}{2} \int d^3x \operatorname{Tr} J^{\mu}(x) A_{\mu}(x).$$

We shall now briefly discuss the Lagrangian formalism for a point particle with internal symmetry[7]. It is closely related to the Kostant-Sourieau approach to finding the UIR's of a Lie group using suitable coadjoint orbits as phase spaces[8], and to the Borel-Weil-Bott method for finding UIR's of semisimple compact Lie groups[9].

We are interested in particles associated with unitary irreducible representations of the internal symmetry group G as sources, just as in the abelian case we are interested in particles of fixed electric charge as sources. As explained in Ref. 7., this means that one can regard the internal vector $I(t)$ of a particle as belonging to a fixed orbit of G in the adjoint representation. The orbit in question determines the UIR. Representing the elements of G by matrices as usual, and denoting by K a fixed fiducial point on the orbit, we may write

$$I(t) = I_j(t)\lambda_j = s(t)Ks(t)^{-1}, s(t) \in G \tag{15}$$

where λ_j are hermitean generators of the Lie algebra $\underline{G}$ of G normalized according to

$$\mathrm{Tr}\, \lambda_j \lambda_k = 2\delta_{jk}. \tag{16}$$

The "free" Lagrangian describing a particle associated with a UIR of G (and no other degree of freedom) is

$$L_{INTERNAL} = i\, \mathrm{Tr}\, K s(t)^{-1} \dot{s}(t) \quad , \quad \dot{s}(t) = \frac{ds(t)}{dt}.$$

As discussed in Ref. 10., the variation of L_{INT} gives the equation of motion

$$\dot{I}(t) \equiv \frac{dI(t)}{dt} = 0, \quad I(t) = s(t)Ks(t)^{-1}.$$

It is not possible to quantize $L_{INTERNAL}$ unless K is subject to certain quantization conditions[7]. The precise nature of these conditions depends on the rules of quantization adopted. Here, we adopt the following:

Let $\widetilde{I}_j^R$ be the generators of $\underline{G}$ acting on s from the right, the nontrivial PB's involving $\widetilde{I}_j^R$ and s being

$$\{\tilde{I}^R_j, s\} = -i\, s\, \frac{\lambda_j}{2}, \tag{17}$$

$$\{\tilde{I}^R_j, \tilde{I}^R_k\} = C_{ijk}\, \tilde{I}^R_l \qquad \text{if} \quad [\frac{\lambda_j}{2}, \frac{\lambda_k}{2}] = iC_{jkl}\frac{\lambda_l}{2}. \tag{18}$$

Define also

$$\tilde{I}^R = \tilde{I}^R_j \lambda_j. \tag{19}$$

Then,

$$\tilde{I}^R = -s^{-1} \tilde{I}^L s. \tag{20}$$

where

$$\begin{aligned} \{\tilde{I}^L_j, s\} &= i\, \frac{\lambda_j}{2}\, s, \\ \{\tilde{I}^L_j, \tilde{I}^L_k\} &= C_{jkl}\, \tilde{I}^L_l \end{aligned} \tag{21}$$

The Lagrangian $L_{INTERNAL}$ implies the constraints

$$\tilde{I}^L_j \approx I_j \tag{22}$$

or

$$\tilde{I}^L \approx I = s\, K\, s^{-1} \quad , \quad \tilde{I}^L = \tilde{I}^L_j \lambda_j \tag{23}$$

relating these generators and sKs^{-1}. The constraints in terms of $\tilde{I}^R$ are then

$$\tilde{I}^R + K \approx 0. \tag{24}$$

Consider the Lie algebra $\underline{C}_K$ of the stability group of K. It is spanned[6] by elements K , $\overline{K}(\rho)$ where $\mathrm{Tr}\, K\, \overline{K}(\rho) = 0$ and $\overline{K}(\rho)$ span a Lie algebra $\underline{D}_{\overline{K}}$. It is shown in Ref. 7. that the components of the constraints (24) in the directions lying in $\underline{C}_K$ are first class. The first class constraints are thus

$$\operatorname{Tr} K\widetilde{I}^R + \operatorname{Tr} K^2 \approx 0, \qquad \operatorname{Tr} \overline{K}(\rho)\widetilde{I}^R \approx 0. \tag{25}$$

The components of the constraints in directions orthogonal to $\underline{C}_K$ are second class. The constraints to be finally applied as conditions on the physical states turn out to be the following[6] (calling $\widehat{I}_j^R$ the quantum operator for the classical variable $\widetilde{I}_j^R$ with $\widehat{I}^R = \widehat{I}_j^R \lambda_j$):

$$[\operatorname{Tr} K\ \widehat{I}^R\ ]\psi = -\operatorname{Tr} K^2\ \psi \quad , \quad \operatorname{Tr} K^2 = \text{ an integer} \times \kappa \ , \tag{26}$$

$$[\operatorname{Tr} \overline{K}(\rho)\widehat{I}^R]\psi = 0, \tag{27}$$

$$[\operatorname{Tr} E_\alpha\ \widehat{I}^R]\psi = 0\ ,\ \alpha < 0. \tag{28}$$

where E_α are simultaneous eigenvectors of all the elements in $\underline{C}$ (a Cartan subalgebra of $\underline{C}_K$) under the adjoint action[6].(If $\underline{C}$ is of dimension r [$\underline{G}$ is of "rank" r] and has a basis H_i, then $\alpha = (\alpha_1, ..., \alpha_r)$ is an r dimensional vector called a "root" and $[H_i, E_\alpha] = \alpha_i\ E_\alpha$. The "lowering" operators are defined as those E_α for which $\alpha < 0$.)

The above conditions admit the following interpretation. The wavefunctions ψ can be regarded as functions on the group $\{s\}$. The group $G = \{g\}$ has the following right action on the functions χ of s (whether or not they fulfill (26,27,28):

$$g\ :\ \chi \rightarrow\ R(g)\ \chi\ ,$$

$$[R(g)\ \chi](s) \equiv \chi(s\ g). \tag{29}$$

The infinitesimal form of this equation defines the action of $\widehat{I}_j^R$ on these functions. (29) means that the physical states ψ transform as a highest weight state for this right action of the group on $\{s\}$. This highest weight state furthermore is an eigenstate of $[\operatorname{Tr} K\ \widehat{I}^R]$ for eigenvalue $-\operatorname{Tr} K^2$ and a $\underline{D}_{\overline{K}}$ singlet by (26) and (27). It is well known that there is a unique UIR with such a highest weight state.

It is possible to display the physical states ψ explicitly. Let ρ label all the inequivalent UIR's of G and let $D^\rho(s)$ be the image of s in the UIR with label ρ. Consider the matrix elements $D^\rho_{\alpha' \alpha}(s)$ of $D^\rho(s)$ between suitable basis states $|\alpha>, |\alpha'>$, $D^\rho_{\alpha' \alpha}(s) = <\alpha'|D^\rho(s)|\alpha>$. It is well known that these matrix elements span the space of functions on s. The constraints state that the physical states are spanned by those $D^\rho_{\alpha' \alpha}$ for which $|\alpha>$ is a highest weight state fulfilling (26,27,28). These requirements can be fulfilled in only one UIR and that too by a unique $|\alpha>$. This means in particular that the UIR describing the particle is uniquely determined.

It is worth noting that the generators of the internal symmetry transformations in quantum theory are $\widehat{I}_j$ and not $\widehat{I}_j^R$ where $\widehat{I}_j$ are the quantum operators associated with the classical variables $\widehat{I}_j^L$. The action of $\widehat{I}_j$ on ψ is obtained from the infinitesimal version of the left action

$$g : \psi \to L(g)\,\psi ,$$

$$[L(g)\,\psi](s) = \psi(g^{-1}s).$$

of G on ψ. Since $D^\rho(g^{-1}s) = D^\rho(g^{-1})D^\rho(s)$, the wavefunctions fulfilling (26,27,28) are seen to transform irreducibly under internal symmetry transformations.

The correct scalar product for physical states is given by group theory: If ψ and ψ' are physical states, then

$$(\psi', \psi) = \int_G d\mu(s)\psi'(s)^* \psi(s),$$

$d\mu(s)$ being the invariant measure on G.

A final remark concerning the quantization : in our approach here, the constraints are not eliminated at the classical level. They are imposed as suitable constraints on the quantum states. For example, take the case of $G = SU(2)$, s(t) can be regarded as an element of its 2×2 UIR and one can assume $K = C\tau_3$, C being a constant. Then the method described above requires $C^2 = j^2$, j being a non-negative integer or half integer (the UIR being associated with the Lagrangian for a given j is $(2j+1)$ dimensional). However, if one quantizes the classical system after first eliminating all constraints[7], the conditions on K are different and become: $C^2 = j(j+1)$.

A "BRAID" QUANTIZATION CONDITION

Let's go back to Eq.(11) which gives L_{EFF} after elimination of the gauge potential:

$$L_{EFF} = L_{EFF}^{(\alpha)}$$

$$= \sum_{\alpha} \frac{1}{2} m[\dot{z}^{a(\alpha)}(t)]^2 + j\,\mathrm{Tr}[K^{(\alpha)} S^{(\alpha)}(t)^{-1} D_t S^{(\alpha)}(t)]$$

We must now check that (11) leads to reasonable equations of motion. In particular we must verify that there is free motion in the coordinates $z^{a(\alpha)}$.

As stated before, the solution (5) is valid only if no two tails $\triangle^{(\alpha)}$ and $\triangle^{(\beta)}$ $(\alpha \neq \beta)$ cross. That is L_{EFF} is correct only if source α never crosses the common tail of two sources β and γ , that is if $\triangle^{(\alpha)} \cap \triangle^{(\beta)} \cap \triangle^{(\gamma)} = \emptyset$. In other words, L_{EFF} is correct (in the limit of infinitely thin $\triangle^{(\alpha)'}s$) if no three particles have the same first coordinate. Excluding these source configurations , one can easily show that we get free equations of motion in the particle coordinates. To extend L_{EFF} to configurations with three or more overlapping $\triangle^{(\alpha)'}s$, note that the coordinates of the N particle configuration space Q can be written as a $2N$ dimensional vector $\xi = (\xi^1, \xi^2, ... \xi^{2N})$. The interactions terms involving ξ in L_{EFF}(for $\triangle^{(\alpha)} \cap \triangle^{(\beta)} \cap \triangle^{(\gamma)} = \emptyset$) can be written as

$$iB_\mu \dot{\xi}^\mu$$

where the sum is from 0 to $2N$, ξ^0 is identified with time t and

$$B_\mu = -\frac{1}{2}\frac{i}{k}\partial_\mu[\sum_{\beta \neq \alpha} \mathrm{Tr}\, I^{(\alpha)} K^{(\beta)} \psi^{(\beta)}(z^{(\alpha)})].$$

It does not affect the equation of motion of ξ_μ because

$$\partial_\mu B_\nu - \partial_\nu B_\mu = 0. \tag{30}$$

Now B_μ is defined only on a subset of $Q \times \mathrm{I\!R}^1$ [points of $\mathrm{I\!R}^1$ labelling instants of time]. We must now extend its definition to all of $Q \times \mathrm{I\!R}^1$ maintaining the condition (30) so that the $x^{(\alpha)}$ motion continues to be free:

Consider the paths in Fig. 2 : these are paths in Q (at an instant of time) (here, for the case of three particles) associated with the left- and right-hand sides of the following braid relation: $\sigma_\alpha \sigma_{\alpha+1} \sigma_\alpha = \sigma_{\alpha+1} \sigma_\alpha \sigma_{\alpha+1}$, σ_α being the element associated with the exchange of particles α and $\alpha + 1$ in the braid group B_N[11]. The path P_1 corresponds to $\sigma_1 \sigma_2 \sigma_1$ where we first exchange 1 and 2 , then 2 and 3 and finally 1 and 2 again. The path P_2 corresponds to $\sigma_2 \sigma_1 \sigma_2$ where 2 and 3 are exchanged first, then 1 and 2 and finally 2 and 3. These two paths in Q are homotopic and along these paths, $\triangle^{(1)} \cap \triangle^{(2)} \cap \triangle^{(3)} = \emptyset$. So, if ∂D is the loop $P_1 \cup P_2{}^{-1}$ in Q, then any disc D which it bounds contains a point q with $\triangle^{(1)} \cap \triangle^{(2)} \cap \triangle^{(3)} = \emptyset$. If the potential B is extendible as a flat connection to all of D including q, then it is necessary that

$$exp - \int_{P_1 \cup P_2{}^{-1}} B_a d\xi^a = 1$$

or

$$exp - \int_{P_1} B_a d\xi^a = exp - \int_{P_2} B_a d\xi^a$$

Evaluating both sides of these equation in quantum theory, we find the quantization condition:

$$\begin{aligned} & exp\{\frac{\pi i}{k} \mathrm{Tr}[\hat{I}^{(1)} K^{(2)}]\} exp\{\frac{\pi i}{k} \mathrm{Tr}[\hat{I}^{(1)} K^{(3)}]\} \\ & \qquad = exp\{\frac{\pi i}{k} \mathrm{Tr}[\hat{I}^{(1)} K^{(3)}]\} exp\{\frac{\pi i}{k} \mathrm{Tr}[\hat{I}^{(1)} K^{(2)}]\}. \end{aligned} \tag{31}$$

The result can be generalized for $N > 3$, where we get more than one such condition.

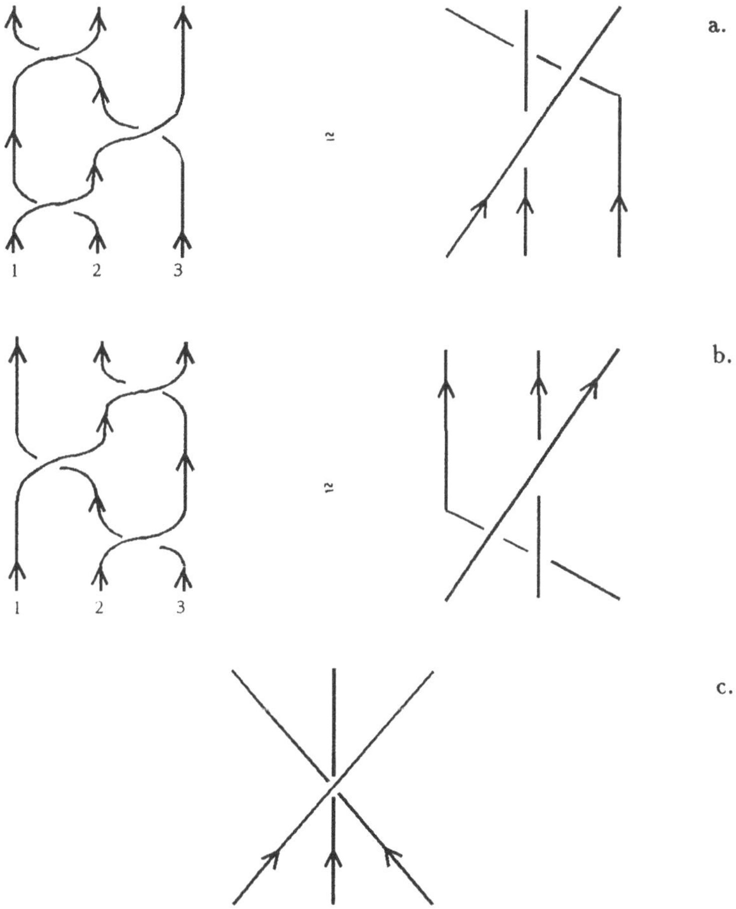

Fig.2. a. shows the motion of the three particles corresponding to the path $\mathcal{P}_1$ for $\sigma_1\sigma_2\sigma_1$ in Q. b. shows the motion of three particles corresponding to the path $\mathcal{P}_2$ for $\sigma_2\sigma_1\sigma_2$ in Q. The homotopy from $\mathcal{P}_1$ to $\mathcal{P}_2$ involves a point where all three particles have same first coordinate as shown in c.

STATISTICS OF IDENTICAL PARTICLES

We can write Eq.(11) as:

$$L_{EFF} = L_{EFF}^{(\alpha)}$$

$$= \sum_{\alpha} \frac{1}{2} m[\dot{z}^{a(\alpha)}(t)]^2 + j\,\mathrm{Tr}[K S^{(\alpha)}(t)^{-1} D_t S^{(\alpha)}(t)]$$

where we have now rotated all the $K^{(\alpha)}$ to K. Indeed, we have seen that the internal symmetry representation of the particle α is determined by the orbit of $K^{(\alpha)}$ under G. It follows that for identical particles, $K^{(\alpha)}$ are in the same orbit of G and can be written as conjugates of a fixed K on this orbit:

$$K^{(\alpha)} = \Theta^{(\alpha)} K \Theta^{(\alpha)^{-1}} \ , \ \Theta^{(\alpha)} \in G. \tag{32}$$

(We can choose $\Theta^{(\alpha)}$ to be time independent in (32), since $K^{(\alpha)}$ and K are time independent.) Now, set $S^{(\alpha)}(t) = \overline{S}^{(\alpha)}(t)\Theta^{(\alpha)^{-1}}$ in (12), and call $\overline{S}^{(\alpha)}(t)$ as $S^{(\alpha)}(t)$ again, then we get the above equation. The elements $K^{(\beta)}$ in (10) are not affected by this transformation, but they of course fulfill (32). Remember here that only a global transformation will preserve the form (10) of $A^{(\alpha)}$.

The statistical properties of sources will be studied by looking at the symmetries of L_{EFF} under exchanges. L_{EFF} is not always invariant under the exchanges

$$x^{(\alpha)} \leftrightarrow x^{(\beta)} \ , \ S^{(\alpha)} \leftrightarrow S^{(\beta)}. \tag{33}$$

We can however restore exchange symmetry by following up such an exchange with internal transformations $S^{(\sigma)} \rightarrow T^{(\sigma)^{-1}} S^{(\sigma)} \ , \ T^{(\sigma)} \in G$. In order that L_{EFF} is exchange invariant, $T^{(\sigma)'}s$ must exist such that L_{EFF} is invariant under the combination of both these transformations. Such $T^{(\sigma)'}s$ don't always exist.

For the case when $N = 3$, the exchanges of α and β , including possible internal transformations, can be written as:

$$x^{(\alpha)} \leftrightarrow x^{(\beta)} \ , \ x^{(\gamma)} \leftrightarrow x^{(\gamma)} \ , \ \gamma \neq \alpha \text{ or } \beta,$$

$$S^{(\alpha)} \rightarrow G_{\alpha\beta}{}^{-1} S^{(\beta)} \ , \ S^{(\beta)} \rightarrow G_{\beta\alpha}{}^{-1} S^{(\alpha)} \ , \ S^{(\gamma)} \rightarrow H_{\alpha\beta}{}^{-1} S^{(\gamma)}. \tag{34}$$

Furthermore, for L_{EFF} to be exchange invariant, $G_{\alpha\beta}$, $G_{\beta\alpha}$ and $H_{\alpha\beta}$ must fulfill:

$$\begin{aligned} G_{\alpha\beta} K^{(\beta)} G_{\alpha\beta}^{-1} &= K^{(\alpha)} \ , \ G_{\alpha\beta} K^{(\gamma)} G_{\alpha\beta}^{-1} = K^{(\gamma)} \ , \\ G_{\beta\alpha} K^{(\alpha)} G_{\beta\alpha}^{-1} &= K^{(\beta)} \ , \ G_{\beta\alpha} K^{(\gamma)} G_{\beta\alpha}^{-1} = K^{(\gamma)} \ , \\ H_{\alpha\beta} K^{(\beta)} H_{\alpha\beta}^{-1} &= K^{(\alpha)} \ , \ H_{\alpha\beta} K^{(\alpha)} H_{\alpha\beta}^{-1} = K^{(\beta)}. \end{aligned} \tag{35}$$

One can assume that $G_{\beta\alpha} = G_{\alpha\beta}{}^{-1}$, $H_{\alpha\beta} = H_{\beta\alpha}$.

If such $G_{\alpha\beta}$, $H_{\alpha\beta}$ don't exist, L_{EFF} is not exchange invariant. Physically, this means the following: The potential seen by particle α is $A^{(\alpha)}$ (as given in (10)); consider an exchange as in (33); the internal transformations must be such that (33) effectively exchanges the potentials $A^{(\alpha)}$ and $A^{(\beta)}$ in L_{EFF} and leaves $A^{(\gamma)}$ invariant. If these internal transformation do not exist, then the potentials seen by the particles are not exchanged by any $\alpha \leftrightarrow \beta$ exchange. When L_{EFF} is not invariant under an exchange σ_α of particles α and $\alpha+1$, σ_α transforms L_{EFF} to a new Lagrangian $\sigma_\alpha L_{EFF}$. All the Lagrangians L_{EFF} , $\sigma_\alpha L_{EFF}$, $\sigma_\alpha\sigma_\beta L_{EFF}$,... must be considered together to restore exchange invariance of the system and identity of the particles.(see Ref. 6.)

From the above, we thus expect that the statistics of the sources (as defined by the representations of B_N on the quantum states) depends on the symmetry properties of $A^{(\alpha)}$. Indeed, distinct B_N representations occur for suitable potential choices. Therefore the nonabelian sources are not uniquely associated with a particular statistics, rather the latter depends on the choice of $A^{(\alpha)}$.

The space of states for an exchange invariant L_{EFF} is spanned by $|x^{(1)}x^{(2)}x^{(3)}>$ $|m^{(1)}m^{(2)}m^{(3)}>$ where the internal group of particle α acts on the index $m^{(\alpha)}$. The exchange σ_1 of particles 1 and 2 acts as follows on the quantum states[6]:

$$\begin{aligned}\widehat{\sigma}_1|x^{(1)}x^{(2)}x^{(3)} > |m^{(1)}m^{(2)}m^{(3)} > = \\ |x^{(2)}x^{(1)}x^{(3)} > |lmn > [D(G_{21})]_{lm^{(2)}}[D(G_{12})]_{mm^{(1)}}[D(H_{12})]_{nm^{(3)}}.\end{aligned} \tag{36}$$

where $D(.)$ are unitary representations. Remembering the conditions (35) and the braid relation $\widehat{\sigma}_\alpha\widehat{\sigma}_{\alpha+1}\widehat{\sigma}_\alpha = \widehat{\sigma}_{\alpha+1}\widehat{\sigma}_\alpha\widehat{\sigma}_{\alpha+1}$, one can find solutions for the $H's$ and $G's$ such that $\widehat{\sigma}'_\alpha s$ generate a representation of the braid group.

Applying σ_1 twice gives:

$$\begin{aligned}\widehat{\sigma}_1^2|x^{(1)}x^{(2)}x^{(3)} > |m^{(1)}m^{(2)}m^{(3)} > = \\ |x^{(1)}x^{(2)}x^{(3)} > |m^{(1)}m^{(2)}n > [D(H_{12}^2)]_{nm^{(3)}}.\end{aligned} \tag{37}$$

We will now give an example of all this for the gauge group $SU(2)$ for the case of three particles ($N = 3$): We regard here $S^{(\alpha)'}s$, $K^{(\alpha)'}s$ etc. as associated with the two-dimensional UIR of $SU(2)$. A first condition that must be satisfied is the braid quantization condition (31). Let's assume that $K = j\tau_3$ with $j \geq 0$. Using (18),(21) and (24), we can write a solution of (31) as

$$\frac{j}{k} \in \{0, \pm 1, \pm 2, ...\}. \tag{38}$$

Conditions (35) simplify to

$$\begin{aligned} \operatorname{Tr} K^{(1)^2} &= \operatorname{Tr} K^{(2)^2} , \\ \operatorname{Tr} K^{(2)^2} &= \operatorname{Tr} K^{(3)^2} , \\ \operatorname{Tr} K^{(1)} K^{(3)} &= \operatorname{Tr} K^{(2)} K^{(3)} , \\ \operatorname{Tr} K^{(2)} K^{(1)} &= \operatorname{Tr} K^{(3)} K^{(1)}. \end{aligned} \tag{39}$$

A simple choice for j , k , K and $K^{(\alpha)}$ consistent with (26,27,28) and the above and leading to $\hat{\sigma}_1^2 = 1$ is $j = 1$, $k = \pm 1$ and $K = K^{(\alpha)} = \tau_3$ $(\alpha = 1, 2, 3)$. In this case the B_3 representation becomes also a representation of S_3. We have used here only the simplest solution of (28). Other solutions could lead to nontrivial representations of the braid group. For examples with other gauge groups, see Ref. 6.

FINAL REMARKS

We have neglected self-interactions of the sources while deriving L_{EFF}. However, the sources acquire intrinsic spin because of the Chern-Simons term . By using the form of the connection[6] as the source is approached in the angular direction φ, the intrinsic spin of a source turns out to be $\frac{1}{2k} \operatorname{Tr} K^2$. It is quantized in the nonabelian case in units of $\frac{\kappa}{2k}$ ($\operatorname{Tr} K^2$ being quantized in units of κ (26)).

As we have seen, in the nonabelian case, there is no unique UIR of B_N we can unambiguously associate with the states of N identical particles. As this UIR determines the statistics of the particle, there is no obvious correlation in this case between the spin and statistics of a particle.

A final note: in this work, we have considered open trajectories of particles (always going forward in time) as sources. This implies that we did not consider antiparticles. Preliminary results on the possible solutions of the field equations when sources trace out space-time loops show that the allowed gauge equivalence classes of potentials become restricted, therefore possibly restricting the statistics available.

ACKNOWLEDGEMENTS

I thank Ted Allen for helping me with TEX.

REFERENCES

1. See references 1., 2. and 3. of : A.P. Balachandran, M. Bourdeau and S. Jo, Syracuse University preprint SU-4228-406 (1989) and Int. J. Mod. Phys. A (in press).

2. A. Schwarz, Lett. Math. Phys. 2 (1978) 247; W. Siegel, Nucl. Phys. *B*156 (1979) 135; R. Jackiw and S. Templeton, Phys. Rev. *D*23 (1981) 2291; J. Schonfeld, Nucl. Phys. *B*185 (1981) 157; S. Deser, R. Jackiw and S. Templeton, Phys. Rev. Lett. 48 (1982) 975; Ann. Phys. 140 (1982) 372; C.R. Hagen, Ann. Phys. (N.Y.) 157 (1984) 342; J. Fröhlich and P.A. Marchetti, Lett. Math. Phys. 16 (1988) 347; G.V. Dunne, R. Jackiw and C.A. Trugenberger, MIT preprint CTP-1711 (1989); G. Zemba, MIT preprint CTP-1721 (1989); X.G. Wen and A. Zee, Phys. Rev. Lett. 61 (1988) 1025; J. Grundberg, T.H. Hansson, A. Karlhede and U. Lindström, Phys. Lett. *B*218 (1989) 321; A.S. Goldhaber, R. MacKenzie and F. Wilczek, Harvard preprint HUTP-88/A044 (1988); A.P. Polychronakos, Florida U. preprint UFIFT-89-7 (1989).

3. R.B. Halperin, Phys. Rev. Lett. 50 (1983) 1395.

4. G. Baskaran and P.W. Anderson, Phys. Rev. *B*37 (1988) 580; P.W. Anderson, Varenna Lectures, Proceedings of the International School of Physics, Enrico Fermi (1987); I. Affleck, Z. Zou, T.Hsu and P.W. Anderson, Phys. Rev. *B*18 (1988) 745; E. Dagotto, E. Fradkin and A. Moreo, Phys. Rev. *B*38 (1988) 2926; K. Wu, L. Yu and C.-J. Zhu, Mod. Phys. Lett. *B*2 (1988) 979; Z. Zou, Phys. Lett. *A*131 (1988) 197 and Stanford preprint (1988); A.P. Balachandran, M.J. Bowick, K.S. Gupta and A.M. Srivastava, Mod. Phys. Lett. *A*3 (1988) 1725; G. Baskaran and R. Shankar, Mod. Phys. Lett. *B*2 (1988) 1211; C. Aneziris, A.P. Balachandran and A.M. Srivastava, Syracuse University preprint SU-4228-422 (1989); I.J.R. Aitchison and N.E. Mavramotos, Phys. Rev. *B*39 (1989) 6544; Mod. Phys. Lett. *A*4 (1989) 521; Oxford preprints OUTP-89-12P and OUTP-89-28P (1989) and references therein.

5. S.K. Wong, Nuovo Cim. 65*A* (1970) 689.

6. A.P. Balachandran, M. Bourdeau and S. Jo, Syracuse University preprint, SU-4228-406 (1989) and Int. J. Mod. Phys. A (in press), and references therein.

7. A.P. Balachandran, S. Borchardt and A. Stern, Phys. Rev. D17 (1978) 3247; A.P. Balachandran, G. Marmo, B.-S. Skagerstam and A. Stern, "Gauge Symmetries and Fibre Bundles", Springer-Verlag Lecture Notes in Physics 188 (Springer-Verlag, Berlin and Heidelberg, 1983).

8. Cf. N. Woodhouse, "Geometric Quantization" [Clarendon Press, 1980].

9. E. Witten, Commun. Math. Phys. 113 (1988) 529.

10. P.O. Horvathy, Phys. Rev. D33 (1986) 407.

11. Cf. J. Birman, "Braids, Links and Mapping Class Groups", Annals of Math. Studies #82 (Princeton University Press, Princeton, 1973).

THE QUANTUM GROUP METHODS OF QUANTISING THE SPECIAL LINEAR GROUP SL(2,C)

Nigel Burroughs

Department of Applied Mathematics and Theoretical Physics
Silver Street Cambridge England CB3 9EW

ABSTRACT: There are two main methods for obtaining a description of a quantised version of the group Sl(2,*C*) : a one parameter deformation of the universal enveloping algebra of the Lie algebra sl(2,*C*), and a non-commutative deformation of functions on Sl(2,*C*). These two methods are discussed and their equivalence proved.

§1. INTRODUCTION

Quantum groups have been mentioned in this school a number of times in relation to vertex models by De Vega, Conformal field theory by Seiberg, 3-dimensional field theory by Fröhlich, and knots and links by Wu and Lee. These fields are all intimately connected, knots and links, 3-dimensional Chern Simons theory and 2-dimensional conformal field theory are all related in the work by Witten [23], and vertex models have conformal field theories as their continuum limit. Quantum groups are connected with all these areas, most explicitly in knots and links [16], [18] and vertex models [6]. The connection with conformal field theory is still incomplete.

Quantum groups and quantised algebras can be considered from two different view points, these leading to the two alternative approaches for quantising the Lie groups. Either one deforms the universal enveloping algebra of the corresponding Lie algebra [8], or one quantises the functions on the group [9]. These approaches are in fact equivalent, and related by duality [4]. The occurrence of quantum groups in Physics can be expressed in either of these formulations, in particular the enveloping algebra approach occurs in knot theory, and the function approach in vertex models. The suggested connection between quantum groups

Physics, Geometry, and Topology
Edited by H. C. Lee
Plenum Press, New York, 1990

and CFT are expressed by a combination : chiral vertex operators appear to be a quantum group comodulc and the braiding matrices of CFT are identical to the R-matrix of some quantised algebra in representation form, [2], [10], [15]. Presumably there is also a quantum group generalisation of the Chern-Simons interpretation of the Jones polynomial, reproducing the quantised algebra valued knot invariants [18], and giving quantised algebra valued 3-manifold invariants. This paper connects these two approaches to quantum groups, demonstrating their equivalence. Hence quantum group symmetries may be expressed in either form, as is convenient.

Only the Lie group Sl(2,C) is considered in this work. However this does not restrict the validity of the analysis to this particular case, the extension to other Lie groups being in most cases obvious. This paper is organised as follows. In §2, the methods for obtaining some type of a quantisation for the topological group Sl(2,C) are considered, thus motivating the quantisation of the universal enveloping algebra of sl(2,C) and the functions on Sl(2,C). Section 3 outlines the quantisation of the functions as defined by Faddeev et al [9]. The corresponding quantisation of sl(2,C), ie the one parameter deformation of the enveloping algebra Usl(2,C), is discussed in §4. This is the definition as given by Drinfel'd [8]. Starting from the quantised algebra (§4), §5 reproduces the quantisation of the functions as formulated by Faddeev et al [9], by considering the fundamental representation of the deformed enveloping algebra. The rest of the paper is dedicated to obtaining the quantised algebra as formulated by Faddeev et al [9], a formulation that is very natural from the quantised function point of view. An explicit isomorphism between the generators used in this description of the quantised algebra, and that in §4, Drinfeld's [8], is obtained. In order to accomplish this we will exploit the quantum double construction [8], which is summarised in §6. Finally the desired reconstruction of the quantised algebra can be done in §7.

This paper is based on the work in [4]. However the emphasis is different, the analysis in [4] emphasising the role of representations, and only using representation logic in the proofs. Here the same results will be obtained, but in a more algebraic manner. Hence this paper is to some extent complementary to [4].

§2. QUANTISING CONSIDERATIONS

In order to provide a motivation for the quantisation programs presented in §3 and §4, we shall consider the problem of quantising the classical group Sl(2,C). This is a topological group, so there is a topology that must also be considered in any quantisation process. A direct quantisation of Sl(2,C) involves some generalisation of topology, in particular the concept of a point would be lost. However the procedure for obtaining the required quantum generalisation of topology is unknown, and the issues one faces are difficult and as yet unresolved [12]. So in order to proceed it is necessary to use indirect methods, methods that are more algebraic and hence closer to the conventional quantisation process, ie replacing some commutative algebra and Poisson structure, with a non commutative algebra. The obvious choice is to consider the functions on Sl(2, C), Fun(Sl(2,C)), a commutative algebra with a Poisson structure. The Gelfand spectral theorems assert that we can rebuild Sl(2,C) as a topological group purely from the knowledge of the function algebra. Thus quantising the functions will produce a description of the quantised group. However, there is also a theorem that relates the algebra Fun(Sl(2,C)) to the universal enveloping algebra of sl(2,C), the Lie algebra of the original group. This states that the linear dual to Usl(2,C), denoted $\mathrm{Usl}(2,C)'$ is isomorphic to Fun(Sl(2,C))† :

$$\mathrm{Fun}(\mathrm{Sl}(2,C)) \cong \mathrm{Usl}(2,C)'. \tag{1}$$

This isomorphism can be understood by considering the derivatives of a function at the identity. An element of $\mathrm{Usl}(2,C)'$ specifies all derivatives of some function on Sl(2,C), and hence the function can be reconstructed by a Taylor series expansion about the identity. Due to the duality in (1), a quantisation of the universal enveloping algebra Usl(2,C) will also produce a description of quantised Sl(2,C). Since Sl(2,C) can be considered as embedded in the universal enveloping algebra as the exponentials of the Lie algebra††, we in fact obtain the triangle of relationships depicted in figure one.

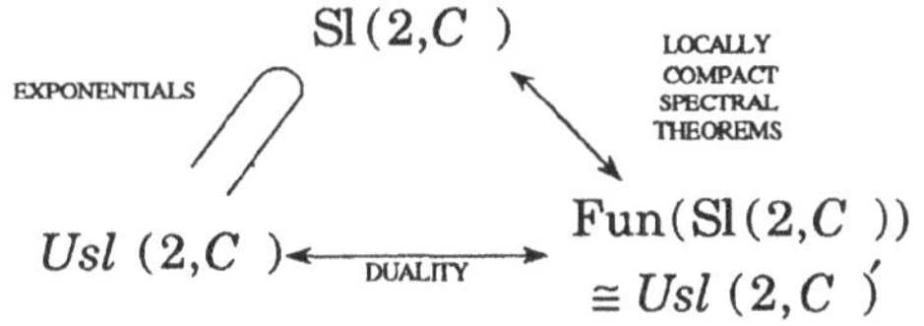

FIGURE 1

† More precisely, Usl(2,C) corresponds to the distributions on Sl(2,C) with support at the identity.
†† In particular, the exponentials of sl(2) form a group in Usl(2) isomorphic to Sl(2), are group like [1] and the C-span is Usl(2).

So a quantisation of Sl(2,C) can be achieved by starting from the classical system positioned at any corner of this triangle. However the only two viable methods at present are via the function space or the enveloping algebra. The quantisation of the function space is the path pursued by Faddeev et al [9] and Manin [19], [20] while Drinfel'd [8] and Jimbo [13], [14] consider a one parameter deformation of the universal enveloping algebra. These two constructions are equivalent, as claimed in [9] and proved in [4]. The constructions that allow their equivalence to be deduced are the subject of this paper. The two quantisation processes are summarised in the following diagram; a quantisation of figure 1.

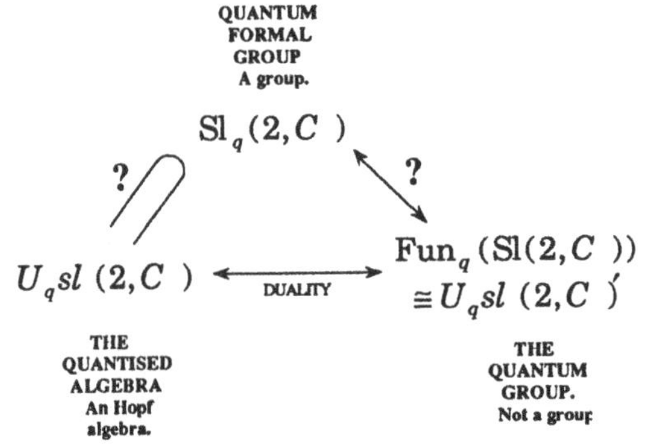

FIGURE 2

Here $U_q sl(2)$ is a one parameter deformation of the universal enveloping algebra of sl(2,C), Fun$_q$(Sl(2,C)) is a quantisation of the functions on Sl(2,C), and Sl$_q$(2, C) is a quantisation of the topological group Sl(2,C). The last object is to some extent speculative, as the author does not know to what extent this object can be described. From the category duality point of view [24], the quantised functions Fun$_q$(Sl(2,C)) can be identified with the 'continuous functions' on Sl$_q$(2, C) :

$$\text{Fun}_q(\text{Sl}(2, C)) \cong \text{Fun}(\text{Sl}_q(2, C)).$$

The actual quantum group, as opposed to the generic term for this subject, is Fun$_q$(Sl(2,C)); the linear dual to the quantised algebra $U_q sl(2)$. This definition is logical in relation to the work of Woronowicz on compact matrix pseudogroups [25], [26].

§3. QUANTISING THE FUNCTION SPACE Fun(Sl(2,C))

In this section the function space of Sl(2,*C*) is considered. The commutative algebra structure and coalgebra structure are demonstrated, however the Poisson structure is left until §4, where it is discussed in the context of the universal enveloping algebra. The quantisation of the functions, as presented by Faddeev et al in [9], is then reproduced, and the duality structure considered, pointing out that the linear dual to the quantised functions should be the one parameter deformation of the universal enveloping algebra of Drinfel'd [8]. This is in fact the case, as proved in §7.

The Function Space of SL(2,C)

The most common picture of Sl(2,*C*) involves two by two matrices, the fundamental representation of Sl(2,*C*). Thus we can consider an element of Sl(2,*C*) as the unit determinant matrix :

$$A = \begin{pmatrix} a & c \\ b & d \end{pmatrix}, \quad ad - bc = 1. \tag{2}$$

If the determinant condition is dropped, we obtain the a representation of Gl(2,C). From this representation, we deduce that the functions on Gl(2,*C*) are then isomorphic to C^4. If the functions are expanded in a Taylor series, then a system of generators for the function space can be deduced :

$$f(A) = \sum_{rstu \geq 0} \alpha_{rstu}\, a^r b^s c^t d^u, \quad \alpha_{rstu} \in C, \quad \forall\, A \in Gl(2,C).$$

By the observation that $a = \rho_{11}(A)$, $b = \rho_{21}(A)$, etc, this can rewritten as :

$$f(A) = \sum_{rstu \geq 0} \alpha_{rstu}\, \rho_{11}^r \rho_{21}^s \rho_{12}^t \rho_{22}^u (A).$$

Since this holds for all elements of Gl(2,C), an expression for functions in terms of four generators is obtained :

$$f = \sum_{rstu \geq 0} \alpha_{rstu}\, \rho_{11}^r \rho_{21}^s \rho_{12}^t \rho_{22}^u .$$

Note that this expression involves the commutative multiplication in Fun(Sl(2,*C*)), defined in the point-wise manner :

$$fg(\mathrm{A}) = f(\mathrm{A})g(\mathrm{A}),$$
$$\forall\ f, g \in \mathrm{Fun}(\mathrm{Sl}(2,\mathrm{C})),$$
$$\mathrm{A} \in \mathrm{Sl}(2,\mathrm{C}).$$

The generator expansion of the functions is valid for the general linear group Gl(2). In the case of Sl(2), a determinant condition must be imposed as a constraint on the generators ρ_{ij}, a condition deduced by analysing the determinant condition in (2) :

$$\rho_{11}\rho_{22}(\mathrm{A}) - \rho_{12}\rho_{21}(\mathrm{A}) = 1, \quad \forall \mathrm{A} \in \mathrm{Sl}(2,C),$$

or as a constraint on the generators :

$$\det(\rho) = \rho_{11}\rho_{22} - \rho_{12}\rho_{21} = 1. \tag{3}$$

The identity map, 1, is the map that takes all elements of the group to 1. Thus the functions on Sl(2,C) are generated by $\{1, \rho_{ij}\}$ with the determinant constraint holding (3).

The fact that the fundamental representation generates the continuous functions is only valid for the special linear groups. Additional representations must be included for other groups, for instance the spinor representation is also needed for the orthogonal groups.

Due to the group structure of Sl(2,C), there is an additional structure on Fun(Sl(2,C)), a coalgebra map† :

$$\Delta:\mathrm{Fun}(\mathrm{Sl}(2, C)) \to \mathrm{Fun}(\mathrm{Sl}(2, C)) \otimes \mathrm{Fun}(\mathrm{Sl}(2, C))$$
$$= \mathrm{Fun}(\mathrm{Sl}(2,C) \times \mathrm{Sl}(2, C)).$$

This is given by :

$$\Delta f(\mathrm{A},\mathrm{B}) = f(\mathrm{AB}), \quad \forall\ \mathrm{A},\mathrm{B} \in \mathrm{Sl}(2,C), \qquad f \in \mathrm{Fun}(\mathrm{Sl}(2,C)).$$

It can be verified that the axioms of a Hopf algebra [1] are satisfied by this definition, with an antipodal mapping induced by the group inverse :

$$\mathrm{S}:\mathrm{Fun}(\mathrm{Sl}(2, C)) \to \mathrm{Fun}(\mathrm{Sl}(2,C)),$$
$$\mathrm{S}(f)(\mathrm{A}) = f(\mathrm{A}^{-1}).$$

---

† Strictly speaking Fun(Sl(2,C))⊗Fun(Sl(2,C)) only contains finite sums, so the equality is not exact.

The generators ρ_{ij} possess an extremely simple coalgebra property (sum over k) :

$$\Delta\, \rho_{ij}(A,B) = \rho_{ij}(AB) = \rho_{ik}(A)\rho_{kj}(B)$$
$$= \rho_{ik} \otimes \rho_{kj}(A,B).$$

Here the representation property has been exploited. We deduce that :

$$\Delta\, \rho_{ij} = \rho_{ik} \otimes \rho_{kj}. \qquad (4)$$

This form will recur throughout this paper. We note again that it is purely a result of the representation property. With this coalgebra, it is possible to prove that :

$$\Delta(\det(\rho)) = \det(\rho) \otimes \det(\rho),$$

ie the determinant is group like [1], and hence it generates a coideal. Since the functions are a commutative algebra the determinant trivially generates an ideal, and so there exists a quotient structure. This is the reduction of the function space Fun(Gl(2,C)) to Fun(Sl(2,C)) :

$$\mathrm{Fun}(\mathrm{Sl}(2,C)) \cong \frac{\mathrm{Fun}(\mathrm{Gl}(2,C))}{\langle \det(\rho) - 1 \rangle}. \qquad (5)$$

Quantising the functions on Sl(2,C) will involve finding a non commutative Hopf algebra with a coalgebra as in (4), that reproduces Fun(Sl(2,C)) in the classical limit. Note that the coalgebra in (4) is only a result of the choice of generators for the function space; generators derived from a representation, but the commutative algebra is a result of the point-wise multiplication. Thus destroying the commutativity of the function space corresponds to the loss of the concept of points in a topological interpretation of quantisation.

Quantising the Function Space [9]

Assume the existence of a matrix R (not to be confused with the universal R-matrix introduced in §4), that is valued in $\mathrm{End}(V \otimes V)$, where V is some n dimensional vector space over the ring $C[[h]]$† . R is assumed to satisfy the Quantum Yang Baxter equation (QYBE) :

$$R_{12}R_{13}R_{23} = R_{23}R_{13}R_{12} \qquad (6)$$

† The formal infinite power series in Planck's constant.

Here R_{13} means: the matrix R in positions 1 and 3, with the identity at position 2 of $\mathrm{End}(V\otimes V\otimes V)$. Similarly for R_{23} etc.

An Hopf algebra A(R) is then defined with generators $\{1,t_{ij}\}$ satisfying the following relations :

$$R\,T_1T_2 = T_2T_1R\,,$$
$$\Delta(t_{ij}) = \sum_k t_{ik}\otimes t_{kj}\,. \tag{7}$$

Here the matrix T is a matrix of generators : $(T)_{ij} = t_{ij}$ with $T_1 = T\otimes 1$ and $T_2 = 1\otimes T$. For this algebra, the QYBE (6) corresponds to an associativity condition.

For example, the following matrix satisfies the QYBE :

$$R = \sum_{r=0}^{\infty}\frac{1}{r!}\left(\frac{h}{4}\right)^r\begin{pmatrix}1 & 0\\ 0 & (-1)^r\end{pmatrix}\otimes\begin{pmatrix}1 & 0\\ 0 & (-1)^r\end{pmatrix} + q^{\frac{-1}{2}}(q-q^{-1})\begin{pmatrix}0&1\\0&0\end{pmatrix}\otimes\begin{pmatrix}0&0\\1&0\end{pmatrix} \tag{8}$$

If this is considered as a 4 by 4 matrix it takes the form :

$$R = q^{\frac{-1}{2}}\begin{pmatrix}q & 0 & 0 & 0\\ 0 & 1 & (q-q^{-1}) & 0\\ 0 & 0 & 1 & 0\\ 0 & 0 & 0 & q\end{pmatrix} \tag{9}$$

The q factors are so arranged for later convenience. This R-matrix will in fact generate the quantum group corresponding to $Sl(2,C)$ when the Hopf structure in (7) is imposed. The algebra has the form :

$$[k',k\,] = (q-q^{-1})W^+W^-$$
$$kW^{\pm} = q^{-1}W^{\pm}k\,,\quad k'W^{\pm} = qW^{\pm}k'$$
$$W^+W^- = W^-W^+.$$

Here the T matrix has been written as : $t = \begin{pmatrix}k & W^+\\ W^- & k'\end{pmatrix}$.

There is in fact a family of R-matrices that produce an identical Hopf structure to (9). The following family of R-matrices is easily verified to give an identical algebra :

$$R = \begin{pmatrix} 1 & & & \\ & a & 1-q^{-1}a & \\ & 1-qa & a & \\ & & & 1 \end{pmatrix},$$

where $a \in C[[h]]$ is arbitrary. There are two special members of this family, the R-matrix in (9) obtained from the quantum double construction, ($a = q^{-1}$), and the triangular R-matrix as used in comodule constructions [15], [16] ($a = 2/(q-q^{-1})$) :

$$R = \begin{pmatrix} 1 & & & \\ & 2/(q+q^{-1}) & (q-q^{-1})/(q+q^{-1}) & \\ & -(q-q^{-1})/(q+q^{-1}) & 2/(q+q^{-1}) & \\ & & & 1 \end{pmatrix}.$$

From the algebra A(R), an Hopf algebra U(R) is defined, a subalgebra of the dual to A(R). U(R) is generated by $\{1, 1^{(\pm)}_{ij}\}$, which are defined by the following evaluations :

$$(L^{(\pm)}, T_1 \ldots T_k) = R_1^{(\pm)} \ldots . R_k^{(\pm)} \tag{10}$$

where the matrices of generators $L^{(\pm)}$ are defined as $(L^{(\pm)})_{ij} = 1^{(\pm)}_{ij}$. The two matrices $R^{(\pm)}$ are : $R^{(+)} = PRP$, $R^{(-)} = R^{-1}$, with P being the transposition matrix on the two factors $V \otimes V$. Equation (10) is valued in End($V^{\otimes(k+1)}$), with the labeling in the order :

$$V \otimes V_1 \otimes V_2 \otimes \ldots \otimes V_k$$

By manipulating duality and the evaluation structure, it can be shown that the Hopf structure :

$$R_{21} L_1^{(\pm)} L_2^{(\pm)} = L_2^{(\pm)} L_1^{(\pm)} R_{21}, \quad R_{21} L_1^{(+)} L_2^{(-)} = L_2^{(-)} L_1^{(+)} R_{21},$$
$$\Delta(1^{(\pm)}_{ij}) = \sum_k 1^{(\pm)}_{ik} \otimes 1^{(\pm)}_{kj}, \tag{11}$$

is obtained for the generators $1^{(\pm)}_{ij}$ [9]. All the generators $\{ 1^{(\pm)}_{ij} \}$ are not independent, since the evaluation structure in (9) is degenerate under the following† :

† The quantum determinant condition reduces to a product over diagonal elements when, as is usually the case, the matrices L are triangular.

$$1_{ii}^{(+)} = (1_{ii}^{(-)})^{-1}, \qquad \det_{q^{-1}}(L^{(\pm)}) = 1. \tag{12}$$

A degeneracy on the algebra A(R) is also observed :

$$(l, \det_q(t) - 1) = 0 \quad \forall\ l \in U(R),$$

giving the quantum determinant condition :

$$\det_q t = 1. \tag{13}$$

Here $\det_q t$ is the quantum determinant, a q-analogue of the determinant condition (3). This constraint is consistent with the algebra and coalgebra structure, since the quantum determinant lies in the centre of A(R) and is also group like [1]. Hence the quotient algebra is well defined as an Hopf algebra. This quotient structure is given in (14).

As in classical matrix algebra, the determinant characterises invertibility. The relations (12) and (13) imply that the $L^{(\pm)}$ and T matrices are invertible [4]. In fact it is necessary for the quantum determinant to be equal to the identity for this to be the case, in contrast to the matrix algebra case where the determinant is only required to be non-zero. The quantum determinant conditions are intimately linked with the possibility of defining an antipode for these bialgebras ('Hopf' algebras without antipode [1]) [4].

If we consider the two Hopf algebras A(R) and U(R) in the light of the previous classical structure of Fun(Gl(2,C)), we observe that the generators t_{ij} are q-analogues of the generators ρ_{ij}, the generators of the function space of Gl(2,C). Hence A(R)† corresponds to $\mathrm{Fun}_q(\mathrm{Gl}(2, C))$ with the following quotient structure (cf (5)) :

$$\frac{\mathrm{Fun}_q(\mathrm{Gl}(2, C))}{\langle \det_q(t) - 1 \rangle} \cong \mathrm{Fun}_q(\mathrm{Sl}(2, C)). \tag{14}$$

Lifting the isomorphism in (1) to the quantum regime, the dual elements $\{1, 1_{ij}^{(\pm)}\}$ should generate a quantisation of the universal enveloping algebra. This paper and [4] prove that this is in fact the case. The classical limit of the Hopf algebra A(R) is easily seen to be Fun(Gl(2,C)). However to be a quantisation of this algebra, the Poisson structure, [8], [17]

† The quotient algebra corresponding to the determinant condition holding, (14) will also be referred to as A(R).

must also be reproduced. In this paper, this requirement is only discussed in the context of the quantised universal enveloping algebra, §4.

Note that in both the Hopf algebras A(R) and U(R), only the algebra is q-dependent, (7), (11). Thus the coalgebra is independent of the deformation parameter. This is provided that both algebras are interpreted as quantum formal series Hopf algebras, and not as the quantum universal enveloping algebra equivalent [8]. This difference in interpretation is especially important when the classical limit is taken.

§4. A ONE PARAMETER DEFORMATION OF THE UNIVERSAL ENVELOPING ALGEBRA Usl(2,C)

In this section the quantisation of the universal enveloping algebra of sl(2,C) is discussed, as defined in [8]. The universal enveloping algebra is initially defined and the Lie coalgebra structure of sl(2) is given. This Lie coalgebra structure corresponds to a coboundary in the Lie algebra cohomology, being the coboundary of the classical r-matrix. It can also be interpreted as a copoisson structure on the universal enveloping algebra, the analogue of the poisson structure in classical systems. The reproduction of the copoisson structure will be part of the conditions determining the quantisation.

The universal enveloping algebra Usl(2,C) is the tensor algebra of sl(2,C) with the commutation relations holding. Thus Usl(2,C) has generators $\{1, H, X^{\pm}\}$. This algebra is in fact a cocommutative Hopf algebra [7]. The Hopf structure is summarised below.

$$\begin{aligned}
&[H, X^{\pm}] = \pm 2X^{\pm},\ [X^{+}, X^{-}] = H,\\
&\Delta 1 = 1\otimes 1,\ \Delta H = H\otimes 1 + 1\otimes H,\ \Delta X^{\pm} = X^{\pm}\otimes 1 + 1\otimes X^{\pm},\\
&\mathrm{S}(1) = 1,\ \mathrm{S}(H) = -H,\ \mathrm{S}(X^{\pm}) = -X^{\pm}.
\end{aligned} \tag{15}$$

(The coalgebra map is Δ, and S is the antipode)

There is an additional structure on this Hopf algebra that can be considered as the corresponding notion to the Poisson structure on the function space [8]. This is the copoisson structure [8]. A copoisson structure is a mapping :

$$\mu : \mathrm{Usl}(2) \to \mathrm{Usl}(2) \otimes \mathrm{Usl}(2)$$

that satisfies the co-Jacobi identity and a certain compatibility condition with the multiplication on the Hopf algebra Usl(2,*C*) [8]. This structure is equivalent to a 1-cocycle on the Lie algebra, and our considerations are based on this interpretation. For a fuller account of copoisson structures we refer to [8].

The copoisson structure, when restricted to the Lie algebra corresponds to a 1-cocycle in the $C^*(sl(2), sl(2) \otimes sl(2))$ cohomology, the cohomology of sl(2) valued in sl(2)⊗sl(2) [11]. In fact it is a coboundary, and for this reason sl(2) is called a coboundary Lie bialgebra. The coboundary takes the form :

$$\phi = \mu|_{sl(2)} = \delta(r) \in B^1(sl(2), sl(2) \otimes sl(2)), \tag{16}$$

where δ is the coboundary operator [11], $B^*(sl(2), sl(2) \otimes sl(2))$ the space of coboundaries, and r is the classical r-matrix :

$$r = \tfrac{1}{4} H \otimes H + X^+ \otimes X^- \in C^0(sl(2), sl(2) \otimes sl(2)), \tag{17}$$

a 0-cochain. There is a generalisation to cases where equation (16) still holds, but the classical r-matrix is not valued in $sl(2) \otimes sl(2)$, but the direct product of an embedding algebra of sl(2). These Lie bialgebras are the pseudo coboundary Lie bialgebras [8].

The cocycle ϕ is also called a Lie coalgebra structure, an interpretation that comes from the Manin triple construction of sl(2) [8].

In the case of sl(2), the copoisson/cocycle structure is as follows :

$$\begin{aligned} &\phi(H) = 0, \\ &\phi(X^{\pm}) = \tfrac{1}{2}\left(X^{\pm} \otimes H - H \otimes X^{\pm}\right). \end{aligned} \tag{18}$$

The cocycle condition, $\delta\phi=0$, is the compatibility of the map ϕ with the Lie structure, ie it expresses the interaction between the commutator bracket and the map ϕ :

$$\phi([a,b]) = [\phi(a), b \otimes 1 + 1 \otimes b] + [a \otimes 1 + 1 \otimes a, \phi(b)].$$

The commutator is the ordinary Lie bracket on the appropriate sl(2) factor.

The quantisation of the Lie algebra sl(2) is defined as a one parameter deformation of the universal enveloping algebra of sl(2) [8], [13]. The deformation parameter is Planck's constant, denoted by h. The combination $q = e^{\frac{h}{2}}$ will prove useful, due to its frequent occurrence. The one parameter deformation of the universal enveloping algebra is denoted by $U_q sl(2)$, and is generated by $\{1, H, X^{\pm}\}$ with the following Hopf structure :

$$[H, X^{\pm}] = \pm 2X^{\pm},\ [X^{+}, X^{-}] = \frac{\sinh(\frac{h}{2}H)}{\sinh(\frac{h}{2})},$$

$$\Delta(H) = 1 \otimes H + H \otimes 1,\ \Delta(X^{\pm}) = X^{\pm} \otimes q^{\frac{H}{2}} + q^{\frac{-H}{2}} \otimes X^{\pm},$$

$$S(H) = -H,\ S(X^{\pm}) = -q^{\pm 1} X^{\pm}. \tag{19}$$

It can be verified that this is indeed an Hopf algebra. It is termed the quantisation of sl(2), and in the classical limit it reproduces Usl(2,C) as a Hopf algebra, and the Lie coalgebra structure (18). The Lie coalgebra is reproduced from the observation that the original Hopf algebra is cocommutative, and hence there is an induced structure in the classical limit :

$$\mu = \left.\frac{(\Delta - T \circ \Delta)}{h}\right|_{h=0},$$

the first deviation from cocommutativity. This induced structure is just the copoisson structure/ Lie coalgebra structure (18). It is the analogue of the more traditional requirement that the non commutative algebra describing the quantised system reproduces the commutative algebra, and poisson structure of the classical system in the classical limit.

The quantised algebra $U_q sl(2)$ possesses two important subalgebras, the Borel subalgebras. These are generated by $\{H, X^{+}\}$ for $U_q b_{+}$ and $\{H, X^{-}\}$ for $U_q b_{-}$. Consider the mapping $H \to H, X^{\pm} \to X^{\mp}$ of $U_q sl(2)$ to itself. By considering this transformation on the algebra and coalgebra in (19), it is observed that this is an algebra anti-isomorphism and coalgebra isomorphism, ie preserves the coalgebra but reverses the multiplication. This is an example of a general morphism $\vartheta : U_q sl(2) \to U_q sl(2)$ that exchanges the Borel subalgebras and is an algebra anti-isomorphism and coalgebra isomorphism [4]. A morphism of this type is required in the quantum double construction [4], however it is not determined canonically by the quantised algebra, ie it is not inherent to the quantised algebra $U_q sl(2)$, as are, for instance, the antipodal maps.

There is a choice of coalgebra structure, $T \circ \Delta$ also being a suitable coalgebra, where T is the transposition operator on $U_q sl(2) \otimes U_q sl(2)$. These two coalgebras are related through the universal R-matrix R, the subject of the next definition :

DEFINITION 1 : The universal R-matrix of $U_q sl(2)$, denoted R, is an element of $U_q sl(2) \otimes U_q sl(2)$, that satisfies the following properties† :

1. R has an inverse.
2. $T \circ \Delta(a) R = R\, \Delta(a), \forall\, a \in U_q sl(2)$.
3. $\Delta \otimes \mathrm{Id}(R) = R_{13} R_{23}$, $\mathrm{Id} \otimes \Delta(R) = R_{13} R_{12}$. These equations are valued in $U_q sl(2)^{\otimes 3}$.
4. In the classical limit : $R \to 1 \otimes 1$.

For a further discussion of notation, nomenclature and properties, we refer to [8]. The universal R-matrix is only non-trivial since the Hopf algebra is not cocommutative. An universal R-matrix for a cocommutative Hopf algebra being $1 \otimes 1$, although not necessarily the only possible R-matrix. Condition 4 ensures the continuity of the universal R-matrix to this solution. To first order in Planck's constant it reproduces the classical r-matrix, (17) :

$$r = \left. \frac{R - 1}{h} \right|_{h \to 0}$$

again demonstrating the reproduction of the copoisson structure of the enveloping algebra Usl(2,*C*) and the symplectic structure of Fun(Sl(2,*C*)) [17].

We note that the above properties of definition 1 imply that the universal R-matrix satisfies the quantum Yang Baxter relation on the algebra level :

$$R_{12} R_{13} R_{23} = R_{23} R_{13} R_{12}, \tag{20}$$

an equation valued in $U_q sl(2)^{\otimes 3}$.

A further condition that is often required is the triangular condition [8] : $(R_{12})^{-1} = R_{21}$. This is a type of normalisation, all universal R-matrices can be made triangular [8]. These

† An Hopf algebra with an universal R-matrix satisfying the properties 1-3 is called a quasi-triangular Hopf algebra [8].

triangular universal R-matrices are important in the comodule structures of quantised function spaces, [20].

For $U_q sl(2)$ the universal R-matrix has a form :

$$R = q^{\frac{1}{2}H \otimes H} \sum_{r=0}^{\infty} \frac{(1-q^{-2})^r}{[r; q^{-2}]!} \left(q^{\frac{H}{2}} X^+\right)^r \otimes \left(q^{\frac{-H}{2}} X^-\right)^r \tag{21}$$

where $[r; q^{-2}]!$ is a q-analogue of a factorial :

$$[x;q]! = \prod_{s=1}^{x} \frac{(1-q^s)}{(1-q)} \quad \text{, for x integral.}$$

There exists a construction that enables the universal R-matrices corresponding to the quantisation of any Lie or Kac-Moody algebra to be calculated. This is the quantum double machine as defined in [8]. Given any representation ρ of the corresponding quantised algebra, we obtain a solution to the quantum Yang Baxter equation in $\mathrm{End}(V^{\otimes 3})$ (6), where $R = \rho \otimes \rho(R) \in \mathrm{End}(V) \otimes \mathrm{End}(V)$. Thus the quantum double construction can be used to systematically derive matrix solutions to the quantum Yang Baxter equation. However, all known solutions can not be reproduced, only the algebraic solutions [6]. This is probably because the quantum group is not yet known.

The existence of the universal R-matrix and its explicit form, as obtained by the quantum double method [5], [8] and summarised in §6 will allow us to reconstruct the quantisation program of Faddeev et al [9] from that of Drinfel'd [8] and Jimbo [13]. The importance of the antipodal map in this construction cannot be overemphasised!

§5. THE QUANTUM GROUP

In §3, the function space of Sl(2,*C*) is constructed from the fundamental representation of the Lie group. This representation can also be considered as a representation of the universal enveloping algebra, via the exponential correspondence implied in figure 1. Hence, by analogy, the fundamental representation of the quantised algebra $U_q sl(2)$ should be considered in order to reconstruct the quantised function space $\mathrm{Fun}_q\,(\mathrm{Sl}(2, C))$. The hope is that this will generate the quantised function space construction of §3.

Although the fundamental representation is used in this section, all the manipulations are valid for other representations. However the final construction of the quantised function space, $\mathrm{Fun}_q\,(\mathrm{Sl}(2, C\,))$ depends on the use of the fundamental representation, as is to be expected by analogy with the classical situation where the fundamental representation generates the continuous functions. The fundamental representation of $U_q sl(2)$ is identical to the classical case of sl(2,C) :

$$\rho(H) = \begin{pmatrix} 1 & 0 \\ 0 & -1 \end{pmatrix}, \ \rho(X^+) = \begin{pmatrix} 0 & 1 \\ 0 & 0 \end{pmatrix}, \ \rho(X^-) = \begin{pmatrix} 0 & 0 \\ 1 & 0 \end{pmatrix}. \tag{22}$$

The individual matrix elements ρ_{ij} define a mapping $\rho_{ij}{:}U_q sl(2) \to C\,[[h\,]]$ via the evaluation map $a \to (\rho(a))_{ij}$, and hence $\rho_{ij} \in U_q sl(2)'$, the dual of $U_q sl(2)$. The ρ_{ij} will generate an Hopf subalgebra of $U_q sl(2)'$ which will be denoted by A(R) following §3, R denoting the universal R-matrix in the ρ representation. The Hopf structure of A(R) is induced from the Hopf structure of $U_q sl(2)$ by duality, the coalgebra being given by :

$$(\Delta\rho_{ij}, a \otimes b) \underset{\text{def}}{=} (\rho_{ij}, ab) = \sum_k \rho_{ik}(a)\rho_{kj}(b)$$

$$\forall \quad a,b \in U_q sl(2).$$

Compare this to the similar calculation in §3, which also exploits the representation property. Since a and b are arbitrary, we deduce that :

$$\Delta\rho_{ij} = \sum_k \rho_{ik} \otimes \rho_{kj}\,. \tag{23}$$

To derive the algebra structure we shall exploit the fact that the universal R-matrix relates the two coalgebras of $U_q sl(2)$, the very structures that induce the commutation relations of the dual† :

$$\begin{aligned} \rho_{ij}\,\rho_{kl}(a) &\underset{def}{=} \rho_{ij} \otimes \rho_{kl}(\Delta a) \\ &= \rho_{kl} \otimes \rho_{ij}(\mathrm{T}\circ \Delta a) \\ &= \rho_{kl} \otimes \rho_{ij}\left(R\,\Delta a R^{-1}\right). \end{aligned}$$

Now use the coalgebra structure (23) to expand the multiplications :

† Since the multiplication relations on some algebra A induce the coalgebra relations on the dual of A (ie algebras and coalgebras are dual), the algebra of the ρ_{ij} generators will be induced from all the relations satisfied by the coalgebra. These are encoded in the universal R-matrix coboundary property, number 2 of definition 1.

$$\rho_{ij}\rho_{kl}(a) = \rho_{ka}\otimes\rho_{ic}\otimes\rho_{ab}\otimes\rho_{cd}\otimes\rho_{bl}\otimes\rho_{dj}\left(R\otimes\Delta a\otimes R^{-1}\right)$$
$$= \left(\mathrm{R}^{\rho}\right)_{ka,ic}\rho_{ab}\rho_{cd}(a)\left(\mathrm{R}^{\rho}\right)^{-1}_{bl\,dj},$$

where $\mathrm{R}^{\rho} = \rho\otimes\rho(R)$, (an $n^2\times n^2$ matrix of End$(V\otimes V)$). A summation on the indices *a,b,c,d* is implied. A rearrangement of the tensor products is necessary to obtain this result.

On defining a matrix ρ valued in the dual algebra $U_q sl(2)'$:

$$\rho\in \mathrm{Mat}\left(n, U_q sl(2)'\right) \quad \text{by} \quad (\rho)_{ij} = \rho_{ij},$$

this can be written as a matrix equation, using the notation of (7) :

$$\rho_2\rho_1 = \mathrm{R}^{\rho}\,\rho_1\rho_2(\mathrm{R}^{\rho})^{-1} \quad \text{or} \quad \mathrm{R}^{\rho}\,\rho_1\rho_2 = \rho_2\rho_1\mathrm{R}^{\rho}. \tag{24}$$

This equation in fact fixes all the commutation rules if $R\in U_q b_+\otimes U_q b_-$, a condition that holds for all the universal R-matrices constructed from the quantum double construction [8].

If the above construction is carried out for $U_q sl(2)$, using the universal R-matrix for $U_q sl(2)$, (21) and the representation (22), the R-matrix in (9) is obtained :

$$\mathrm{R} = q^{\frac{-1}{2}}\begin{pmatrix} q & 0 & 0 & 0 \\ 0 & 1 & (q-q^{-1}) & 0 \\ 0 & 0 & 1 & 0 \\ 0 & 0 & 0 & q \end{pmatrix}$$

Thus the Hopf algebra A(R) as defined by the relations (7) with this R-matrix, generates the dual $U_q sl(2)'$. As discussed in §3, there is in fact a constraint condition on the generators, the quantum determinant condition (13). In the context of generating the quantum double from the fundamental representation, this follows from a quantum determinant condition satisfied by the representation ρ, a consequence of the fact that this representation was deduced from an Sl(2,*C*) representation. The condition reads :

$$\det{}_q\,\rho = \rho_{11}\rho_{22} - q^{-1}\rho_{12}\rho_{21} = 1, \tag{25}$$

and may be verified to hold explicitly on all elements of $U_q sl(2)$ by using the representation in (22). The most satisfactory derivation of the quantum determinant with its accompanying properties is with the quantum plane constructions of Manin [19], [20].

The generators of $U_q sl(2)'$ will be denoted by $\{t_{ij}\}$ with ρ being reserved for the fundamental representation. The t_{ij} generators satisfy no more relations if the representation is the fundamental one. This is because the representation satisfies no more relations. Compare to the case of the orthogonal groups where the corresponding orthogonality condition must be included.

The quantum group as defined in [9] and reproduced in (7) has been obtained. It is desired to obtain the quantised algebra in the format of equations (11), this form being very natural from the quantised function space point of view. The similarity of the Hopf structures encoded in (7) and (11) implies that this process can be achieved by a similar technique to that used for the quantum group construction in this section. Thus it is necessary to have two structures available, as can be observed from the previous calculation of the quantum group Hopf structure :

1. An algebra representation. This gave the coalgebra (23).
2. An universal R-matrix. This allowed us to derive the algebra (24).

Thus to obtain a similar formulation of the Hopf structure of the quantised algebra $U_q sl(2)$ it is necessary to have a representation of the dual algebra $U_q sl(2)'$, and an universal R-matrix for $U_q sl(2)'$. But there is a problem : it can be proved that the quantum group $U_q sl(2)'$ does not possess an universal R-matrix [4].

The solution to this problem is based upon the following two theorems [4] :

THEOREM 1: *Given an Hopf subalgebra B of an Hopf algebra A, then the annihilator space :*

$$B^{\perp} = \{\chi \in A' : \chi(B) = 0\} \subset A'$$

is an ideal and coideal with quotient :

$$\frac{A'}{B^{\perp}} \cong B'.$$

THEOREM 2: *The Borel subalgebras are self-dual* :

$$U_q b'_\pm \cong U_q b_\pm.$$

Proofs can be found in [4].

These theorems imply that the dual $U_q sl(2)'$ can be mapped into the original quantised algebra $U_q sl(2)$ by exploiting the Borel subalgebra structure :

$$U_q sl(2)' \rightarrow \frac{U_q sl(2)'}{U_q b_\pm^\perp} \cong U_q b'_\pm \cong U_q b_\pm \rightarrow U_q sl(2) \tag{26}$$

Then the fundamental representation of $U_q sl(2)$ can be used to obtain two representations of the dual $U_q sl(2)'$, the very representation used to describe the quantum group A(R) in this section. This should then reproduce the duality structure in (10). Both morphisms in (26) are in fact required, since by taking the quotient, the possibility of describing all of the quantised algebra with a single representation has been lost [4]. The problem is to construct canonically the two homomorphisms in (26), ie only by use of the structure available from the quantised algebra. This is achieved by exploiting the quantum double construction, the subject of the next section.

§6. THE QUANTUM DOUBLE CONSTRUCTION

The quantum double construction is the central feature of the analysis in [4], that relates the two quantisations of the Lie group Sl(2,*C*), as outlined in §3 and §4. From the quantum double, the two homomorphisms (26), that are the possible resolution of the difficulties in meeting the two requirements in §5, can be constructed. These homomorphisms are constructed only from structure that is intrinsic to the quantised algebra, ie from the universal R-matrix of $U_q sl(2)$ and the antipodes. This becomes obvious from the formulae for the elements of $U_q sl(2)$ that correspond to the matrix element generators $\sigma^\pm_{ij}$ (30). In this section the main details of the construction of the two morphisms (26) will be omitted. This is for simplicity, but unfortunately loses the emphasis felt desirable in [4], mainly that the reconstruction of the quantised algebra follows due to the possibility of constructing two representations of the quantum group, (26). All resulting properties are then a result of these representations, and their structure.

This section summarises the properties of the quantum double needed for our purposes, and is not meant to be an introduction or thorough exposition, these being found in [5] and [8], while the necessary adaptations for our purposes are described in [4]. The necessary Hopf subalgebra properties and the existence of an homomorphism onto the quantised algebra $U_q sl(2)$ are simply stated. Proofs can be found in [4] and [8].

The crucial point is that there exists an Hopf algebra called the quantum double of $U_q b_+$, and denoted by D($U_q b_+$), that has the following three properties :

1. D($U_q b_+$) possesses an universal R-matrix, denoted R.
2. D($U_q b_+$) has $U_q b_+$ and $U_q b'_-$ as Hopf subalgebras. The quantum double is in fact constructed to be the minimal Hopf algebra containing these two Hopf algebras as disjoint Hopf subalgebras.
3. There is a Hopf homomorphism π, from D($U_q b_+$) onto $U_q sl(2)$. Thus an universal R-matrix structure, definition 1, is induced on the quantised algebra $U_q sl(2)$. The induced universal R-matrix on $U_q sl(2)$ is given by $\pi \otimes \pi(R)$. This is in fact the method of construction used to calculate the universal R-matrix in (21), [8].

We note that the existence of an antipode is important in the ability to construct this object [8]. The duality structure in theorem 2 is necessary in order that the quotient structure of property 3 exists [8]. An arbitrary morphism ϑ is also required, the quotient mapping π being ϑ dependent. This morphism ϑ is the general algebra anti-isomorphism and coalgebra automorphism of $U_q sl(2)$ discussed in §4, that exchanges the Borel subalgebras. Our interest is in the Borel subalgebras, so we restrict ϑ to $U_q b_-$, thus :

$$\vartheta : U_q b_- \to U_q b_+. \tag{27}$$

There is an algebra isomorphism, coalgebra anti-isomorphism induced on the dual Hopf algebras :

$$\vartheta' : U_q b'_+ \to U_q b'_- \tag{28}$$

defined by : $\vartheta' \zeta(a) = \zeta(\vartheta a)$ for all $a \in U_q b_-$, $\zeta \in U_q b'_+$. Thus, by the use of these morphisms, the Borel algebras $U_q b'_+$ and $U_q b_-$ can be mapped into the quantum double, ie by mapping to the Hopf subalgebras required by property 2. Note however, that only one of these morphisms is an algebra homomorphism, it is necessary to compose ϑ with another anti-algebra homomorphism in order to construct a representations.

The linear dual to the quantum double, denoted $D(U_q b_+)'$, is also needed. This has the quantum group $U_q sl(2)'$ as an Hopf subalgebra [4], a consequence of the fact that there exists a Hopf homomorphism from $D(U_q b_+)$ onto $U_q sl(2)$. This is the converse to theorem 2 [4]. Thus $D(U_q b_+)'$ can be considered instead of $U_q sl(2)'$ in the construction of the morphisms of $U_q sl(2)'$ into $U_q sl(2)$, (26). Further use of theorem 2 implies that the dual to the quantum double, $D(U_q b_+)'$ possesses a quotient structure induced from the Borel subalgebras of $D(U_q b_+)$ [4] :

$$D(U_q b_+)' \xrightarrow{\pi^+} \frac{D(U_q b_+)'}{U_q b_+^{\perp}} \cong U_q b'_+$$
$$D(U_q b_+)' \xrightarrow{\pi^-} \frac{D(U_q b_+)'}{U_q b'^{\perp}_-} \cong U_q b_- \qquad (29)$$

§7. RECONSTRUCTION OF THE QUANTISED ALGEBRA $U_q sl(2)$ FROM THE QUANTUM GROUP

The two necessary requirements of §5 for the reconstruction of the Hopf structure of $U_q sl(2)$ with a q-independent comultiplication have now been met. The two representations $\sigma^{\pm} = \rho \circ \phi^{\pm}$ are defined, with $\phi^{\pm}$ given by the following sequence of mappings :

$$D(U_q b_+)' \xrightarrow{\pi^+} \frac{D(U_q b_+)'}{U_q b_+^{\perp}} \cong U_q b'_+ \xrightarrow{\vartheta'} U_q b'_- \xrightarrow{1} D(U_q b_+) \xrightarrow{\pi} U_q sl(2)$$
$$D(U_q b_+)' \xrightarrow{\pi^-} \frac{D(U_q b_+)'}{U_q b'^{\perp}_-} \cong U_q b_- \xrightarrow{S\circ\vartheta} U_q b_+ \xrightarrow{1} D(U_q b_+) \xrightarrow{\pi} U_q sl(2)$$

Thus : $\phi^+ = \pi \circ \vartheta'$, $\phi^- = \pi \circ S \circ \vartheta$, the various morphisms being defined in (27), (28), (29). Note the occurrence of the antipode S with the map ϑ, this being required in order that ϕ^- is an algebra homomorphism (otherwise σ^- is an anti-representation). In this construction ϑ is arbitrary, the only requirements being that it is an anti-algebra isomorphism, coalgebra isomorphism and exchanges the Borel subalgebras. A change in the morphism ϑ produces a compensating change in the quotient map π, ie the ϑ dependence cancels. This is necessary as ϑ is not canonically associated to the quantised algebra $U_q sl(2)$.

In [4] it is shown that with these representations the elements of $U_q sl(2)$ to which the matrix elements $\sigma^{\pm}_{ij}$ correspond (recall that $\sigma^{\pm}_{ij}$ gives a linear map from $U_q sl(2)'$ to $C[[h]]$, and hence is an element of the linear dual, ie $U_q sl(2)$) are given by :

$$\sigma^{+}_{ij} = \mathrm{Id} \otimes \rho_{ij}\left(R_{U_q sl(2,C)}\right) \in U_q b_+ ,$$
$$\sigma^{-}_{ij} = \rho_{ij} \otimes \mathrm{Id}\left(R^{-1}_{U_q sl(2,C)}\right) \in U_q b_- . \tag{30}$$

The analysis in [4] continues by only exploiting the representations $\sigma^{\pm}$, emphasising how these two representations allow a reconstruction of the desired Hopf structure in (11) (the matrices $L^{\pm}$ being denoted here by $\sigma^{\pm}$). A different approach will be followed here, the expressions (30) for the elements $\sigma^{\pm}_{ij}$ and the universal R-matrix properties in definition 1 and (20), will be exploited to attain the same results. This has the advantage of greater simplicity, but tends to understate the fact that the structure in (11) follows purely from representation theoretic properties.

First note that the desired evaluation structure (10) between the T matrix and $\sigma^{\pm}$ follow directly from (30), because the T matrix is constructed from the representation ρ, §5. For example :

$$\left(t_{kl}, \sigma^{+}_{ij}\right) = \rho_{kl}\left(\sigma^{+}_{ij}\right) = \rho_{kl} \otimes \rho_{ij}\left(R_{U_q sl(2,C)}\right)$$
$$\text{or} \quad (t, \sigma) = R^{\rho},$$

as in (10) (k=1). The other k values follow from the coalgebra relation (23).

The coalgebra relation in (11), $\Delta\sigma^{\pm}_{ij} = \sum \sigma^{\pm}_{ik} \otimes \sigma^{\pm}_{kj}$ follows immediately since $\sigma^{\pm}$ are representations, but also from the property 3 of the universal R-matrix, definition 1 :

$$\Delta\sigma^{+}_{ij} = \Delta \otimes \rho_{ij}\left(R_{U_q sl(2)}\right) = \sum_k \mathrm{Id} \otimes \rho_{ik} \otimes \mathrm{Id} \otimes \rho_{kj}(R \otimes R)$$
$$= \sum_k \sigma^{+}_{ik} \otimes \sigma^{+}_{kj} ,$$

and similarly for σ^{-}_{ij}. Note that these coalgebra relations imply that the Borel subalgebras are in fact compact matrix pseudo groups† [26], or in this context quantum groups themselves. Compare to the self duality in theorem 2.

† Strictly, a pseudogroup has a C* structure; hence it is necessary to restrict $U_q sl(2)$ to $U_q su(2)$ in order to attain a *-morphism [9].

Consider mapping to the representation space in the second and third positions of the Quantum Yang Baxter relation (20) :

$$\mathrm{Id} \otimes \rho \otimes \rho(R_{12}R_{13}R_{23}) = \mathrm{Id} \otimes \rho \otimes \rho(R_{23}R_{13}R_{12})$$
$$\text{or} \quad \sigma_2^+\sigma_3^+\mathrm{R}_{23}^{\rho} = \mathrm{R}_{23}^{\rho}\sigma_3^+\sigma_2^+.$$

Relabeling the vector spaces gives the desired relation (11).

The other two algebra expressions in (11) follow from the two similar calculations (for + + and + − respectively) :

$$\rho \otimes \rho \otimes \mathrm{Id}\left(R_{13}^{-1}R_{23}^{-1}R_{12}\right) = \rho \otimes \rho \otimes \mathrm{Id}\left(R_{12}R_{23}^{-1}R_{13}^{-1}\right)$$
$$\rho \otimes \mathrm{Id} \otimes \rho\left(R_{13}R_{23}R_{12}^{-1}\right) = \rho \otimes \mathrm{Id} \otimes \rho\left(R_{12}^{-1}R_{23}R_{13}\right)$$

The two constraints (12) on the generators have differing origins. The relation $\sigma_{ii}^{(+)} = (\sigma_{ii}^{(-)})^{-1}$ follows because the diagonal elements only depend on the purely H-dependent part of the universal R-matrix, ie only on the $q^{\frac{1}{2}H \otimes H}$ prefactor in (21), which is symmetric, and because $S(H)=-H$. In contrast, the quantum determinant condition is a result of a property of the representation. It follows from the quantum determinant condition satisfied by the fundamental representation (25).

In order to derive the quantum determinant condition in (12), it is necessary to introduce the q-analogue Levi-Civita symbol [4]. Define :

$$\varepsilon^q = \begin{pmatrix} 0 & 1 \\ -q^{-1} & 0 \end{pmatrix}, \tag{31}$$

which in this 2-dimensional form also occurs in [22]. Then the quantum determinant can be expressed as [4] :

$$\varepsilon_{ij}^q \, t_{ri} t_{sj} = \varepsilon_{rs}^q \det{}_q t\,, \tag{32}$$

an obvious generalisation of the more familiar classical expression. Then we have :

$$\mathrm{Id} \otimes \det{}_q \rho(R\ ) = \mathrm{Id} \otimes 1(R\ ) = 1 \quad \in \ U_q sl(2, C\ )$$

or upon manipulating the left hand side using (32) :

$$\mathrm{Id} \otimes \varepsilon^q_{ij}\, \rho_{1i}\, \rho_{2j}(R) = \varepsilon^q_{ij}\, \mathrm{Id} \otimes \rho_{1i} \otimes \rho_{2j}(\mathrm{Id} \otimes \Delta(R))$$
$$= \varepsilon^q_{ij}\, \mathrm{Id} \otimes \rho_{1i} \otimes \rho_{2j}(R_{13} R_{12})$$
$$= \varepsilon^q_{ij}\, \sigma^+_{2j} \sigma^+_{1i} = \varepsilon^{q^{-1}}_{ij}\, \sigma^+_{1i} \sigma^+_{2j}.$$

Here we have used the easily verified result that : $\varepsilon^q_{ij} = -\, q^{-1} \varepsilon^{q^{-1}}_{ji}$. Thus we obtain the constraint : $\det_{q^{-1}} \sigma^{\pm} = 1$.

There are no other relations between the $\sigma^{\pm}$ generators, since the universal R-matrix and fundamental representation satisfy no further relations.

For $U_q sl(2)$ the following forms for $\sigma^{\pm}$ are obtained from (30), (21) and (22) :

$$\sigma^+ = \begin{pmatrix} q^{\frac{H}{2}} & 0 \\ q^{\frac{-1}{2}}(q - q^{-1})X^+ & q^{\frac{-H}{2}} \end{pmatrix}, \quad \sigma^- = \begin{pmatrix} q^{\frac{-H}{2}} & -\, q^{\frac{1}{2}}(q - q^{-1})X^- \\ 0 & q^{\frac{H}{2}} \end{pmatrix} \tag{33}$$

It can be verified that the Hopf structure of these generators, as encoded in (11) is indeed that of $U_q sl(2)$ (19). The two constraints on the generators (12) :

$$\sigma^{(+)}_{ii} = (\sigma^{(-)}_{ii})^{-1}, \quad \det_{q^{-1}} \sigma^{\pm} = 1,$$

obviously hold by direct observation of the form of the matrices in (33).

Note that the algebra generated by $\{1, \sigma^{\pm}_{ij}\}$ is only a Hopf subalgebra of $U_q sl(2)$ as defined in §4. However these Hopf algebras only differ in the treatment of the Cartan subalgebra, U(R) only containing the combination $q^{\frac{H}{2}}$, and not H itself. This is not an important difference. The quantised algebra $U_q sl(2)$ as defined by Jimbo in [13] is just our U(R). Thus we deduce, modulo this difference in the treatment of the Cartan subalgebra, that U(R) generates the quantised algebra $U_q sl(2)$, and A(R) generates the linear dual, ie the quantum group $\mathrm{Fun}_q(\mathrm{Sl}(2,C))$, as claimed in §3 and [9]. The two approaches to quantising the Lie group Sl(2,*C*) are thus equivalent.

CONCLUSION

The discussion in §2 on the problem of obtaining some quantisation prescription of the Lie group Sl(2,*C*) lead us to a consideration of the function space of Sl(2,*C*) and the universal enveloping algebra Usl(2,*C*). The two methods in the literature for performing this quantisation, [8], [9], [13], [14] are then related and their equivalence proved (§5, §7). The discussion is restricted to Sl(2,*C*) for convenience and simplicity of exposition, but all the definitions and proofs can be applied directly to all the other Lie groups, or at least require only small changes. Due to the presence of the constraint relations in the construction of Faddeev et al [9], the deformation of the universal enveloping algebra approach [8] is simpler (but more abstract). However we have proved that starting from the quantised algebra as defined by Drinfel'd [8], we may equivalently define the quantum group and quantised algebra by :

$$\left.\begin{array}{l} RT_1T_2 = T_2T_1R \\ \Delta(t_{ij}) = \sum_k t_{ik} \otimes t_{kj} \end{array}\right\} \quad \mathrm{Fun}_q(\mathrm{Gl}(2, C))$$

$$\det{}_q\, t = 1 \qquad \text{Quotients to } \mathrm{Fun}_q(\mathrm{Sl}(2, C)).$$

$$\left.\begin{array}{l} R_{21}L_1^{(\pm)}L_2^{(\pm)} = L_2^{(\pm)}L_1^{(\pm)}R_{21} \\ R_{21}L_1^{(+)}L_2^{(-)} = L_2^{(-)}L_1^{(+)}R_{21} \\ \Delta(l_{ij}^{(\pm)}) = \sum_k l_{ik}^{(\pm)} \otimes l_{kj}^{(\pm)} \end{array}\right\} \text{Generates } U_q sl(2) \left\{\begin{array}{l} l_{ii}^{(+)} = (l_{ii}^{(-)})^{-1}, \\ \det_{q^{-1}}(L^{(\pm)}) = 1. \end{array}\right.$$

with an evaluation structure given in (10), expressing the duality of these Hopf algebras :

$$\mathrm{Fun}_q(\mathrm{Sl}(2,C)) \cong U_q sl(2,C)',$$

the analogue of the classical expression (1).

The problem is to obtain an explicit description of the $L^{(\pm)}$ matrices in terms of the generators used in the quantised algebra description of [8] and [14]. The expressions (30) require a knowledge of the universal R-matrix, this only being known for the A_n Lie algebra series [3], [21]. The construction of the matrices $L^{(\pm)}$ in this case is carried out in [4]. However a similar analysis for the other Lie algebra series has not been done. In these cases there will be a greater number of constraints on the $L^{(\pm)}$ generators corresponding to the orthogonality etc (cf (12)), and it will be necessary to consider more than just the fundamental representation in the construction of the function spaces. The extent of the applicability of similar calculations for the quantised Kac-Moody algebras is unknown.

The third method of quantisation, via a quantum topology is still open, and is perhaps a testing ground for quantum topology.

ACKNOWLEDGEMENTS

I would like to thank the organisers of the Banff Summer School for a very interesting conference and allowing me to talk. I would like to thank C. Isham for discussing the question of topology in relation to $Sl_q(2, C)$, and the spectral theorems. I am indebted to C. Yastremiz for helping with the preparations of my talk, and Dr. A. Macfarlane for his useful advice in the preparations of this script.

REFERENCES

[1] Abe, E. : Hopf Algebras. Cambridge University Press 1977.
[2] Alvarez-Gaumé, L. Gomez, C. Sierra, G. : Duality and Quantum Groups. CERN preprint. TH. 5369/89.
[3] Burroughs, N.J. : The Universal R-matrix for $U_q\, sl\,(3)$ and Beyond! DAMTP/R-89/4. CMP to be published.
[4] Burroughs, N.J. : Relating the approaches to Quantised Algebras and Quantum Groups. DAMTP/R-89/11. CMP to be published.
[5] Burroughs, N. : In preparation.
[6] De Vega, H.J. : Quantum Groups (YBZF algebras), integrable Field Theories and statistical models. Paris preprint LPTHE 87-54. Also see the lectures at this summer school.
[7] Dixmier, J. : Enveloping Algebras. North Holland Mathematical Library. Vol 14, 1977.
[8] Drinfel'd, V.G. : Quantum Groups. ICM Berkeley 1986, Ed. A.M. Gleason, AMS, Providence (1987) 798.
[9] Faddeev, L. Reshetikhin, N. Takhtajan, L. : Quantisation of Lie groups and Lie algebras. Leningrad preprint LOMI E-14-87.
[10] Felder, G. Fröhlich, J. Keller, G. : Braid Matrices and Structure Constants for Minimal Conformal Models. CMP 124, 647-664 (1989).
[11] Fuks, D.B. : Cohomology of infinite dimensional Lie algebras. Contemporary Soviet Mathematics, 1986.

[12] Isham, C. : See for example the lectures in this summer school.

[13] Jimbo, M. : A q-Difference Analogue of U(g) and the Yang-Baxter Equation. LMP 10 (1985) 63-69.

[14] Jimbo, M. : A q-analogue of the $U_q(gl(N+1))$, Hecke Algebra, and the Yang-Baxter Equation. LMP 11 (1986) 247-252.

[15] Kanie, Y. Tsuchiya, A. : Vertex Operators in Conformal Field Theory on P^1 and Monodromy Representations of Braid Group. Adv. Studies in Pure Math. Vol. 16, 1988, Jimbo, M. Miwa, T. Tsuchiya, A. (Eds) Conformal Field Theory and Solvable Lattice Models. pp 297-372.

[16] Kirillov, A.N. Reshetikhin, N. Yu. : Representations of the Algebra $U_q sl(2)$, q-Orthogonal Polynomials and Invariants of Links. Leningrad Preprint LOMI E-9-88.

[17] Kosmann-Schwarzbach, Y. Poisson Drinfeld groups. Appears in Topics in Soliton theory and exactly solvable nonlinear equations. Ablowitz, M. Fuchssteiner, B. and Kruskal M. (Eds.) : World Scientific, Singapore 1987.

[18] Lawrence, R.J. : A Universal Link Invariant Using Quantum Groups. Oxford preprint. (Mathematical Institute.)

[19] Manin, Yu. I. : Some remarks on Koszul algebras and quantum groups. Ann. Inst. Fourier, Tome XXXVII, F.4, 191-205 (1987).

[20] Manin, Yu. I. : Quantum groups and non-commutative geometry. Prcprint Montreal University, CRM-1561, 1988.

[21] Rosso, M. : An analogue of P.B.W. Theorem and the Universal R-matrix for $U_h sl(N+1)$. Palaiseau preprint (1989). CMP to be published.

[22] Vokos, S. Wess, J. Zumino, B. : Properties of Quantum 2×2 Matrices. LAPP-TH-253/89.

[23] Witten, E. : Quantum Field Theory and the Jones Polynomial. CMP 121,351-399 (1989).

[24] Woronowicz, S.L. : Pseudospaces, pseudogroups and Pontriagin duality, Proceedings of the International Conference on Mathematics and Physics Lausanne 1979, Lecture Notes in Physics, 116 (1980).

[25] Woronowicz, S.L. : Twisted $SU(2)$ Group. An Example of a Non-Commutative Differential Calculus. Publ. RIMS 23, 117-181, 1987.

[26] Woronowicz, S.L. : Compact Matrix Pseudogroups. CMP 111, 613-665 (1987).

2+1 DIMENSIONAL QUANTUM GRAVITY AND THE BRAID GROUP

Steven Carlip

The Institute for Advanced Study
Princeton, NJ 08540 U.S.A.

Elsewhere in these proceedings [1], Jackiw has described an interesting calculation of scattering amplitudes in 2+1 dimensional gravity. His approach [2], like that of 't Hooft [3], starts from the observation that there is no classical gravitational radiation in 2+1 dimensions: solutions of the field equations are locally flat outside sources, and the gravitational interaction depends only on the global geometry. In the absence of propagating gravitational degrees of freedom, one way to quantize the theory is to use the field equations to eliminate the metric, and to then quantize the remaining matter degrees of freedom. One then has the physically intuitive picture of quantized particles each moving in the background conical geometry induced by the others. For two-particle scattering, this picture is especially simple, since one may use center of mass coordinates [3] to reduce the problem to one of solving the Schrödinger equation on a cone.

While this approach is attractive, it is not obvious that it is equivalent to ordinary canonical quantization. Further, since the total deficit angle is determined classically by the energy, we are forced to work in a fixed eigenstate of the Hamiltonian, and the superposition of energy eigenstates becomes difficult to understand [4]. In addition, by assuming the form of the background geometry, we evade some of the important conceptual issues of quantum gravity, such as the role of time in a theory in which time translations are symmetries [5] and problem of constructing diffeomorphism-invariant observables [6]; while this simplifies calculations, it makes the generalization to 3+1 dimensions less clear.

A conventional canonical treatment of 2+1 dimensional quantum gravity is therefore of interest. Such a quantization is now possible, thanks to the recent observation of Witten [7] that 2+1 dimensional gravity can be treated as a Chern-Simons gauge theory. By comparing such an exact quantization to the physically clear picture of Jackiw *et al.*, we may hope to gain some insight into the quantization of gravity in 3+1 dimensions, while at the same time testing the 2+1 dimensional canonical framework.

Physics, Geometry, and Topology
Edited by H. C. Lee
Plenum Press, New York, 1990

1. 2+1 GRAVITY AS A CHERN-SIMONS THEORY

Let us begin by reviewing Witten's treatment [7] of $2+1$ dimensional gravity as a Chern-Simons theory for the $2+1$ dimensional Poincaré group $ISO(2,1)$. This group is characterized by two sets of generators, the local Lorentz generators $\mathcal{J}^a$ and the translation generators $\mathcal{P}_a$, obeying the usual algebra

$$\begin{aligned} [\mathcal{P}_a, \mathcal{P}_b] &= 0 \\ [\mathcal{P}_a, \mathcal{J}_b] &= \epsilon_{abc}\mathcal{P}^c \\ [\mathcal{J}_a, \mathcal{J}_b] &= \epsilon_{abc}\mathcal{J}^c \end{aligned} \tag{1.1}$$

To construct a Chern-Simons theory, we associate gauge fields $A_\mu = \{e^a{}_\mu, \omega_{a\mu}\}$ to the symmetry generators $\{\mathcal{P}_a, \mathcal{J}^a\}$, and form the standard Chern-Simons action $S = \int \mathrm{Tr}\left(A\wedge dA + \frac{2}{3}A\wedge A\wedge A\right)$. To make sense of this expression, we must define "trace" for $ISO(2,1)$, i.e., we must specify a nondegenerate group-invariant pairing of the generators. Such a pairing is almost unique [7]: it is

$$\begin{aligned} &\mathrm{Tr}\,(\mathcal{J}^a\mathcal{P}^b) = \eta^{ab} \\ &\mathrm{Tr}\,(\mathcal{J}^a\mathcal{J}^b) = \mathrm{Tr}\,(\mathcal{P}^a\mathcal{P}^b) = 0 \end{aligned} \tag{1.2}$$

It is then easy to check that

$$S = \frac{1}{2}\int_M \epsilon^{\rho\mu\nu} e^a{}_\rho\left(\partial_\mu\omega_{a\nu} - \partial_\nu\omega_{a\mu} + \epsilon_{abc}\omega^b{}_\mu\omega^c{}_\nu\right) \tag{1.3}$$

which can be recognized as the standard first order form of the Einstein action if we identify $\omega_{a\mu}$ with the spin connection $\frac{1}{2}\epsilon_{abc}\omega_\mu{}^{bc}$.

S is invariant under the gauge transformations

$$\begin{aligned} \delta e^a{}_\mu &= -\partial_\mu\rho^a - \epsilon^{abc}\omega_{b\mu}\rho_c - \epsilon^{abc}e_{b\mu}\tau_c \\ \delta\omega^a{}_\mu &= -\partial_\mu\tau^a - \epsilon^{abc}\omega_{b\mu}\tau_c \end{aligned} \tag{1.4}$$

The τ^a parameterize the local Lorentz transformations, while the ρ^a transformations are equivalent on shell to the diffeomorphisms. Note that this means we can either gauge-fix the diffeomorphisms or the ρ transformations: their difference is a trivial invariance of the kind discussed by Henneaux at this school [8].

We now have a Chern-Simons description of $2+1$ dimensional gravity without sources. There are various methods of incorporating point particles: by adding source terms to the action, by including timelike Wilson lines [9], or, most simply, by deleting particle world lines from our spacetime manifold and investigating the surrounding geometry. Let us choose this latter approach: we shall study quantum gravity on the manifold $\mathbb{R}\times(\Sigma - \{x_\alpha\})$, where Σ is a two-surface (which we will often take to have the topology $\mathbb{R}^2$) and $\{x_\alpha\}$ are the locations of sources.

One slight complication must be addressed if we wish to consider scattering problems. The classical solution of the $2+1$ dimensional Einstein equations in the presence of a point source contains a conical singularity at the location of the source. At such a point, the group of rotations of the tangent space changes: if we denote the deficit angle of a cone by β, a vector at the apex of the cone returns to its original position after a rotation by $2\pi - \beta$ rather than 2π. To allow for this kind of singular behavior, we should replace $ISO(2,1)$ by its universal covering space $\widetilde{ISO}(2,1)$. We can always recover the original $ISO(2,1)$ theory by an appropriate projection, but, as we shall see below, by allowing the projection to vary at chosen points we can also obtain the appropriate singular behavior.

Unfortunately, the passage from $ISO(2,1)$ to $\widetilde{ISO}(2,1)$ is not unique, and we will also find a number of ambiguities of 2π in the final theory. For example, the deficit angle of the conical singularity for a classical point source is equal to the mass of the source; but deficit angles are periodic, while masses are not, so the geometry alone is insufficient to determine the physics. A similar ambiguity occurs in the Regge calculus approach to gravity, in 2+1 or 3+1 dimensions [10]. While this ambiguity can be resolved classically by an appeal to the low-mass limit, its role in the quantum theory is not well-understood.

2. QUANTIZATION

Our goal is to quantize the system described by the action (1.3) on a manifold $\mathbb{R} \times (\Sigma - \{x_\alpha\})$. Decomposing our fields into space and time components, we find that

$$S = \int dt \int_\Sigma \left(-\epsilon^{ij} e^a{}_i \frac{d}{dt}\omega_{aj} + e^a{}_0 \tilde{N}_a + \omega_{a0} N^a\right) \tag{2.1}$$

where

$$\begin{aligned} N^a &= \frac{1}{2}\epsilon^{ij}\left(\partial_i e^a{}_j - \partial_j e^a{}_i + \epsilon^{abc}(\omega_{bi}e_{cj} - \omega_{ci}e_{bj})\right) \\ &\qquad + \textit{boundary terms} \\ \tilde{N}^a &= \frac{1}{2}\epsilon^{ij}\left(\partial_i \omega^a{}_j - \partial_j \omega^a{}_i + \epsilon^{abc}\omega_{bi}\omega_{cj}\right) \end{aligned} \tag{2.2}$$

Since no time derivatives of $e^a{}_0$ or ω_{a0} appear in the action, the N^a and $\tilde{N}^a$ are constraints. From the $e\dot{\omega}$ term, on the other hand, we can read off the commutators

$$\begin{aligned} [\omega_{ai}(x), e^b{}_j(x')] &= i\epsilon_{ij}\delta^b_a\delta^2(x - x') \\ [\omega_{ai}(x), \omega_{bj}(x')] &= [e^a{}_i(x), e^b{}_j(x')] = 0 \end{aligned} \tag{2.3}$$

It may then be checked that the constraints obey the $ISO(2,1)$ algebra (1.1) and that the generator of gauge transformations (1.4) is

$$G = -\int_\Sigma (\tau_a N^a + \rho_a \tilde{N}^a) \tag{2.4}$$

We must next construct a Hilbert space $\mathcal{H}$ on which operators e and ω act. Because of the structure of the constraints, it turns out to be easiest to work in a "momentum representation," in which the Hilbert space is built out of functionals of the ω_{ai}. To impose the constraints $N^a = \tilde{N}^a = 0$, we have two options: as Seiberg has explained in his lectures here [11], such constraints may either be solved before quantization or imposed as operator equations after quantization.

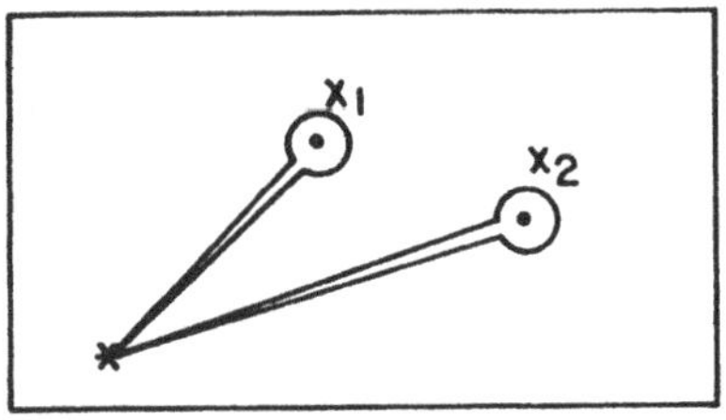

Figure 1. A flat connection on a punctured plane is determined by its holonomies around the punctures.

Let us choose the former approach. To solve the constraints, we observe that the condition $\tilde{N}^a = 0$ is simply the statement that ω_{ai} is a flat $\widetilde{SO}(2,1)$ connection on $\Sigma - \{x_\alpha\}$. N^a then generates the $\widetilde{SO}(2,1)$ transformations of ω, so the constraint $N^a = 0$ requires us to identify connections which are gauge-equivalent. Hence the physical Hilbert space is the space of square integrable functions on the moduli space $\mathcal{N}$ of flat $\widetilde{SO}(2,1)$ connections on $\Sigma - \{x_\alpha\}$.

More concretely, a flat connection is determined up to gauge transformations by its holonomies around the nontrivial loops in $\Sigma - \{x_\alpha\}$. To specify these holonomies, we must associate an element of the gauge group with each homotopy class of loops; in other words, we must give a group homomorphism $\pi_1(\Sigma - \{x_\alpha\}, *) \to \widetilde{SO}(2,1)$, where $*$ is a base point for the fundamental group of $\Sigma - \{x_\alpha\}$. Under a gauge transformation $\mathcal{G}$: $\Sigma - \{x_\alpha\} \to \widetilde{SO}(2,1)$, the holonomies are conjugated at $*$ by $\mathcal{G}(*)$. Hence the moduli space of flat connections may be written as

$$\mathcal{N} = \mathrm{Hom}\big(\pi_1(\Sigma, *), \widetilde{SO}(2,1)\big)/ \sim \tag{2.5}$$

where two homomorphisms are identified under $\sim$ if they differ by conjugation.

If Σ is not compact, we must modify this result slightly, since the action (1.3) is then only invariant under gauge transformations which fall off sufficiently fast at infinity. The simplest approach is to choose the base point $*$ to be at infinity, so gauge transformations are the identity at $*$, and to omit the equivalence relation $\sim$ from the definition of $\mathcal{N}$. In particular, if Σ has the topology $\mathbb{R}^2$, a point in $\mathcal{N}$ is determined by a set of holonomies

$$\Lambda_\alpha = e^{-ip_{\alpha a}\mathcal{J}^a} \in \widetilde{SO}(2,1) \tag{2.6}$$

around the punctures $\{x_\alpha\}$ (see figure 1).

We are thus led to consider a Hilbert space of square integrable functions $\psi(p_{\alpha a})$. One cautionary note is needed, however. As noted above, we must allow our holonomies to be elements of $\widetilde{SO}(2,1)$, not just $SO(2,1)$. This means in effect that the wave functions ψ must be multivalued: the 2π rotation operator $\exp\{2\pi i \mathcal{J}_\alpha{}^0\}$ need not act as the identity, but may change the phase of ψ. We therefore obtain a quantum theory described by a Hilbert space $\mathcal{H}$ of multivalued square integrable functions of the parameters $p_{\alpha a}$ which label the holonomies of ω_a.

3. DYNAMICS

With this understanding of the kinematics of 2+1 dimensional quantum gravity, can turn to the question of the dynamics, that is, time translation and the Hamiltonian. If space is compact, this is can be a confusing task: time translations are gauge symmetries, the Hamiltonian is a constraint, and even the definition of time becomes ambiguous. If space is asymptotically locally flat, however, the Hamiltonian includes a nonvanishing boundary contribution, and there is a sensible definition of time translation based on the behavior of the metric at infinity.

To derive this boundary contribution, Regge and Teitelboim [12] give a simple general argument. Let G be the generator of a symmetry such as time translation. For the Poisson bracket of G with the fields to be well-defined, we must be able to take functional derivatives of G; this will be possible only if variations δG contain no boundary contributions. While boundary contributions which do occur can sometimes be eliminated by restricting the allowed variations of the fields, certain variations — those corresponding to asymptotic symmetries — must be permitted on physical grounds, and the unwanted boundary variations must instead be cancelled by adding terms to G.

For $2+1$ dimensional gravity, the classical metric for isolated sources is asymptotically conical [13]:

$$ds^2 \sim \left(dt - \frac{\alpha}{2\pi - \beta} d\phi\right)^2 - dr^2 - r^2 d\phi^2 \qquad 0 \leq \phi < 2\pi - \beta \tag{3.1}$$

where β is the total deficit angle and α is proportional to the total angular momentum. Such a metric admits two asymptotic $ISO(2,1)$ symmetries, time translation

$$\rho^0 \sim -1 \tag{3.2}$$

and spatial rotation

$$\begin{cases} \tau^0 \sim 1 \\ \rho^2 \sim -r \end{cases} \tag{3.3}$$

For the generator G of gauge transformations defined by equation (2.4),

$$\delta G = -\int_{\partial\Sigma} \rho_a \delta\omega^a{}_{\parallel} + \mathit{volume\ terms} \tag{3.4}$$

where the subscript $\parallel$ denotes the component tangent to $\partial\Sigma$. (Because of the boundary terms in (2.2), there is no τ_a contribution.) Hence for an asymptotic time translation (3.2), we must cancel the boundary variation by adding a term

$$H = -\int_{\partial\Sigma} \omega^0{}_{\parallel} \tag{3.5}$$

to G.

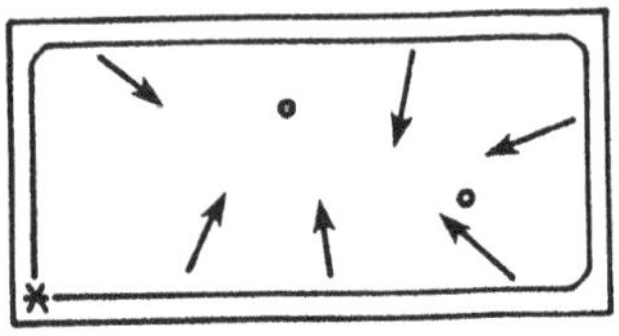

Figure 2. The contour defining H can be deformed to give a product of holonomies.

Classically, $H = \beta$, the total deficit angle. When $\Sigma = \mathbb{R}^2 - \{x_\alpha\}$, there is a useful trick for evaluating this quantity. In the frame for which (3.1) holds, the only component of ω at infinity lies in the $a = 0$ direction. Hence

$$\mathrm{Tr}\, e^{-iH\mathcal{J}_0} = \mathrm{Tr}\ P\exp\{i\int_{\partial\Sigma} \omega^a{}_i dx^i \mathcal{J}_a\} = \mathrm{Tr}\,(\prod_\alpha \Lambda_\alpha) \tag{3.6}$$

where the holonomies Λ_α are given by (2.6), and the last equality comes from deforming the contour of integration through regions of zero curvature (see figure 2). On the other hand, in the fundamental representation of $SO(2,1)$,

$$\mathrm{Tr}\, e^{-iH\mathcal{J}_0} = 1 + 2\cos H \tag{3.7}$$

For the case of two Wilson lines, for instance, this means that

$$\cos\frac{H}{2} = \cos\frac{m_1}{2}\cos\frac{m_2}{2} - \sin\frac{m_1}{2}\sin\frac{m_2}{2}(\frac{p_1\cdot p_2}{m_1 m_2}) \tag{3.8}$$

where

$$m_\alpha^2 = p_\alpha^2 \tag{3.9}$$

This expression agrees with the result of Deser *et al.* [14] for the addition of deficit angles. From the classical solutions described in [14], it may be checked that the symbols p and m, which we introduced to parameterize the holonomies, really do have interpretations as momenta and masses. Note also that in the fundamental representation of $SO(2,1)$, the Hamiltonian H — and for that matter the masses m_α — are only determined mod 2π. This is an early sign of the ambiguity in the lifting from $ISO(2,1)$ to $\widetilde{ISO}(2,1)$; it will reappear when we try to evaluate scattering amplitudes.

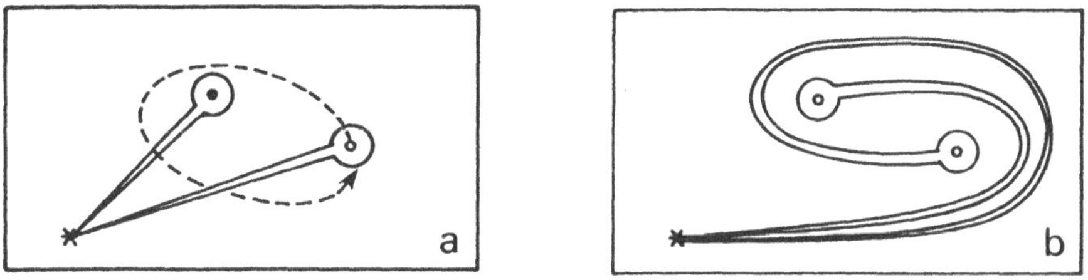

Figure 3. The holonomies around a pair of punctures (a) change under the action of the braid group (b).

4. BRAIDS

We now have a Hilbert space and a Hamiltonian for 2+1 dimensional gravity. The Hamiltonian is essentially the same as that of Deser and Jackiw [2] and 't Hooft [3], and it gives the same differential equations for scattering amplitudes as those proposed by Jackiw. But we have not yet seen any sign of the conical boundary conditions so crucial to these other analyses.

To understand the origin of these boundary conditions, we must examine more carefully the symmetries of the theory. We saw above that diffeomorphisms in 2+1 dimensional gravity are equivalent on shell to ρ gauge transformations, and thus are factored out when we gauge-fix. However, the proof of this equivalence starts with the infinitesimal form (1.4) of the transformations, and, strictly speaking, only holds for those transformations which can be reached by exponentiating infinitesimal ones. There remain the diffeomorphisms which are not isotopic to the identity, which must still be factored out.

For a manifold $\mathbb{R} \times (\mathbb{R}^2 - \{x_\alpha\})$, such diffeomorphisms are generated by Dehn twists around the punctures x_α. At first sight, such transformations appear to have no effect on the holonomies Λ_α, which are defined in terms of coordinate-independent integrals. If we look more carefully, however, we see that the definition (2.6) of Λ_α depends implicitly on a choice of generators for the fundamental group $\pi_1(\Sigma - \{x_\alpha\})$. A Dehn twist has the effect of changing this choice of generators, as is apparent in figure 3a-b: when one puncture is wrapped around another, the relevant holonomies change by conjugation. This is the well-known action of the braid group on the fundamental group of a punctured surface, and it gives the relevant action of the mapping class group on the holonomies.

If we wish to consider the full diffeomorphism group to be a symmetry of our theory, we must demand that the wave functions ψ be invariant under this action. Note that the generalization from $\mathbb{R}^2$ to an arbitrary two-manifold Σ is straightforward; there will be new elements of the braid group coming from the possibility of wrapping punctures around handles, but the properties of the resulting group are well-understood [15,16].

For concreteness, let us specialize to the case of two punctures. Our Hilbert space then consists of square integrable functions of two momenta p_1 and p_2, the Hamiltonian is given by (3.8), and it is apparent from figure 3 that the action of the braid group is

$$\begin{aligned} \Lambda_1 &\mapsto \Lambda_2\Lambda_1\Lambda_2^{-1} = (\Lambda_2\Lambda_1)\Lambda_1(\Lambda_2\Lambda_1)^{-1} \\ \Lambda_2 &\mapsto (\Lambda_2\Lambda_1)\Lambda_2(\Lambda_2\Lambda_1)^{-1} \end{aligned} \tag{4.1}$$

The momenta p after the transformation will thus differ by an $SO(2,1)$ matrix $\Lambda_2\Lambda_1$ from the initial momenta. This matrix, in turn, determines an operator $\widehat{\Lambda_2\Lambda_1}$ on the Hilbert space $\mathcal{H}$, and states must be invariant under its action.

Note that once again a 2π ambiguity has appeared. The action of the braid group on $SO(2,1)$ holonomies is determined geometrically, but its lift to an operator acting on $\mathcal{H}$ is not unique. Instead, $\widehat{\Lambda_2\Lambda_1}$ generates an $\widetilde{SO}(2,1)$ transformation characterized by a rotation by an angle $2\pi n - H$, with the integer n undetermined. Indeed, a simple calculation shows that

$$\widehat{\Lambda_2\Lambda_1} = e^{-iP_a(J_1+J_2)^a} \tag{4.2}$$

where

$$\begin{aligned} \sin\frac{H}{2}\Big(\frac{P_a}{|P|}\Big) = \Big(\cos\frac{m_2}{2}\sin\frac{m_1}{2}\Big)\frac{p_{1a}}{m_1} + \Big(\cos\frac{m_1}{2}\sin\frac{m_2}{2}\Big)\frac{p_{2a}}{m_2} \\ + \Big(\sin\frac{m_1}{2}\sin\frac{m_2}{2}\Big)\frac{(p_1\times p_2)_a}{m_1m_2} \end{aligned} \tag{4.3}$$

and

$$|P| = H - 2\pi n \tag{4.4}$$

Roughly speaking, P_a is the momentum of the center of mass, and in a suitable frame at infinity — i.e., after an appropriate $SO(2,1)$ transformation at $*$ — the condition of invariance under the braid group can be shown to be [9]

$$e^{i(2\pi n - H)\hat{L}}\psi = \psi \tag{4.5}$$

where p is a suitably defined relative momentum and $\hat{L}$ is the angular momentum operator for p in the center of mass frame.

We can now determine the integer n from physical arguments. In the limit of small masses, equation (4.5) should reduce to the ordinary projection from $\widetilde{SO}(2,1)$ to $SO(2,1)$ in order to recover the correct flat space limit. We must therefore choose $n = 1$; any other value will give a wave function defined not on the plane, but on an n-fold branched cover of the plane.

If we now take ψ to be an eigenstate of H with eigenvalue E, we have precisely recovered 't Hooft's description [3] of the Hilbert space and the Hamiltonian for two-particle scattering. Equivalently, viewing $\hat{L}$ as a generator of rotations, equation (4.5) is the condition of invariance under rotation by $2\pi - E$; this is precisely Deser and Jackiw's conical boundary condition. At the same time, we have avoided some of the ambiguities of the earlier approaches: ψ need not be an energy eigenstate, and the generalization to more than two particles and to more complicated topologies is clear. The price we have paid is in 2π ambiguities; other choices of n in (4.5) may eventually be of interest, but for now their significance is not understood.

5. THE EMERGENCE OF SPACE AND TIME

One striking feature of this analysis may have implications for quantum gravity in 3+1 dimensions. A long-standing problem in quantum gravity is the "breaking of general covariance," that is, the appearance of an approximately classical spacetime structure in a theory which is initially diffeomorphism-invariant and which involves a sum over all metrics. The problem can be thought of as one of symmetry-breaking: the metric must acquire an expectation value which is not diffeomorphism-invariant. It is evident that such symmetry-breaking has occurred here, since our scattering amplitudes have clear spacetime interpretations; it is of interest to understand exactly how this has happened.

We can recover the spatial structure more explicitly from our Hilbert space by Fourier transforming the wave functions ψ:

$$\psi(\mathbf{p}) = \int \frac{d^2\mathbf{y}}{(2\pi)^2} e^{i\mathbf{p}\cdot\mathbf{y}} \phi(\mathbf{y}) \tag{5.1}$$

The condition of invariance under the braid group then becomes

$$\phi(e^{i(2\pi - E)\hat{L}}\mathbf{y}) = \phi(\mathbf{y}) \tag{5.2}$$

which means that $\mathbf{y}$ acts as a coordinate on a cone with deficit angle E. Recall, however, that $\mathbf{p}$ originated not as a local spatial variable, but as a parameter characterizing a holonomy. We have thus built a spatial geometry out of a set of variables which had no obvious local geometric interpretation.

To understand this better, consider the classical version of equation (5.2). The relevant classical holonomies are $\widetilde{ISO}(2,1)$ matrices

$$\Xi = P\exp\{i \int (\omega_a \mathcal{J}^a + e_a \mathcal{P}^a\} = \exp\{-i(p_a \mathcal{J}^a + J_a \mathcal{P}^a)\} \tag{5.3}$$

where for spin zero sources we can take $J = q \times p$. Conjugation of one such a holonomy by another represents parallel transport in $\widetilde{ISO}(2,1)$, that is, translation and rotation of p and q. In particular, for a classical solution with two holonomies Ξ_1 and $|Xi_2$, conjugation by Ξ_1 corresponds to the transformation

$$\begin{aligned} q_2 &\rightarrow q_1 + \Xi_1(q_2 - q_1) \\ p_2 &\rightarrow \Xi_1 p_2 \end{aligned} \tag{5.4}$$

Invariance under such a transformation means that although q_2 looks like a flat coordinate, it is really only locally flat; in fact, (5.4) is a typical matching condition for a coordinate in a classical solution for $2+1$ dimensional gravity with point sources [14]. The requirement of invariance under the braid group thus allows us to reconstruct the classical geometry from the parameters q which label $\widetilde{ISO}(2,1)$ holonomies. In the quantum system, the translations $\mathcal{P}^a$ no longer appear in the holonomies, and the conical geometry becomes fuzzy, but its presence can still be felt in the boundary condition (4.5).

Finally, let us turn to the role of time in our quantization of $2+1$ dimensional gravity. Our construction of the Hamiltonian, and thus our interpretation of dynamics, was possible because we worked in an asymptotically locally flat space, which allowed us to define time translations as translations at infinity. But $2+1$ dimensional gravity is a topological theory, and infinity is not really such a special point. In a compact space Σ, we can instead choose a preferred puncture x_0 to play the same role.

In particular, x_0 can serve as a basepoint for the holonomies of ω_{ai} around the remaining x_α. Wave functions will now depend on the holonomies Λ_α, $\alpha \neq 0$, and not on Λ_0, but there is no inconsistency here: for a closed space Σ, the holonomies are not all independent, but satisfy one relation which can be used to eliminate Λ_0. We can then define time translations as translations at x_0. Equation (3.6) will again determine the generator of such translations, provided that Λ_0 is omitted from the trace on the right hand side. The world line of x_0 thus serves as a clock with which we can measure time evolution of the rest of the universe. Similar ideas have been proposed in $3+1$ dimensions, but it has never been clear how to isolate a "clock" variable.

Canonical quantization of $2+1$ dimensional gravity thus successfully reproduces the physically intuitive results of Deser, Jackiw, and 't Hooft for point particle scattering. In principle, it provides a framework in which those results can be unambiguously extended to more complicated topologies and larger numbers of particles. At the same time, it provides a simple model in which some of the deeper conceptual questions of quantum gravity can be addressed.

Acknowledgements

This work was supported in part by the U.S. Department of Energy under grant DE-AC02-76ERO2220.

REFERENCES

1. R. Jackiw, lectures in these proceedings
2. S. Deser and R. Jackiw, *Commun. Math. Phys.* **118**(1988), 495
3. G. 't Hooft, *Commun. Math. Phys.* **117**(1988), 685
4. R. Jackiw, "Quantum Gravity in Flatland," MIT preprint CTP-1622 (1988)
5. B. DeWitt, *Phys.Rev.* **160**(1967), 1113; W. G. Unruh and R. M. Wald, "Time and the Interpretation of Canonical Quantum Gravity," Santa Barbara preprint ITP-88-190 (1988)
6. L. Smolin, "Nonperturbative Quantum Gravity via the Loop Representation," Syracuse preprint, to appear in *Proceedings of the 1988 Osgood Hill Meeting on Quantum Gravity*, A. Ashtekar and J. Stachel, eds.
7. E. Witten, *Nucl. Phys.* **B311**(1988), 46
8. M. Henneaux, lectures in these proceedings
9. S. Carlip, *Nucl. Phys.* **B324**(1989), 106

10. M. Rocek and R. Williams, to appear

11. N. Seiberg, lectures in these proceedings

12. T. Regge and C. Teitelboim, *Ann. Phys.* **88**(1974), 276

13. M. Henneaux, *Phys. Rev.* **D29**(1984), 2766

14. S. Deser, R. Jackiw, and G. 't Hooft, *Ann. Phys.* **152**(1984), 220

15. J. S. Birman, *Braids, Links, and Mapping Class Groups*, Annals of Mathematics Studies 82 (Princeton University Press, Princeton, 1975)

16. J. S. Birman, *Comm. Pure Appl. Math.* **22**(1969), 213

FINITE RENORMALIZATION OF CHERNS-SIMONS GAUGE THEORY

Wei Chen[1,2], Gordon W. Semenoff[3], and Yong-Shi Wu[1]

[1]Department of Physics, University of Utah, Salt Lake City, Utah U.S.A.; supported in part by U.S. NSF grant #PHY-8706501
[2]On leave from Institute of High Energy Physics, Academia Sinica Beijing, People's Republic of China
[3]Department of Physics, University of British Columbia, Vancouver British Columbia, Canada; supported in part by NSERC Canada

ABSTRACT

We investigate perturbation theory for a non-abelian Chern-Simons gauge field theory. Three regularization methods are examined. It is shown that regularization by introducing a conventional Yang-Mills action makes the theory finite to all orders and yields a finite integer-valued renormalization of the coefficient k of the Chern-Simons term at one loop, and we conjecture no renormalization of k at any higher order. We have found that dimensional regularization does not respect gauge invariance at two loops, where logarithmic divergences first appear. A variant of this approach called regularization by dimension reduction is shown to obey both the Ward identity and quantization condition and gives zero renormalization of the theory to three loops. By explicit calculation, We also demonstrate vanishing of the beta-function for k and therefore absence of a scale anomaly to three loops and no general covariance anomaly in two- and three-point functions to two loops.

INTRODUCTION

In this school, people have been talking about three dimensional Chern-Simons (CS) gauge theory. The topics that have been discussed in previous talks$^{(1)}$ are mainly concerned with the topological aspects of the theory. Here we shall take a different perspective of conventional perturbative quantum field theory.

The action that describes the non-abelian CS gauge theory is a three-form integrated over a three-manifold M:

$$S_{cs} = \int_M L, \qquad L = -i\ \mathrm{Tr}(A\wedge dA+\frac{2}{3}gA\wedge A\wedge A), \tag{1}$$

Written version of the talk given by Wei Chen at the Workshop in the NATO/Banff Summer School, August 1989.

Physics, Geometry, and Topology
Edited by H. C. Lee
Plenum Press, New York, 1990

where d is the exterior derivative operator on M; and the one-form A takes value in the Lie algebra of the gauge group G. Instead of an overall coefficient $k/4\pi$ in action (1), we give the AAA interaction term a coupling constant g. However, as we shall demonstrate, the theory has finite renormalization, so the two expressions with k and g may be transformed into each other by a rescaling

$$A\to gA \quad \text{and} \quad 4\pi/k\to g^2 \tag{1'}$$

or the inverse without any singularity. The expression (1) is obviously convenient for perturbative expansion.

We begin with a brief review of some novel features of the theory. The first is diffeomorphism invariance. To see this, let us perform a transformation over S_{cs} through a Lie derivative,

$$\mathcal{L}_\xi S_{cs} = \int_M [d(L(\xi))+(dL)(\xi)], \tag{2}$$

where ξ is a vector field on the tangent space of M. The last term above is zero since L is a top-form on M, while the first term vanishes if $\partial M = 0$ or $\xi|_{\partial M} = 0$. The very fact shows that CS gauge theory is a topological field theory since the action takes the same value in any coordinate system of the manifold M. Physically, this implies general covariance. Since the Lagrangian (1) is a three-form, it admits no spacetime metric $g_{\mu\nu}$. (1) is invariant under any local deformation of the metric of the spacetime manifold M. As a result, the Hamiltonian of the system vanishes, $\delta S_{cs}/\delta g_{\mu\nu} = 0$.

Secondly, the Lagrangian (1) is linear in derivatives. The variation of (1) with respect to A^a_μ gives the "equation of motion" of A^a_μ:

$$F^a_{\mu\nu} = \partial_\mu A^a_\nu - \partial_\nu A^a_\mu + gf^{abc}A^b_\mu A^c_\nu = 0. \tag{3}$$

Actually (3) is a flat connection condition but not a wave equation of motion. Therefore, the gauge field A_μ is not a propagating field.

Taking into account the above features, perhaps one can say that pure topological CS theory has no local dynamics, so it has no classical physics.

Moreover, although action (1) is invariant under an infinitesimal gauge transformation, $\delta A = D\lambda$, where D is the covariant derivative and λ is an arbitrary zero-form taking value on the Lie algebra of group G, it is not invariant under the "large" gauge transformations that are associated with the non-trivial elements of the third homotopy group of the gauge group $G^{(3)}$:

$$S_{cs} \xrightarrow[\text{transformations}]{\text{"large" gauge}} S'_{cs} = S_{cs} + i\times\text{Const.}\times n \tag{4}$$

where n is the winding numbers of the map $g: x \to g(x)$.

However, the story of quantum theory is different. Since an action appears in an exponential in the partition function, $\int[dA]\exp(-S_{cs}[A])$ (henceforward we use Euclidean spacetime), those actions will describe the same quantum theory only if they differ by an integer multiple of 2π. Therefore, large gauge invariance will survive with the quantization condition

$$4\pi/g^2 = \text{integer (or } k = \text{integer)}. \tag{5}$$

Eq.(5) is a topological condition.

Topological quantum CS gauge field theory has been known(2,1,4) as an exactly soluble model and its contents are extremely rich. For example, it provides a three dimensional quantum field theory description for knot theory by associating link invariants with the expectation values of Wilson loops, $W_R(C) = Tr_R P\exp(\int_C A)$, which is a main (if not only) sort of topological quantities in the field theory besides the Lagrangian itself. On the other hand, it has been related to 1 + 1 dimensional conformal field theory by establishing a correspondence between the conformal blocks and the quantum states.

When coupled with matter, as a "background field", Chern-Simons (abelian or nonabelian) gauge field endows the matter fields with remarkable features(5), such as fractional statistics and Fermi-bose transmutations, which are argued to have something to do with high T_c superconductivity, superfluids, and the quantum Hell effect.

In this talk we shall consider perturbative non-abelian CS gauge field theory. A question arises naturally — since CS theory is a topological theory, why does one need a perturbative analysis? Instead of answering the question, we like to raise a relevant question — does CS gauge theory define a sensible quantum field theory? To have a definite answer, one has to understand the theory better. An incomplete list of what one perhaps wants to know about is as following:

1. calculability of correlation functions of A_μ's.
2. renormalizations of coupling constant g and gauge field A_μ.
3. scaling behavior.
4. anomalies.

We shall see that perturbation theory can tell us something about these.

Although the gauge field A_μ in CS gauge theory is not a propagating field as we mentioned above, an A-propagator — the lowest order two-point function of A field — is still well-defined (we shall work in the Landau gauge):

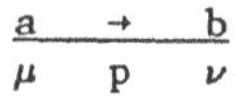

$$-\delta^{ab}\epsilon^{\mu\nu\lambda}p_\lambda/p^2 \ . \qquad (6)$$

And from action (1), an AAA-vertex — the lowest order three-point function of A field — is

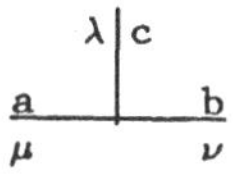

$$igf^{abc}\epsilon^{\mu\nu\lambda}. \qquad (7)$$

(The Faddeev-Popov ghosts are temporarily ignored.) As we shall see, the antisymmetric structure of the two- and three-point functions in spacetime indices leads to very interesting results.

As in conventional quantum field theories, the two- and three-point functions will get corrections from loop diagrams in general. Since the only parameter, the coupling constant g, is dimensionless, the theory is obviously renormalizable and it is necessary to regularize divergent Feynman diagrams in loops calculations.

The miracle here is that, as we shall see, at least to three loops there are no singularities in either individual diagrams or ones combining pertinent regularized Feynman diagrams when regularization parameters go to limits. This implies that one has neither to introduce counterterms to absorb singularities (technically), nor refer to "bare" quantities as unobservable or unphysical (conceptually). What we have in the perturbative CS theory will be finite renormalizations at least to the order we have studied. It provides the possibility that both the "bare" coupling constant, g, which appears in the tree-level action (1) and the renormalized one, g_r, are physically meaningful: the former accords to the lowest order observable and the latter the quantum corrected one to proper order.

Furthermore, we shall show that the renormalized quantity depends on the regularization scheme. Therefore, in the finite renormalization case, perhaps different (reasonable) regularization methods define different quantum field theories, since they give different quantum corrections to the same classical theory.

In this talk, three types of regularization will be examined. We shall discover that dimensional regularization fails at the two-loop level since it does not respect the Ward identities. And regularization by dimensional reduction gives zero remormalization to at least three-loop order in perturbative theory. It seems that the theory is perturbatively trivial in the calculation of two- and three-point functions with this regularization. While F^2 regularization[6] leads to a finite renormalization. At one-loop order the coupling constant g gets a finite correction which satisfies the quantization condition. It implies a shift of k:

$$k \rightarrow k_r = k + \frac{1}{2}c_2(G), \tag{8}$$

where $c_2(G)$ is the value of the quadratic Casimir operator of the group G in the adjoint representation, such as $c_2[SU(N)] = 2N$. We argue that the higher order corrections should not modifies the shift (8). Otherwise the corresponding quantization condition could not be obeyed.

Since the coupling constant g is quantized, the beta function for g should be zero, i.e. the quantum theory must be scale invariant.

However, this point is not perturbatively obvious so needs to be verified. Also, it remains to make clear whether $\beta(g)$ is identically zero or whether it vanishes only for quantized values of g. As is known, the latter case occurs in WZW model. But since the regularized CS field theory has only finite renormalization, $\beta(g)$ of this theory must be zero. Hence there is no scale anomaly in CS quantum gauge field theory. We shall demonstrate this by explicit calculation to three-loop order.

Aside from scale invariance, the classical action (1) has two other symmetries — general covariance and gauge invariance. Therefore it is important to check whether they survive quantization. Now suppose these symmetries remain, what should the renormalized CS action look like?

First of all, it can not be spacetime metric dependent. Otherwise general covariance will be violated. Therefore only local three forms containing $A_r \wedge dA_r$ and $A_r \wedge A_r \wedge A_r$ are allowed in the renormalized action, where A_r is renormalized gauge field.

Secondly, the relative coefficient of these two terms must coincide with the one in the tree-level action so that the small gauge transformations remain a symmetry. This means that if there are neither local gauge nor diffeomorphism anomalies, the renormalized action can only take the same form with action (1) with the bare quantities A and g replaced by the renormalized ones A_r and g_r.

Furthermore, if the large gauge transformations remain a symmetry, the renormalized coupling constant must satisfy the quantization condition (5).

However, we do not know if this is the case before completing the renormalization procedure. As an example, we can not exclude a finite nonlocal term like

$$\int_M d^3x \mathrm{Tr}\{F_{\mu\nu}[1/\surd(D_\lambda D^\lambda)]F^{\mu\nu}\}, \tag{9}$$

(where $F_{\mu\nu}$ is the field strength) which would imply a diffeomorphism anomaly. Loop diagrams can contribute to the action (9) if they contain even numbers of the three order antisymmetric tensor $\epsilon^{\sigma\tau\lambda}$. This is the situation of any odd-loop gauge field self-energy diagram. The contraction of even number of $\epsilon^{\sigma\tau\lambda}$ will give rise to symmetric spacetime tensors.

Nevertheless, we shall find that the cancellations between the gauge field loops and the ghost loops are so complete that no term like (9) appears. And no other noninvariant terms have been found in the two- and three-point functions to two loops.

This Lecture will be arranged as follows. In sect.(2) we point out the subtlety of an axial gauge choice for CS theory in Euclidean spacetime, introduce covariant gauge fixing first, and discuss powercounting. It is shown that the actual ultraviolet behavior of the theory is much better than what is expected by naive powercounting. Then in sect.(3), we describe the three regularization procedures. The most interesting one is the F^2 regularization, which actually makes the theory finite! In this section, the renormalization constant and the Ward identities are also defined. Explicit calculations for one-loop and two-loop are given in sect.(4) and (5), respectively. We find that F^2 regularization gives finite renormalization at one loop; dimensional regularization breaks the gauge symmetry of the theory at two loops; and the regularization by dimension reduction respects all Ward identities but gives no renormalization up to three loops. Finally, sect.(6) gives a summary of our result.

2. Gauge Fixing and Powercounting

It seems that the simplest gauge fixing of the theory at hand is the temporal gauge, say $A_0 = 0$. In this case, the ghosts decouple and the interactions in action (1) disappear. The action now reads

$$S'_{cs} = i\int_M \mathrm{Tr}(\epsilon^{0ij}A_i\partial_0 A_j). \tag{10}$$

Action (10) looks like a free theory. Of course it is supplemented with a gauge constraint $F_{ij} = 0$. However this is unseen perturbatively. Therefore the perturbation theory in this gauge seems trivial.

However, since we work in the Euclidean spacetime S^3, the gauge transformation to the temporal gauge is illegal: If the spacetime manifold is compact, it can not be done without introducing singularities in the gauge connection A. If the spacetime manifold is open, the generator of gauge transformation is identity at infinity and the change in the Chern-Simons term in action (10) will not be proportional to a correctly quantized integer. Therefore, the theories described by actions (1) and (10) are not identical. However, their perturbative structure must be very similar.

Hence we like to have a linear covariant gauge fixing, $\partial^\mu A_\mu = 0$, as this sort of gauge condition can be realized on a compact spacetime. Meanwhile we also introduce Faddeev-Popov ghosts through the standard procedure. The corresponding action is

$$S_{g.f.} = \int_M d^3x\sqrt{g}\,Tr[-\frac{1}{\beta}(\partial^\mu A_\mu)^2 + g^{\mu\nu}\partial_\mu \bar{c}D_\nu c]. \tag{11}$$

The action (1) and (11) is BRST invariant, and the BRST charge is nilpotent, $Q^2 = 0$. The BRST invariance is very useful for deriving the Ward identities.

Adopting a covariant quantization, one unavoidably introduces a metric. It obviously breaks the general covariance of the gauge fixed classical action. Furthermore, it is impossible to regulate ultraviolet divergences without a metric, i.e. using any cutoff implies use of a distance scale which only makes sense when there is a metric. Fortunately, we shall see that the renormalized action is cutoff independent. Thus the metric introduced in regularization does not appear in the final results. On the other hand, there have been arguments that the metric introduced in gauge fixing does not affect the final results either. The first is that in path integral approach, it has been shown that[2] the combining result of the two metric dependent terms in (11) is a topological invariant upto a metric dependent phase factor which is controllable at least at one-loop order. Secondly, if we consider the BRST canonical quantization of CS gauge field theory, we find that (also see ref. (10)) although the energy momentum tensor, $T^{\mu\nu}$, is no longer zero due to the action (11), it is a BRST commutator: $T^{\mu\nu} = [Q, X^{\mu\nu}]$, where $X^{\mu\nu}$ is a curtain operator. It turns out that the energy momentum is not measurable, i.e. $\langle phys|T^{\mu\nu}|phys\rangle = 0$, because $Q|phys\rangle = 0$. Furthermore, in this talk we provide new evidence by demonstrating perturbatively that the metrics introduced by either gauge fixing or regulating do not bring new terms, especially like (9), into the renormalized CS action.

The Feynman rules for the ghost propagator and the $\bar{c}cA$ vertex are as usual:

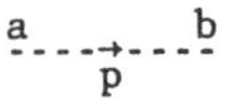

$$\delta^{ab}(-i)/p^2, \tag{12}$$

and

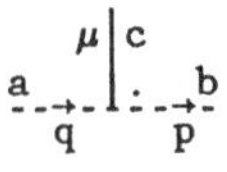

$$gf^{abc}p_\mu, \tag{13}$$

where we have chosen the flat metric, $g^{\mu\nu} = \delta^{\mu\nu}$, and the Landau gauge, $\beta = \infty$. The advantage of this gauge is that the calculations are greatly simplified and it avoids potential infrared singularities[7].

Let us consider the ultraviolet behavior of the theory. From the Feynman rule, we know that the propagators (6) and (12) have dimensions of momentum p^{-1} and p^{-2}, respectively; the AAA vertex is dimesionless, and the $\bar{c}cA$ one is linear in p. If we denote the superficial degree of divergence of a diagram Γ by $D(\Gamma)$, naive power counting shows that it is

$$D(\Gamma) = 3 - E, \tag{14}$$

where $E = E_A + E_c$, while E_A and E_c are the number of external gauge fields and ghost lines, respectively. (Eq.(14) is different from the one, $D'(\Gamma) = 3 - E_A - \frac{1}{2}E_c$ (see reference (11), say) by standard power counting. The reason is that the ghost external lines in any diagram Γ always appear in pairs and the two external ghost lines in each pair are connected by internal

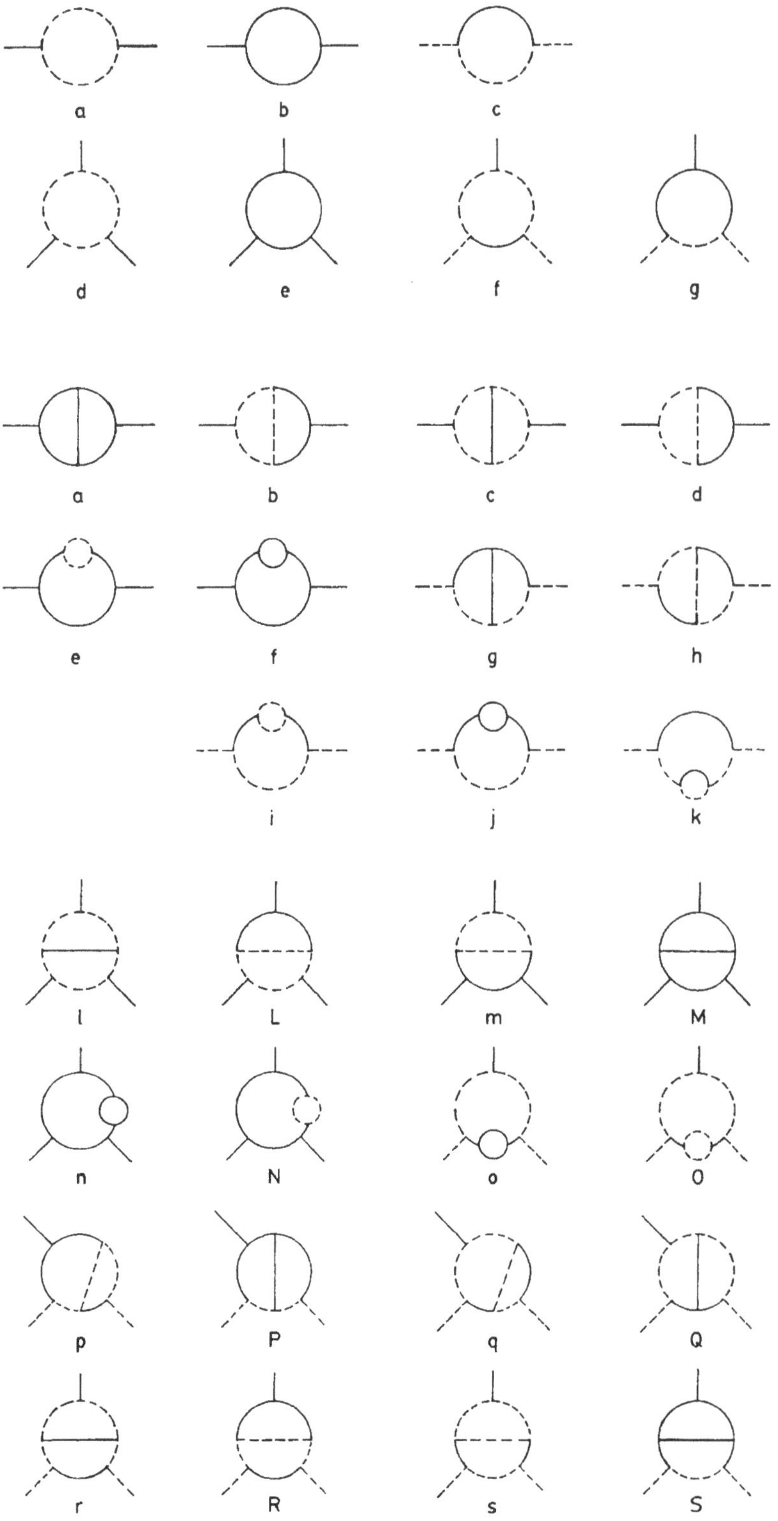

Fig. 1

ghost line(s), then one of the two $\bar{c}cA$ vertices associated with the two external ghost lines in a pair has a factor of external momentum (by the Feynman rule (13)). This means that the two $\bar{c}cA$ vertices actually contribute to the superficial degree of divergence for a given diagram by only one but not two. It turns out that each pair of external ghost lines makes $D'(\Gamma)$ one less. This results our formula (14).) Eq.(14) tells us that only two- and three-point functions in this theory are superficially divergent.

But, remember that the non-negative superficial degree of divergence of a given diagram does not mean that the diagram is necessarily divergent. Symmetries may render the ultraviolet behavior of some diagrams better. We should look at it more carefully.

First of all, we shall show that all two- and three-point functions are convergent at the one-loop level (Fig.1), although they have superficial degrees of divergence $D = 1$ and 0, respectively. 1) As in any three-point diagram the divergent terms (according to $D = 0$) are three powers of internal momentum, the actual degree of divergence for them is $D = -1$ (under the integrand over the internal momentum). Namely, they are actually convergent. 2) The one diagram for the ghost self-energy vanishes because in that diagram there is one antisymmetric tensor $\epsilon^{\sigma\tau\lambda}$ coming from the internal gauge field line, and its three indices have to contract with one internal and one external momentum; then the contraction gives zero. 3) In the one-loop correction to the gauge field self-energy, each diagram contains an even number of $\epsilon^{\sigma\tau\lambda}$. Since the gauge field two-point function is transverse in the external momentum p, it is in proportion to $(\delta^{\mu\nu}p^2-p^{\mu}p^{\nu})$. On the other hand, the gauge field inverse propagator only has the dimension of mass. Therefore, the gauge field two-point function is ultraviolet convergent at one loop. Let us emphasize this point with an explicit calculation. Consider a typical Feynman integral in the one loop gauge field self-energy diagrams:

$$I(\omega) = (2\pi)^{-3}\int d^{\omega}k\; k_{\mu}(k+p)_{\nu}/k^2(k+p)^2. \tag{15}$$

Above we have analytically continued the integral to $\omega = 3-\epsilon$ dimension (ϵ is a small positive quantity) so that the integral is precisely defined. Then introduce Feynman parameters and integrate over the internal momentum k as well as the Feynman parameters through a standard procedure, we get

$$I(\omega) = \frac{1}{64\sqrt{\pi}}\,\frac{1}{p}\,[\tfrac{1}{2}\Gamma(1-\tfrac{\omega}{2})\delta^{\mu\nu}p^2-\Gamma(2-\tfrac{\omega}{2})p^{\mu}p^{\nu}]. \tag{15'}$$

We have no singularity when $\omega\to 3$. This completes the proof of the convergence of one loop.

Now we turn to two-loop diagrams (Fig.2). At this level, every gauge field self-energy diagram contains an odd number of $\epsilon^{\sigma\tau\lambda}$ therefore the gauge

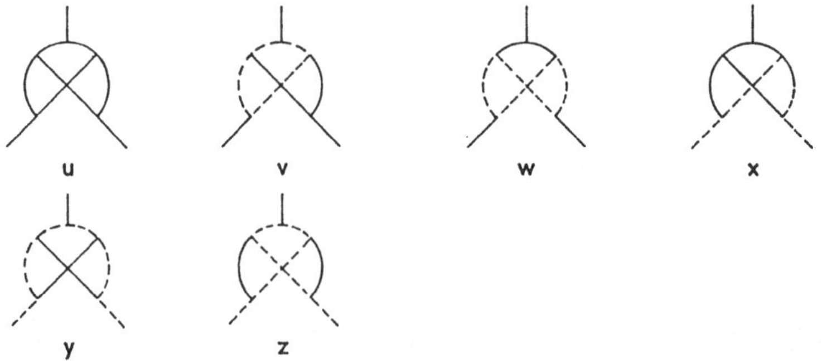

Fig. 2

field two-point function takes the form of $\Pi_0(p)\epsilon^{\mu\nu\lambda}p_\lambda$; while every ghost self-energy diagram contains an even number of $\epsilon^{\sigma\tau\lambda}$ so it takes the form of $\tilde{\Pi}(p)p^2$. Remembering that the gauge field and the ghost two-point functions have dimensions of p and p^2, respectively, $\Pi_0(p)$ and $\tilde{\Pi}(p)$ therefore the two point functions at two-loop level should be logarithmically divergent, a bit better than the linear divergence of naive powercounting. We shall also verify this directly from the diagram calculation.

On the other hand, the two types of three-point functions at the two-loop level are actually convergent. To see this, we notice that in every two-loop three-point diagram there exists at least one $\epsilon^{\sigma\tau\lambda}$ that must contract over all its three antisymmetric spacetime indices with three momenta carried by internal gauge field line(s) or $\bar{c}cA$ vertex(es). On the other hand, the diagram has only two independent internal momenta. Therefore the contraction must contain the external momentum. In this way, the superficial degree of divergence D = 0 is decreased by at least one and the three-point functions at two-loop level are certainly ultraviolet convergent.

For higher loops, a similar analysis is available but much more complicated. Generally speaking, the ultraviolet behaviors of odd-(even-)loop diagrams are similar to those of one-(two-)loop ones, because their antisymmetic tensor structures are similar. Particularly, like one-loop diagrams, we have demonstrated that all three-loop diagrams are convergent.

Anyway, since there are divergences in, e.g. two-loop two-point vertices, regularization is needed.

3. Regularization, Renormalization, and Ward Identities

We shall consider three types of regularization. The first, called F^2 regularization, we supplement the action (1) and (11) with a Yang-Mills term

$$S_{Y\text{-}M} = -\int_M \frac{1}{2\theta}\mathrm{Tr}(F^{\mu\nu}F_{\mu\nu}). \tag{16}$$

The cutoff θ above has dimension of mass and will go to infinity at the end of calculation. At first sight, this regularization makes the theory powercounting super-renormalizable at the price of introducing the metric and complicating the Feynman rules. It respects BRST symmetry of the theory. The Feynman rules for gauge field two- and three-point vertices now contain a symmetric part:

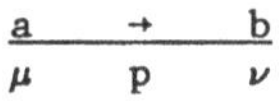

$$\delta^{ab}\frac{\theta}{p^2(p^2+\theta^2)}(p^2\delta^{\mu\nu}-p^\mu p^\nu-\theta\epsilon^{\mu\nu\lambda}p_\lambda) \tag{17}$$

and

$$\frac{i}{\theta}gf^{abc}[\theta\epsilon^{\mu\nu\lambda}-(r-q)_\mu\delta_{\nu\lambda}-(q-p)_\lambda\delta_{\mu\nu}-(p-r)_\nu\delta_{\lambda\mu}]. \tag{18}$$

With this regularization, we shall still work in the Landau gauge. According to action (16), a dimensionless four-gauge field vertex is also involved. However, since it has good ultraviolet behavior and any contributions associated with the vertex will go to zero as θ goes to infinity, we can simply ignore it.

Let us make a powercounting analysis about the regularization. With θ finite, the two-point vertex (17) behaves at large momentum like p^{-2} for the symmetric part and p^{-3} for the antisymmetric one. While the three-point vertex (18) behaves like p and constant for the two parts, respectively. The ghost propagator and $\bar{c}cA$ vertex are not affected. Since in the most divergent parts, we lose one power in each AAA-vertex but gain one in each AA-propagator with this regularization, we expect that the ultraviolet behavior of higher-loop diagrams will be improved.

Precisely, powercounting shows that all diagrams are superficially divergent except for two-point functions at one-loop which are superficially linear divergent and three-point functions at one-loop and two-point functions at two-loop which are superficially logarithmically divergent.

Surprisingly, these superficially divergent diagrams turns out to be convergent! The F^2 regularization makes the theory finite. The reasons for this have been discussed in the last section. Let us see how it works. First of all, in any one-loop three-point diagram, the superficially (logarithmically) divergent terms are odd-number powers of the integral momentum, so three-point functions are actually convergent under integrals over internal momenta. Secondly, due to the regulator (19), the antisymmetric part of the gauge field propagator behaves at large momentum like p^{-3}, much better than p^{-1} as before. Therefore the antisymmetric parts of two-point functions have good ultraviolet behavior, i.e. they are convergent by naive powercounting. Finally, for the symmetric part of two-point functions, although the degree of superficial divergence is $D = 1$ at one-loop order and $D = 0$ at two-loop, the tensor structures is of form $i\tilde{\Pi}(p)p^2$ for the ghost and $\frac{1}{\theta}\Pi_e(p)(\delta^{\mu\nu}p^2 - p^\mu p^\nu)$ for the gauge field. As a result, $\tilde{\Pi}(p)$ and $\Pi_e(p)$ are actually convergent. Now we have completed the statement that with F^2 regularization the loop corrections of two- and three-point functions therefore the theory turns out to be finite. We shall show later that the finiteness remains when $\theta \to \infty$.

We now consider dimensional regularization. It has the obvious advantage that it does not complicate the Feynman rules so higher orders in perturbation theory are more accessible. According to this approach, all tensors should be defined on the regularization dimension $\omega = 3 - \epsilon$. But there is an ambiguity in the dimensional continuation of antisymmetric tensor $\epsilon^{\sigma\tau\lambda}$. We do not know how to analytically continue $\epsilon^{\sigma\tau\lambda}$ to ω. However one may try to define the contraction between $\epsilon^{\sigma\tau\lambda}$ and the dual $\epsilon_{\sigma\theta\eta}$ in ω dimension:

$$\epsilon^{\sigma\tau\lambda}\epsilon_{\sigma\theta\eta} = (\tilde{\delta}^{\tau}{}_{\theta}\tilde{\delta}^{\lambda}{}_{\eta} - \tilde{\delta}^{\tau}{}_{\eta}\tilde{\delta}^{\lambda}{}_{\theta})\Gamma(\omega-1), \tag{19}$$

with

$$\tilde{\delta}^{\sigma}{}_{\sigma} = \omega. \tag{20}$$

It certainly works at one-loop, since one-loop diagrams are actually convergent as we have discussed in the last section. We shall find that dimensional regularization leads to finite corrections to the two-point functions at the two-loop level where logarithmical divergences first appear. Unfortunately, the resulting renormalization constants satisfy neither the quantization condition nor the Ward identity. It means the failure of applying dimensional regularization in CS gauge theory.

An alternative approach is the regularization by dimensional reduction. This approach has been used to regulate supersymmetric field theories[13]. The tensor algebra is performed in three dimensions before the Feynman integrals are worked out and only the dimension of the integrations is analytically continued. Namely, we have

$$\epsilon^{\sigma\tau\lambda}\epsilon_{\sigma\theta\eta} = (\delta^{\tau}{}_{\theta}\delta^{\lambda}{}_{\eta}-\delta^{\tau}{}_{\eta}\delta^{\lambda}{}_{\theta}), \quad \text{and } \delta^{\sigma}{}_{\sigma}=3. \tag{21}$$

This procedure appears to preserve the symmetries of the theory and give no relative correction at all to at least two loops. We conjecture that in this scheme there is no renormalization to all orders in perturbative theory.

Next let us consider renormalization. Notice that the inverse A-propagator should take the form:

$$\Delta^{-1}_{\mu\nu}(p) = Z_A(p)\epsilon_{\mu\nu\lambda}p^{\lambda}+Z'_A(p)(p^2\delta_{\mu}-p_{\mu}p_{\nu}), \quad (\Delta^{ab}_{\mu\nu} = \delta^{ab}\Delta_{\mu\nu}). \tag{22}$$

And the gauge field three-point function is

$$\Gamma^{abc}_{\mu\nu\lambda}(p,q,r) = igf^{abc}\{Z_g(p,q,r)\epsilon_{\mu\nu\lambda}-Z'_g(p,q,r)[(r-q)_{\mu}\delta_{\nu\lambda}+\text{cycle}]+\cdots\}. \tag{23}$$

There exist two possible sources for the symmetric terms in eqs.(22) and (23). The first is the contractions of even number of $\epsilon^{\sigma\tau\lambda}$ brought in by the A-propagators and the AAA-vertices. (By the way, it is the case for all odd-loop diagrams in dimensional regularization and thus dimensional reduction.) In the F^2-regularization, even the Feynman rules themselves contain symmetric terms. In the latter case, we have

$$Z_A(p) = 1+\Pi_o(p), \qquad Z'_A(p) = \frac{1}{\theta}Z''_A(p) = \frac{1}{\theta}[1+\Pi_e(p)], \tag{22'}$$

$$Z_g(p,q,r) = 1+T_g(p,q,r), \qquad Z'_g(p,q,r) = \frac{1}{\theta}[1+T'_g(p,q,r)]. \tag{23'}$$

The exact A-propagator is then

$$\Delta_{\mu\nu}(p) = \frac{\theta}{Z''_A(p)p^2\{p^2+[Z_A(p)/Z''_A(p)]^2\theta^2\}}[-\frac{Z_A(p)}{Z''_A(p)}\theta\epsilon_{\mu\nu\lambda}p^{\lambda}+p^2\delta_{\mu\nu}-p_{\mu}p_{\nu}]. \tag{24}$$

It looks like a propagating gauge field[3,7] with mass $\theta Z_A(p)/Z'_A(p)$. With $\theta\to\infty$, the propagating components of A-field are infinitely massive and the A-field does not propagate.

The renormalized ghost-propagator and $\bar{c}cA$-vertex are

$$\tilde{\Delta}^{ab}(p) = -i\delta^{ab}/Z_{gh}(p)p^2, \qquad Z_{gh}(p) = 1+\tilde{\Pi}(p), \tag{25}$$

and

$$\tilde{\Gamma}^{abc}{}_{\lambda}(p,q;r) = -igf^{abc}\tilde{Z}_g(p,q;r)p_{\lambda}+\cdots, \qquad \tilde{Z}_g(p,q;r) = 1+T_g(p,q;r), \tag{26}$$

where we have denoted the ghost-wave function and $\bar{c}cA$-vertex renormalization constants with $Z_{gh}(p)$ and $\tilde{Z}_g(p)$. In the theory at hand, the $\bar{c}cA$-vertex (26) at even-loop levels probably contains terms that involves $\epsilon^{\sigma\tau\lambda}$ (the propagator (25) has no the problem since it has only one external momentum and no spacetime index). Nevertheless, such terms in (26) as well as the symmetric parts in gauge field propagator (22) and vertex (23), if not vanishing, would bring into renormalized action new terms.

Fortunately, it is not the case. The explicit calculations to the order we have studied show that after removing the regulator $Z_A'(p) = Z_g'(p,q,r) = 0$ and no extra term appears in Eq.(26). Particularly, the gauge field propagator takes the form

$$\Delta_{\mu\nu}(p) = -\frac{1}{Z_A(p)p^2}\epsilon_{\mu\nu\lambda}p^{\lambda}. \tag{27}$$

Eq.(27) shows that it is $Z_A(p)$ that plays the role of the wave function renormalization constant of gauge field.

Furthermore, we shall see that the renormalization constants are independent of external momenta, i.e.

$$Z_A(p) = Z_A = Z_A(0),\ Z_{gh}(p) = Z_{gh} = Z_{gh}(0), \text{ and so on.} \tag{28}$$

From the wave function and the vertex renormalization constants, it is easy to define the renormalized coupling constant

$$g_r = g\, Z_g Z_A^{-3/2} = g\tilde{Z}_g Z_{gh}^{-1} Z_A^{-1/2}. \tag{29}$$

In the last equality we have applied the Ward identity for the two- and three-point functions, which must be satisfied if gauge invariance survives quantization.

Finally, the invariance under large gauge transformations requires the renormalization coupling constant g_r obey the quantization condition

$$4\pi/g_r^2 = \text{integer}, \quad (\text{or } k_r = \text{integer}), \tag{30}$$

which is called the topological Ward identity[7].

In the next two sections, we shall examine whether The three regularization schemes described here preserve eqs.(29) and (30).

4. One Loop Structure

Since, as we have demonstrated, at one loop CS gauge field theory is actually finite, it seems that the result at this order should be independent of regularization schemes although a regularization is necessary so that Feynman integrals are precisely defined.

The exception is F^2-regularization. Although the Yang-Mills regulator is so effective that all two- and higher-loop ultraviolet divergences are cured, it is unable to solve the similar problem at one loop; therefore a further treatment such as dimensional continuation of Feynman integral is necessary. More interesting, the existing F^2 term at one loop dramatically leads to a finite quantum correction to the coupling constant g. In other words, it provides a shift $\frac{1}{2}c_2(G)$ to the overall coefficient k when the cutoff $\theta\to\infty$. So perturbation theory in this case is not trivial as it is in the other two regularization methods we are examining, which produce no correction at all.

Thus let us first consider the F^2 regularization. With F^2 term, the Feynman rules (see eqs.(17), (18), (12), and (13)) therefore the calculations are much more complicated. Fortunately, most calculations for one loop correction could be found in ref.(3,7,8) where the authors studied the three dimensional massive Yang-Mills theory with the Chern-Simons action as a mass term.

The simplest diagram is the ghost self-energy, (c) of Fig.1. The contribution from the antisymmetric term in the gauge field propagator is trivially zero and we have (see eq.(25))

$$ip^2\delta^{ab}\tilde{\Pi}(p) = -ig^2 f^{adc}f^{cdb}\theta\int\frac{d^3k}{(2\pi)^3}\,\frac{(k+p)_\rho p_\sigma(k^2\delta_{\rho\sigma}-k_\rho k_\sigma)}{k^2(k+p)^2(k^2+\theta^2)}. \tag{31}$$

Thus

$$\tilde{\Pi}(p) = \frac{g^2\theta\, c_2(G)}{2p^2}\int\frac{d^3k}{(2\pi)^3}\,\frac{k^2p^2-(k\cdot p)^2}{k^2(k+p)^2(k^2+\theta^2)}, \tag{32}$$

where $\delta^{ab}\frac{1}{2}c_2(G) = f^{adc}f^{dcb}$. It is obviously ultraviolet convergent. Performing the integration and taking the limit $\theta\to\infty$, we get

$$\tilde{\Pi}(p) = \tilde{\Pi}(0) = -\frac{1}{12\pi}g^2c_2(G). \tag{33}$$

Fig.(1.a) and (1.b) give the gauge field self-energy corrections. The contribution will be divided into two parts: the terms with an odd number of $\epsilon^{\sigma\tau\lambda}$ go to $\Pi_o(p)$ and others to $\Pi_e(p)$ (see eq.(22) and (22')). For $\Pi_e(p)$, the exact value is not important. What we need to know is whether it is finite or it goes infinity more slowly than θ does when $\theta\to\infty$ so that $Z_A'|_{\theta\to\infty} = 0$. We have verified that it is the first case. The value of $\Pi_o(p)$ is important for renormalization, after some tensor algebra we have (also see ref.(7))

$$\Pi_o(p) = \frac{g^2\theta c_2(G)}{2p^2}\int\frac{d^3k}{(2\pi)^3}\,\frac{[k^2p^2-(k\cdot p)^2](5k^2+5k\cdot p+4p^2+2\theta^2)}{k^2(k^2+\theta^2)(k+p)^2[(k+p)^2+\theta^2]}. \tag{34}$$

The integral is also convergent. Performing standard Feynman integration and taking $\theta\to\infty$, it turns out that

$$\Pi_o(p) = \Pi_o(0) = \frac{7}{24\pi}g^2c_2(G). \tag{35}$$

By eq.(33) and (35) we have the wave function renormalization constants

$$Z_A(p) = Z_A(0) = 1+\frac{7}{24\pi}g^2c_2(G), \tag{36}$$

and

$$Z_{gh}(p) = Z_{gh}(0) = 1-\frac{1}{12\pi}g^2c_2(G). \tag{37}$$

Next we consider three-point functions. The net result of corrections to the $\bar{c}cA$ vertex from (f) and (g) of Fig.1 is zero once $\theta\to\infty$: The one-$\epsilon^{\sigma\tau\lambda}$ term of (1.f) cancels against the three-$\epsilon^{\sigma\tau\lambda}$ term of (1.g); The one-$\epsilon^{\sigma\tau\lambda}$ terms of (1.g) go to zero and the non-$\epsilon^{\sigma\tau\lambda}$ terms cancel against each other between (1.f) and (1.g) when $\theta\to\infty$. This is in agreement with the arguments by Taylor[14] that to any order one has

$$\tilde{Z}_g = 1. \tag{38}$$

The AAA-vertex correction is given by (d) and (e) of Fig.1. The purely $\epsilon^{\sigma\tau\lambda}$ part of gauge field loop (1.e) cancels against the ghost loop 1.(d) and the odd-number $\epsilon^{\sigma\tau\lambda}$ part of (1.e) tends to zero when $\Theta\to\infty$. The remaining part of (1.e) thus is ultraviolet convergent. Since the theory with the regularization at this order is finite, the Ward identity (29) must be satisfied. Therefore, instead of doing the tedious Feynman diagram calculation, we are able to simply use eqs.(36-38) and Ward identity (29) to get the coupling constant renormalization constant Z_g. The result is

$$Z_g = \frac{\tilde{Z}_g}{Z_{gh}Z_A} = 1 + \frac{3}{8\pi}g^2c_2(G). \tag{39}$$

Now we come to the crucial point — checking quantization condition (30). By definition (29),

$$\frac{4\pi}{g_r^2} = \frac{4\pi Z_A^3}{Z_g^2g^2} = \frac{4\pi Z_{gh}^2Z_A}{\tilde{Z}_g^2g^2} = \frac{4\pi}{g^2} + \frac{1}{2}c_2(G). \tag{40}$$

(40) shows the renormalized coupling constant g_r satisfies the quantization condition (30) since $c_2(G)$ is always an integer. (40) also implies the shift of k:

$$k_r = k + \frac{1}{2}c_2(G), \tag{8}$$

which has first been given by Witten[2,12].

However, the shift can not be obtained in the other two regularization methods, which we shall consider next.

Actually, there is no difference between dimensional regularization and regularization by dimensional reduction for one loop calculations since the theory is finite at this order. Therefore it is unnecessary to discuss them separately.

The Feynman rules are given by eqs. (6), (7), (12), and (13) in the present case.

By inspecting the Feynman rules and one-loop diagrams of Fig.1, we know that at this level one must have (see eqs.(22), (23), (25), and (26)):

$$Z_A = Z_{gh} = Z_g = \tilde{Z}_g = 1. \tag{41}$$

The reason is that every one-loop two- or three-A vertex diagram includes factors of an even number of $\epsilon^{\sigma\tau\lambda}$ so it will not contribute Z_A or Z_g; while every one-loop diagram of the $\bar{c}c$ or $\bar{c}cA$ vertex includes factors of an odd number of $\epsilon^{\sigma\tau\lambda}$ therefore it will not contribute to Z_{gh} or $\tilde{Z}_g$.

Thus the only possibility for any one-loop diagram of Fig.1 is to contribute to Z'_A or Z'_g, the symmetric part of two- or three-A vertex, or add new terms to eq.(24) or (25). Any non-zero contribution as such would mean an anomaly or non-renormalizability. Below we shall find that such a case does not occur.

First, it is easy to see that the ghost self-energy (c) of Fig.1 is trivially zero since it contains an $\epsilon^{\sigma\tau\lambda}$, all three spacetime indices of which must be contracted, but there are only two momenta — an internal and an external one, i.e.

$$(c) \sim \int d^{\omega}k \frac{\epsilon^{\sigma\tau\lambda} k_{\lambda}(k+p)_{\sigma} k_{\tau}}{k^2 (k+p)^2} \equiv 0. \qquad (42)$$

Then let us show that each of remaining six diagrams in Fig.1 is finite. Furthermore there is cancellation between the pairs, (a) and (b), (d) and (e), and (f) and (g):

$$(1.a) = -(1.b) = -\frac{1}{2}\delta^{ab} g^2 c_2(G) \int \frac{d^{\omega}k}{(2\pi)^3} \frac{(k+p)_{\mu} k_{\nu} + k_{\mu}(k+p)_{\nu}}{k^2 (k+p)^2}, \qquad (43)$$

$$(1.f) = -(1.g) = T^{abc} g^3 \int \frac{d^3k}{(2\pi)^3} \frac{k_{\lambda} \epsilon^{\sigma\tau\eta} r_{\sigma} q_{\tau} k_{\eta}}{K^2 (k+q)^2 (k-r)^2}, \qquad (44)$$

where $T^{abc} = f^{ead} f^{dbg} f^{gce}$ and $r+p+q = 0$, and

$$(1.d) = -(1.e) = -iT^{abc} g^3 \int \frac{d^{\omega}k}{(4\pi)^3} \frac{(k-r)_{\mu}(k+q)_{\nu} k_{\lambda} + (k+q)_{\mu} k_{\nu} (k-r)_{\lambda}}{k^2 (k+q)^2 (k-r)^2}. \qquad (45)$$

To complete this section, we have seen that the F^2-regularization leads to a finite renormalization and dimensional regularization as well as dimensional reduction give zero renormalization. No anomaly has been found in the calculations of the two- and three-point functions at one-loop level with any of the three regularization schemes we have used.

5. Two Loop Considerations

Logarithmic divergences first appear at the two-loop order. And the F^2 regularization makes the theory finite by powercounting and symmetry analysis. Therefore at this level especial interest is in dimensional regularization and dimensional reduction. We shall find in this section that dimensional regularization is not suitable to CS gauge field theory because it does not satisfy Ward identities, Dimensional reduction is free of such problems.

The two-loop two- and three-point vertices are collected in Fig.2. From the diagram structures and the Feynman rules, we see that, contrary to the one-loop case, these two-loop diagrams only give corrections to the renormalization constants Z_A, Z_g, Z_{gh}, and $\tilde{Z}_g$ but no others. This is because that each gauge field two- or three-point vertex has factors of an odd number of $\epsilon^{\sigma\tau\lambda}$ and each ghost-ghost or ghost-ghost-gauge vertex has an even number of $\epsilon^{\sigma\tau\lambda}$.

According to our analysis in section 2, all two-loop three-point vertices are essentially convergent, therefore the different regularization schemes used to regulate superficially divergent Feynman integrals should not give different final results. In fact, as we shall show, the two-loop corrections to three-point functions turn out to be zero.

First of all, the planar three-leg diagrams in Fig.2 are canceled in pairs. To see this, we pair the planar diagrams of Fig.2, denoted by (l) and (L), (m) and (M), and so on. It can be seen, in the two diagrams of each pair, some parts are the same but the remaining different parts are just these that form one of the one-loop pairs of Fig.1 with the expression given by one of eqs.(43-45), which cancel within the pair.

Secondly, the nonplanar three-leg diagrams of Fig.2 vanishe individually because of the symmetry of gauge group indices. To see this, we pick one such diagram and cut any one of two crossing internal propagators, then we get a one-loop four-point vertex connecting with a bare three-point vertex through a propagator. A four-point vertex has a gauge group factor denoted by T^{abcd}, which is equal to $f^{eag}f^{gbh}f^{hci}f^{ide}$ for an one-loop diagram. It is easy to verify that T^{abcd} is symmetric under the exchange of two indices which are not neighbor, i.e. $T^{abcd} = T^{adcb} = T^{cbad}$. On the other hand, the bare three-point vertex has a factor of the gauge group structure constant, and two of its three indices will contract with two non-neighboring indices of T^{abcd} due to the Feynman diagram structure. Each nonplanar two-loop three-point vertex has such a group factor $T^{abcd}f^{jbd}$ and is therefore zero.

Then we come to two-point functions. It seems that the cancellation mechanism used in the planar three-point vertices is also suitable to the two-point vertices of Fig.2 since they have similar sub-diagram structures. However we must be very careful since we are dealing with logarithmically divergent diagrams. In fact, we shall find that surprisingly the two-loop contributions to two-point functions are without singularities and that the final results are different in the different regularization schemes we have used. Particularly, the complete cancelation mentioned above only occurs in the dimensional reduction approach.

Let us first consider dimensional regularization. Here we do not know how to continue $\epsilon^{\sigma\tau\lambda}$ into noninteger spacetime dimension $\omega=3-\epsilon$. Instead of attempting it let us try to define the contraction between $\epsilon^{\phi\tau\lambda}$ and its dual $\epsilon_{\phi\sigma\eta}$ as $\epsilon^{\phi\tau\lambda}\epsilon_{\phi\sigma\eta}=(\tilde{\delta}^{\tau}{}_{\sigma}\tilde{\delta}^{\lambda}{}_{\eta}-\tilde{\delta}^{\tau}{}_{\eta}\tilde{\delta}^{\lambda}{}_{\sigma})\Gamma(\omega-1)$ with $\tilde{\delta}^{\phi}{}_{\phi}=\omega$, eqs.(19,20). The regularization procedure will be as follows. We contract all antisymmetric tensors, according to eqs.(19,20) first, and then perform Feynman integrals, the dimension of which will have been analytically continued to ω.

In this approach, two-point functions will get finite corrections. We shall first work out the radiative correction of the ghost propagator. The corresponding diagrams are (g), (h), (i), (j), and (k) of Fig.2. The last one, Fig. (2.k) is zero since it contains a sub-diagram Fig. (1.c) which we have shown vanishes. And the first two give:

$$(2.g) = i\Gamma(\omega-1)g^4R^{ab}I_1(\omega), \tag{46}$$

$$(2.h) = -ig^4R^{ab}I_1(\omega), \tag{47}$$

where

$$R^{ab} = R\delta^{ab} = f^{ade}f^{dcg}f^{ghe}f^{hcb}. \tag{48}$$

The factor $\Gamma(\omega-1)$ in (2.g) is from the contraction of two of its four $\epsilon^{\phi\tau\sigma}$, and the Feynman integral is

$$I_1(\omega) = \int\frac{d^{\omega}k}{(4\pi)^3}\frac{d^{\omega}q}{(4\pi)^3}\frac{\epsilon^{\phi\sigma\lambda}q_{\phi}p_{\sigma}k_{\lambda}\epsilon^{\tau\eta\kappa}k_{\tau}p_{\eta}q_{\kappa}}{k^2q^2(k+q)^2(k-p)^2(q+p)^2}. \tag{49}$$

The primary divergence of (49) is logarithmic. Through the standard Feynman integral procedure, a simple pole $\Gamma(3-\omega)$ can be isolated:

$$I_1(\omega) = -\Gamma(3-\omega)\frac{\Gamma(\omega)\Gamma(\omega-1)}{4(4\pi)^{\omega}}p^2 I_1'(\omega), \tag{49'}$$

where the Feynman parameter integral $I_1'(\omega)$ is a finite number when $\omega\to3$:

$$I_1'(3) = \int_0^1 dx_1 dx_2 dx_3 \theta(1-x_1-x_2-x_3)\frac{1-x_1-x_2-x_3}{[(1-x_1-x_2)(x_1+x_2+x_3)-x_3^2]^{5/2}}. \tag{50}$$

From (2.g) plus (2.h) we can extract a factor $1-\Gamma(\omega-1)$, which is zero as $\omega\to3$, apart from the factor $\Gamma(3-\omega)$. A simple calculation gives

$$[1-\Gamma(\omega-1)]\Gamma(3-\omega) = 1-\gamma, \quad \text{when } \omega\to3, \tag{51}$$

where $\gamma = 0.5772...$ is the Euler constant. Surprisingly, the singularity is gone when the two singular diagrams are added together.

The other couple of diagrams to the radiation correction of the ghost propagator gives

$$(2.i) = -i\frac{1}{4}\delta^{ab}[c_2(G)]^2 g^4 I_2(\omega), \tag{52}$$

$$(2.j) = i\frac{1}{4}\delta^{ab}[c_2(G)]^2 g^4 [\Gamma(\omega-1)]^2(\omega-2)I_2(\omega), \tag{53}$$

where

$$I_2(\omega) = \int\frac{d^{\omega}k}{(2\pi)^3}\frac{p_\sigma \epsilon^{\sigma\tau\lambda}k_\lambda(\frac{1}{32}\surd k^2\delta_{\tau\eta})\epsilon^{\eta\xi\upsilon}k_\upsilon p_\xi}{k^2k^2(k+p)^2}$$

$$= \frac{1}{32}\Gamma(\omega-1)\int\frac{d^{\omega}k}{(4\pi)^3}\frac{(k\cdot p)^2-k^2p^2}{k^3(k+p)^2}. \tag{54}$$

The bracket in the first line in Eq.(54) is the contribution from the one-loop two-point vertex insertion. Like $I_1(\omega)$, the integral (54) is primarily logarithmically divergent and it gives a simple pole:

$$I_2(\omega) = \frac{(1-\omega)}{32(4\pi)^{3/2}}\Gamma(\frac{3-\omega}{2})p^2. \tag{54'}$$

A relation like eq.(51) is

$$\{1-(\omega-2)[\Gamma(\omega-1)]^2\}\Gamma(\frac{3-\omega}{2}) = 2(3-\gamma) \text{ as } \omega\to\infty$$

therefore (2.i) plus (2j) is not divergent either.

Adding together (2.g, h, i, j), out of divergent diagrams, we have a finite correction to the ghost self-energy at two-loop:

$$\tilde{\Pi} = \frac{g^4 R}{2(4\pi)^3}(1-\gamma)I_1'(3) + \frac{g^4[c_2(G)]^2}{32(4\pi)^{3/2}}(3-\gamma). \tag{55}$$

The result is finite and independent of external momentum p.

Next we calculate the radiative correction to the gauge field at the two-loops. Contracting $\Pi_{\mu\nu}(p)$ with $\epsilon^{\mu\nu\lambda}p_\lambda$, we have $\Pi_o(p) = \frac{1}{2p^2}\epsilon^{\mu\nu\lambda}p_\lambda\Pi_{\mu\nu}(p)$ (see eq.(22)). The contributions to $\Pi_o(p)$ from (a-f) of Fig.2 are as followings:

$$(2.a) = \frac{1}{2p^2}g^4R^{ab}[\Gamma(\omega-1)]^4(\omega-1)I_1(\omega), \tag{56}$$

$$(2.b) = -\frac{1}{p^2}g^4R^{ab}\Gamma(\omega-1)I_1(\omega), \tag{57}$$

$$(2.c) = \frac{1}{p^2}g^4R^{ab}I_1(\omega), \tag{58}$$

$$(2.d) = (2.b), \tag{59}$$

$$(2.e) = -\frac{1}{8p^2}g^4[c_2(G)]^2\delta^{ab}I_3(\omega) \tag{60}$$

$$(2.f) = \frac{1}{8p^2}g^4[c_2(G)]^2\delta^{ab}[\Gamma(\omega-1)]^2(\omega-2)I_3(\omega) \tag{61}$$

where $I_1(\omega)$ is given by eq.(48) and

$$I_3(\omega) = \frac{-1}{32}\int\frac{d^\omega k}{(2\pi)^3}\frac{(k\cdot p)^2}{k^3(k+p)^2} = -\frac{\Gamma(\frac{3-\omega}{2})}{64(4\pi)^{3/2}}p^2. \tag{62}$$

After some algebra, we finally obtain

$$\Pi_o = \frac{g^4R}{4(4\pi)^3}(7-6\gamma)I_1'(3) + \frac{g^4[c_2(G)]^2}{256(4\pi)^{3/2}}(3-\gamma). \tag{63}$$

It is remarkable that without invoking any counterterms the singularities cancel between ghost loops and gauge field loops and a finite result is obtained. Unfortunately, either the Ward identity (29) nor the quantization condition (30) are satisfied by these results, since $Z_g = \tilde{Z}_g = 1$ but $Z_A \neq Z_{gh}$ by Eqs.(55) and (63). The fact forces us to abandon dimensional regularization in CS gauge theory. We attribute this to the ambiguity in the dimensional continuation of the antisymmetric tensor $\epsilon^{\sigma\tau\lambda}$.

On the other side, let us show that if one insist on defining antisymmetric tensors on three dimensions, the ambiguity may be avoided. Especially, the contraction of $\epsilon^{\sigma\tau\lambda}$ and its dual is defined as $\epsilon^{\phi\tau\lambda}\epsilon_{\phi\sigma\eta} = \delta^\tau_\sigma\delta^\lambda_\eta - \delta^\tau_\eta\delta^\lambda_\sigma$, with $\delta^\phi_\phi = 3$, eq.(21). The strategy we shall take is that when regularizing a potentially divergent Feynman integral, all $\epsilon_{\mu\nu\lambda}$ will be extracted from the integral first, and then the rest will be analytically continued to ω-dimension. This method is so-called the regularization by dimensional reduction. Now we shall see how it works at this level.

Based on the contraction (21), we simply fix all ω that come from the contractions between $\epsilon_{\mu\nu\lambda}$ to 3 in (2.a) to (2.j) and find that the complete cancellations happen between (2.a) and (2.b), (2c.) and (2.d), and so on.

As a result, with the regularization by dimension reduction we have obtained no corrections to both two- and three-point functions. Therefore we have zero renormalization up to two-loop order. It will be correct also at three loops because at this order (and any odd-loop order) the diagrams will not contribute to the renormalization constants. It is obvious that the regularization by dimensional reduction does respect all symmetries, although trivially, at least to two loops.

To conclude this section, we would like to make an comment on F^2-regularization at two loops and beyond. Although there is no reason to exclude non-zero corrections to renormalization constants in F^2-regularization at higher-loop order, it is unlikely to add new term to the shift of $k \to k_r = k + \frac{1}{2}c_2(G)$, eq.(8). In other words, the renormalized coupling constant

$$g_r = \frac{Z_g}{Z_A^{3/2}} g = (1 - \frac{g^2 c_2(G)}{16\pi}) g \tag{64}$$

should not be changed at two-loop order and beyond, if the quantization condition (30) remains. To see this, suppose g_r get a correction at two loops, $\frac{-Bg^4}{2(4\pi)^2}$, where B is a non-singularity number. Then the quantization condition requires $B\frac{g^2}{4\pi}$ be an integer for a fixed constant B but an arbitrary integer $\frac{4\pi}{g^2}$. The only possibility is B=0.

6. Conclusions

Now we return to the questions we asked at the beginning of the Lecture. By the above analysis of two- and three-point functions, we can say that the CS gauge theory does define a sensible quantum field theory in the sense of perturbation theory.

In the explicit calculation of two- and three-point functions up to two loops, we have found no gauge, general covariance, or scale anomalies with either the F^2 regularization or the regularization by dimensional reduction, while dimensional regularization is ruled out since it is not gauge invariant at two loops. Furthermore, the metric introduced in regularization does not break the general covariance of the theory so that the renormalized action is independent on the metric but takes the same form as the bare action with the bare quantities replaced by renormalized ones.

More interesting, the perturbation theory which is defined locally respects the requirement of topological considerations — the quantization condition of k or g.

The new lesson here is about finite renormalization. We have demonstrated that dimensional reduction leads to no renormalization up to three loops. And to higher orders, we have reasons to believe that the theory is finite, even though it probably does not continue to have zero renormalization. On the other hand, we have shown that the F^2-regularization essentially makes the theory finite to any order and gives a finite shift of the overall coefficient k by $\frac{1}{2}c_2(G)$ at one loop order.

The advantage of a theory with finite renormalization is that we do not need to invoke ambiguous counterterms to absorb singularities coming form loop corrections or attribute the bare quantities as unobservable or physically meaningless infinities. A reasonable explanation in this case seems to be that both bare and renormalized quantities are physically meaningful: the former are classical quantities and the latter are quantum ones.

Since CS gauge field theory is of this sort, it faces another problem. Starting from a given classical CS action (a fixed classical coupling constant), with regularization by dimensional reduction we do not get quantum correction as we have seen, but with the F^2-regularization we do. Then the problem is which procedure should we trust? A priori, we have no reason to prefer either since both of them obey all symmetries of the theory. A possibility is that different proper regularization schemes define different reasonable quantum theories.

Finally, since the renormalized theory is finite, it is obvious that the beta-function for the only parameter of the theory is exactly zero, i.e. the theory is free of scale anomalies.

This talk is based on the work in ref.(9). Recently, perturbative Chern-Simons theory has been discussed also by other authors[(10,11)].

References

1. The lecture notes in this school by J. Frohlich, R. Jackiw, N. Seiberg, G. Semenoff, and Y.-S. Wu.
2. E. Witten, Comm. Math. Phys. **121** (1989) 351.
3. S. Deser, R. Jackiw and S. Templeton, Phys. Re. Lett. **48** (1982) 975; Ann. Phys. (N.Y.) **140** (1982) 372.
4. G. Moore and N. Seiberg, Phys.Lett. **B220** (1989) 422; M. Bos and V.P. Nair, Phys. Lett. **B223** (1989) 61; Y. Hosotani, Phys. Rev. Lett. **62** (1989) 2785; H. Murayama, "Explicit Quantization of the Chern-Simons Action", Univ. of Tokyo preprint. UT-542, March 1989; K. Yamagishi, M.L. Ge and Y.-S. Wu, Univ. Utah preprint UU-HEP-89/1.
5. G. Baskaran and P.W. Anderson, Phys. Rev. **B17** (1988) 580; P.W. Anderson, Varenna Lectures, Proc. of the Internaltional School of Physics, Enrico Fermi (1987); I. Affleck, Z. Zou, T. Hsu, and P.W. Anderson, Phys. Rev. **B18** (1988) 754; K. Wu, L. Yu, and C.-J. Zhang, Mod. Phys. Lett. **B2** (1988) 979; Z. Zou, Phys. Lett. **A131** (1988) 197; A.P. Balachandran, M.j. Bowick, K.S. Gupta, and A.M. Srivastava, Mod. Phys. Lett. **A3** (1988) 1725; A.M. Polyakov, Mod. PHys. Lett. **A3** (1988) 325; G.W. Semenoff, Phys. REv. Lett. **61** (1989) 517.
6. G.W. Semenoff, P. Sodano, and Y.-S. Wu, Phys. Rev. Lett. **62** (1989) 751.
7. R.D. Pasarski, and S. Rao, Phys. REv. **D32** (1985) 2081.
8. Y.-C. Kao, and M. Suzuki, Phys. Rev. **D31** (1985) 2137, Y.-C. Kao, J. Koller, and H. Yamagishi, Phys. Rev. Lett. **58** (1987) 1077.
9. W. Chen, G.W. Semenoff, and Y.-S. Wu, Utah Uni. Preprint, May 1989.
10. E. Guadanini, M. Martellini, and M. Mintchev, CERN-TH.5324/89 & IFUP-TH6/89, CERN-TH.5420/89.
11. L. Alvarez-Gaume, J.M.F. Labastida, and A.V. Ramallo, CERN-TH.5480/89.
12. The shift of k has been got also by the authors of ref.(7), where they cosidered the three dimensional massive Yang-Mills perturbation theory.
13. S.J. Gates Jr., M.T. Grisaru, M.Rocek, and W. Siegal, "Superspace or One Thousand and one Lessons in Supersymmetry", (Benjamin/Cummings,1983). The method has also been used in two dimensional σ model, see say M. Bos, Phys. Lett. **B189** (1987) 435.
14. J.C. Taylor, Nucl. Phys. **B33** (1971) 436.

A NEW FAMILY OF N-STATE REPRESENTATIONS OF THE BRAID GROUP

M. Couture*, **H.C. Lee***□ *and* **N.C. Schmeing***†

* *Theoretical Physics Branch, Chalk River Nuclear Laboratories, Atomic Energy of Canada Limited, Research Company, Chalk River, Ontario, Canada K0J 1J0*

□ *Department of Applied Mathematics, University of Western Ontario, London, Ontario, Canada N6A 5B9*

† *Permanent address: 17 Beach Avenue, Deep River, Ontario Canada K0J 1P0*

Abstract

For the Artin braid group B_n, we give two types of N-state representations for N=2,3,4,5,6.

1. Introduction

The theory of braids and knots is related to many topics in physics. Recently a connection between knot theory and the theory of exactly solvable models in statistical mechanics and many body systems in two dimensions has been discovered [1-5]. Central to this is the Yang-Baxter relation which is a sufficiency condition for the transfer matrices to commute in those models. Exploiting the similarity between the defining relation of Artin's braid group B_n and the Yang-Baxter relation, various representations of B_n have been obtained from statistical models at criticality. The Markov trace defined on B_n is used to construct topological invariants for knots and links. State models associated to link diagrams have been shown to be partition functions of certain statistical models. For instance, Kauffman [6] has defined a state model for the Alexander-Conway polynomial which can be seen as the low temperature limit of the partition function of a generalized Potts model. Methods for constructing link polynomials via the braid group or a diagrammatic approach rely on a two-dimensional definition of link polynomials. Recently within the context of a solvable topological quantum field theory in 2+1 dimensions, Witten [7] has proposed an intrinsically three-dimensional definition of the Jones polynomials. This idea has been carried further and a new hierarchy of link polynomials has been derived from a topological Chern-Simons gauge theory [8]. The braid group is of interest in its own right, as the fun-

---

Presented by M. Couture at Workshop on "Physics, Braids & Links", Banff NATO Summer School, August 14-25, 1989

damental group of the configuration space of n indistinguishable particles in two dimensions. Braid group representations are related to strange statistics (neither Bose or Fermi) in quantum mechanics in 2+1 dimensions [9-12] and quantum field theory in 1+1 dimensions [13] and to the monodromy of multipoint correlation functions in two dimensional conformal field theory [14,15].

In an earlier paper [12], we proposed an algorithm for constructing N-state representations of B_n of the maximally symmetric type (defined in section 2); from this family of representations one can derive an infinite sequence of link polynomials. The N=2 case in this family corresponds to the famous Jones polynomial. The method consisted of solving a set of equations of the Yang-Baxter type. The resulting family of solutions had been first discovered by Akutsu and Wadati [1,2]; their method consisted of extracting representations of B_n from exactly solvable N-state vertex models in statistical mechanics at criticality. Recently our method led to the discovery of a new family of N-state representations of B_n; solutions for N=2, 3 and 4 were given [16,17].

The object of my talk (M.C.) is to give a detailed account of those representations. In section 2 solutions to both families of representations are given up to N=6.

2. N-State Representations of B_n

Artin's braid group B_n [18,19] is generated by a set of (n-1) generators (elementary braids) $g_1, g_2, \dots, g_{n-1}$ and their inverses subject to the following necessary and sufficient defining relations

$$g_i g_j = g_j g_i \qquad |i-j| \geq 2 \tag{2.1a}$$

$$g_i g_{i+1} g_i = g_{i+1} g_i g_{i+1} \tag{2.1b}$$

An element β in B_n, called a braid, is a word in the g_i's

$$\beta = g_{i_1}^{\sigma_1} g_{i_2}^{\sigma_2} \dots, \qquad \sigma_i \in \mathbb{Z}$$

Let V be an N-dimensional vector space and $R \in End(V \otimes V)$ be an $N^2 \times N^2$ matrix that has an inverse. The mapping

$$\rho : B_n \to End(V^{\otimes n}) \tag{2.2}$$

with

$$\rho(g_i) = I_1 \otimes \dots \otimes I_{i-1} \otimes R \otimes I_{i+2} \otimes \dots \otimes I_n \tag{2.3}$$

is a representation of B_n, where $I \in End(V)$ is the identity matrix, the subscript i means the ith vector space in $V^{\otimes n}$, and R acts on the ith and $(i+1)^{th}$ vector spaces. The form of (2.3) insures that (2.1a) holds. Let us now consider (2.1b). Component wise :

$$[\rho(g_i)]^{b_1,\ldots b_n}_{a_1,\ldots a_n} = \delta^{b_1}_{a_1}\cdots\delta^{b_{i-1}}_{a_{i-1}} (R)^{b_i,b_{i+1}}_{a_i,a_{i+1}} \delta^{b_{i+2}}_{a_{i+2}} \cdots\delta^{b_n}_{a_n},$$

$$a_i,b_i=1,..,N \tag{2.4}$$

and therefore

$$\left[\rho(g_i)\rho(g_{i+1})\rho(g_i)\right]^{b_1,\ldots b_n}_{a_1,\ldots a_n}$$

$$= \delta^{b_1}_{a_1}\ldots\delta^{b_{i-1}}_{a_{i-1}} \delta^{b_{i+3}}_{a_{i+3}}\ldots\delta^{b_n}_{a_n} \sum_{k,\ell,m} (R)^{k,m}_{a_i,a_{i+1}} (R)^{\ell,b_{i+2}}_{m,a_{i+2}} (R)^{b_i,b_{i+1}}_{k,\ell} \tag{2.5a}$$

$$\left[\rho(g_{i+1})\rho(g_i)\rho(g_{i+1})\right]^{b_1,\ldots b_n}_{a_1,\ldots a_n}$$

$$= \delta^{b_1}_{a_1}\ldots\delta^{b_{i-1}}_{a_{i-1}} \delta^{b_{i+3}}_{a_{i+3}}\ldots\delta^{b_n}_{a_n} \sum_{k,\ell,m} (R)^{m\ell}_{a_{i+1},a_{i+2}} (R)^{b_i,k}_{a_i,m} (R)^{b_{i+1},b_{i+2}}_{k,\ell} \tag{2.5b}$$

It is understood that a summation will always be from 1 to N. It follows from (2.1b) and (2.5) that ρ is a representation of B_n provided R satisfies the Yang-Baxter relation

$$\sum_{k,\ell,m} R^{km}_{rs} R^{\ell c}_{mt} R^{ab}_{k\ell} = \sum_{k,\ell,m} R^{m\ell}_{st} R^{ak}_{rm} R^{bc}_{k\ell} \tag{2.6}$$

Our approach in constructing representations of B_n consists of solving (2.6) directly. The solutions we seek are of the "charge conserving" type meaning that

$$R^{ab}_{cd} = 0 \quad \text{if} \quad a+b \neq c+d\ . \tag{2.7}$$

For N=2, this condition constrains R to those appropriate to the six-vertex state models. Because of (2.7), the R-matrix is block diagonal

$$R_{(N)} = \begin{bmatrix} A_1 & & & & & & \mathbf{O} \\ & A_2 & & & & & \\ & & \ddots & & & & \\ & & & A_N \equiv A'_N & & & \\ & & & & A'_{N-1} & & \\ & \mathbf{O} & & & & \ddots & \\ & & & & & & A'_1 \end{bmatrix} \tag{2.8}$$

where the submatrices A_m and A'_m are m×m and $R_{(N)}$ is an $N^2 \times N^2$ matrix. We further demand that all elements in the upper left triangle in the submatrices A_m and A'_m be identically zero, and all other elements in those submatrices be non zero. We now note certain symmetries of the YB relations (2.6).

Consider the diagonal $N^2 \times N^2$ matrices

$$(S_1)^{cd}_{ab} = \delta^{cd}_{ab} \exp[\alpha_1(a+b)]$$

$$(S_\ell)^{cd}_{ab} = \delta^{cd}_{ab} \exp[\alpha_\ell(\phi^\ell(a) + \phi^\ell(b))] \quad \ell = 2,3,... \tag{2.9a}$$

where the α_ℓ are arbitrary constants and the ϕ^ℓ arbitrary functions. It is easy to show that if $R_{(N)}$ satisfies the YB relations, so does SRS^{-1}, where

$$S = \prod_\ell S_\ell . \tag{2.9b}$$

The algorithm used to solve $(2.6)_{cd}$ consists in solving for solutions that are invariant under transposition ($R^{cd}_{ab} = R^{ab}_{cd}$). More general solutions are then obtained by using the transformation (2.9). For the N-component model, only N-1 of the S_ℓ's, including S_1, are effective; they introduce N-1 asymmetry parameters. See [12] for example.

Let us use

$$[YB] = (r,s,t,a,b,c) \tag{2.10}$$

to denote the set of 6 indices in (2.6) that specifies a particular YB relation. Consider the matrix element $(R_{(N)})^{cd}_{ab}$ and the following transformations

$$\begin{array}{lll} \text{"charge conjugation"} & (a,b,c,d) \xrightarrow{C} (N\text{-}a+1,N\text{-}b+1,N\text{-}c+1,N\text{-}d+1) \\ \text{"parity change"} & (a,b,c,d) \xrightarrow{P} (b,a,d,c) \\ \text{"time reversal"} & (a,b,c,d) \xrightarrow{T} (c,d,a,b) \end{array} \tag{2.11}$$

The set of YB relations are invariant under a CP transformation

$$(\alpha,\beta,\gamma,\lambda) \xrightarrow{CP} (N\text{-}\beta+1,N\text{-}\alpha+1,N\text{-}\lambda+1,N\text{-}\gamma+1) . \tag{2.12}$$

To every YB equation there is a corresponding CP conjugate

$$\begin{aligned}[YB] &= (r,s,t,a,b,c) \rightarrow [YB]^{CP} \\ &= (N\text{-}t+1,N\text{-}s+1,N\text{-}r+1,N\text{-}c+1,N\text{-}b+1,N\text{-}a+1)\end{aligned} \tag{2.13}$$

Under a CP transformation

$$A_m \underset{\longleftarrow}{\overset{CP}{\longrightarrow}} A'_m . \tag{2.14}$$

This CP symmetry will manifest itself in the solutions of (2.6).

We have solved (2.6) for N = 2,3,4,5 and 6; throughout we normalize to $A_1=1$. Under the constraints described above, there exists two families of solutions. These will be referred to as the maximally symmetric (MS) and the non-maximally symmetric (NMS) families. We now describe some of their properties:

(a) A_m and A'_m are symmetric under a T transformation;

(b) Denote the submatrices of the MS and NMS families by $\mathcal{X}_m(t)$ and $A_m(t;\omega)$ respectively. NMS solutions depend on N only through the parameter ω; for N=3,4,5 and 6 and arbitrary values of t, $\omega=\omega_N^{\pm 1}$ where $\omega_N=e^{2\pi i/N}$. The MS and NMS solutions differ in their CP properties. Under CP transformation

$$\mathcal{X}_m(t) \xrightarrow{CP} \mathcal{X}'_m(t) = \mathcal{X}_m(t) \qquad (MS) \qquad (2.15a)$$

for the MS solutions, and

$$A_m(t;\, \omega) \xrightarrow{CP} A'_m(t;\, \omega) = \omega t^{N-1}\, A_m(\omega^2 t^{-1};\, \omega) \qquad (NMS) \qquad (2.15b)$$

for the NMS solutions. It is easy to show that in both cases (trivially in (2.15a)) the CP transformation is unipotent. Note that (2.15b) imposes that $A_N(t;\, \omega)$ be CP invariant. The NMS solutions satisfy the following relations (tr $\equiv$ trace)

$$tr(A_i) + tr(A'_{N-i}) = tr(A'_i) + tr(A_{N-i}) = tr\,(A_N) \quad i=1,2,\ldots,[\tfrac{N}{2}] \qquad (2.15c)$$

and therefore

$$tr(R_{(N)}) = N\, tr(A_N)\ ; \qquad (2.15d)$$

where [X] is the largest integer not greater than X. The MS solutions do not satisfy (2.15c) and (2.15d).

(e) The two families of solutions are related by (see Appendix A)

$$\mathcal{X}_m(t= \omega^{-1}) = A_m(t=\omega;\, \omega) \qquad (2.16)$$

The submatrices for N = 2,3,4,5 and 6 are

$$(A_m)_{ij} = c_{ij} t^{\sigma_{ij}} (a_m)_{ij} \qquad m = 2,\ldots,N$$

and

$$a_2 = \begin{bmatrix} 0 & \\ 1 & X_0 \end{bmatrix}$$

$$a_3 = \begin{bmatrix} 0 & & \\ 0 & 1 & \\ 1 & (Y_1 X_{0,1})^{1/2} & X_{0,1} \end{bmatrix}$$

$$
a_4 = \begin{bmatrix}
0 & & & \\
0 & 0 & & \\
0 & 1 & Y_1X_1 & \\
1 & (Y_2X_{0,2})^{1/2} & X_1(Y_2X_{0,2})^{1/2} & X_{0,1,2}
\end{bmatrix}
$$

$$
a_5 = \begin{bmatrix}
0 & & & & \\
0 & 0 & & & \\
0 & 0 & 1 & & \\
0 & 1 & (Y_{1,2}X_{1,2})^{1/2} & Y_2X_{1,2} & \\
1 & (Y_3X_{0,3})^{1/2} & (Y_1^{-1}Y_{2,3}X_{0,1,2,3})^{1/2} & X_{1,2}(Y_3X_{0,3})^{1/2} & X_{0,1,2,3}
\end{bmatrix}
\qquad (2.17)
$$

$$
a_6 = \begin{bmatrix}
0 & & & & & \\
0 & 0 & & & & \\
0 & 0 & 0 & & & \\
0 & 0 & 1 & Y_2X_2 & & \\
0 & 1 & (Y_{1,3}X_{1,3})^{1/2} & X_2(Y_{2,3}X_{1,2})^{1/2} & Y_3X_{1,2,3} & \\
1 & (Y_4X_{0,4})^{1/2} & (Y_1^{-1}Y_{3,4}X_{0,1,3,4})^{1/2} & X_2(Y_1^{-1}Y_{3,4}X_{0,1,3,4})^{1/2} & Z_1 & Z_2
\end{bmatrix}
$$

where

$$Z_1 = X_{1,2,3}\,(Y_4X_{0,4})^{1/2}$$

$$Z_2 = X_{0,1,2,3,4}$$

$$X_{s,t\ldots} = X_sX_t \ldots.$$

$$Y_{s,t\ldots} = Y_sY_t \ldots.$$

for the NMS solutions we have that

$$
\begin{aligned}
c_{ij} &= \omega^{(m-i)(m-j)} \\
\sigma_{ij} &= (2m-i-j)/2 \\
X_\ell &= 1 - \omega^\ell t \\
Y_\ell &= \sum_{k=0}^{\ell} \omega^k
\end{aligned}
\qquad \text{(NMS family)} \qquad (2.18)
$$

and for the MS solutions

$$c_{ij} = 1$$

$$\sigma_{ij} = [(2m-i-j)(N-m+1)+(m-i)(m-i-1)+(m-j)(m-j-1)]/2$$

$$X_\ell = 1 - t^{N-\ell 1} \qquad \text{(MS family)}$$

$$Y_\ell = \sum_{k=0}^{\ell} t^k \tag{2.19}$$

In Appendix A the NMS solutions are given explicitly up to N=5. For the MS solutions, $R_{(N)}$ satisfy the following reduction formula

$$\prod_{s=1}^{N} (\check{R} + (-)^s t^{\rho_s}) \equiv \check{R}^N - \sum_{s=1}^{N} \Upsilon_s(t)\, \check{R}^{N-s} = 0; \quad \rho_s \equiv \frac{1}{2}(2N-s)(s-1) \tag{2.20}$$

For the NMS family the reduction formula is

$$\prod_{s=1}^{N} (R + (-)^s \omega^{\sigma_s} t^{s-1}) \equiv R^N - \sum_{s=1}^{N} f_s(t) R^{N-s} = 0; \; \sigma_s \equiv \frac{1}{2}(s-1)(s-2) \tag{2.21}$$

from which the recursion relation for generating the set of functions f_s for N from the set for N-1 can be derived, provided ω is treated as a free parameter (but not a root of 1):

$$f_s^{(N)} = (1-\delta_{N,s}) f_s^{(N-1)} + (-)^N \left[(1-\delta_{s,1}) f_{s-1}^{(N-1)} - \delta_{s,1}\right] \omega^{\sigma_N} t^{N-1}; \; s=1,..,N \tag{2.22}$$

The pattern of the reduction formulas and the algorithm used to obtain the MS and NMS representations lead us to suspect that, although (2.16), (2.17), (2.20) and (2.21) have been established only up to N=6 they may be generally true for all N.

The existence of the NMS family of braid group representation raises many interesting questions. First the MS family of braid group representations has been shown to be associated to a family of link polynomials; the N=2 case being the Jones polynomial. Is there also a family of link polynomials associated with the NMS family? By letting the spectral parameter of a family of N-state vertex model tend to infinity, Akutsu and Wadati obtained the MS family of braid group representations. Is there a family of statistical models whose limit would give us the NMS family? Finally, it is well known that the MS family can be derived from the quantized universal enveloping algebra of SL(2,C). Is there a quantum group corresponding to the NMS family?

Akutsu and Wadati [1,2] obtained the MS family (up to N=4) of R-matrices from a series of vertex models which included the 6-vertex model of Lieb and Wu and the 19-vertex model by Zamolodchikov and Fateev; the procedure consisted in letting the spectral parameters tend to infinity. It is easily verified that the N=2 case of the NMS family can be obtained from the free fermion model of Fan and Wu [20]; the cases N>2 remain an open question. The relation of the Alexander-Conway polynomial with the N=2 case of the NMS family has been discussed by Lee and Couture [16] and Kauffman [21]. See the contribution by H.C. Lee in these proceedings for a detailed discussion on the quantum group structure associated to the NMS family of solutions.

APPENDIX A

In this appendix we give the NMS solutions for N = 2, 3, 4 and 5 explicitly. In all cases

$$A_1 = 1 \quad ; \qquad A_2 = \begin{bmatrix} 0 & \\ t^{1/2} & 1-t \end{bmatrix} . \tag{A.1}$$

N = 2

$$A'_1 = -t \tag{A.2}$$

N = 3

$$A'_1 = \omega t^2 \quad ; \qquad A'_2 = \begin{bmatrix} 0 & \\ \omega^2 t^{3/2} & -tX_1 \end{bmatrix}$$

$$A_3 = \begin{bmatrix} 0 & & \\ 0 & \omega t & \\ t & i\,\omega(tX_{0,1})^{1/2} & X_{0,1} \end{bmatrix} \tag{A.3}$$

N = 4

$$A'_1 = \omega t^3 \quad ; \qquad A'_2 = \begin{bmatrix} 0 & \\ \omega^2 t^{5/2} & -\omega^3 t^2 X_1 \end{bmatrix}$$

$$A_3 = \begin{bmatrix} 0 & & \\ 0 & \omega t & \\ t & (tY_1 X_{0,1})^{1/2} & X_{0,1} \end{bmatrix} \quad ; \quad A'_3 = \begin{bmatrix} 0 & & \\ 0 & t^2 & \\ \omega^3 t^2 & -\omega^3(\omega^3 t^3 Y_1 X_{1,2})^{1/2} & \omega^2 t X_{1,2} \end{bmatrix}$$

$$A_4 = \begin{bmatrix} 0 & & & \\ 0 & 0 & & \\ 0 & \omega^2 t^{3/2} & \omega t Y_1 X_1 & \\ t^{3/2} & t(Y_2 X_{0,2})^{1/2} & X_1(tY_2 X_{0,2})^{1/2} & X_{0,1,2} \end{bmatrix} \tag{A.4}$$

N = 5

$$A'_1 = \omega t^4 \quad ; \qquad A'_2 = \begin{bmatrix} 0 & \\ \omega^2 t^{7/2} & -\omega^3 t^3 X_3 \end{bmatrix}$$

$$A_3 = \begin{bmatrix} 0 & & \\ 0 & \omega t & \\ t & (tY_1 X_{0,1})^{1/2} & X_{0,1} \end{bmatrix} \quad ; \quad A'_3 = \begin{bmatrix} 0 & & \\ 0 & \omega^4 t^3 & \\ \omega^3 t^3 & \omega^2 t^2 (tY_1 X_{2,3})^{1/2} & \omega t^2 X_{2,3} \end{bmatrix}$$

$$A_4 = \begin{bmatrix} 0 & & & \\ 0 & 0 & & \\ 0 & \omega^2 t^{3/2} & \omega t Y_1 X_1 & \\ t^{3/2} & t(Y_2 X_{0,2})^{1/2} & X_1 (t Y_2 X_{0,2})^{1/2} & X_{0,1,2} \end{bmatrix}$$

$$A'_4 = \begin{bmatrix} 0 & & & \\ 0 & 0 & & \\ 0 & \omega t^{5/2} & -\omega^2 t^2 Y_1 X_2 & \\ \omega^4 t^{5/2} & -\omega t^2 (Y_2 X_{1,3})^{1/2} & \omega^3 X_2 (t^3 Y_2 X_{1,3})^{1/2} & -t X_{1,2,3} \end{bmatrix}$$

$$A_5 = \begin{bmatrix} 0 & & & & \\ 0 & 0 & & & \\ 0 & 0 & \omega^4 t^2 & & \\ 0 & \omega^3 t^2 & \omega^2 t^{3/2} (Y_{1,2} X_{1,2})^{1/2} & Y_2 X_{1,2} \omega t & \\ t^2 & (t^3 Y_3 X_{0,3})^{1/2} & t (Y_1^{-1} Y_{2,3} X_{0,1,2,3})^{1/2} & X_{1,2} (t Y_3 X_{0,3})^{1/2} & X_{0,1,2,3} \end{bmatrix}$$

with

$$X_{s,t,\ldots} = X_s X_t \ldots .$$

$$Y_{s,t,\ldots} = Y_s Y_t \ldots .$$

and

$$X_\ell = 1 - \omega^\ell t \quad ; \qquad Y_\ell = \sum_{k=0}^{\ell} \omega^k \quad ;$$

where

$$\omega = \omega_N^{\pm 1} .$$

References

[1] Akutsu Y and Wadati M 1987 J. Phys. Soc. Jpn. **56**, 839-842.

[2] Akutsu Y and Wadati M 1987 J. Phys. Soc. Jpn. **56**, 3039.

[3] Akutsu Y, Deguchi T and Wadati M 1987 J. Phys. Soc. Jpn. **56**, 3464-3479.

[4] Deguchi T, Akutsu Y and Wadati M 1988 J. Phys. Soc. Jpn. **57**, 757-776.

[5] Akutsu Y, Deguchi T and Wadati M 1988 J. Phys. Soc. Jpn. **57**, 1173-1185.

[6] Kauffman LH 1988 *State Models for Link Polynomials.* Univ. of Illinois preprint.

[7] Witten E 1988 *Quantum Field Theory and the Jones Polynomial*, "Braid Group, Knot Theory, and Statistical Mechanics", *Advanced Series in Mathematical Physics, Vol.* **9**, *239-329, World Scientific*, C.N. Yang and M.L. Ge - Editors.

[8] Yamagishi K, Ge ML and Wu YS 1989 *New Hierarchies of Knot Polynomials from Topological Chern-Simons Gauge Theory*, Univ. of Utah preprint UU-HEP-89/1.

[9] Wilczek F 1982 Phys. Rev. Lett. **48**, 1144.

[10] Wu YS 1984 Phys. Rev. Lett. **52**, 2103.

[11] Wilczek F and Zee A 1983 Phys. Rev. Lett. **51**, 2250.

[12] Lee HC, Ge ML, Couture M and Wu YS 1989 Int. J. Mod. Phys. A, Vol. 4, 9, 2333.

[13] Fröhlich J 1987 *Statistics of Fields, the Yang-Baxter Equation and the Theory of Knots and Links*, in "Nonperturbative Quantum Field Theory" (Cargèse 1987). G.'t Hooft *et al. (eds.)*, Plenum Pub., 1988.

[14] Tsuchiya A and Kanie Y 1988 Adv. Studies Pure Math. **16**, 297.

[15] Khono T 1988 *Linear Representations of Braid Groups and Classical Yang-Baxter Equations*, Univ. of Nagoya (Japan) preprint.

[16] Lee HC and Couture M 1988 *A Method to Construct Closed Braids from Links and a New Polynomial for Connected Links*, Chalk River (Canada) preprint CRNL-TP-88-1118R.

[17] Lee HC, Couture M and Schmeing NC 1988. *Connected Link Polynomials*, Chalk River (Canada) preprint CRNL-TP-88-1125R.

[18] Artin E 1947 Ann. of Math. **48**, 101-126.

[19] Birman JS 1974 *Braids, Links and Mapping Class Groups*, Princeton Univ. Press.

[20] Fan C and Wu FY 1970 Phys. Rev. B**2**, 723

[21] Kauffman 1989 preprint - *Knots, abstract tensors and the Yang-Baxter equation.*

Link Polynomials and Solvable Models

Tetsuo Deguchi

Institute of Physics, College of Arts and Sciences, [1]
University of Tokyo, Komaba, Meguro-ku, Tokyo 153, Japan

Abstract

Through a general method we construct link polynomials from exactly solvable models in statistical mechanics. Various examples are explicitly shown. From the crossing symmetry we derive link polynomials with the graphical calculation. By use of transformations we obtain different link polynomials from a solvable model.

1 Introduction

The Yang-Baxter relation is a sufficient condition for the solvability of models in statistical mechanics and field theories such as 1-dimensional quantum spin chains, 2-dimensional lattice systems, many body systems in (1+1)-dimensions, etc.. [1,2,3,4,5,6,7] For various models this relation is written in terms of the operators $X_i(u)$ as [1,3,13,14],

$$\begin{aligned} X_i(u)X_{i+1}(u+v)X_i(v) &= X_{i+1}(v)X_i(u+v)X_{i+1}(u), \\ X_i(u)X_j(v) &= X_j(v)X_i(u), |i-j| \geq 2. \end{aligned} \tag{1}$$

[1]Address after April 1, 1990: Department of Physics, Faculty of Science, University of Tokyo, Hongo 7-3-1, Bunkyo-ku, Tokyo 113, Japan.

Physics, Geometry, and Topology
Edited by H. C. Lee
Plenum Press, New York, 1990

The Yang-Baxter relation in this form has an advantage that we can easily see connection of solvable models to the braid group.

Recently, the Yang-Baxter relation has been found to be a key to several fields in mathematical physics. Various link polynomials [8,9,10,11,12] and their extensions are obtained from exactly solvable models through a general method. [13,14,15,17,18,19,20,21,23,24,25,26, 27] The purpose of this paper is to show a general theory for construction of link polynomials from exactly solvable models in statistical mechanics.

The outline of this paper is given in the following. In §2, vertex models, IRF models and factorized S-matrices are introduced. In §3, the braid group and the mothod for construction of the representations are explained. In §4, link polynomials are constructed. The crossing symmetry is used for the graphical calculation of the link polynomials. In §5, some examples are shown. In §6, transformations of solvable models are explained. In §7 we give concluding remarks.

2 Exactly solvable models

2.1 Solvable models in statistical mechanics

Let us explain solvable models in two-dimensional statistical mechanics. [3,24] There are two types of solvable models, vertex models and IRF models. (Fig.1) Let us first consider vertex

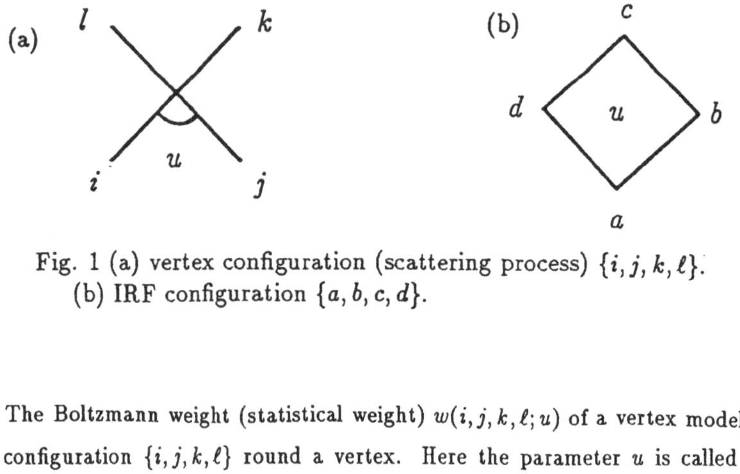

Fig. 1 (a) vertex configuration (scattering process) $\{i, j, k, \ell\}$.
(b) IRF configuration $\{a, b, c, d\}$.

models. The Boltzmann weight (statistical weight) $w(i, j, k, \ell; u)$ of a vertex model defined is for a configuration $\{i, j, k, \ell\}$ round a vertex. Here the parameter u is called spectral parameter which controls the anisotropy (and strength) of the interactions for the model.

The Yang-Baxter relation is a sufficient condition for the commutativity of the transfer matrices of the model. In this sense it gives the solvability of the model. There are various methods to calculate physical quantities (free energy, one-point function, etc.) for the solvable models, such as Bethe ansatz method, corner transfer method, inversion method, etc.. [3,28] Models whose Boltzmann weights (or matrix elements) satisfy the Yang-Baxter relation are called to be solvable. For vertex models the Yang-Baxter relation is given by

$$\begin{aligned}
&\sum_{abc} \quad w(b,c,q,r;u)w(a,k,p,c;u+v)w(i,j,a,b;v) \\
&=\sum_{abc} \quad w(a,b,p,q;v)w(i,c,a,r;u+v)w(j,k,b,c;u). \qquad (2)
\end{aligned}$$

Let us consider IRF models. The Boltzmann weight of an IRF model $w(a,b,c,d;u)$ is defined on a configuration $\{a,b,c,d\}$ round a face (Fig.1).

IRF models have constraints on the configurations. The symbol $b \sim a$ denotes that the "spin" b is admissible to the "spin" a under the constraint of the model. If the conditions $b \sim a, a \sim d, b \sim c$ and $c \sim d$ are all satisfied, then the configuration $\{a,b,c,d\}$ in Fig.1 is called to be allowed. The Boltzmann weights for not-allowed configurations are set to be 0. For IRF models the Yang-Baxter relation is written as

$$\begin{aligned}
&\sum_{c} w(b,d,c,a;u)w(d,e,f,c;u+v)w(c,f,g,a;v) \\
= &\sum_{c} w(d,e,c,b;v)w(b,c,g,a;u+v)w(c,e,f,g;u) \qquad (3)
\end{aligned}$$

The IRF configuration a,b,c,d in Fig.1 corresponds to the vertex configuration in Fig.1 by $i=a-d, j=b-a, k=b-c$ and $\ell=c-d$. We refer to this correspondence as Wu-Kadanoff-Wegner transformation [34,18]. In general, we can transform any (unrestricted) IRF model into a vertex model by taking the Wu-Kadanoff-Wegner transformation and taking a limit [18] which brings "the base point" ω_0 of the IRF spin states into infinity:$\omega_0 \to \infty$.

2.2 Factorized S-matrices

Let us introduce factorized S-matrices. We write the amplitude of the scattring process: $i \to k$, $j \to \ell$ as $S_{j\ell}^{ik}(u)$ (Fig.1), where u is the rapidity difference. In general, the "charge" variables i,j,k and ℓ of $S_{j\ell}^{ik}(u)$ take vector values (weight vectors). The factorized S-matrices represent the elastic scattering of particles where only the exchanges of momenta and the phase shifts occur. The rapidity difference of the scattering particles can be depicted by the angle in the diagram. It is known that factorized S-matrices are mathematically equivalent to corresponding solvable vertex models. [29]

When $S_{j\ell}^{ik}(u)$ is non-zero only for the case $i+j=k+\ell$, we say that the model has "charge conservation" property. [13,14,23,24]

The Yang-Baxter relation for the S-matrices reads as

$$\sum_{abc} S_{cr}^{bq}(u) S_{kc}^{ap}(u+v) S_{jb}^{ia}(v) = \sum_{abc} S_{bq}^{ap}(v) S_{cr}^{ia}(u+v) S_{kc}^{jb}(u). \tag{4}$$

This relation is often referred to as the factorization equation. [1,6,5,7]

2.3 Basic relations

The Boltzmann weights for most of solvable models satisfy the following basic relations in addition to the Yang-Baxter relation. [13,14,18,19,23,24] In this subsection we write the relations in terms of the factorized S-matrices.

1) standard initial condition

$$S_{j\ell}^{ik}(u=0) = \delta_{i\ell}\delta_{jk}. \tag{5}$$

2) inversion relation (unitarity condition)

$$\sum_{mp} S_{p\ell}^{mk}(u) S_{jm}^{ip}(-u) = \rho(u)\rho(-u)\delta_{i\ell}\delta_{jk}, \tag{6}$$

where $\rho(u)$ is a model-dependent function.

3) second inversion relation (second unitarity condition)

$$\sum_{pm} S_{p\ell}^{im}(\lambda-u) S_{mj}^{kp}(\lambda+u) \cdot \left(\frac{r(m)r(p)}{r(i)r(j)r(k)r(\ell)}\right)^{1/2} = \rho(u)\rho(-u)\delta_{ij}\delta_{k\ell}. \tag{7}$$

We call the parameter λ crossing parameter (crossing point) and $\{r(i)\}$ crossing multipliers.

4) crossing symmetry (Fig.2)

$$S_{j\ell}^{ik}(u) = S_{\bar{k}\bar{i}}^{j\ell}(\lambda-u)\left(\frac{r(i)r(\ell)}{r(j)r(k)}\right)^{\frac{1}{2}}, \tag{8}$$

Here, we have used the notation $\bar{j}$ for the "antiparticle" of j. We assume that $r(\bar{j}) = 1/r(j)$. Note that the second inversion relation and the crossing symmetry define the crossing multipliers.

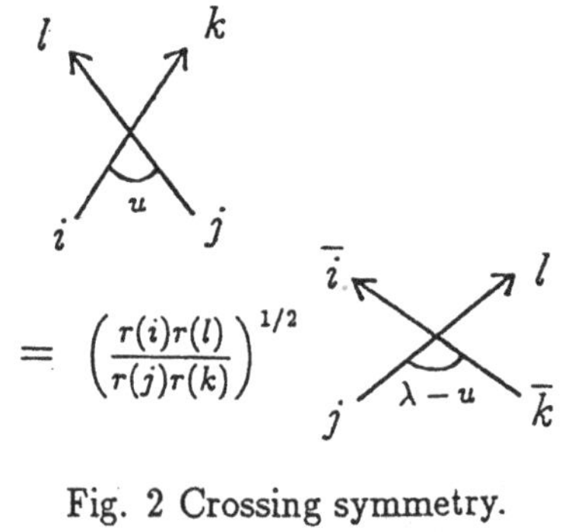

Fig. 2 Crossing symmetry.

The Boltzmann weights for most of IRF models satisfy the basic relations corresponding to (5)-(8). For example, the crossing symmetry is

$$w(a,b,c,d;u) = w(b,c,d,a;\lambda-u)\left(\frac{\psi(a)\psi(c)}{\psi(b)\psi(d)}\right)^{1/2}, \tag{9}$$

where $\{\psi(\ell)\}$ are the crossing multipliers for the IRF model. Crossing multipliers $\{\psi(\ell)\}$ for an IRF model are related to those for the corresponding vertex model by $r^2(j) = \psi(b)/\psi(a)$, when $j = b - a$ and $b \sim a$.

The above relations have the following physical meanings. [13,14,18,19,24] The standard initial condition indicates that there is no scattering between two particles with zero relative velocity. The crossing symmetry is a relation between s-channel and t-channel scatterings. For the 2-dimensional lattice systems the symmetry describes the invariance of the system under 90 degree rotation. Note that from the standard initial condition and the crossing symmetry, the inversion relation and the second inversion relation are derived. We shall see the basic relations and the Yang-Baxter relation are related to the local moves on link diagrams, known as the Reidemeister moves in knot theory.

2.4 Yang-Baxter operator

In order to see the connection of exactly solvable models to the braid group we introduce Yang-Baxter operator $X_i(u)$. [13,14,18,23,24] The operator is, in statistical mechanics, a unit constituent of the diagonal-to-diagonal transfer matrix. [3] For factorized S-matrices we define Yang-Baxter operator by

$$\begin{aligned} X_i(u) &= \sum_{abcd} S_{da}^{cb}(u) I^{(1)} \otimes \cdots \otimes e_{ac}^{(i)} \otimes e_{bd}^{(i+1)} \\ &\quad \otimes I^{(i+2)} \otimes \cdots \otimes I^{(n)}. \end{aligned} \tag{10}$$

Here $I^{(i)}$denotes the identity matrix and e_{ab} a matrix such that $(e_{ab})_{jk} = \delta_{ja}\delta_{kb}$. The Yang-Baxter operators $\{X_i(u)\}$ satisfy the following relations (Yang-Baxter algebra),

$$X_i(u)X_{i+1}(u+v)X_i(v) = X_{i+1}(v)X_i(u+v)X_{i+1}(u), \tag{11}$$

$$X_i(u)X_j(v) = X_j(v)X_i(u), \qquad |i-j| \geq 2. \tag{12}$$

In terms of the Yang-Baxter operators, the Yang-Baxter relation for vertex models and IRF models is written in the same form.

3 Braid group

3.1 Braids and closed braids

We introduce braids and the braid group. [30] The braid group B_n is defined by a set of generators, $b_1, \cdots, b_{n-1}$ which satisfy

$$b_i b_{i+1} b_i = b_{i+1} b_i b_{i+1},$$

$$b_i b_j = b_j b_i, \qquad |i-j| \geq 2. \tag{13}$$

It is known that any oriented link can be expressed by a closed braid. The equivalent braids expressing the same link are mutually transformed by a finite sequence of two types of operations, Markov moves I and II. The Markov trace $\phi(\cdot)$ is a linear functional on the representation of the braid group which have the following properties (the Markov properties):

$$I.\ \phi(AB) = \phi(BA), \qquad A, B \epsilon B_n, \tag{14}$$

$$\begin{aligned} II.\ \phi(Ab_n) &= \tau\phi(A), \\ \phi(Ab_n^{-1}) &= \bar{\tau}\phi(A), \\ & A\epsilon B_n, \quad b_n \epsilon B_{n+1}, \end{aligned} \tag{15}$$

where

$$\tau = \phi(b_i), \quad \bar{\tau} = \phi(b_i^{-1}), \text{ for all } i. \tag{16}$$

From the Markov trace we obtain a link polynomial $\alpha(\cdot)$ as [13,14,23,24,25]

$$\alpha(A) = (\tau\bar{\tau})^{-\frac{n-1}{2}}(\frac{\bar{\tau}}{\tau})^{\frac{1}{2}e(A)}\phi(A), A\epsilon B_n \tag{17}$$

Here $e(A)$ is the exponent sum of b_i's in the braid A, which is equivalent to the writhe of the link diagram. For instance, if $A = b_1^4 b_2^{-2} b_3 b_1^{-1}$, then $e(A) = 4 - 2 + 1 - 1 = 2$.

3.2 Construction of the braid operator

The braid operator $G(+)_i$, the inverse operator $G(-)_i$ and the identity I are given by [13,14]

$$G(\pm)_i = \lim_{u\to\infty} X_i(\pm u)/\rho(\pm u), \tag{18}$$

$$I = X_i(0). \tag{19}$$

The limit $u \to \infty$ (more precisely, an infinite limit in a certain direction in the complex u-plane) requires that the Boltzmann weights be parametrized by hyperbolic (trigonometric) functions. Hereafter we write the matrix elements of the braid operator as

$$G_{cd}^{ab}(\pm) = \lim_{u\to\infty} S_{da}^{cb}(\pm u)/\rho(\pm u). \tag{20}$$

Then we can express the braid operator (18) constructed from the Yang-Baxter operator as

$$G(\pm)_i = \Sigma_{abcd} G_{cd}^{ab}(\pm) I^{(1)} \otimes \cdots \otimes e_{ac}^{(i)} \otimes e_{bd}^{(i+1)} \otimes I^{(i+2)} \otimes \cdots \otimes I^{(n)}. \tag{21}$$

It is sometimes convenient to write the matrix elements of the braid operator

$$[G_i]_{b_1\cdots b_n}^{a_1\cdots a_n} = \prod_{j=1}^{i-1} \delta_{b_j}^{a_j} \cdot G_{b_i b_{i+1}}^{a_i a_{i+1}} \cdot \prod_{j=i+2}^{n} \delta_{b_j}^{a_j}, \tag{22}$$

where δ_b^a is the Kronecker delta. We can also construct braid operators for IRF models by the formula (18). [18,20,22,23,24,25]

4 Construction of link polynomials

4.1 Construction of the Markov trace

We shall obtain link polynomials by constructing the Markov trace on the representations of the braid group derived from the solvable models. The Markov trace takes the follwing form [13,14,23,24,25]

$$\phi(A) = \frac{\hat{T}r(H(n)A)}{\hat{T}r(H(n))}, A \epsilon B_n,$$
$$[H(n)]^{a_1 a_2 \cdots a_n}_{b_1 b_2 \cdots b_n} = \prod_{j=1}^{n} r^2(a_j)\delta^{a_j}_{b_j}. \tag{23}$$

For the models with the crossing symmetry (and the second inversion relation), $r(p)$ is nothing but the crossing multiplier of the model. We present sufficient conditions for the Markov properties explicitly. We can show that the trace $\phi(\cdot)$ defined in (23) is the Markov trace by proving for the Markov property I the "charge conservation" property and for the Markov property II the following conditions:

$$\Sigma_b G^{ab}_{ab}(\pm) r^2(b) = \chi(\pm) \qquad \text{(independent of } a). \tag{24}$$

The τ-factors are related to $\chi(\pm)$ as $\tau/\tau = \chi(-)/\chi(+)$.

We can prove the extended Markov property, [18,20,23,24,25] which is an extension of the Markov property with finite spectral parameter.

$$\sum_b X^{ab}_{ab}(u) h(b) = H(u;\eta)\rho(u) \qquad \text{(independent of } a), \tag{25}$$

where the function $H(u;\eta)$ is called characteristic function.

For IRF models we introduce a "constrained trace" $\tilde{T}r(A)$ [18,20,23,24,25]:

$$\tilde{T}r(A) = \tilde{\sum_{\ell_1 \ell_2 \cdots\cdots\ell_n}} A^{\ell_0 \ell_1 \cdots \ell_n}_{\ell_0 \ell_1 \cdots \ell_n} \frac{\psi(\ell_n)}{\psi(\ell_0)}, \qquad (\ell_0 : fixed) \tag{26}$$

where the symbol $\tilde{\Sigma}$ represents the summataion over admissible multi-indices $\ell_i : \ell_{i+1} \sim \ell_i$for $i = 0, \cdots, n-1$ with ℓ_0 being fixed. Then the Markov trace $\phi(\cdot)$ is written as

$$\phi(A) = \frac{\tilde{T}r(A)}{\tilde{T}r(I(n))}, \qquad A \epsilon B_n, \tag{27}$$

where $I(n)$ is the "identity" operator for n strings.

We can prove the extended Markov property also for IRF models. [18,20,23,24,25]

4.2 Graphical calculation

The crossing symmetry is significant in algebraic and graphical aspects of the knot theory. For solvable (vertex and IRF) models with the crossing symmetry, the Yang-Baxter operator becomes the Temperley-Lieb operator at the point $u = \lambda$. [19] In fact, setting

$$E_i = X_i(\lambda), \tag{28}$$

we find that the operators $\{E_i\}$ satisfy the following relations [31]

$$\begin{aligned} E_i E_{i\pm 1} E_i &= E_i, \\ E_i^2 &= q^{\frac{1}{2}} E_i, \\ E_i E_j &= E_j E_i, \quad |i-j| \geq 2, \end{aligned} \tag{29}$$

where the quantity $q^{1/2}$is related to the crossing multipliers $r(a)$ (or $\psi(i)$) by [13,14,18,19]

$$q^{\frac{1}{2}} = \sum_j r^2(j), \qquad \text{for S-matrix (vertex model),} \tag{30}$$

$$= \sum_{b\sim a} \frac{\psi(b)}{\psi(a)}, \qquad \text{for IRF model,} \tag{31}$$

where in (31) the summation is over all states b allowable to a. The relations (29) are the defining relations of the Temperley-Lieb algebra.

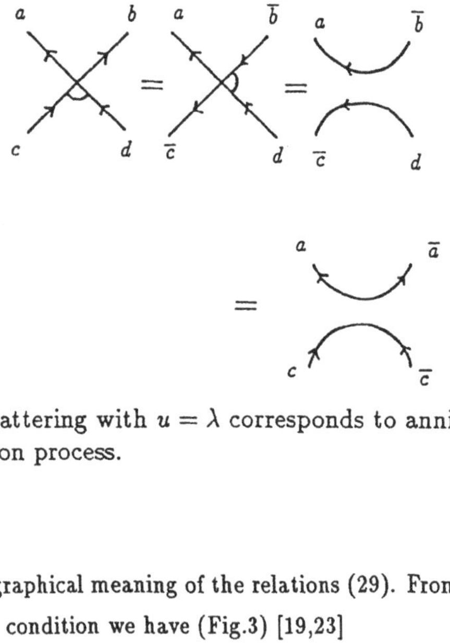

Fig. 3 Scattering with $u = \lambda$ corresponds to annihilation-creation process.

Let us consider the graphical meaning of the relations (29). From the crossing symmetry and the standard initial condition we have (Fig.3) [19,23]

$$\begin{aligned} S_{da}^{cb}(\lambda) &= \left(\frac{r(a)r(c)}{r(b)r(d)}\right)^{\frac{1}{2}} S_{b\bar{c}}^{da}(0) \\ &= r(a)\delta(a,\bar{b}) \cdot r(c)\delta(c,\bar{d}), \end{aligned} \tag{32}$$

where $\delta(a,c) = \delta_{ac}$ is the Kronecker delta. We can regard the elements $r(c)\ \delta(c,\bar{d})$ and $r(a)\ \delta(a,\bar{b})$ as the weights for the pair-annihilation diagram and the pair-creation diagram, respectively(Fig.4). Then, the Yang-Baxter operator at $u = \lambda$ is depicted as the monoid diagram, by which the Temperley-Lieb algebra is explained. This interpretation is consistent with a fact that the energy at the point λ is related to the pair-creation energy.

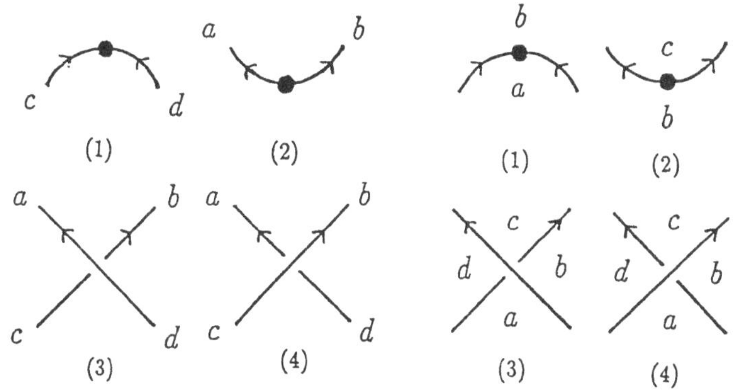

Fig. 4 Elements of link diagram.
(1) pair-annihilation diagram: $r(c)\delta_{c,d}$; $(\psi(a)/\psi(b))^{1/2}$.
(2) pair-creation diagram: $r(a)\delta_{a,b}$; $(\psi(c)/\psi(b))^{1/2}$.
(3) braid diagram with $\epsilon = -1$: $G^{ab}_{cd}(+)$; $G(a, b, c, d; +)$.
(4) braid diagram with $\epsilon = 1$: $G^{ab}_{cd}(-)$; $G(a, b, c, d; -)$.

For IRF models, the weights $\{\psi(a)/\psi(b)\}^{1/2}$ and $\{\psi(c)/\psi(b)\}^{1/2}$ correspond to the pair-annihilation and pair-creation diagrams, respectively (Fig.4).

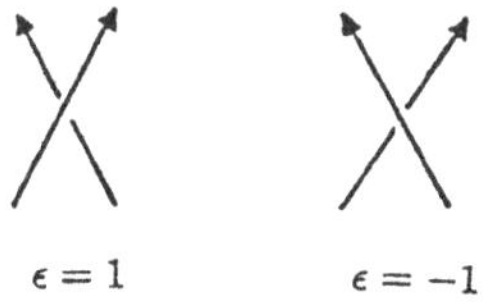

Fig. 5 Sign $\epsilon(C)$.

We can formulate link polynomials with the crossing symmetry directly on link diagrams. Link diagram $\hat{L}$ is a 2-dimensional projection of a link L. The writhe $w(\hat{L})$ is the sum of signs for all crossings C_i in the link diagram (Fig.5):

$$w(\hat{L}) = \Sigma_{C_i}\epsilon(C_i), \tag{33}$$

We calculate statistical sum $Tr(\hat{L})$ on the diagram $\hat{L}$ by the rules given in Fig.4. The link polynomial for the link L is calculated as

$$\alpha(L) = c^{-w(\hat{L})}\frac{Tr(\hat{L})}{Tr(K_0)}, \tag{34}$$

where $\hat{K}_0$ is the trivial knot diagram (a loop) and the constant c is defined by a relation

$$G_i E_i = c E_i, \tag{35}$$

or by

$$c = \left(\frac{\chi(-)}{\chi(+)}\right)^{\frac{1}{2}}. \tag{36}$$

It is easy to see that $\alpha(L)$ is invariant under the Reidemeister moves (Fig.6), and therefore $\alpha(L)$ is a topological invariant of the link L.

I

II

III

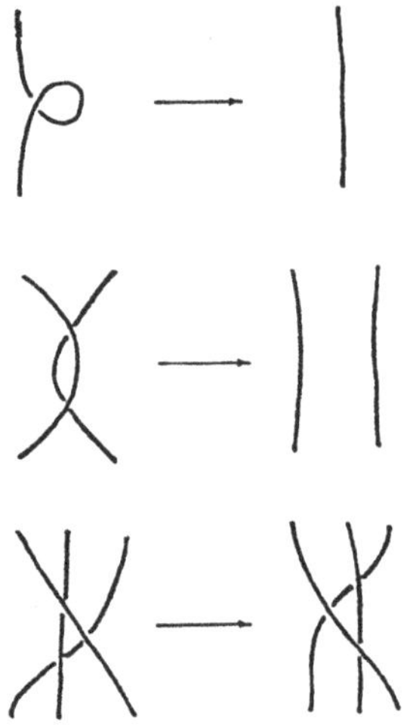

Fig. 6 Reidemeister moves.

Thus we have shown that the link polynomials constructed from solvable models with the crossing symmetry can be graphically formulated. The monoid diagram and the weights for the creation and annihilation diagrams were used by L.H. Kauffman for the Bracket polynomial which gives a graphical calculation of the Jones polynomial. [32] The graphical calculation is named "state model". We have remarks. The graphical formulation applied to closed braids yields the Markov trace (Fig.7). For the link polynomials with the crossing symmetry, the formulation based on the Markov trace is equivalent to the graphical formulation. This viewpoint is consistent with the braid-plat correspondence [33].

The link diagrams are considered as the Feynman diagrams for the high energy processes of charged particles and the link polynomials as the scattering amplitudes. At the lowest point in the diagram there occurs a pair creation and at the highest point a pair annihilation. Further, if we regard the link diagrams as distorted 2-dimensional lattices, the link polynomials are considered as the partition functions.

To conclude this section, we put emphasis on the fact that the crossing symmetry has the algebraic and graphical meanings. Algebraically, the symmetry leads to the Temperley-Lieb algebra (and the braid-monoid algebra).

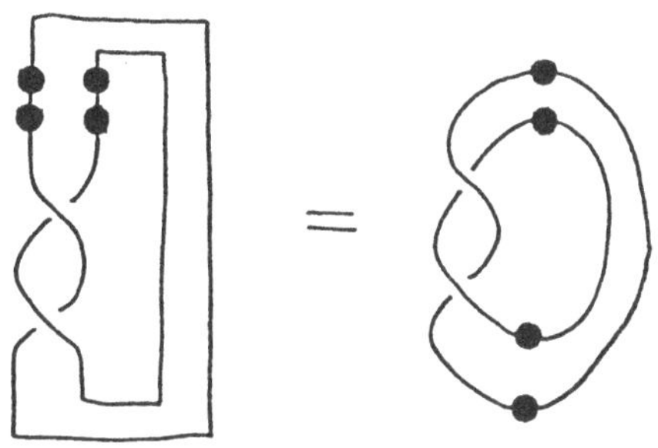

Fig. 7 Equivalence of the Markov trace and the graphical calculation.

Graphically, the pair-creation and pair-annihilation diagrams are introduced through the crossing symmetry.

5 Various Examples

5.1 N-state vertex model

From the N-state vertex models a hierarchy of link polynomials are obtained by the general method presented in §3 and §4. [13,14] The model corresponds to the factorized S-matrices with spin s particles, where $N = 2s + 1$. For the case $N = 3$, there are 19 vertex configurations. [35] The Boltzmann weights of the N-state vertex model can be systematically calculated by using recursion relations. [34] Therefore, an algorithm for construction of the hierarchy of link polynomials has been established. [13,14]

From the N-state vertex model (asymmetrized by the symmetry breaking transformation) we get the braid operator which satisfies an N-th order relation: [13,14]

$$(G_i - C_1)(G_i - C_2)\cdots(G_i - C_N) = 0 \tag{37}$$

where for $j = 1, 2, \cdots, N$

$$C_j = (-1)^{j+N} t^{\frac{1}{2}N(N-1)-\frac{1}{2}j(j-1)}, \quad t = e^{2\lambda}. \tag{38}$$

We call a relation for G_i such as the relation (37) reduction relation of the braid operator. The crossing multiplier for the asymmetrized N-state vertex model is [13,14]

$$r(k) = e^{-\lambda k} = t^{-k/2}, \qquad k = -s, -s+1, \cdots, s, \tag{39}$$

where

$$s = (N-1)/2. \tag{40}$$

The extended Markov property [18,24] is satisfied with the characteristic function given as [23,24]

$$H(u;\lambda) = \frac{\sinh(N\lambda - u)}{\sinh(\lambda - u)}. \tag{41}$$

The constants τ and $\bar{\tau}$ are

$$\tau = 1/(1+t+\cdots+t^{N-1}), \tag{42}$$

$$\bar{\tau} = t^{N-1}/(1+t+\cdots+t^{N-1}). \tag{43}$$

It is remarkable that there exists an infinite sequence of link polynomials corresponding to the N-state vertex models ($N = 2,3,4,5,\cdots$). [13,14,24] The $N = 2$ case corresponds to the Jones polynomial. [9] In the $N \geq 3$ cases we have new link polynomials. From the reduction relation, we obtain the skein relations (the Alexander-Conway relations) for the link polynomials:

$$\alpha(L_+) = (1-t)t^{\frac{1}{2}}\alpha(L_0) + t^2\alpha(L_-), \quad (N=2) \tag{44}$$

$$\begin{aligned} \alpha(L_{2+}) &= t(1-t^2+t^3)\alpha(L_+) + (t^4-t^5+t^7)\alpha(L_0) \\ &\quad -t^8\alpha(L_-), \qquad (N=3) \end{aligned} \tag{45}$$

$$\begin{aligned} \alpha(L_{3+}) &= t^{3/2}(1-t^3+t^5-t^6)\alpha(L_{2+}) + t^6(1-t^2+t^3+t^5-t^6+t^8)\alpha(L_+) \\ &\quad +t^{25/2}(-1+t-t^3+t^6)\alpha(L_0) - t^{20}\alpha(L_-), \qquad (N=4). \end{aligned} \tag{46}$$

In (44), by L_+, L_0 and L_- we have denoted links which have the configuration of b_i, b_i^0 and b_i^{-1}, at an intersection. Similarly, L_{2+}, L_+, L_0 and L_- in (45) and L_{3+}, L_{2+}, L_+, L_0 and L_- in (46) should be understood.

We can also present a general expression for the braid matrix derived from the N-state vertex model. The symbol $\sigma^{(N,c)}$ denotes the charge submatrix acting in the sector of the total charge c ($c = i+j = k+\ell$).

$$\begin{aligned} (\sigma^{(N,c)})_{mn} &= (-1)^{m+n}\left(Q_{n-1,N-|c|-m}(t^{|c|})Q_{m-1,N-|c|-n}(t^{|c|})\right)^{\frac{1}{2}}, \\ &\quad \text{for } m,n = 1,\cdots,N-|c|, \end{aligned} \tag{47}$$

where

$$Q_{mn}(z) = \frac{(t;m)(tz;m)}{(t;m-n)(t;n)(tz;n)} z^n t^{n^2}, \tag{48}$$

$$\begin{aligned} (z;n) &= (1-z)(1-zt)\cdots(1-zt^{n-1}) \text{ for } n \geq 1, \\ &= 1 \text{ for } n = 0, \\ &= \infty \text{ for } n \leq -1. \end{aligned} \tag{49}$$

The general expression of the braid matrix was obtained in the following way. We first obtained recursively the general expression of the regular representation matrices [17] of the composite braid operator in the composite string representation (the operator is the composite operator constructed from the generators of the Hecke algebra.). By comparing the regular representation matrices with the braid matrices derived from the (asymmetrized) N-state vertex model, we found the expression of the matrix elements. We can also check (43) from the knowledge of the composite Yang-Baxter operator.

5.2 Graph state IRF model

We can construct solvable IRF models corresponding to arbitrary graphs in any dimensions. [36,18] Let us express the constraint of the model by a graph. In the graph each point represents the spin state. When a spin c is admissible to d then the point for c is connected to the point for d. For ADE type graphs the models are called ADE models. [37] There also exist solvable models with elliptic parametrization for extended Dynkin diagrams [36,38].

Let us construct the graph state IRF models. [18] We solve the eigenvalue equation for the graph

$$\sum_{b\sim a} \psi(b) = \Lambda\psi(a), \tag{50}$$

where the summation is over all spin state b admissible to a. For example square lattice graph we have

$$\psi(\vec{a}) = \sin(\vec{a}\cdot\vec{n} + \omega_0), \tag{51}$$

where $\vec{a} = (a_1, a_2)$ and $\vec{n} = (n_1, n_2)$. Constructing the Temperley-Lieb operator

$$[E_i]^{p_1\cdots p_n}_{k_1\cdots k_n} = \prod_{j=0}^{i-1} \delta^{k_{i+1}}_{k_{i-1}} \frac{\psi(p_i)\psi(k_i)}{\psi(p_{i-1})} \prod_{j=i+1}^{n} \delta^{p_j}_{k_j}. \tag{52}$$

we have the Yang-Baxter operartor

$$X_i(u) = \frac{\sinh(\lambda - u)}{\sinh(\lambda)}\left(I + \frac{\sinh u}{\sinh(\lambda - u)}E_i\right). \tag{53}$$

From the models we have braid operator by taking the limit $u \to \infty$ and the Markov trace on the braid group representation by using the crossing multipliers. The link polynomial satifies the second degree skein relation.

We can consider vertex models corresponding to the graph state IRF models under the Wu-Kadanoff-Wegner transformation and the base-point-infinity limit . [27] We call them vertex models in TL class. [18,19,27] From these vertex and IRF models we have multi-variable braid matrices.[27]

5.3 ABCD IRF models

The IRF model corresponding to affine Lie algebra $A^{(1)}_{m-1}$ ($B^{(1)}_m$, $C^{(1)}_m$, $D^{(1)}_m$) is called $A^{(1)}_{m-1}$ ($B^{(1)}_m$, $C^{(1)}_m$, $D^{(1)}_m$) model. [39] The crossing parameter λ and the sign factor σ are defined as

$$\lambda \;=\; m\omega/2, \quad \sigma = 1 \text{ for } A^{(1)}_{m-1}, \tag{54}$$

$$\lambda = (2m-1)\omega/2,\ \sigma = 1 \text{ for } B_m^{(1)}, \tag{55}$$

$$\lambda = (m+1)\omega,\ \sigma = -1 \text{ for } C_m^{(1)}, \tag{56}$$

$$\lambda = (m-1)\omega,\ \sigma = 1 \text{ for } D_m^{(1)}, \tag{57}$$

where ω is a parameter. The reduction relations are

$$(G_i - 1)(G_i + \gamma^2) = 0 \text{ for } A_{m-1}^{(1)}, \tag{58}$$

$$(G_i - 1)(G_i - \beta)(G_i + \gamma^2) = 0 \text{ for } B_m^{(1)}, C_m^{(1)} \text{ and } D_m^{(1)}, \tag{59}$$

with

$$\gamma = e^{-i\omega} \text{ for } A_{m-1}^{(1)},\ B_m^{(1)},\ C_m^{(1)} \text{ and } D_m^{(1)}, \tag{60}$$

$$\beta = \sigma e^{-i[2\lambda+\omega(1+\sigma)]} \text{ for } B_m^{(1)},\ C_m^{(1)} \text{ and } D_m^{(1)}. \tag{61}$$

The extended Markov property is proved and the characteristic functions are calculated as

$$H(u) = \frac{\sin(m\omega - u)}{\sin(\omega - u)} \text{ for } A_{m-1}^{(1)}, \tag{62}$$

$$H(u) = \frac{\sigma \sin(2\lambda - u)\sin(\sigma\omega + \lambda - u)}{\sin(\lambda - u)\sin(\omega - u)} \text{ for } B_m^{(1)},\ C_m^{(1)} \text{ and } D_m^{(1)}, \tag{63}$$

(The explicit forms of the crossing multipliers are given in [17]). Using the reduction relations and the Markov traces, we obtain the (generalized) skein relations:

$$\alpha(L_+) = (1-t)t^{(m-1)/2}\alpha(L_0) + t^m \alpha(L_-) \text{ for } A_{m-1}^{(1)}, \tag{64}$$

$$\begin{aligned}\alpha(L_{2+}) = {} & (1-t+\beta)e^{-i(2\lambda+\omega(\sigma-1))} \cdot \alpha(L_+) \\ & +(t+\beta t-\beta)e^{-2i(2\lambda+\omega(\sigma-1))} \cdot \alpha(L_0) \\ & -t\beta e^{-3i(2\lambda+\omega(\sigma-1))} \cdot \alpha(L_-), \\ & \text{for } B_m^{(1)},\ C_m^{(1)} \text{ and } D_m^{(1)},\end{aligned} \tag{65}$$

where

$$t = e^{-2i\omega}. \tag{66}$$

For $A_{m-1}^{(1)}$ model, the Alexander polynomail is obtained by the limit $m \to 0$, while $m = 2$ corresponds to the Jones polynomial.

Link polynomials thus obtained are one-variable invariants for each fixed m. It is noted that m is independent of t. We now have two variables t and m. The link polynomial constructed from $A_{m-1}^{(1)}$ model corresponds to the two-variable extension [10,11] of the Jones polynomial. The link polynomails from $B_m^{(1)}$, $C_m^{(1)}$, $D_m^{(1)}$ models correspond to the Kauffman polynomial [12]. We thus have explicit realizations of the Kauffman polynomial and the two-variable extension of the Jones polynomial (HOMFLY polynomial). The braid matrices constructed by Turaev [40,41] correspond to the vertex-model analog of the present braid matrices constructed from $A_{m-1}^{(1)}$, $B_m^{(1)}$, $C_m^{(1)}$, $D_m^{(1)}$ IRF models. From the IRF models we can construct braid matrices and the Markov trace for the vertex models by the Wu-Kadanoff-Wegner transformation and the base-point-infinity limit. [18] For example, from A-type IRF models we derive the multi-state vertex models [42] related to SU(n). From the Markov trace [20] for the IRF model we have that [40] for the vertex model.

6 Transformations of solvable models

6.1 Transformations

The solvable models with charge conservation condition are invariant under several transformations. [34,24,27] We introduce symmetry breaking transformations (or gauge transformations) [34] for factorized S-matrices (equivalently, for the Boltzmann weights of the vertex models):

$$S_{j\ell}^{ik}(u) \to \tilde{S}_{j\ell}^{ik}(u) = \alpha_{ij,k\ell}(u)\beta_{ij,k\ell}\gamma_{ij,k\ell}\delta_{ij,k\ell}S_{j\ell}^{ik}(u), \tag{67}$$

$$\alpha_{ij,k\ell}(u) = \exp[\vec{\mu}\cdot(\vec{k}-\vec{i}-\vec{\ell}+\vec{j})u] \tag{68}$$

$$\beta_{ij,k\ell} = \exp[\vec{\nu}\cdot(\vec{\ell}-\vec{i}-\vec{k}+\vec{j})], \tag{69}$$

$$\gamma_{ij,k\ell} = \exp[\omega(\vec{k}\cdot\vec{\ell}-\vec{i}\cdot\vec{j})], \tag{70}$$

$$\delta_{ij,k\ell} = \exp[\pi\sqrt{-1}(\vec{j}+\vec{k})\cdot\vec{\epsilon}], \tag{71}$$

where ω is a free parameter, $\vec{\mu}$ and $\vec{\nu}$ are arbitrary vectors, and $\vec{\epsilon}$ is a vector such that $(\vec{j}+\vec{k})\cdot\vec{\epsilon}$ is an integer for any weight vectors $\vec{j}$ and $\vec{k}$. Using these transformations C, P and T invariances for the S-matrices can be broken. There are symmetry breaking transformations for IRF models corresponding to those for S-matrices. [18,23,24]

The transformed matrix elements satisfy the crossing symmetry with

$$\tilde{r}(\vec{k}) = r(\vec{k})\exp[-2\vec{k}\cdot\vec{\mu}\lambda]. \tag{72}$$

Under the transformations the standard initial condition and the first inversion relation are invariant. The second inversion relation holds for the transformed matrix elements with the crossing multipliers modified as (72).

6.2 Deformation of solvable models

We shall show that the symmetry breaking transformation changes algebraic structure of the Yang-Baxter operator. Let us take the 6-vertex model. We set $\beta=\gamma=\delta=1$ and consider only the transformation $\alpha_{ij,k\ell}(u)$ in the following discussion. We assume that the vector $\vec{\mu}$ is parallel to the weight vectors and write it simply as μ. The Boltzmann weights of the 6-vertex model are given by

$$S_{1/2\ 1/2}^{1/2\ 1/2}(u) = S_{-1/2\ -1/2}^{-1/2\ -1/2}(u) = \frac{\sinh(\lambda-u)}{\sinh\lambda}, \tag{73}$$

$$S_{-1/2\ 1/2}^{1/2\ -1/2}(u) = S_{1/2\ -1/2}^{-1/2\ 1/2}(u) = 1, \tag{74}$$

$$S_{1/2\ 1/2}^{-1/2\ -1/2}(u) = S_{-1/2\ -1/2}^{1/2\ 1/2}(u) = \frac{\sinh u}{\sinh\lambda}. \tag{75}$$

They satisfy the standard initial condition and have the crossing symmetry with the trivial crossing multiplier: $r(j) = 1$ for $j = \pm 1/2$. We see that the Yang-Baxter operator of the 6-vertex model satisfies a cubic relation:

$$\left(X_i(u) - \frac{\sinh(\lambda - u)}{\sinh\lambda} I\right) \quad \left(X_i(u) - (1 - \frac{\sinh u}{\sinh\lambda}) I\right) \times \left(X_i(u) - (1 + \frac{\sinh u}{\sinh\lambda}) I\right) = 0. \tag{76}$$

If we apply the symmetry breaking transformation $\alpha_{ij,k\ell}(u)$ with $\mu = \pm 1/2$ to the 6-vertex model, then we have an "asymmetrized" 6-vertex model with the nontrivial crossing multipliers:

$$\tilde{r}(k) = \exp(-k\lambda), \quad k = \pm\frac{1}{2}, \text{ for } \mu = \frac{1}{2}, \tag{77}$$

$$\tilde{r}(k) = \exp(k\lambda), \quad k = \pm\frac{1}{2}, \text{ for } \mu = -\frac{1}{2}, \tag{78}$$

The transformed Yang-Baxter operator $\tilde{X}_i(u)$ satisfies a quadratic relation:

$$\left(\tilde{X}_i(u) - \frac{\sinh(\lambda - u)}{\sinh\lambda} I\right) \left(\tilde{X}_i(u) - \frac{\sinh(\lambda + u)}{\sinh\lambda} I\right) = 0. \tag{79}$$

Further, we can decompose the Yang-Baxter operator $\tilde{X}_i(u)$ as

$$\tilde{X}_i(u) = \rho(u)(I + f(u)E_i), \tag{80}$$

where

$$\begin{aligned} \rho(u) &= \frac{\sinh(\lambda - u)}{\sinh\lambda}, \\ f(u) &= \frac{\sinh\lambda}{\sinh(\lambda - u)}, \end{aligned} \tag{81}$$

and E_i is the Temperley-Lieb operator. For the symmetry breaking transformation $\alpha_{ij,k\ell}(u)$ with arbitrary μ, we find that the transformed Yang-Baxter operator $\hat{X}_i(u)$ satisfies a cubic relation. The Yang-Baxter operator satisfies a quadratic relation only when $\mu = \pm 1/2$.

By changing the value of the parameter μ, we get different representations of the braid group from the Yang-Baxter operator. Hereafter in this sub-section, we set

$$t = \exp(2\lambda). \tag{82}$$

The braid matrices are given in the following. (i) $\mu = \frac{1}{2}$

$$G^{ab}_{cd}(+) = \begin{pmatrix} 1 & 0 & 0 & 0 \\ 0 & 0 & -t^{\frac{1}{2}} & 0 \\ 0 & -t^{\frac{1}{2}} & 1-t & 0 \\ 0 & 0 & 0 & 1 \end{pmatrix}. \tag{83}$$

(ii) $-\frac{1}{2} < \mu < \frac{1}{2}$

$$G^{ab}_{cd}(+) = \begin{pmatrix} 1 & 0 & 0 & 0 \\ 0 & 0 & -t^{\frac{1}{2}} & 0 \\ 0 & -t^{\frac{1}{2}} & 0 & 0 \\ 0 & 0 & 0 & 1 \end{pmatrix}. \tag{84}$$

(iii) $\mu = -\frac{1}{2}$

$$G^{ab}_{cd}(+) = \begin{pmatrix} 1 & 0 & 0 & 0 \\ 0 & 1-t & -t^{\frac{1}{2}} & 0 \\ 0 & -t^{\frac{1}{2}} & 0 & 0 \\ 0 & 0 & 0 & 1 \end{pmatrix}. \tag{85}$$

The braid matrices for the cases (i) and (iii) are equivalent if we interchange up-spin and down-spin. They have the Markov traces. It is remarked that they satisfy the defining relations of the Hecke algebra [9]:

$$\begin{aligned} G_i G_{i+1} G_i &= G_{i+1} G_i G_{i+1}, \\ G_i G_j &= G_j G_i, \text{ for } |i-j| \geq 2, \\ G_i^2 &= (1-t)G_i + tI. \end{aligned} \tag{86}$$

The operator G_i can be decomposed into the Temperley-Lieb operator as [9]

$$G_i = I - t^{\frac{1}{2}} E_i. \tag{87}$$

In the case (ii) , the operator G_i satisfies a cubic relation:

$$(G_i - I)(G_i^2 - tI) = 0. \tag{88}$$

6.3 6-vertex model and link polynomials

By making use of the general method presented we shall construct the Markov trace for the braid group representations derived from the 6-vertex model. From the asymmetrized 6-vertex model with $\mu = \pm 1/2$, we obtain the Jones polynomial by using the transformed crossing multipliers in the trace. The Jones polynomial has a quadratic skein relation corresponding to the quadratic reduction relation.

For the symmetric 6-vertex model with $\mu = 0$, the crossing multiplier is equal to 1 and then the Markov trace is a trace on the braid matrix (the case (ii)) with the trivial matrix just ($H = I$). We thus obtain a link polynomial with a cubic skein relation:

$$\alpha(L_{3+}) = \alpha(L_{2+}) + t\alpha(L_{+}) - t\alpha(L_0), \tag{89}$$

where L_{n+} is the link which has n twist at a crossing point in the link diagram. This link polynomial is also obtained by soving directly the defining relation of the braid group with the assumption that the braid matrix does not satisfy the charge conservation condition. [43]

The link polynomial (89) for a link with two strings is determined by the linking number of the link. Thus, from the 6-vertex model we obtain two different link polynomials, the Jones polynomial and the link polynomial related to the linking number. [27] It is remarked there are many multi-variable link polynomials related to the linking number. [27]

We have a comment. In the case of the symmetric 6-vertex model, the Markov trace is given by the ordinary trace: $\phi(A) = Tr(A)$. Therefore the partition function for the 6-vertex model on a lattice automatically becomes the Markov trace and also the link polynomial.

7 Concluding Remarks

We have shown that various link polynomials are systematically constructed from exactly solvable models.

The existence and properties of the link polynomials [13,14] constructed from the N-state vertex model [34] can be proved also by the construction of composite models (fusion method) in terms of the Temperley-Lieb algebra and the graphical formulation derived from the crossing symmetry. [19] Note that the combination of the crossing symmetry and the Temperley-Lieb algebra characterize the link polynomials.

We can construct composite solvable models from the graph-state IRF models and vertex models in TL class. [18,19,27] From these composite models we obtain the link polynomials constructed from the N-state vertex models. [19,27]

Due to the limited space we have omitted the discussion for construction of two-variable link invariants [16,17,23,24,25] which may be regarded as two-variable extension of the link polynomials constructed from A type composite vertex and IRF models. In the papers [16,17] an algorithm for calculation of the two-variable link invariants for any links has been established, and some examples have been given.

Some class of braid matrices obtained from vertex models related to Lie algebras can be reconstructed by using the knowledge of q-analogue of universal enveloping algebra of the Lie algebra. [41,44] For example, the matrix elements of the braid operator obtained from the N-state vertex model are also calculated by using the knowledge of $su(2)$ [45].

Recently, there are some attempts to obtain braid matrices by solving directly the defining relation of the braid group. [43,46] Connection of these braid matrices to solvable models is an interesting problem.

There are several problems in physics related to the braid group. [47,48,49,50,51] Interestingly, solvable models and conformal field theories share many mathematically similar points in common. [52,53,54,55,56,57,58] Through the fusion rule, mathematical structures analogous to IRF models appear in conformal field theories. [58,59,60] It may be instructive to compare the viewpoints of field thoeries and statistical mechanics.

It seems that there are many interesting problems concerning applications of link polynomials to physics, chemistry and biology. We hope that the knowledge exhibited in this paper will be helpfull for studying those applications of the link polynomials.

Acknowledgements

The author would like to express his sincere thanks to Prof. M. Wadati and Dr. Y. Akutsu for continuous encouragements, critical reading of the manuscript and fruitful collaborations on which this paper is based. He would also like to thank Prof. H.C. Lee for inviting him to NATO Advanced Study Institute, Banff, August 13 ~ 25, 1989.

References

[1] C.N. Yang: Phys. Rev. Lett. **19** (1967) 1312.

[2] R.J. Baxter: Ann. of Phys. **70** (1972) 323.

[3] R.J. Baxter: *Exactly Solved Models in Statistical Mechanics* (Academic Press, 1982).

[4] L.A. Takhtadzhan and L.D. Faddeev, Russian Math. Surveys **34** (1979) 11.

[5] A. B. Zamolodchikov and A.B. Zamolodchikov, Ann. of Phys. **120** (1979) 253.

[6] M. Karowski, H.J. Thun, T.T. Truong and P.H. Weisz: Phys. Lett. **67B** (1977) 321.

[7] K. Sogo, M. Uchinami, A. Nakamura and M. Wadati: Prog. Theor. Phys. **66** (1981) 1284.

[8] J.W. Alexander, Trans. Amer. Math. Soc. **30** (1928) 275.

[9] V.F.R. Jones: Bull. Amer. Math. Soc. **12** (1985) 103.

[10] P. Freyd, D. Yetter, J. Hoste, W.B.R. Lickorish, K. Millett and A. Ocneanu: Bull. Amer. Math. Soc. **12** (1985) 239.

[11] J.H. Przytycki and K.P. Traczyk: Kobe J. Math. **4** (1987) 115.

[12] L.H. Kauffman, *On Knots* (Princeton University Press, 1987);
Trans. Amer. Math. Soc. (to appear).

[13] Y. Akutsu and M. Wadati: J. Phys. Soc. Jpn. **56** (1987) 839.

[14] Y. Akutsu and M. Wadati: J. Phys. Soc. Jpn. **56** (1987) 3039.

[15] Y. Akutsu, T. Deguchi and M. Wadati: J. Phys. Soc. Jpn. **56** (1987) 3464.

[16] Y. Akutsu and M. Wadati: Commun. Math. Phys. **117** (1988) 243.

[17] T. Deguchi, Y. Akutsu and M. Wadati: J. Phys. Soc. Jpn. **57** (1988) 757.

[18] Y. Akutsu, T. Deguchi and M. Wadati: J. Phys. Soc. Jpn. **57** (1988) 1173.

[19] T. Deguchi, M. Wadati and Y. Akutsu: J. Phys. Soc. Jpn. **57** (1988) 1905.

[20] T. Deguchi, M. Wadati and Y. Akutsu: J. Phys. Soc. Jpn. **57** (1988) 2921.

[21] M. Wadati and Y. Akutsu: Prog. Theor. Phys. Suppl. **94** (1988) 1.

[22] M. Wadati, T. Deguchi and Y. Akutsu: in *Nonlinear Evolution Equations, Integrability and Spectral Methods*, ed. A. Fordy (Manchester University Press, 1989).

[23] Y. Akutsu, T. Deguchi and M. Wadati: in *Braid Group, Knot Theory and Statistical Mechanics*, ed. C.N. Yang and M.L. Ge (World Scientific Pub., 1989) p. 151.

[24] M. Wadati, T. Deguchi and Y. Akutsu: Phys. Reports **180** (1989) 427.

[25] T. Deguchi, M. Wadati and Y. Akutsu: Adv. Stud. in pure Math. **19** (1989), Kinokuniya-Academic Press, p. 193.

[26] M.Wadati, Y. Akutsu and T. Deguchi, Link Polynomials and Exactly Solvable Models, to appear in *Lecture Notes in Mathematics* (Springer-Verlag, Berlin, Heidelberg, 1989).

[27] T. Deguchi : Braids, Link Polynomials and Transformations of Solvable Models, preprint UT-Komaba 89-11 July 1989.

[28] H.J. de Vega, Adv. Stud. in pure Math. **19** (1989), Kinokuniya-Academic Press, p. 567.

[29] A.B. Zamolodchikov, Commun. Math. Phys. **69** (1979) 165.

[30] J.S. Birman: *Braids, Links and Mapping Class Groups* (Princeton University Press, 1974).

[31] Temperley H.N.V. and Lieb E.H., 1971 Proc. Roy. Soc. London **A322** 251.

[32] L. H. Kauffman, Statistical Mechanics and The Jones polynomial, preprint 1987 (to appear in Proceedings of 1986 Santa Cruz Conference on the Artin Braid Group); State Models for Link Polynomials, IHES/M/88/46 Septembre 1988, preprint.

[33] J.S. Birman and T. Kanenobu, Proc. Amer. Math. Soc. **102** (1988) 687.
T. Kanenobu, Math. Ann. **285** 115.

[34] K. Sogo, Y. Akutsu and T. Abe, Prog. Theor. Phys. **70** (1983) 730,739.

[35] A.B. Zamolodchikov and V.A. Fateev, Sov. J. Nucl. Phys. **32** (1980) 293.

[36] Y. Akutsu, A. Kuniba and M. Wadati, J. Phys. Soc. Jpn. **55** (1986) 1466.

[37] V. Pasquier, J. Phys. A: Math. Gen. **20** (1987): L217, L221.

[38] A. Kuniba and T. Yajima, J. Phys. A: Math. Gen. **21** (1988) 519;
J. Stat. Phys. **50** Nos. 3/4, (1988) 829.

[39] M. Jimbo, T. Miwa and M. Okado, Commun. Math. Phys. **116** (1988) 353.

[40] V.G. Turaev: Invent. Math. **92** 1988 527.

[41] N. Yu. Reshetikhin, LOMI preprint E-4-87,E-17-87, Leningrad 1988.

[42] I.V. Cherednik, Theor. Math. Phys. **43** (1980) 356.
O. Babelon, H. J. de Vega and C. M. Viallet, Nucl. Phys. **B190** (1981) 542.
Cherie L. Schultz, Phys. Rev. Lett. **46** (1981) 629; J.H.H. Perk and C.L. Schultz, Phys. Lett. **84A** (1981) 407.

[43] Z.Q. Ma and B.H. Zhao, J. Phys. A: Math. Gen. **22** (1989) L49.

[44] A.N. Kirillov and N.Yu. Reshetikhin, Representations of the algebra $U_q(sl(2))$, q-orthogonal polynomials and invariants of links, LOMI preprint E-9-88, Leningrad 1988.

[45] M. Nomura, J. Math. Phys. **30** (10) (1989) 2397.

[46] H.C. Lee, M. Couture and N.C. Schmeing, Chalk River preprint CRNL-TP-1125R December 1988.

[47] Y.S. Wu, Phys. Rev. Lett. **52** (1984) 2103.

[48] G.W. Semenoff, Phys. Rev. Lett. **61** (1988) 517.

[49] A.M. Polykov, Mod. Phys. Lett. **3A** (1988) 325.

[50] C. Rovelli and L. Smolin, Phys. Rev. Lett. 61 (1988) 1155.

[51] E. Witten, Commun. Math. Phys. **121** (1989) 351.

[52] A. Kuniba, Y. Akutsu and M. Wadati, J. Phys. Soc. Jpn. **55** (1986) 3285.

[53] E. Verlinde, Nucl. Phys. **B300** [FS22] (1988) 360.

[54] G. Moore and N. Seiberg, Phys. Lett. **B212** (1988) 451; Classical and Quantum Conformal Field Theory, IASSNS-HEP-88/39, Princeton preprint 1988.

[55] K.H. Rehren and B. Schroer, Einstein Causality and Artin braids, preprint 1988.

[56] J. Fröhlich, Statistics of fields, the Yang-Baxter equation, and the theory of knots and links, preprint 1988.
G. Felder, J. Fröhlich and G. Keller, ETH preprint 1989.

[57] H.C. Lee, M.L. Ge, M. Couture and Y.S. Wu, Int. J. Mod. Phys. A, **4** (1989) 2333.

[58] M. Wadati, Y. Yamada and T. Deguchi: J. Phys. Soc. Jpn. **58** (1989) 1153

[59] P.Di Francesco and J.B. Zuber, Saclay preprint S.Ph-T/89/92.

[60] P. Ginsparg, Harvard University preprint HUTP-89/A027.

INTEGRABLE RESTRICTIONS OF QUANTUM SOLITON THEORY AND MINIMAL CONFORMAL SERIES

André LeClair

Newman Laboratory
Cornell University
Ithaca, New York 14853, U.S.A.

For special values of the sine-Gordon theory coupling, we restrict the Hilbert space of the theory in a way that preserves the integrability by using the underlying quantum group structure. We argue that the new theories renormalize to the $c < 1$ minimal conformal series. We discuss generalizations to affine Toda theories.

1. Introduction

Conformal symmetry in two-dimensional quantum field theory is generically a property of a massless theory, or of the renormalization group fixed point of a massive theory. The general principles of conformal symmetry are well understood [1], and for example, lead to a complete classification of unitary theories with central extension of the Virasoro algebra $c \leq 1$ [2].

A physically well motivated problem is to find a massive quantum field theory whose behavior at a renormalization group fixed point is described by a given conformal field theory. It is this problem we wish to address in the case of the minimal $c < 1$ series. Given a conformal field theory, there is no reason to believe that there exists a unique quantum field theory that flows to it. We therefore significantly limit the number of possibilities by requiring the field theory to be integrable. Complete integrability as it is usually defined in field theory requires an infinite number of commuting and conserved quantities. A conformal field theory is integrable in this sense. Loosely speaking, the theory can be solved exactly because there are as many constants of the motion as there are degrees of freedom. Further, we require that the theory be local and relativistically invariant. A well-known example occurs at $c = 1/2$, where the massive theory is a free Majorana fermion. Here, the integrability is trivial since the theory is free. Generalizations to the rest of the minimal series were until recently unknown.

Talk delivered at workshop on "Physics, Braids, and Links", August 1989 Banff, Alberta, Canada

Physics, Geometry, and Topology
Edited by H. C. Lee
Plenum Press, New York, 1990

There is good reason to expect that massive models that flow to the minimal series will possess novel features from the point of view of local quantum field theory. We will see some of these features in the course of this talk. The reason is that the minimal models have an interesting new superselection structure. By this we mean the structure of charge sectors and the intertwining fields between them. For a description of this, I refer you to J. Fröhlich's lectures at this school[3]. In the context of conformal field theory, this is the structure of conformal blocks and chiral vertex operators [4]. The existence of novel superselection structures is a possibility in lower dimensional quantum field theory, and it should be stressed that conformal symmetry is not a requirement[3][5]. Thus, we expect that the superselection structure of a minimal model will not be broken when the conformal invariance is broken by making it massive. As we will see our construction is in this spirit, since it relies on techniques relevent for descriptions of the superselection rules of conformal field theory. For applications to particle theory, it is the manifestation of the superselection rules on multiparticle asymptotic states that is of interest.

Perhaps the most celebrated integrable field theory in the particle physics literature is the sine-Gordon (SG) model. It has the Minkowski space action

$$S = \frac{1}{\beta^2} \int dx dt \left[\frac{1}{2} \partial_\mu \phi \partial^\mu \phi + m^2 (cos(\phi) - 1) \right] . \tag{1.1}$$

As we will see, in the deep ultraviolet this theory is a free boson with $c = 1$. We will argue that for

$$\beta^2 / 8\pi = p/(p+1), \tag{1.2}$$

where p is an integer ≥ 3, there exists a coupling-dependent restriction of the SG Hilbert space that preserves the integrability, and these new theories flow to

$$c = 1 - \frac{6}{(p+1)p} . \tag{1.3}$$

We will propose a novel type of exact S-matrix for the new theories that is a sort of confinement of the original SG S-matrix [6]. Our method centers on the Yang-Baxter equation and its associated quantum group structure, and parallels recent results in conformal field theory [7][8][9][10]and Chern-Simons theory [11][12].

The remainder of this lecture is organized as follows. In section 2, we will review some results from the SG model. In section 3 we will analyze the conformal structure and derive the relation between β^2 and c. Section 4 will describe the relation to some classical lattice statistical mechanics models, from which we derived some inspiration. In section 5 the exact method of Quantum Inverse Scattering (QISM) will be reviewed. The restricted sine-Gordon (RSG) model will be described in section 6. In section 7, the relation to the method of Zamolodchikov for perturbing a conformal model to obtain a massive integrable model will be studied.

2. Sine-Gordon Theory

We summarize here some known exact results concerning the SG theory [13][14][15]. The renormalization of the theory was studied by Coleman where he also showed that the theory is equivalent to the Massive Thirring Model (MTM). For $0 < \beta^2/8\pi < 1$, the coupling β^2 is unrenormalized. All infinities can be removed by normal ordering the $cos\phi$ interaction, and absorbing them into the mass parameter m. The renormalized mass is

$$m_r^2 = m^2(m^2/\Gamma^2)^{\beta^2/8\pi}, \tag{2.1}$$

where Γ is a cuttoff. The spectrum consists of solitons and antisolitons of mass $M_s = 8m/\gamma'$, where γ' is a loop corrected coupling:

$$\gamma' = \beta^2/(1-\beta^2/8\pi). \tag{2.2}$$

The solitons are the Thirring fermions. There are also N soliton-antisoliton bound states of mass

$$M_n = (16m/\gamma')sin(n\gamma'/16), \tag{2.3}$$

$n = 1, 2, \cdot\cdot N$, where N is the largest integer less that $8\pi/\gamma'$. The S-matrices of these particles were found by exploiting the topological $U(1)$ symmetry, crossing symmetry, and unitarity. Sklyanin, Takhtajan, and Faddeev developed the QISM to solve the model in a way that preserves the integrability.

3. Conformal Analysis

From the form of the renormalized mass (2.1) , we see that m_r goes to zero in the deep ultraviolet. This is a fixed point of the renormalization group equation for m_r. Thus at this fixed point the $cos\phi$ interaction disappears, and we are left with a free boson. Define Euclidean coordinates $z = (t+ix)/2, \bar{z} = (t-ix)/2$. For convenience rescale the SG field ϕ in (1.1) to $\phi = \phi'\beta/\sqrt{4\pi}$ so that the Euclidean propagator is

$$\phi'(z,\bar{z})\phi'(w,\bar{w}) \sim log|z-w|^2. \tag{3.1}$$

The traceless energy-momentum tensor at $m_r = 0$ is then

$$T_{zz} = \frac{1}{2}\partial_z\phi'\partial_z\phi'. \tag{3.2}$$

(The $\bar{z}$ sector is identical). The central charge c is defined by the operator product expansion

$$T(z)T(w) \sim c/2(z-w)^4. \tag{3.3}$$

Thus we recover the well-known result that the SG theory flows to $c = 1$ for all $0 < \beta^2/8\pi < 1$.

In [6] the following consistency argument was given for c as a function of β^2 for the restricted SG model. Expand $cos\phi$ in terms of $\exp(\pm i\beta\phi'/\sqrt{4\pi})$. Suppose that one of the operators $\exp(\pm i\beta\phi'/\sqrt{4\pi})$ has anomalous dimension (1,1) at the fixed point. Then m would be dimensionless, and the theory would not necessarily flow to $c = 1$, due to the presence of the remaining operator in the action. This anomalous dimension can be accomplished if at the fixed point T_{zz} has an additional background charge term

$$T_{zz} = \frac{1}{2}\partial_z\phi'\partial_z\phi' - i\sqrt{2}\alpha_0\partial_z^2\phi'. \tag{3.4}$$

The anomalous dimension d of an operator $O(w)$ is defined by the operator product expansion

$$T_{zz}(z)O(w) \sim -dO(w)/(z-w)^2. \tag{3.5}$$

Take $O(w)$ to be $\exp(i\beta\phi'/\sqrt{4\pi})$. (Taking instead $\exp(-i\beta\phi'/\sqrt{4\pi})$ is related to the following by a Z_2 symmetry.) Then α_0 is fixed to be a solution of

$$1 = \beta^2/8\pi - 2\alpha_0\beta/\sqrt{8\pi}. \tag{3.6}$$

For $T(z)$ of the form (3.4) , by (3.3) ,

$$c = 1 - 24\alpha_0^2, \tag{3.7}$$

where α_0 is the function of β defined in (3.6) . Inserting the value (1.2) for the coupling β^2, we get (1.3) .

To summarize, we have assumed that the RSG model has a background charge at the fixed point, and we have found a consistency condition for this charge as a function of the SG coupling β^2. There are two refinements of this argument. First, note that our consistency condition implies that $\exp(-i\beta\phi'/\sqrt{4\pi})$ disappears at the fixed point leaving a Liouville action

$$S = \frac{1}{\beta^2}\int dxdt\left[\frac{1}{2}\partial_\mu\phi\partial^\mu\phi + \frac{m^2}{2}e^{i\phi}\right]. \tag{3.8}$$

Liouville theory is known to have a background charge classically, i.e. there exists an improved traceless energy-momentum tensor for the action (3.8) , with background charge $\alpha_0 = -\sqrt{2\pi}/\beta$. By taking into account normal ordering, the authors of [16] have computed the quantum corrections to the classical background charge. The result is given by the relation (3.6) . Secondly, in [17] renormalization group flows for the SG model with an explicit background charge term $R\phi$ in the action, where R is the scalar curvature, were studied, and are consistent with the above picture.

Energy momentum tensors with the background charge term are familiar from the Feigin-Fuchs description of the minimal models [18]. The Feigin-Fuchs construction begins with the Hilbert space of a free massless boson and reaches the minimal model Hilbert space by removing the null vectors. This essential truncation of the Hilbert space can be formulated as a cohomology problem [19]. It is therefore clear that the restriction of the Hilbert space in the RSG model (to be described in section 6) is analagous to the Feigin-Fuchs truncation, and must reduce to it in the massless limit.

4. Classical Lattice Statistical Mechanics

A classical lattice statistical mechanics model in two spacial and zero time dimensions consists of a set of defined Boltzman weights for a given configuration of degrees of freedom on a two dimensional lattice. The degrees of freedom are spins or their generalizations. There is a well-known correspondence between classical statistical mechanics and quantum mechanics, where Boltzman's constant corresponds to Planck's, and partition functions become path integrals. In this way a statistical mechanics model may be related to a 1+1 dimensional lattice Hamiltonian system. There are a class of statistical models that are exactly solvable. In the case of the 8-Vertex model, solved by Baxter, the corresponding Hamiltonian is the XYZ Heisenberg spin chain [20]. It has the Hamiltonian

$$\mathcal{H}^{XYZ} = -\frac{1}{2}\sum_{i=1}^{M}(J_x\sigma_i^x\sigma_{i+1}^x + J_y\sigma_i^y\sigma_{i+1}^y + J_z\sigma_i^z\sigma_{i+1}^z), \tag{4.1}$$

where $J_{x,y,z}$ are free parameters, $\sigma^{x,y,z}$ are Pauli matrices, and M is the number of lattice sites.

There is a well established connection to sine-Gordon theory. Luther has shown that in the continuum limit the XYZ model becomes the massive Thirring model [21]. Using the relation between SG and MTM couplings, Luther has shown that

$$cos(\pi(1-\beta^2/8\pi)) = -J_z/J_x. \tag{4.2}$$

The 8-Vertex model at criticality is known to be $c = 1$, consistent with our discussion of the fixed point of SG. Andrews, Baxter, and Forrester (ABF) [22], discovered that for special values of the 8-Vertex parameters, the allowed configurations of spins could be truncated and the resulting model could still be solved exactly. Huse [23] in turn realized that the new models constitute the $c < 1$ minimal series at the critical point. See also [24]. We relate the special values of the 8-Vertex parameters to the SG coupling. In Baxter's parametrization

$$J_z/J_x = -\frac{cn(2\eta)dn(2\eta)}{1-ksn^2(2\eta)}, \tag{4.3}$$

where cn, dn, and sn are Jacobi elliptic functions [20] . The special values of η are $\eta = K/(p+1)$, where K is a complete elliptic integral, and p is an integer ≥ 3 [22] . The elliptic modulus of the above functions measures the distance from criticality, where $k = 0$ is critical. In the continuum limit, k and the lattice spacing are taken to zero simultaneously while maintaining a finite mass. For this reason $k = 0$ is not necessarily the critical regime in the continuum. Taking this limit in (4.3) one finds the expected values of the SG coupling (1.2) .

A wide generalization of the ABF result has been obtained by the Kyoto school [25][26], and Pasquier [27][28]. The classification of models parallels the classification of simple Lie algebras. As we will see in the next section there is an analagous generalization of SG to affine Toda theories, where SG corresponds to SU(2). We will return to the affine Toda generalizations of the RSG model.

Pasquier has described the truncation of the statistical models at criticality in terms of special properties of quantum groups. It is this technique that we will use to restrict the SG theory, to be described below. The reason the techniques at criticality in the statistical model are applicable to the off-critical continuum theory is that one takes k to zero in going to the continuum. In principle, one can use the identification of Luther to deduce the implications of the ABF restriction on the Bethe-ansatz states of MTM. We do not chose to pursue this line of work. Henceforth, we will be working directly in the continuum.

5. QISM for Affine Toda Systems

In this section we review some classic results on QISM as applied to affine Toda theories. For a review see [29][30]. The classical Toda theories are studied in [31][32]. To motivate the discussion, recall how we diagonalize the Hamiltonian for a single free boson. The field is expanded in terms of Fourier components

$$\phi(x,t) = \int dk \left[a(k)e^{-ikx} + a^*(k)e^{ikx}\right]. \tag{5.1}$$

With the creation-annihilation operators $a(k)$ and $a^*(k)$ one can construct a particle Fock space. The inverse scattering method of solving classical soliton equations is a generalization of Fourier analysis for integrable non-linear equations of motion. The aim of the QISM is to find the analog of the creation-annihilation operators.

The integrability of a classical soliton equation is most clearly revealed through its zero curvature representation and its relation to affine Kac-Moody algebras [33]. Let $\hat{g}$ be a Kac-Moody algebra , and $\{e_i, f_i, h_i; i = 0, \cdot\cdot, r\}$ a Chevalley basis of generators of $\hat{g}$, satisfying

$$\begin{aligned} [h_i, h_j] &= 0, \qquad & [e_i, f_j] &= \delta_{ij} h_i \\ [h_i, e_j] &= K_{ji} e_j, \qquad & [h_i, f_j] &= -K_{ji} f_j. \end{aligned} \tag{5.2}$$

The generalized Cartan matrix K_{ij} is normalized to be 2 down the diagonal. To simplify the discussion, we consider only the case where $\hat{g}$ is untwisted, in which case $r = \text{rank}(g)$. For zero central extension of $\hat{g}$, there exists a representation of $\hat{g}$ with generators $T^a\lambda^n$, where T^a is a generator of the simple Lie algebra g, and λ is a parameter (in integrable systems theory, λ is called the spectral parameter). A representation of the Chevalley basis in terms of generators of the simple Lie algebra g is

$$\begin{aligned} h_i &= \frac{2}{|\alpha^{(i)}|^2}\alpha^{(i)} \cdot H \qquad & e_i &= E_{\alpha^{(i)}} \qquad & f_i &= E_{-\alpha^{(i)}} \qquad & i = 1, \cdot\cdot, r \\ h_0 &= -\frac{2}{|\psi|^2}\psi \cdot H \qquad & e_0 &= \lambda E_{-\psi} \qquad & f_0 &= \lambda^{-1} E_{+\psi}, \end{aligned} \tag{5.3}$$

where $\alpha^{(i)}$ are simple roots, ψ is the highest root, and $E_{\alpha^{(i)}}$,H_i are Cartan-Weyl generators of g. The gradation of $\hat{g}$ is only meaningful up to automorphisms of g. The above gradation is called the homogeneous one. Another useful gradation is the principle gradation, where the Chevalley basis is as above except that for $i = 1,\cdot\cdot,r$, $e_i = \lambda E_{\alpha^{(i)}}$, $f_i = \lambda^{-1}E_{-\alpha^{(i)}}$. The relation between the two gradings for an element $a \in \hat{g}$ is

$$a^{prin}(\lambda) = \sigma a^{homo}(\lambda^h)\sigma^{-1}, \tag{5.4}$$

where $\sigma = \lambda^t$, t measures the length of a root: $[t, E_\alpha] = length(\alpha)E_\alpha$, and $h = length(\psi)+1$ is the Coxeter number of g.

For the generalized Toda theories, we define the gauge fields

$$\begin{aligned} \partial_+ + A_+ &= e^{-\phi/2}\partial_+ e^{\phi/2} + m\, e^{\phi/2}\Lambda e^{-\phi/2} \\ \partial_- + A_- &= e^{\phi/2}\partial_- e^{-\phi/2} + m\, e^{-\phi/2}\bar{\Lambda} e^{\phi/2}, \end{aligned} \tag{5.5}$$

where $\partial_\pm = \partial_x \pm \partial_t$, $\phi = \sum_{i=1}^r \phi_i h_i$, and $\Lambda = \sum_{i=0}^r e_i$, $\bar{\Lambda} = \sum_{i=0}^r f_i$. The field equations for the Toda fields ϕ_i are then written as $F_{+-} = [\partial_+ + A_+, \partial_- + A_-] = 0$. These field equations follow from the action

$$S = \frac{2}{\beta^2}\int dxdt \left[\sum_{i,j=1}^r \frac{1}{|\alpha^{(i)}|^2}\partial_+\phi_i K_{ij}\partial_-\phi_j + m^2\sum_{i=0}^r \frac{2}{|\alpha^{(i)}|^2}e^{K_{ij}\phi_j}\right]. \tag{5.6}$$

The SG theory corresponds to $g = su(2)$, and $\phi \to i\phi/2$.

That the Toda field equations have an infinite number of conserved currents follows from: i) There exists a gauge transformation $\omega(\partial_\pm + A_\pm)\omega^{-1} = \partial_\pm + a_\pm$ such that $a_\pm \in Ker(Ad_\Lambda)$. ii) $Ker(Ad_\Lambda)$ is infinite dimensional and abelian. Since the zero-curvature condition is gauge invariant, this implies $\partial_+ a_- - \partial_- a_+ = 0$. The infinity of conserved currents are generated by $j^\mu(\lambda) = \varepsilon^{\mu\nu}a_\nu(\lambda)$. The dimensions of the conserved densities modulo the coxeter number of $\hat{g}$ are given by the exponents of $\hat{g}$.

In the quantum theory, a fundamental role is played by the monodromy matrix, which is a Wilson line of the gauge field:

$$T(\lambda) = \mathcal{P}\exp\left(-\int_{-L}^{L} A_x dx\right). \tag{5.7}$$

$\mathcal{P}$ denotes path ordering, and the fields are taken to be periodic, with period $2L$. In (5.7) the gauge field is taken to be in a representation of g, with generators acting on vector space V. The aforementioned gauge transformation implies

$$T(\lambda) = \omega^{-1}(L)\exp(-\int a_x dx)\omega(-L). \tag{5.8}$$

Thus the classical integrals of motion can be recovered by expanding $tr\ T(\lambda)$ in powers of λ . It is these integrals of motion that are promoted to the quantum theory. Quantum integrability requires that the integrals of motion commute as quantum operators. Suppose that upon imposing the canonical commutation relations on the fields ϕ , the monodromy matrix satisfies the following quantum operator equation:

$$\Re(\lambda/\mu)T_1(\lambda)T_2(\mu) = T_2(\mu)T_1(\lambda)\Re(\lambda/\mu), \tag{5.9}$$

where $T_1(\lambda) = T(\lambda)\otimes 1$, $T_2(\mu) = 1\otimes T(\mu)$. $\Re$ is an ordinary c-number matrix acting on $V\otimes V$. Then $trT(\lambda)$ generates quantum commuting integrals of motion, i.e. $[trT(\lambda), trT(\mu)] = 0$.

Using the regularization techniques in [15] one can show that (5.9) reduces to the following conditions on $\Re$:

$$\begin{aligned}
\Re(\lambda/\mu)(h_i \otimes 1 + 1 \otimes h_i) &= (h_i \otimes 1 + 1 \otimes h_i)\Re(\lambda/\mu) \\
\Re(\lambda/\mu)(e_i \otimes K_i^{-1} + K_i \otimes e_i) &= (e_i \otimes K_i + K_i^{-1} \otimes e_i)\Re(\lambda/\mu) \\
\Re(\lambda/\mu)(f_i \otimes K_i^{-1} + K_i \otimes f_i) &= (f_i \otimes K_i + K_i^{-1} \otimes f_i)\Re(\lambda/\mu)
\end{aligned} \tag{5.10}$$

for all $i = 0, \cdot\cdot r$, where $K_i = exp((i\beta^2/16)\alpha^{(i)2}h_i)$. The authors of [34][35] have found the $\Re$ matrices satisfying (5.10) for arbitrary $\hat{g}$ and V defined by the fundamental representation. The result for SU(n) in the homogeneous gradation is

$$\begin{aligned}
\Re(x,q) = (x - q^{-2})\sum_{\alpha} E_{\alpha\alpha} \otimes E_{\alpha\alpha} + q^{-1}(x-1)\sum_{\alpha\neq\beta} E_{\alpha\alpha} \otimes E_{\beta\beta} \\
+(1-q^{-2})(\sum_{\alpha<\beta} + x\sum_{\alpha>\beta})E_{\alpha\beta} \otimes E_{\beta\alpha},
\end{aligned} \tag{5.11}$$

where $x = \lambda/\mu$, $q = \exp(-i\beta^2/4)$, and $E_{\alpha\beta}$ are unit matrices for the fundamental representation, i.e. $(E_{\alpha\beta})_{\gamma\sigma} = \delta_{\alpha\gamma}\delta_{\beta\sigma}$, $i = 1, \cdot\cdot n$. Furthermore, it can be shown that $\Re$ satisfies the Yang-Baxter (YB) equation

$$\Re_{12}(\lambda/\mu)\Re_{13}(\lambda/\nu)\Re_{23}(\mu/\nu) = \Re_{23}(\mu/\nu)\Re_{13}(\lambda/\nu)\Re_{12}(\lambda/\mu). \tag{5.12}$$

The above relation is an equation in $V \otimes V \otimes V$, and the subscripts refer to the vector spaces where the matrix is not unity.

Let us now specialize to the SG theory. We take V to span the two dimensional spin 1/2 representation of su(2). $\Re$ is then the 4×4 matrix given in (5.11) . The monodromy matrix is a 2×2 matrix of operators

$$T(\lambda) = \begin{pmatrix} A(\lambda) & B(\lambda) \\ C(\lambda) & D(\lambda) \end{pmatrix} \tag{5.13}$$

whose commutation relations are given by (5.9) . We have seen that $A(\lambda) + D(\lambda)$ generates the infinite number of integrals of motion. Remarkably, it was found in [15] that the operator $B(\lambda)$ serves as a creation operator for the states that diagonalize the Hamiltonian, thereby providing an algebraic Bethe ansatz solution of the model.

6. Restricted Sine-Gordon Theory

We will now use the Yang-Baxter structure of the last section, especially its associated quantum group structure, to restrict the SG Hilbert space. The main requirement is to preserve the integrability.

A continuation of our line of discussion would necessarily involve a description of the restriction at the level of the Bethe ansatz states in the QISM. The main technical difficulty in pursuing this is the filling of the Dirac sea. We have not solved this important problem. Instead we describe a restriction of the SG S-matrix. Physically the restriction then becomes less obscured by technical problems. Furthermore, since the algebraic Bethe ansatz provides a derivation of the usual S-matrix, it is clear that our restriction implies a corresponding restriction of the Bethe ansatz. In fact, we use the fact that the Yang-Baxter structure for the S-matrix follows from that for $\Re$, making this argument very plausible.

Complete integrability leads to factorization of the S-matrix, i.e the N-body S-matrix can be expressed in terms of 2-body S-matrices [14]. This is because the conservation laws imply the complete momentum distribution is conserved. We first describe the unrestricted SG S-matrix for the values of the coupling (1.2) . At this coupling there is a repulsive force between the solitons, thus they don't bind to form bound states. There is a $U(1)$ symmetry in the theory underwhich solitons and antisolitons have charge $+1$ and -1 respectively. Let us arrange the soliton and antisoliton states into an isovector $|\alpha = \pm 1/2\rangle$, where h is the $U(1)$ charge and $h|\alpha\rangle = 2\alpha|\alpha\rangle$. This two dimensional vector space will be denoted as V. We parametrize the energy and momentum of a particle in terms of the rapidity θ :

$$p^0 = m cosh\theta, p^1 = m sinh\theta. \tag{6.1}$$

From the explicite form of the soliton-antisoliton S-matrix given in [14] and equation (5.11) one finds

$$S(\theta) = s(\theta)\Re^{prin}(x = e^{8\pi\theta/\gamma'}, q = -e^{-i8\pi^2/\gamma'}). \tag{6.2}$$

In (6.2) $s(\theta)$ is a scalar function of the rapidity difference $\theta = \theta_1 - \theta_2$:

$$s(\theta) = \frac{-i}{\pi} sh(\frac{8\pi}{\gamma'}(i\pi - \theta))\Gamma(\frac{8\pi}{\gamma'})\Gamma(1 + i\frac{8\theta}{\gamma'})\Gamma(1 - \frac{8\pi}{\gamma'} - i\frac{8\theta}{\gamma'}) \prod_{n=1}^{\infty} \frac{R_n(\theta)R_n(i\pi - \theta)}{R_n(0)R_n(i\pi)}, \tag{6.3}$$

$$R_n(\theta) = \frac{\Gamma(2n\frac{8\pi}{\gamma'} + i\frac{8\theta}{\gamma'})\Gamma(1 + 2n\frac{8\pi}{\gamma'} + i\frac{8\theta}{\gamma'})}{\Gamma((2n+1)\frac{8\pi}{\gamma'} + i\frac{8\theta}{\gamma'})\Gamma(1 + (2n-1)\frac{8\pi}{\gamma'} + i\frac{8\theta}{\gamma'})}. \tag{6.4}$$

$\Re^{prin}$ is the $\Re$ matrix of (5.11) in the principle gradation

$$\Re^{prin}_{12}(x) = \sigma_{12}\Re(x^h)\sigma_{12}^{-1}, \tag{6.5}$$

where $\sigma_{12} = \lambda^t \otimes \mu^t$, and x is again λ/μ. (In (6.5) , $h = 2$, the coxeter number of $su(2)$). The relation (6.2) it is not at all obvious from what we have said so far. It can be heuristically explained as follows: $\Re$ determines the commutation relations of the creation operators of the Bethe ansatz and moreover, in 1+1 dimensions the interchange of particles is an interaction.

The multiparticle S-matrix is computed as follows. In the far past there are N particles (solitons or antisolitons) located at $x_1 < x_2 < \cdots < x_N$, with momenta $p_1 > p_2 > \cdots > p_N$.

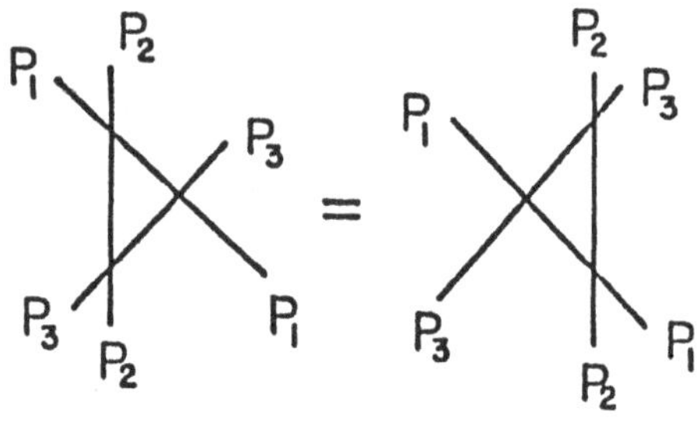

figure 1

After $N(N-1)/2$ pair collisions the out state again consists of N particles (no particle creation). Since the set of momenta is unchanged in each pair collision, the outgoing particles have the same set of momenta but are now located at $x_1 > x_2 > \cdots > x_N$. One can assign a space-time picture to this process. For example, 3-particle scattering is shown in figure 1. A fixed momentum is assigned to a straight line with slope proportional to $-p^1$. The successive 2-body scatterings are denoted by a crossing, as in figure 2. Sums

$$\underset{\alpha_2' \quad \alpha_1'}{\overset{\alpha_1 \qquad \alpha_2}{X}} \quad = \quad \langle \alpha_1' | \otimes \langle \alpha_2' | \; S \; | \alpha_1 \rangle \otimes | \alpha_2 \rangle$$

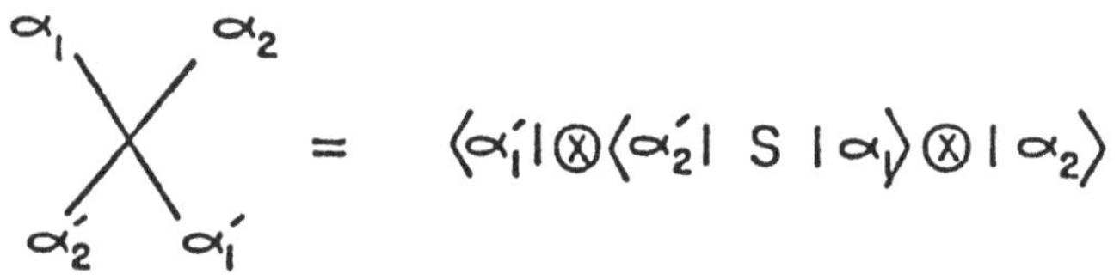

figure 2

over intermediate states are assumed. Thus, 3-particle S-matrices are matrix elements of $S_{23}(\theta_{23})S_{13}(\theta_{13})S_{12}(\theta_{12})$, where $\theta_{ij} = \theta_i - \theta_j$. The consistency relation displayed in figure 1 is ensured by the YB equation (5.12) for $\Re$.

For convenience, we define a braid-type S-matrix $\hat{S} = PS$, where P is the permutation matrix: $Pu \otimes v = v \otimes u$, $u \otimes v \in V \otimes V$. The 2-body S-matrix diagram of figure 2 is now $\langle \alpha_2' | \otimes \langle \alpha_1' | \hat{S}(\theta_{12}) | \alpha_1 \rangle \otimes | \alpha_2 \rangle$. In terms of $\hat{S}$ a diagram for 3-body scattering for example is shown in figure 3. The horizontal lines in figure 3 will be given meaning shortly. For the process displayed in figure 3, the S-matrix is computed as matrix elements of $\hat{S}_{12}(\theta_{12})\hat{S}_{23}(\theta_{13})\hat{S}_{12}(\theta_{23})$, where a vertical line corresponds to a single vector space V.

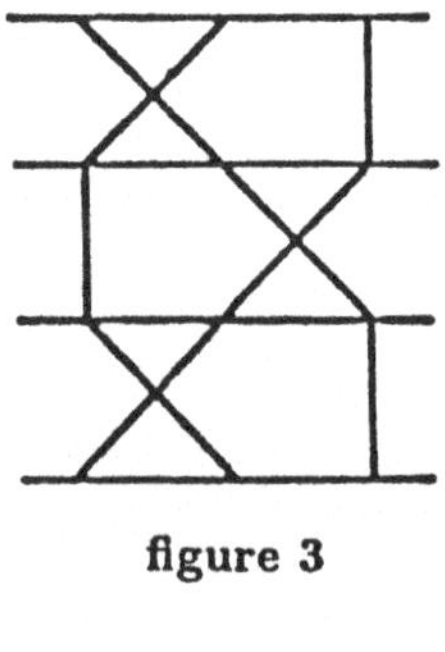

figure 3

Due to (6.2) and (6.5) , the multiparticle S-matrix elements can be computed from the matrix elements of $P\Re$ up to a computable multiplicative constant, and a gauge transformation. For convenience we rescale $P\Re$ to

$$\hat{\Re}(x) \equiv \frac{q}{\sqrt{x}} P\Re(x) = \sqrt{x}\hat{R} - \frac{1}{\sqrt{x}}\hat{R}^{-1}, \tag{6.6}$$

$$\hat{R} = q\sum_{\alpha=1}^{2} E_{\alpha\alpha}\otimes E_{\alpha\alpha} + \sum_{\alpha\neq\beta=1}^{2} E_{\beta\alpha}\otimes E_{\alpha\beta} + (q-q^{-1})\sum_{\alpha>\beta=1}^{2} E_{\beta\beta}\otimes E_{\alpha\alpha}. \tag{6.7}$$

$P\hat{\Re}$ satisfies the YB equation by virtue of the Hecke algebra relations satisfied by $\hat{R}$ [36]

$$\begin{aligned}(\hat{R}-q)(\hat{R}+q^{-1}) &= 0 \\ \hat{R}_{21}\hat{R}_{32}\hat{R}_{21} &= \hat{R}_{32}\hat{R}_{21}\hat{R}_{32}.\end{aligned} \tag{6.8}$$

We will now define a restriction of the multisoliton Hilbert space that preserves the factorizability of the S-matrix and thus preserves the integrability. The technique we will

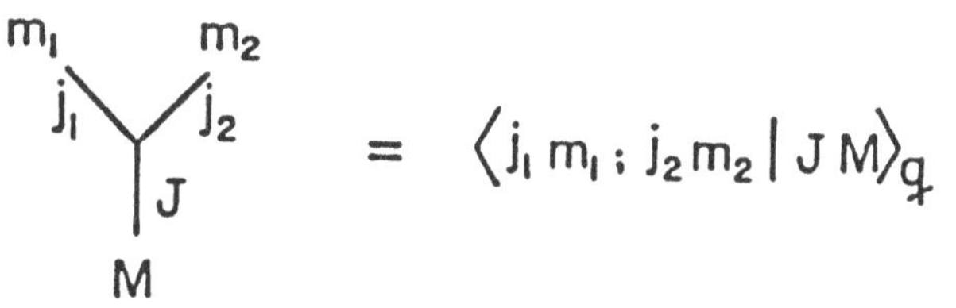

figure 4

use to restrict the model relies extensively on some results from the theory of quantum groups [37][38][28]. The technique was used by Pasquier in his study of the ABF models at criticality. It is the so-called Vertex/RSOS correspondence. For a discussion in the context of conformal field theory, see [8] [10] .

In order to describe the restriction, we first make a change of basis in the space of states. Ordinary $su(2)$ is not a symmetry of our system, thus organization of states into irreducible $su(2)$ multiplets is not useful. A meaningful change of basis is provided by the quantum-$su(2)$ algebra $\mathcal{U}_q(SU(2))$ [39][40][41]. $\mathcal{U}_q(SU(2))$ is generated by $J_\pm, h$ satisfying

$$[J_+, J_-] = \frac{q^h - q^{-h}}{q-q^{-1}}, \qquad [h, J_\pm] = \pm 2J_\pm. \tag{6.9}$$

The distinguishing feature of the relations (6.9) is that they allow the analog of addition of angular momentum: given two representations of the algebra, one can find a new representation on the tensor product space. This in turn implies that $\mathcal{U}_q(SU(2))$ has a well defined representation theory. Formally it is said that $\mathcal{U}_q(SU(2))$ admits a comultiplication Δ . Δ provides a representation of $\mathcal{U}_q(SU(2))$ on $V\otimes V$ via

$$\begin{aligned}\Delta(h) &= 1\otimes h + h\otimes 1 \\ \Delta(J_\pm) &= q^{h/2}\otimes J_\pm + J_\pm \otimes q^{-h/2}.\end{aligned} \tag{6.10}$$

Similarly, one can define $\Delta^{(N)}$ on $V^{\otimes N}$. Note that as $q \to 1$, we recover the usual $su(2)$ algebra and comultiplication. The comultiplication allows one to generalize many of the ordinary group theory constructions such as Clebsch-Gordon decompositions.

Consider an incoming N-soliton-antisoliton state whose $U(1)$ quantum numbers are specified by some vector in $V^{\otimes N}$. A meaningful change of basis for $V^{\otimes N}$ is to decompose it

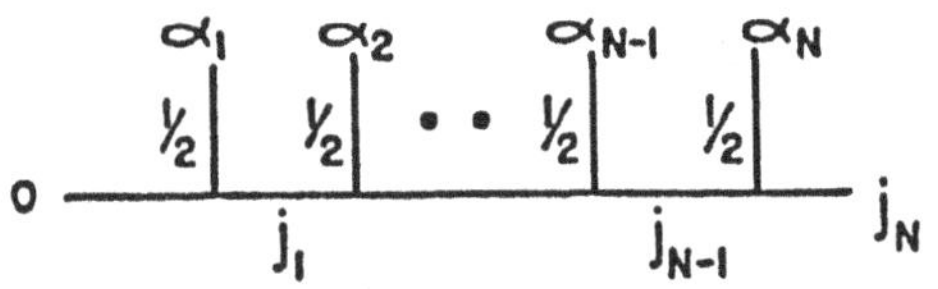

figure 5

into irreducible representations of $\mathcal{U}_q(SU(2))$. This is because the S-matrix commutes with this decomposition. In other words, the S-matrix does not mix different representations of $\mathcal{U}_q(SU(2))$. That the S-matrix commutes with the comultiplication $\Delta^{(N)}$ follows from the defining relations (5.10) for $\Re$ and the form of the comultiplication (6.10). More precisely, we write

$$V^{\otimes N} = \oplus_{\mathcal{J}} V^{j_N}, \tag{6.11}$$

where V^{j_N} is an irreducible representation space of $\mathcal{U}_q(SU(2))$ of dimension $2j_N + 1$, and $\mathcal{J}$ denotes the history or 'path'of the decomposition: $\mathcal{J} = \{j_1 = 1/2, j_2, \cdots j_N\}$, $j_{i+1} = j_i \pm 1/2 \geq 0$. The new basis of states is $|\mathcal{J}; M\rangle$, $-j_N \leq M \leq j_N$, and is related to the previous basis with q-Clebsch-Gordon (q-CG) coefficients $\langle j_1 m_1; j_2 m_2 | J, M\rangle_q$. Graphical techniques have proven to be very useful in quantum group theory. Denote a q-CG coefficient as in figure 4. Then a state $|\mathcal{J}; M\rangle$ is represented as in figure 5. The other objects we need are the q-6j symbols (q-analogues of Wigner-Racah coefficients) defined in figure 6. The q-CG and q-6j symbols were computed in [38]. They are given by

$$\begin{aligned}\begin{Bmatrix} a & b & e \\ d & c & f \end{Bmatrix}_q &= (-1)^{-a-b+c+d+2e}([2e+1][2f+1])^{1/2}\Delta(abe)\Delta(dce)\Delta(acf)\Delta(dbf) \\ &\quad (\sum_{z\geq 0}(-1)^z[z+1]!([z-a-b-e]![z-d-c-e]![z-a-c-f]! \\ &\quad [z-d-b-f]![a+b+c+d-z]![a+e+d+f-z]![b+e+c+f-z]!)^{-1}).\end{aligned} \tag{6.12}$$

Above, in the sum over z only terms with positive argument within the bracket are kept, and

$$[n] = \frac{q^n - q^{-n}}{q - q^{-1}}, \tag{6.13}$$

$[n]! = [n][n-1]\cdot\cdot[1]$, $[0] \equiv 1$,

$$\Delta(abc) = \left(\frac{[-a+b+c]![a-b+c]![a+b-c]!}{[a+b+c+1]}\right)^{1/2}. \tag{6.14}$$

Let us return to the braid-type S-matrix diagrams, as in figure 3. The q-CG decomposition before and after a pairwise collision is signified by horizontal lines. To compute the S-matrix in this basis, we need only the matrix elements of $\hat{R}$, which are

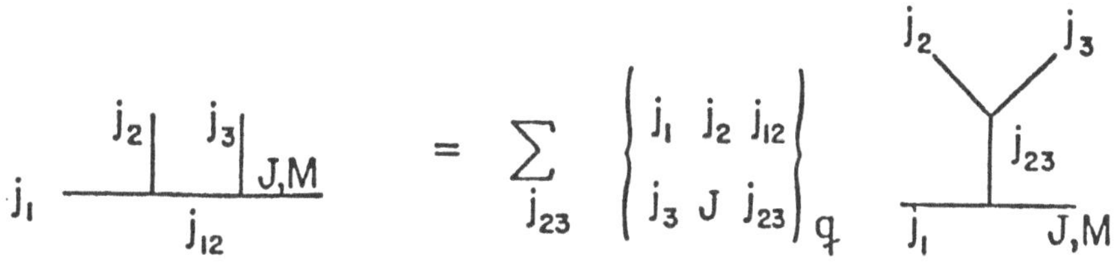

figure 6

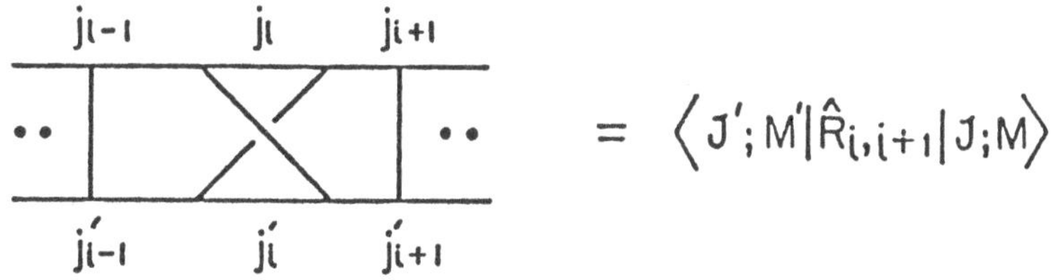

figure 7

$$\langle \mathcal{J}'; M' | \hat{R}_{i,i+1} | \mathcal{J}; M \rangle = \delta_{MM'} \left(\prod_{k \neq i} \delta_{j_k j_k'} \right) (-1)^{j_i' + j_i - j_{i+1} - j_{i-1}} \tag{6.15}$$
$$q^{2(c_{j_{i+1}} + c_{j_{i-1}} - c_{j_i'} - c_{j_i})} \begin{Bmatrix} 1/2 & j_{i-1} & j_i' \\ 1/2 & j_{i+1} & j_i \end{Bmatrix}_q ,$$

where $c_j = j(j+1)$. This matrix element is depicted graphically in figure 7. $\hat{R}^{-1}$ is represented graphically as in figure 7 with an inverted central crossing.

Special properties of representations of $\mathcal{U}_q(SU(2))$ for q a root of unity are the key to restricting the model. Here, $q^{2(p-1)} = 1$. For q a root of unity the q-CG and q-6j symbols become singular ($\to \infty$) unless one restricts the allowed spins: $j \leq j_{max} = (p-1)/2 - 1$. That is, for q a root of unity, there are only a finite number of representations of $\mathcal{U}_q(SU(2))$, labeled by j with j belonging to the set $\{0, 1/2, 1, \cdots j_{max}\}$. I refer you to Keller's lecture for a discussion of this point [42]. This leads us to restrict the N-body SG Hilbert space as follows: in the q-CG decomposition (6.11) we require all spins j_i in $\mathcal{J}$ to be less than j_{max}. The RSG S-matrix is then defined by the braid-type diagrams as before using (6.15) , with the restricted Hilbert space assumed. Remarkably, it turns out this new S-matrix still satisfies the YB relations, since $\hat{R}$ in this restricted basis continues to satisfy the Hecke algebra relations. The reason is connected to the fact that there are only a finite number of representations of $\mathcal{U}_q(SU(2))$. This is a non-trivial result, and we do not prove this here but merely point out that by modifying the innerproduct, one can project out the unwanted states, i.e.

$$tr' \left(1_{v^{j_1}} \otimes 1_{v^{j_2}} \right) = \sum_{j=|j_1 - j_2|}^{min(j_1+j_2, j_{max})} tr' \left(1_{v^j} \right), \tag{6.16}$$

where $tr'(a) = tr(aq^h \otimes q^h)$, $j_1, j_2 \leq j_{max}$, and $h|j,m\rangle = 2m|j,m\rangle$.

Due to the fact that $\hat{R}_{i,i+1}$ is diagonal in the M quantum number in (6.15) , one can effectively mod it out, and consider the RSG S-matrix as acting only on the j-quantum numbers that define a path. Thus the $U(1)$ quantum numbers of the original soliton states becomes confined in the RSG model. The S-matrix scattering diagrams then have quantum numbers in the regions between lines taking values in the set $\{0, 1/2, 1, \cdots j_{max}\}$ and correspond to the 'shadow world' diagrams in [38].

We point out that the q-6j symbols that define the RSG S-matrix are precisely the same as those that describe the braid matrices for the minimal model it flows to. These braid matrices are induced by the monodromies of the conformal blocks. We do not know the general principle that would inply this, but see it as further evidence for our conjecture.

7. Perturbations of Conformal Field Theory

Zamolodchikov has developed a scheme for generating massive integrable quantum field theories as perturbations of conformal field theories by relevent operators. By examining the field content implied by the Virasoro characters he can demonstrate some non-trivial integrals of motion for the perturbed theory [43][44]. Let us set up some definitions. There exists the minimal model $M_{p/q}$ with

$$c = 1 - 6\frac{(p-q)^2}{pq}, \tag{7.1}$$

where $q = p+1$ corresponds to the unitary minimal series. There are primary fields $\Phi_{m,n}$ with dimensions

$$\Delta_{m,n} = 1/4\left[\frac{p}{q}(n^2-1) + \frac{q}{p}(m^2-1) + 2(1-nm)\right], \tag{7.2}$$

where m and n are integers over a finite range.

Let us return to the conformal analysis of section 3. If $exp(i\beta\phi'/\sqrt{4\pi})$ develops anomalous dimension 1,1 due to the background charge, then using (3.5) we can compute the dimension of the other operator in the action $exp(-i\beta\phi'/\sqrt{4\pi})$. It turns out to have dimension $(p-1)/(p+1)$, the dimension of the $\Phi_{1,3}$ primary field. Thus we conclude that the RSG model is a perturbation of the minimal series by the $\Phi_{1,3}$ operator.

An explicit result relating our work with the Zamolodchikov method was found by Eguchi and Yang [45]. These authors invoked the Feigin-Fuchs description of the minimal models, then studied the effect of the perturbation. They found that the Feigin-Fuchs field then became a solution of the SG equation of motion. The relation between β^2 and c is the same as derived in section 3. It must be stressed that this does not imply a direct connection between SG and perturbed $M_{p/p+1}$, since SG does not have a background charge, nor is its Hilbert space restricted at is fixed point. Equations of motion do not reveal the structure of the Hilbert space. Of course the above is entirely consistent with RSG theory.

We now discuss generalizations of our results to affine Toda models. For the same reason as in the su(2) or SG case, the action (5.6) by itself has no direct connection with a minimal series; it must be restricted. Recall from our conformal analysis of section 3 that at the fixed point RSG was described by a Liouville type equation of motion. The generalization for $\hat{g}$ affine Toda field theory is the simple g Today field theory, which is known to have improved traceless energy-momentum tensor. Furthermore, for the simple g Toda field theories there exists an analagous **Feigin-Fuchs** construction of the resulting W-algebra symmetry [46][47], leading to the minimal series with

$$c = r\left(1 - \frac{h(h+1)}{p(p+1)}\right), \tag{7.3}$$

where $r = rank(g)$, and h is the Coxeter number of g. The analog of the perturbing field $\Phi_{1,3}$ is a field of dimension $(p-h+1)/(p+1)$. Eguchi and Yang have extended their result to the W-algebra series, and found the generalized **Feigin-Fuchs** fields satisfy the equation of motion resulting from the action (5.6) . Again, this result does not indicate the spectrum. The study of this question is complicated by the fact that at the coupling β^2 that relate to the W-series, the action is not Hermitian. To see this, note that β^2 is positive, thus the kinetic term is made positive by $\phi_i \to i\phi_i$, taking the potential to $exp(iK_{ij}\phi_j)$. Only for su(2) is this potential real, due to the symmetry $\phi \to -\phi$. Thus for the affine Toda models we cannot strictly speaking describe a restriction, since the unrestricted model is sick. Nevertheless, we conjecture that the W-series perturbed by the analog of the $\Phi_{1,3}$ operator has S-matrix of the RSOS form, and follows from the restricted $\Re$ matrix of the affine Toda models. The necessary q-group theory is easily generalized.

For the lowest element of the series (7.3) , i.e. $p = h + 1$, Fateev and Zamolodchikov have proposed exact S-matrices. They are the minimal solution to the S-matrix with r particles with the masses that follow from the action (5.6) . See also [48][49][50]. Thus it is clear that the lowest theory in the series is special, to the extent that the spectrum follows simply from the Lagrangian. In the case $0 < c < 1$, which occurs for A_1, E_8, A_2, E_6 and E_7, at $c = 1/2, 1/2, 4/5, 6/7$, and $7/10$ respectively, there is no contradiction with our results since the minimal model is now being perturbed by a different operator. For example, for E_8, the $c = 1/2$ theory is perturbed by $\Phi_{1,2}$ rather than $\Phi_{1,3}$.

8. Conclusion

Though we have not proven that the S-matrices of section 6 describe models that flow to the minimal series, we have given a number of arguments supporting this conjecture. At the critical point, the structure of the Hilbert space follows from projecting null vectors, i.e. it follows from Virasoro algebraic structure. Our work provides intriguing evidence for an algebraic structure that plays this role in the massive model.

A study of the spectrum and its physical interpretation will be reported elsewhere in work done with D. Bernard [51].

Independently, Smirnov [52] has considered restrictions of the SG model at different values of coupling than the ones considered here. He looks at

$$\beta^2/8\pi = 2/(2n+3). \tag{8.1}$$

At this coupling the unrestricted model has both bound states and solitons. His restriction is of a different nature than ours in this case, in that he simply removes the solitons from the asymptotic states, and does not rely on the quantum group structure. Again this changes c from 1 to the value (7.1) with $q/p = 2/(2n+3)$. The relation between c and β^2 is the same as we derived in section 3. Smirnov's technique is to examine the form factors of certain local fields and demand that upon restriction the theory remains local. He argues that the background charge is thereby induced. The resulting S-matrices were arrived at independently by using Zamolodchikov methodology without reference to SG theory in [53]. Though the restriction at our values of the coupling is more complicated since it removes states from the multiparticle Hilbert space rather than simply the single particle Hilbert space, perhaps his techniques will generalize. The combination of our results with Smirnov's suggests there is a general result for every model $M_{p/q}$.

Acknowledgements

I would like to thank the organizers of this workshop for the opportunity to present this work here, and J. Frölich for some helpful discussions during this school. I would also especially like to thank Denis Bernard for his recent collaboration. This work was supported in part by the National Science Foundation.

References

[1] A. A. Belavin, A. M. Polyakov, and A. B. Zamolodchikov, Nucl. Phys. B241 (1984) 333.

[2] D. Friedan, Z. Qiu, and S. Shenker, Phys. Rev. Lett. 52 (1984) 1575.

[3] J. Fröhlich, Zurich preprint, lectures given at Banff Summer School in Theoretical Physics, August 1989.

[4] G. Moore and N. Seiberg, Phys. Lett. B 212 (1988) 451; IAS preprint IASSNS-HEP-88/39.

[5] K. Fredenhagen, K.H. Rehren, and B. Schroer, Berlin preprint, *Superselection Sectors with Braid Group Statistics and Exchange Algebras, I: General Theory*, Sept 1988.

[6] A. LeClair, Princeton preprint PUPT-1124, *Restricted Sine-Gordon Theory and the Minimal Conformal Series*, April 1989, to appear in Physics Letters B.

[7] A. Tsuchiya and Y. Kanie, Lett. Math. Phys. B (1987) 303.

[8] L. Alvarez-Gaumé, C. Gomez, and G. Sierra, Phys. Lett. B220(1989) 142; CERN preprint CERN-TH 5369/89.

[9] G. Felder, J. Fröhlich and G. Keller, 'Braid matrices and Structure Constants for Minimal Conformal Models', preprint.

[10] G. Moore and N. Reshetikhin, IAS preprint , *A Comment on Quantum Group Symmetry in Conformal Field Theory* , IASSNS-HEP-89/18, March 1989.

[11] E. Witten, IAS preprint, *Gauge Theories and Integrable Lattice Models*, IASSNS-HEP-89/11 February 1989.

[12] E. Witten, IAS preprint, *Gauge Theories, Vertex Models, and Quantum Groups*, IASSNS-HEP-89/32, May, 1989.
[13] S. Coleman, Phys. Rev. D 11 (1975) 2088.
[14] A. B. Zamolodchikov and A. B. Zamolodchikov,Annals Phys. 120 (1979) 253.
[15] E. K. Sklyanin, L. A. Takhtadzhyan, and L. D. Faddeev, Theor. Math. 40 (1980) 688.
[16] T. L. Curtright and C. B. Thorn, Phys. Rev. Lett. 48(1982) 1309.
[17] M. T. Grisaru, A. Lerda, S. Penati, and D. Zanon, MIT preprint CTP no. 1790, Sept. 1989.
[18] V. Dotsenko and V. Fateev, Nucl. Phys. B240 (1984) 312.
[19] G. Felder, 'BRST Approach to Minimal Models', preprint.
[20] R. Baxter, *Exactly Solved Models in Statistical Mechanics*, Academic Press, London, 1982.
[21] A. Luther, Phys. Rev. B 14 (1976) 2153.
[22] G. Andrews, R. Baxter, and P. J. Forrester, Jour. Stat. Phys. 35 (1984) 193.
[23] D. A. Huse, Phys. Rev. B 30 (1984) 3908.
[24] V. Bazhanov and N. Reshetikhin, *Critical RSOS Models and Conformal Field Theory*, Serpukhov preprint 1987, to appear in Int. Jour. of Mod. Phys.
[25] E. Date, M. Jimbo, A. Kuniba, T. Miwa, and M. Okado, Nucl. Phys. B 290 (1987) 231.
[26] E. Date, M. Jimbo, T. Miwa, and M. Okado, Lect. at AMS summer inst. 'Theta Functions'.
[27] V. Pasquier, Nucl. Phys. B295 (1988) 491.
[28] V. Pasquier, Comm. Math. Phys. 118 (1988) 355.
[29] L. Faddeev, Les Houches Lectures 1982, Elsevier Science Publishers (1984).
[30] V. V. Bazhanov, Comm. Math. Phys. 113 (1987) 471.
[31] V. Drinfel'd and V. Sokolov, Jour. Sov. Math. 30 (1984) 1975.
[32] D. Olive and N. Turok, Nucl. Phys. B 257 (1985) 277.
[33] For a review see: P. Goddard and D. Olive, Int. Jour. Mod. Phys. A 1 (1986) 303.
[34] M. Jimbo, Comm. Math. Phys. 102 (1986) 537.
[35] V. V. Bazhanov, Phys. Lett. B159 (1985) 321.
[36] M. Jimbo, Lett. Math. Phys. 11 (1986) 247.
[37] N. Reshetikhin, LOMI preprint E-4-87, E-17-87
[38] A. N. Kirillov and N. Reshetikhin, LOMI preprint E-9-88
[39] P. P. Kulish and N. Yu. Reshetikhin, J. Soviet Math. 23 (1983) 2435.
[40] V. G. Drinfel'd, Doklady Akad. Nauk. SSSR 283 (1985) 1060.
[41] M. Jimbo, Lett. Math. Phys. 10 (1985) 63.
[42] G. Keller, lecture at this workshop.
[43] A. B. Zamolodchikov, Int. Journ. of Mod. Phys. A4 (1989) 4235.
[44] V. A. Fateev and A. B. Zamolodchikov, preprint 1989, *Conformal Fiel d Theory and Purely Elastic S-matrices*.
[45] T. Eguchi and S-K. Yang, *Deformations of Conformal Field Theories and Soliton Equations*, Kyoto preprint RIFP-797, March 1989.
[46] A. Bilal and J-L. Gervais, Phys. Lett. B 206 (1988) 412.
[47] V. Fateev and S. Lykanov, Int. Jour. Mod. Phys. A3 (1988) 507.

[48] P. Christe and G. Mussardo, *Integrable Systems away from Criticality: the Toda Field Theory and S-matrix of the Tricritical Ising Model*, Santa Barbara preprint UCSBTH-89-19, May 1989.

[49] T. Hollowood and P. Mansfield, *Rational Conformal Field Theories at and Away from Criticality as Today Field Theories*, Oxford preprint 89-17P, May 1989.

[50] H. W. Braden, E. Corrigan, P. E. Dorey, and R.Sasaki, Durham Preprint UDCPT-89-23.

[51] D. Bernard and A. LeClair, in preparation

[52] F. A. Smirnov, Leningrad preprint, *Reductions of Quantum Sine-Gordon Model as Perturbations of Minimal Models of Conformal Field Theory*, E-4-89, February 1989.

[53] P. G. O. Freund, T. R. Klassen, and E. Melzer, Chicago preprint, *S-Matrices for Perturbations of Certain Conformal Field Theories*, EFI 89-29, June 1989.

TANGLES, LINKS AND TWISTED QUANTUM GROUPS

H.C. Lee

Theoretical Physics Branch, Chalk River Nuclear Laboratories
Atomic Energy of Canada Limited, Research Company
Chalk River, Ontario, Canada K0J 1J0
and
Dept. of Applied Mathematics, University of Western Ontario
London, Ontario, Canada N6A 5B9

1. Introduction

In this paper we report three main results: (a) An algebraic-geometric construction of universal link invariants in quantum groups based on Reidemeister's theorem; (b) A similar construction of universal tangle invariants -- diagrammatically a tangle is a link diagram with one external edge cut -- and a proof showing that, with a certain restriction on the quantum group, the set of all tangle invariants forms a subset of the centre of the quantum group; (c) A demonstration that a classical simple Lie (and Kac-Moody) algebra can be at least twice deformed to give a "twisted" quantum group with an associated twisted Hopf structure -- the classical Alexander-Conway link Polynomial is the simplest link invariant constructed from the twisted quantum group of $sl(2)$. These results are used to derive a number of theorems on the quantum group invariants of tangles, knots and links and their representations. They are also used to give insight to our understanding of the relation between quantum groups

* *Based on presentation given at Workshop on Physics, Braids & Links, Banff NATO ASI on Physics, Geometry & Topology, Banff, Alberta, Canada, 1989 August 14-25. Work supported in part by a Canadian NSERC Grant and a NATO Coll. Res. Grant.*

Physics, Geometry, and Topology
Edited by H. C. Lee
Plenum Press, New York, 1990

and conformal and topological field theories. It appears that, effectively, only the maximum Abelian subalgebra of the Hopf structure of the quantum group acts in these theories. It is shown that a tangle to a link in knot theory is what a "Wilson tangle" is to a Wilson line in the Chern-Simons-Witten topological field theory. The tangle theorem derived in (b) asserts that the set of all Wilson tangles forms a U(1) group, which is a natural holonomy group of the CSW theory.

In the last few years evidence has been accumulating to show that link invariants and quantum groups[1,2,3] provide an underlying structure common to a large and diverse array of topics in mathematics and low-dimension physics including braid group representations,[3-6] knot and link theory,[3-9] fractional statistics,[10-13] exactly solvable[14] statistical and lattice models,[2,2a,4,6,15] conformal field theory in two dimensions[16] and quantum theory[17] and topological field theory[18] in three dimensions. Most constructions of link invariants in quantum groups have been based on representation theory. In this scheme, a representation of the universal $\mathcal{R}$-matrix of the quantum group is used to generate a representation for Artin's braid group,[19] from which a link invariant is then constructed using Markov's theorem.[20] Recently Lawrence,[9] still utilizing the braid group and Markov's theorem, constructed a link invariant that is a universal (*i.e.*, representation independent) invariant of the quantum group. Our construction is based on Reidemeister's theorem[21]; it makes reference to neither the braid group nor Markov's theorem. While our approach is very geometric, we also make full use of the algebraic properties of the quantum group.

Our construction is similar to the state-model construction of link invariants (not universal) by Kauffman[8], also based on Reidemeister's theorem (but not in the context of a quantum group). It is closest in spirit to Witten's construction in the context of a topologically invariant field theory in three dimensions -- in our construction one may view elements of the Hopf algebra as particles moving in a three-dimensional manifold acted on by the topological Chern-Simons action.

Another aspect that sets the present approach apart from previous ones is the identification of *three* elements in the quantum group as fundamental for the construction of a universal link invariant. For terminology, by quantum group we shall mean the Hopf algebra, or the universal enveloping algebra $\mathcal{A}$ of the q-analogue g' of the the Lie algebra g. The three fundamental elements are the well-known universal matrix $\mathcal{R} \in g' \otimes g'$, the universal element $h \in g'$, and the central element $\lambda \in$ centre of g'. A careful study of the properties of these elements and their relation to the link invariant yields the second and perhaps most important result of this paper: the set of all tangle invariants in a quasitriangular quantum group whose h is not unipotent in g'/centre forms a subset of the centre of the quantum group. We believe all quantum groups of simple Lie algebras belong to the above restricted type. The relation between the tangle invariant $\mathcal{V}$ and the associated link invariant P is simple: since $\mathcal{V}$ is in the centre of g', $\mathcal{V} = \tilde{\mathcal{V}} e_0$, where e_0 is the identy element in g'; then $P = \tilde{\mathcal{V}}\, \mathrm{Tr}(h)$ to within a normalization (given explicitly in the text), where Tr maps g' to $\mathbb{C}$ and is invariant under cyclic permutations of its argument. In matrix representation, Tr is just the trace of the matrix. One may view the universal tangle invariants as an infinite set of Casimir operators of the quantum group, and think of P as their h-weighted traces. There is however an important difference in the relations between a Casimir operator and its trace in a Lie algebra and between $\mathcal{V}$ and P in a quantum group. In a Lie algebra, the ratio of the eigenvalues of a Casimir operator and its trace is just the dimension of the representation. In a quantum group the ratio, proportional to the eigenvalue of Tr(h), is a nontrivial functional of the representation and may also be a function of the deformation parameter q, such that for many representations Tr(h) vanishes for certain values of q. There are also known representations for twisted quantum groups in which the eigenvalues of Tr(h) vanish identically. The point is that tangle invariants are the more fundamental invariants, and that it is possible always to normalize $\mathcal{V}$, but not always P, such that the eigenvalue for the unknot is, say, unity. We show that the Alexander-Conway polynomial is precisely such a $\mathcal{V}$ whose corresponding P is identically zero.

From the tangle theorem one can deduce a number of interesting consequences, some of which are: the group of all (quantum group) tangles is homomorphic to a U(1) group; any link invariant is the commuting product of invariants of irreducible tangles (an irreducible tangle cannot be separated into two tangles by cutting any one of its strings); invariants of links that are nonisotopic but whose tangles factorize into the same set of irreducible tangles are degenerate. These results apply to universal invariants, but it is straightforward to take representations of such invariants, including the most general case of one representation for each component of a link. Except for accidental ones, the degeneracy mentioned above is generally removed when all the representations for the components in a link are distinct.

A very current and far from completely understood subject is the relation between quantum groups and the CSW and conformal field theories. Whereas much previous discussions on quantum group invariants have focussed on their representation theory and have restricted q to be *not* an integral root of unity (probably because for such values of q the Hopf algebra has a complicated and not fully understood ideal), many properties of the CSW and conformal field theories can be identified with those of representations of the quantum group *only* if it is assumed that the value of q *is* some integral root of unity. These two points of view are reconciled in a universal link (or tangle) invariant whose representations are well-defined for any value of q. It turns out that the effect of q being equal to a certain integral root of unity is to set to zero the representation for the noncommuting sector of the quantum group. In other words, the effect is equivalent to restricting the quantum group such that only its maximum Abelian Hopf subalgebra acts in these theories; the fact that q is an integral root of unity also makes the algebra of finite order. Since an Abelian group is homomorphic to a U(1) group, this explains why all the field intertwinings in conformal field theory and all the Wilson lines in the CSW theory give simple phase factors. (Of course, the fact that these are simple phase factors does not necessitate the restriction mentioned above.)

It appears that in conformal field theory the restriction on q can be traced to the requirement that a two-point correlation function has a U(1) monodromy group. Translated into the language of quantum group, it is a requirement that $\mathcal{R}(\mathcal{T}\cdot\mathcal{R})$ is diagonal in $g' \times g'$; $\mathcal{T}$ is the transposition operator. For all the representations of nontwisted quantum groups that we know, this restriction on q also assures that the representation of h is proportional to the identity matrix. This is consistent with the observation that, whereas h is indispensible in the construction of quantum group universal link invariants, its role, if any, is invisible in Witten's construction of the representations of presumably the same link invariants in the CSW theory. For representations of twisted quantum groups, the monodromy restriction on q does not make the representation of h prpoportional to the identity matrix. Our feeling is that link invariants of the twisted quantum groups, such as the Alexander-Conway polynomial[22], as well as invariants of nontwisted quantum groups with arbitrary values of q, cannot be computed as a Wilson line without some modification to the CSW theory.

An interesting application of the tangle theorem to the CSW theory is the recognition that if one neglects to take the trace of the representation on one of the contours in Witten's Wilson line, then the resulting object -- we call it a Wilson tangle -- is a link invariant times the identity matrix. The U(1) group of Wilson tangles can be viewed as the natural generalized holonomy group of the CSW theory. It is generalized because not only can the closed loops defining the holonomy be knotted, they can also be multicomponent links. In fact the U(1) nature of tangles is more persistent than what is revealed in the CSW theory which, as mentioned earlier, sees only the Abelian structure of the quantum group. Another related application that will be developed elsewhere is the generalization of fractional statistics to nonscalar anyons, the representations of whose exchange algebra generate a nonunitary representations of the braid group.

2. Quantum Group and Hopf Algebra

Notation

$\mathfrak{g}$	:	simple Lie algebra or bialgebra
$\mathfrak{g}'$	:	q-analoque of $\mathfrak{g}$
q	:	deformation parameter $q \equiv \exp(\hbar)$
$\mathcal{A}$	:	Hopf algebra or quantum group of $\mathfrak{g}$
m	:	multiplication map $\mathfrak{g}' \otimes \mathfrak{g}' \to \mathfrak{g}'$
Δ	:	comultiplication map $\mathfrak{g}' \to \mathfrak{g}' \otimes \mathfrak{g}'$
ε	:	counit map $\mathfrak{g}' \to \mathfrak{g}'$
S	:	antipode map $\mathfrak{g}' \to \mathfrak{g}'$ (antiautomorphic)
e_σ	:	basis for Borel subalgebra $\mathscr{C}_-$ of $\mathfrak{g}'$
e^σ	:	basis dual to e_σ for Borel subalgebra $\mathscr{C}_+$ of $\mathfrak{g}'$
$\mathscr{T}$	:	transposition operator in $\mathfrak{g}'$
$\mathscr{R}$	:	invertible universal R-matrix in $\mathcal{A}$; $\mathscr{R} \equiv e_\sigma \otimes e^\sigma$
$h, \bar{h}$	:	invertible universal element in $\mathfrak{g}'$; $h \equiv e_\sigma S(e^\sigma)$, $\bar{h} = S(e^\sigma) e_\sigma$
λ	:	central element in $\mathfrak{g}'$; $\lambda \equiv h\bar{h}$

For quantum group we adopt the definition of Drinfel'd.[1] Simply, given a simple Lie bialgebra $\mathfrak{g}$ with commutation bracket [,], we q-deform it to yield $\mathfrak{g}'$ and then "quantize" $\mathfrak{g}'$ by demanding that its commutation bracket {,} has as the small $\hbar$ limit [,]. The quantized $\mathfrak{g}'$, which we shall refer to as the q-analogue of $\mathfrak{g}$, has an associative multiplication m: $\mathfrak{g}' \otimes \mathfrak{g}' \to \mathfrak{g}'$: and a coassociative comultiplication $\Delta : \mathfrak{g}' \to \mathfrak{g}' \otimes \mathfrak{g}'$. These induce a unique counit $\varepsilon : \mathfrak{g}' \to \mathfrak{g}'$ which is a homomorphism, and a unique antipode $S : \mathfrak{g}' \to \mathfrak{g}'$ which is an antiautomorphism (*i.e.* S(ab) = S(b)S(a); a,b $\in \mathfrak{g}'$). An algebra equipped with the set $(m, \Delta, \varepsilon, S)$ is a Hopf algebra $\mathcal{A}$. So a quantum group is a Hopf algebra. We shall use these two terms in the same sense. The presence of comultiplication implies that the Hopf algebra contains $(\mathfrak{g}')^{\infty}$, therefore the Hopf algebra is also referred to as the quantized universal enveloping algebra $\mathscr{U}_q(\mathfrak{g})$ of $\mathfrak{g}$. As a simple example consider the q-analogue of the Lie algebra $s\ell(2)$, with generators H, X^- and X^+, and brackets

$$[\,H, X^{\pm}] = \pm 2X^{\pm} \tag{2.1}$$

$$[\,X^{+}, X^{-}] = (q^{H}-q^{-H})/(q-q^{-1}) \tag{2.2}$$

which in the limit $\hbar \to 0$ are just the brackets of $\mathcal{A}(2)$ (We shall actually use the brackets [,] instead of {,} for $\mathcal{g}'$). The Borel subalgebra $\mathcal{C}_-$ is generated by H and X^-, and its dual $\mathcal{C}_+$ by H and X^+. Define

$$k \equiv q^{H/2} \tag{2.3}$$

Then the Hopf algebra $\mathcal{A}$ is generated by $\{k^{\pm 1}, X^+, X^-\}$ with

$$\Delta(k) = k \otimes k; \qquad \Delta(X^{\pm}) = X^{\pm} \otimes k + k^{-1} \otimes X^{\pm} \tag{2.4}$$

$$S(k) = k^{-1}; \qquad S(X^{\pm}) = -kX^{\pm}k^{-1} \tag{2.5}$$

$$\varepsilon(k) = 1; \qquad \varepsilon(X^{\pm}) = 0 \tag{2.6}$$

It is clear that the bases e_σ and e^σ contain the infinite sets

$$\{e_\sigma\} = \left\{k^p (X^-)^q \ ; \quad p \in \mathbb{Z},\ q \in |\mathbb{Z}|\right\} \tag{2.7}$$

$$\{e^\sigma\} = \left\{k^p (X^+)^q ; \quad p \in \mathbb{Z},\ q \in |\mathbb{Z}|\right\} \tag{2.8}$$

However, a representation of $\mathcal{g}'$ may be finite. It has been shown[3] that for any simple Lie or Kac-Moody algebra $\mathcal{g}$ there is a one-to-one correspondence between representations of $\mathcal{g}$ and (the untwisted; *see* §7) $\mathcal{g}'$.

If $\mathcal{g}$ is of rank r, meaning that it has r simple roots α_i, i=1,.., r, then $\mathcal{g}'$ is generated by $\{H_i, X_i^+, X_i^-;\ i = 1,.., r\}$ with brackets

$$[\ H_i, H_j\ ] = 0 \tag{2.9}$$

$$[\ H_i, X_j^{\pm}\ ] = \pm 2(\alpha_i \bullet \alpha_j / \alpha_j^2) X_j^{\pm} \tag{2.10}$$

$$[\ X_i^+, X_j^-\ ] = \delta_{ij}(k_i^2 - k_i^{-2})/(q-q^{-1}) \tag{2.11}$$

and maps for $k_i \equiv q^{H_i/2}$ and $X_i^{\pm}$

$$\Delta(k_i) = k_i \otimes k_i; \qquad \Delta(X_i^{\pm}) = X_i^{\pm} \otimes k_i + k_i^{-1} \otimes X_i^{\pm} \tag{2.12}$$

$$S(k_i) = k_i^{-1}; \qquad S(X_i^{\pm}) = -q^{H_\rho}\, X_i^{\pm}\, q^{-H_\rho} \tag{2.13}$$

$$\varepsilon(k_i) = 1; \qquad \varepsilon(X_i^{\pm}) = 0 \tag{2.14}$$

where the subscript ρ in (2.13) refers to the half sum of the positive roots of $\mathcal{g}$. Given any root α, with the expansion in simple roots α_i

$$\alpha = n_i \alpha_i \tag{2.15}$$

in the Chevalley basis, we define

$$H_\alpha = n_i(\alpha \bullet \alpha_i / \alpha^2) H_i\ . \tag{2.16}$$

There is a complicated relation[2,3] between $X_i^{\pm}$ and $X_j^{\pm}$, $i \neq j$, which we shall not give, since it will not be needed in this paper.

We now summarize the important properties of $\mathcal{A}$. Because e_σ and e^σ are dual bases, their multiplication and comultiplication are related by

$$m(e_\sigma \otimes e_\rho) \equiv e_\sigma e_\rho = m^\tau_{\sigma\rho} e_\tau ; \qquad \Delta(e^\tau) = m^\tau_{\sigma\rho}\, e^\sigma \otimes e^\rho \tag{2.17}$$

$$\Delta(e_\tau) = \mu^{\sigma\rho}_\tau\, e_\sigma \otimes e_\rho; \qquad m(e^\rho \otimes e^\sigma) \equiv e^\rho e^\sigma = \mu^{\sigma\rho}_\tau e^\tau \tag{2.18}$$

where m and μ are c-numbers. Comultiplication is coassociative,

$$(\mathrm{id} \otimes \Delta)\Delta(a) = (\Delta \otimes \mathrm{id})\Delta(a), \qquad \forall\ a \in \mathcal{g}' \tag{2.19}$$

The counit is uniquely determined by Δ and S by

$$m(\mathrm{id} \otimes S)\Delta(a) = m(S \otimes \mathrm{id})\Delta(a) = \varepsilon(a)e_0, \qquad \forall\ a \in \mathcal{g}' \tag{2.20}$$

e_0 is the identity element in $\mathcal{g}'$. Define the skew antipode S_0 by

$$m(\mathrm{id} \otimes S_0)\mathcal{T}\bullet\Delta(a) = m(S_0 \otimes \mathrm{id})\mathcal{T}\bullet\Delta(a) = \varepsilon(a)e_0 \ . \tag{2.21}$$

Then S_0 is the inverse of the antipode; $SS_0 = S_0S = \mathrm{id}$, with the properties

$$S(e_\sigma) \equiv S^\rho_\sigma e_\rho; \qquad S_0(e^\rho) = S^{-1}(e^\rho) = S^\rho_\sigma e^\sigma \tag{2.22a}$$

$$S_0(e_\sigma) = S^{-1}(e_\sigma) \equiv (S^{-1})^\rho_\sigma e_\rho; \qquad S(e^\rho) = (S^{-1})^\rho_\sigma e^\sigma \tag{2.22b}$$

The three most important elements in $\mathcal{A}$ are the universal $\mathcal{R}$-matrix

$$\mathcal{R} \equiv e_\sigma \otimes e^\sigma \in \mathcal{g}' \otimes \mathcal{g}' \tag{2.23}$$

and

$$h \equiv e_\sigma S(e^\sigma) = S^{-1}(e_\sigma)e^\sigma \in \mathcal{g}' \tag{2.24}$$

$$\bar{h} \equiv S(e^\sigma)e_\sigma = e^\sigma S^{-1}(e_\sigma) \in \mathcal{g}' \tag{2.25}$$

The following properties are well known:[3,23,24] $\mathcal{R}$ is invertible with

$$\mathcal{R}^{-1} = (S \otimes \mathrm{id})\mathcal{R} = (\mathrm{id} \otimes S_0)\mathcal{R} \tag{2.26}$$

It then follows that

$$\mathcal{R} = (S \otimes S)\mathcal{R} = (S_0 \otimes S_0)\mathcal{R} \tag{2.27}$$

The Hopf algebra $\mathcal{A}$ is *quasitriangular* if it skew-commutes with $\Delta(a)$

$$(\mathcal{T}\bullet\Delta(a))\mathcal{R} = \mathcal{R}\Delta(a), \qquad \forall\ a \in \mathcal{g}' \tag{2.28}$$

It is possible to use eqs. (2.17,18,20,22,28) as constraints to construct[29] a finite dimensional representation of a Hopf algebra.

If $\mathcal{A}$ is quasitriangular then h and $\bar{h}$ commute and their product lie in the centre of $\mathcal{g}'$,

$$\mathcal{h}\bar{\mathcal{h}} = \bar{\mathcal{h}}\mathcal{h} \equiv \lambda \in \text{centre of } g' \qquad (2.29)$$

This means that in any representation π of g' (or $\mathcal{A}$) λ is a c-number and and $\pi(\mathcal{h})$ and $\pi(\bar{\mathcal{h}})$ are mutual reciprocals to within a factor of $\pi(\lambda)$. Eq. (2.30) and the invertibility of $\mathcal{h}$ and $\bar{\mathcal{h}}$ are derivable from (2.21,22,26,28)

$$\mathcal{h}^{-1} = S^2(e_\sigma)e^\sigma = e_\sigma S^{-2}(e^\sigma) \qquad (2.30)$$

$$\bar{\mathcal{h}}^{-1} = S^{-2}(e^\sigma)e_\sigma = e^\sigma S^2(e_\sigma) \qquad (2.31)$$

Conjugation with respect to $\mathcal{h}$ and $\bar{\mathcal{h}}$ are given respectively by

$$\mathcal{h}a\mathcal{h}^{-1} = S^{-2}(a), \qquad \bar{\mathcal{h}}a\bar{\mathcal{h}}^{-1} = S^2(a) \qquad \forall\, a \in g' \qquad (2.32)$$

For quantum groups of simple Lie algebras S is not an involution: $S^2(k_i) = k_i$ but $S^2(X_i^{\pm}) \neq X_i^{\pm}$. This means that $\mathcal{h}$ and $\bar{\mathcal{h}}$ (do not) commute with (any of) all the $(X_i^{\pm})$ k_i generators of g'. Therefore $\mathcal{h}$ and $\bar{\mathcal{h}}$ lie in the invariant subalgebra g' generated by the k_i's. It then follows that $\mathcal{h}$ has the expansion

$$\mathcal{h} = c_{[p]}\, e_{[p]} \equiv c_{[p]} \left(\prod_{i=1}^{r} k_i^{p_i} \right), \qquad \{e_{[p]}\} \subset \{e_\sigma\} \qquad (2.33)$$

summed over the integer sets $[p] = [p_1,\dots,p_r]$. Similarly for $\bar{\mathcal{h}}$.

The universal $\mathcal{R}$-matrix is the key link between quantum groups and braid group representations, solutions of the Yang-Baxter equations, and exactly solvable lattice models. The universal elements $\mathcal{h}$, $\bar{\mathcal{h}}$ and λ are the essential additional ingredients needed for the construction of universal link invariants and tangles, and for understanding why the action of quantum groups is restricted to their maximum Abelian Hopf subalgebra in conformal and topological field theories. The *operator* relations in the direct product $g' \otimes g' \otimes g'$

$$(\Delta \otimes \mathrm{id})\mathcal{R} = \mathcal{R}_{13}\mathcal{R}_{23}(\Delta \otimes \mathrm{id}), \qquad (\mathrm{id} \otimes \Delta)\mathcal{R} = \mathcal{R}_{13}\mathcal{R}_{12}(\mathrm{id} \otimes \Delta) \qquad (2.34)$$

follow immediately from (2.17,18 and 28), where $\mathcal{R}_{13} = e_\sigma \otimes e_0 \otimes e^\sigma$, and so on. The relations are valid when acting on any element $a \otimes b \in g' \otimes g'$. They are the universal versions of the fusion-braiding relations in conformal field theories in two dimensions.[16] A customarily given, but weaker version of these relations is

$$((\Delta \otimes \mathrm{id})\mathcal{R}) = \mathcal{R}_{13}\mathcal{R}_{23}, \qquad ((\mathrm{id} \otimes \Delta)\mathcal{R}) = \mathcal{R}_{13}\mathcal{R}_{12} \qquad (2.34')$$

Another important consequence of quasitriangularity is the quantum Yang-Baxter,[25] or braid group relation[19]

$$\mathcal{R}_{12}\mathcal{R}_{13}\mathcal{R}_{23} = \mathcal{R}_{23}\mathcal{R}_{13}\mathcal{R}_{12} \qquad (2.35)$$

The following is a dictionary between quantum group, field theory and diagrammatics.

Hopf Algebra	**Field Theory**	**Diagrammatics**
element	field	
m,Δ	fusion rule	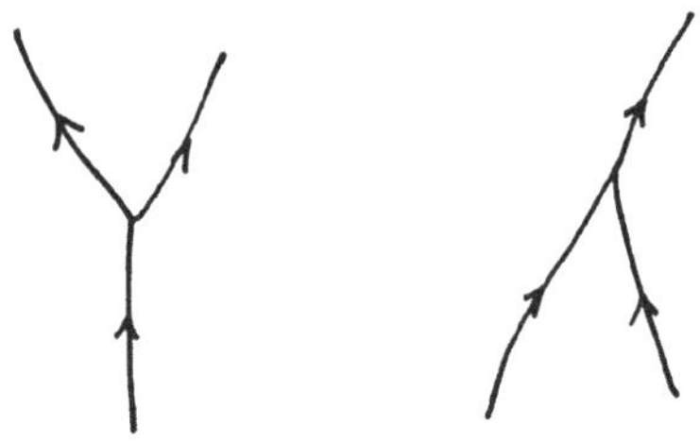
$\mathcal{R}$	intertwining of fields	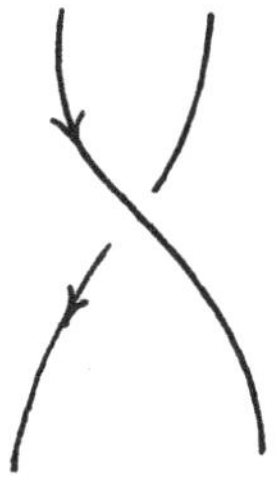

The following relations involving the universal elements h, $\bar{h}$ and λ are crucial for the construction of link invariants.

$$e_\sigma h e^\sigma = e^\sigma \bar{h} e_\sigma = e_0 \tag{2.35}$$

$$e^\sigma h S(e_\sigma) = S(e_\sigma)\bar{h}e^\sigma = \lambda \tag{2.36}$$

$$\begin{aligned} e_\sigma S(e_\tau) \otimes e^\tau h e^\sigma &= S(e_\sigma)e_\tau \otimes e^\tau h e^\sigma \\ &= e^\sigma e^\tau \otimes e_\tau h S(e_\sigma) = e^\sigma e^\tau \otimes S(e_\tau) h e_\sigma = e_0 \otimes h \end{aligned} \tag{2.37}$$

$$\begin{aligned} e_\sigma S(e_\tau) \otimes e^\tau \bar{h} e^\sigma &= S(e_\sigma)e_\tau \otimes e^\tau \bar{h} e^\sigma \\ &= e^\sigma e^\tau \otimes e_\tau \bar{h} S(e_\sigma) = e^\sigma e^\tau \otimes S(e_\tau) \bar{h} e_\sigma = e_0 \otimes \bar{h} \end{aligned} \tag{2.38}$$

We demonstrate the derivation of some of these relations:

(2.35): $$e^\sigma \bar{h} e_\sigma = m[(S \otimes \mathrm{id})\ \mathcal{T}\cdot(\mathcal{R}\mathcal{R}^{-1})] = e_0 \tag{2.39}$$

(2.35): $$e^{\sigma} h S(e_{\sigma}) = h S^2(e^{\sigma}) S(e_{\sigma}) = h S(e^{\sigma}) e_{\sigma} = h\bar{h} = \lambda \tag{2.40}$$

(2.37): $$e_{\sigma} S(e_{\tau}) \otimes e^{\tau} h e^{\sigma} = e_{\sigma} S(e_{\tau}) \otimes h S^2(e^{\tau}) e^{\sigma}$$

$$= e_0 \otimes h(\mathrm{id} \otimes S)\mathcal{R}^{-1}\mathcal{R} = e_0 \otimes h \tag{2.41}$$

The derivation of (2.32) is somewhat more complicated. Consider $\Delta(a) = u_i \otimes v_i$, $\forall\, a \in \mathcal{g}'$. From (2.21) and (2.28)

$$m(\mathrm{id} \otimes S)[\mathcal{R}^{-1}(\mathcal{T}\cdot\Delta(a))\mathcal{R}] = m(\mathrm{id} \otimes S_0)[\mathcal{T}\cdot\Delta(a)]$$

$$\text{left-hand side} = e_{\sigma} v_i e_{\rho} S(e^{\rho}) S(u_i) e^{\sigma}$$

$$\text{right-hand side} = v_i S^{-1}(u_i) = \varepsilon(a) e_0$$

left multiply by e_{τ} and right multiply by $S(e^{\tau})$ on both sides and use $e_{\tau}e_{\sigma} \otimes e^{\sigma}S(e^{\tau}) = \mathcal{R}^{-1}\mathcal{R} = e_0 \otimes e_0$ to obtain

$$v_i(h S(u_i) h^{-1} - S^{-1}(u_i)) = 0 \tag{2.42}$$

Let $a = k_i$ and substitute $\Delta(k_i) = k_i \otimes k_i$ into (2.42) to establish

$$h S(k_i) h^{-1} = S^{-1}(k_i) \tag{2.43}$$

Let $a = X_i^{\pm}$ and substitute $\Delta(X_i^{\pm})$ into (2.42) and use (2.43) to establish

$$h S(X_i^{\pm}) h^{-1} = S^{-1}(X_i^{\pm}) \tag{2.44}$$

This proves the first part of (2.32). The proof of the second part follows a similar route.

3. Universal Link Invariant

By an ℓ-component link[19] we mean the disjoint union of ℓ closed curves without self intersection in R^3. A knot is a one-component link. Link invariants are maps of links classified by ambient isotopy. We consider link diagrams which are two-dimensional projections of links. In these diagrams curves still do not intersect, but they do cross. According to a theorem by Reidemeister,[21] two unoriented links are equivalent (*i.e.*, ambient isotopic) if their corresponding diagrams can be deformed to each other by a sequence of moves composed of the three Reidemeister moves :

Reidemeister move I (3.1)

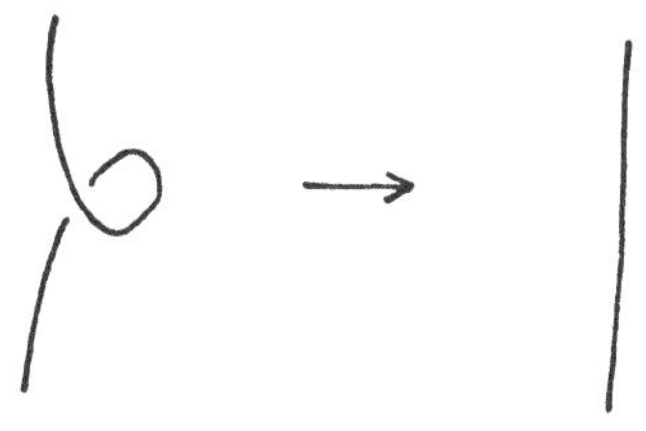

Reidemeister move II (3.2)

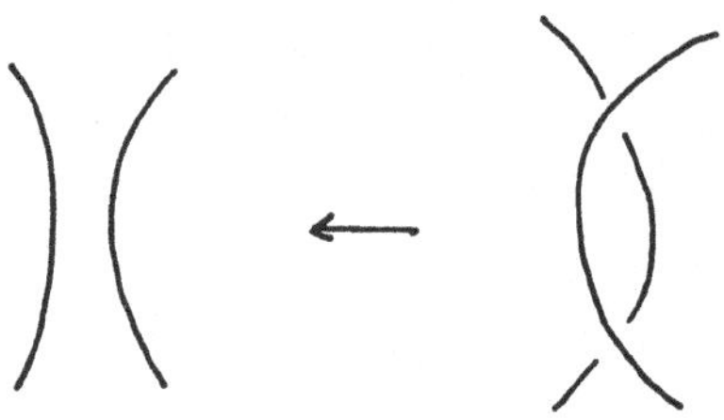

Reidemeister move III (3.3)

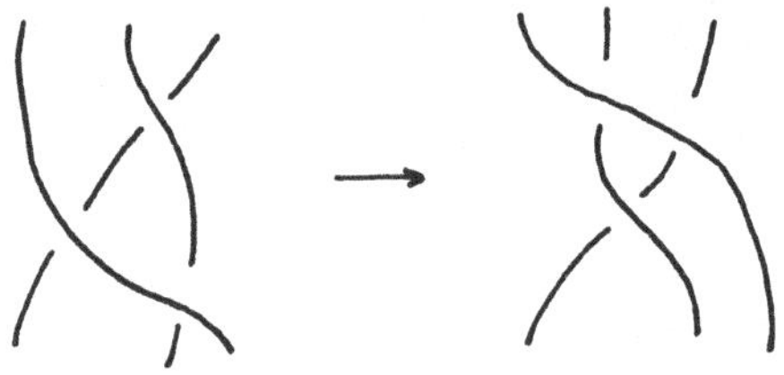

We use the words link and link diagram interchangably. The strategy of using the Reidemeister moves to construct link invariants in an algebraic context is to assign algebraic meanings to the curves and crossings in the diagrams above and show that the two sides in each of the moves are algebraically equivalent. For quantum groups there is a crucial complication: because the q-analogue $\mathcal{g}'$ (as well as $\mathcal{g}$) is noncommutative, the links are oriented, impliying that the curves in (3.1-3) must be directional, or have arrows attached to them. A sufficient generalization of the Reidemeister moves to cover the case of oriented links is the set[8]

Reidemeister move Ia (3.4a)

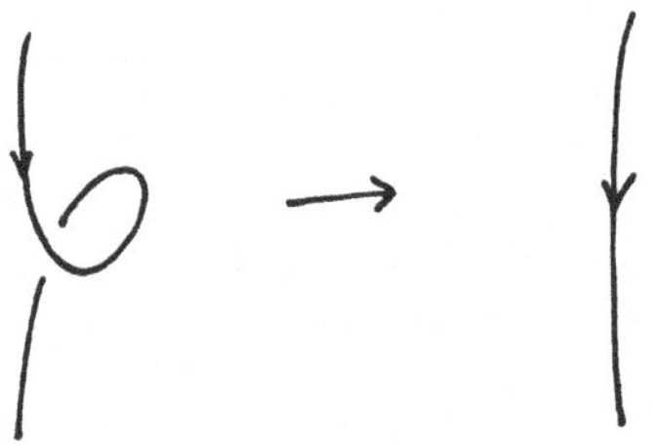

Reidemeister move Ib (3.4b)

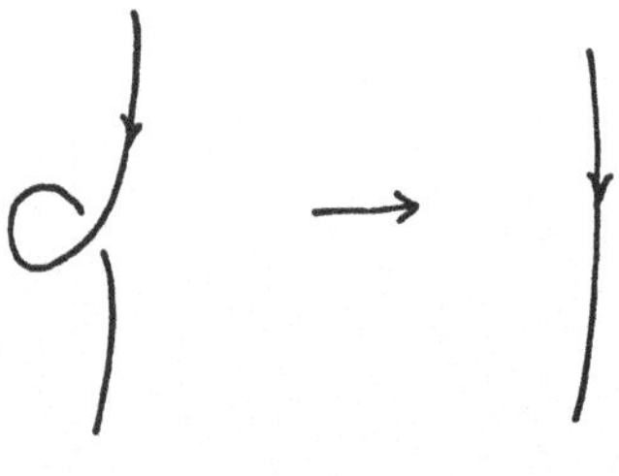

Reidemeister move IIa (3.5a)

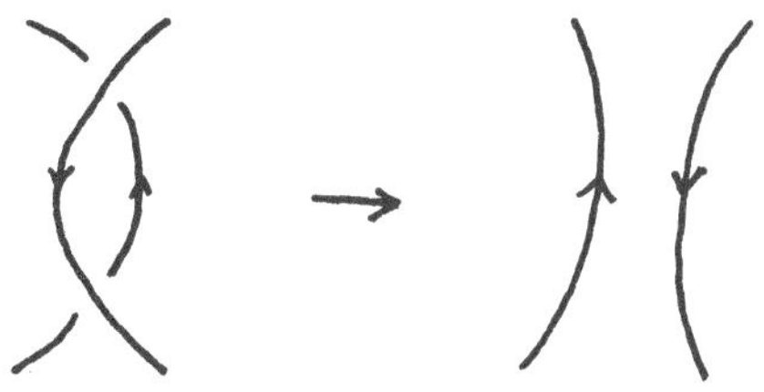

Reidemeister move IIb (3.5b)

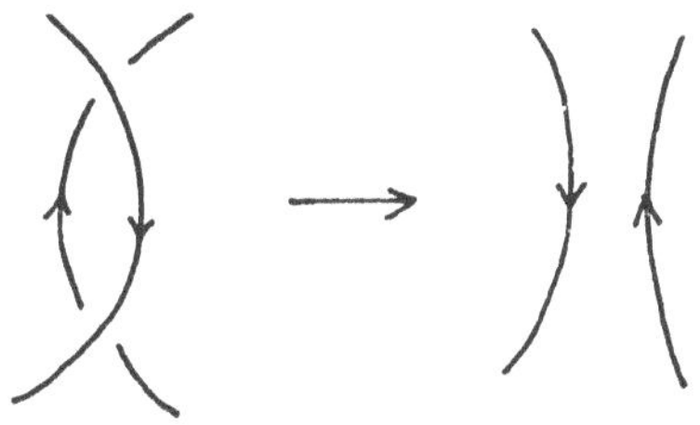

Reidemeister move III (3.6)

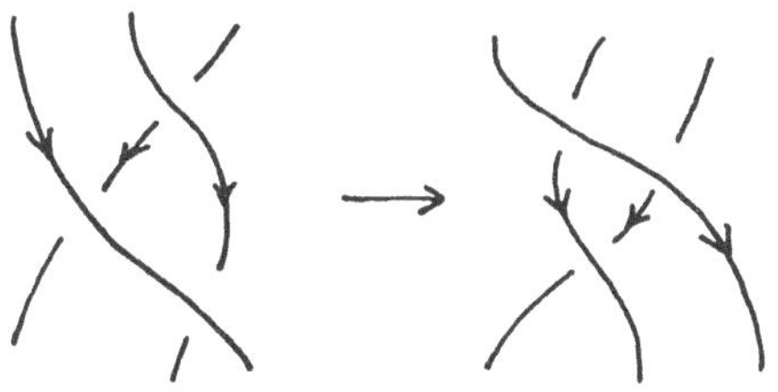

Kaufmann[8] has shown that given Ia, Ib, IIa, IIb and III, it is sufficient to deduce invariance under other type III Reidemeister moves in which one or two of the curves point upward.

We begin by making the following rules.

-- An arrowed curve is to be considered as an element in $\mathscr{g}'$.

-- A curve that does not cross with any other curve is assigned the identity element $e_0 \in \mathscr{g}'$.

-- A positive crossing is one in which the curve crossing from above is counterclockwise relative to the curve crossing from below; at a negative crossing the top curve is clockwise relative to the bottom curve.

-- A crossing is labelled by a dummy Greek index (ρ, σ, τ,..) and considered as an element in $\mathscr{g}' \otimes \mathscr{g}'$.

-- A positive (negative) crossing is assigned a value $\mathscr{R}$ ($\mathscr{T} \cdot \mathscr{R}^{-1}$); more precisely, at a positive (negative) crossing the top curve is given a value e_σ ($S(e_\sigma)$) and the bottom curve a value e^σ (e^σ).

-- Given a link diagram, deform it, without causing any new crossings to be generated or any existing ones to be eliminated, into one in which no two crossings are on the same lattitude; a section of a curve between two distinct crossings is an edge, and is said to have the right (wrong) direction if it is pointing generally downward (upward).

-- A right-direction edge is assigned the value e_0; a wrong-direction edge is assigned the element h ($\bar{h}$), if it is a section of an anticlockwise (clockwise) closed curve in the spliced diagram of the link.

-- If a curve self crosses, then the section of the curve between the crossing is considered to be a wrong-direction edge and is assigned h ($\bar{h}$) if it is anticlockwise (clockwise).

-- Every wrong-direction edge is to be marked by a solid triangle pointing to its right (left) if it is assigned the value h ($\bar{h}$).

-- A spliced diagram is obtained from a link diagram by replacing all crossings in the manner given below;

a spliced diagram is thus composed of a set of noncrossing closed curves. Two examples are

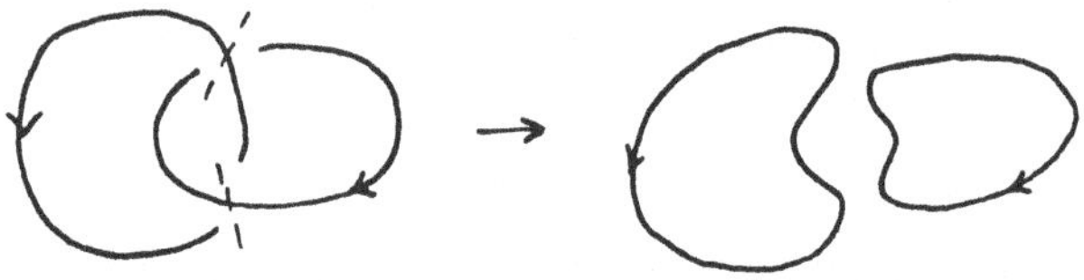

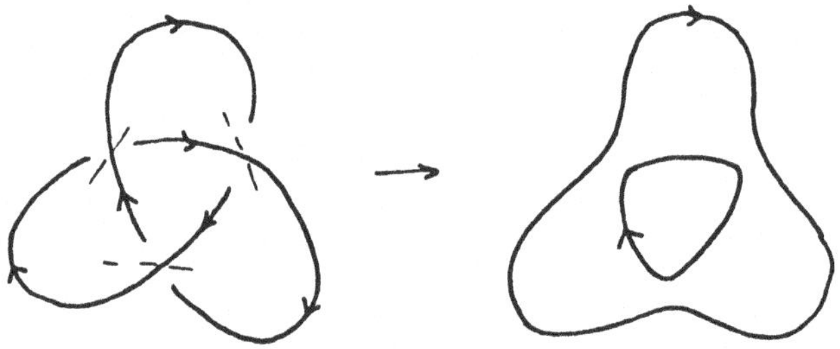

A completely labelled and marked trefoil is

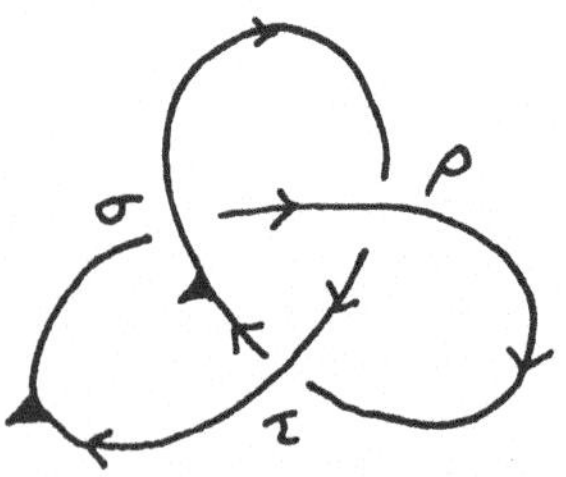

The above rules for assignment, which are completely specified by the arrows, labels and triangular marks on a link diagram, map (but not uniquely so) each component of a link into an element of $\mathscr{g}'$ composed of an ordered product of e_σ, $S(e_\sigma)$, e^σ, h and $\bar{h}$. The assignment thus maps an ℓ-component link into a direct product in $(\mathscr{g}')^\ell \in \mathscr{A}$. A possible result for the trefoil given above is

$$e_\sigma e^\rho e_\tau \bar{h}\, e^\sigma e_\rho e^\tau \bar{h}$$

Other possible results are obtained by cyclic permutations.

We now show that the map is invariant under the oriented Reidemeister moves. First we note that it is invariant under move IIa by definition, since $\mathscr{R}\mathscr{R}^{-1} = e_0 \otimes e_0$. Note that because of intertwining, the negative crossing is assigned $\mathscr{T}\bullet\mathscr{R}^{-1}$, not $\mathscr{R}^{-1}$. That the map is invariant under move III follows from the braiding relation (2.35). Invariance under Ia, Ib and IIb, which involve wrong-direction edges, come from the properties of h and $\bar{h}$:

$$e_\sigma h e^\sigma = e^\sigma \bar{h} e_\sigma = e_0 \qquad (3.7)$$

$$S(e_\sigma)\bar{h} e^\sigma = e^\sigma h S(e_\sigma) = \lambda \qquad (3.8)$$

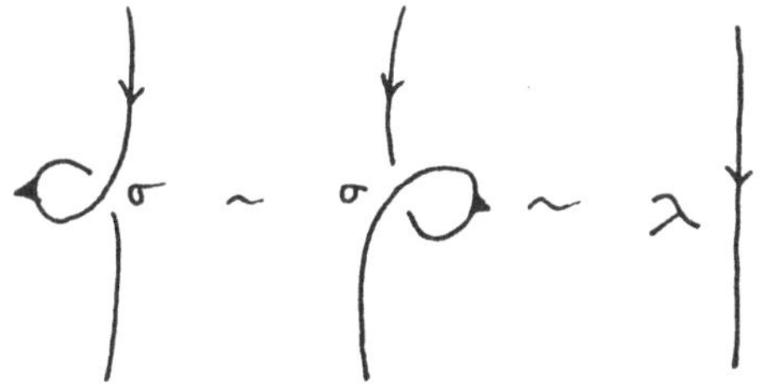

$$e_\rho h S(e_\sigma) \otimes e^\sigma e^\rho = e^\rho h e^\sigma \otimes e_\sigma S(e_\rho) = h \otimes e_0 \qquad (3.9)$$

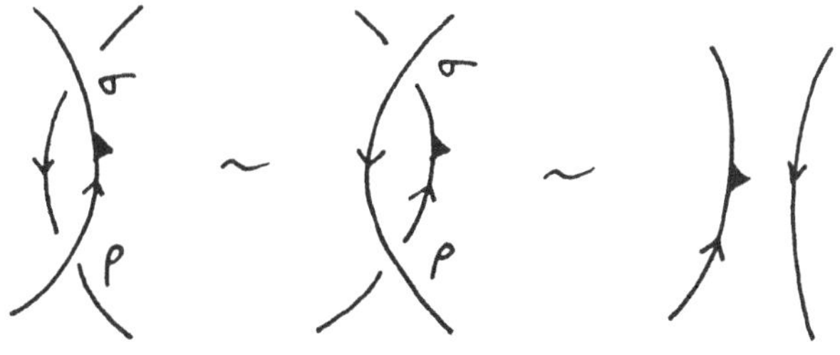

$$S(e_\sigma)e_\rho \otimes e^\rho \bar{\lambda} e^\sigma = e^\sigma e^\rho \otimes S(e_\rho)\bar{\lambda} e_\sigma = e_0 \otimes \bar{\lambda} \tag{3.10}$$

The right-hand side of (3.8) has a factor of λ, a central element in $\mathcal{g}'$ which in general is not equal to the identity element. For example, under the rules given above,

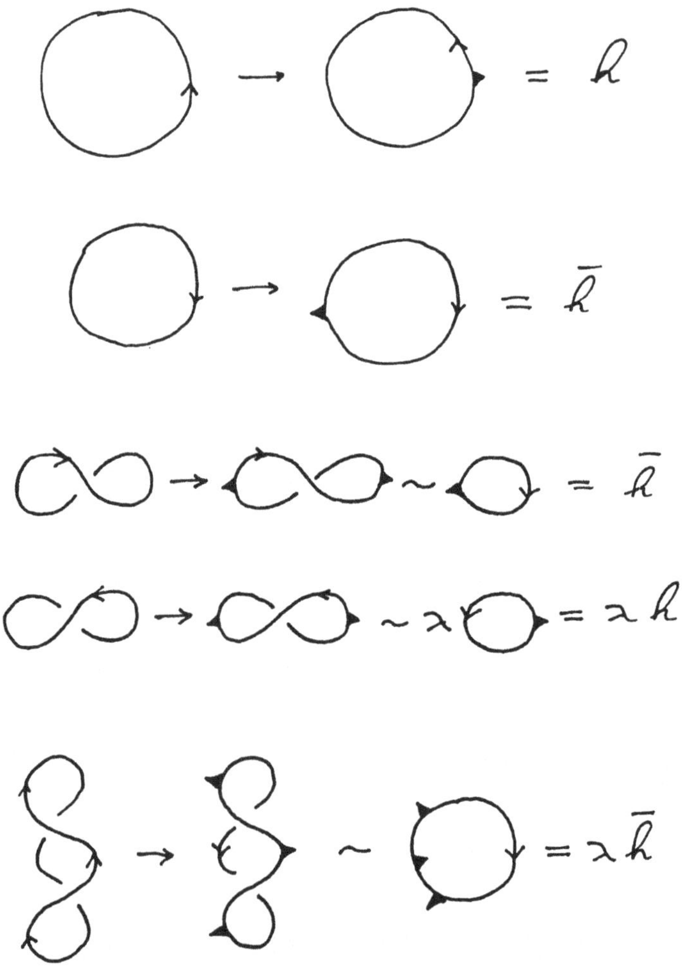

This suggests that the map needs to be properly normalized. The correct normalization factor, one for each component of the link, is

$$\prod_{m=1}^{\ell} \lambda^{(w_m - 2N_m)/4} \tag{3.11}$$

where $\lambda \equiv \chi e_0$, w_m is the total number of times the m*th* component curve passes through a positive crossing minus the number of times it passes through a negative crossing, and N_m is the total number of h's and $\bar{h}$'s on the component. Since each crossing is the intersection of two curves, the w's summed over all components is twice the writhe number, an invariant of the link. With this normalization, the examples given above become

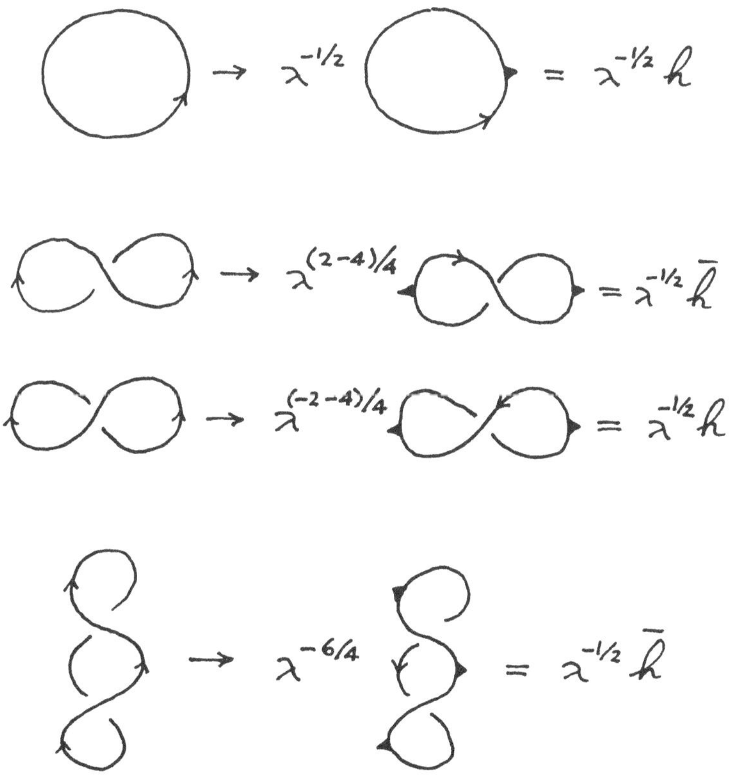

Finally, the normalized map is not unique unless each component is made invariant with respect to cyclic permutation of its ordered factors of e_σ's, e^σ's, h's and $\bar{h}$'s. For this purpose we define a map

$$\mathrm{Tr} : \mathcal{g}' \to \mathbb{C} \tag{3.12}$$

that is invariant under cyclic permutation. The notation suggests itself: for each matrix representation π of $\mathcal{g}'$, we take Tr to be the trace in π, denoted by Tr_π. Denote by a_m the element into which the m*th* component of the link is mapped. We now have:

<u>Link Invariant Theorem</u>. *The map P*

$$P : \text{Link} \to \mathbb{C}$$

by

$$P[L] = \prod_{m=1}^{\ell} \chi^{(w_m - 2N_m)/4} \, Tr(a_m) \qquad (3.13)$$

is a universal link invariant.

For the unknot, w=0, N=1 and $a=\bar{h}$ (h) if it is closed (counter)clockwise. The link invariant is therefore $\chi^{-1/2}$ times $Tr(\bar{h})$ or $Tr(h)$. But $Tr(\bar{h}) = Tr(h)$ because of the cyclic property of Tr. We therefore define

$$\tau = \chi^{-1/2} Tr(h) = \chi^{-1/2} Tr(\bar{h}) \qquad (3.14)$$

Note that τe_0 is an element in the centre of g'. Thus we have

The Unknot. *The link invariant for the unknot is equal to the element τ in g'/e_0,*

$$P[\text{unknot}] = \tau \in g'/e_0 \qquad (3.15)$$

It is important to realize that τ is representation-dependent, but not a c-number. Thus one may not normalize a universal link invariant such that its value for the unknot is always, say, unity. The implication of this becomes clear when one considers a multirepresentation, which we shall also refer to as a multicolour, link invariant. Such an invariant is trivially obtained from (3.13). We have the following,

Multicolour Link Invariant. *Let $L\{\pi_m;\ m=1,..,\ell\}$ be ℓ (possibly) distinct representations of g', then a (possibly) ℓ-coloured link invariant is*

$$(\pi_1 \otimes \cdots \otimes \pi_\ell) P[L] = \prod_{m=1}^{\ell} \chi_{\pi_m}^{(w_m - 2N_m)/4} \, Tr_{\pi_m}(a_m) \qquad (3.16)$$

where χ_π is value of χ in the π-representation.

4. Comparison with other Constructions

There are a number of constructions of link invariants based on quantum groups. We mention some of them for comparison; the list is not intended to be complete. Several authors use Alexander's theorem[26] and Markov's[20] theorem to construct the invariant via the braid group. Wadati and coworkers[4] work with solutions of the Yang-Baxter equation (but do not explicitly mention the quantum group), Reshetikhin[3] and Turaev[5] work in representation theory of the quantum group, whereas Lawrence[9] works in the algebra g'. All employ an enhanced Yang-Baxter set $\{\mathcal{R},\mathcal{M}\}$ where $\mathcal{R}$ is the universal $\mathcal{R}$-matrix, (for convenience we shall use the language of algebra, although all references except Lawrence speak of explicit representations) and $\mathcal{M} \in g'$ (for which Turaev uses the notation μ, instead of $\mathcal{M}$, Wadati, h, and Lawrence, X) satisfies the property

$$[\, \mathcal{M}\otimes\mathcal{M},\ \mathcal{R}\,] = 0 \qquad (4.1)$$

$$m\big((id \otimes \mathcal{M})\mathcal{R}\big) = z e_0; \qquad m\big((id \otimes \mathcal{M})\mathcal{T}\cdot\mathcal{R}\big) = z^{-1} e_0 \qquad (4.2)$$

In our notation, $\mathcal{M}$ is just the univeral element $\lambda^{-1/2} h$ and z is the square root of the element $\mathcal{X}^{-1}$. Lawrence, the only one of the above authors who constructed a universal link invariant, does not however identify X as a universal element in $\mathfrak{g}'$.

In addition to being universal, another significant difference between our method and those mentioned above is our use of Reidemeister's theorem instead of Markov's. Unlike the other methods, it is not necessary in our method to first reduce a link to the equivalent closed braid; the invariant can be directly constructed on the link. At the same time, in our method it is straightforward to compute a link invariant from a closed braid, as illustrated below:

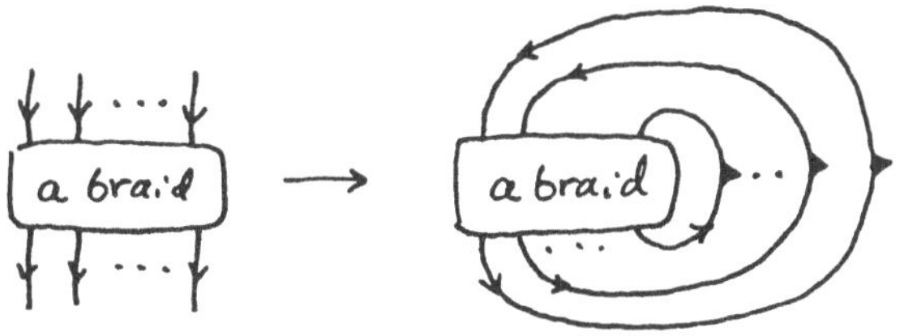

(4.3)

By definition all edges inside the box have the right direction, so they take values only at crossings. The only wrong-direction edges are those outside the box closing the braid counterclockwise, which is the standard convention. Therefore each wrong-direction edge is be assigned a value λ, as indicated in the diagram abovc. What if one chooses to close one or more of the edges clockwise? Our rule says that each one of such edges be assigned a value $\bar{\lambda}$. Since the resulting link is isotopic to the original one, the link invariant must not change. The diagram below suffices to show that this is indeed the case.

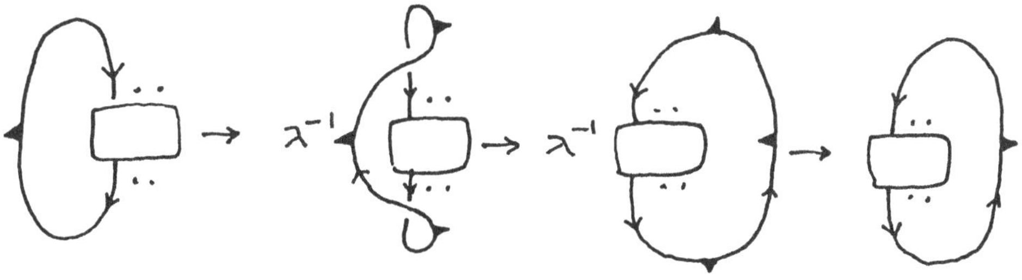

(4.4)

The first equivalence is from Reidemeister move I, and commutation of the wrong-direction edge with the braid is from moves II and III.

Our construction is closer in spirit to the bracket or state model construction of Kaufmann,[8] also based on invariance under the Reidemeister moves. We view it as being closest to a field theory construction.[18] The point is that in our construction the curves in a link are given primary attention -- crossings are treated as *interactions* which change the *values* of the curves, as opposed to the Markov construction where preeminence is given to the *order* of curves and their *braidings* at crossings.

A final remark that will be expanded later: because our construction is completely universal, its validity depends explicitly neither on the properties of any specific representation of $\mathfrak{g}'$ nor on the value of the deformation parameter q. In particular there is no intrinsic difficulty with our link invariant when q is equal to any rational root of unity.

5. Universal Tangle Invariant

We describe how to get a tangle diagram from a link diagram L. Our convention, as before, is that an edge with an arrow pointing downward has the right direction. First, use the rules given in section 3 to attach either h or $\bar{h}$ to all the wrong-direction edges in L. Then draw a box that completely encloses L. Now take any edge in L and pull it horizontally outside of the box. If the edge is wrong-directioned and has an h ($\bar{h}$) attached to it, pull it to the right (left) of the box. If the edge is right-directioned pull it either to the right or to the left of the box. If it is pulled to the right (left) of the box then give the edge an extra counterclockwise (clockwise) writhe and attach an h ($\bar{h}$) to the resulting loop, which by definition is wrong-directioned. It does not matter whether the writhe has a positive or a negative crossing. The above manouver only involves a number of Reidemeister moves, so the new link diagram L′ is isotopic to the original one. In other words, P[L′] = P[L]. Now remove all the attached h or $\bar{h}$ from L′, cut the pulled-out edge and pull the top (bottom) end of the cut edge towards $z=+\infty$ ($-\infty$). Remove the box. We shall call the resulting diagram a tangle T cut from the link L. It may be viewed as the two-dimensional projection of a tangle in three dimensions. It should have a right-directioned open string entering (leaving) the bulk of the tangle from (to) $z=+\infty$ ($-\infty$). Clearly there can be as many tangles cut from a link as there are edges in the link. Below are two examples:

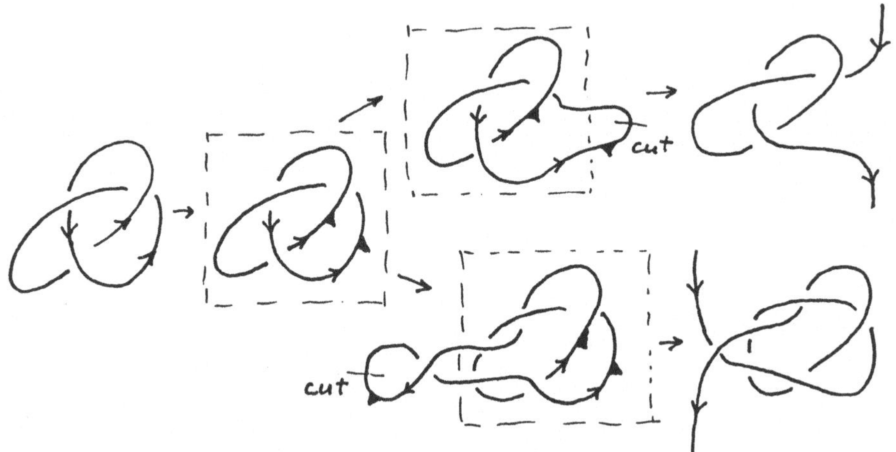

By isotopy of (oriented) tangles we shall mean the invariance of a tangle under the (oriented) Reidemeister moves, with the restriction that no curve is allowed to move beyond either end of the tangle. It is clear that if we assign values to various parts of a tangle in exactly the manner that was described in section 3 for a link, then, before taking the Tr map as defined in (3.12), the tangle is also mapped into an element in $(g')^{\otimes \ell}$ which is invariant under isotopy,

$$T \rightarrow a_1 \otimes a_2 \cdots \otimes a_\ell \tag{5.1}$$

where, as before, the element a_m corresponds to the m*th* component of the tangle. In our convention the first component refers to the cut string. To construct a tangle invariant, each element except a_1 -- it has a natural order by virtue of the cut -- needs to be made cyclically symmetric with respect to its product factors. This is again achieved with the map Tr in (3.12). A tangle invariant $\mathcal{V}$ is an element in $\mathfrak{g}'$

$$\mathcal{V}[T] = \mathcal{X}^{(w_1-2N_1)/4} a_1 \prod_{m=2}^{\ell} \mathcal{X}^{(w_m-2N_m)/4} \mathrm{Tr}(a_m) \in \mathfrak{g}' \tag{5.2}$$

where w and N are defined as before.

If the tangle in (5.1) is cut from the link L, then the latter is mapped to either the element

$$L \to a_1 h \otimes a_2 \otimes \cdots \otimes a_\ell \tag{5.3a}$$

if the cut edge was originally closed counterclockwise, or to the element

$$L \to a_1 \bar{h} \otimes a_2 \otimes \cdots \otimes a_\ell \tag{5.3b}$$

if the cut edge was originally closed clockwise. Eq. (4.4) shows that the link invariant does not depend on which way the cut edge is closed. Therefore

$$P[L] = \mathcal{X}^{-1/2}\mathrm{Tr}(\mathcal{V}[T]h) = \mathcal{X}^{-1/2}\mathrm{Tr}(\mathcal{V}[T]\bar{h}) \tag{5.4}$$

It follows that

$$\mathrm{Tr}\{\, (\mathcal{V}[T])(\, h-\bar{h}\,)\,\} = 0 \tag{5.5}$$

for any tangle T.

For convenience, we label tangles by the subscripts s, t,.. and write $\mathcal{V}[T_s]$ as $\mathcal{V}_s$. Given two tangles T_s and T_t, define the product tangle $T_{st} = T_sT_t$ as the tangle obtained by connecting the bottom end of T_s to the top end of T_t. Then T_{st} and T_{ts} are isotopic.

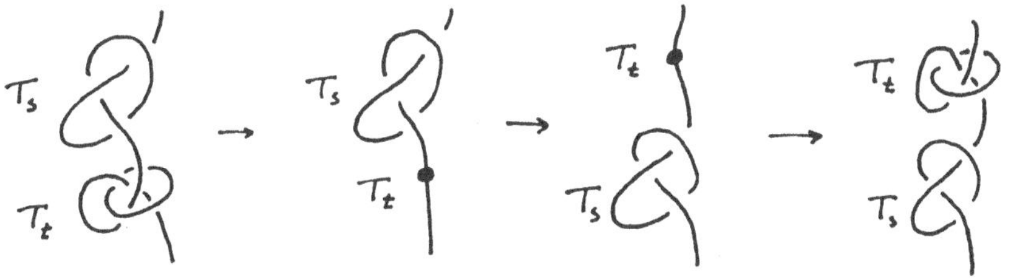

Equivalently, the links obtained by closing the two product tangles T_{st} and T_{ts} are isotopic. Since $\mathcal{V}$ is an isotopic homorphism of tangles, the set of all tangle invariants Z(T) forms an abelian subalgebra of $\mathfrak{g}'$,

$$[\mathcal{V}_s, \mathcal{V}_t] = 0, \qquad \mathcal{V}_s, \mathcal{V}_t \in \{\text{set of all tangles}\} \tag{5.6}$$

We now show that under a certain restriction on the quantum group, Z(T) is an invariant subalgebra of $\mathfrak{g}'$. First note that h commutes with all the commuting generators k_j of $\mathfrak{g}'$. Therefore h can be spanned by the commuting subset $\{e_{[p]}\}$ in the basis e_σ, (see (2.33)), which coincides with the commuting subset in e^σ,

$$e_{[p]} = \prod_{j=1}^{r} k_j^{p_j} , \qquad p_1,\ldots, p_r \in \mathbb{Z} \tag{5.7}$$

The identity is $e_0 = e_{[0]}$, and clearly,

$$e_{[p]}e_{[q]} = e_{[p+q]} \tag{5.8}$$

The map Tr in (3.12) and (5.2) is just an inner product in this basis. Write

$$\mathscr{h} - \bar{\mathscr{h}} = c_{[p]}e_{[p]} \tag{5.9}$$

where the c's are expansion coefficients and summation over [p] is understood. Then for the untangle (whose closure is the unknot)

$$\mathrm{Tr}(\mathscr{h}) - \mathrm{Tr}(\bar{\mathscr{h}}) = c_{[p]}\mathrm{Tr}(e_{[p]}) = 0 \tag{5.10}$$

Similarly, since the tangle invariants live in the commuting sector of $\mathscr{g}'$, we write

$$\mathscr{V}[T_s] = d^{(s)}_{[q]}\, e_{[q]} \tag{5.11}$$

From (5.5,8), for any tangle T_s

$$d^{(s)}_{[q]}c_{[p]}\mathrm{Tr}(e_{[p]}e_{[q]}) = d^{(s)}_{[q]}c_{[p]}\mathrm{Tr}(e_{[p+q]}) = 0 \tag{5.12}$$

Since this equation must be satisfied by the complete set of infinite number of distinct tangles, at least one of the conditions

(i) $\quad c_{[p]}\mathrm{Tr}(e_{[p+q]}) = 0, \qquad$ for every [q]

(ii) $\quad d^{(s)}_{[q]} = 0, \qquad$ for all $[q] \neq [0]$

must be true. Suppose (i) is true. Consider the set $\{[q]\} = \{[-p]\}$. Then a necessary condition that all c's do not vanish is that the determinant of the $\mathrm{Tr}(e_{[p+q]})$'s vanishes. Since e[p] span a (albeit infinite dimensional) linear vector space, and Tr is just the inner product, the determinant is, with proper normalization, proportional to the dimension of the vector space, which certainly does not vanish. Therefore we conclude that all c's must vanish. Or equivalently, $\mathscr{h}=\bar{\mathscr{h}}$. Since $\bar{\mathscr{h}}=\lambda\mathscr{h}^{-1}$, (i) is not true unless

$$\mathscr{h}^2 = \lambda \in \text{centre of } \mathscr{g}' \quad \text{(if (i) true)} \tag{5.13}$$

Note that (5.13) is a universal constraint on the algebra. A sufficient condition for it to be false is if it is false in *any* representation. As far as we know (5.13) is false for quantum groups of simple Lie algebras.

We therefore consider the case when (5.13) is not true . Then condition (ii) above must hold. In this case only $d^{(s)}_{[0]} \neq 0$, therefore $\mathscr{V}[T]$ for every tangle is proportional to the identity element in $\mathscr{g}'$. Note however, that although $\mathscr{V}[T]$ commutes with every other element in $\mathscr{g}'$, it is not a c-number times the identity element, but is an element in the centre of $\mathscr{g}'$; it is representation-dependent. We therefore have

Tangle Theorem: *Let $\mathcal{A}$ be the quantum group of $\mathcal{g}$ that has a quasitriangular Hopf algebra structure equipped with a noninvolutive antipode, $\mathcal{g}'$ the q-analoque of $\mathcal{g}$, $\hbar$ the universal element and λ the central element of $\mathcal{g}'$ as defined in (2.24,29), and $\hbar^2$ is not in the centre of $\mathcal{g}'$. The tangle invariant $\mathcal{V}$ defined in (5.2) is a universal isotopic map of tangles to the centre of $\mathcal{g}'$,*

$$\mathcal{V}[T] \equiv \mathcal{V}[T]\, e_0 \in \text{centre of } \mathcal{g}'. \tag{5.14}$$

Eq. (5.14) may be viewed as a generalization of the invariance under the Reidemeister move I. In what follows we restrict our considerations to quasitriangular quantum groups with noninvolutive antipodes and whose $\hbar^2$ are not in the centre of $\mathcal{g}'$. As far as we know, all quantum groups of simple Lie algebras $\mathcal{g}$ belong to this category. We then have,

Corollary: $\mathcal{V}[T]$ *is a universal tangle invariant.*

Corollary: Let {T; L} *be the set of tangles cut from the link L,* P[L] *be the universal link invariant defined by (3.13),* $\mathcal{V}[T]$ *be the universal tangle invariant defined by (5.2,14), then*

$$P[L] = \tau\, \mathcal{V}[T_s]; \qquad \forall\, T_s \in \{T;\, L\} \tag{5.15}$$

Corollary: *The element τ factors out from every universal link invariant.*

Recall that $\tau = \chi^{-1/2}\mathrm{Tr}(\hbar)$ is an element in the centre of $\mathcal{g}'/e_0$. One needs to be careful in the interpretation of (5.15). For in a representation of P, the colour (that is, representation) of τ and $\mathcal{V}$ must be the same. This is not a problem when L is a knot. In this case, the $\mathcal{V}$'s for all the T's cut from the knot are equal. This means $\mathcal{V}$ maps the isotopy of knots, which is greater then the isotopy of tangles. Thus we may define, for the equivalence class of knots [K]

$$Q[K] = \mathcal{V}\,[\text{any T cut from K}] \tag{5.16}$$

and have

Knot Theorem: Q[K] *is a universal knot invariant.*

The discussion above applies also to monocolour links. Therefore we have

Monocolour Link Theorem: Q[L] *is a universal invariant for mono colour links.*

The representation versions of (5.14 and 15) were first conjectured in ref. 26. A consequence of (5.14) was mentioned by Kauffman.[33] Note that for the unknot, P[unknot] = τ, while Q[unknot] = 1. This difference between these two invariants is significant in representation theory. If rep(τ) $\neq$ 0, then the representations of the two invariants differ by only a trivial normalization. But there are representations of quantum groups for which rep(τ) vanishes,[26,28] then rep(P) becomes a trivial map -- all links mapped to zero, while rep(Q) is not trivial. The well-known Alexander-Conway link polynomial[22] is precisely of this type.

We make a few remarks on the consequences of (5.14-16). We shall use the notion of "decorating (a component of) a link with T" to mean cutting the link at some point (on the component) and reconnecting it after inserting a tangle T at that point. If the undecorated component has value $\mathscr{a}$ in $\mathscr{g}'$, then after decoration it has value $\mathscr{a}\mathscr{V}[T]$. We refer to the link before (after) decoration as the skeleton (decorated) link. Since $\mathscr{V}[T]$ is in the centre of $\mathscr{g}'$, the invariant of the decorated link depends only on *which component* of the skeleton link is decorated, but does not depend on *where* on that component T is inserted.

Extended Casimir Operators of Quantum Group. Since Tr is invariant under cyclic permutation of its argument, $\mathscr{V}$ and P are invariant under similarity transformations of the basis. This, together with (5.14), means that the tangle invariants $\mathscr{V}$ can be viewed a set of infinite *extended Casimir operators* of the quantum group. Like $\mathscr{V}$, the Casimir operators $\mathscr{C}$ of a classical simple Lie (or Kac Moody) algebra $\mathscr{g}$ also lie in the centre of $\mathscr{g}$. Some significant difference between $\mathscr{V}$ and $\mathscr{C}$ is reflected in the nonunipotency of $\mathscr{h}$. For example in $\mathscr{g}'$, $e_\sigma \mathscr{h} e^\sigma = e_0$ and $S(e_\sigma)\mathscr{h}e^\sigma = \lambda$ are invariants but $e_\sigma e^\sigma$ is not. In Lie algebras a Casimir operator often refers to $\mathrm{Tr}(\mathscr{C})$ instead of $\mathscr{C}$. It is simply related to $\mathscr{C}$ by $\mathrm{Tr}(\mathscr{C})e_0 = \mathscr{C}\,\mathrm{Tr}(e_0)$. In a representation $\mathrm{Tr}(e_0)$ is just the dimension of the representation. In quantum groups, the quantity analogous to $\mathrm{Tr}(\mathscr{C})$ is P, but P and $\mathscr{V}$ are related by $Pe_0 = \lambda^{-1/2}\mathscr{V}\mathrm{Tr}(\mathscr{h})$, where $\mathrm{Tr}(\mathscr{h})$ is not simply related to the dimension of the representation unless $\mathscr{h}$ is unipotent. In fact, unlike $\mathrm{Tr}(e_0)$, $\mathrm{Tr}(\mathscr{h})$ can sometimes vanish.

U(1) Group of Tangles. Consider the function

$$\phi[T] = \mathscr{V}[T]/|\mathscr{V}[T]| \tag{5.17a}$$

and let G′ be the group of tangles equipped with the multiplication of $\mathscr{g}'$, then G′ *is an Abelian group, and*

$$\phi\colon G' \to [0,2\pi] \tag{5.17b}$$

is a U(1) *representation of* G′.

Chains. By an undecorated positive ℓ-chain we mean a union of ℓ circles, with the *kth* circle linked to the k+1st circle by two positive crossings, k=1 to ℓ-1.

A Hopf link is an undecorated positive 2-chain. Write the invariant of the tangle obtained by cutting the Hopf link as

$$\mathscr{V}[\mathrm{Hopf}] \equiv \mathscr{S}$$

$$= (\lambda^{-1/2} e_{\sigma} e^{\rho})\ \mathrm{Tr}(e^{\sigma} h e_{\rho}) \equiv \mathscr{S}_{o} =$$

(5.18)

Then *the link invariant for the Hopf link is*

$$\mathrm{P[Hopf]} = \mathscr{S}\,\tau \tag{5.19}$$

The tangle invariant for an ℓ-chain is therefore

$$\mathscr{V}[\ell\text{-chain}] = \mathscr{S}^{\ell_1} \tag{5.20}$$

and its link invariant is

$$\mathrm{P}[\ell\text{-chain}] = \mathscr{S}^{\ell_1}\,\tau \tag{5.21}$$

The above generalizes straightforwardly to chains with a combination of positive and negative crossings.

Chain of Tangles. *Let* L *be a decorated ℓ-chain whose mth component is decorated with the tangle* T_m, *then*

$$\mathrm{P[L]} = \left[\prod_{m=1}^{\ell} \mathscr{V}[T_m]\right] \mathscr{S}^{\ell_1}\,\tau \tag{5.22}$$

Link of Tangles. *Let the skeleton* L *be an ℓ-component link, and the decorated* L′ *be obtained by decorating* L *with a set of tangles* $\{T_s\}$ *anywhere, then*

$$\mathrm{P[L']} = (\prod_{s} \mathscr{V}[T_s])\mathrm{P[L]} \tag{5.23}$$

Some tangle and link invariants calculated using (5.2) are given in ref. 27.

6. Factorization, Degeneracy, Representation and Classification

From (5.15) and (5.23), the link invariant P for any link has the form

$$\mathrm{P[L]} = (\text{product of } \mathscr{V}\text{'s})\tau \tag{6.1}$$

We shall call a tangle nontrivial if it is not the cut unknot. Define an irreducible tangle as a tangle that cannot be seperated into two nontrivial tangles by cutting a single line in the original tangle. A reducible tangle is always the propduct of two or more commuting irreducible tangles. If the tangle T cut from a link L is irreducible, then from (5.15) all tangles cut from L must also be irreducible. This notion defines an irreducibles link. It follows that

Factorization Theorem. *The link invariant* P *of any link is factorizable to a product of invariants of irreducible tangles and at least one factor of* τ.

Completeness Theorem. *The universal invariants* $\mathcal{V}$[T] *for the set of all irreducible tangles T and the element* τ *are commuting generators of universal invariants for all links.*

If L is a split link with M disjoint parts, then clearly P[L] will have M factors of τ. Thus

Corollary for Split Links. *Let L be a split link with* M *disconnected parts. Then* P[L] *is a product of invariants of irreducible tangles and* M *factors of* τ.

Consider now the partition of the set $\{T_s\}$ of irreducible tangles in the decoration of an ℓ-component skeleton link L. By a partition of $\{T_s\}$ we mean a division of $\{T_s\}$ into ℓ nonintersecting subsets X_m whose union is $\{T_s\}$. Two subsets are equivalent if they differ only by permutation of member tangles (because tangles commute). A partition u is distinct from a partition v if not all their corresponding subsets are equivalent. Let L_u be obtained from L by decorating it with the u partition of $\{T_s\}$, where the m*th* component of L is decorated by all the tangles in the subset X_m anywhere on the component. Similarly for L_v, where v is a partition distinct from u. Then by (5.23)

$$P[L_u] = P[L_v] \tag{6.1}$$

Since the partitions u and v are distinct, L_u and L_v are not necessarily isotopic -- whether they are isotopic or not also depends on L. If they are not isotopic, then (6.1) expresses a class I degeneracy. We define class I degeneracy as follows. Let L_1 and L_2 be two nonisotopic links. Then L_1 and L_2 are class I degerate under P if rep(P[L_1]) = rep(P[L_2]) when all the representations on L_1 and L_2 are the same. Below is an example.

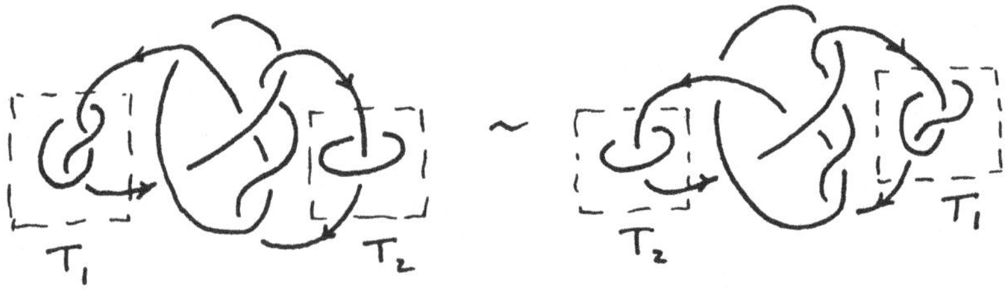

We therefore have

Degeneracy Theorem. *Let P be a universal link invariant constructed from a quasitriangular quantum group whose* h^2 *is not in the centre of* g', *and let* L_u *and* L_v *be two nonisotopic links obtained from two distinct partitions u and v of tangle decorations on* L. *Then* L_u *and* L_v *are class I degenerate under P.*

From this theorem it is easy to contruct any number of sets of nonisotopic links that are class I degenerate under P. The theorem also suggests that class I degeracy may be resolved by making the links *multicolour*. For this purpose we consider the representation of P. First consider the representation of $\mathscr{V}$. We say a tangle has ℓ components if the link from which it is cut has ℓ components. We always designate the cut component as the first component of a tangle. We shall use the notion that the $(\pi_1 \otimes \pi_2 \otimes \cdots \otimes \pi_\ell)$ representation of the universal link invariant P[L] is the link invariant of the $(\pi_1, \pi_2, .., \pi_\ell)$-colour L. Thus we write

$$(\pi_1 \otimes \pi_2 \otimes \cdots \otimes \pi_\ell) P[L] = P[L; \pi_1, .., \pi_\ell] \tag{6.2}$$

Similarly for $\mathscr{V}$ and $\mathscr{V}$. Thus

$$P[L; \pi_1, .., \pi_\ell] = \mathscr{V}[T; \pi_1, .., \pi_\ell] \, \tau_{P_1} \tag{6.3}$$

where T is obtained by cutting the component (with the colour) π_1. We call the colours $\pi_2, .., \pi_\ell$ of T its internal colours.

Now consider the representation of an ℓ-component irreducible skeleton link L decortated with the partition $\{X_m\}$ of the set of irreducible tangles $\{T\}$. If the invaraint of the skeleton is $P[L; \pi_1, .., \pi_\ell]$, then the invariant of the decorated link L′ is

$$P[L'; \pi_1, ..., \pi_\ell, \text{ internal colours of inserted tangles}]$$

$$= P[L; \pi_1, .., \pi_\ell] \prod_{X_m \in \{T\}} \Big(\prod_{T_s \in X_m} \mathscr{V}(T_s; \pi_m, ..) \Big)$$

where, except π_m, the internal colours of the inserted tangles are not given explicitly. Thus, if L_u and L_v are two coloured links obtained by decorating the skelton L with two distinct partitions, then their respective invariants are not in general degenertate provided all colours in L are distinct. Conversely, if the two colours π_i and π_j of L are identical, then the invariants of all the decorated L's with partitions that differ only in the two subsets X_i and X_j are degenerate. In general, if colours $\pi_i, .., \pi_k$ are identical, then P[L] is insensitive to differences in the partition of the union of the subsets $(X_i, .., X_k)$.

Classification. The discussion above shows that *coloured links are classified by the set: the irreducible ℓ-component skeleton* L; *its colour* $\pi_1, \pi_2, .., \pi_\ell$; *the decorating tangles* $\{T\}$; *its partition* $(X_1, X_2, .., X_\ell)$.

7. Twisted Quantum Groups

Here we briefly show that the Lie algebra A_1 can be twice deformed to obtain a quantum group that has a Hopf structure distinct from the Hopf algebra of standard quantum group of A_1 as described in section 2. Let $\{H, X^+, X^-\}$ be the generators of the q-analogue of A_1, and call the deformation parameter q_1 instead of q. That is, $[X^+,X^-] = (q_1^H-q_1^{-H})/(q_1-q_1^{-1})$. Now introduce H_0 which commutes with $\{H,X^\pm\}$ and another deformation parameter q_2 and define†

$$k_1 \equiv (q_1q_2)^{H/2} \; ; \qquad k_2 \equiv q_2^{H_0}(q_1/q_2)^{H/2} \; ; \tag{7.1}$$

$$Y^\pm \equiv q_2^{-H/2} X^\pm \tag{7.2}$$

Then we have a Hopf algebra generated by $\{k_1^{\pm 1}, k_2^{\pm 1}, Y^+, Y^-\}$ with relations and maps

$$k_1 Y^\pm k_1^{-1} = (q_1q_2)^{\pm 1} \; Y^\pm \tag{7.3}$$

$$k_2 \; Y^\pm \; k_2^{-1} = (q_1/q_2)^{\pm 1} \; Y^\pm \tag{7.4}$$

$$Y^+Y^- - q_2^2 \; Y^-Y^+ = [(q_1/q_2)^H - (q_1q_2)^{-H}]/[q_1/q_2 - (q_1q_2)^{-1}] \tag{7.5}$$

$$\Delta(k_m) = k_m \otimes k_m \; ; \qquad m = 1,2 \tag{7.6}$$

$$\Delta(Y^\pm) = Y^\pm \otimes k_2 + k_1^{-1} \otimes Y^\pm \tag{7.7}$$

$$S(k_m) = k_m^{-1} \; ; \qquad S(Y^\pm) = - \; k_1 Y^\pm k_2^{-1} \tag{7.8,9}$$

$$\varepsilon(k_m) = 1 \; ; \qquad \varepsilon(Y^\pm) = 0 \tag{7.10}$$

This gives a twice deformed quantum group, which we shall call a twisted quantum group. The Hopf algebra of the twisted quantum group specializes to the standard one when $q_2=1$, $q_1=q$. For a generalization of this structure to A_N see ref. 34.

Examples of representations of the twisted quantum group for the case $q_1q_2 = q$, $q_1/q_2 = q^{-1}\omega = q^{-1}\exp(2\pi i/N)$ are known.[26-29] In this case the pair (k_1,k_2) can be reexpressed in terms of

$$k \equiv q^{H/2} \; ; \qquad z \equiv (q\omega^{-1/2})^{H_0} \; \omega^{H/2} \tag{7.11}$$

by

$$k_1 = k \; ; \qquad k_2 = k^{-1}z \tag{7.12}$$

Note that

$$z_i^N \in \text{centre of Hopf algebra} \tag{7.13}$$

---

† *I am grateful to Nigel Burroughs for suggesting this particular presentation.*

so z is the Hopf algebra valued generators of Z_N. In this case the twisted quantum group can be viewed as a deformation of a *spontaneously broken* $A_1 \times Z_N$. In terms of k and z, the Hopf structure of this twisted quantum group is[27]

$$kY^{\pm} k^{-1} = q^{\pm 1} Y^{\pm} \tag{7.14}$$

$$zY^{\pm} z^{-1} = \omega^{\pm 1} Y^{\pm} \tag{7.15}$$

$$Y_i^+ Y_j^- - q^2\omega^{-1} Y^- Y^+ = k^{-2} q(z^2 - 1)/(\omega-1) \tag{7.16}$$

$$\Delta(k) = k \otimes k \; ; \qquad \Delta(z) = z \otimes z \tag{7.17}$$

$$\Delta(Y^{\pm}) = Y^{\pm} \otimes k^{-1}z + k^{-1} \otimes Y^{\pm} \tag{7.18}$$

$$S(k) = k^{-1} \; ; \qquad S(z) = z^{-1} \tag{7.19}$$

$$S(Y^{\pm}) = -(\omega/q)^{\pm 1} k^2 z^{-1} Y^{\pm} \tag{7.20}$$

$$\varepsilon(k) = \varepsilon(z) = 1 \; ; \qquad \varepsilon(Y^{\pm}) = 0 \tag{7.21}$$

Actually this structure is valid even when ω is not an integral root of unity. In that case the relation (7.13) will not be satisfied. However, if the $N \times N$ matrix representation of $\mathcal{R}$ is required to be quasitriangular *(see (2.28))*, then ω must be an *Nth* root of unity. Comparing (7.16) with the Poisson bracket for the untwisted quantum group suggests that it is ω that plays the role of deformation parameter $e^{\hbar}$ for which the limit $\hbar \to 0$ can be taken. In a twisted quantum group q is the extra deformation parameter.

We give a simple example in the case of $\mathcal{g} = s\ell(2,\mathbb{C})$, with N=2. That is with $\omega=-1$. The twisted $\mathcal{g}'$ has a 2×2 matrix representation with basis

$$\pi(e_1) = \begin{pmatrix} q^{-1/2} & 0 \\ 0 & q^{1/2} \end{pmatrix}, \qquad \pi(e_2) = \begin{pmatrix} q^{1/2} & 0 \\ 0 & -q^{3/2} \end{pmatrix}, \qquad \pi(e_3) = \begin{pmatrix} 0 & 0 \\ \eta^{1/2} & 0 \end{pmatrix},$$

$$\pi(e^1) = \begin{pmatrix} 1 & 0 \\ 0 & 0 \end{pmatrix}, \qquad \pi(e^2) = \begin{pmatrix} 0 & 0 \\ 0 & 1 \end{pmatrix}, \qquad \pi(e^3) = \begin{pmatrix} 0 & \eta^{1/2} \\ 0 & 0 \end{pmatrix}, \tag{7.22}$$

($\eta \equiv (1-q^2)q^{-1/2}$) whose representations under antipode are

$$\pi(S(e_1)) = \begin{pmatrix} q^{1/2} & 0 \\ 0 & q^{-1/2} \end{pmatrix}, \; \pi(S(e_2)) = \begin{pmatrix} q^{-1/2} & 0 \\ 0 & -q^{-3/2} \end{pmatrix}, \; \pi(S(e_3)) = -\begin{pmatrix} 0 & 0 \\ q^{-1}\eta^{1/2} & 0 \end{pmatrix},$$

$$\pi(S(e^1)) = \frac{1}{2}\begin{pmatrix} q+q^{-1} & 0 \\ 0 & 1-q^{-2} \end{pmatrix}, \; \pi(S(e^2)) = \frac{1}{2}\begin{pmatrix} 1-q^{-2} & 0 \\ 0 & q^{-1}+q^{-3} \end{pmatrix}, \; \pi(S(e^3)) = +\begin{pmatrix} 0 & q^{-1}\eta^{1/2} \\ 0 & 0 \end{pmatrix} \tag{7.23}$$

The representations of $\mathcal{R}$, h and $\mathcal{X}$ are

$$(\pi \otimes \pi)\mathcal{R} = \begin{pmatrix} q^{-1/2} & 0 \\ 0 & q^{1/2} \end{pmatrix} \otimes \begin{pmatrix} 1 & 0 \\ 0 & 0 \end{pmatrix} + \begin{pmatrix} q^{1/2} & 0 \\ 0 & -q^{3/2} \end{pmatrix} \otimes \begin{pmatrix} 0 & 0 \\ 0 & 1 \end{pmatrix} + \eta \begin{pmatrix} 0 & 0 \\ 1 & 0 \end{pmatrix} \otimes \begin{pmatrix} 0 & 1 \\ 0 & 0 \end{pmatrix} \quad (7.24)$$

$$\pi(h) = q^{1/2} \begin{pmatrix} 1 & 0 \\ 0 & -1 \end{pmatrix} ; \qquad \pi(\mathcal{X}) = q^{-1} \quad (7.25)$$

Note that $\pi(\tau) = \pi(\mathcal{X}^{-1/2}\mathrm{Tr}(h))$ vanishes identically, so the corresponding link invariant P[L] is trivial -- all links are mapped to zero, but the link invariant Q[L] is not. It is easy to see that Q[L] satisfies the skein relation

$$Q[L_+] - Q[L_-] = (q^{-1}-q)Q[L_0] \quad (7.26)$$

where, relative to L_+, one of the positive crossings in L_+ is spliced in L_0, and is replaced by a negative crossing in L_-. (7.26) is just the Skein relation for the Alexander-Conway link polynomial.[22] This shows that, for the Alexander-Conway polynomial, it would be incorrect to say that the link polynomial P for the unknot can be normalized to unity. Rather, it is only the tangle polynomial Q for the the unknot that can be normalized to unity. Note that Q[L] still vanishes for all split links (see corollary for split links, section 6).

A comparison with the 2×2 representation for the untwisted $\mathcal{g}'$ of $s\ell(2,\mathbb{C})$ is instructive. In this representation one has

$$(\pi \otimes \pi)\mathcal{R} = \begin{pmatrix} q^{-1/2} & 0 \\ 0 & q^{1/2} \end{pmatrix} \otimes \begin{pmatrix} 1 & 0 \\ 0 & 0 \end{pmatrix} + \begin{pmatrix} q^{1/2} & 0 \\ 0 & q^{-1/2} \end{pmatrix} \otimes \begin{pmatrix} 0 & 0 \\ 0 & 1 \end{pmatrix} + \eta \begin{pmatrix} 0 & 0 \\ 1 & 0 \end{pmatrix} \otimes \begin{pmatrix} 0 & 1 \\ 0 & 0 \end{pmatrix} \quad (7.27)$$

$$\pi(h) = q^{1/2} \begin{pmatrix} 1 & 0 \\ 0 & q^{2} \end{pmatrix} ; \qquad \pi(\mathcal{X}) = q^{3} \quad (7.28)$$

Here $\pi(\tau)$ does not vanish unless $q = e^{i\pi/2}$. So when $q \neq e^{i\pi/2}$, P[L] and Q[L] differ only by a L-independent nonzero factor. The skein relation for P (and for Q) is

$$q^{-2}P[L_+] - q^{2}P[L_-] = (q^{-1}-q)P[L_0] \quad (7.29)$$

So we identify P as the Jones polynomial.[7] When $q = e^{i\pi/2}$, P vanishes identically, but one still has the Jones polynomial, now given only by Q. In this case, just like the Alexander-Conway polynomial, $Q[L;\ q=e^{i\pi/2}]$ vanishes for all split links, except that the Alexander-Conway Q is still a function of q, whereas the Jones Q is a c-number. This illustrates the remark following (7.21) that the deformation parameter ω in the twisted $\mathcal{g}'$ corresponds to the parameter q in the untwisted $\mathcal{g}$, while the parameter q in the twisted $\mathcal{g}'$ is extra. In the literature it is sometimes said that the link polynomials constructed in quantum groups are not well-defined when q is a rational root of unity. We see that this is not the case, at least for monocolour links, provided one always uses Q instead of P.

Several other finite dimensional representations of the twisted quantum groups of broken $s\ell(2,\mathbb{C})\times Z_N$, N=3,4,5 and $s\ell(n,\mathbb{C})\times Z_2$ are given in refs. 28,34. For examples of link invariants see ref. 27. The construction of representations of the twisted quantum groups where neither q nor ω is a rational root of unity is an open problem. It is likely that such quantum groups only have infinite dimensional representations. A more detailed discussion of twisted quantum groups will be given elsewhere.

8. Conformal and Topological Field Theories

We make a few remarks on the connection between quantum groups and topological and conformal field theories. See also the articles by Seiberg,[16] Fröhlich,[17] DeVega,[13] Boudreau,[30] Carlip,[31] LeClair[32] in this volume.

Quantum group has a universal fusion-braiding relation (see (2.33)). Thus the relation is true for any representation at any value(s) of the deformation parameter(s).

Monodromy and the restriction on q to be a rational root of unity. Often derivations of link invariants based on representation theory of quantum groups restrict q *not* to be a rational root of unity. This is related to the fact that the Hopf algebra of the quantum group may have a nontrivial ideal for such values of q. The tangle and link invariants derived in this paper are universal, so they are valid for any value of q. On the other hand, quantum groups appear to manifest themselves in conformal[16] and topological[18] field theories only for values of q that *are* rational roots of unity. One of the properties of conformal field theory is that two-point correlation functions have a well defined covering space. This implies that if the two fields defining the correlation function are intertwined, the correlation function changes only by a phase. Such phases define the monodromies of the two-point correlation function. Translated into the language of quantum group, this property means that the element $\mathscr{R}(\mathscr{T}\bullet\mathscr{R}) \in \mathscr{g}'\otimes\mathscr{g}'$ must be diagonal in both algebras of the direct product. To examine what this constraint means we divide the bases e_σ and e^σ of $\mathscr{g}'$ into two parts,

$$\{e_\sigma\} = \{e_k, e_x\} \qquad \{e^\sigma\} = \{e^k, e^x\} \tag{8.1}$$

where e_k (e^k) is a product of k_i's but not of the raising and lowering elements $X_i^{\pm}$'s, while e_x (e^x) has factors involving the X_i^+'s (X_i^-'s). We call $\{e_k, e^k\}$ the Cartan sector. It generates an Abelian subalgebra $\mathscr{g}'_C$, any of whose matrix representation can be diagonalized. We call $\{e_x, e^x\}$ the X-sector; any subalgebra generated by generators involving a nonvanishing subset of the X-sector is nonAbelian; its matrix represenation cannot be diagonalized. Write

$$\mathscr{R} = e_k\otimes e^k + \eta_x\, e_x\otimes e^x \tag{8.2}$$

where for convenience we have factored out a q-dependent central factor η_x from the second term. Then

$$\mathscr{R}(\mathscr{T}\bullet\mathscr{R}) = e_k e^j\otimes e^k e_j + \eta_x(e_k e^x\otimes e^k e_x + e_x e^k\otimes e^x e_k) + \eta_x\eta_y e_x e^y\otimes e^x e_y \tag{8.3}$$

This would lie in the direct product of two Cartan sectors only if the nonAbelian parts in the fourth term could cancel completely the nonAbelian second and third terms. This cancellation is not a universal property of $\mathscr{g}'$. Cancellation can occur, however, in representations, at least for specific values of the deformation parameter q. It is known that for the fundamental representations[5] of $\mathscr{g} = A_n$, B_n, C_n, D_n and $A^{(2)}_n$, and for higher representations of untwisted[4,35] as well as twisted[28,34] quantum groups, η_x has an x-independent common factor of the form

$$\pi(\eta) = q^a (1 - q^b) \tag{8.4}$$

where a and b are some rational number that depends on the algebra. For example, for the fundamental representation of $\mathscr{g} = A_{n-1} \sim s\ell(n)$,

$$a = -(n-1)/n, \qquad b = 2; \qquad \mathscr{g} = s\ell(n) \tag{8.5}$$

In these cases, $\mathscr{R}(\mathscr{T}\bullet\mathscr{R})$ will be in the Cartan sector provided that

$$q = e^{m\pi i}; \qquad m = \text{some intger} \tag{8.6}$$

Note from (8.2), however, that (8.6) is a sufficient condition for $\mathscr{R}$ itself to be restricted to the Cartan sector. This suggests that, at least in these cases, the monodromy condition in conformal field theory restricts the quantum group to be effectively Abelian for which all representations are homomorphic to some unitary representations.

For the fundamental representations π of (untwisted) $s\ell(n)$, it can be shown that the universal element h and the central element τ are

$$\pi(h) = q^{-a} \begin{bmatrix} 1 & & & \\ & q^b & & \\ & & \ddots & \\ & & & q^{nb} \end{bmatrix} \tag{8.7}$$

$$\pi(\tau) = q^{-nb/2} (1-q^{(n+1)b})/(1-q^b) \tag{8.8}$$

where a and b are given by (8.5). Thus, under the monodromy constraint of (8.6), $\pi(h)$ is just the identity operator to within a phase, and $\pi(\tau)$ is just a phase to within a normalization factor equal to the dimension of the representation. This is another indication that the monodromy condition forces the standard quantum group to act trivially in conformal field theory. Note that, because of the tangle theorem and (5.18), writhing and knotting of a field gives only a phase, for any value of q.

The situation in twisted quantum groups is somewhat more complicated. Here the condition restricting $\mathscr{R}(\mathscr{T}\bullet\mathscr{R})$ to the Cartan sector still appears -- we do not have a theorem for this -- to be given by (8.4) such that $q^b=1$ for some rational b. This again restricts R itself to the Cartan sector, but now $\pi(h)$ as a rule is not proportional to the identity matrix. For the N-dimensional representations of the spontaneously broken $s\ell(2)\times Z_N$,

$$\pi(h) \sim \begin{bmatrix} 1 & & & & \\ & \omega & & & \\ & & \omega^2 & & \\ & & & \ddots & \\ & & & & \omega^{N-1} \end{bmatrix} \tag{8.9}$$

where $\omega = e^{2\pi i/N}$. Note that for these cases $\pi(\tau)=0$. Although having $\pi(\tau)=0$ guarantees monodromy, and it is known that for N=2 (which gives the Alexander-Conway polynomial) this system corresponds to an exactly solvable free fermion lattice model,[36] the identification of conformal field theories -- if such exist -- corresponding to the representations of the twisted quantum groups is still an open problem.

Holonomy and the Wilson Tangle. Witten[18] has argued that link invariants defined in quantum groups are just the expectation values of Wilson lines

$$W[L;\rho_1,\rho_2..] = \prod_{C_i} (\mathrm{Tr}_{\rho_i} \mathscr{P} \exp(\oint_{C_i} A_{\rho_i} dx)) \tag{8.10}$$

given by a topological field theory in a three-dimension manifold M with a pure Chern-Simons Lagrangian

$$\mathscr{L}_{CS} = k \int_M (AdA + \frac{2}{3} A^3) \tag{8.11}$$

The gauge field A is valued in some Lie algebra $\mathscr{g}$, ρ_i are representations of $\mathscr{g}$, the link L is the union of the contours C_i, $\mathscr{P}$ means path ordering, and Tr is the trace of the representations. If we do not take, say, the trace of the representation ρ_1 on the contour C_1, then the quantity

$$\mathscr{W}[L;\rho_1,\rho_2..] = \mathscr{P} \exp\left[\oint_{C_1} A_{\rho_1} dx\right]\left(\prod_{C_i \neq C_1} \mathrm{Tr}_{\rho_i} \mathscr{P} \exp(\oint_{C_i} A_{\rho_i} dx)\right) \tag{8.12}$$

is matrix-valued in the ρ_1 representation of $\mathscr{g}$. If (8.10) is indeed a link invariant, then from the tangle theorem in section 5 we expect the quantity $\mathscr{W}$ defined by (8.12), which we shall call a Wilson tangle, to be the invariant of the tangle T_1 obtained by cutting C_1 in L, and therefore be proportional to the identity matrix in ρ_1. Actually there is a difference between Witten's link invariant and link invariants in quantum groups: the role of the universal element $\mathscr{k}$ in $\mathscr{g}'$, which is crucial for the construction of link invariants in quantum groups, is obscure in Witten's link invariant. For example, if the Wilson tangle is indeed proportional to the identity matrix,

$$\mathscr{W}[L;\rho_1,\rho_2..] = (\mathscr{W}[T_1])1_{\rho_1} \tag{8.13}$$

then a comparison of (8.10) and (8.12) require that

$$W[L;\rho_1,\rho_2..] = N_1 \mathscr{W}[T_1] \tag{8.14}$$

where $N_1 = \mathrm{Tr}_{\rho_1}(1)$ is the dimension of ρ_1. However, as we have mentioned several times, in quantum groups the quotient of the link invariant and the tangle invariant is proportional to rep(Tr($\mathscr{k}$)), not to the dimension of the representation. The discrepancy is removed only if rep($\mathscr{k}$) is proportional to the identity matrix. Earlier we have seen that for non-twisted quantum groups the constraint

$$\mathrm{rep}(\mathscr{k}) \sim \text{unit matrix} \tag{8.15}$$

is the same as the constraint requiring $\mathscr{R}(\mathscr{T}\bullet\mathscr{R})$ (and indeed $\mathscr{R}$ itself) to be in the Cartan sector, which is to restrict q to be some integral root of unity. (For $\mathscr{sl}_k(2)$, the requirement[18] is $q = \exp(2\pi i/(2+k))$ which is consistent with (8.6) with m=1, k=0.) We shall assume that Witten's link invariant indeed corresponds to a link invariant of the (untwisted) quantum group in some representation for which rep(h)$\sim$1. In this case, tangle theorem guarantees that (8.13) is satisfied. It was shown in (5.17) that for any quasitriangular quantum group whose h is not unipotent the set of all tangles is homomorphic to a U(1) group. This implies that the Wilson tangle of (8.12) should be interpreted as a generalized holonomy of the Chern-Simons-Witten theory, whose holonomy group is just the U(1) group of tangle invariants. Thus in the Chern-Simons-Witten theory, when a particle is adiabatically transported around a closed curve in M, even if the curve is knotted, or is linked with other closed curves, the tangle theorem assures that the particle will acquire at most a phase change. From our discussion in section 5, representations of the holonomies are equivalent to eigenvalues of Casimir operators of the quantum group.

Our discussion seems to point to the notion that untwisted quantum groups are restricted to acting unitarily - in the sense that rep($\mathscr{R}$) is restricted to the Cartan sector and rep(h)$\sim$1 - in conformal and topological field theories. The restriction, realized by setting q equal to some integral root of unity, can be traced to the requirement that in such theories both the holonomy and monodromy groups be U(1) groups. But the tangle theorem insures the homomorphism between tangles and some U(1) group for *any*, including nonunitary, representation of the quantum group. This suggests that if we can find a way to incorporate the appropriate action of h in the definition of the Wilson tangle (and the Wilson line), then we may be able to remove the restriction on conformal and topological field theories having only unitary representations of quantum groups. In this case, the holonomy group would still be U(1) - incidentally, this also implies that anyons would not be restricted to unitary representations of the braid group, but the monodromy group would not be reducible to U(1).

I have benefitted from communications and discussions with C.N. Yang, J. Birman, Michel Couture, Mo-Lin Ge, Louis Kauffman, Yong-Shih Wu, Jürg Fröhlich and Ruth Lawrence. I especially thank Peter Leivo for helping me understand Hopf algebra, Shahn Majid for showing me the proof of (2.30), Vaughan Jones for encouraging me to use the Reidemeister theorem to construct link invariants, Hosein Hooshangi for testing by computation some of the ideas in sections 3 and 5, Nigel Burroughs for helping me get the twisted Hopf algebra right, and Wei-Dong Zhao for making several suggestions that have been incorporated into sections 6 and 8.

References

1. V.G. Drinfel'd, *Proc. Int'l Cong. Math., (Berkeley, 1986), p.798; DAN SSSR, 5(1985)1060.*
2. M. Jimbo, *Lett. Math. Phys. 11(1986)247; Comm. Math. Phys. 102(1986)537.*

2a. V.F.R. Jones, *Pacific Jour. Math. 137(1989)311.*

3. N.Y. Reshetikhin, *Quantized Enveloping Algebra, the Yang-Baxter Equa tion and the Invariants of Links, LOMI preprint E-4-87, I & II (Leningrad, 1988).*
4. Y. Akutsu, T. Deguchi and M. Wadati, *J. Phys. Soc. Jap. 56(1987)3464, 59(1987)3034.*
5. V.G. Turaev, *Invent. Math. 92(1988)527.*
6. T. Deguchi, *in these proceedings.*
7. V.F.R. Jones, *Bull. Amer. Math. Soc. 12(1985)103.*
8. L.H. Kauffman, *Topology 26(1987)395; Am. Math. Monthly 95(1988)195.*

9. R. Lawrence, *A Universal Link Invariant Using Quantum Groups, Proc. Int. Conf. Diff. Geom. Meth. Theo. Phys. 1988 (World Scientific).*
10. F. Wilczek, *Phys. Rev. Lett. 48(1982)1144.*
11. Y.S. Wu, *Phys. Rev. Lett. 52(1984)2103.*
12. J. Fröhlich, *Statistics of Fields, the Yang-Baxter equation and the Theory of Knots and Links, Cargese lectures (1987).*
13. H.C. Lee, M.L. Ge, M. Couture & Y.S. Wu, *Int. J. Mod. Phys. 4(1989)2333*
14. R.J. Baxter, *Exactly Solved Models in Statistical Mechanics, (Acad. Press, 1982).*
15. H.J. De Vega, *Yang-Baxter Algebra, Integrable Models and Quantum Groups in these Proceedings.*
16. G. Moore and N. Seiberg, *Rational Conformal Field Theory, in these proceedings; Nucl. Phys. B313(1989)16; Comm. Math. Phys. 123(1989)177.*
17. J. Fröhlich, *Braid Statistics in Three-Dimenasional Local Quantum Theory, in these proceedings.*
18. E. Witten, *Comm. Math. Phys. 121(1989)351.*
19. J.S. Birman, *Braids, Links and Mapping Class Groups, (Princeton U. Press, 1974).*
20. A.A. Markov, *Recueil Math., Moscou 1(1935)73; see also ref. 19.*
21. K. Reidemeister, *"Knoten Theorie" (Chelsea Pulishing, 1948).*
22. J.W. Alexander, *Trans. Am. Math. Soc. 30(1923)275;* J.H. Conway, *Computational Problems in Abstract Algebra, Ed. J. Leech (Pergamon Press, 1969*
23. S. Mahjid, *Quasitriangular Hopf Algebras and Yang-Baxter Equations (U. Swansea Preprint, 1989).*
24. N.Y. Reshetikhin and V.G. Turaev, *Ribbon Graphs and their Invariants derived form Quantum Groups (LOMI prepreint 1989).*
25. C.N. Yang, *Phys. Rev. 168(1967)1920;* R.J. Baxter, *ref. 14.*
26. H.C. Lee and M. Couture, *A Method for Constructing Closed Braids from Links and a New Polynomial for connected Links (Chalk River preprint 1988).*
27. H.C. Lee, *Q-deformation of* $\mathcal{sl}(2)\times Z_N$ *and Link Invariants, Proc. NATO ARW on Phys. and Geometry, Lake Tahoe 1989 July (Plenum Press, 1990).*
28. H.C. Lee, M. Couture and N.C. Schmeing, *Connected Link Invariants (Chalk River preprint, 1989)*; M. Couture et al., *A New Family of N-State Representations of the Braid Group,* in these proceedings.
29. H.C. Lee, Hopf Algebra, *Complexification of* $\mathcal{U}_2(\mathcal{sl}(2,\mathbb{C}))$ *and Link Invariants, Proc. Symp. on Fields, Fields and Quantum Gravity, Beijing 1989 June (Gordon & Breach, 1990).*
30. M. Bourdeau, *The NonAbelian Chern-Simons term with Sources and Braid Source Statistics, in these proceedings.*
31. A. LeClair, *Integrable Restrictions of Quantum Soliton Theory and Minimal Conformal Series, in these proceedings.*
32. S. Carlip, *2+1 Dimensional Quantum Gravity and the Braid Group, in these proceedings.*
33. L.H. Kauffman, *Knots, Abstracts Tensors and the Yang-Baxter Equation, preprint 1989.*
34. H.C. Lee, *Twisted Quantum Groups of* A_N *and the Alexander-Conway Polynomial* (submitted to Pac. J. Math.);
H.C. Lee, M. Couture and M.L. Ge, *Twisted Quantum Groups of* A_N*: II, Non Quasi-Classical* $\mathcal{R}$*-Matrix Ribbon Links and Graded Vertex Models, Yang-Baxter Equation,* in preparation.
35. M.L. Ge, *private communication.*
36. C. Fan and F.Y. Wu, *Phys. Rev. B2(1970)723;* K. Sogo, M. Uchinami, Y. Akutsu and M. Wadati, *Prog. Theo. Phys. 68(1982)508.*

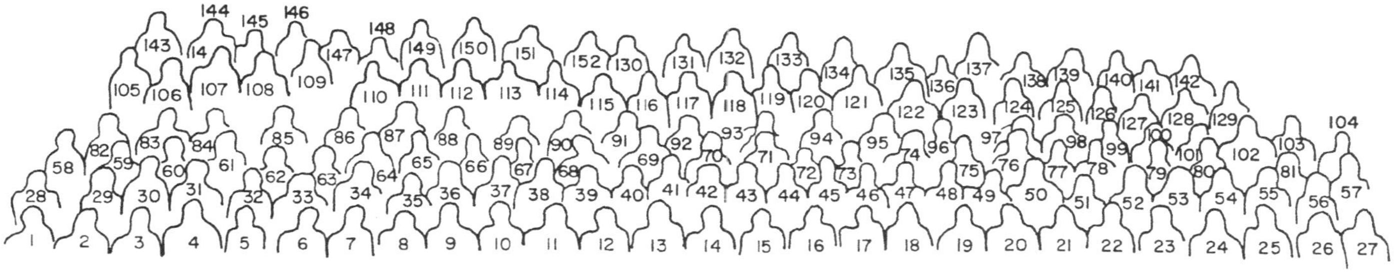

1. F. Ruiz; 2. R. Laflamme; 3. A. Shapere; 4. M. Ortiz; 5. T. del Rio Gaztelurrutia; 6. S.-J. Rey; 7. Y.-S. Wu; 8. R. Kobes; 9. D.J. Toms; 10. G. Kunstatter; 11. J. Patera; 12. H.C. Lee; 13. D. Olive; 14. C.J. Isham; 15. G.'tHooft; 16. H.J. de Vega; 17. L. Vinet; 18. J. Fröhlich; 19. N. Seiberg; 20. G. Semenoff; 21. I. Affleck; 22. C. Viallet; 23. M. Henneaux; 24. M. Rocek; 25. A. LeClair; 26. M. Bourdeau; 27. D.E. Freed; 28. M.J. Rodriguez; 29. T.T. Chia; 30. J. Gauntlett; 31. T. Nakanishi; 32. D. Bar-Natan; 33. S.Q. Chen; 34. S. Iso; 35. T. Deguchi; 36. K. Isler; 37. C.A. Trugenberger; 38. G. Dunne, 39. H.P. Leivo; 40. J.G. Williams; 41. R. Picken; 42. K.W. Mak; 43. C.C. Chen; 44. H.T. Cho; 45. A. Dabholkar; 46. L. Vinet; 47. G. Siopsis; 48. K. Tas; 49. D. Fivel; 50. A. Vercin; 51. G. Bhamathi; 52. R. Loll; 53. J. Cohn; 54. G.J. Ni; 55. H.L. Yu; 56. M.C.B. Abdalla; 57. S.M. Fournier; 58. Y. Saint-Aubin; 59. R. Mendel; 60. M. Bergeron; 61. A. Anderson; 62. Y.C. Kao; 63. N. Ishibashi; 64. R. Bluhm; 65. M. Groot; 66. D. Murray; 67. N. Hill; 68. J. Dixon; 69. G. Cleaver; 70. J. Paterson; 71. P. Van Driel; 72. A. Polychronakos; 73. J. Zuk; 74. J. Gegenberg; 75. *unidentified;* 76. A. Schirrmacher; 77. R. MacKenzie; 78. J-M. Lina; 79. J. Martinez; 80. J. Wang; 81. R. Tzani; 82. B. Rosenstein; 83. L. Benoit; 84. A. Rebhan; 85. D. Lancaster; 86. P. Bouwknegt; 87. J. Segert; 88. S. Cordes; 89. J. Sniatycki; 90. L. Bombelli; 91. J. Talman; 92. P. Griffin; 93. J. Horne; 94. H. Riggs; 95. A. Eastaugh; 96. G. Harris; 97. G. Grignani, 98. R. Floreanini; 99. R. Percacci; 100. S. Pesquale; 101. M. Pieralberto; 102. G. Giavarini; 103. T.R. Klassen; 104. S. Carlip; 105. S.C. Lee; 106. C. Yastremiz; 107. M.E. Carrington; 108. C. Rovelli; 109. M. Blencowe; 110. D. Rohrlich; 111. M.B. Paranjape; 112. P. Roche; 113. E. Piard; 114. M. Picco; 115. D. Gurarie; 116. T. Kuramoto; 117. T. Onogi; 118. A. Rutherford; 119. I. Adjali; 120. N. Manojlovic; 121. H. Frahm; 122. E.C.G. Sudarshan; 123. R. Douglas; 124. M. Fowler; 125. G. Esposito-Farèse; 126; F. Thuillier; 127. W. Chen; 128. J. Avan; 129. G. Papadopoulos; 130. G. McKeon; 131. M. Légaré; 132. R. Link; 133. D. Aubin; 134. L. Tarasov; 135. L. Culumovic; 136. K.L. Chang; 137. M. Carreau; 138. R. Plesser; 139. S. Dasmahapatra; 140. K. Intriligator; 141. D. Chang; 142. A. Kumar; 143. P. Fuchs; 144. G. Gat; 145. S.H. Park; 146. A. Kovner; 147. N. Burroughs; 148. N. Schmeing; 149. M. Couture; 150. *unidentified;* 151. M. Leblanc; 152. R. Mann - **Not in Picture:** A. Berkovich; R. Jackiw; S.Y. Pi; J. Miller

Lecturers

I. Affleck	*Physics Dept., Univ. of British Columbia, Vancouver, B.C., Canada V6T 2A3*
J. Fröhlich	*Theoretical Physics, ETH-Hönggerberg, CH-8093, Zürich, Switzerland*
M. Henneaux	*Phys. Théorique, Univ. Libre de Bruxelles, Boulevard du Triomph, CP 225, B-1050 Bruxelles, Belgium*
G. 't Hooft	*Inst. voor Theor. Fysica, Univ. of Utrecht, Princetonplein 5, P.O.Box TA Utrecht, The Netherlands*
C.J. Isham	*The Blackett Lab., Prince Consort Road, London SW7 2BZ, U.K.*
R. Jackiw	*Centre for Theor. Physics, Dept. of Phys., Mass. Inst. of Technology, Cambridge, Mass. 02139, U.S.A.*
D. Olive	*The Blackett Lab., Prince Consort Road, London SW7 2BZ, U.K.*
J. Patera	*CRM, Univ. de Montreal, C.P. 6128, Montreal, Quebec, Canada H3C 3J7*
N. Seiberg	*Inst. for Advanced Study, Princeton, New Jersey 08540, U.S.A.*
G. Semenoff	*Physics Dept., Univ. of British Columbia, Vancouver, B.C. Canada V6T 2A3*
H.J. DeVega	*Lab. de Physique Theor. et Hautes Energies, Université Paris VII, Tour 16 - ler étage, 4 Place Jussieu, 75230 Paris Cedex 05, France*
C. Viallet	*Lab. de Physique Theor. et Hautes Energies, Université Paris VII, Tour 16 - ler étage, 4 Place Jussieu, 75230 Paris Cedex 05, France*
L. Vinet	*Physics Dept., Univ. of Montreal, C.P. 6128, Montreal, Quebec, Canada H3C 3J7*
Y.S. Wu	*Physics Dept., Univ. of Utah, Salt Lake City, Utah 84112, U.S.A.*

Organizing Committee

H.C. Lee	*Chalk River Nuclear Laboratories, Chalk River, Ont., Canada K0J 1J0*
R. Kobes	*Dept. of Physics, Univ. of Winnipeg, Winnipeg, Manitoba, Canada R3B 2E9*
G. Kunstatter	*Dept. of Physics, Univ. of Winnipeg, Winnipeg, Manitoba, Canada R3B 2E9*
D.J. Toms	*Univ. of Newcastle Upon Tyne, Newcastle, United Kingdom*
Y.S. Wu	*Dept. of Physics, University of Utah, Salt Lake City, Utah 84112, U.S.A.*

LIST OF ASI STUDENTS (ASI 870/88)

Abdalla, M.C.	Inst. de Fisica Teorica da UNESP, Sao Paulo, Brazil
Adjali, M.I.	Univ. of Oxford, Oxford, England
Anderson, A.	Univ. of Utah, Salt Lake City, Utah, U.S.A.
Aubin, D.	McGill Univ., Montreal, Quebec, Canada
Avan, J.	Lab. de Physique Theorique, Paris Cedex 05 75252, France
Bar-Natan, D.	Princeton Univ., New Jersey, U.S.A./Israel
Bellucci, S.	INFN-Lab. Nazionali Di Frasscati, Frascati, Italy
Benoit, L.	Lab. de Physique Nucleaire, Univ. de Montreal, Montreal, Quebec, Canada

Bergeron, M.	Univ. of British Columbia, Vancouver, B.C., Canada
Berkovich, A.	Inst. of Theoretical Physics, Stony Brook, New York, U.S.A.
Bhamathi, G.	Univ. of Madras, Madras, India
Blencowe, M.	Theory Group, Blackett Lab., Imperial College, London, England
Bluhm, R.	Indiana Univ., Bloomington, Indiana, U.S.A.
Bombelli, L.	Univ. of Calgary, Calgary, Alberta, Canada
Bourdeau, M.	Syracuse Univ., Syracuse, New York, U.S.A.
Bouwknegt, P.	Massachusetts Inst. of Tech., Cambridge, Mass., U.S.A./Holland
Bugajska, C.	Univ. of Western Ontario, London, Ontario, Canada
Burroughs, N.	DAMTP, Cambridge, England
Campbell, B.	Univ. of Alberta, Edmonton, Alberta, Canada
Carlip, S.	Inst. for Advanced Study, Princeton, New Jersey, U.S.A.
Carreau, M.	Massachusetts Inst. for Technology, Cambridge, Mass., U.S.A.
Carrington, M.	Univ. of Minnesota, Minneapolis, Minnesota, U.S.A.
Cespedes, J.	Univ. Autonoma Barcelona, Bellaterra 08193, Spain
Chang, D.	Northwestern Univ. Boston, Massachusetts, U.S.A.
Chang, Kow-Lung	National Taiwan Univ., Taipeo, Taiwan, R.O.C.
Chen, Chia-Chu	Texas A&M Univ., College Station, Texas, U.S.A.
Chen, S.Q.	Fudan Univ., Shanghai, China
Chen, W.	Univ. of Utah, Salt Lake City, Utah, U.S.A./China
Chia, T.T.	National Univ. of Singapore, Kent Ridge, Singapore 0511
Cho, H.T.	Ohio State Univ., Columbus, Ohio, U.S.A.
Cleaver. G.	California Inst. of Technology, California, U.S.A.
Cohn, J.	Inst. for Advanced Study, Princeton, New Jersey, U.S.A.
Cordes, S.	Texas A&M Univ., College Station, Texas, U.S.A.
Couture, M.	Chalk River Nuclear Labs., Chalk River, Ontario, Canada
Culumovic, L.	Univ. of Western Ontario, London, Ontario, Canada
D'Hoker, E.	Univ. of California at Los Angeles, Los Angeles, California, U.S.A.
Dabholkar, A.	Princeton Univ., Princeton, New Jersey, U.S.A.
Dasmahapatra, S.	State Univ. of New York, Stony Brook, New York, U.S.A./India
De Groot, M.	Oxford Univ., Oxford, OX1, 3NP, England
Deguchi, T.	Univ. of Tokyo, Komaba, Japan
Del Rio Gaztelu, T.	DAMTP, Cambaridge, England/Spain
Dixon, J.	Univ. of Victoria, Victoria, British Columbia, Canada
Douglas, R.	Univ. of British Columbia, Vancouver, B.C., Canada
Dunne, G.	Massachusetts Inst. of Technology, Cambridge, Mass., U.S.A./Australia
Eastaugh, A.	Univ. of Chicago, Enrico Fermi Inst., Chicago, Illinois, U.S.A.
Esposito-Farese, G.	Centre de Phys. Theor. CNRS Luminy Case 907, Marseille Cedex 9, France
Fivel, D.	Univ. of Maryland, College Park, Maryland, U.S.A.
Floreanini, R.	INFN, c/o Dip. di Fisisca Teorica, Strada Costiera 11, Trieste, Italy
Fournier, S.	Univ. of Waterloo, Waterloo, Ontario, Canada
Fowler, M.	Univ. of Virginia, McCormich Road, Charlottesville, Virginia, U.S.A.
Frahm, H.	Univ. of Virginia, McCormich Road, Charlottesville, Virginia, U.S.A.
Freed, D.	Jadwin Hall, Princeton Univ., Princeton, New Jersey, U.S.A.
Fuchs, P.	TRIUMF, 4004 Wesbrook Mall, Vancouver, British Columbia, Canada/Germany
Gat, G.	Univ. of British Columbia, Vancouver, B.C., Canada
Gauntlett, J.	DAMTP, Silver St., Cambridge, U.K./Australia
Gegenberg, J.	Univ. of New Brunswick, Fredericton, New Brunswick, Canada
Giavarini, G.	Universita di Parma, Parma 43100, Italy
Green, B.	Lyman Lab. of Physics, Harvard Univ., Cambridge, Mass., U.S.A.
Griffin, P.	Fermi National Accelerator Lab., Batavia, Illinois, U.S.A.

Grignani, G.	Centre for Theor. Phys., Mass. Inst. for Tech., Cambridge, Mass., U.S.A.
Gurarie, D.	CWR Univ., Cleveland, Ohio, U.S.A.
Harnard, J.	Math. Science Research Inst., Centennial Drive, Berkeley, California, U.S.A.
Harris, G.	James Franck Inst., Univ. of Chicago, Chicago, Illinois, U.S.A.
Hill, N.	Univ. of Western Ontario, London, Ontario, Canada
Horne, J.	Princeton Univ., Princeton, New Jersey, U.S.A.
Intriligator, K.	Jefferson Lab., Harvard Univ., Cambridge, Massachusetts, U.S.A.
Ishibashi, N.	KEK, National Lab. for High Energy Physics, Ibaraki-ken 305, Japan
Isler, K.	Lab. de Phys. Nucl., Univ. de Montreal, Montreal, Quebec, Canada
Iso, S.	Univ. of Tokyo, 7-3-1 Hongo Bunkyo ku, Tokyo 113, Japan
Kao, Y.C.	National Tsing-Hua Univ., Taiwan
Karliga, B.	G.U. Fen-Ed. Fak. Matematik, Bolümü Ögretim Üyesi, Ankara, Turkey
Keller, G.	Inst. fur Theor. Physik, ETH - Honggerberg, Zurich, CH-8093, Switzerland
Keung, W.Y.	Univ. of Illinois, Chicago, Illinois, U.S.A.
Klassen, T.	Enrico Fermi Inst., Univ. of Chicago, Chicago, Illinois, U.S.A.
Kostelecky, V.A.	Indiana Univ., Bloomington, Indiana, U.S.A.
Kovner, A.	School of Phys. & Astronomy, Tel Aviv Univ., Ramat Avid, Tel Aviv, Israel
Kumar, A.	Northwestern Univ. Boston, Massachusetts, U.S.A.
Kuramoto, T.	Queen Mary College, Mile End Road, London E1 4NS, England
Laflamme, R.	Univ. of British Columbia, Vancouver, B.C.
Lancaster, D.	Kyoto Univ., Kitashirakawa, Kyoto 606, Japan
Leblanc, M.	Univ. of Waterloo, Waterloo, Ontario, Canada
LeClair, A.	Jadwin Hall, Princeton Univ., Princeton, New Jersey, U.S.A.
Lee, S.C.	Inst. of Phys., Academia Sinica, Nankang, Taipei, Taiwan, R.O.C.
Legare, M.	Univ. of Alberta, Edmonton, Alberta, Canada
Leivo, P.	Univ. of Winnipeg, Winnipeg, Manitoba, Canada
Lina, J-M.	Lab. de Phys. Nucl., Univ. de Montreal, Montreal, Quebec, Canada
Link, R.	Univ. of British Columbia, Vancouver, B.C., Canada
Loll, R.	Theor. Phys. Dept., Imperial College, London SW7 2BQ, England/Germany
Mackenzie, R.	Lab. de Phys., Univ. de Montreal, Montreal, Quebec, Canada
Madore, J.	Univ. de Paris Sud, Paris, France
Mak, K.	Univ. of Manitoba, Winnipeg, Manitoba, Canada
Mann, R.	Univ. of Waterloo, Waterloo, Ontario, Canada
Manojlovic, N.	Blackett Lab., Imperial College, London SW7 2BZ, England
Marchetti, P.	Univ. di Padova, Via Marzola 8, Padova 351131, Italy
Martinez, J.	Univ. of Illinois, Chicago, Illinois, U.S.A.
Matsuo, Y.	Enrico Fermi Inst., Univ. of Chicago, Chicago, Illinois, U.S.A.
McKeon, G.	Univ. of Western Ontario, London, Ontario, Canada
Mendel, R.	Univ. of Western Ontario, London, Ontario, Canada
Miller, J.	California Inst. of Tech., Pasadena, California, U.S.A.
Murray, D.	Univ. of Guelph, Guelph, Ontario, Canada
Nakanishi, T.	Univ. of Tokyo, Komaba, Meguro-ku, Tokyo 153, Japan
Ni, G.J.	Fudan Univ., Shanghai, China
Onogi, T.	KEK, National Lab. for High Energy Physics, Ibaraki-ken 305, Japan
Ortiz, M.	DAMTP, Silver St., Cambridge CB3 9EW, U.K./Spain
Papadopoulos, G.	King's College London, Strand, London WC2R 2LS, England/Greece
Paranjape, M.	Lab. de Phys. Nucl., Univ. de Montreal, Montreal, Quebec

Park, S.	Univ. of Texas at Austin, Austin, Texas, U.S.A.
Paterson, J.	Glasgow Univ., Glasgow G12 ORY, Great Britain
Percacci, R.	SISSA, Trieste, Italy
Pfeffer, D.	New York University, New York, New York, U.S.A.
Pi, S.Y.	Boston Univ., Boston, Massachusetts, U.S.A.
Piard, E.	215 Williamson Hall, Univ. of Florida, Gainesville, Florida, U.S.A./France
Picco, M.	Lab. de Physique Theorique, 4 Place Jussieu, Tour 16, Paris Cedex, France
Picken, R.	Grupo Teo'rico des Altas Energias, Av. Prof. Gama Pinto, 2, 1699 Lisboa Codex, Portugal
Plesser, M.R.	Harvard Univ., Cambridge, Massachusetts, U.S.A.
Polychronakos, A.	Univ. of Florida, Gainesville, Florida 32611, U.S.A./Greece
Rebhan, A.	TechnisheUniversität Wien, Wiedner Hampst. 8-10, Vienna A-1040, Austria
Rey, S.J.	Univ. of California, Santa Barbara, California, U.S.A.
Riggs, H.	Enrico Fermi Inst., Univ. of Chicago, Chicago, Illinois, U.S.A.
Rocek, M.	State Univ. of New York, Stony Brook, New York, U.S.A.
Roche, P.	Ecole Polytechnique, France
Rodriguez, M.	DAMTP, Silver St., Cambridge CB3 9EEW, U.K./Spain
Rohrlich, D.	Tel Aviv Univ., Ramat Avid, Tel Aviv 69978, Israel
Rosenstein, B.	Univ. of Texas at Austin, Austin, Texas, U.S.A.
Rovelli, C.	Univ. di Roma, p. Moro 2, Roma 00185, Italy
Ruiz, F.	DAMTP, Silver St., Cambridge, CB3 9EW, U.K./Spain
Rutherford, A.	Univ. of British Columbia, Vancouver, B.C., Canada
Saint-Aubin, Y.	Univ. de Montreal, C.P. 6128, Montreal, Quebec, Canada
Schirrmacher, A.	Univ. of Oxford, Oxford, England/Switzerland
Schmeing, N.	Chalk River Nuclear Labs., Chalk River, Ont., Canada
Segert, J.	California Inst. of Tech., Pasadena, California, U.S.A.
Shapere, A.	Inst. for Advanced Study, Princeton, New Jersey, U.S.A.
Siopsis, G.	Texas A&M Univ., College Station, Texas, U.S.A./Spain
Smith, J.W.	Univ. of British Columbia, Vancouver, B.C., Canada
Sniatycki, J.	Univ. of Calgary, Calgary, Alberta, Canada
Sodano, P.	Univ. di Perugia, Italy
Sudarshan, E.C.G.	Univ. of Texas, Austin, Texas, U.S.A.
Talman, J.	Univ. of Western Ontario, London, Ontario, Canada
Tarasov, L.	Univ. of Toronto, Toronto, Ontario, Canada
Tas, K.	Hacettepe Univ., Beytepe-Ankara 6532, Turkey
Thuillier, F.	LAPP, B.P. 110, Annecy le Vieux, Cedex 74941, France
Trugenberger, C.	Massachusetts Inst. of Tech., Cambridge, Massachusetts, U.S.A./Switzerland
Tanzi, R.	City College of CUNY, New York, New York, U.S.A./Greece
Van Driel, P.	Univ. of Amsterdam, Holland
Vercin, A.	Ankara Univ., Batikent/Ankara, Turkey
Viswanathan, K.	Simon Fraser Univ., Burnaby, British Columbia, Canada
Wang, J.	Univ. of Illinois, Chicago, Illinois, U.S.A.
Williams, J.	Brandon Univ., Brandon, Manitoba, Canada
Woloshyn, R.	TRIUMF, 4004 Wesbrook Mall, Vancouver, British Columbia, Canada
Yastremiz, C.	DAMTP, Silver St., Cambridge, CB3 9EW, U.K./Argentina
Yu, H.L.	Academia Sinica, Nankang, Tapiep, Taiwan
Zuk, J.	Max-Planck Inst. fur Kernphysik, Postfach 103980, Heidelberg 1, D-6900, Fed. Rep. Germany

INDEX

Abelian braid statistics, 15, 23, 63
Abelian Chern-Simons term, 493-494, 498, *see also* Abelian Chern-Simons theory
Abelian Chern-Simons theory, *see also* Abelian Chern-Simons term
  in planar physics, 204, 212
    point-particles with, 214-219
    with sources, 205-207
  spin-statistics and, 374-375
Abelian gauge theory, 23
ADE models, 595
Affine commutation relations, 143
Affine Toda theory, 609, 610-612, 619, *see also* Toda field theory
Aharonov-Bohm analysis, *see also* Bohm-Aharonov phase
  braid statistics and, 19, 23
  in FQHE, 482, 484
  in planar physics, 207, 232, 234
Alexander-Conway polynomial, 573
  braid groups and, 579
  quantum groups and, 623, 625, 643, 650, 652
  solvable models and, 594
Alexander polynomial, 596
Alexander's theorem, 425, 638
Angular momentum, not quantized, 216
Anomalous statistics, 212-213
Anomalous theory, *see under* Symmetry
Anti-atomic lattices, 173, 184
Antibrackets, 97, 99
Antiferromagnetism, 3
Antifields, 97, 98, 100, 101, 102
Antighost numbers, 87, 92, 95, 96, 99
Anti-holomorphic fields, 308, 310, 311
Antipodal maps, 525, 527
Antipode
  in Sl(2,C), 522, 531, 532, 533
  in YBA, 396, 397
Anyons, 19, 20, 21, 486*n*
Artin's braid group, 624
  N-state representation of, 573, 574-579
Atiyah-Witten axioms, 323
Atomic lattices, 173, 174, 184
Automorphism
  in quantum topology, 180, 184
  in RCFT, 311, 312, 323, 343, 351
  in Sl(2,C), 532
  in symmetry, 437

BA, *see* Bethe Ansatz
BAE, *see* Bethe Ansatz equation
Banach space, 145
Base, for topology, 161
Becchi-Rouet-Stora (BRS) operator, 435, 445-446, 455
Belavin-Knizhnik theorem, 264
Bell's inequalities, 123
Bethe Ansatz (BA)
  coordinate, 416
  of MTM, 610
  nested, 411, 417
  in planar physics, 225
  QISM and, 612, 613
  solvable models and, 585
  in YBA, 387, 390, 402, 407, 415, 429
Bethe Ansatz equation (BAE)
  IFT and, 428, 429-430
  light-cone lattices and, 416, 417-418, 420
  nested, 418
  six vertex model and, 409-410, 411, 415
Bianchi identity, 236, 377
Biedenharn sum-rule, 300

Bijective maps, 165-166
Black holes, 105-128
  description of, 107-109
  discrete physics and, 122-124
  eternal, 121
  Hawking effect and, 111-114
  horizon of, *see* Horizon
  realistic, 109
  Rindler space and, 109-112, 113, 114, 118
  string theory and, 107, 120, 125-126
  thermodynamics and, 115-117
  white holes and, 119-122
Bloch theorem, 473
Bogolyubov transformation, 112
Bohm-Aharonov phase, 364, 369, 379, *see also* Aharonov-Bohm analysis
Bohr correspondence, 397
Boltzmann's constant, 609
  black holes and, 105, 114, 115
Boltzmann weight
  in solvable models, 584, 585, 586, 587, 588, 593
  in statistical mechanics, 608-609
Boolean algebra, 150, 151
Borel subalgebra
  quantum groups and, 626
  in Sl(2,C), 525, 531, 532, 533, 534
Borel-Weil-Bott method, 500
Bose condensation, 484
Bose-Einstein statistics, 16
Bose statistics, 17, 62, 367, 368
Bosonization, *see* Non-abelian bosonization
Boundary conditions, twisted
  in QHE, 475-476
  in YBA, 399, 402, 425, 428
BPZ model, 265, 269
Bracket polynomial, 592
Braid groups
  in exotic spin-statistics, 367
  fractional statistics and, 626
  link invariants and, 624
  link polynomials and, *see under* Link polynomials
  N-state representations of, 573-581
  quantum gravity and, 547-548, 549
  quantum groups and, 626, 629, 639
  in solvable models, 583, 587-588, 595, 600
  from YBA, 387, 421-428
Braiding/fusing, 15, 629
  in RCFT, 276, 279-281, 282, 284, 291, 294, 329, 331
Braid matrix, 51, 72
  in CFT, 514
  in RCFT, 331
  in SG theory, 618
  in solvable models, 598-599
    link polynomials and, 594-595, 596-597, 600
Braid operator, 588, 593, 595
Braid quantization condition, 505-507
Braid statistics, 15-74, *see also* Three-dimensional local quantum theory
  abelian, 15, 23, 63
  algebraic formulation in, 27-43
  Chern-Simons term in, *see* Non-abelian Chern-Simons term
  Chern-Simons theory in, 15, 21, 23-27, 73-74
  intertwiners and, *see under* Intertwiners
  non-abelian, 15, 24, 26, 64
  physical realizations of, 21-23
  spin-statistics and, 16, 17, 51-52, 59-64
Brillouin zone, 474, 475
BRS operator, *see* Becchi-Rouet-Stora operator
BRST generator, 94, 95
BRST invariance, 96
BRST symmetry, 81-103
  algebraic topology of, 83-87
  geometric application of, 88-94
  Hamiltonians and, 81, 82, 94-96, 99
  Lagrangians and, 81, 82, 97-100, 101
  non-abelian Chern-Simons theory and, 558, 561

Canonical commutation relations, 143
Canonical formalism, 441, 448
Canonical quantization
  in CSW theory, 334
  of general relativity, 144
  of quantum gravity, 131, 132, 550
  in RCFT, 346, 356
Canonical quantum field theory (QFT), 376-382

Canonical transformation, 95, 99, 101
Cardy's postulate, 247-249
Cartan algebra
in CFT, 250, 256
non-abelian Chern-Simons term and, 503
in RCFT, 336-337, 346
in Sl(2,C), 536
YBA and, 394, 416, 428
Cartan-Weyl generators, 610
Cartesian polarization, 199, 203, 206
Casimir operator
in CFT, 252
in non-abelian Chern-Simons theory, 556
in planar physics, 219
of quantum groups, 624, 644, 654
in RCFT, 339
YBA and, 427
CBA, *see* Coordinate Bethe-Ansatz
Cellular automation, 123
Cerenkov radiation, 121
CFT, *see* Conformal field theory
Charge-transport operators, 31
Chebyshev polynomials, 430
Chern-Simons action, 624
in planar physics, 194, 227, 237
in RCFT, 352, 355
Chern-Simons term, *see also* Chern-Simons theory
abelian, *see* Abelian Chern-Simons term
non-abelian, *see* Non-abelian Chern-Simons term
in planar physics, 193, 194, 195, 196-197, 213, 216, 220, 221, 237
in RCFT, 352
Chern-Simons theory, *see also* Chern-Simons term
abelian, *see* Abelian Chern-Simons theory
braid statistics and, 15, 21, 23-27, 73-74
in CFT, 251, 255
link polynomials and, 573
non-abelian, *see* Non-abelian Chern-Simons theory
pure, 197-198
QHE and, 461
quantum gravity as, 541, 542-543
quantum groups and, 513, 514
in RCFT, 290
spin-statistics connection in, 363-384, *see also under* Spin-statistics
Chern-Simons-Witten (CSW) theory
in CFT, 266
quantum groups and, 624, 625, 626
in RCFT, 263, 265, 330, 342-353, 354, 355, 356, 357
coset models and, 342, 348-350
extended algebras and, 342, 343-348
orbifolds and, 342, 350-353
quantization and, 334-342
Wilson lines/tangles in, 624, 625, 626, 653, 654
Chiral algebra
in braid statistics, 51
in RCFT, 307-313, 342, 343, 345, 346
Chiral anomaly, 447
Chiral fermion model, 416, 421
Chiral Gross-Neveu model, 420
Chiral vertex operators (CVOs), 606
quantum groups and, 514
in RCFT, 266-277, 285, 307, 308, 309, 313, 342, 357
Classical lattice statistical mechanics, 606, 608-610
Classical R-matrix, 523
Clifford algebra, 258
Closed algebras, 82, *see also* specific types
Closed braids, 587-588
Closed sets, 141, 158, 162, 166-167
Closure on-shell, 97
Coboundaries, in planar physics, 201
Cocycle relation
in Lie algebra, 524
in planar physics, 195, 196, 199, 201
in symmetry, 452-453
Cofinite topology, 161, 167, 169, 184
Cohomology
in BRST symmetry, 84, 85, 87, 91, 100-103
in RCFT, 351
of Sl(2,C), 523, 524
in symmetry, 446-447, 453
Coistropic surface, 96
Co-Jacobi identity, 524
Coleman-Mandelstam construction, 364
Coleman's theorem, 8
Collision (scattering) theory, 71
Colored braids, 48

Compact braided monoidal category, 305, 331
Compact space, 162
Comultiplication, 274
Configuration space, 170, 179, 183, 365
Conformal analysis, 606, 607-608
Conformal blocks, 256, 606
  left-moving, 266
  non-abelian Chern-Simons theory and, 555
  in RCFT, 266, 268, 269, 275, 278, 286-287, 288, 296, 297, 307, 311, 313, 354
  right-moving, 266
  in SG theory, 618
  in YBA, 425
Conformal field theory (CFT), 65, 264, 275, *see also* Rational conformal field theory
  bosonization and, 1, 2
  braid groups and, 574
  braid statistics in, 51
  Chern-Simons theory in, 251, 255
    non-abelian, 555
  CSW theory and, 266
  IFT and, 430-431
  infinite dimensional algebras in, 241-259, *see also* Virasoro algebra
  perturbations of, 606, 618-619
  QFT and, 605, 606
  quantum groups and, 513, 514, 623-624, 625, 651-654
  SG theory and, 606, 618-619
  two-dimensional, 15, 18, 68
  YBA in, 403, 430-431
Conformal Lie algebra, 219
Conformal symmetry, 241, 242, 245, 605
Conformal transformation, 217
Conformal weight, 244, 246, 247, 248, 249
Continuous maps, 135, 163-164, 165, 168, 169
Convergence, 153, 154, 155, 157, 158, 162, *see also* Convergent sequences
Convergent sequences, 135, 138, 142, 154, 157, 164
Coordinate Bethe-Ansatz (CBA), 416
Copoisson structure, 523-524, 525
Cosets
  in CFT, 254-255, 256, 257
  in RCFT, 265, 342, 348-350, 351, 356, 357
Cotton tensor, 236, 237
Coulomb gauge, 368, 370, 377
Coulomb interaction, 379, 481
Coulomb potential, 479
Coulomb repulsion, 2, 3
Crossing symmetry, 587, 589-590, 591, 592, 593, 598
CSW theory, *see* Chern-Simons-Witten theory
CVOs, *see* Chiral vertex operators

Dehn twist
  in quantum gravity, 547
  in RCFT, 292, 294, 295, 297
Deligne's condition, 302-304, 306, 316
Differential geometry, 129, 132, 134, 184-185, *see also* specific types
Differential modulo delta, 85
Differentials, in BRST symmetry, 84
2+1 Dimensional quantum gravity, 541-550
  braid groups in, 547-548, 549
  as Chern-Simons theory, 541, 542-543
  dynamics in, 545-546
  quantization in, 543-544
  space-time in, 543, 545, 549-550
3+1 Dimensional quantum gravity, 549
Dimension of phase space, on surface of genus *g*, 338
Dirac analysis, 131, 441-443
Dirac bracket, 142, 375, 379, 383
Dirac constraints, 181, 183
Dirac delta function, 144
Dirac energy momentum tensor, 258
Dirac equation, 232, 234, 235
Dirac fermion theory, 5, 454
Dirac matrix, 196
Dirac operator, 449
Dirac quantization, 499
  weak, 103
Dirac sea, 410, 415, 613
Directed sets, 162
Discrete physics, 122-124
Discrete topology, 160, 161, 167, 186
Distance function, 137
Distributional topology, 171, 175
Dual Coxeter number
  in CFT, 252, 257, 258, 259
  in RCFT, 288, 289, 339
Dual ideals, 156, 157

Duality
in braid statistics, 29, 32
two-dimensional, 266, 313-330
Duality identity, 278-290
Duality matrix, 266-277, 285, 354
Duality transformation, 296
Dual theory, 356-357
Dynkin diagram, 250
link polynomials and, 595
in RCFT, 312, 343
in YBA, 417

Edge states, gapless current-carrying, 489-490
Eight vertex model
in SG theory, 609
in YBA, 401, 404, 411, 429
Einstein-Hilbert action, 236, 237
Einstein Podolski Rosen paradox, 123
Einstein's equation, 107, 109, 227-228, 236, 542
Einstein's theory, 226, 227, 241
Einstein tensor, 226, 236
Energy-momentum tensor
in CFT, 248, 253, 258
Dirac, 258
fractional spin and, 382-383
in planar physics, 213, 228, 230
in SG theory, 607-608, 619
Equivalence relation, 147, 165, 167
Euler-Lagrange equation, 198, 214
Euler-Maclaurin formula, 430
Exactly solvable models, 573, 584-587, 600
Exotic fractional statistics, 478
Exotic spin-statistics, 363, 365-371, 384
canonical QFT with, 376-382
Exotic statistics, 18
Extended algebras, 342, 343-348, *see also* specific types
Exterior derivative operator, 89, 95

Factorization equation, *see* Yang-Baxter equation
Faddeev-Popov determinant, 364, 435, 447, 449, 450-451
Faddeev-Popov ghosts, 453, 555, 558
Feigin-Fuchs construction, 608, 618, 619
Fermi statistics, 17, 62, 367, 368
Field theory, *see also* specific types
in Hubbard model, 1-13, *see also* One-dimensional Hubbard model
in the light cone lattice approach, 390
three-dimensional, 513
two-dimensional, 192
in YBA, 398
Filling factor, 463, 465, 472, 485
Filter base, 157, 158, 159, 163, 164
Filters, 157, 158, 159, 162, 167
Filtration degree, 87
First-class constraints
in BRST symmetry, 94-95, 96
exotic spin-statistics and, 379
in RCFT, 349
in topology, 131
Fixed sets, 170
Fock modules, 357
Fock representation, 103
Fock space, 121, 610
Fourier transforms, 177, 549
black holes and, 111, 112, 125
FQHE, *see* Fractional quantum Hall effect
FQS theorem, *see* Friedan-Qiu-Shenker theorem
FRA, *see* Fusion rule algebra
Fractional quantum Hall effect (FQHE), 461, 464, 465
braid statistics and, 15, 21, 74
exotic spin-statistics and, 363
fractional statistics and, 461, 478, 481-486
ground state degeneracy and, 461, 462-463, 478, 479-480, 486-490
impurities and, 481, 487
IQHE compared with, 478
Laughlin's wave function for, 462, 478-481, 483, 484-485, 489, 493
Fractional spin, 382-384
Fractional statistics, 16-21, 23, *see also* specific types
abelian Chern-Simons term and, 493
braid groups and, 626
Chern-Simons theory and, 365, 368
exotic, 478
in FQHE, 461, 478, 481-486
Frames, 164, 168-169, 187
Free energy, thermodynamic limit in, 399

Free fermion model, 579, 652
Free Majorana fermion, 605
Frenkel-Kac construction, 350
Friedan-Qiu-Shenker (FQS) theorem
in CFT, 243-244, 255
in RCFT, 265
Friedan-Shenker modular geometry, 276, 289, 295
Functional integrals, 130
Fundamental selection rule, 483-484
Fusion matrix
in braid statistics, 57, 59, 64
in RCFT, 307
Fusion rule algebra (FRA), *see also* specific types
in braid statistics, 34
in CFT, 248
in RCFT, 268, 270, 271, 287, 291, 307, 311-312

Galileo boosts, 219
Gauge choice
non-abelian Chern-Simons theory and, 557
in RCFT, 285
in symmetry, 454
in topology, 155, 156
Gauge field elimination, 499-505
Gauge-fixed BRST cohomology, 100-103
Gauge-fixed stationary surfaces, 102
Gauge fixing
Gribov ambiguity in, 444
non-abelian Chern-Simons theory and, 557-561
in symmetry, 438, 439, 444
Gauge groups, 174, 183
Gauge invariance, 101, 319
Gauge symmetry, 82, 191
Gauge systems, 94-96
Gauge theory, *see also* specific types
abelian, 23
black holes and, 110
braid statistics and, 72, 73
in BRST symmetry, 94
non-abelian, 23, 197
in planar physics, 192-226, *see also* Planar gauge theory
in topology, 155
Gauge transformation
in IQHE, 471, 472
non-abelian Chern-Simons term and, 498
in non-abelian Chern-Simons theory, 554, 556, 557
Gauge transformation (continued)
quantum gravity and, 542, 543, 544, 547
SG theory and, 611, 615
symmetry and, 437-438, 439, 445, 454
in YBA, 394-395
Gauss condition, 442
Gauss' law
Chern-Simons modified, 196, 198, 200, 201, 202, 203, 204, 205
fractional spin and, 383
in planar physics, 194, 195, 196, 198, 200, 201, 202, 203, 204, 205, 212
in RCFT, 335-337, 338, 339
spin-statistics and, 375, 379, 380
Gel'fand spectral theorem, 176, 177, 515
General relativity, 129, 133, 143, 144, 159, 226, 227
General topology, 129, 183-187
lattices in, *see under* Lattices
metric spaces in, *see* Metric spaces
partially ordered sets in, 135, 146-149, 161, 181
topological spaces in, 153-169, *see also* Topological spaces
Ghost field, 435, 445-446
Ghost numbers, *see also* Antighost numbers
in BRST symmetry, 84, 85, 86, 87, 91, 93, 95, 98, 99, 101, 103
Faddeev-Popov, 453, 555, 558
in non-abelian Chern-Simons theory, 555, 557, 558, 560, 561, 562, 563, 565, 566, 567, 568, 569, 570
pure, 87
in symmetry, 453
Ghosts of ghosts, 91
Ginsburg-Landau theory, 488
GKO coset construction, 265
Gluing axiom, 296, 324, 325-326
Godel universes, 237
Graded commutators, 83
Graded derivations, 83
Gravity
black holes and, 105, 111
canonical quantization of, 131
planar, *see* Planar gravity
quantum, *see* Quantum gravity
symmetry and, 435, 439, 449

Gravity (continued)
topology of, 131, 132-133
Green's function
black holes and, 119
exotic spin-statistics and, 367
in planar physics, 208, 210
in QHE, 478
Gribov ambiguity, 435
Gross-Neveu model, 416
Ground state degeneracy, 461, 462-463, 478, 479-480, 486-490
Ground states
antiferroelectric, 410
in CFT, 242, 246, 247, 249, 250-251, 252
extreme, 250-251
in YBA, 410, 429-430

Haag-Ruelle theory, 71
Haldane's generalization, 489
Half-filling, 1, 2, 3, 9-12
Hall conductance, 461, 462, 465, 478
electron motion and, 466, 469
ground state degeneracy and, 486, 488
as topological invariant, 470-477
Hall current, 463, 466, 472, 484
Hamiltonian formalism, 436, 447, 452
Hamiltonians
black holes and, 109, 117
in braid statistics, 24, 74
BRST symmetry and, 81, 82, 94-96, 99
Dirac, 235
equivalently constrained, 435
in Hubbard model, 2-3, 6, 10
light-cone, 421
in non-abelian Chern-Simons theory, 554
non-local Dzialozhinski-Moriya interaction, 407
in planar physics, 198, 202, 204, 205, 206, 212, 213, 215, 217, 219, 220, 221, 235
in QHE, 471, 472, 475, 476, 480, 487
quantum, 401
in quantum gravity, 541, 545, 546, 547, 548, 550
in quantum mechanics, 22
in quantum topology, 184
in RCFT, 338, 351
spin-statistics and, 365*n*, 368, 369, 371, 374, 375, 378
Hamiltonians (continued)
in statistical mechanics, 609
in symmetry, 435, 436, 441-442, 447, 452
in Toda theory, 610
XXZ Heisenberg, 405
in YBA, 390, 401, 405, 407, 421
Hausdorff space, 166, 167, 168, 169, 176, 177, 184
Hawking effect, 111-114
Hawking radiation, 106
Hawking temperature, 114, 115, 116
Hecke algebra
link polynomials and, 595, 599
in RCFT, 331
in SG theory, 615
Heegaard splitting, 330
Heisenberg algebra, 488
Heisenberg model, 1, 3
Hexagon identity, 283-284
Higgs model, 365
High-Tc superconductivity
braid statistics and, 15
exotic spin-statistics and, 363
Hubbard model and, 1, 2, 10
non-abelian Chern-Simons theory and, 555
planar physics and, 192
QHE and, 461
two-dimensional, 15
Hilbert space
black holes and, 113, 114, 121-123, 124, 126, 127, 128
braid statistics and, 19, 20, 25, 27, 30, 35, 71, 73
in BRST symmetry, 103
in CFT, 242, 245, 249, 256
in partially ordered lattices, 152
in QHE, 489, 490
for quantum gravity, 543, 544, 547, 548, 549
in quantum topology, 170, 171, 176, 178, 181
in RCFT, 264-265, 273, 274, 308, 309, 335, 341, 351, 352, 354, 356
in SG theory, 605, 608, 612, 615, 617, 618, 619, 620
in YBA, 397
Holomorphic fields, 308, 310, 311
Holomorphic polarization, 199, 203, 207
Holonomy
in quantum gravity, 544, 546, 547, 549, 550
Wilson tangles and, 653-654

Holons, 1
Homeomorphism
  in general topology, 165, 166, 168
  in quantum topology, 170, 176, 179, 180, 183
HOMFLY polynomial, 596
Homological perturbation theory, 82, 85, 102
Homomorphism
  in quantum gravity, 544
  in RCFT, 298
  in Sl(2,C), 531, 532, 533
  in topology, 164, 169
Homotopy, contracting, 92
Hopf algebra
  braid statistics and, 68, 69
  quantum groups and, 624, 625, 626-631, 643, *see also* Hopf structure, twisted
  quasitriangular, 628, 643
  in RCFT, 330-331
  Sl(2,C) and, 518, 519, 520, 521, 522, 523, 524, 525, 526, 529, 530, 532, 533, 534, 536
  YBA and, 393, 397, 426
Hopf links, 644-645
Hopf structure, twisted, 623, 647-649
Horizon, 107, 109, 115
  displacements of, 118-119
  operator algebra for, 124-128
Hubbard model
  one-dimensional, 1-13, *see also* One-dimensional Hubbard model
  two-dimensional, 1, 12

Ideals, 156-157, 175-177
Identification topology, 165, 186, 187
Identity matrix, 626, 653
IFT, *see* Integrable field theory
Incompressible quantum fluid states, 478
Indiscrete topology, 161
Induced maps, 174
Induced topology, 164
Infinite dimensional algebras, 241-259, *see also* Virasoro algebra
Infinite distributive law, 169
Injective maps, 169
Integrable field theory (IFT)
  CFT and, 430-431
  YBA and, 387, 390, 397, 419-420, 422-423, 428-431
Integral quantum Hall effect (IQHE), 461, 464, 465
  FQHE compared with, 478
  impurities and, 487
  Laughlin's argument for, 462, 470-472, 486, 489
  non-interacting electrons and, 472-475
Internal time, 134
Intertwiners
  braid statistics and, 34, 35, 40-52, 53-59, 71, 72
  in RCFT, 274, 297-299, 302, 303, 304, 309
Invariant Thirring model, 6
Inverse set maps, 163
IQHE, *see* Integral quantum Hall effect
IRF models
  basic relations in, 587
  link polynomials and, 589, 590, 591, 595-597, 600
  in statistical mechanics, 584, 585
Irreducible representation
  in CFT, 242, 245, 246, 247, 248, 249, 253, 255, 256, 259
  in RCFT, 299, 300
  in SG theory, 616
  unitary, *see* Unitary irreducible representation
Ising model
  in CFT, 246
  in RCFT, 267, 271, 291-294
  two-dimensional, 18
Isomorphism
  in RCFT, 298, 299, 303, 305, 323
  in Sl(2,C), 514, 515, 517, 522, 525, 533
  in topology, 165, 166

Jacobians
  in quantum topology, 181-182
  in symmetry, 452
Jacobi identity, *see also* Co-Jacobi identity
  in BRST symmetry, 83, 84, 86
  in CFT, 241, 258
Jastrow form, 479, 480
Jones polynomial
  braid groups and, 573, 574, 579
  Chern-Simons interpretation of, 514
  quantum groups and, 650
  solvable models and, 592, 594, 596, 599

Jordan-Wigner transformation, 246

Kac-Moody algebra
in CFT, 242, 249-250, 252, 254, 255, 256
quantum groups and, 623, 627, 644
in RCFT, 288-289, 312, 337, 341, 343-348
in Sl(2,C), 527, 538
in Toda theory, 610
Kac-Peterson formula, 289
Kaufman polynomial, 596
Kerr-Newmann solution, 108
Kerr solution, 228
Klein-Gordon equation, 232, 234
Knizhnik-Zamolodchikov equation, 24, 339-340
Knots, 323, 513, 514
Chern-Simons theory and, 364
non-abelian, 555
crossing symmetry in, 589
exactly solvable models and, 573
framed, 314
link/tangle relationship in, 624
in RCFT, 313-319, 331
in three-dimensional local quantum theory, 65-71
universal link invariant and, 631, 638
universal tangle invariant and, 642, 643
in YBA, 422-423, 425
Kostant-Sourieau approach, 500
Kosterlitz-Thouless transition, 404
Koszul complex, 102
Koszul-Tate resolution, 92, 95, 98
Kronecker delta, 209, 588, 590
Kruskal coordinates, 108, 109, 111, 118
Kubo formula, 473-474, 476

Lagrange density, 192, 193, 197
Lagrangian formalism
non-abelian Chern-Simons term and, 499-505
in symmetry, 449-450
Lagrangians
black holes and, 110, 114
braid statistics and, 72, 73, 74
BRST symmetry and, 81, 82, 97-100, 101
exotic spin-statistics and, 376
Lagrangians (continued)
non-abelian Chern-Simons term and, 493, 494, 495, 499-505, 509
non-abelian Chern-Simons theory and, 554, 555
in planar physics, 202-204, 205, 206, 213, 214, 215, 216, 217, 218, 237
in RCFT, 341
singular, 435
in symmetry, 435, 438, 441-443, 449-450
Yang-Mills gauge, 435
in YBA, 420, 421
Landau gauge
in Chern-Simons theory, 555, 558, 561
in QHE, 467-468
Landau-Ginzburg theory, 372
Landau level
exotic spin-statistics and, 371
in QHE, 463, 465, 468, 469, 471, 473, 478
fractional statistics and, 484
Laughlin wave function and, 479
Langevin equation, 467
Lattices, 186, 187
anti-atomic, 173, 184
atomic, 173, 174, 184
complemented, 150, 152
complete, 150
continuum limit of, 415
distributive, 150
in general topology, 140, 156-158, 163, 164, 166, 168, 169, 183, 184
partially ordered type, 135, 149-152
light-cone, *see* Light-cone lattices
modular, 150
non-distributive, 151
in quantum topology, 172-177, 178, 180
in solvable models, 583, 587, 592, 599
Laughlin's argument, 462, 470-472, 486, 489
Laughlin's wave function
in braid statistics, 20
for FQHE, 462, 478-481, 483, 484-485, 489, 493
Left-movers
in chiral algebra, 308-313
on lattices, 412
Leibnitz rule, 83, 84

Lie algebra
in CFT, 249, 250, 251, 252, 254, 256, 258, 259
conformal, 219
non-abelian Chern-Simons term and, 500, 501, 502
in non-abelian Chern-Simons theory, 554
in planar physics, 193, 219
quantum groups of, 623, 624, 626, 627, 629, 642, 643, 644
in RCFT, 267, 304, 331, 334, 336, 353, 354
Sl(2,C) and, 513, 514, 515, 523, 524-525, 527, 531, 536, 537
in solvable models, 595, 600
in symmetry, 436, 437, 439, 446
in Toda theory, 610
YBA and, 387, 394, 401, 410, 416, 419-420, 426, 428
Light-cone hamiltonians, 421
Light-cone lattices, 390, 411-421, 429
Limit points, 142
Linear maps, 534
Link invariants
non-abelian Chern-Simons theory and, 555
quantum groups and, 513, 623, 624, 625, 640, 645, 646-647, 650, 651, 653, 654
comparison with other constructions in, 638-639
in three-dimensional local quantum theory, 65, 67
universal, *see* Universal link invariant
in YBA, 422-423
Link polynomials, *see also* specific types
braid groups and, 573, 574, 579, *see also under* solvable models, this section
solvable models and, 583-600
braid groups in, 583, 587-588, 595, 600
examples of, 593-597
Markov trace in, 588, 589, 592, 593, 595, 596, 597, 599
N-state vertex model in, 593-595, 600
six vertex model in, 599
statistical mechanics in, 583, 584-585
Link polynomials (continued)
in symmetry, 453
in YBA, 425
Liouville theory, 608, 619
Locales, 164, 168-169, 187
Local observable algebra, 29, 31, *see also* specific types
Lorentz invariants, 241
black holes and, 114
in Hubbard model, 5, 6, 10
in symmetry, 449-450
Lorentz transformation
black holes and, 110
braid statistics and, 27
quantum gravity and, 542

MacLane coherence theorem, 300
Mandelstam string operators, 25, 27, 28
Manin triple construction, 524
Maps
antipodal, 525, 527
bijective, 165-166
continuous, *see* Continuous maps
induced, 174
injective, 169
inverse set, 163
linear, 534
surjective, *see* Surjective maps
in topology, 163-166, 168, 169, 174, 176, 179, 186
Marginal operators, in Hubbard model, 7
Markov's theorem, 624, 638, 639
Markov trace
braid groups and, 573
link polynomials and, 588, 589, 592, 593, 595, 596, 597, 599
Massive Thirring model (MTM)
SG theory and, 607, 609, 610
in YBA, 416, 420, 429
Matrix pseudo groups, 534
Maxwell's equation, 241
in braid statistics, 23
in exotic spin-statistics, 376-377
Maxwell term
in planar physics, 197, 210
spin-statistics and, 364, 376, 378, 384
Metric functions, 142-146
Metrics
bounded, 140
equivalent, 137-138
inequivalent, 138-139
isometric, 137
operations on, 139-140
symmetry and, 438-440

Metric spaces, 134-135, 136-142, 153, 158, 159, 160, 162, 164, 167
Minkowski space
in planar physics, 212
three-dimensional, 43
two-dimensional, 17, 18
Minkowski space-time, 411
Modular functor, 295-297
Modular group
representation of, 291
S-matrix in, 276, 323
Modular invariants
in CFT, 256-259
Cardy's postulate of, 247-249
in RCFT, 312, 313, 342
Modular lattices, 150
Modular tensor category (MTC), 297-308, 311, 330, 331, 332, 350-353, 354, 355-356
Monodromy
braid groups and, 574
braid statistics and, 58, 59, 60, 62
in Ising model, 18
QISM and, 611
in RCFT, 269, 271, 272, 278, 286-287, 288, 302, 305, 309, 321-322
for two point function on torus, 329
Morphism
braid statistics and, 31, 35, 36, 37, 39, 43-44, 48, 49, 53, 55, 70, 72, 73
in RCFT, 304, 305
in Sl(2,C), 531, 533
in topology, 163-166, 169*n*
Mott-Hubbard insulator, 2, 3
MTC, *see* Modular tensor category
MTM, *see* Massive Thirring model

NBA, *see* Nested Bethe Ansatz
NBAE, *see* Nested Bethe Ansatz equation
Neighborhood spaces, 154-158, 161, 170
defined, 159
Nested Bethe Ansatz equation (NBAE), 418
Nested Bethe Ansatz (NBA), 411, 417
Newtonian attraction, 230
Newtonian gravity, 227
Newton-Lorentz equation, 466
Newton's constant, 105, 226
N-identical particles, 493, 494-495, 507-510
Nineteen vertex model, 579, 593
Noether's energy-momentum tensor, 213
No ghost theorem, 243
Non-abelian bosonization, 1, 2, 7, 8, 9, 10, 11, 12
Non-abelian braid statistics, 15, 24, 26, 64
Non-abelian Chern-Simons term, 493-510, *see also* Non-abelian Chern-Simons theory
field equation solutions in, 496-498
Langrangian formalism and, 499-505
model description in, 494-496
Non-abelian Chern-Simons theory, 197-204, *see also* Non-abelian Chern-Simons term
finite renormalization of, 553-572
gauge fixing and, 557-561
at one loop, 553, 556, 557, 558, 560, 561, 562, 564-567, 568, 569, 571
regularization and, 553, 556, 557, 561-564, 567, 571, 572
at three loops, 555, 556, 571
at two loops, 553, 557, 560-561, 562, 563, 567-571
Ward identity and, 553, 557, 558, 561, 566, 567
in planar physics, 197-204, 206
spin-statistics and, 364
Non-abelian gauge theory, 23, 197
Non-abelian Thirring model, 420
Non-chiral primary field, 245
Non-linear delta model, 129, 170
Non-local Dzialozhinski-Moriya interaction Hamiltonians, 407
Non-Lorentz invariants, 10
N-state representations, 573-581
N-state vertex model
braid groups and, 579
link polynomials and, 593-595, 600
Null vector *L*, 267

Odd-denominator rule, 465, 478, 483-484
One-dimensional Hubbard model, 1-13
continuum limit of, 4-8

One-dimensional Hubbard model (continued)
half-filling in, 1, 2, 3, 9-12
phase diagram for, 11
strong coupling limit in, 2-4
Onsanger's result, 246
Open algebras, 82, *see also* specific types
Open sets, 186, 187
in general topology, 141, 158, 159, 160, 161, 162, 164, 166, 168, 169
in quantum topology, 171, 173
Orbifold theory, 265, 342, 350-353
Orbits
in BRST symmetry, 89, 89-91, 95, 96, 98
p-forms along, 89, 90
in symmetry, 438, 439, 440-441, 443-444, 451
functional measures on, 447-449
O(4) symmetry, 1, 2

Parafermions, 255, 265
Parastatistics, 16-17
Partially ordered lattices, 135, 149-152
Partially ordered sets, 135, 146-149, 161, 181
Partition function of CFT, 247
Pauli matrix, 196, 259
Pauli principle, 17, 365
Pauli-Villars regularization, 196
PCM, *see* Principal chiral model
Pentagon identity, 279
Permutation groups, 179-183, 184
Permutation statistics, 20, 26, 51, 60, 61, 62, 73
Perturbation theory, 556, 571
Perturbative renormalizability, 132
Planar gauge theory, 192-226
Chern-Simons theory in, *see under* Abelian Chern-Simons theory; Non-Abelian Chern-Simons theory
quantum dynamics in, 220-226
quantum holonomy in, 207-212
spin-statistics in, 212-213
topologically massive, 192-197
wave function in, 225
Planar gravity, 192, 222, 226-237
quantum dynamics in, 231-235
space-time in, 226, 227-230
topological elaborations in, 236-237
Planar physics, 191-237, *see also* Planar gauge theory; Planar gravity
Planck length
black holes and, 106, 121, 126
in topology, 129, 133, 134
Planck's constant
Boltzman's constant and, 609
in planar physics, 204
in Sl(2,C), 525, 526
Planck's law of black body radiation, 16
Poincare algebra
in braid statistics, 27, 29
fractional spin and, 383
in planar gravity, 237
in quantum gravity, 542
Pointless topology, 169
Poisson bracket, 94, 545, 649
Poisson structure, 515, 517, 523, 525
Polarization, 198-199, 201, 203, 205, 206-207, 208, 211
Cartesian, 199, 203, 206
holomorphic, 199, 203, 207
rotationally invariant, 205, 206
Polynomial equations, *see also* specific types
braid statistics and, 57, 69
in RCFT, 330-334
Potts model, 259, 573
Power sets, 148
Primary field
assignment of, 247-249
concept of, 244
correlation functions of, 244-245
in extension of algebra, 251-252
finite number in, 249
fundamental property of, 246-247
non-chiral, 245
Principal chiral model (PCM), 421
Principal ideal, 156
Pseudo-metrics, 137, 143, 153, 160, 167, 172

QFT, *see* Quantum field theory
QHE, *see* Quantum Hall effect
QISM, *see* Quantum Inverse Scattering method
Quantization of coupling constant, 194, 202
Quantum double construction, 514, 527, 529, 531-533
Quantum dynamics, 220-226, 231-235

Quantum field theory (QFT), 170, 177
black holes in, 105, 109-111
braid groups and, 573, 574
canonical, 376-382
CFT and, 605, 606
integrable, *see* Integrable field theory
local relativistic, 16
many particle systems and, 372-376
non-abelian Chern-Simons term in, 493, 504
three-dimensional local, *see* Three-dimensional local quantum theory
YBA in, 387, 402, 403, 410, 411, 416, 418, 421
Quantum gravity, 111, 129, *see also* 2+1 Dimensional quantum gravity
black holes and, 111
topology of, 129, 130, 132, 133, 144, 185
Quantum groups, 623-654
CFT and, 513, 514, 623-624, 625, 651-654
CSW theory and, 624, 625, 626
Hopf algebra and, *see under* Hopf algebra
link invariants in, *see under* Link invariants
in RCFT, 330-334
Sl(2,C) quantization in, 513-538, *see also* Sl(2,C) special linear group
in three-dimensional local quantum theory, 15, 65-71
topological field theory and, 623-624, 651-654
twisted, *see* Twisted quantum groups
universal link invariant and, *see* Universal link invariant
universal tangle invariant and, *see* Universal tangle invariant
from YBA, 387, 421-428
Quantum Hall effect (QHE), 461-490
Chern-Simons theory and, 461
non-abelian, 555
electron motion in magnetic field of, 466-470
exotic spin-statistics and, 371
experimental facts in, 462-466
Quantum Hall effect (continued)
fractional, *see* Fractional quantum Hall effect
impurities and, 475
integral, *see* Integral quantum Hall effect
planar physics and, 192
Quantum Hamiltonians, 401
Quantum holonomy, 207-212
Quantum Inverse Scattering method (QISM), 606, 607, 610-612, 613
Quantum mechanics
braid groups and, 574
braid statistics in, 16, 19, 22
in QHE, 467-470
statistical mechanics and, 609
two-dimensional, 363-384
Quantum norm theory, 146
Quantum topology, 129-135, 167, 170-183, 184-185, 187
complementary variables in, 177-179
eigenstates of, 135
lattices in, 172-177, 178, 180
metric functions in, 142-146
metric spaces in, 135
permutation groups in, 179-183, 184
wave functions in, 135
Quantum Yang-Baxter equation (QYBE), 629
Sl(2,C) and, 519, 520, 526, 527, 535
Quasiparticles
fractional statistics for, 481-483
ground state degeneracy and, 487, 489
Laughlin wave function for, 478, 480-481, 483, 484-485, 493
Quasitriangular Hopf algebra, 628, 643
QYBE, *see* Quantum Yang-Baxter equation

Racah coefficients, 285, 299, 303, 307, 320, 333
Racah matrix, 331
Racah's sum-rule, 300
Ramond-Neveu-Schwarz string theory, 246
Random topology, 184*n*
Rational conformal field theory (RCFT), 263-357
chiral algebra in, 307-313, 342, 343, 345, 346
completeness in, 290-297

Rational conformal field theory (RCFT) (continued)
CSW theory in, *see under* Chern-Simons-Witten theory
CVOs in, 266-277, 285, 307, 308, 309, 313, 342, 357
2D duality vs. 3D invariance in, 266, 313-330
duality identity in, 278-290
duality matrix in, 266-277, 285, 354
MTC in, *see* Modular tensor category
Tannaka-Krein theory in, 266, 297-308
Rational torus chiral algebra, 342, 346
R-degree, *see* Resolution degree
Reducible case, 91
Reeh-Schlieder theorem, 31, 74
Regge calculus, 543
Regularization, 553, 556, 557, 561-564, 567, 571, 572
Reidemeister moves
in RCFT, 316-317
in solvable models, 587, 592
in YBA, 422
Reidemeister's theorem
universal link invariant and, 623, 624, 631-633, 635, 639, 640
universal tangle invariant and, 640, 643
Relativity, *see* General relativity; Special relativity
Renormalization group equations, 12
Resolution degree (R-degree), 84, 85, 86, 87, 92, 100, 101
Restricted sine-Gordon (RSG) model, 606, 608, 609, 612-618
CFT perturbations and, 618, 619
Riemann tensor, 236
Riemannian geometry, 226
in exotic spin-statistics, 381
in QHE, 488
in RCFT, 264, 276, 277, 311, 334
in symmetry, 435, 439, 442-443, 444
in topology, 130, 132, 138, 142-144, 145, 153, 172, 185
Right-movers
in chiral algebra, 308-313
on lattices, 412
Rindler space, 109-112, 113, 114, 118
R-matrix
braid groups and, 575, 579, *see also under* YBA, this section
classical, 523
in Sl(2,C), 514, 519, 520-521
triangular, 521
universal, *see* Universal R-matrix
in YBA, 389, 391, 393, 395, 397-398, 399, 403, 404, 407
braid and quantum groups and, 422, 425
IFT and, 429
light-cone lattices and, 416, 417, 420, 421
Rotationally invariant polarization, 205, 206
RSG, *see* Restricted sine-Gordon model

Scattering matrix, 119-122
Schrodinger equation
black holes and, 117, 123
in planar physics, 194, 195, 198, 206, 220, 221, 224, 232
in QHE, 467-468, 473
quantum gravity and, 541
in topology, 131, 132*n*, 184, 185
Schur's lemma, 45, 51-52, 54
Schwarzschild solution
black holes and, 108, 109, 114, 118-119
planar gravity and, 228
Self-linking numbers, 369-370
Semi-classical limit, 132, 134
Semigroups, 150, 177
Separation axioms, 166-167
Sequences, 162, *see also* Convergent sequences
SG theory, *see* Sine-Gordon theory
Simple domain, 28, 31, 43, 48, 53, 55
Sine-Gordon (SG) theory, 605, 619-620
CFT perturbations and, 606, 618-619
conformal analysis of, 606, 607-608
Hubbard model and, 9, 10
known results concerning, 607
QISM for, 606, 607, 611
restricted, *see* Restricted sine-Gordon model

Sine-Gordon (SG) theory (continued)
statistical mechanics in, 606, 608-610
Toda theory and, 609
in YBA, 416
Singularity, in black holes, 107
Six vertex model
braid groups and, 425, 427, 579
deformation of, 597-598
link polynomials and, 599
in YBA, 387, 401, 403, 404-411, 417, 425, 427
light-cone lattices and, 413, 415, 416, 417, 421
Skein relations, 594
Sl(2,C) special linear group, 513-538
braid groups and, 579
function space quantization in, 515, 516, 517-523, 526, 527-528, 536, 537
quantizing considerations in, 514-516
quantum double construction in, 514, 527, 529, 531-533
universal enveloping algebra of, *see under* Universal enveloping algebra
S-matrix
factorized, 585-586, 587, 593
in RCFT, 276, 323
in SG theory, 606, 613-615, 616-618, 619, 620
in solvable models, 590
in YBA, 398, 413, 416, 420, 429, 431
Soliton phenomenon, 191
Solvable models, *see also* Exactly solvable models; specific types
deformation of, 597-599
link polynomials and, 583-600, *see also under* Link polynomials
transformations of, 597
SO(4) symmetry, 3, 4, 6, 7, 10
Space-like cones, 26, 28, 30, 31, 32, 34, 35, 36, 39, 43, 44, 48, 49, 53, 55, 70, 72
Space-time
Minkowski, 411
non-abelian Chern-Simons term and, 510
in non-abelian Chern-Simons theory, 555, 556, 557, 561, 568
in planar gravity, 226, 227-230
Space-time (continued)
in planar physics, 191, 192
in quantum gravity, 543, 545, 549-550
symmetry and, 436
topology of, 129, 130, 185-186, 187
in YBA, 411
Special relativity, 241
Spectral sequence, 87
Spectral topology, 171, 175, 176, 177, 178
Spin
braid statistics and, 23, 39, 40, 50, 51, 72, 73
in Hubbard model, 4, 5, 6, 10, 12
non-abelian Chern-Simons term and, 510
planar gravity and, 228, 229, 230, 232, 234
QHE and, 462
in RCFT, 331, 332
SG theory and, 617
in YBA, 401-402, 410, 421, 427, 429
Spin addition rules, 59-64
Spin chains, 430, 583, 609
Spinless test particles, 231, 232, 235
Spin network theory, 186
Spinons, 1
Spin spectrum, 59-64
Spin-statistics
in CFT, 245
Chern-Simons theory and, 363-384
abelian, 374-375
fractional spin in, 382-384
non-abelian, 364
QFT in, 372-376
exotic, *see* Exotic spin-statistics
non-abelian Chern-Simons term and, 494, 510
in planar physics, 212-213
in QFT, 16, 17, 51-52, 59-64, *see also under* Chern-Simons theory, this section
Stationary surface, 97, 102
Statistical mechanics, 583, 584-585
classical lattice, 606, 608-610
Statistics, *see also* Spin-statistics
anomalous, 212-213
exotic, 18
Statistics matrix, 44, 46

Stoke's theorem, 211, 474, 497
String theory
Belavin-Knizhnik theorem of, 264
black holes and, 107, 120, 125-126
CFT and, 243, 246, 252, 264
fractional statistics and, 18
in planar physics, 192, 230
Ramond-Neveu-Schwarz, 246
symmetry in, 435, 439, 449, 450
in topology, 129
Strongly correlated electrons, 1-13, *see also* One-dimensional Hubbard model
Subbase, for topology, 161
Sublattices, 168, 184, 187
Subspace topology, 165
Sugawara construction, 252-254, 256, 258
Supergravity, 132-133
Superselection sectors, 130
Superstrings, 133
Supersymmetry, 242, 562
Surgery, 324, 326, 332
ambiguity in, 327
calculations with, 327-328
Verlinde's formula from, 328-329
Surjective maps, 165, 168, 169, 176, 186
SU(2) symmetry, 3, 4, 6, 7, 8, 9, 10, 11, 12
Symmetry, 435-455
anomalous theory in, 435
as cohomological problem, 446-447, 453
measure with, 452-453
measure without, 450-451
problem of, 449-450
BRS operator in, 445-446, 455
BRST, *see* BRST symmetry
conformal, *see* Conformal symmetry
Dirac analysis of Lagrangian in, 441-443, *see also* *under* Lagrangians
gauge, 82, 191
gauge fixing in, 438, 439, 444
gauge transformation in, 437-438, 439, 445, 454
ghost field in, 435, 445-446
ghost numbers in, 453
Gribov ambiguity in, 444
integral over all fields in, 453-454
metric and connections in, 438-440
Symmetry (continued)
notations and basic objects in, 436-437
O(4), 1, 2
orbit space in, *see under* Orbits
Riemannian geometry in, 435, 439, 442-443, 444
SO(4), *see* SO(4) symmetry
SU(2), *see* SU(2) symmetry
U(1), *see* U(1) symmetry
Z2, 9, 10

Tangles, 645, 651, 652
CSW theory and, 626
degeneracy in, 647
factorization of, 646
links and, 623, 624, 625, 640
Tannaka-Krein theory, 266, 297-308
Taylor series, 515, 517
Teichmuller modular group, 295
Teichmuller space, 276, 277
Temperley-Lieb algebra, 590, 591, 593, 595, 598, 599, 600
TFT, *see* Toda field theory
Thermodynamics
of black holes, 115-117
in free energy, 399
Thirring model
invariant, 6
massive, *see* Massive Thirring model
non-Abelian, 420
in YBA, 420, 429
Three-dimensional general coordinate invariance, 266, 313-330
Three-dimensional local quantum theory, 15-74, *see also* Braid statistics
algebraic formulation of, 27-43
knot theory and, 65-71
quantum group theory and, 15, 65-71
Time, 132, 134, 184, *see also* Space-time
TLJ algebra, 331
Toda field theory (TFT), 431, *see also* Affine Toda theory
Topological field theory, 129-187, *see also* General topology; Quantum topology
quantum groups and, 623-624, 651-654
in RCFT, 323-324
Topologically massive gauge theory, 192-197

Topologically massive gravity, 236-237
Topological spaces, 137, 153-169
defined, 159
frames and locales in, 164, 168-169
morphism in, 163-166, 169*n*
neighborhood spaces as, *see* Neighborhood spaces
non-metric convergence in, 153-154
separation axioms in, 166-167
Torus
ground state degeneracy on, 487-489
monodromy for two point function on, 329
proof of equation on, 320-322
Totally ordered sets, 146
Transfer matrix, 390, 394, 395, 399
light-cone lattices and, 414
six vertex model and, 406
twisted, 401
Triangle inequality, 142, 145
Triangular relation, *see* Yang-Baxter equation
Triangular R-matrix, 521
Triangular universal R-matrix, 527
Twisted quantum groups, 623, 624-625, 647-651, 652, 653
Two-dimensional quantum mechanics, 363-384, *see also* Spin-statistics, Chern-Simons theory and

UIR, *see* Unitary irreducible representation
Ultra-filters, 174
Unitary irreducible representation (UIR)
non-abelian Chern-Simons term and, 493, 495, 500-501, 503-505, 510
of Vir, 242-243
Universal enveloping algebra, *see also* specific types
in RCFT, 331
of Sl(2,C), 513, 514, 515, 522-523, 527-528, 537, 579
deformation of, 513, 514, 516, 523-527, 537
quantization of, 514, 515, 522-523, 537
reconstruction of, 533-536
Universal link invariant, 623, 624, 625, 629, 631-638, 640
Universal link invariant (continued)
degeneracy in, 646-647
Universal R-matrix
in quantum groups, 624, 626, 628, 629
in Sl(2,C), 519, 526-527, 529, 530, 531, 532, 534, 535, 536, 537
triangular, 527
Universal tangle invariant, 623, 625, 629, 640-645
Unknots, 638, 642, 645, 650
Upper sets, 155-156, 161
U(1) symmetry, 3, 5, 6, 9, 10, 11

Vacuum
in CFT, 243, 246, 247, 248, 249
in planar physics, 226
Vector spaces
in partially ordered lattices, 152
in RCFT, 303, 305
in YBA, 387, 388
Verlinde's formula, 270, 286, 287-288, 289-290, 306-307, 311, 312
from surgery, 328-329
Vertex models, *see also* specific types
braid groups and, 579
link polynomials and, 589, 595, 597, 600
N-state, *see* N-state vertex model
quantum groups and, 513
spin-statistics and, 364
in statistical mechanics, 584-585
in TL class, 595, 600
in YBA, 399, 402
Vertex-sos transformation, 69
Virasoro algebra (Vir)
in CFT, 241-259
cosets in, 254-255, 256, 257
extreme ground states in, 250-251
FQS theorem in, 243-244, 255
Ising model in, 246
Kac-Moody algebra and, *see* Kac-Moody algebra, in CFT
need for the extension of, 249
primary field in, *see* Primary field
Sugawara construction in, 252-254, 256, 258
UIR of, 242-243

Virasoro algebra (Vir) (continued)
in RCFT, 265, 266, 273, 274, 276, 343, 349
in SG theory, 618, 619

W-algebras, 265, 342
Ward identity, 553, 557, 558, 561, 566, 567
Wave functions, 135, 225
Wess-Zumino action, 453, 454
Wess-Zumino consistency condition, 446
Wess-Zumino functional, 341
Wess-Zumino-Witten (WZW) model
braid statistics and, 24
in CFT, 253, 255
in Hubbard model, 7, 8, 9, 10, 12
non-abelian Chern-Simons theory and, 556
in RCFT, 265, 271, 288, 289, 309, 334, 337, 341, 342, 348, 351, 357
Weyl alcove, 337, 343
Weyl anomaly, 450
Weyl fermions, 446, 449
Weyl gauge, 195, 198, 204
Weyl group, 346
Weyl-Kac characters, 288, 337
Weyl tensor, 236
Wheeler-DeWitt equation, 131, 132
White holes, 119-122
Wigner's formula, 235
Wilson lines
in CSW theory, 624, 625, 626, 653, 654
QISM and, 611
quantum gravity and, 542, 546
in RCFT, 336, 338, 339, 341-342, 344, 345, 346
Wilson loops, 21, 27, 28
Wilson tangles
in CSW theory, 624, 626, 653, 654
holonomy and, 653-654
Witten's link invariant, 654
Witten's triple cosets, 350
WKB approximation, 338, 355
Wong's equation, 495
Wu-Kadanoff-Wegner transformation, 585, 595, 597
WZW model, *see* Wess-Zumino-Witten model

XXZ Heisenberg Hamiltonians, 405
XYZ Heisenberg spin chain, 609

Yang-Baxter algebra (YBA), 387-431
Yang-Baxter algebra (YBA) (continued)
braid groups from, 387, 421-428
in braid statistics, 18
description of, 387-398
gauge transformation in, 394-395
group invariance of, 392
IFT and, *see under* Integrable field theory
Lie algebra and, *see under* Lie algebra
light-cone lattices and, 390, 411-421, 429
physical realizations of, 398-403
quantum groups from, 387, 421-428
reproduction property of, 393
shift invariance of, 393
six vertex model in, *see under* Six vertex model
trigonometric/hyperbolic, 387, 392, 402, 419-420, 421-423, 425-426, 428
Yang-Baxter equation (YBE), 388-389, 391, 396
braid groups from, 422, 423, 427, 574
in braid statistics, 48
classical, 397-398
quantum, *see* Quantum Yang-Baxter equation
quantum groups from, 422, 423, 427, 629, 638
in RCFT, 278, 279, 284, 294
SG theory and, 612-613, 614, 615
six vertex model and, 404
trigonometric/hyperbolic, 404
universal link invariant and, 638
Yang-Baxter generator, 388, 389, 393, 394, 405, 422
Yang-Baxter matrix, 23, 68
Yang-Baxter operator, 387-388, 389-390, 392
braid groups from, 422, 423, 425, 427, 588
elliptic, 392, 402
quantum groups from, 422, 423, 425, 427
rational, 392, 402
in solvable models, 587, 588, 589-591, 595, 598
trigonometric/hyperbolic, 392, 402
Yang-Baxter relation
braid groups and, 573, 575-576
in SG theory, 617

Yang-Baxter relation (continued)
in solvable models, 583-585, 586, 587, 597
Yang-Baxter set, enhanced, 638
Yang-Mills action, 553
Yang-Mills density, 192, 194
Yang-Mills field, 442
Yang-Mills gauge Lagrangians, 435
Yang-Mills potential, 495
Yang-Mills regulator, 564
Yang-Mills term, 197, 364, 561
Yang-Mills theory
in symmetry, 435, 447, 449, 450, 452
in topology, 129, 131, 148
YBA, *see* Yang-Baxter algebra
YBE, *see* Yang-Baxter equation

Z_2 symmetry, 9, 10
Z-variance, 389

The manufacturer's authorised representative in the EU is Springer Nature Customer Service Centre GmbH, Europaplatz 3, 69115 Heidelberg, Germany. If you have any concerns regarding our products, please contact ProductSafety@springernature.com

Printed and bound by CPI Group (UK) Ltd, Croydon, CR0 4YY
29/04/2026
02099472-0015